The NASA Psyche Mission

James F. Bell III • Carol A. Polanskey •
Linda T. Elkins-Tanton
Editors

The NASA Psyche Mission

The First Up-Close Exploration of a Metal-Rich World

Previously published in *Space Science Reviews, Topical Collection: The NASA Psyche Mission: Science Instruments and Investigations*

Springer

Editors
James F. Bell III
Arizona State University
Tempe, AZ, USA

Carol A. Polanskey
Jet Propulsion Laboratory
California Institute of Technology
Pasadena, CA, USA

Linda T. Elkins-Tanton
Space Sciences Laboratory
University of California
Berkeley, CA, USA

ISBN 978-94-024-2376-1

Cover illustration: A SpaceX Falcon Heavy rocket with the Psyche spacecraft onboard is launched from Launch Complex 39A, Friday, Oct. 13, 2023, at NASA's Kennedy Space Center in Florida. NASA's Psyche spacecraft will travel to a metal-rich asteroid by the same name orbiting the Sun between Mars and Jupiter to study it's composition and origin.
Photo Credit: (NASA/Aubrey Gemignani)

This Springer imprint is published by the registered company Springer Nature B.V.
The registered company address is: Van Godewijckstraat 30, 3311 GX Dordrecht, The Netherlands

Contents

Space Science Reviews (2026) 222:5
https://doi.org/10.1007/s11214-025-01255-6

The NASA Psyche Mission: The First up-Close Exploration of a Metal-Rich World

James F. Bell III[1] · Lindy Elkins-Tanton[2] · Carol A. Polanskey[3]

Received: 21 November 2025 / Accepted: 24 November 2025 / Published online: 3 January 2026

The main belt asteroid (16) Psyche was discovered in 1852 at the Astronomical Observatory of Capodimonte in Naples, Italy by Annibale de Gasparis, who named the new minor planet after the eponymous Greek goddess. Over the more than 220 years since, astronomers have discovered that Psyche is the largest and most massive of the so-called "M Class" asteroids, with average visible to near-IR colors and spectral properties similar to those of many of the metallic meteorites in the world's collections. Indeed, Psyche is now thought to singularly comprise roughly 1% of the mass of the main belt, and to have among the highest densities of any known asteroid ($\sim$4.1 g/cm^3).

Some of the most recent telescopic observations of Psyche as well as new laboratory measurements of a variety of meteorites and mixtures of meteorites and typical asteroidal silicate minerals suggest that this object is likely not simply composed of only typical FeNi minerals found in metallic meteorites. For example, modeling of the available density and spectroscopic properties allow for the potential presence of tens of percent (or more) of non-metallic components with densities comparable to those of typical silicates found in meteorites or on other asteroids. There is also minor evidence that some of the silicate component could be hydrated. Regardless, the most preferred models still suggest that (16) Psyche is ultimately likely to be "metal rich" in its bulk properties.

But what is this small ($\sim$220 km average diameter), distant ($\sim$3 AU from the Sun), enigmatic world really like? No one has ever seen a world made mostly of metal rather than of rock, ice, or gas. The best current models of the shape of Psyche come from telescopic imaging, which reveals a roughly triaxial ellipsoid shape of dimensions $\sim 290 \times 245 \times 170$ km as well as several large concavities suggestive of impact craters or basins. But it's impossible to constrain the asteroid's surface geology, topography, or elemental/mineral composition further using existing telescopic capabilities. And if it is indeed metal-rich, is its origin related to that hypothesized for metallic meteorites? Is Psyche a metallic core of a terrestrial planet or a precursor planetesimal? If so, studying this object might be the best way to study planetary cores and the core formation process, directly. To answer these kinds of questions and take the next steps, we need to get a much closer look, not only by taking high-resolution images but also by collecting diagnostic compositional, mineralogic, and other data.

✉ J.F. Bell III
Jim.Bell@asu.edu

1 Arizona State University, School of Earth and Space Exploration, Tempe, AZ, USA

2 University of California Berkeley, Space Sciences Laboratory, Berkeley, CA, USA

3 Jet Propulsion Laboratory, California Institute of Technology, Pasadena, CA, USA

In the early 2010s a team of scientists and engineers, led by Prof. Lindy Elkins-Tanton, began to formulate plans for a dedicated robotic space mission – named after the asteroid itself – that would do just that. Taking advantage of the opportunity to propose Principal Investigator-led proposals to NASA's Discovery program of lower cost deep space missions, the Psyche mission team's initial bid was selected as a Concept Study finalist in 2015 and then, eventually, approved for flight in 2017. After several years of design, reviews, fabrication, and testing – punctuated by the extreme challenges imposed by the global COVID-19 pandemic – the Psyche spacecraft and its mission of the same name were finally launched into space from Kennedy Space Center's Launch Pad 39A on October 13, 2023, aboard a SpaceX Falcon Heavy rocket.

At a high level, the Psyche mission is essentially a mission of exploration and discovery, designed to reveal for the first time the details of a class of solar system objects not yet seen up close – a metal-rich world. Still, three very specific science goals have been identified for the mission: (1) Understand a previously-unexplored building block of planet formation: iron-rich planetesimals; (2) Look inside terrestrial planets, including Earth, by directly examining the interior of a differentiated body, which otherwise could not be seen; and (3) Explore a new type of world that is rich in metal rather than in rock or ice. To achieve these goals, five very specific primary science and measurement objectives have been identified for the mission: (i) Determine whether Psyche is a core, or if it is primordial unmelted material; (ii) Determine the relative ages of regions of Psyche's surface; (iii) Determine whether small metal bodies incorporate the same light elements into the metal phase as are expected in the Earth's high-pressure core; (iv) Determine whether Psyche was formed under conditions more oxidizing or more reducing than Earth's core; and (v) Characterize Psyche's morphology.

To achieve these objectives, the team outfitted a novel spacecraft – which uses a Xe-fueled ion thruster system rather than a traditional chemical propulsion system – with a highly capable and high-heritage payload of remote sensing science instruments. These instruments include: (a) a pair of redundant Multispectral Imagers designed to characterize the asteroid's geology, topography, and surface metal to silicate mineral ratio globally at scales from a few to a few hundred meters resolution; (b) a combined Gamma-Ray and Neutron Spectrometer instrument designed to measure the abundance of key surface elements (e.g., Fe, Si, K, Ni, Ca, Al) to further constrain the metal-silicate ratio, the composition of surface silicates, the redox conditions of Psyche's formation, and the relationship of the asteroid's observed surface compositions with those of achondritic and/or chondritic meteorites; (c) a Magnetometer with a pair of sensors designed to determine if Psyche preserves evidence of a magnetic field with an intensity in the range of natural remanent magnetization found in iron meteorites – potentially providing the strongest evidence that the asteroid is a planetary core; and (d) a Gravity Science investigation that will use the spacecraft's telecommunications subsystem to determine the mass and internal mass distribution of the asteroid.

After a just under 6-year cruise through the inner solar system, the science phase of the Psyche mission will involve initial capture into a high circular orbit around the asteroid in mid-2029, for initial reconnaissance of the asteroid's geology, topography, shape, spin axis, and gravity/magnetic fields. Over the course of the following 27 months, the spacecraft will be commanded to a series of different orbital altitudes designed to acquire higher resolution global imaging, topography, and gravity/magnetic field data and, at the lowest altitude, km- to tens-of-km-scale elemental compositional data. Because the obliquity of Psyche is already known to be quite high ($\sim$100°), only the southern hemisphere will be illuminated for the first set of orbital altitude maps. Thus, a second set of later observations, after the asteroid's seasons have advanced accordingly, will be acquired to map the northern hemisphere. The Prime Mission will nominally be completed in November of 2031.

The Psyche mission team is organized into a series of five thematic science-focused Working Groups that are aligned with the five primary science and measurement objectives noted above: (i) Psyche as a Planetary Core; (ii) Relative Ages of Surface Units; (iii) Light Elements; (iv) Oxidizing/Reducing Conditions; and (v) Topography/Geomorphology. Each Working Group will be analyzing and interpreting data and generating results and data products gleaned from multiple payload investigations, emphasizing the interdisciplinary and synergistic nature of the science instruments and team. Importantly, the team also incorporates a substantial engineering and management workforce that designed, built, tested, and now operates the spacecraft for NASA. A hallmark of the Psyche mission is the close-knit integration of scientists, engineers, and managers – with experience spanning students to senior researchers – into a single unit, the Psyche team.

This topical collection of papers is intended to serve as the primary initial reference and starting point for readers to understand the background, motivation, and goals/objectives of the mission; the high-fidelity instrument investigations that are being used to achieve those goals/objectives; the details of the innovative spacecraft systems and operational processes that are taking those instruments to this interesting asteroid; and the detailed design of the mission that will be carried out there.

Indeed, the collection begins with a detailed description of the mission design, including the fascinating history of the mission's origins and the many challenges encountered on the way to launch. The next five papers then provide background, context, details, and plans from each of the five science-focused thematic Working Groups. Those are then followed by four papers that describe each of the mission's science instrument investigations in detail, and one paper that provides detailed information about the spacecraft and its many subsystems that had to be optimized to conduct this specific mission. Finally, the collection ends with a wonderful description of the mission's novel and ambitious student and public engagement program, which plays a major role in the team's communication and outreach to the many different stakeholders in the mission.

As this editorial is being written, the Psyche mission has completed approximately a third of its long cruise to the asteroid, and is approaching Mars for a May 2026 gravity assist encounter that will slingshot the spacecraft into the main asteroid belt. The spacecraft and its instruments are operating well after initial checkouts and several iterations of periodic "maintenance and calibration" activities, and science operations at Psyche remain on target for August, 2029. While the team has specific hypotheses to test and everyone has their own variety of expectations in mind, what we will discover when we actually get up close to this curious little metal world for the first time is, truly, unknown. Of course, that's what makes exploration – and missions like Psyche – so incredibly exciting!

Declarations

Competing Interests The authors declare no competing interests.

Publisher's Note Springer Nature remains neutral with regard to jurisdictional claims in published maps and institutional affiliations.

Space Science Reviews (2025) 221:95
https://doi.org/10.1007/s11214-025-01218-x

Psyche Mission Description and Design Rationale

Carol A. Polanskey[1] · Linda T. Elkins-Tanton[2,3] · James F. Bell III[2] · Eleanor K. Alonge[1] · Sarah H. Bairstow[1] · Richard P. Binzel[4] · Abhijit Biswas[1] · Luke Bury[1,5] · Ernest Cisneros[2] · Dongsuk Han[1] · Insoo Jun[1] · William M. Klipstein[1] · David J. Lawrence[6] · Timothy J. McCoy[7] · Nickolaos Mastrodemos[1] · José M.G. Merayo[8] · Sarah K. Noble[9] · David Y. Oh[1] · Rona Oran[4] · Carolyn A. Ortega[1] · Ryan S. Park[1] · Patrick N. Peplowski[6] · Thomas H. Prettyman[10] · Marc D. Rayman[1] · Jodie B. Ream[4] · Thomas Roatsch[11] · Timothy M. Weise[1] · Daniel D. Wenkert[1] · Benjamin P. Weiss[4] · Maria T. Zuber[4]

Received: 17 April 2025 / Accepted: 6 September 2025 / Published online: 14 October 2025

Abstract
The Psyche spacecraft launched on October 13, 2023 to journey to the asteroid of the same name. Psyche is the largest M-class asteroid and possibly the remanent core of an early differentiated planetesimal that was disrupted by collisions. The Psyche mission will test that hypothesis as the 14th mission in NASA's Discovery Program. An alternative hypothesis is that the asteroid is unmelted primordial material. We describe the proposal competition process leading to selection of the mission and its context with other small body missions. This paper will briefly introduce the three science instruments, gravity science investigation, and Deep Space Optical Communications technology demonstration, leading into a detailed explanation of the science mission architecture. The orbital science phase is divided into a series of circular mapping orbits at four distinct altitudes, each selected to address specific science objectives. The requirements and objectives for each orbit are accompanied by an assessment of the effectiveness of each phase. We discuss the structure of the Psyche team during the operations phase along with the roles and responsibilities of the science and flight operations teams. Key elements of mission operations that are unique to the Psyche mission are provided. The Science Data Center manages and archives the Psyche mission data. The contents of the archive data sets for each instrument are outlined as well as the interfaces between the Science Data Center, the instrument teams, and the Planetary Data System.

Keywords Asteroids · (16) Psyche · Psyche mission · Science operations · Mission operations

1 Introduction

This paper provides an overview of the Psyche mission, a detailed exploration of the science implementation and mission operations, and context for the remaining articles in this special collection. Section 1 looks back to how the mission began, places it in context with other missions to asteroids, provides a description of the target asteroid (16) Psyche, and outlines

Extended author information available on the last page of the article

the mission science objectives. Section 2 provides a brief overview of the science instruments and investigations along with references to the Psyche instrument papers. Section 3 describes the approach to science planning, the driving constraints, and other considerations that influence the architecture of the science mission. Details of the Psyche science orbits and observation designs are expanded in Sect. 4. Section 5 describes the Psyche science and flight operations teams along with details of science and engineering operations, while Sect. 6 outlines Psyche science data archiving.

1.1 Mission Synopsis

The Psyche spacecraft, launched on October 13, 2023, will visit the metal-rich, main-belt asteroid (16) Psyche (Dibb et al. 2024). The orbital science phase at Psyche, beginning in August 2029, follows a 5.6-year interplanetary cruise.

The spacecraft carries three science instruments: a Magnetometer, two redundant Multispectral Imagers, and a Gamma-Ray and Neutron Spectrometer (GRNS). Gravity science will be accomplished with the X-band high gain antenna (HGA) that also provides communications for the spacecraft, as well as the three low gain antennas (LGAs). A brief description of the instruments and science investigations is provided in Sect. 2 and in other articles in this collection (Weiss et al. 2023; Bell et al. 2025; Lawrence et al. 2025, and Zuber et al. 2022). The spacecraft also carries a high-rate laser technology demonstration, the Deep Space Optical Communications (DSOC) instrument (Biswas et al. 2024), which has already proven the feasibility of optical communications in the deep space environment at the distance of Mars and beyond. DSOC is a separate NASA project and was developed independently but in coordination with the Psyche project. DSOC is described further in Sect. 2.5.

The Psyche spacecraft launched from the Kennedy Space Center on a Falcon Heavy rocket provided by SpaceX. Figure 1 shows the interplanetary trajectory from launch through orbital operations. Each of the spacecraft subsystems and instruments were checked out and calibrated during the first 100 days of the mission. The solar electric propulsion system will continue to operate for most of the cruise phase until rendezvous with the asteroid at approximately 2.7 astronomical units (AU) solar range (see Sect. 5.3). The DSOC technology demonstration has periodic opportunities to communicate with their ground tracking stations during their prime mission, the first two years of Psyche cruise.

The spacecraft trajectory includes a gravity assist from Mars in May 2026. Prior to approaching Mars, there will be approximately 10 months when it is optimal for the trajectory to pause thrusting and coast. The only planned thrusting during this time is for a Mars Trajectory Correction Maneuver and two potential Mars Trim Maneuvers, which adjust the spacecraft's trajectory to ensure optimal arrival at the asteroid using the gravity assist. These maneuvers occur during the first half of the Mars gravity assist (MGA) phase, which spans from 80 days before the spacecraft's closest approach to Mars until 2 days after. The science instruments will be active during the coasting period around Mars, performing routine calibrations and acquiring data during Mars closest approach. None of the instrument activities during this period are required for mission success, but they will serve as a dry run for multi-instrument operations in orbit around Psyche and provide calibration data from a well-characterized target body.

The orbital phase is divided into four science orbit altitudes that accomplish specific mission objectives. Figure 2 illustrates the relationship between the science orbits. The highest altitude phase, Orbit A, provides the first opportunity to characterize Psyche. Imaging campaigns in Orbit B are designed to provide complete geologic and topographic maps of

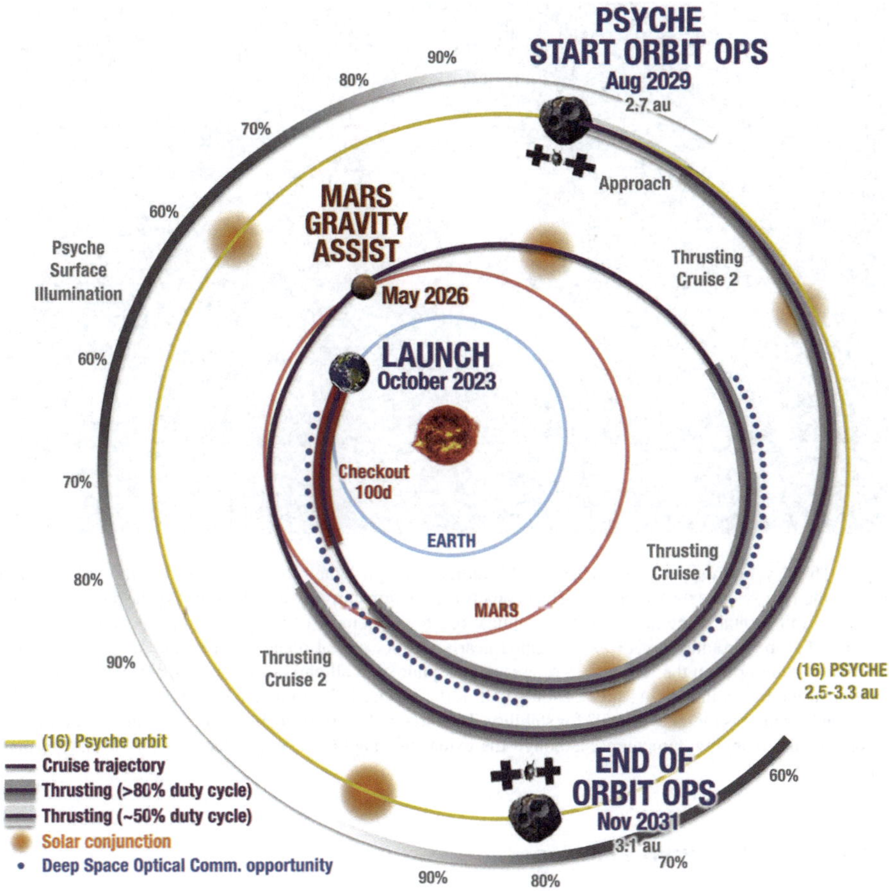

Fig. 1 The Psyche baseline cruise trajectory includes a Mars gravity assist in 2026 with arrival at Psyche approximately three years later. The wide grey lines plotted over the trajectory indicate when the spacecraft is thrusting with its electric propulsion subsystem. The blue dots indicate weekly DSOC opportunities over the first two years of the cruise phase. There are multiple periods in cruise and in orbit around Psyche when the spacecraft is in solar conjunction and communications with the earth are disturbed by the sun. The grey-scale semi-circle beyond Psyche's orbit indicates the percentage of illuminated surface at that time during orbital operations

Psyche's surface. The gravity science objectives are accomplished in Orbit C along with improved topographic models. Orbit D is the lowest altitude planned for the primary mission and enables the mapping of elemental composition with the GRNS. Magnetometry objectives are supported by several aspects of the mission plan including the lower altitudes of Orbits C and D and the unique geometries of transfers to and from Orbit D. The project uses the spacecraft's solar electric propulsion subsystem to transfer between each of the science orbits.

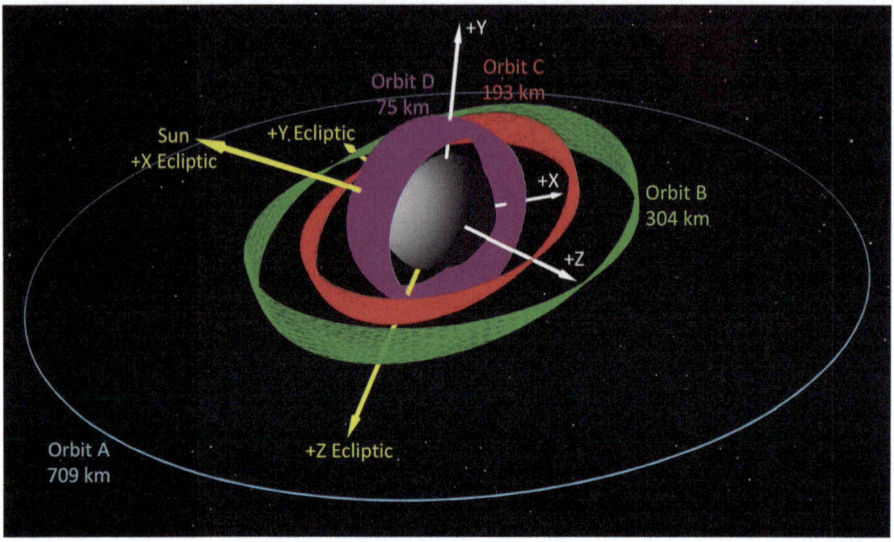

Fig. 2 The Psyche mission's four science orbit altitudes are designed to address the specific objectives of each of the four science investigations. The geometrical relationship between the sun, Psyche's spin axis, and the spacecraft orbital plane has important implications for placement of imaging campaigns on the mission timeline. Psyche's rotation axis (+Z) is tipped nearly into its orbital plane. At the time of this simulation, sunlight is coming from the left, and Psyche is at an equinox, so all parts of its surface are illuminated each rotation around its spin axis. At solstice, only half of the surface is illuminated. Orbits A, B, and C are polar orbits and Orbit D is inclined at 160° for stability. The three polar orbits enable global ground track coverage to support the mapping campaigns, although the extent of imaging coverage depends on Psyche's orbital season

1.2 Origins of the Psyche Mission

1.2.1 Origins

In 2011, Linda Elkins-Tanton and Erik Asphaug received emails from Bruce Bills and Daniel Wenkert at the Jet Propulsion Laboratory (JPL), asking whether they were interested in thinking about a mission proposal to test the hypotheses of a paper Elkins-Tanton, Benjamin Weiss, and Maria Zuber had published that year (Elkins-Tanton et al. 2011). Motivated by paleomagnetic measurements of the Allende carbonaceous chondrite from Weiss' group (Carporzen et al. 2011), the paper proposed the idea of partial differentiation for planetesimals: some planetesimals, in the first 2 Myr of the solar system, may have melted and differentiated into a metal core and rocky mantle in their interiors, but retained a colder, unmelted, primitive crust. The resulting planetesimal structure allows chondrites (primitive unmelted meteorites) to originate on internally differentiated parent bodies, from which they can record a magnetic field from the core dynamo (Weiss and Elkins-Tanton 2013).

Over the rest of 2011 and well into 2012, we discussed possible targets for this mission. We sought asteroid families, with the goal of detecting mantle cumulates, metallic core, and chondritic lid elements in the bodies, but no such family had been definitively identified, and the mission operations challenges of moving from one to another body in a family were too great. In 2012 we settled with some certainty upon the asteroid (16) Psyche, thought to be the largest metal asteroid by far.

Part of the challenge of proposing a mission is convincing a NASA Center to be your partner. We wanted to be included in the JPL portfolio for the upcoming Discovery proposal call, and we negotiated that process until well into 2013. The cost of supporting a proposal is significant, and Centers know they can only adequately support a certain number; strategically, choosing late when more information on each is available and proposals are more developed makes sense from the point of view of the Center but is risky for the proposal team, who has probably lost their chance to court another Center by the time the decision is made.

1.2.2 Preparation for Step 1 Proposal

From 2012 through 2013, the team participated in a series of Innovation Foundry exercises at JPL. The Foundry is JPL's team for ideating new concepts, and they bring to bear all the best practices in innovating and brainstorming, including using a room full of whiteboards, LEGO®, sticky notes, overhead projectors, and room to walk around.

A tension exists between the need to have experts in the room to make decisions about science objectives and which instruments can make measurements to answer them, and the need to wait until these are well-developed before selecting the specific instruments and inviting their leads to be Co-Investigators. Those very instrument leads are the top experts on what instruments can do and how to make the science interpretations, but once they are on the team, the decision is generally made that their instrument is going to fly. We purposefully kept our team as lean as was possible and attempted to avoid inviting any instrument providers before we knew we wanted their instrument.

The first payload concept for the mission was put together in January 2012, and the science team consisted of Elkins-Tanton, Asphaug, Weiss, Richard Binzel, Bills, and Wenkert. Our work was greatly enabled by Foundry members and others from JPL who joined the planning meetings. Later in 2012, as we continued with Foundry and did a first Team X planning session (see below), Bill Bottke, Simone Marchi, and Insoo Jun joined the team.

Finally, in spring 2013, JPL committed to the Psyche project and David Oh, electric propulsion expert and systems engineer, joined us as our Capture Lead, with the job to manage the proposal process.

In 2013, Carol Polanskey joined the team, along with Maria Zuber and Tom Prettyman. In May, we brought on Jim Bell as Deputy Principal Investigator. We led a session at the Lunar and Planetary Science Conference and held a Planetesimal workshop at the Carnegie Institution for Science that resulted in an edited volume of papers, *Planetesimals* (Elkins-Tanton and Weiss 2017), from Cambridge University Press on the state of the art of planetesimal science.

At this point, we had fairly closely defined five driving science objectives and were well into creating a Science Traceability Matrix. We knew we needed a magnetometer and imager, and we debated about the need to fly an ultraviolet/visible (UV/VIS) spectrometer; they were expected by the community at this time because the active missions of the day were all investigating rocky bodies with silicate crystalline minerals, well-suited for that kind of spectrometry. UV/VIS spectrometers are not, however, discriminating for metal. After careful debate we decided to fly a gamma-ray and neutron spectrometer instead, as it would be able to measure the compositions of the metal we expect to find. We were also forced to reconsider instrumentation because of a NASA ruling that required a certain percentage of the payload to be American, which further firmed our conclusion that this mission had to fly the simplest, highest-heritage payload possible to enable the science we had defined. At that time, we invited David Lawrence and Patrick Peplowski to join the team and provide the

GRNS. We eventually settled on a paradigm we have kept ever since: we have no descopes in payload. We have selected exactly the needed instrument suite.

It was clear that accomplishing the science objectives would require entering orbit at Psyche and remaining there for many months. Under the anticipated Discovery cost cap, we thought our only viable mission architecture candidate model was the Dawn mission, which had used its electric propulsion system to travel to and enter orbit around both (4) Vesta and (1) Ceres. However, there were key changes to be made. By 2013, components of Dawn's electric propulsion system had become functionally obsolete. Commercial electric propulsion systems used near-Earth offered the necessary performance and the benefits of strong flight heritage and a strong industrial base but were not formally qualified for interplanetary applications. NASA's follow-on developmental initiative at Glenn Research Center, the NASA Evolutionary Xenon Thruster ion thruster system, remained incomplete. A viable alternative needed to be found for Psyche.

To address this need, JPL issued a Request for Information seeking partners to provide an interplanetary spacecraft that used solar electric propulsion (SEP). A variety of solutions were proposed using both NASA and commercial electric propulsion systems, but one option that stood out above the others came from Space Systems/Loral (SSL, now Maxar Space Systems).

SSL was already an experienced provider of electric propulsion spacecraft for commercial communications satellites. It had flown over a dozen mission-unique electric propulsion spacecraft with solar power systems that generated over 20 kW—the same power level needed for Psyche. SSL had production volume—they regularly launched from four to seven product line spacecraft per year, and based on their commercial experience, they were uniquely willing to work on a firm-fixed-price contract. The potential benefit of such an arrangement for a cost-capped Discovery mission was clear. However, because SSL's commercial satellites all operated in Earth orbit, SSL lacked experience working on interplanetary spacecraft. Psyche would be an opportunity for SSL to benefit from NASA's experience working in deep space, while NASA would gain the benefits of SSL's experience building commercial SEP spacecraft.

At JPL's request, SSL proposed to build a SEP chassis at their production line in Palo Alto and deliver it to JPL for the completion of integration and testing. The SEP chassis would comprise the spacecraft's main composite structure, solar power system, propulsion systems, and thermal subsystem while the balance of the spacecraft, including the main computer, flight software, and deep space telecommunications transponder would be provided by JPL. The proposed architecture leveraged the strengths of SSL's experience building high-power electric propulsion for Earth-orbiting satellites with JPL's experience building autonomous software and fault protection for deep space missions. However, a potential disadvantage was that it would require an unprecedented level of coordination between two organizations which had never before worked together at a spacecraft level. This would be SSL's first time building a spacecraft traveling beyond lunar orbit, and because SSL's design comprised only part of the spacecraft, a further collaborative design iteration with JPL would be needed to fully understand the technical, cost, and management implications for the whole spacecraft.

1.2.3 Team X

This task was ideally suited to JPL's "Team X" concurrent engineering environment. Team X is a long-established part of JPL's Innovation Foundry that provides processes and infrastructure to bring together a diverse team of design specialists to work in a single location, sharing data in real-time to rapidly create an integrated point design of a space mission

(Sherwood and McCleese 2013). In March 2014, we invited SSL to participate in a joint design session at JPL's Team X facility in Pasadena. SSL sent a team of ~20 engineers to participate in two three-hour design sessions, held over a two-day period during which they worked side by side with JPL's engineering experts to complete the spacecraft design. For most participants on both sides, this was the first time they had ever interacted with their colleagues from the other organization. The combined team rapidly identified areas of synergy and created a point design for Psyche that incorporated heritage design elements from both SSL and JPL. The results showed that the SEP chassis architecture was technically feasible and that the cost of the combined spacecraft was appropriate for a Discovery class mission. More importantly, the design sessions showcased the ability of the JPL and SSL teams to work effectively together, which built confidence that the unique partnership could work despite the unique arrangement and lack of prior work experience. The results of this collaborative Team X design study were critical to the final decision to partner with SSL for Psyche's SEP chassis, which was made in April 2014.

1.2.4 Step 1 Proposal

A draft of NASA's Announcement of Opportunity for Discovery selection 13 was released on July 2, 2014, and we went immediately to work. JPL created an Excel spreadsheet of all the sections that were needed, and the team began to populate the sections with writers, editors, reviewers, and documentarians. We reached over 60 drafts of our Science Traceability Matrix before it was complete, and it has changed very little since Step 1. We wrote the 218-page Step 1 proposal and handed it to NASA in January 2015.

1.2.5 NASA Downselect, Phase A Proposal, and Selection

We received notification that Psyche had been selected for Phase A on September 30, 2015, followed by a written and verbal debrief of the Step 1 evaluation by NASA's Science, Technical, Management, and Cost (STMC) review board a month later. The bulk of the work on the Concept Study occurred over a nine-month period between November 2015 and July 2016. We submitted the study report to NASA in August 2016 and then held a site visit with the STMC review board three months later. The debrief report from Step 1 served as the primary roadmap used to prioritize work during Phase A. The report's findings included a list of major and minor "strengths" and "weaknesses," and although the report listed only a handful of major weaknesses, because this was a first-time evaluation for the mission concept it also listed 59 minor weaknesses, each of which had the potential to turn into a major weakness at the end of Phase A. Our key challenge was to maintain the strengths while addressing these weaknesses in the mission concept within the limited time and resources available for the concept study.

The strategy used in Phase A was the same we used in Step 1: support excellent science with an eminently feasible mission concept formulated to be as simple as possible. In addition to addressing weaknesses identified by the STMC and by the team itself, we sought to highlight the strength of the team (particularly in systems engineering) and to develop a straightforward implementation plan for the science mission and payload.

Technical work began with a series of subsystem level heritage reviews applied to all elements of the payload and spacecraft. These on-site reviews lasted two to four hours per subsystem. Face-to-face meetings were a key element of team building throughout the study period. For example, over 25 face-to-face meetings were held between JPL and SSL personnel during Phase A. These meetings built strong relationships that would endure through the balance of the project. Out of the heritage reviews came a checklist of all technical work to be done in Phase A to address known weaknesses and establish the final technical baseline for the Concept Study Report.

In addition to conducting design work, we also conducted a series of risk reduction tests to mature critical elements of the Psyche spacecraft design. These tests included an end-to-end laboratory demonstration of the modifications needed to SSL's power system to operate Hall thrusters in deep space and a testbed demonstration of the compatibility of JPL and SSL's avionics interfaces. In addition, risk reduction testing of a demonstration model pulse tube cryocooler provided confidence that a new cooler design could reliably meet mission requirements for the Gamma-Ray Spectrometer instrument, resulting in a switch from a rotary cryocooler design to the pulse tube cryocooler during Phase A.

After the baseline design was declared "frozen" in April, we turned to the completion of the management and cost plans and the writing of the Concept Study Report itself. This involved multiple drafts, strategic editing sessions to "handcraft" text, and committee reviews conducted by JPL's Team X and by an excellence review committee. The final Concept Study Report, over 1000 pages long, was submitted to NASA for evaluation in August 2016.

Three months later, the evaluation team's site visit was conducted at SSL's facility in Palo Alto. Preparations took place over many weeks and included a mock site visit conducted with a stand-in board of independent reviewers three weeks before the real site visit. The mock site visit was critical to our preparations, as a particular challenge for the Psyche project management team was that none of us had ever previously participated in a site visit. In addition, because SSL had never previously worked for NASA as a prime contractor, most members of the STMC were not familiar with the company or its capabilities. Building NASA's confidence in SSL capabilities was an important goal for the site visit.

The work began in earnest a week before the presentation date, when we received written questions from the STMC and began the process of creating written and oral responses. We used the mock site visit to develop a seven-day workflow that engaged experts from across the team working around the clock to write responses. Each response was generated by a subject matter expert and then reviewed by two independent technical content reviewers and a technical editor before it was cleared for release. The workflow proved to be robust and produced over 150 pages of high-quality written technical material in the seven days prior to the final site visit. The oral presentations were similarly developed and dry run before a group of independent reviewers two days before presentation day and then were edited and cleared for final presentation to the STMC.

In the end, the site visit proved to be a valuable opportunity to demonstrate the strength of the team and to introduce the STMC and the members of the Discovery program office to SSL's manufacturing capabilities. A tour of SSL's spacecraft manufacturing facilities proved to be a highlight of the visit, as the visitors were able to view some 15 spacecraft under various stages of construction on SSL's factory floor. In addition, in feedback received after the site visit, the STMC stated that they could perceive the strength of the team, see the strong cooperation between team members, and appreciate the open communication philosophy embraced by the management team.

The final step of Phase A was for a subset of the team to travel to Washington DC and present for one hour to Thomas Zurbuchen, the NASA Associate Administrator. As part of our strategy for showing how ideal SSL was for prime partnership, the presentation was given primarily by Steve Scott from SSL and Lindy Elkins-Tanton as Principal Investigator.

The announcement that Psyche had won was made in January 2017. Many of the design decisions and outcomes of the build of the spacecraft have been described in other papers

(Oh et al. 2019, 2022, 2025a); the major change that our plan underwent was a launch slip from August 2022 to October 2023, caused by a variety of issues including building the spacecraft entirely during the COVID-19 pandemic of 2020-2021. During 2022, Ben Weiss from the Massachusetts Institute of Technology (MIT) replaced Jim Bell as Deputy PI, and Prof. Bell has stayed on as a highly valued Co-Investigator, lead of the Imagers, and lead of the science data center at Arizona State University (ASU).

1.3 Psyche's Place Amongst Small Body Missions

The minor bodies in our Solar System are a fascinating set of worlds that only become more interesting and varied as we visit them and expand our inventory of explored places. We continually work to categorize them—by size, by color, by location, by comparing them to the meteorites in our collections—but every world we visit has surprised us and often defied our expectations. Psyche will undoubtedly be no different.

Although humankind has trained their telescopes on asteroids for hundreds of years, our first close encounter with them came courtesy of Galileo and its flybys of (951) Gaspra and (243) Ida in 1991 and 1993, respectively (Veverka et al. 1994; Belton et al. 1996). Those images showed us little, rocky worlds, covered in regolith, cratered and weathered. Ida surprised us with a moon, Dactyl, though we have now come to understand that satellites and binaries are not uncommon among asteroid populations, and searching for them is standard procedure for asteroid missions.

The Near Earth Asteroid Rendezvous (NEAR) mission to (433) Eros in 2000 was our first chance to orbit and thoroughly explore an asteroid. We found Eros to be similar in many ways to Gaspra and Ida, all three are S-type asteroids of roughly similar size. But there were differences too. Space weathering effects on Gaspra and Ida appear similar to lunar weathering, with changes to both albedo and color evident in impacts that expose fresh material. Weathering on Eros, in contrast, results in large changes of albedo but almost no color variation, the reason for which is still a mystery (Pieters and Noble 2016).

Five years after NEAR visited Eros, the Japanese Hayabusa mission visited tiny (25143) Itokowa, an asteroid similar in composition to Eros, but vastly different in size, providing an opportunity to explore some of the consequences that size and gravity have on asteroid evolution. Hayabusa also returned a small amount of sample from Itokowa. These samples, along with the detailed chemistry obtained by NEAR, confirmed the link between S-type asteroids and ordinary chondrites (Veverka et al. 2000; Binzel et al. 2001; Nakamura et al. 2011). Despite this and other progress, linking asteroids to their parent bodies remains a challenge. The Psyche mission certainly faces that challenge; we have a large array of metal-rich meteorites in our collections, but none seems to perfectly match our understanding of Psyche so far.

The Dawn mission showed us the other end of the size scale, orbiting the two largest asteroids in the asteroid belt, protoplanet Vesta and dwarf planet Ceres. These two little worlds took very different evolutionary paths. Water-rich Ceres cooled fast and stratified into a rocky, clay-rich mantle and an ice- and hydrate-rich outer shell (Russell et al. 2016). In contrast Vesta, with little or no water, cooled slowly, differentiating into a core, mantle, and crust (Russel et al. 2012). Psyche will be the third largest asteroid we have visited and should further help us understand the different pathways planetesimals and planets follow as they form and evolve.

Our catalog of compositional differences is also growing, with Origins, Spectral Interpretation, Resource Identification, and Security – Regolith Explorer (OSIRIS-REx) and the Japanese Hayabusa2 missions, which have now visited - and successfully sampled - a couple

Table 1 List of asteroids that have been (white) or will soon be (grey) visited by spacecraft

Asteroid	Mission	Mean diameter	Class	Asteroid	Mission	Mean diameter	Class
Ceres	Dawn	946 km	C	Ida	Galileo	16 km	S
Vesta	Dawn	525 km	V	Steins	Rosetta	4.6 km	E
Psyche	Psyche	226 km	M	Donaldjohanson	Lucy	4 km	C
Patroclus	Lucy	113 km	P	Annefrank	Stardust	4 km	S
Menoetius	Lucy	104 km	P	Toutatis	Chang'e 2	2.5 km	S
Lutetia	Rosetta	100 km	M	Braille	Deep Space 1	1.6 km	Q
Eurybates	Lucy	64 km	C	Ryugu	Hayabusa2	0.9 km	C
Matilde	NEAR	58 km	C	Didymos	DART	0.78 km	S
Orus	Lucy	51 km	D	Dinkinesh	Lucy	0.7 km	S
Leucus	Lucy	40 km	D	Bennu	OSIRIS-REx	0.5 km	B
Polymele	Lucy	21 km	P	Apophis	OSIRIS-APEX	0.34 km	S
Gaspra	Galileo	17 km	S	Itokawa	Hayabusa	0.3 km	S
Eros	NEAR	16 km	S	Dimorphos	DART	0.16 km	S

of carbon-rich (C-type) targets, Bennu and Ryugu, respectively. Those samples are already providing a wealth of information and will continue to reveal their secrets for generations to come.

The recent Double Asteroid Redirection Test (DART) mission further added to our collection of imagery of S-type surfaces, including some dramatic close-ups of the surface of tiny Dimorphos (160 m diameter) as the impactor approached. This impactor, alongside results from Hayabusa2 and OSIRIS-REx, has shown that small rubble piles apparently have almost zero cohesive strength (Ferrari and Tanga 2022; Walsh et al. 2022).

Finally, as we look forward, our "sister" mission, Lucy (because it was also selected for flight in January 2017), will help us build up the diversity of our catalog listed in Table 1, conducting flybys of several Jupiter Trojans, a significantly different population from the main belt, and exploring an array of P-, D-, and C-types. Together, Lucy and Psyche, the first mission to orbit an M-class asteroid, will allow us over the next decades to make large strides in understanding the diversity of asteroids across the solar system.

1.4 History and Characteristics of (16) Psyche

Psyche was discovered at the Astronomical Observatory of Naples by Italian astronomer Annibale de Gasparis (1819-1892) on 17 March 1852 (de Gasparis 1852). Gasparis named the asteroid Psyche after the Greek mythological figure, the mortal wife of Cupid, and the personification of the Greek word for "the soul." Psyche is the 16th asteroid spotted by astronomers since the discovery of Ceres some 50 years earlier and is one of eight found by Gasparis (Badolati 2007). It was one of the last minor planets to be assigned a planet-like iconic symbol, which is meant to resemble a butterfly wing under a star-like asterisk ⯘ (Paper Elemental Blog 2021).

Besides its orbital elements, little was known about the asteroid for more than a century. Starting in the 1960s and 1970s, a combination of advancements in telescopes and detector technologies, and renewed interest by space scientists in asteroids and their potential connections to meteorites, fueled new discoveries about Psyche. For example, optical and infrared radiometric, polarimetric, and photometric observations of Psyche revealed quickly that it is a relatively rapidly rotating (spin period just over 4 hours), relatively large object (with an average diameter > 200 km) corresponding to about 1% of the mass of the Main Asteroid Belt (Gehrels 1970; Morrison 1974; Chapman et al. 1975). Early broadband multispectral observations revealed the asteroid to have featureless reddish visible-wavelength color properties that immediately suggested similarities with the colors of iron-nickel meteorites, earning it a prominent status among the M-class asteroids (e.g., Chapman and Salisbury 1973; McCord and Gaffey 1974; Zellner and Gradie 1976; Chapman 1976; Dollfus and Mandeville 1977; Tholen 1984; Bell et al. 1989).

Radar observations of Psyche also date back to the 1980s (e.g., Ostro et al. 1985; Magri et al. 2007; Shepard et al. 2008), with Psyche yielding one of the highest radar albedos among all measured main-belt and near-Earth asteroids with a value of 0.37 +/− 0.09 (Shepard et al. 2017). The high radar albedo is arguably the strongest evidence suggesting that Psyche's surface and regolith is dominated by iron-nickel metal. Radar and ground-based adaptive optics and stellar occultation imaging observations have also helped to significantly refine the asteroid's size and shape, which is now known to be best approximated by Shepard et al. (2021) with a triaxial ellipsoid of approximate dimensions 280 × 240 × 170 km (effective spherical diameter ≈ 222 km). However, significant concavities, likely related to the presence of large impact craters or basins, are also evident in the most current shape models (Shepard et al. 2021; Viikinkoski et al. 2018). Size estimates combined with mass estimates derived from orbital perturbations of the asteroid by other solar system bodies have enabled estimates of the density of Psyche, which is likely between 3400 and 4100 kg/m^3 (see review by Elkins-Tanton et al. 2020), with the most recent estimate at 4172 ± 145 kg/m^3 (Farnocchia et al. 2024), values that could be consistent with a mix of metals, silicates, and porosity rather than just an entirely metallic composition.

Additional evidence for a more complex composition, structure, and origin for Psyche comes from telescopic evidence for the presence of what are likely relatively small abundances of iron-bearing silicates (olivine and pyroxene) as well as hydrated mineral phases on the asteroid's surface, with progress in spatial and spectral resolution revealing surface heterogeneities (e.g., Binzel et al. 1995; Hardersen et al. 2005; Ockert-Bell et al. 2010; Sanchez et al. 2017; Takir et al. 2017).

There are significant remaining uncertainties in the physical properties, composition, and potential origin of Psyche, even though the most powerful modern instruments in ground- and space-based optical, infrared, and radar astronomy have been brought to bear on its study. The Psyche mission can now address these uncertainties with an up-close spacecraft investigation of this important but enigmatic small world.

1.5 Psyche Mission Goals and Science Objectives

The science requirements levied by NASA on the Psyche mission were refined from those developed from the proposal. To summarize at a high level, the mission objectives are to probe the history of Psyche and address the fundamental question whether Psyche is the remnant of a planetary core or has some other origin. Section 4 of this paper describes how the science data acquisition plan provides the means to satisfy these objectives within mission constraints.

The Psyche mission addresses two of the NASA Science Mission Directorate's Planetary Science Division goals and ties into the two most recent National Academies of Science and Engineering Planetary Science Decadal Surveys (National Academies of Sciences, Engineering, and Medicine 2011, 2022) with an investigation that has three broad goals:

Table 2 Relationship between the four science orbits and their science and operational goals. The goals of the Magnetometry Investigation span all orbits and orbit transfers but are called out only in Orbit C here

Orbit	Science and operational goals	Science measurements	Instrument
A	Characterize Psyche's gravity field, shape, and spin axis to enable navigation to lower science orbits Determine the relative ages of regions of Psyche's surface by counting craters with diameters larger than 1 km	Doppler and ranging Global multispectral imaging Off-nadir imaging Magnetic field monitoring	HGA and LGA Imager Magnetometer
B	Characterize Psyche's topography and morphology Identify silicate regions on Psyche's surface Search for oldhamite on Psyche's surface Develop a geologic map of Psyche's surface	Global nadir and off-nadir imaging Global multispectral imaging Magnetic field monitoring	Imager Magnetometer
C	Determine Psyche's gravity field Characterize Psyche's surface density variations Refine Psyche's shape model Determine whether Psyche has a remanent magnetic field (applies to all science orbits)	Doppler and ranging Global nadir and off-nadir imaging Magnetic field monitoring	HGA and LGA Imager Magnetometer
D	Discriminate between achondritic and chondritic silicates on Psyche's surface by measuring Fe, Si, K, Ca, and Al Determine the global average weight percent of nickel over Psyche's surface Determine the metal-silicate ratio of Psyche's surface	Gamma-ray and neutron measurements over 70% of Psyche's surface Magnetic field monitoring	GRNS Magnetometer

1. Understand a previously unexplored building block of planet formation: iron cores.
2. Look inside the terrestrial planets, including Earth, by directly examining the interior of a differentiated body, which otherwise could not be seen.
3. Explore a new type of world. For the first time, examine a world not made of rock or ice, but of metal.

The goals were further refined into the five science objectives for the mission:

1. Determine whether Psyche is a core, or if it is primordial unmelted material
2. Determine the relative ages of regions of Psyche's surface
3. Determine whether small metal bodies incorporate the same light elements into the metal phase as are expected in the Earth's high pressure core
4. Determine whether Psyche was formed under conditions more oxidizing or more reducing than Earth's core
5. Characterize Psyche's morphology.

The science objectives became the basis of the high-level science requirements of the mission, which were then transformed into a set of science measurement requirements. These measurement requirements drove the design of the instruments described in Sect. 2 and the orbital operations architecture of the mission found in Sect. 4. Based on the measurements and the instrument suite, the mission science was divided into four orbital phases. Table 2 provides a summary of how the science requirements map to the science orbit altitudes and measurements. Each science objective is shepherded by a specific working group within the Psyche team, and each is explored in depth in the accompanying papers by Elkins-Tanton

Table 3 List of science instruments and basic attributes

Science instrument	Magnetometer	Multispectral Imagers	Gamma-Ray and Neutron Spectrometer
Provider	Technical University of Denmark (DTU)	Malin Space Sciences Systems, Inc. for Arizona State University (ASU)	Johns Hopkins/Applied Physics Laboratory (APL)
Heritage	European Space Agency (ESA) Swarm	Mars Science Laboratory Mastcam; Mars 2020 Mastcam-Z; Mars Reconnaissance Orbiter MARCI	MESSENGER Lunar Prospector
Components	2 sensors 2 electronics boxes	2 camera heads 1 electronics box	2 instruments 2 electronics boxes
Mounting	On a dedicated 2-m boom	−X direction	On a dedicated 2-m boom
Mass	3.635 kg	5.12 kg	21.42 kg
Power	4.4 W	5.6 W standby and 8.7 W imaging, per camera	69.1 W (peak)
Specifications	Sampling frequency: 50 Hz Measurement uncertainty requirement at system level: < 1.5 nT/axis Range: $\pm$ 80,000 nT Noise: 25 pT_{RMS} Resolution: 11 pT Thermal stability: <0.25 nT in operating temperature range [−125 °C, +60 °C]	1 broadband polychromatic filter and 7 narrowband color filters Field of view (FOV): $4.6° \times 3.4°$ Instantaneous field of view (IFOV): 50 μrad pixel scale Focal ratio: f/2.9 Focal length: 148 mm	Gamma-ray energy range: 60 keV to 9 MeV Energy resolution: 1.92 keV @ 1333 keV (ground); 2.09 keV @ 1333 keV (in space)

et al. (2022), Marchi et al. (2022), Prettyman et al. (2025), McCoy et al. (2022), and Jaumann et al. (2022).

2 Psyche Instrumentation

As described in Sect. 1.2, the minimum set of three instruments, the Magnetometer, the Multispectral Imager, and the GRNS, plus X-band gravity science, were selected to meet mission science requirements. A duplicate Multispectral Imager was added for redundancy as that instrument is also needed for optical navigation. Details on each of the science instruments are provided in subsequent articles of this collection (Weiss et al. 2023; Bell et al. 2025; Lawrence et al. 2025; and Zuber et al. 2022), but a brief description is provided below as context for the flight operations plans. Table 3 summarizes their characteristics and heritage while Fig. 3 shows the layout of the instruments on the Psyche spacecraft.

While our goal was to fly the simplest, highest-heritage payload possible to complete the science portion of the mission, the NASA Discovery 2014 Announcement of Opportunity offered an incentive for adding a technology demonstration opportunity to the payload. DSOC was chosen from the options provided because the Psyche interplanetary trajectory enabled demonstration of laser communications where NASA is intending to use it: in the vicinity of Earth, at the distance of Mars, and beyond Mars orbit. DSOC is not required or intended for use during the Psyche baseline orbital mission; however, the spacecraft flight software supports transmission of both engineering and science downlink telemetry

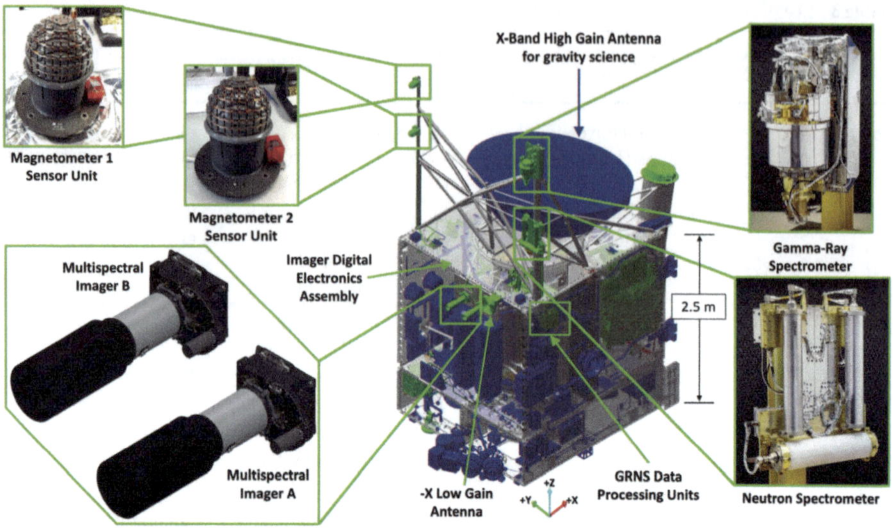

Fig. 3 Placement of science instruments on the spacecraft. The two Magnetometer sensor units are located on a 2-m boom with their electronics units (not shown) in a gradiometer configuration. The Gamma-Ray Spectrometer and Neutron Spectrometer are located on a separate 2-m boom with the Gamma-Ray Spectrometer at the 2 m point and the Neutron Spectrometer at the 1 m point of the boom. Each has its own electronics box mounted inside the spacecraft. The two Multispectral Imagers are mounted on the central cylinder of the spacecraft with views out of the $-X$ direction of the spacecraft. Their electronics are packaged into a single box mounted inside the spacecraft. Gravity science is performed with the X-band HGA as well as three LGAs mounted on the $+X$, $-X$, and $-Z$ decks of the spacecraft (not all are visible here). Blue components on the spacecraft drawing are provided by Maxar. Green components are from JPL or other partner institutions

via DSOC. Transmission of spacecraft engineering telemetry by DSOC was successfully demonstrated during cruise.

2.1 The Magnetometry Investigation

The Magnetometry Investigation (Weiss et al. 2023) is motivated by the mission science objective of determining whether asteroid Psyche is a differentiated body containing a metallic core or is a primordial unmelted aggregate (Elkins-Tanton et al. 2020). It addresses this objective by seeking to constrain the remanent magnetization of Psyche. Sufficiently strong magnetization would provide evidence that Psyche cooled and/or was aqueously altered in the presence of an ancient magnetic field. Remanent magnetization will be constrained by measuring the ambient magnetic field and in particular searching for an intrinsic field from remanent magnetization.

The Magnetometer consists of two identical sensor units positioned at 1.45 m and 2.15 m away from the spacecraft along a boom. Each sensor unit is connected by a harness to its own electronics located on the spacecraft bus. The two sensor units each measure the three components of the magnetic field. In combination, these enable a measurement of the gradient of the field, which will be used to characterize and suppress the contributions from the spacecraft magnetic field (Ream et al. 2022). Each sensor unit acquires measurements over 0-50 Hz, has a dynamic range of $\pm 80{,}000$ nT, and a required 3-σ instrument uncertainty of 0.22 nT per axis. The Magnetometry Investigation is led by the Massachusetts Institute of Technology, and the Magnetometer is provided by the Technical University of Denmark.

Table 4 Psyche Imaging science measurement requirements driving orbit design

Imager physical parameter measured	Pixel scale	Surface coverage	Target orbit	Target pixel scale
Mapping craters > 1 km diameter (1 Imager filter)	200 m	50%	A	35 m/pix
Metal-silicate ratio map (4 Imager filters ≥ 700 nm)	500 m	80%	B	15 m/pix
Mapping oldhamite on the surface (3 Imager filters)	200 m	80%	B	15 m/pix
Mapping topography to support gravity determination (1 Imager filter, ≤ 70 m height scale)	70 m	80%	B	15 m/pix
Mapping for topography (1 filter, ≤ 50 m height scale)	20 m	50%	B	15 m/pix
Geologic mapping (1 Imager filter)	20 m	80%	B	15 m/pix

The Magnetometer derives high heritage from the ESA Swarm Vector Field Magnetometer (Merayo et al. 2008).

The Magnetometer operates using the fluxgate principle in which a high-permeability toroidal core is cycled in and out of magnetic saturation by the application of a time-variable current applied to windings around the core (Primdahl 1979). An ambient magnetic field can be detected because the core will spend a larger fraction of each magnetization cycle saturated in one polarity compared to the other.

2.2 The Multispectral Imager Investigation

The Psyche Multispectral Imager ("Imager"; Bell et al. 2025) includes two redundant camera heads, each paired with their own electronics; the Imager is used to characterize the geology and topography of Psyche's surface and to provide some constraints on the asteroid's surface mineralogy. The cameras protrude from the −X panel of the spacecraft with a toe-out angle of 3.7° between them.

Major goals (Table 4) of the Imager Investigation include:

(1) Characterizing the geology and topography of a metal-rich world to reveal potentially unique impact, erosional, weathering, and possibly even tectonic and/or volcanic processes that could be characteristic of this class of small body
(2) Characterizing the relative ages of Psyche's surface regions by mapping statistically significant numbers of impact craters across the asteroid's surface at diagnostic spatial scales
(3) Deriving topographic maps of the asteroid's surface from digital terrain models to assist with crater counting and shape modeling, and to provide insights into internal structure and homogeneity via comparisons with lower spatial resolution gravity field, magnetic, and neutron/gamma-ray compositional measurements
(4) Assessing the relative spatial abundances of iron-bearing silicates and metals on the surface of the asteroid, to help constrain the evolutionary history of the surface and to provide potential connections to specific meteorite classes
(5) Searching for the presence of specific diagnostic sulfide minerals like oldhamite [(Ca, Mg, Fe)S] and/or troilite (FeS) that could provide information on oxidizing or reducing conditions predicted by differentiated versus primitive hypotheses for the asteroid's origin.

These are first-order exploration goals and questions, and the Imager will directly address them by providing both polychromatic (clear) filter imaging and narrowband multispectral filter imaging (see Bell et al. 2025 and Winhold et al. 2025) at smaller spatial and vertical (topographic) resolutions as the orbital altitudes decrease during the mission. Geologic and topographic imaging goals will primarily be met using the polychromatic filter (540 ± 125 nm), which can acquire high spatial resolution (Table 4), high signal-to-noise ratio (SNR $>$ 100) images at short enough exposure times to avoid target motion smear. Imaging goals related to mapping of iron-bearing silicates and the silicate/metal ratio will be achieved using a combination of short-wave near-infrared filters that provide continuum (722 nm) and in-band (840, 942, and 1014 nm) imaging at SNR $>$ 50 designed to help detect and constrain the abundances of iron-bearing phases like pyroxene and olivine via mapping of the so-called "1-μm band." Imaging goals related to detecting and potentially discriminating between sulfide phases will be achieved using imaging also at SNR $>$ 50 with short-wave visible filters (442, 494, and 548 nm) positioned to provide maximum sensitivity to known sulfide absorptions in that wavelength region.

The Imager camera heads are powered and controlled by a separate Digital Electronics Assembly (DEA) that can compress the images via lossless or lossy compression algorithms and that can store more than 4000 images per camera in its own internal memory. The DEA is powered by and interfaces with the spacecraft's main computer to command the camera heads and to control the transfer of locally-stored images to the spacecraft for eventual downlink.

2.3 The Gamma-Ray and Neutron Spectrometer Investigation

The Psyche GRNS instrument (Lawrence et al. 2025) includes two sensor subsystems—a Gamma-Ray Spectrometer (GRS) and a Neutron Spectrometer (NS)—and measures elemental abundances of Ni, S, K, Si, Fe, Ca, and Al, as well as Psyche's metal-to-silicate fraction. Since the Psyche mission science objectives require measurement of Ni abundances, this measurement requires use of a high-purity Ge (HPGe) sensor due to its excellent energy resolution and detection sensitivity. In particular, the primary Ni gamma-ray line at 1.454 MeV is very close in energy to a ubiquitous K gamma-ray line at 1.46 MeV and therefore needs the energy resolution provided by HPGe technology. The excellent performance of HPGe technology also enables the measurement of other elements such as Al, Ca, and S.

The Psyche GRS is an evolution of the MErcury Surface, Space ENvironment, GEochemistry, and Ranging (MESSENGER) GRS (Goldsten et al. 2007) that incorporates various lessons learned from MESSENGER experience (Burks et al. 2020). The GRS sensor is a 5-cm-long by 5-cm-diameter cylindrical HPGe crystal. To properly operate, the Ge crystal needs to be cooled to cryogenic temperatures. To facilitate this cooling by thermally isolating the sensor, the Ge sensor is enclosed in a cryostat designed and built by Lawrence Livermore National Laboratory. The sensor is cooled to its operating temperature of 85 K using a long-life, pulsed-tube cryocooler built by Lockheed Martin Corporation. As with prior GRS systems, the Ge sensor is surrounded by a boron-loaded anticoincidence shield. The anticoincidence shield serves three functions. First, the anticoincidence shield provides active background rejection for non-gamma-ray signals (e.g., charged particles) that occur in the Ge sensor. Second, the boron loading in the scintillator enables the measurement of fast neutrons, which is a key measurement for characterizing Psyche's metal-to-silicate fraction. Finally, energy deposition measurements in the anticoincidence shield provide a direct monitor of the time varying galactic cosmic-ray flux, which is a normalization factor needed for deriving accurate elemental abundances.

The NS consists of three ^{3}He gas proportional sensors that use different material coverings to discriminate three separate neutron energy ranges. Thermal (neutron energy $E_n < 0.4$ eV) and low-energy ($0.4 < E_n <\sim 10$ keV) epithermal neutrons are measured using bare and Cd-covered sensors. High-energy ($E_n < 100$ keV) epithermal neutrons are measured with a polyethylene-covered ^{3}He sensor. This energy band is used to more effectively characterize the expected metal-rich content of Psyche (Yokley et al. 2019; Lawrence et al. 2020, 2021, 2025).

To reduce background from spacecraft-originating gamma rays and neutrons, both sets of sensors are mounted on a boom to physically separate the sensors from other spacecraft materials.

Finally, each sensor subsystem is operated by separate data processing units, which provide capabilities for low and high voltages, command processing, autonomous fault protection, signal processing, and data packet generation. The data processing units are located in the spacecraft body and are connected to the sensor subsystems via harnesses approximately 1–2 m in length.

2.4 The Gravity Science Investigation

The Gravity Science Investigation aims to map the gravitational variation of asteroid Psyche to provide constraints on its interior structure (Zuber et al. 2022). The gravity field will be measured via the spacecraft's X-band telecommunication system, which is capable of two-way radio link (7.2 GHz uplink and 8.4 GHz downlink) between the Deep Space Network (DSN) on Earth and the spacecraft, allowing high-precision range and Doppler measurements. The spacecraft trajectory is sensitive to gravitational perturbations, thus, accurately tracking the velocity of the spacecraft allows determination of Psyche's gravity field. The performance of the Doppler data is expected to be better than 0.1 mm/s at 60-s integration time and the performance of the range data is expected to be 1-2 m (Konopliv et al. 2014; Park et al. 2016). The Psyche spacecraft is also equipped with a camera that can be used to determine the shape and landmarks via stereophotoclinometry (Park et al. 2019). The Doppler and imaging data are primary for recovering the gravity field and the spacecraft orbit relative to the asteroid while the range data are primary for determining the heliocentric orbit of Psyche (Park et al. 2016, 2020).

The Psyche mapping plan has four orbital phases (see Sects. 3 and 4). For gravity science, Orbit C is the prime phase with the orbit altitude of $\sim$190 km. During the Orbit C phase, radio tracking data will be acquired for the equivalent of at least 35 days, which is expected to be sufficient for a global sampling of Psyche's gravity field (Zuber et al. 2022). Gravity data will also be acquired in the lower-altitude Orbit D phase, during which time measurements of surface composition are prioritized.

The radio and imaging observables will be processed using JPL's MONTE (Mission Analysis, Operations, and Navigation Toolkit Environment) orbit determination software, which employs a least-squares approach to estimate gravity field (Evans et al. 2018). MONTE has been utilized and demonstrated in a similar capacity in numerous other planetary gravity investigations and is capable of processing all of Psyche's gravity observations (Park et al. 2016, 2020).

A detailed gravity simulation based on realistic modeling, measurement scheduling, and expected data performance shows that a degree and order 10 gravity field can be recovered (Zuber et al. 2022), which is equivalent to spatial block size, or half-wavelength, average resolution of $\sim$36 km. Combining the gravity data with the shape, magnetometry, and compositional data will allow probing the interior structure of Psyche. Specifically, techniques

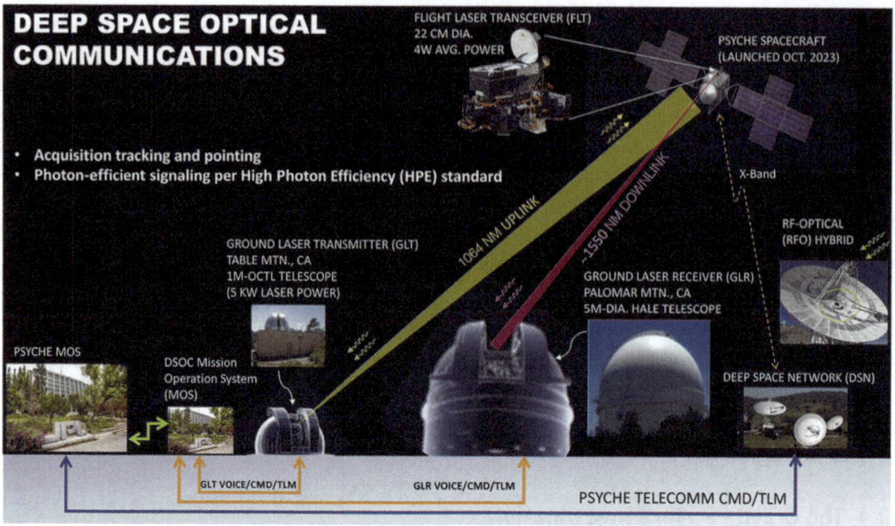

Fig. 4 DSOC technology demonstration architecture showing the FLT on board the Psyche spacecraft, the ground laser transmitter and ground laser receiver on the ground with the DSOC mission operations system for coordinating operations. Coordination between the flight and ground assets in addition to the Psyche project was key to DSOC project success

(Goossens and Smith 2023) developed to analyze gravity data acquired in the NASA GRAIL mission (Zuber et al. 2013) will allow recovery of lateral density variations with depth in spatial areas (cf. Goossens et al. 2025). The mass estimate from gravity data and the shape estimate from imaging data will provide the bulk density of the asteroid. The degree-2 gravity field will provide the gravitational flattening and spin axis precession, related to the radial distribution of mass. Correlation of the gravity field with the surface data (e.g., topography, magnetic field, and composition map) will provide mass, and potentially, structural variations (Park et al. 2020).

2.5 Technology Demonstration: Deep Space Optical Communications

The DSOC project provides the first technology demonstration of free-space optical communications from planetary distances (Biswas et al. 2024). Validating the free-space optical communications technology will enable an order of magnitude increase in data rates from future NASA science and human exploration missions (Deutsch 2014).

Figure 4 shows the operational architecture for the DSOC technology demonstration. The three key elements include: the flight laser transceiver (FLT) accommodated on the Psyche spacecraft; the ground laser transmitter, integrated to the Optical Communications Telescope Laboratory located at JPL's Table Mountain Facility near Wrightwood, CA; and the ground laser receiver integrated to the 5-m diameter Hale telescope at the Palomar Observatory, Palomar Mountain, CA. An optical link is initiated when the ground laser transmitter points a 1064 nm beacon laser at the FLT while the spacecraft "coarsely" points to Earth. The FLT searches out the laser beacon signal by scanning over the spacecraft-pointing uncertainty region. Upon acquisition the FLT actively tracks the beacon signal establishing a spatial reference used to point back the 1550 nm downlink laser, to the ground laser receiver. The DSOC mission operations system shown in Fig. 4 coordinates activities between the DSOC space and ground assets and Psyche mission planning.

The Psyche spacecraft trajectory is shown in Fig. 1 with blue dots to indicate the DSOC contacts during the first two years of cruise. As the Palomar observatory cannot operate in the daytime, DSOC operation must be suspended when the Psyche spacecraft's line of sight to ground laser receiver occurs during that period. Optical link operations occurred in the pre-conjunction phase (November 2023-July 2024) and resumed in the post-conjunction phase in early December 2024.

Pre-scripted DSOC operations sequences are uploaded to the spacecraft in advance. Sequence execution assisted by spacecraft avionics and FLT electronics interaction occurs on a pre-defined schedule. Ground activities at the ground laser transmitter, ground laser receiver, and the mission operations system are coordinated to the schedule unless sequence execution is aborted in advance due to predicted weather outages. On the few occasions when weather outages occur without advance prediction, the FLT continuously searches for the beacon signal until the end of the sequence.

Downlink data-rates to the ground laser receiver vary from 267 Mbps to 8 Mbps, from 0.175 AU to 2.68 AU using 2 W of average 1550 nm laser power. Low-rate uplink communications at 1.8 kbps was achieved out to 3.3 AU with the upper bound of ground uplink 1064 nm laser power being 3 kW. In the spirit of a true technology demonstration, comprehensive system telemetry is gathered in space and on the ground to learn how to operate future planetary optical communications services from deep space.

A significant departure of DSOC from previous free-space optical communications demonstrations is the substantial increase in distance. For example, NASA's Lunar Laser Communication Demonstration Project established 622 Mbps downlink in 2013 (Boroson et al. 2014), from lunar range (0.00267 AU). Defining link difficulty as the product of data-rate in Mbps $\times$ squared-distance, AU^2, indicates more than a 1000-fold increase, or 60 dB increase, in difficulty for achieving 1 Mbps data rate. The DSOC project implemented several new technologies that enabled photon-efficient signaling combined with microradian accuracy laser beam pointing so that link difficulties up to 113 have been achieved to date.

Pulse position modulation and serially concatenated pulse position modulation error correction codes (Moision and Hamkins 2003; CCSDS Blue Book 2019) are used by DSOC for photon-efficient communications. A high peak to average power space laser transmitter (Dailey et al. 2019) coupled to 22 cm diameter silicon carbide optics (Driscoll et al. 2020) is used to transmit the downlink data-modulated laser beam. Flight electronics with the JPL legacy universal space transponder (Pugh et al. 2017) serve as a modem for generating the laser modulation signal and host the flight software and firmware for operating the FLT.

On the ground, a superconducting nanowire single photon detector array (Wollman et al. 2024) at the focus of the Hale telescope is used to detect the laser pulses. The electrical output from the array is time-stamped with a time-to-digital converter front-end signal processing assembly (Rogalin et al. 2021; Srinivasan et al. 2017). Subsequent signal processing of time stamps achieves temporal synchronization, parameter estimation, de-interleaving, and decoding to extract the information codewords.

The multi-beam uplink laser assembly has 8-10 lasers transmitted through sub-apertures of the 1 m Optical Communications Telescope Laboratory telescope. It is blind pointed to the spacecraft using spacecraft ephemeris provided by the Psyche team. Multi-beaming partially mitigates atmospheric turbulence-induced fades through incoherent averaging of the atmospheric turbulence-induced power fluctuations. The mean irradiance delivered at the FLT aperture is typically about 10 pW/m^2. This irradiance suffices for acquisition tracking and uplink communications.

The optical transceiver assembly (OTA) is mounted on struts equipped with magnetic actuators and sensors. When powered, the struts "float" or isolate the OTA from base disturbance. This allows 6-degrees-of-freedom control to search and acquire the beacon signal on

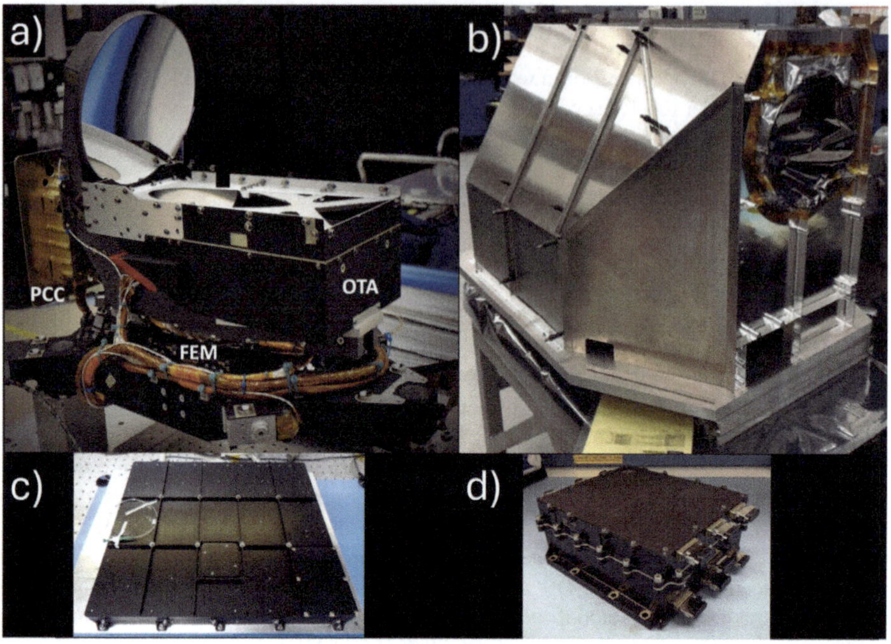

Fig. 5 (a) Optical Transceiver Assembly (OTA) with photon-counting camera and floating electronics module integrated; (b) DSOC accommodation kit for accommodating OTA; (c) Laser transmitter assembly mounted separately on spacecraft panel and fiber coupled to OTA; (d) Stationary electronics module interfaces to spacecraft electronics and houses flight software and firmware

the single-photon-sensitive photon-counting camera (Buck et al. 2020). Upon acquisition of signal, a low bandwidth control loop maintains spatial and temporal "lock" onto the beacon laser spot. The pointing reference established by the "lock" allows the onboard-computed point-ahead angle of the downlink beam to be implemented. An electronically controlled piezo-electric point-ahead mirror in the OTA transmit path implements the point-ahead angle. Figure 5 shows a photograph of the DSOC FLT.

The FLT operating from space is pointed to Earth with relatively small (tens of degrees) angular separation from the sun. The resulting stressful thermal environment is partially mitigated by a DSOC accommodation kit that is comprised of an interface plate that mates to the spacecraft and an enclosure for thermal management of the OTA. One-time deployable launch locks are integrated to the interface plate of the DSOC accommodation kit. A latching mechanism is used to stow the OTA when it is unpowered.

3 Planning, Constraints, and Considerations for Science Orbital Operations

After selecting a planetary target and a suite of instruments, the next step was to design an architecture for the science observations. The observation strategy for the Psyche mission was inherited from the Dawn mission to Vesta and Ceres. Dawn had demonstrated the use of SEP to efficiently observe a massive body with unknown size, gravity, and other physical characteristics by systematically mapping the body's surface without regard to individual surface features or characteristics (Polanskey et al. 2011). Having successfully developed and executed a science observing campaign for Vesta and Ceres (Polanskey et al. 2012, 2014, 2016), the same observation strategy was applied to Psyche, despite the project having a different spacecraft provider, SEP technology, and instrument suite (Polanskey et al. 2018).

Psyche is among the ten most massive objects in the main asteroid belt, so uncertainty as to Psyche's gravity plays a key role in planning. Without data from previous missions or even flybys of Psyche, the project must observe size, shape, and gravity during approach and feed that information into the design of the spacecraft trajectory without disrupting the predefined science observation plans. This is accomplished by maintaining the orbit resonance (the number of Psyche rotations over the number of spacecraft orbits) with emphasis on preserving the orbit period when redesigning the spacecraft orbit for a new asteroid shape model. In addition, hours or days of margin are added to the start time of each science orbit campaign to accommodate uncertainties in the time needed to thrust from one orbit altitude to the next (see Sect. 5.3). Therefore, the observing plan must also be responsive to changes in orbit timing, particularly since the files of sequenced commands that execute the plan on the spacecraft are designed in advance, well before the spacecraft arrives at Psyche. Anchoring each orbit's data acquisition to the geometric epoch when the spacecraft crosses from the dark (night) side to the lit (day) side of the body provides a straightforward mechanism to update observing plans to changing conditions. The strategy outlined in this section is proactive, rather than reactive, to the challenges posed by a low thrust mission orbiting a massive asteroid for the first time, one that may have significant mass density contrasts in its interior.

The observing community on Earth is using the latest techniques to improve our knowledge of the asteroid before the spacecraft's arrival. While the measured dimensions from each study vary, all models agree that Psyche is irregularly shaped. The irregular shape can impact observation strategies and orbit stability. The Psyche flight team has already gained experience adapting to new estimates of Psyche's size and shape over the course of the mission. For the Discovery proposal, the Kaasalainen et al. (2002) shape was used as the baseline model for trajectory planning, but by the time the mission was selected, a new model had been published by Shepard et al. (2017). The Shepard model was adopted as the baseline for Phase B and new science orbits were redesigned to the same resonances, requiring no significant changes to transition the observation plans. This exercise provided an early validation of the team's ability to maintain the original operations plan while adapting to the latest shape model. By 2020, there were additional shape models published by Hanuš et al. (2017) and Viikinkoski et al. (2018), so the project performed a sensitivity study to gravity, shape, and spin axis orientation. Through these bounding cases, the project demonstrated that updated orbit parameters could be derived that met all requirements and preserved the observation plans. The baseline shape model was changed to Shepard et al. (2021) when the mission replanning began for the 2023 launch date. Table 5 lists the current baseline physical characteristics and Fig. 6 shows the selection of recent Psyche shape models evaluated.

The following subsections describe the key aspects that influence the development of the architecture of the Psyche science mission.

3.1 Functional Redundancy and Heritage Approach to Observing

If there is anything given in spacecraft operations, it is that plans will change, and data will be lost. Data can be unavailable either because it was never acquired, transmission to Earth failed, or it was obtained under conditions that make it unsuitable for scientific analysis. To ensure that the science objectives of the mission are met, the Psyche science plan oversamples the dataset needed for each investigation by collecting data that are functionally

Table 5 Psyche baseline characteristics used for mission and science planning

Characteristic	Value (Shepard et al. 2021)
Effective diameter	222 (−1/+4 km)
Dimensions a, b, c	278 (−4/+8 km), 238(−4/+6 km), 171 km (−1/+5 km)
Pole (was EMO2000)	right ascension, declination (36°, 6°) ± 2° EME2000
Optical albedo	0.16 ± 0.01
GM	1.601 ± 0.017 $km^3\ s^{-2}$; bulk density 4172 ± 145 $kg\ m^{-3}$ (Farnocchia et al. 2024)

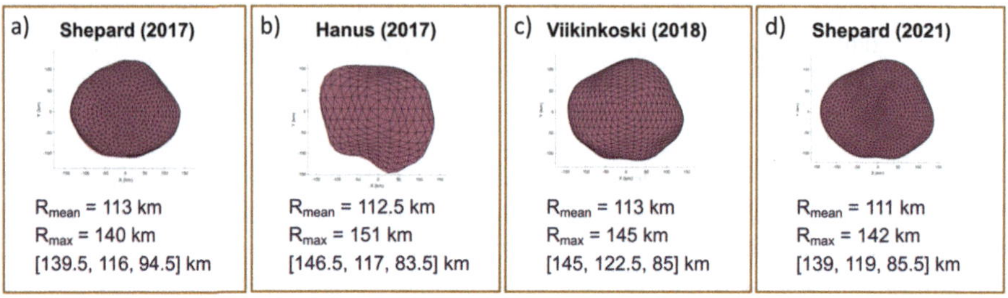

Fig. 6 Four recent Psyche shape models are shown for comparison. (a) Shepard et al. (2017) was the planning baseline for the 2022 launch opportunity. (b) Hanuš et al. (2017) was used to evaluate our plans in the "bounding cases" study. (c) Viikinkoski et al. (2018) was a second model used in the "bounding cases" study. (d) Shepard et al. (2021) is the current planning baseline for the 2023 launch mission. Our mapping campaign was shown to be successful with each of the shape models, demonstrating the resilience of our strategy for designing science orbits

redundant. Functional redundancy in this context means that each dataset designed to satisfy mission science requirements will be acquired more than once, although the circumstances of acquisition can be varied to allow the combination of those two datasets to yield better results than either dataset alone (Polanskey et al. 2011). Even if significant portions of either dataset are lost, there will be sufficient data to meet requirements without altering future plans.

Other strategies are used to ensure we meet and exceed the science requirements. For example, orbit altitudes are designed to provide better imaging resolution than required to allow for minor adjustments to orbit altitude in response to Psyche's physical characteristics. For measurements that need to accumulate signal over time, the accumulation time is extended to provide margin against lost data or periods of reduced data quality. Data volume margin is held to protect against uncertainty in scheduling sufficient DSN tracking to return the data or to accommodate the need to retransmit data. Unlike spacecraft mass and power margins in the development phase that are intended to be used whenever needed, science margin is built into the plan to reduce the workload and time needed to replan the mission timeline if additional resources are needed. The margin built into the science plan is encumbered for the execution phase of the mission and is not available to the team to make use of during the development or design phases.

A series of circular science orbit altitudes with dedicated science requirements for each altitude enables sufficient redundancy and margin-keeping, as demonstrated by Dawn (Polanskey et al. 2016). Figure 2 illustrates the four science orbit altitudes chosen specifically to meet Psyche's requirements. Psyche is a mapping mission that methodically observes the surface and in-situ environment to obtain global measurements. The specific

Table 6 Optical topography requirements for image data set (Polanskey et al. 2018)

Stereo technique	Optimal conditions	Useable conditions
Stereophotogrammetry	5° < Phase Angle < 90°	5° < Phase Angle < 90°
	10° < Incidence Angle < 60°	5° < Incidence Angle < 70°
	Emission Angle < 55°	Emission Angle < 70°
	Illumination Variation < 10°	Illumination Variation < 15°
	Stereo Separation 15°, 15°-65°	Stereo Separation 10°, 15°-65°
Stereophotoclinometry	0° < Phase Angle < 90°	0° < Phase Angle < 90°
	10° < Incidence Angle < 75°	10° < Incidence Angle < 85°
	20° < Emission Angle < 50°	0° < Emission Angle < 70°
	20° < Illumination Variation < 70°	10° < Illumination Variation < 85°
	Stereo Separation 0°, 20°-90°	Stereo Separation 0°, 15°-90°

observation technique with the highest inheritance from Dawn is the optical topography campaign. The Psyche spacecraft does not carry a laser altimeter, so topography must be derived from stereo imaging. While Psyche does carry two Imagers that are offset by 3.7° in their mounting, only one Imager is needed for the primary topographic data set. The second Imager is reserved as a backup. There can be many approaches to designing stereo observations, but Psyche will use the approach developed by the Dawn mission for both Vesta and Ceres. Psyche will image a series of complete maps of the surface with a different body-relative fixed attitude for each map. Modeling and experience have shown that a minimum of four maps of the surface (one acquired with a nadir attitude and three with different off-nadir attitudes) are necessary to derive a high-fidelity shape model for an asteroid (Raymond et al. 2011; Jaumann et al. 2022). The topographic solution improves with each additional off-nadir view. In addition to the number of different viewing angles, the quality of the image dataset is critical as well. Table 6 lists the criteria necessary to acquire an imaging set that will maximize the quality of the topography solution for each of the two stereo techniques (stereophotogrammetry and stereophotoclinometry) that the science team plans to use. The beta angle, the angle between the spacecraft orbit plane and the direction to the sun, is an additional constraint used to achieve optimal observing conditions for topography.

3.2 Science Orbit Requirements

The science orbits are designed to characterize Psyche both to enable navigation from one orbit to the next and to efficiently accomplish the science objectives and requirements of the mission. Fortunately, the observing requirements of the Psyche instrument suite are synergistic, so it was straightforward to select a series of orbit altitudes that met both navigation and science requirements. Since Psyche's rotation axis is tipped nearly into its orbital plane, the surface is only fully illuminated during the spring and autumn equinox every 2.5 years. Based on the baseline spin axis model, peak lighting of the surface is achieved in June 2029 as well as July 2031.

Creating the orbital operations timeline was like assembling a puzzle in which all the differently shaped pieces need to fit together to complete the picture. The first step was to select a series of orbit altitudes that meet the science and navigation requirements. The mission designers determined the maximum change between orbit altitudes that could be supported with the gravity knowledge available from the current orbit(s). For example, a

Table 7 Science orbit requirements are listed with the current baseline estimates noted in italics below each requirement

Orbit attribute	Orbit A	Orbit B	Orbit C	Orbit D
Science and Operational Objectives	Characterize gravity and shape, multispectral mapping	Multispectral mapping and topography	Gravity field mapping, improved topography, and magnetometry	Elemental mapping, improved gravity, and magnetometry
Averaged Altitude *(using shape model)*	600–2000 km *709 km*	< 400 km *304 km*	$< 1.7\ R_{Psyche}$ *193 km*	$< 1\ R_{Psyche}$ *75 km*
Inclination (deg)	90 ± 5 *90*	90 ± 5 *90-90.7*	90 ± 5 *89.7*	70% coverage *160*
% Surface Illumination (<70° incidence angle)	> 80% *81-75%*	> 80% *B1=68%* *B2=98%*	No requirement	No requirement
|Beta angle| (deg)	10–30 *[29–30]*	30–40 *[31–34], [36-38]*	≤ 45 *[35–38]*	No requirement *[55-85]*
Duration	≥ 40 days *56 days*	≥ 6 cycles *21 cycles*	≥ 100 days *100 days*	≥ 100 days *100 days + 63 days margin*

transition directly from approach to the altitude needed for the required imaging resolution was not feasible. Therefore, the original mission timeline that supported launches in 2021 and 2022, both arriving at Psyche in early 2026, included a four-orbit set:

- Orbit A was insensitive to Psyche's gravity while providing the first opportunity to characterize Psyche's gravity and confirm its spin axis orientation.
- Orbit B provided sufficient imaging resolution to satisfy mapping requirements, utilizing the best surface illumination conditions available.
- Orbit C was low enough to meet gravity science and magnetometry requirements but high enough to support a stable polar orbit with global ground track coverage.
- Orbit D provided close-up sensing of elemental composition.

This orbit plan provided a methodical platform for meeting science and navigation objectives. Evolution of this plan to the 2023 launch opportunity is discussed in Sect. 3.4.

Next, orbit resonances were selected, providing a ground track that would efficiently accumulate the surface coverage required at each altitude. The selected orbit resonance specifies the length of a global mapping cycle (the time for the ground track to begin repeat coverage), which is a critical building block of the science plan. The unit of a ground track repeat cycle is used to plan repeatable templates for imaging and other spacecraft operations. The science objectives dictated the number of mapping cycles needed at each orbit, which in turn defined the length of each orbit phase. Next, the orbit phases were placed on the timeline to assure that the surface coverage requirements could be met when paired with illumination conditions and that sufficient time was allocated to transfer between each orbit altitude. A summary of the orbit requirements is found in Table 7.

All orbits must also meet a stability requirement that the spacecraft does not crash or escape within 28 days, the maximum time allocated to recover from a significant anomaly and perform orbit maintenance maneuvers. This requirement drove the selection of a highly inclined orbit for Orbit D.

3.3 Data Budget and Margins

One of the more challenging issues on many flight projects is the allocation of data volume (the amount of data that can be sent back to Earth) between engineering activities and the various science instruments or scientific objectives. The Psyche mission data volume is also constrained by virtue of a fixed HGA that can only be pointed towards Earth when the spacecraft is not oriented for science observations. An articulating HGA was out of scope, which drove two important decisions during mission concept development. The spacecraft data storage capacity was increased to allow for a large accumulation of images during periods of high-frequency acquisition relative to the downlink intervals. In addition, the range of telemetry rates implemented in spacecraft flight software included the highest rates achievable at a DSN station during orbital operations rather than the average rate. This allows the flight team to command the downlink rates to rise and fall during each communications pass to take maximum advantage of the DSN antenna's capability as the spacecraft rises and sets during the antenna's viewing period.

Rather than have each instrument team competing for the data to support their own observations, the project allocates data volume in a structured manner. The first element of the data budget is the routine engineering housekeeping data that must be returned to operate the spacecraft safely, diagnose anomalies, and monitor trends. The total volume of spacecraft housekeeping data can be estimated for a given science orbit phase and subtracted from the total available data that can be transmitted to Earth per that science orbit's observation schema. Early in project development, the undefined telemetry needs of the spacecraft make it difficult to estimate the housekeeping data volume, but this can be measured and adjusted after launch during the interplanetary cruise phase. Next, the constant rate of science instrument housekeeping data is computed and allocated.

The Magnetometer and GRNS produce a constant stream of science data products, although the GRNS can customize that rate by command. The fixed rate science data volume is estimated and allocated, leaving the imaging data volume as the only discrete and variable dataset. Since Psyche is a mapping mission, our planning tools can model the number of image footprints to cover the surface and compute total imaging data requirements by multiplying the number of image stations by the required number of filters for that objective, taking into account the expected data compression. This approach was used during the proposal phase to specify the required average data rate of the spacecraft and correspondingly the size of the HGA.

The total data volume available to downlink mission data is computed from the time allocated to HGA tracking per the observing strategy of each science orbit phase adjusted for expected liens. These liens are accounted for by encumbering margin on the overall downlink capability. It is well known that the DSN resources are oversubscribed, and the Psyche project can only take advantage of DSN tracks that occur during specific times when the instruments are not collecting data so the antenna can point to Earth. We subtract a margin of 27% from the theoretical data return and encumber it as DSN-allocation margin for orbital operations. Remaining margin, if any, is unencumbered and can be used to address new issues or opportunities at Psyche. Each science orbit has a different estimated margin, but the overall margin available to the mission is roughly 30%. The data margin must be managed within each orbit phase because unused margin in Orbit A is not available to be used later in Orbit B. The Psyche project does not plan to use DSOC to transmit any of the required orbital science or engineering data for the mission.

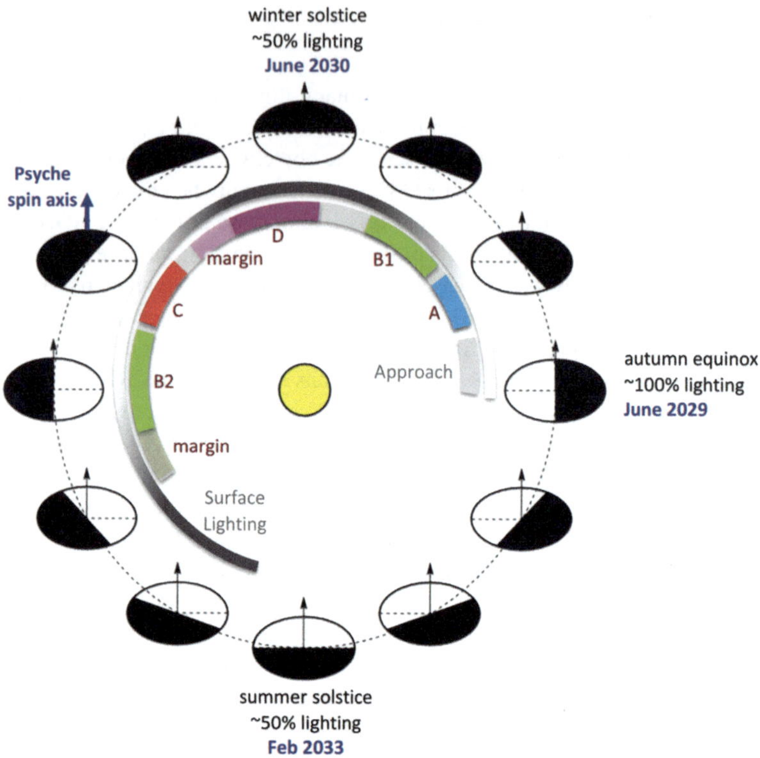

Fig. 7 Simplified graphic of Psyche's surface illumination as viewed from the north ecliptic pole. Psyche spin axis is the "positive pole" and that hemisphere is the basis of the seasonal references. The science orbit phases are shown on the inside circle. Orbits A through C are polar orbits and Orbit D is a near equatorial orbit. The best surface illumination is available during the two equinoxes. Orbit B2 is the only orbital phase that occurs during optimal illumination conditions

3.4 Restructuring the Mission Timeline for the 2023 Launch Date

Following the launch slip from 2022 to October 2023, the mission design team conducted a study of different interplanetary trajectories to deliver the spacecraft into a Psyche orbit. The preferred trajectory begins orbital operations at Psyche in August 2029, after the peak in surface illumination as shown in Fig. 7. The next time Psyche will be fully illuminated is in July 2031. The most demanding imaging requirement of obtaining greater than 80% surface coverage at less than 20 m/pixel dictates that Orbit B should be scheduled during periods of optimal illumination, but Orbit B imaging is also necessary for the updated gravity and shape knowledge required to transition to lower orbits. The only solution to these constraints is to conduct two mapping (Orbit B) campaigns. Orbit B1 follows Orbit A with insufficient surface illumination to meet science requirements but sufficient shape model updates to support the transfer to lower orbit altitudes. Orbit B2 occurs in mid 2031 to complete the imaging needed to satisfy the science requirements.

Orbit C does not have surface illumination requirements, but the gravity science investigation requires a Sun-Earth-Probe angle greater than 30° for at least 70 days, although greater than 40° is preferred. In the scenario where Orbit C followed Orbit B1, the Sun-Earth-Probe angle would be less than 40° for most of Orbit C. In addition, Orbit C would

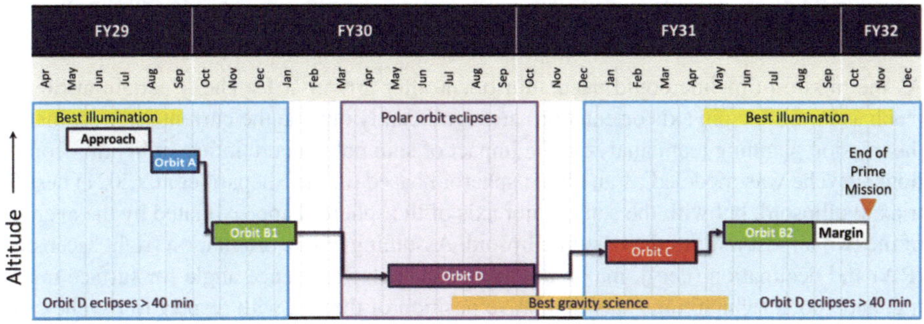

Fig. 8 The Psyche Approach phase, science orbits, and transfer orbits are shown on the mission timeline for the 2023 launch planning baseline. Illumination conditions are based on a model using a triaxial ellipsoid. The eclipse season for equatorial orbits is shown in pale blue and for polar orbits in pale violet. The period for optimal gravity science when the Sun-Earth-Probe angle is > 40° is shown in orange. Avoiding eclipse seasons while satisfying surface illumination and other observation requirements drove the design of the mission profile

occur during a period when Psyche would be blocking sunlight from reaching the spacecraft. While the spacecraft is qualified to go into eclipse, it is constrained to avoid eclipses longer than 65 minutes occurring as often as once every 6.6 hours. The Orbit C eclipse season would begin in March 2030 and continue through September 2030, and then by the time the spacecraft could complete the orbit transfer to begin Orbit D, the period when equatorial orbits experience eclipses would have begun. This disadvantageous set of geometric relationships drove the project to change the order of the science orbits.

An alternate consideration was to put the spacecraft in a high parking orbit and wait for the eclipse season for polar orbits to end, but that would add many months to the orbital operations timeline. Fortunately, Psyche is closer to the sun in its orbit during the 2029 scenario, providing more power to the solar arrays and enabling more efficient thrusting during the transfer periods between science orbits. This enabled the design of a transfer trajectory between Orbit B1 and Orbit D without having to stop at Orbit C first. This novel solution avoids eclipses in both Orbit D and C and has other advantages. In the previous mission timeline, the elemental composition measurements would be acquired at the end of the mission and could take advantage of any remaining operations margin; this is no longer possible under the new timeline. Therefore, margin was added to the duration of Orbit D to mitigate data loss from any solar energetic particle events that would mask Psyche-relevant data collection and to avoid starting Orbit C too early before the end of its eclipse season. Orbit C was then placed following Orbit D in a period where the Sun-Earth-Probe angle is > 40° for the entire duration, allowing the final transfer to Orbit B2 to complete at an opportune time to begin imaging the surface in the northern hemisphere as it becomes illuminated.

The result illustrated in Fig. 8 is a mission duration of 26.5 months that meets all science observing requirements, avoids eclipses, maximizes conditions for gravity science, and still has timeline margin to handle unexpected interruptions. The adaptability of the mission profile to a completely new set of constraints speaks to the advantages of having science orbit phases that can be treated as mostly independent building blocks in the timeline. If there are no delays in the interplanetary trajectory that impact the arrival date at Psyche, then the lower resolution multispectral imaging objectives could be satisfied in Orbit A, leaving only the higher resolution geologic mapping and optical topography objectives to Orbit B2.

3.5 Lighting Variability

As the mission timeline could need adjustment after arrival at Psyche if certain attributes (such as Psyche's spin axis orientation) are significantly outside the current error estimates, the mission planning team analyzed the impact of spin pole uncertainties on lighting conditions. Psyche was modeled as an oblate spheroid based on the Shepard et al. (2021) best fit triaxial ellipsoid, but with the semi-major axis of the spheroid approximated by the average of the two largest radii in the best fit ellipsoid. Assuming a pole orientation [right ascension (RA) and declination (dec)], maximum acceptable solar incidence angle for surface imaging, and epoch (which translates to relative position of the sun with respect to Psyche), the model determines the geodetic latitude at which the limiting incidence angle occurs. The lighting percentage is calculated as the percentage of surface below that latitude compared against the total surface area of the spheroid. Both hemispheres are considered separately, because the areas around both poles can simultaneously experience unacceptable lighting near the equinox for incidence angle limits of less than 90°. This simple model, intended for use in quick what-if studies, was compared against a detailed time-varying simulation incorporating more complex shape modeling (Shepard et al. 2021 shape model), body rotation, and self-shadowing, and was found to estimate the lit surface percentage to within 2%. Figure 9 shows the percentage of the surface that is illuminated during the nominal timeline of the Psyche mission. A 3-standard deviation worst case pole error shows the impact of variation in optimal lighting conditions and how the strategic placement of Orbits B1 and B2 provide the conditions to meet science surface coverage objectives under any possible illumination in this study. While our plans are designed to be robust to the expected uncertainty in Psyche's physical characteristics, we also acknowledge that Psyche's real shape and spin pole orientation may require adjustments to our plans.

4 Science Implementation by Orbit Phase

The following subsections describe, in mostly chronological order, the details of each science orbit and orbit transfer phase that includes science data collection. A summary of the science goals that informed each orbit can be found in Table 2. The Psyche science data in each phase builds on data collected earlier in the mission. All efforts were made to modularize and standardize the spacecraft and instrument activities to reduce complexity of operations while simultaneously designing in the ability to be flexible to changes in the timeline and other conditions.

Low-thrust missions allow timing of each phase to fluctuate with routine re-optimization of the trajectory. The date of capture into orbit around Psyche is allowed to drift in response to the effectivity of cruise thrusting and any unplanned loss of thrust. The time between science orbit phases may also change once the characteristics of Psyche are known and realistic spacecraft transfer sequences are developed. The Psyche mission operations concept embraces these uncertainties, and the flight team operational processes are designed to respond nimbly to expected timing changes. However, long-range planning tasks can benefit from a stable planning baseline. Table 8 contains the dates currently assumed for the planning baseline, knowing that the actual dates are guaranteed to change. The following subsections describe the layout of the science observing plans independent of the actual start and end dates of each science orbit phase, so it is expected that certain geometric traits of the asteroid (such as surface illumination) will vary based on actual execution times as well as actual tilt of the asteroid spin axis as discussed above.

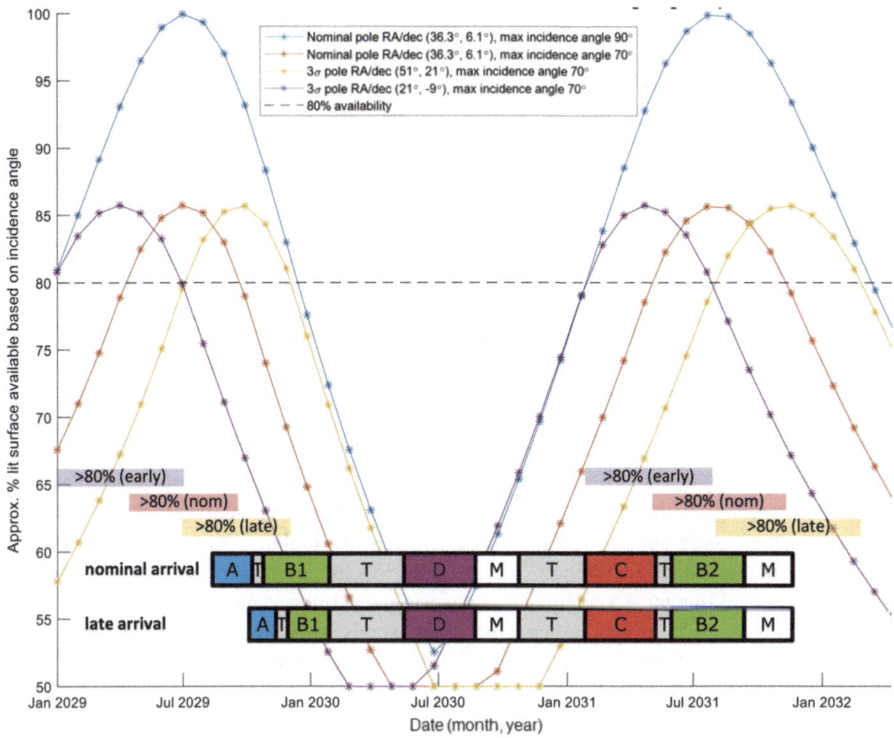

Fig. 9 Variability in illumination conditions for Psyche assuming different spin axis orientations. The blue curve is for Psyche's nominal pole RA/dec (36.3°, 6.1°) and maximum incidence angle of 90°. The salmon curve is for the nominal pole with maximum incidence angle of 70°. The yellow curve shows a worst case 3-sigma variation in Psyche's pole RA/dec (51°, 21°) and the purple curve is the opposite worst case 3-sigma pole RA/dec (21°, −9°), both with maximum incidence angle of 70°. The science orbit phases are shown with respect to the lighting curves for both the nominal date for injection into Orbit A and a worst-case injection date. Imaging in Orbit B2 will satisfy the requirement to provide surface coverage of greater than 80% even with a worst-case injection date

Table 8 Orbital operations phases

Operation phase	Duration (days)	Start date (baseline for planning)
Approach	100	May 9, 2029
Orbit A	56	August 17, 2029
Transfer to Orbit B1	17	October 12, 2029
Orbit B1	92	October 29, 2029
Transfer to Orbit D	98	January 29. 2030
Orbit D	163	May 7, 2030
Transfer to Orbit C	90	October 17, 2030
Orbit C	100	January 15, 2031
Transfer to Orbit B2	23	April 25, 2031
Orbit B2	100	May 18, 2031
Operations Margin	67	August 26, 2031
End of Mission	n/a	November 1, 2031

The spacecraft attitude modeling is performed with the Science Opportunity Analyzer science planning tool (Ortega et al. 2021). This is the same tool that was used to create the mission architecture in the proposal phase. To uniquely define spacecraft attitude, the primary pointing vector (i.e. Imager pointed to Psyche nadir) and secondary pointing vector (sometimes referred to as twist) must be specified. For missions with solar arrays, the twist is often controlled to keep the solar arrays facing the sun although the arrays may be slightly off-pointed to control angular momentum of the spacecraft.

4.1 Observations and Navigation on Approach to Psyche

Once the spacecraft reaches a distance where imaging Psyche will provide improvements on ground-based observations, the project will embark on an imaging campaign to support optical navigation (OpNav). Since the navigation team relies on the same Multispectral Imager used for the science observations to perform this mission critical navigation function, redundant imagers were built and mounted to the −X axis of the spacecraft to protect the mission from a single point failure. While there is a 3.7° offset in the mounting orientation of the imager boresights, the mission requires only one camera to complete all engineering and scientific goals, and a single Imager will be operated throughout the completion of imaging requirements. Both Imagers are calibrated during the approach phase, so the OpNavs alternate between Imager A and Imager B ensuring that Imager B is ready if needed during orbital operations. As a project policy, OpNav and science imaging are shared between both teams. The labeling of the activity as OpNav or science is based on which team requires and designs the observation.

The primary objective of the approach imaging is to refine estimates of Psyche's spin axis orientation and shape to enable the design of and delivery to Orbit A. These attributes of Psyche are also important science requirements that feed into the shape derivation and gravity modelling. This objective requires images taken with the polychromatic (clear) filter. Additional observations using both polychromatic and/or all narrowband filters are planned to calibrate the image exposure times to be used during the orbital operations phase and to search for evidence of any secondary objects in orbit around Psyche. Analysis has demonstrated that the existence of orbiting secondaries would not pose a threat to the Psyche spacecraft, so the search for satellites is purely of scientific interest.

The approach phase is a transition from cruise operations that are focused primarily on thrusting to orbital operations with long coasts during the science orbits and thrusting between those orbits. Thrusting is still a critical part of this phase, and the imaging periods need to either be reserved in advance or scheduled during periods when it is optimal to coast. During this period there will be a change from a Sun-centered ephemeris to a Psyche-centered ephemeris. Science operation processes specific to the orbital operations phase will be integrated with the processes still being used from the cruise phase in a way that is unique to this period.

4.1.1 Optical Navigation Plan

On approach to Psyche, the key objectives for OpNav are to provide optical observations for the day-to-day orbit determination activities and to update key Psyche physical properties, such as the spin axis, spin rate, and shape model. The goal of these objectives is an accurate insertion into the reference Orbit A science orbit. During the first seven weeks of approach, with Psyche less than 50 pixels diameter, OpNav is based on body center-finding techniques, such as cross-correlation and limb scans. The only geometric requirement for these techniques is a phase angle less than 100°–110°, to allow part of the body to be illuminated. Once Psyche exceeds ∼50 pixels diameter, the main OpNav observables are landmarks, constructed with stereophotoclinometry. Landmarks are body-centered vectors determined via small digital terrain models, or landmark maps. The construction and maintenance of such a network of landmarks allows the estimation of the rotational properties of Psyche and the large and small-scale shape. These landmark maps also enable terrain-relative navigation when Psyche exceeds the field of view (FOV) of the Imager.

Stereophotoclinometry places several geometric constraints on the imaging and illumination geometry (i.e., on the emission and incidence angles). Ideally the emission angle will vary both in elevation and azimuth, with elevation angles ranging from nadir to 60° off-nadir and azimuth from 0°–360°. In practice, on approach to Psyche such an emission range is accomplished by pointing the Imager in the nadir and off-nadir Psyche directions and by imaging for a full rotation period over a range of sub-spacecraft latitudes, enabling east-west and north-south azimuth camera directions. Ideally the incidence angle will have an elevation range of 20°–70° and a 0°–360° azimuth. Such a variation in the incidence angle is difficult to accomplish. In practice we rely on the beta angle being less than 15° to avoid very low incidence angle observations. The rotation of the body allows for a variation in the incidence angle, but the solar azimuth is restricted to morning and evening observations rather than a wider range. The range of solar elevation also imposes a similar constraint in the phase angle in the range of 20°–70°. In addition, to maximize the surface coverage and to allow more favorable observing conditions for the rotational frame estimation, there is a need to image, for at least a limited time, at sub-spacecraft latitudes as close to Psyche's equator as possible.

The approach low-thrust trajectory design optimizes xenon fuel usage and arrival time and tends to produce trajectories that reach Orbit A from a high phase angle while keeping the sub-spacecraft latitude mostly near the north pole of Psyche, which is largely not illuminated. As a result, several state constraints had to be added to allow, in a statistical sense, the trajectory to meet the OpNav imaging geometry needs. These constraints are discussed below.

At two different fixed epochs of 48.015 days and 12.358 days prior to Orbit A insertion, or approximate range of 93,000 km and 3800 km respectively, we impose a phase angle constraint in the 0°–90° range, which keeps the phase angle from growing too large during the period where landmarks are the key OpNav observations. In the current reference trajectory these constraints keep the phase angles in the range of 35°–69°.

At a range of 10,000 km, or approximate epoch of 16.5 days to Orbit A, a constraint of a sub-spacecraft latitude between 0°–30° North latitude has been added. This constraint forces the trajectory to move closer to Psyche's equator, while staying always at north sub-spacecraft latitudes. In the current reference this constraint keeps the trajectory between 30° and 17° North latitudes during key OpNav sessions. Figure 10 provides examples of several simulated OpNav images during the approach phase, showing the increasing resolution and differing illumination conditions and including a simulated surface texture for context.

Operational considerations for thrust cycles create another constraint. The mission design divides the approach phase into seven thrust design cycles. Design cycles 5–7 target insertion into Orbit A based on the Psyche team's knowledge of physical parameters. To improve the spin axis knowledge in design cycles 5 and 6, a constraint has been added to slow the spacecraft down relative to Psyche, starting at a range of 22,000 km, or an epoch of 28.504 days relative to Orbit A insertion, so as to increase the number of OpNav imaging sessions and ground processing time, during which landmark observables are being constructed.

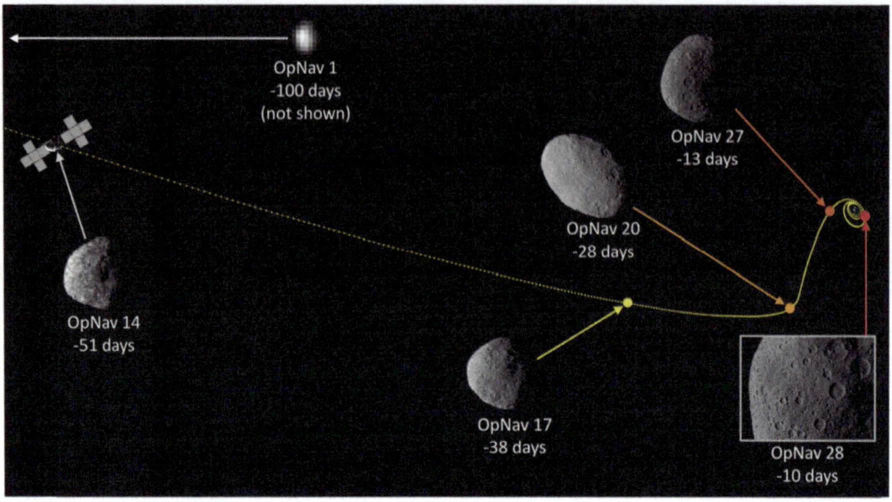

Fig. 10 Six simulated OpNav images shown on approach to Psyche. Psyche's diameter changes from 6.7 pixels to 2900 pixels between OpNav 1 and OpNav 28. For this approach trajectory, the phase angles range between 35° to 72° and all but OpNav 28 are taken from a northern sub-spacecraft latitude. Psyche surface texture is simulated

4.1.2 Multispectral Imager Calibrations

Ten of the OpNav opportunities will include additional science imaging with all seven narrowband filters (five opportunities for each Imager) to allow the Imager team to refine the image exposures planned for orbital operations. The first pair are acquired early during the first week of the approach phase, and the remaining four pairs are during the last five weeks of approach when Psyche fills a larger portion of the camera FOV. Two straylight tests for each Imager are also planned towards the end of the approach phase.

4.1.3 Psyche Satellite Search Observations

Two opportunities are planned during the approach phase to search for natural satellites by imaging the star field within Psyche's Hill sphere, where any bound satellite orbits are expected to reside. To date, no moons have been detected within Psyche's Hill sphere by ground-based observations. Viikinkoski et al. (2018) established an upper limit of undetected satellites of 1.5 km diameter within 150 km of Psyche and 800 m diameter within 2000 km of Psyche. The primary challenge of these observations is to design image exposures that will provide sufficient signal to detect small moving objects without excessive smear from spacecraft jitter. The details for these observations will be planned during the cruise phase. While McFadden et al. (2015, 2018) conclude that large, intact protoplanets do not have satellites, the fact that Psyche's spin pole is approximately in its orbit plane argues that the asteroid was disrupted early in its history and thus could have been subject to different or unique circumstances regarding formation and retention of satellites, compared to "normal" large asteroids. Therefore, if satellites are observed around Psyche, they could contribute additional information to discriminate between models of Psyche's origin.

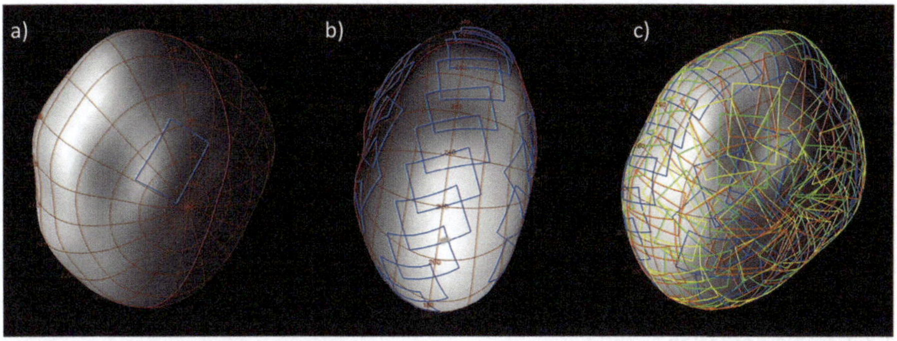

Fig. 11 Imaging in Orbit A taken from the perspective of the spacecraft as modeled by the Science Opportunity Analyzer software. (a) One Imager FOV is shown positioned near the dark-to-lit terminator in Orbit A. (b) One orbit of nadir imaging. The spiral shape of the orbit ground track and imaging swath results from the spacecraft orbit period being greater than the asteroid rotation period. (c) Four orbits of imaging, less than half of the 9-orbit cycle, provide nearly global surface coverage of the illuminated surface

4.2 Orbit A: Characterizing Psyche

Insertion of the spacecraft into Orbit A marks the beginning of the orbital operations phase of the mission. Unlike spacecraft with chemical propulsion, SEP missions do not utilize discrete orbit insertion maneuvers. Instead, all thrusting from the beginning of the cruise phase through the approach phase comprise an extended maneuver designed to gradually enter Orbit A. The interplanetary thrusting changes the spacecraft's heliocentric trajectory to slowly become more and more like (16) Psyche's. Thrusting does not target gravitational capture by the target body. Rather capture occurs simply as part of normal electric propulsion thrusting during the approach phase to Orbit A.

4.2.1 Objectives

While some imaging science requirements may be achieved in Orbit A, the primary motivation for this orbit is to characterize Psyche by obtaining gravity measurements and deriving a shape model sufficient to design the transfer trajectory to Orbit B1. Imaging with all filters in this cycle will contribute to the global mapping of silicates and metals, if not intimately mixed within the rock, and the search for oldhamite. Coverage requirements for these objectives are difficult to satisfy with the predicted illumination conditions. However, all imaging objectives can be satisfied in Orbit B2, so the data acquired in Orbit A simply provides an opportunity to begin mapping early and enables fine tuning the observation parameters and techniques for subsequent orbits. Mapping of craters with diameters greater than 1 km over 50% of the surface can be fully accomplished during Orbit A. Figure 11 illustrates the scale of the Imager FOV on Psyche's surface in Orbit A and how imaging coverage builds up over the course of a nine-orbit mapping cycle.

To determine Psyche's gravity sufficiently to design and build the spacecraft sequences that command thrusting to the Orbit B altitude, the navigation and science team will use X-Band Doppler data from the HGA and LGA to create the first global gravity model of Psyche. One of the most important scientific results to come from the gravity investigation during approach and/or Orbit A is a precise measurement of Psyche's mass, which will finally give us an accurate density and tell us whether Psyche really is metal rich.

Table 9 Orbit A planned characteristics (*values expected to change after arrival). $R_{Psyche} = 111$ km

Orbit A start date*	8/17/2029	Average orbit period	32.6 hours
Orbit A end date*	10/12/2029	Orbit resonance	70:9
Duration	56 days	Orbits per repeat cycle	9
Average radius*	811 km (7.3 R_{Psyche})	Ground track repeat cycle duration	12.2 days
Average altitude*	709 km	Number of ground track repeat cycles	4.5
Beta angle range*	[29-30]°	Number of orbits in phase	41
Maximum ground speed*	61 m/s	Average Imager pixel scale*	35 m

4.2.2 Implementation

The orbit altitude was constrained to be greater than 600 km by the need to limit perturbations from Psyche's gravity to less than 1% and lower than 2000 km to ensure that the images acquired satisfy the mineralogy resolution requirement of 200 m/pixel with margin. The current baseline altitude of 709 km provides an image pixel scale of 35 m/pixel, substantially better than the required 500 and 200 m/pixel for the metal to silicate ratio determination and oldhamite mapping objectives.

The 56-day duration of Orbit A was driven by operational considerations of the timeline for planning and building the thrust sequences to Orbit B1, so it is longer than needed to map the surface with all Imager filters with 100% redundancy. Table 9 lists the attributes of the current Orbit A planning baseline.

We use global image maps created with polychromatic-filter imaging to improve the Psyche shape model. To efficiently accommodate both imaging and gravity objectives, the spacecraft will be commanded to point the prime Imager towards Psyche when the spacecraft is traversing the lit portion of the orbit, then turn to point the HGA towards Earth to return accumulated data and collect X-band Doppler data for gravity science while the spacecraft is traversing the dark side of the orbit. Figure 12 is a cartoon of this operations schema including the expected spin axis (+Z) orientation of Psyche.

After 12.2 days and nine orbits, the spacecraft will have completed its first ground track repeat cycle, and the following orbit will begin to cover terrain that has previously been mapped. This repeated coverage provides redundancy and opportunities to observe the surface differently. The allocation of time between the spacecraft pointing to acquire imaging and pointing the HGA to Earth for communications is illustrated in Fig. 13.

Over the first 12-day mapping cycle, imaging will be acquired at a strictly nadir attitude. The orbit resonance (70:9) was selected to accommodate measurements of Psyche's potential magnetic signature, although the field would need to be strong to be detected at this altitude. The ground track repeat pattern is designed so that each point on Psyche's surface lies within 10° of arc of the ground track while providing overlapping swaths of image footprints to obtain complete coverage of the illuminated surface. The ground tracks are spaced closely enough that duplicate imaging coverage can be acquired in one mapping cycle.

The second mapping cycle will be a repeat of the first, offering an opportunity for the Imager team to update image exposures based on results from the first cycle and redundant coverage for any terrain that was missed.

The third mapping cycle is allocated to off-nadir imaging with the polychromatic filter to improve the Psyche shape model for optical navigation and science planning. This also allows more time for the imaging operations team to evaluate and update parameters for their next full-color imaging campaign in the fourth cycle. Improving the shape model with stereo imaging early in the orbit phase allows time to feed those results into the thrust planning for

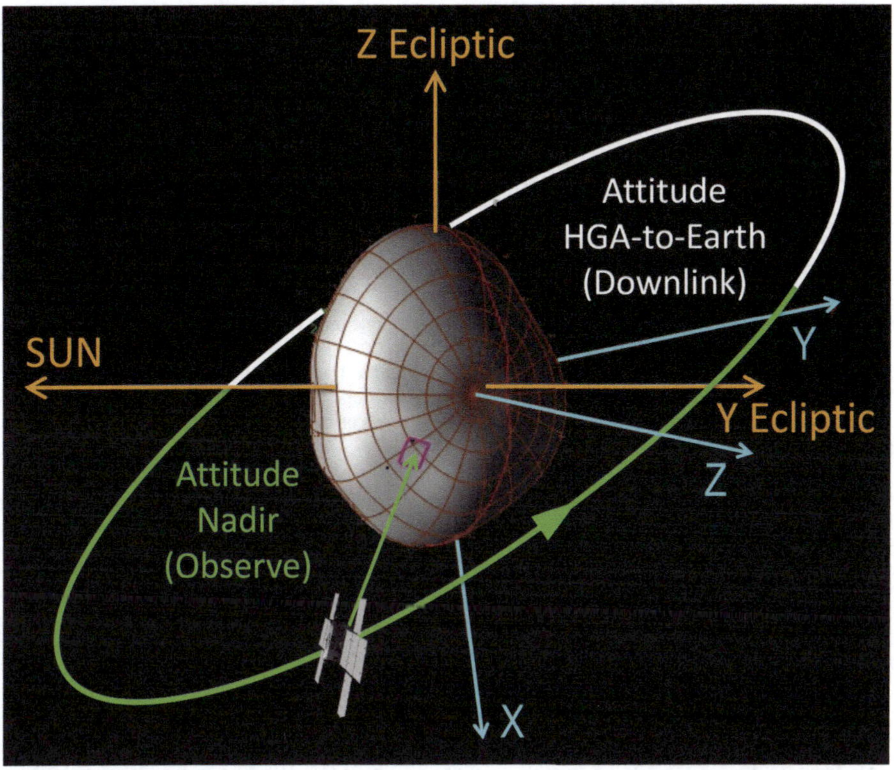

Fig. 12 The spacecraft transitions between data downlink as it passes over the dark side of the asteroid during its orbit and science observing while passing over the lit side of the asteroid. The spacecraft enters the hit hemisphere in the south and travels to the north. Activities are planned relative to the terminator crossings of each orbit. This scheme allows the imaging sequences to be developed as a repeating template that can be adjusting if the timing of the terminator crossing changes. The magenta rectangle is a projection of the Imager field of view on the surface. The Psyche body-fixed axes are turquoise. The spacecraft is not to scale

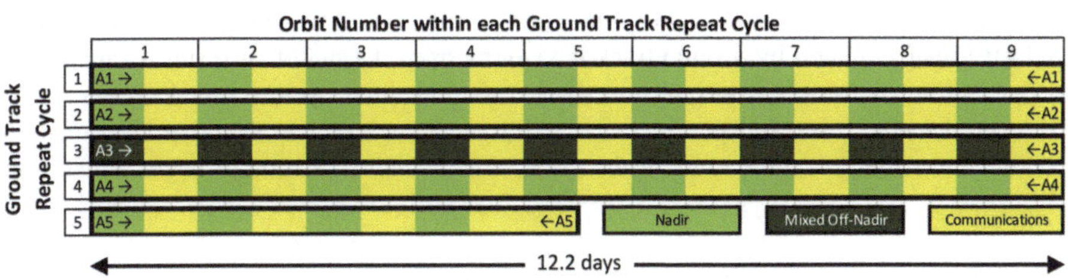

Fig. 13 The operations time in Orbit A naturally divides into 4.5 ground track repeat cycles. Each 9-orbit cycle provides the opportunity to completely map the surface with the Imager and map gravity using the telecommunications subsystem. The time spent pointing the Imager nadir or off nadir is colored green and the communications periods are yellow. Cycle 3 is dedicated to off-nadir OpNav imaging

the transfer to the next orbit and has little impact on the final multispectral coverage. Off-nadir pointing in Orbit A attempts to control the camera azimuth and elevation, relative

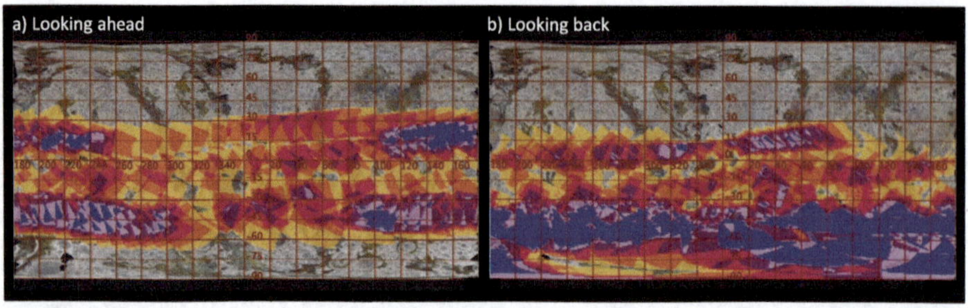

Fig. 14 Examples of different off-nadir imaging coverage for one orbit of Orbit A plotted on a latitude/longitude map. The background texture map is an artist rendering. (a) The spacecraft is pointing the Imager ahead of the orbital velocity vector. (b) The spacecraft is pointing the Imager behind the orbital velocity vector. The lightest orange indicates coverage with a single image and duplicate coverage is shown with increasingly dark colors until blue, which indicates coverage with six images or more covering the same area

to the local surface, by using a specific imaging strategy in each of the nine orbits of the mapping cycle.

For each orbit the camera follows a specific azimuth direction relative to the spacecraft, scanning either left, right, ahead, or behind the spacecraft. While in each of these four azimuth directions, we execute a 1-dimensional scan pattern to map as much of the surface as possible. In the first four of these orbits, the turn angle is "large" to image the surface at local emission angles in the range of 40°–60° and in next four of these orbits the turn angle is "small" to image the local surface at emission angles in the range of 15°–40°. The key assumption with this plan is that at the limit of complete surface coverage, each patch of surface will have been imaged in four distinct and widely separated directions in camera azimuth and two distinct directions in emission angle. The ninth orbit aims at imaging the illuminated limb of Psyche by positioning the Imager boresight to $\sim 20\%$ off the body.

Figure 14 demonstrates two examples of off-nadir attitude mapping. Looking ahead and looking back shows some overlap in coverage where the same patch of surface is imaged in these two azimuth directions, but also areas where, due to the irregular body shape, there is no overlap.

The fourth and partial fifth cycles are a repeat of the first two with optimized imaging exposures. These cycles are primarily intended to improve on the images taken previously and fill in gaps created by ground track deviations from the plan that result from delivery errors and small forces perturbing the spacecraft. The final cycle is not long enough to complete all nine orbits but provides one additional coverage map. Multispectral imaging continues for the first four orbits of the cycle, after which it is time to begin thrusting to Orbit B1.

4.2.3 Illumination Constraints and Mapping Coverage

Peak illumination conditions occur during the approach to Psyche, so orbital science operations begin with the north polar region in shadow down to about 60° north latitude. While 81% of the surface is theoretically illuminated with incidence angles below 70° at the start of Orbit A, northern latitudes will become less illuminated throughout this phase. These conditions may be sufficient to satisfy multispectral coverage requirements in the first cycle, but it is likely that parts of the northern hemisphere will need to be imaged again towards the end of the mission. However, these early maps in combination with lower resolution

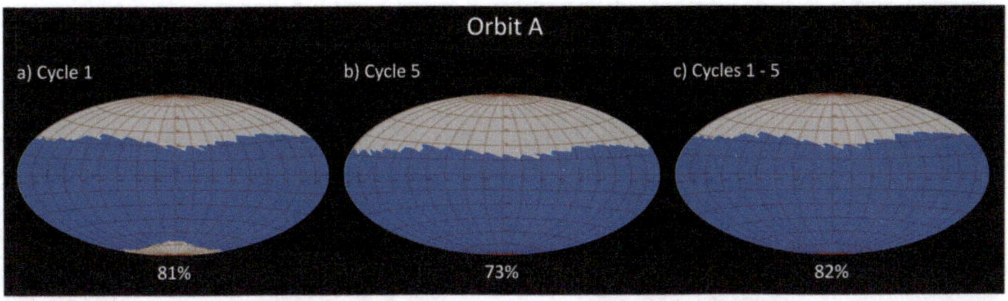

Fig. 15 (a) The first cycle of Orbit A provides 81% surface coverage for incidence angles < 70°. (b) As the sun moves towards the southern hemisphere, overall coverage decreases by the end of the phase although additional terrain in the south becomes visible with good illumination. (c) The cumulative coverage is slightly above the required 80% for multispectral imaging, but this could change with small variations in spin axis or arrival date. Coverage is shown using a Mollweide projection

approach imaging will provide the multispectral baseline for future planning and help the project better prepare to complete the multispectral mapping much later in the mission when illumination improves in the northern hemisphere. Figure 15 illustrates the surface coverage that can be achieved in Cycle 1 and Cycle 5 of Orbit A. The combination of the two cycles provides a minor increase in overall mapping coverage.

Following completion of Orbit A activities, 17 days of thrusting are allocated to transfer the spacecraft to the Orbit B altitude to begin Orbit B1. A short trajectory correction maneuver with the electric propulsion system may be required several days after nominal thrusting completes to fine-tune the orbital parameters to achieve the required reference orbit.

4.3 Orbit B: Geologic and Topographic Mapping of Psyche's Surface

4.3.1 Objectives

Many of the driving requirements for imaging will be completed at the altitude defined for Orbit B. The most challenging requirement is to obtain the images needed to create a geologic map of at least 80% of the surface at a resolution $\leq$ 20 m/pixel. Global multispectral imaging is also required to search for evidence of the sulfide-bearing mineral oldhamite that would indicate that Psyche was formed under reducing conditions, and evidence of large regions of achondritic or cumulate silicates on Psyche's surface that could indicate that Psyche is a stripped core. Data sets sufficient to derive a global shape model using stereo imaging techniques are also planned for this orbit phase and comprise the majority of the timeline. As noted earlier, the surface illumination during the Orbit B1 is insufficient to meet imaging coverage requirements; however, the spacecraft will return to the same altitude to conduct Orbit B2 towards the end of the mission. Most attributes of Orbits B1 and B2 are similar, so the two phases will be described together in this section.

4.3.2 Implementation

Designed to meet all imaging requirements for topography and multispectral imaging, Orbit B is required to have at least six mapping cycles in each phase, two at a nadir attitude and four with off-nadir attitudes. The baseline plan accommodates ten cycles in Orbit B1 and 11 cycles in Orbit B2, providing significant margin. The mean orbit altitude must be <400 km

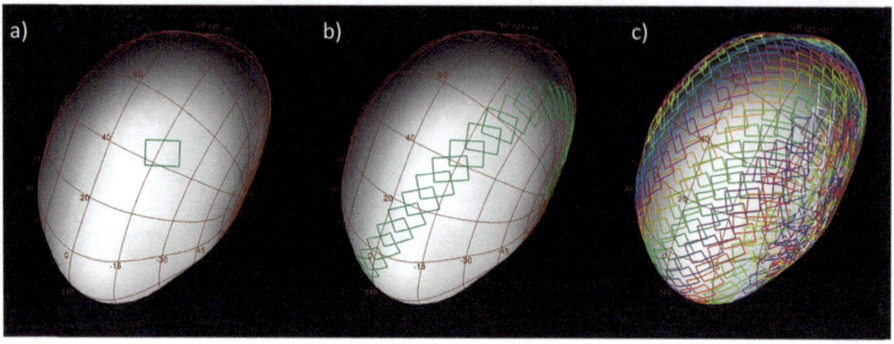

Fig. 16 Imaging in Orbit B1 taken from the perspective of the spacecraft. (a) One Imager FOV. (b) One orbit of nadir imaging. The spiral shape of the orbit ground track is evident. The twisting of the FOV is driven by the secondary pointing axis that strives to keep the solar arrays oriented towards the sun. (c) One complete cycle of 17 orbits of imaging provides global coverage of the illuminated portion of the surface

Table 10 Orbit B planned characteristics

Orbit B1 start date	10/29/2029	Average period	11.6 hours
Orbit B1 end date	1/29/2030	Orbit resonance	47:17
Orbit B1 duration	92 days	Orbits per ground track repeat cycle	17
Average radius	406 km (3.7 R_{Psyche})	Ground track repeat cycle duration	8.2 days
Average altitude	304 km	Operational repeat cycle duration	9.2 days
Average inclination	90.7°	Number of operational repeat cycles	10
Orbit beta angle range	[31-34]°	Total number of orbits	190
Maximum ground speed	66 m/s	Average Imager pixel scale	15 m
Orbit B2 start date	5/18/2031	Average period	11.6 hours
Orbit B2 end date	8/26/2031	Orbit resonance	47:17
Orbit B2 duration	100 days	Orbits per ground track repeat cycle	17
Average radius	406 km (3.7 R_{Psyche})	Ground track repeat cycle duration	8.2 days
Average altitude	304 km	Operational repeat cycle duration	9.2 days
Average inclination	90.0°	Number of operational repeat cycles	10
Orbit beta angle range	[36-38]°	Total Number of orbits	206
Maximum ground speed	65 m/s	Average Imager pixel scale	15 m

to satisfy the imaging requirements for the 20 m pixel scale, and the baseline altitude provides 25% margin on the requirement. Over 17 orbits (one ground track repeat cycle), the spacecraft ground track provides complete imaging coverage. Figure 16 illustrates the projection of the Imager's footprints from the vantage point of Orbit B1. Each FOV drawn on the surface is called an "image station" because it could correspond to a single image acquisition with one filter or a complete set of all eight filters.

Because the ground track coverage is the driving requirement for this orbit, the final altitude in operations will be selected to preserve the planned orbital period and resonance. This selection will be made once the gravity field and shape knowledge are improved with data from Orbit A. These parameters influence the design of the orbit transfer, so the science and mission design teams collaborate on the final orbit selection. Table 10 lists the attributes of the current baseline design for both Orbit B phases.

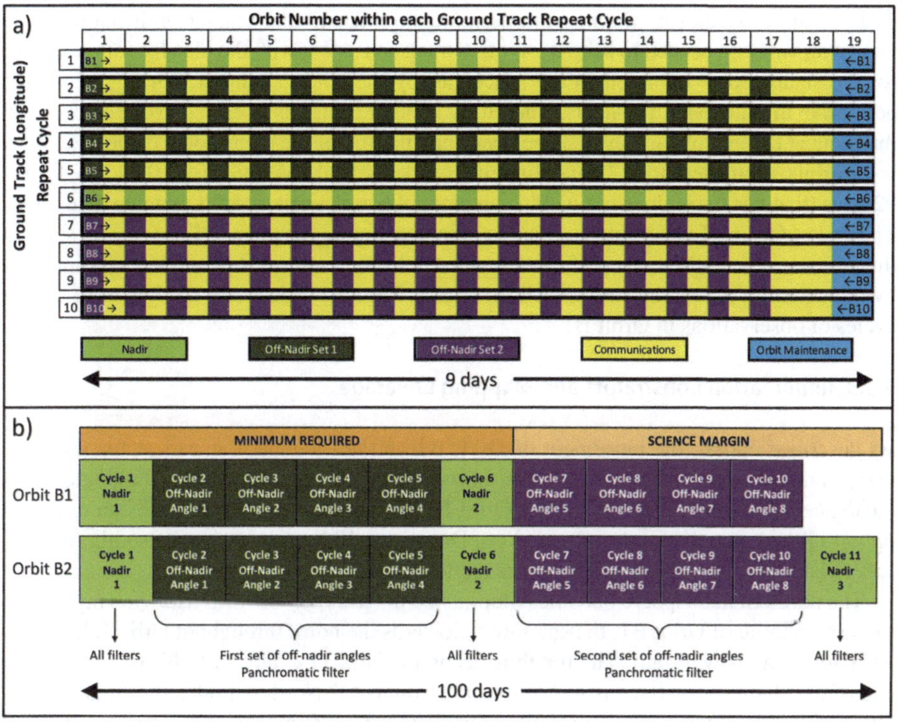

Fig. 17 (a) Graphical depiction of the 10 sequence cycles for Orbit B1. Each ground track repeat cycle is 17 orbits with additional orbits added for extra downlink and orbit maintenance maneuvers. Yellow indicates periods when the HGA is pointed to Earth for communications and shades of green and purple indicate when the Imager is pointed nadir or off nadir to observe. The illuminated surface is fully mapped 10 times in Orbit B1. (b) Summary timeline of all cycles in Orbits B1 and B2 demonstrating the ample science margin in the plan

Like Orbit A, the spacecraft will be pointed towards Psyche on the illuminated side of each orbit with LGA tracking for gravity science, and the HGA will be pointed towards Earth to return the imaging data and collect two-way Doppler tracking data on the dark side of the orbit. However, with an 11.6-hour orbit period, the 40-minute duration of the spacecraft turns consumes a significant portion of the dark-side pass.

Images acquired with all filters during nadir cycles have a high enough data volume to preclude returning all images with this downlink strategy alone. An extra orbit devoted only to data return will occur at the end of each 17-orbit ground track repeat cycle. The current timelines include multiple opportunities to acquire complete nadir maps with all eight camera filters in Orbit B (two in Orbit B1 and three in Orbit B2) as shown in Fig. 17. Despite the extra downlink-only orbit, the number of images acquired in the multispectral cycles cannot be returned to Earth with only the downlink available during that cycle. Therefore, the nadir cycles are separated in time to allow some of those images to carry over to the next polychromatic-filter cycle and trickle down after the data from those orbits are returned. This separation of nadir cycles also provides an opportunity for the illuminated regions to change, increasing coverage, and for the Imager team to evaluate the data quality and make any adjustments to exposures or other parameters prior to the subsequent multispectral imaging campaign.

In addition, studies by the mission design team indicate that regular orbit maintenance maneuvers may be needed to keep the ground track consistent with the plan by compensating for the small forces imparted to the spacecraft by the cold gas thruster firings needed to desaturate the momentum wheels. Therefore, another orbit has been reserved for potential thrusting at the end of each cycle. The resulting duration of the operational cycle in Orbit B is 19 orbits, equivalent to 9.2 days. Imaging of the surface continues for 10 cycles, lasting 92 days (that last cycle does not include an extra playback orbit or orbit maintenance orbit). The end of Orbit B1 observing is timed to complete before the polar eclipse season starts at this altitude. While the spacecraft could withstand short eclipses, operations are simplified by transitioning before the start of eclipse season. Figure 17 illustrates the layout of the ten cycles of observations in Orbit B1.

4.3.3 Illumination Constraints and Mapping Coverage

As described earlier, the placement of Orbit B1 on the timeline is not ideal for illumination conditions. It is understood that at most 67% of the surface will be illuminated with incidence angles $< 70°$ at the start of Orbit B1. However, Orbit B1 provides an excellent opportunity to map the south polar regions. During the 10th cycle, the entire southern hemisphere is fully observable with incidence angles $< 70°$.

The northern hemisphere becomes illuminated in Orbit B2. Starting with viewing similar to the beginning of Orbit B1, the sun moves towards the north throughout Orbit B2, continually revealing new terrain. The timeline accommodates 11 cycles of Orbit B2, and by the end of the phase, the surface is well illuminated up to 45° North latitude. The final cycle of Orbit B2 will be imaged at a nadir attitude to allow for multispectral coverage to the highest extent towards the north pole. All imaging requirements could be met in Orbit B2 alone with a cumulative coverage of 91% of the surface, but when combined with the south polar views, the coverage increases to 92%. The operations margin of the mission is allocated to the end of Orbit B2. If any operations margin is consumed during the earlier orbit phases, then Orbit B2 will be delayed and execute during periods with better lighting without risk of introducing a gap between B1 and B2 coverage. If the operations margin is not consumed, then Orbit B2 observing can extend to the nominal end of prime mission and provide coverage up to 75° North latitude. There is a solar conjunction towards the end of the nominal observing period when the spacecraft cannot communicate with the DSN due to interference from the sun, but there is nothing to preclude the spacecraft from continuing to image the surface during conjunction. The images would be stored onboard until conjunction has ended, and the data transmission can resume. Figure 18 illustrates the illuminated surface available for observing in both Orbits B1 and B2.

4.3.4 Accommodation of Optical Topography Imaging Requirements

The imaging science requirements for Orbit B can be operationally categorized into two separate but compatible campaigns: multispectral nadir mineralogy mapping described above, and single-filter off-nadir topography. To acquire an off-nadir imaging dataset that supports use of optical topography, the imaging plan must meet a set of lighting and stereo separation requirements as described previously in Table 6. The two numerical optical topography techniques used by the Psyche project, stereophotogrammetry and stereophotoclinometry, benefit from imaging datasets with slightly different attributes. For more details on these techniques see Jaumann et al. (2022). The goal of the observing campaign is to select off-nadir pointing angles that will satisfy both sets of stereo criteria. Selecting off-nadir angles

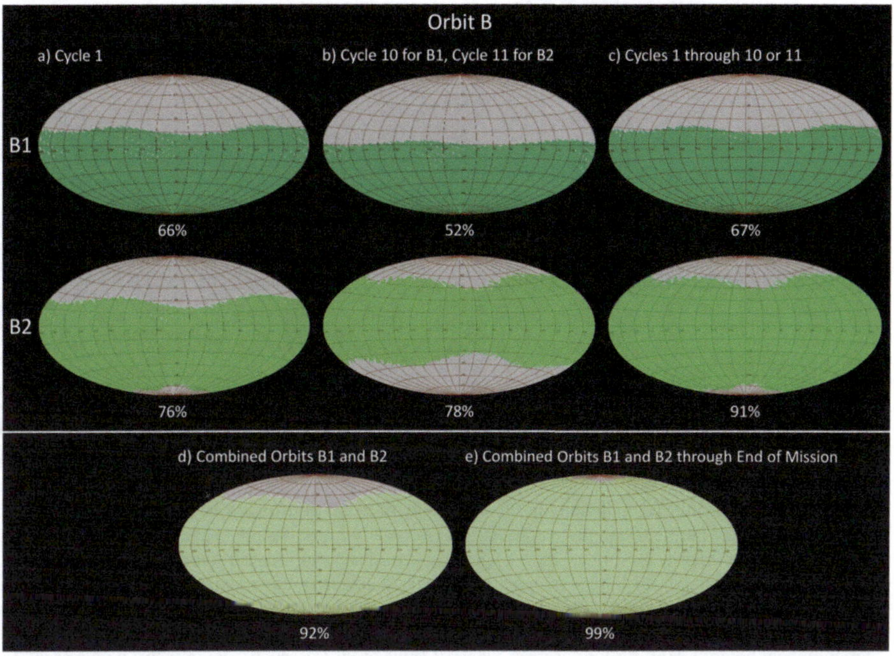

Fig. 18 The illuminated surface available with incidence angles $< 70°$ at the beginning and end of Orbits B1 and B2 is shown here. (a) Cycle 1 coverage for both Orbits B1 and B2 are shown. The improved surface illumination in Orbit B2 is evident in the coverage results. (b) Coverage results for the final cycle for Orbits B1 (Cycle 10) and B2 (Cycle 11) are shown. (c) The accumulated coverage over each science orbit phase for Orbits B1 and B2. (d) Coverage results from combining Orbits B1 and B2. (e) The addition of bonus mapping in the operations margin at the end of Orbit B2 brings the total mapping coverage to 99% for the mission

in groups of four that form a rectangle centered on nadir when plotted in terms of look direction provides a more robust stereo dataset that accommodates the requirements for both stereophotoclinometry and stereophotogrammetry.

The current baseline that satisfies our requirements is the set of eight fixed off-nadir attitudes shown in Fig. 19 (a) where "Ahead" refers to looking ahead of or behind the spacecraft velocity vector and "Cross" is looking perpendicular to the velocity vector. On the Psyche spacecraft we command these attitudes in terms of the "local vertical" and "local horizontal" directions. With eight off-nadir cycles, we were able to construct two different sets of complementary off-nadir attitudes. The additional cycle at the end of Orbit B2 returns to the nadir attitude. The imaging plan was modeled using these attitudes and then scored against the topography criteria in Table 6 with the CKVIEW multi-mission planning tool (Matz and Roatsch 2018). CKVIEW evaluates the attributes of all images of each surface element against the criteria and provides a score of the number of contributing views, enabling science planners to predict how well the observation plan will support stereo image processing. Four or more views of each surface element are desired for robust topography extraction although three views can be sufficient. The results shown in Fig. 19 (c) and (d) for the "useable" criteria demonstrate that greater than 80% of the surface satisfies the topography criteria for both techniques using the complete datasets of Orbits B1 and B2. The combination of the extensive south polar mapping in Orbit B1 with the optimal equatorial surface illumination in Orbit B2 creates a robust imaging dataset for topography. Stereopho-

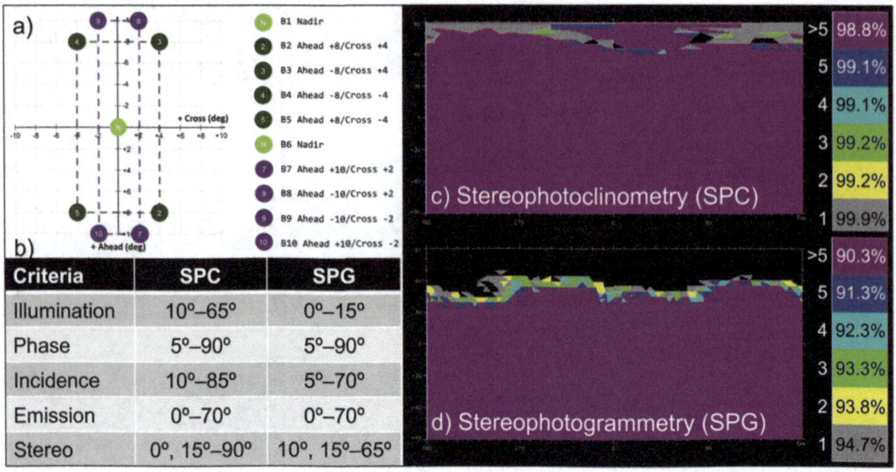

Criteria	SPC	SPG
Illumination	10°–65°	0°–15°
Phase	5°–90°	5°–90°
Incidence	10°–85°	5°–70°
Emission	0°–70°	0°–70°
Stereo	0°, 15°–90°	10°, 15°–65°

Fig. 19 Summary of Orbit B topography results. (a) Representation of off-nadir angles planned for Orbit B. The dashed green and purple rectangles emphasize the two sets of four off-nadir angles chosen. (b) Definition of "useable" stereo criteria used to evaluate expected performance in reconstructing topography from the planned imaging campaign for both stereophotoclinometry and stereophotogrammetry. (c) Stereo coverage on a latitude/longitude body map using the stereophotoclinometry criteria. Colors defined in legend show how many stereo views are available for each element of the surface. Coverage results are cumulative with the requirement being > 80% for four or more views. Results of 99.1% satisfy requirements with margin. (d) Stereo coverage using the stereophotogrammetry criteria. Results of 92.3% satisfy requirements with margin

toclinometry techniques can incorporate images with incidence angles above 70° resulting in higher coverage results than shown in Fig. 18.

Orbit B1 provides topography for only the southern hemisphere, but that is sufficient for us to design and implement the thrust sequences for the complex orbit transfer to Orbit D.

4.4 Transferring to and from Orbit D

4.4.1 Objectives

Between each science orbit, the spacecraft must thrust with the electric propulsion subsystem to achieve the next set of orbit requirements. The objectives of these orbit transfers are strictly driven by navigation and engineering to safely transport the spacecraft to the next science orbit. The transfers from Orbit B1 to D and from Orbit D to C are unique because they include a plane change between polar and equatorial orbits plus additional objectives for the GRNS and Magnetometry Investigations.

The GRNS Investigation benefits from this fortuitous opportunity to bookend their primary science phase, Orbit D, with two high-altitude calibrations where the spacecraft is far from gamma rays and neutrons originating from Psyche's surface. GRNS requires 40-days of background calibration at an altitude greater than 3.5 body radii after capture into Psyche orbit and in the vicinity of Orbit D. This requirement is satisfied during these two large orbit transfers as long as the GRS is configured in its full science mode, cooled to cryogenic temperatures.

The Magnetometry Investigation is searching for evidence of Psyche's remanent magnetic field by sampling the region around Psyche at different geometries with respect to the solar wind and with respect to the body. The additional viewing points provided by transfer orbits, which allow for continuous magnetic field observations over a large range of distances, will augment science opportunities for the Magnetometry Investigation to capture structures like a magnetopause or magnetic tail, whose actual distance from the body is not known. Data collected during the transfer orbits will aid in characterizing the shape of the interaction region between the asteroid and the solar wind, which is critical for extracting the fields caused by any remanent magnetization of the asteroid.

4.4.2 Transferring Between Polar and Equatorial Orbits

The orbit plane change required to transition from a polar orbit to a nearly equatorial orbit is most efficiently accomplished at a higher altitude than Orbit B1. Orbital velocity decreases with altitude and rotating a smaller velocity vector requires less change in spacecraft velocity, or delta-V (ΔV), than a larger one. Consequently, it can be more efficient to temporarily raise and lower the orbit to perform an inclination change than to execute the maneuver at the original altitude. This is especially true for a low-thrust spacecraft that cannot quickly perform maneuvers at optimal locations. Therefore, before spiraling down to Orbit D, the spacecraft will spiral up to altitudes significantly higher than Orbit A (see Fig. 20). These orbit transfers are significantly longer than the other transfers, at 98 days for transfer into Orbit D and 90 days to transfer into Orbit C. These duration estimates currently contain 50% margin, and the transfer to Orbit D also must accommodate a period of solar conjunction when the sun is between the spacecraft and Earth and blocks our ability to communicate with the spacecraft. Finally, the orbit transfer from Orbit B1 to Orbit D needs to achieve the high-altitude plane change before eclipses are predicted to occur for polar orbits but arrive in Orbit D after the period when eclipses are expected in equatorial orbits.

The specific implementation details of these orbits transfers are still being designed as they depend on resolution of other trades that the spacecraft team is working during the cruise phase.

4.4.3 Instrument Activities During These Orbit Transfers

While all aspects of the orbit transfer design are driven by the need to navigate the spacecraft between the polar orbits Orbit B1 and C on either side of the equatorial Orbit D, activities for GRNS and the Magnetometer can be accommodated without impacting the engineering objectives. Aside from these orbit transfers, Orbit D is the only phase where GRS will be operating its HPGe detector at cryo-temperatures. Prior to configuring for Orbit D, the HPGe crystal will be annealed near the end of Orbit B1 operations. The detector will be cooled and the instrument configured for its high-rate science collections. This allows GRS to collect calibration data during the transfer up to high altitudes before and after collecting its primary dataset in Orbit D. The requirement to accumulate 40 days of calibration data can be achieved by combining the data from both orbit transfers.

No additional commanding is required for the Magnetometer during the orbit transfers; however, the quality of the magnetic field measurements is degraded while the Hall thrusters are operating. There are frequent coast periods for communication with the spacecraft when the thrusters are off, and data collected during these coast periods is an important contribution to this study. It is possible that the thruster signatures can be calibrated out of the data if it would appear to be of high value. However, even after calibration, the noise floor would still be about twice as high during thrusting intervals than during coasting intervals, so the quality of the fully calibrated data would be degraded for these intervals.

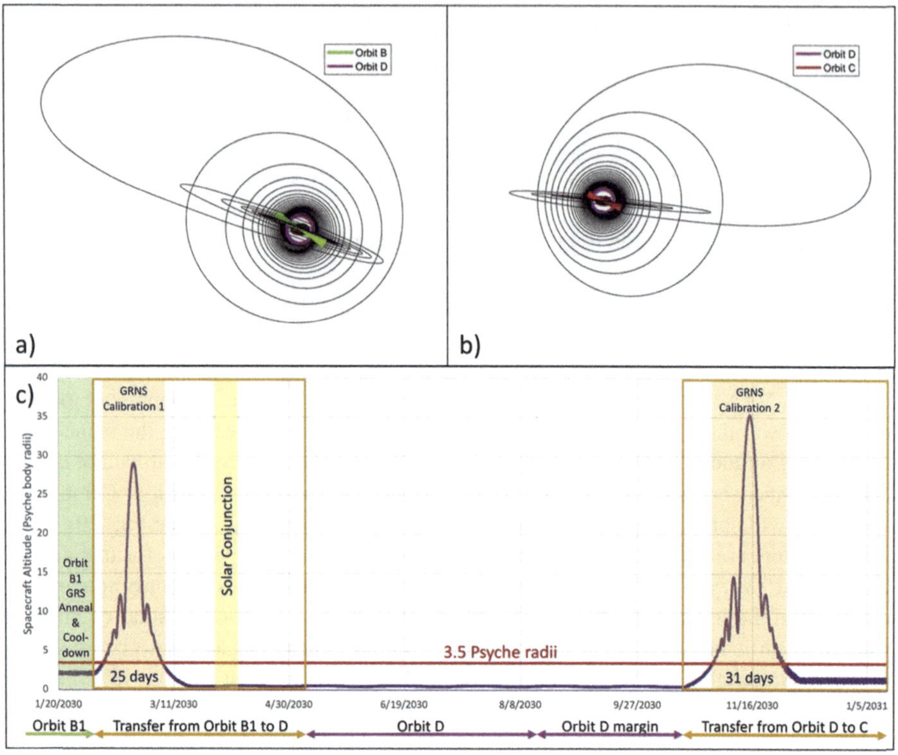

Fig. 20 Illustrations of an example (a) orbit transfer from Orbit B1 to Orbit D and (b) orbit transfer from Orbit D to Orbit C showing the plane change at high altitude. Note that these transfer orbits reach altitudes of 29 and 35 Psyche radii, respectively, far exceeding the highest science orbit altitude. (c) This graphical timeline demonstrates the placement of the GRNS calibrations during the high-altitude periods of the orbit transfers before and after Orbit D. The purple curve is the spacecraft altitude in units of Psyche body radii. Thicker portions of the line are due to altitude fluctuations. The horizontal red line at 3.5 body radii is the altitude above which the GRNS can obtain useful background calibration data. The 40-day calibration requirement is satisfied by combining the 25 days from the first transfer with the 31 days from the second transfer. The orbit transfer durations are designed to have 50% margin as they are expected to increase in duration once all constraints are accommodated. The transfer to Orbit D includes a 10-day solar conjunction period

4.5 Orbit D: Determining Psyche's Elemental Composition

4.5.1 Objectives

Orbit D differs significantly from the other science orbits in the mission. Its design is driven by the requirement to orbit less than one Psyche body radius from the surface to provide gamma-ray and neutron data collection with sufficient signal from Psyche's surface compared to the background counts from galactic cosmic rays. Data from GRNS will provide global abundances of key elements that could answer the question of whether Psyche is a core. Identifying the major elements in silicate compositions may help discriminate between achondrites, cumulates, and chondritic compositions. A global average nickel abundance measurement of less than 4 wt% indicates Psyche formed under reducing conditions, whereas abundances greater than 12 wt% indicate highly oxidizing conditions. Lastly, mea-

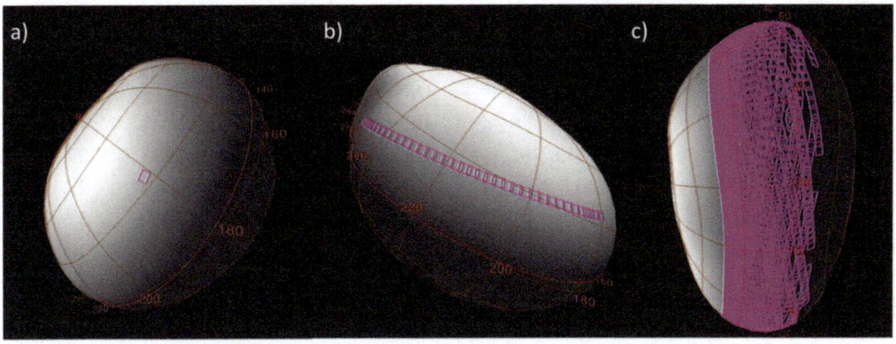

Fig. 21 Imaging in Orbit D taken from the perspective of the spacecraft. (a) One Imager FOV. (b) One nadir swath of 37 images covering about 25% of the lit-side pass. The trajectory hugs a band around the terminator for the entire phase. (c) The equivalent of 25 days (171 orbits) of imaging with 37 images/orbit. Placement of the images over features of interest viewable from a nadir attitude will be planned during Orbit B1 execution and the transfer to Orbit D

surements of the average abundance of light elements that appear to be in the metal phase can be compared to models of their abundances in Earth's core.

Gravity science requires the equivalent of 35 days of DSN tracking with the Sun-Earth-Probe angle greater than 40°. While Orbit D begins with very low Sun-Earth-Probe angles and a solar conjunction, there are 118 calendar days where the Sun-Earth-Probe angle meets requirements corresponding to 59 cumulative days after accounting for a more conservative 50% DSN duty cycle. With high quality gravity tracking at low altitude, we expect to recover the gravity field to degree and order 12 for Psyche's equatorial band, the only region sampled from this high inclination orbit (Zuber et al. 2022). The global gravity dataset is acquired later in Orbit C.

OpNav requires a minimum of 24 images per lit side pass to support orbit determination; however, any science imaging acquired in Orbit D can be used to meet the OpNav requirement as long as it is distributed across the observing period. The data budget limits acquisition to 37 images per orbit which is about a quarter of the images needed to cover each lit-side swath; therefore global mapping is not possible at this altitude.

Opportunistic imaging is planned to capture higher resolution imaging of features of interest identified during Orbits A and B1 that are visible along the spacecraft ground track in Orbit D. The high ground speed will limit useful imaging to the polychromatic filter to reduce exposure time and avoid smear. The equatorial orbit constrains the terrain accessible for imaging as shown in Fig. 21. Off-nadir pointing is restricted to 10° around nadir. This allowance is intended to provide relief from accumulated pointing errors but can also be used operationally if needed.

4.5.2 Implementation

As described in the previous section, an almost equatorial orbit (160° inclination) is required to achieve the required orbit stability for the current Psyche shape model. While the baseline orbit is sufficient to provide good coverage for the GRNS instrument, the final orbit inclination will be re-evaluated once the spacecraft is in Orbit A and Psyche's gravity field is mapped. If possible, the inclination will be reduced to provide better coverage for most science investigations. The equatorial inclination benefits the Magnetometry Investigation

Table 11 Orbit D planned characteristics

Orbit D start date	5/7/2030	Average orbit period	3.6 hours
Orbit D end date	10/17/2030	Orbit resonance	153:196
Duration	163 days	Orbits per ground track repeat cycle	196
Average radius	198 km (1.8 R_{Psyche})	Ground track repeat cycle duration	29.12 days
Average altitude	75 km (0.8 R_{Psyche})	Operational repeat cycle duration	29.7 days
Average inclination	160°	Number of operational repeat cycles	5.6
Orbit beta angle range	[55–85] deg	Number of orbits	1086
Maximum ground speed	125 m/s	Imager pixel scale	4 m/pixel

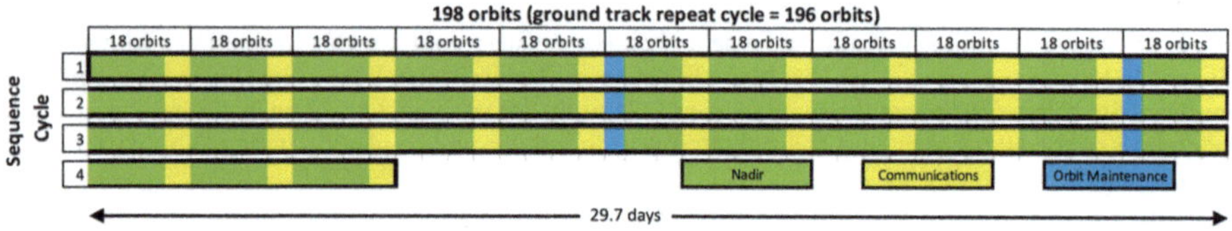

Fig. 22 Graphical depiction of the four cycles for Orbit D. Each ground repeat cycle is 196 orbits divided into roughly 75% nadir observing (green) for GRNS and 25% with HGA pointed to Earth for communications (yellow). The 18-orbit template was selected to provide the opportunity to evaluate spacecraft health every two days

by allowing sampling of new orientations of Psyche with respect to the solar wind. Details of Orbit D are listed in Table 11.

An advantage of the new mission profile is that eclipses are no longer expected in Orbit D. The spacecraft has been qualified to withstand eclipses up to 65 minutes in duration if circumstances change due to updates in Psyche's physical parameters or a shift in operational timeline, as originally included in the plan for the 2022 mission.

Unlike Orbits A and B1, the short 3.6-hour orbit period precludes turning the HGA to the DSN every orbit, so the spacecraft is pointed towards Psyche for 13.5 orbits to collect science data and then the HGA is pointed towards Earth for 4.5 orbits of telecommunication and Doppler data for gravity science. The turns can occur on the lit or dark side of the orbit because imaging is very limited in this phase. There is a 3-orbit period allocated to orbit maintenance at the beginning of the sixth and eleventh science periods. One ground track repeat cycle in Orbit D is 196 orbits. The operational repeat cycle shown in Fig. 22 is 11 sets of 18 orbits for a total of 198 orbits.

The GRNS will already be configured in its full science collection mode coming into Orbit D and is expected to require little commanding during this phase. GRNS does not need the surface to be illuminated to collect gamma rays and neutrons, but the spacecraft is required to point within +/− 10° of nadir during the data collection periods with a reconstructed Psyche-relative pointing knowledge uncertainty of 1° per spacecraft axis.

Modeling of GRNS instrument performance indicates that 52 days of data collection while pointing nadir is sufficient to accumulate enough signal to detect calcium (Ca) (Lawrence et al. 2025). The Ca requirement was used because it is the longest accumulation time of any of the required elements. Nickel, a key element in the science requirements, requires only 36 days of accumulation time and iron can be detected in less than a day. The 52 days of nadir accumulation will take 72 calendar days to acquire in Orbit D assuming a 75% nadir duty cycle and an orbit maintenance period. This provides 91 calendars days of margin in Orbit D. These values are based on an assumed orbit altitude of 0.8 Psyche radii. There

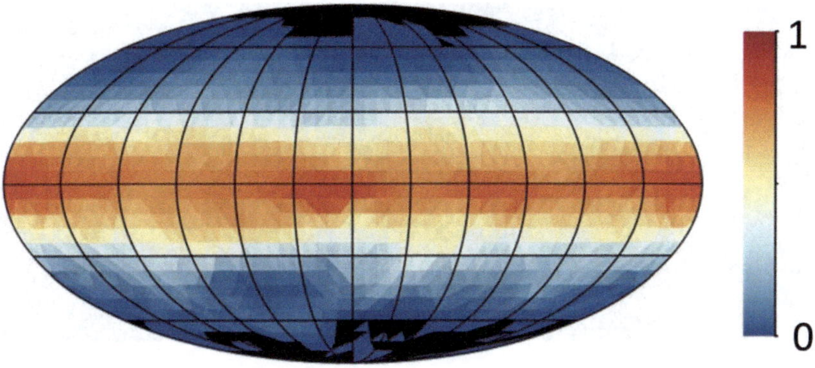

Fig. 23 Surface sampling. The map shows the relative contribution of surface regions to the response of the GRNS while in the Orbit D with the Shepard et al. (2017) shape model. Since the orbit has high inclination, the equatorial band between +/−20° latitude contributes more than the mid-to-high latitude regions. This is primarily due to the solid angle of surface parcels at higher latitudes being lower than those at the equatorial sub-spacecraft points. When averaged over all orbital measurements, 76.5% of the surface contributes >5% of the maximum signal to the instrument response. This threshold criterion is used to evaluate coverage for different orbits. The sampling map is from an Orbit D "bounding cases" study for which several permutations of Psyche's shape and corresponding orbits were considered. The map was normalized so that the maximum sampling intensity across all permutations considered was 1. The baseline orbit provides ample sampling of the equatorial region

may be additional time at other attitudes where GRNS would have an unobstructed view of Psyche by virtue of the boom. GRNS can sense Psyche even when not nadir pointed if the spacecraft is not blocking its view of the asteroid, but this time is considered bonus data and not part of the formal requirement. Events that would consume Orbit D margin are strong solar energetic particle events that can dominate the Psyche-originating gamma rays and neutrons for days to weeks, or a spacecraft safe mode entry that would ramp down the GRS high voltage. To avoid warming the HPGe crystal during safing events, which would then require a lengthy anneal of the crystal to recover, spacecraft fault protection does not always power off the GRS cyrocooler during transition to spacecraft safe mode (see Sect. 5.3.8 for details).

4.5.3 GRNS Surface Coverage

Due to the large field of view of GRNS, pointing the instrument nadir in Orbit D provides greater surface coverage than what would be expected from inspecting the spacecraft ground track alone. The instrument response was modeled for the Orbit D scenario to validate the requirement that greater than 70% of the surface is sampled. The results of this analysis are shown in Fig. 23 for the Shepard et al. (2017) shape model and 2022 launch trajectory. The current Orbit D trajectory is similar, and the timeline now includes an additional 63 days of margin at this altitude.

To investigate the signal intensity and spatial resolution of GRNS in Orbit D, Fig. 24 provides a histogram of solid-angle-equivalent altitudes (h_{eq}) for a series of orbital locations sampled uniformly in time, analogous to GRNS data acquisition. The equivalent altitude has units of body radii and is the altitude above a sphere of unit radius that gives the same solid angle as that of the Psyche shape model (Prettyman et al. 2019). As such, h_{eq} has an inverse relationship with signal intensity. For comparison, low altitude elemental mapping orbits by Dawn at Vesta and Ceres had mean altitudes of about 0.8 body radii (Prettyman

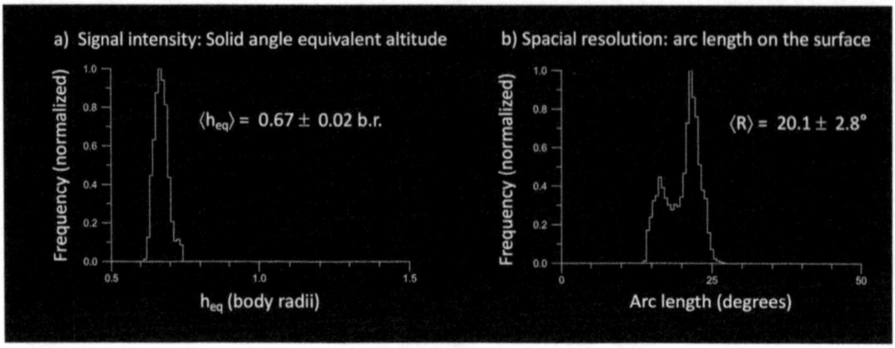

Fig. 24 Signal intensity and spatial resolution for the Orbit D baseline. (a) A histogram of solid-angle-equivalent altitudes (h_{eq}) for a series of orbital locations sampled uniformly in time. (b) A histogram of the width of the GRNS instrument response (1-sigma arc length on the surface) determined for the same time series of locations

et al. 2012, 2017). The measurement geometry of the Psyche's baseline Orbit D, with a mean h_{eq} of 0.67 body radii, would have higher signal intensity than achieved by Dawn. The variability of h_{eq} indicated by the histogram reflects variations in orbital altitude and the shape of Psyche. Also shown is a histogram of the width of the GRNS instrument response (1-sigma arc length on the surface) determined for the same time series of locations. Surface regions with different compositions can be resolved by the instrument if they are separated by about 2-sigma (Prettyman et al. 2019). The average 1-sigma width is 20°, which implies a full width half maximum resolution of about 40 degrees arc length on the surface. This would enable elemental analyses of roughly 10 separate regions in the equatorial band.

4.6 Orbit C: Gravity Science, Improved Topography, and Magnetometry

4.6.1 Objectives

Orbit C is designed to provide the conditions necessary to meet the gravity science requirements that require 35 days of tracking with the DSN at a Spacecraft-Earth-Probe angle greater than 40° in a polar orbit with global ground track coverage at an altitude less than 1.7 Psyche radii. The Orbit C baseline altitude of 174 km is approximately 1.7 Psyche radii, satisfying the gravity science requirement. To support gravity modeling, additional imaging from nadir and off-nadir attitudes provides a higher resolution shape model for the illuminated southern hemisphere to supplement the model produced with data from Orbit B. The smaller projected FOV on the surface increases the number of image stations needed for coverage (Fig. 25) and precludes imaging with all narrowband filters within the available data budget, although some filters may be added once in orbit if the data budget allows. There are no imaging surface coverage requirements for Orbit C.

The altitude is also sufficiently low to meet the magnetometry requirement to determine whether Psyche has a remanent magnetic field, although that requirement is expected to have been satisfied in Orbit D. Magnetometer requirements are compatible with and less driving than gravity science and imaging so specify no special attributes of Orbit C.

4.6.2 Implementation

A polar orbit is required to provide the necessary ground track coverage to meet the gravity science requirements. The 100-day duration of this phase is driven by the gravity science

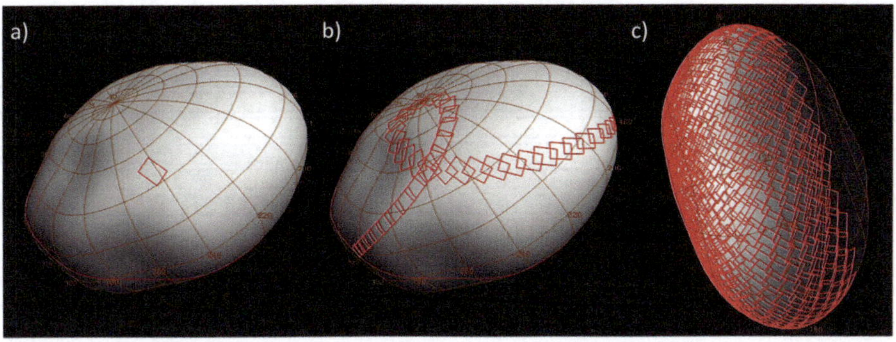

Fig. 25 Imaging in Orbit C taken from the perspective of the spacecraft. (a) One Imager FOV projected on the surface in the southern hemisphere; (b) One orbit of nadir imaging; (c) One cycle of nadir imaging coverage if every orbit was able to be imaged. The operations plan skips multiple orbits for communications passes as shown in a later figure. The northern hemisphere is not well illuminated

Table 12 Orbit C planned characteristics

Orbit C start date	1/15/2031	Average orbit period	7.2 hours
Orbit C end date	4/25/2031	Orbit resonance	77:45
Duration	100 days	Orbits per ground track repeat cycle	45
Average radius	295 km (2.7 R_{Psyche})	Ground track repeat cycle duration	13.46 days
Average altitude	193 km (1.7 R_{Psyche})	Operational repeat cycle duration	15.6 days
Average inclination	89.7°	Number of operational repeat cycles	6.4
Orbit beta angle range	[35–38] deg	Number of orbits	333
Maximum ground speed	71 m/s	Imager pixel scale	9 m

objective to acquire the equivalent of 35 days of tracking with the DSN at a Spacecraft-Earth-Probe angle greater than 40°. Conveniently, Orbit C begins with the Sun-Earth-Probe angle of 136° and ends at 56°, clearly exceeding that requirement. This provides high confidence that the requirement to resolve Psyche's gravity field to degree and order 10 can be satisfied by the operational plans (for details see Zuber et al. 2022). Details of Orbit C are listed in Table 12.

The architecture of Orbit C is a blend of the approach used for Orbits A and B with that of Orbit D. The short 6.7-hour orbit period motivates grouping multiple orbits when the spacecraft is pointed for imaging followed by grouping orbits pointing HGA-to-Earth for improved gravity science and communications. We allocate 75% of each cycle to imaging with LGA tracking and 25% to HGA communications, neglecting the small amount of time lost to spacecraft turns between those attitudes. Twice per cycle, two orbits are reserved for orbit maintenance maneuvers. As a result, some lit side passes each ground track repeat cycle must be sacrificed while the Imager is turned away from Psyche. The 52-orbit operational cycle shown in Fig. 26 was developed for the original 49-orbit Orbit C resonance and has been adapted to the current 45-orbit resonance. The resulting operational cycle is seven orbits longer than the spacecraft ground track repeat cycle, which intentionally moves the downlink passes to occur over different longitudes each cycle, avoiding systematic gaps in coverage. This strategy also benefits gravity science by providing evenly distributed HGA Doppler tracking over all longitudes. Scanning the chart vertically, each longitude band (as represented by an orbit number in the cycle), has at least two nadir observing passes.

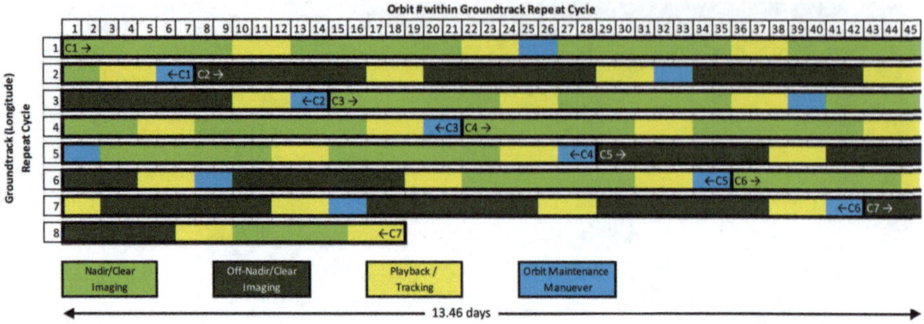

Fig. 26 Orbit C architecture showing observing periods (green), HGA communications (yellow), and orbit maintenance maneuver windows (blue). Each ground track repeat cycle is 45 orbits in duration and each operational repeat cycle is offset from the previous ground track repeat cycle by seven orbits to shift the HGA tracks

Most have at least one HGA pass as well. Future optimization of the template could yield increased off-nadir opportunities for some cycles.

To improve upon the Psyche shape model, four observation cycles are performed on a best-effort basis with the spacecraft at off-nadir angles. The lower altitude drives the selection of larger off-nadir attitudes (dark green periods in Fig. 26) than used in Orbit B. Most portions of the surface will only have two off-nadir angles in addition to nadir.

4.6.3 Illumination Constraints and Mapping Coverage

Orbit C begins with 56% of the surface illuminated, primarily the southern hemisphere, but more of the northern surface is illuminated as the cycle continues. Figure 27 shows how the imaging swaths of the illuminated surface are combined throughout the orbit phase.

The magnetic field mapping coverage provided to the Magnetometry Investigation is computed differently than for the Imager. To retrieve the magnetic field and address the Magnetometry Investigation objectives, the spacecraft trajectory must provide good coverage not only of the surface but also of the three-dimensional magnetic topology around the body. For a given magnetized feature within the body, the magnetic topology extending into space changes continuously due to its interaction with the highly time-variable solar wind. To mitigate this variability, the data are binned for different solar wind conditions and in two reference frames: the Psyche-body-fixed frame, which rotates with the body, and the Psyche-solar-orbit frame, which is tied to the solar wind-body interaction (see Weiss et al. 2023 for details).

To assess the quality of the science orbits for magnetic field retrieval, the body's surface is divided into $10^\circ \times 10^\circ$ bins in longitude and latitude in the Psyche-body-fixed frame. First, we count how many total measurements would be taken in each bin, assuming the Magnetometer operates in its nominal science mode at a frequency of 50 Hz. This can yield up to 10^5 measurements per bin. However, the total number of measurements in a given Psyche-body-fixed frame bin is distributed over several Psyche-solar-orbit frame locations as the spacecraft orbits the body and the sun vector changes, resulting in different magnetic topologies extending above the bin. We estimate the number of different topologies by counting the number of unique Psyche-solar-orbit frame views of the same Psyche-body-fixed frame bin (where each Psyche-solar-orbit frame view is defined as the spacecraft being within a $10^\circ \times 10^\circ$ bin in longitude and latitude around a central value).

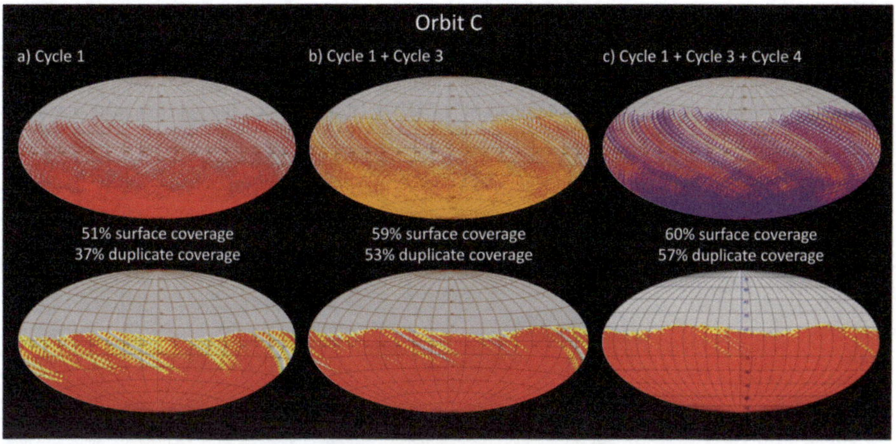

Fig. 27 Buildup of nadir coverage with incidence angle < 70° in Orbit C following the operations schema shown in Fig. 26. (a) Cycle 1 coverage: Top map shows individual Imager FOV and bottom map shows total coverage. Yellow indicates areas imaged only once and red indicates areas imaged more than once. (b) Combined coverage from Cycles 1 and 3. Top map shows individual FOV for Cycle 1 in red and Cycle 3 in orange. Bottom map shows total coverage. (c) Combined coverage from nadir Cycles 1, 3 and 5. Top map shows individual FOV with Cycle 5 in purple, and bottom map is total coverage. By Cycle 4, gaps from downlink passes and maneuvers in earlier cycles are filled in with robust redundant coverage. The northern hemisphere is not well illuminated during this period

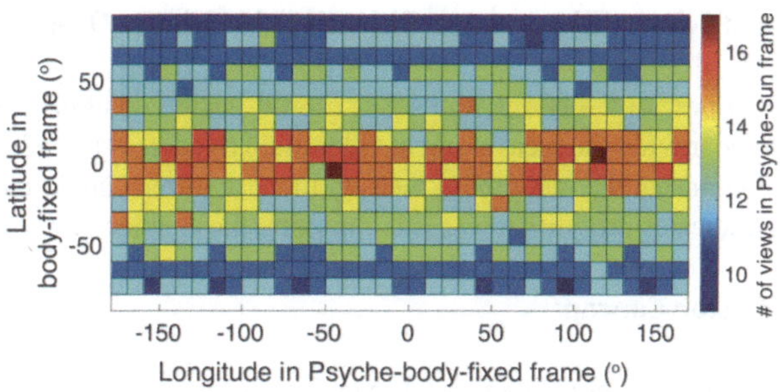

Fig. 28 Colored boxes show the number of unique views in the Psyche-solar-orbit frame from which a given area on the body would be observed by the spacecraft during Orbit C. Each view corresponds to a different interaction between the solar wind and the same region of Psyche's surface. Data are binned over longitude and latitude in the Psyche-body-fixed frame. Each bin is 10° × 10°, and the spacecraft position is sampled at 1-s intervals. The views in the Psyche-solar-orbit frame are also binned at 10° × 10°. Each view comprises tens of thousands of measurements taken during different segments of the orbit

The results for Orbit C are shown in Fig. 28. We find that each possible magnetic feature would be observed from more than 10 different topologies (and likely more, due to solar wind variability), with each configuration measured many thousands of times. This provides confidence that statistical analysis and related techniques will enable detection of the underlying signal.

Following completion of operations in Orbit C the spacecraft has a 23-day orbit transfer to begin Orbit B2. Activities in Orbit B2 were previously described in Sect. 4.3.

Table 13 The completion of each instrument science measurement requirement can be tracked across the science orbits. Yellow cells indicate significant progress is made towards that objective. Green cells indicate the measurement requirement is met or exceeded. Blue cells indicate additional measurements provide margin against requirements or science margin toward meeting that objective. Purple cells indicate enhanced resolution over limited areas. The "to be determined" (TBD) in Orbit C reflects the plan to update the topography campaign prior to arrival

Observation Type	Science Measurement Requirement	Capability in Baseline Mission				
		Orbit A	Orbit B1	Orbit D	Orbit C	Orbit B2
Imaging nadir (color filters)	**Metal to silicate fraction:** 80% of surface at pixel scale <500m.	82% @ <37m	67% @ <17m			~99% @ <17m
	Oldhamite mapping: 80% of surface at pixel scale <200m.	82% @ <37m	67% @ <17m			~99% @ <17m
Imaging nadir (clear filter)	**Crater counting:** 50% of surface at pixel scale <200m.	82% @ <37m	67% @ <17m	Limited @ <5m	60% @ 11m	~99% @ <17m
	Geologic mapping: 80% of surface at pixel scale <20m.		67% @ <17m	Limited @ <5m	60% @ 11m	~99% @ <17m
Imaging off-nadir (clear filter)	**Spin axis determination:** accuracy to 0.5° 1σ. *	Continued OpNav imaging in Orbit A	*Measurement requirement complete by mid-Approach. Accuracy to 0.001° 1σ achieved by end of Orbit A.			
	Topography: 50% of surface at <50m vertical scale.		~56% @ <34m		TBD% @ <26m	~99% @ <34m
	Topography: 80% of surface at <70m vertical scale		~56% @ <34m		TBD% @ <26m	~99% @ <34m
	Topography for gravity field: Provide a minimum of 4 off-nadir mapping cycles in Orbit C.				4 cycles Significant mapping overlap provides margin	
Magnetometer observations	**Magnetic dipole:** Global dataset with ground track designed so that each point on Psyche's surface lies within 10° of arc of the ground track in Orbits A, B, C.	4 global datasets <3° arc	10 global datasets <4° arc	3 partial datasets between +/-20° latitude only <2° arc	7 global datasets <3° arc	9 global datasets <4° arc
Gravity science tracking	**Gravity Field:** HGA tracking (Doppler + ranging) in Orbits A & B for at least 40% of each orbit when Sun-Earth-Probe angle > 20°; 35 cumulative days of tracking in Orbits C/D with Sun-Earth-Probe angle > 40°.	~43% HGA tracking	~37% HGA tracking plus >10% LGA tracking	74 days between +/-20° latitude only	100 days global coverage	
GRNS observations	**Element measurements** *(driving case - Calcium)*: 52 cumulative days at nadir in Orbit D.			71 days		

4.7 Completion of Science Observation Objectives by Science Orbit

As described in the previous sections, the science orbit design combined with the science observation strategy provide the measurements identified to address the mission science objectives. Table 13 connects the instrument measurement requirements to the corresponding science orbit and indicates progress towards completing those objectives.

4.8 Extended Mission Concept

The Psyche team is aware that the elemental mapping campaign would be enhanced by reducing the orbit radius to 1.4 Psyche radii (~44 km altitude) or lower. There will undoubtedly be regions of Psyche's surface of elevated scientific interest that could benefit from higher resolution imaging at a lower altitude. The primary mission will improve knowledge of Psyche's gravity and shape as well as understanding how the flight system performs in orbit and what consumables are remaining. This would enable planning for lower orbit operations; details of any specific orbit altitude and other science requirements would be developed during the primary mission and proposed to NASA for consideration.

5 Mission Operations

5.1 The Psyche Team

Psyche mission operations is a joint venture of the Psyche science team and flight operations team along with other partners such as the DSOC project, the DSN, the Small Bodies Node

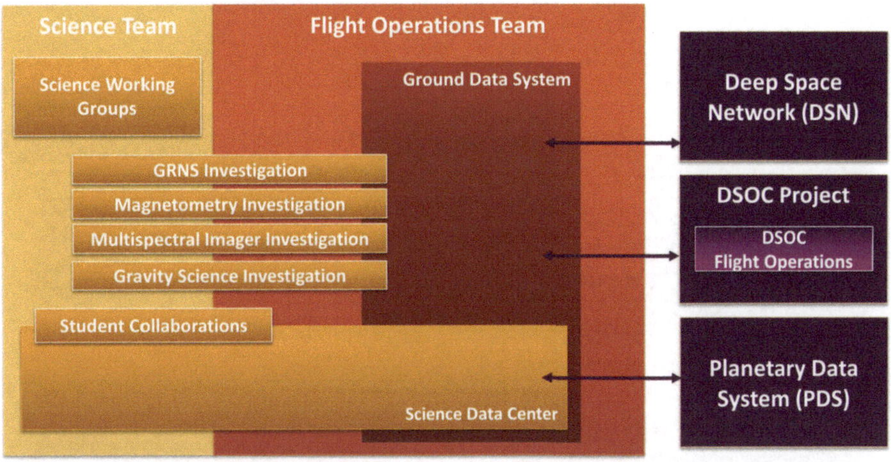

Fig. 29 Project decomposition and external interfaces. The instrument investigation teams and the Science Data Center have responsibilities to both the science team and the flight operations team. While all NASA projects interface to the DSN and PDS, the interface to the DSOC project is unique to Psyche

of the Planetary Data System, and sponsors at NASA. Interfaces between the Psyche project and other close partners are shown in Fig. 29. A vital component of the team not discussed elsewhere in this article is the Student Collaborations Program headquartered at ASU that engages hundreds of undergraduate students each year in capstone projects and in the Psyche Inspired program as described by Bowman et al. (2023). Beyond students, this group also develops the Psyche Innovation Toolkit online courses and leads the Psyche Outreach Interns and Docents as well as assists with other public engagement and media outreach.

Despite the delineations in Fig. 29, the project has never entertained a divide between the science and engineering components of the team. All members of the Psyche project and many external partners are invited to our twice-yearly project team meetings, which feature a mixture of speakers from all disciplines. Prior to orbital operations, these meetings are used to educate the entire project on each other's work.

Perhaps the most critical aspect of the team culture the Psyche leadership team attempts to foster is the idea that every voice needs to be heard. This is an explicit invitation for any person working on any part of the team to speak up with issues or concerns at any time. We have an unofficial motto that "the best news is bad news brought early," allowing enough time for the issue to be resolved. Concerns need to be met with support and inquiry and their bearers embraced. This approach has allowed, again and again over the years, the most junior people on the team, those with their hands on the hardware or their fingers into the code, to raise issues in time to avoid bad outcomes. In fact, the times that problems surprised us were almost always times when someone had not been listened to attentively enough.

We have also led a team norms workshop to realign the team after initial checkout and the stress of making the launch was complete. The team came up with the following norms:

- Behaviors of Individuals: We work to build an environment of mutual trust.
- Ways we Interact: Each person's opinions and questions matter and are heard and valued.
- Actions of Leadership: Leaders recognize each team member's professional excellence and their importance on the team.

The team will continue to revisit and stress-test these norms statements by working through dilemmas and case studies in practice for orbital operations and its attendant higher stress.

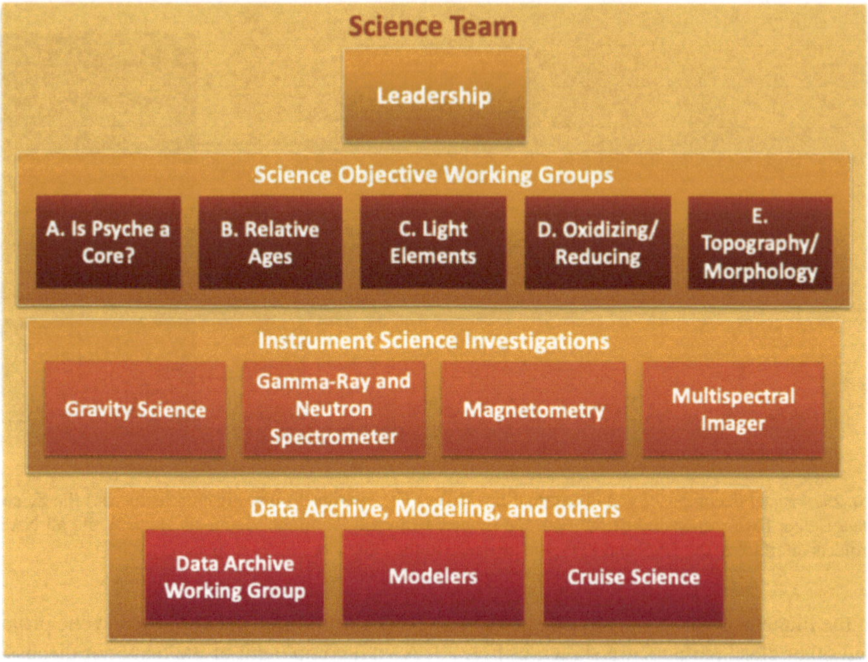

Fig. 30 The working group organization within the science team. In practice, there is significant overlap in the membership between science working groups so it is the responsibility of the working group leads to ensure that each science objective receives dedicated attention and analysis. Members of the instrument science investigations and other groups are also included in the science objective working groups

5.1.1 Science Team and Working Group Structure

The science team, led by the Principal Investigator, consists of the Co-Investigators approved by NASA, scientific Collaborators sponsored by the Co-Investigators, and Participating Scientists selected by NASA prior to Psyche arrival. All members of the Psyche team, both science and engineering, are bound by the Psyche Team Guidelines document and Psyche Communications Plan. The science team is organized into three tiers of working groups (Fig. 30). The top tier is the five Science Objective working groups, one for each of the science objectives. These groups synthesize the relevant data analysis from multiple instruments to address that specific science objective and facilitate the science by ensuring that all relevant data, people, and ideas are involved. By design, the Science Objective working groups are led by Co-Investigators who are not also instrument investigation leads. This encourages a multidisciplinary approach to exploring the key science questions of the mission. While one of the team's leading hypotheses for Psyche's origin is that it is a remnant of a differentiated planetesimal, all of the many possibilities (Elkins-Tanton et al. 2022) must be considered. The science team's organization and philosophy are to keep an open mind and avoid formulating strong conclusions until all mission science data has been collected, synthesized, and evaluated. This discipline is embedded in the ongoing discussions within the working groups and the team at large. Details of the science framework for each working group are described later in this volume (Elkins-Tanton et al. 2022; Marchi et al. 2022; Prettyman et al. 2025; McCoy et al. 2022; Jaumann et al. 2022).

The next tier contains the four instrument science investigations that assure calibrated, performant data for each of the three science instruments plus the gravity science investigation. These working groups also ensure that their data is interpreted properly by the other members of the team. The working groups in the third tier are not connected by a theme. The Data Archive working group provides the interface between the Psyche project and the Small Bodies Node of the Planetary Data System. This group works closely with the Science Data Center. There is another loosely defined group of Co-Investigators who are primarily modelers. They have no formal leadership but coordinate their modeling efforts. Finally, when topics expected to be of broad interest to the Psyche team arise, ad hoc working groups can be formed to meet specific needs.

One example of the last category is the Cruise Science working group that promotes and advocates for opportunistic science analysis activities during the interplanetary cruise phase using data collected for purposes of calibrating the science instruments (e.g., solar proton measurements by GRNS or interplanetary magnetic field data) or from engineering subsystems (e.g., single event upset data from memories used in the spacecraft avionics). Potential scientific topics include heliophysics, emphasizing space weather such as validating/improving the physics-based space weather forecasting tools for solar energetic particles and solar wind, understanding the propagation of coronal mass ejections and solar energetic particle events in interplanetary space using multi-point observations (e.g., Geostationary Operational Environmental Satellite, Europa Clipper, Jupiter Icy Moons Explorer, BepiColombo, etc.), and extrapolating the solar modulation index from the GRNS data.

5.1.2 Flight Operations Team

The spacecraft is operated at JPL in Pasadena, CA. The flight operations team is made up of the following teams: Spacecraft, Mission Planning and Sequencing, Mission Design and Navigation, Science Operations Support, Ground Data System, and Science Data Center at ASU. These teams are supported by project leadership/management, mission management, and mission operations assurance management. The Psyche project organization is shown in Fig. 31, with the flight operations team being the group of teams below mission management.

These teams work together to operate the mission—everything that needs to happen to get the spacecraft to Psyche and then acquire the science data. A brief description of each team's role follows. The functions of the Spacecraft team include the monitoring and analysis of spacecraft health and performance, planning of future spacecraft activities, and generation of spacecraft commands (both sequenced and real-time) to execute planned events. The Science Operations Support team is responsible for coordinating with the instrument teams on instrument health, as well as developing the instrument activities during cruise and at the asteroid. The Mission Planning and Sequencing team is primarily responsible for sequence/command generation, modeling, review and verification, and DSN scheduling. The Mission Design and Navigation team is responsible for tracking and predicting the spacecraft's position, designing the trajectory and maneuvers required to get to Psyche, as well as adjusting the orbits while there. The Ground Data Systems Operations team is responsible for the execution and monitoring of data processing of Psyche mission data during passes, monitoring and maintaining the Ground Data System software and interfaces used to perform operations, and for providing user support. In addition, outside of DSN passes the Ground Data System Operations team performs data accountability for downlink products received and processed by the telemetry processing system, archives and restores (when necessary) data produced by the telemetry processing system, and produces the data

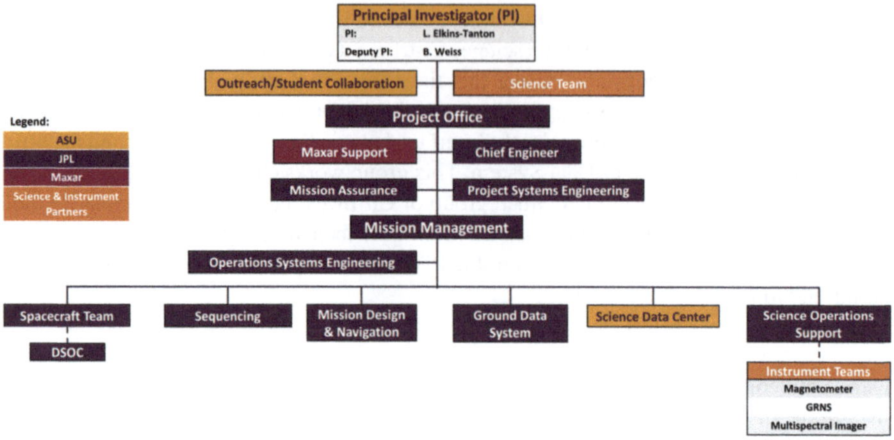

Fig. 31 Psyche Phase E project organization for operations. Many aspects of the Psyche organization are common to deep space flight projects. Unique to the Psyche project is the partnership with Maxar. While the spacecraft is operated from JPL, Maxar provides on-call support for anomaly investigation and other studies as needed

accountability report. Finally, the Science Data Center serves as the central clearing house of PDS products from the Psyche mission, generated by instrument teams and the science team. The Science Data Center is also tasked with being a source of instrument data and telemetry, used by the instrument teams to generate the RAW PDS4 data products that feed into downstream processing to generate CALIBRATED and DERIVED products. In addition to storing and disseminating project PDS4 data products, the Science Data Center team is responsible for the generation, validation, and delivery of mission PDS4 archives to the Small Bodies Node, which curates data relevant to comets, asteroids, and interplanetary dust (see Sect. 6).

5.2 Science and Instrument Operations

Science and instrument planning and sequencing are performed by the Science Operations Support team. This team has responsibility for science system engineering, science planning, science instrument sequencing, telemetry monitoring for instrument health and safety, instrument anomaly resolution, and tracking progress towards meeting the science observing (Level 2) requirements. Science operations support includes staff at JPL as well as the instrument operations teams at the home institutions of the instrument science investigations. Science data processing and archiving are the responsibility of the instrument science investigations that span both the science team and the flight operations team.

5.2.1 Science Planning and Sequencing

The Psyche science operations concept and processes were developed to support implementation of the mission's science goals and objectives that were defined early during the project's proposal phase. These science goals and objectives set the foundation for every aspect of the mission's design, planning, and execution (see Fig. 32). At the start, the initial proposal team developed the mission concept, which included specifying the target destination, selecting the science instruments, outlining the navigation timeline, and identifying

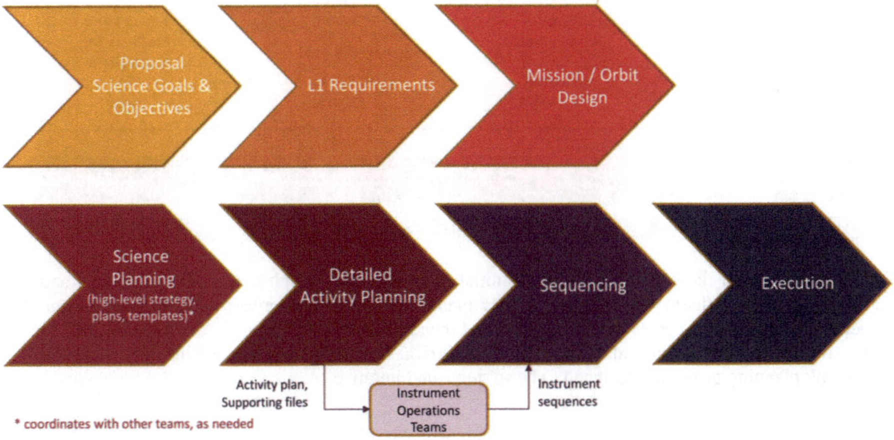

Fig. 32 Top-down approach drives science planning. Activity plans and sequences flow from top-level science goals rather than instrument-driven requests. This approach eliminates time-consuming negotiations between instrument teams for resources and ensures that the science data collected addresses the primary science objectives

the required engineering flight system capabilities. During the project's early development years, these high-level concepts were then translated into detailed requirements, which provided guidelines for spacecraft design, operations, and scientific data collection. As these requirements cascaded downward, they became progressively more detailed, extending to all subsystems and teams involved in the mission.

The mission's science objectives also led to requirements that informed the design of the science orbits, which aim to ensure that the spacecraft will be positioned in a way that allows optimal observations of Psyche. Orbit design needs to account for factors such as the spacecraft's distance from Psyche and the time available for scientific observations verses communication with Earth and downlink of data. All orbits need to be stable to allow consistent data collection from the Magnetometer and GRNS while also allowing imaging of the asteroid at various angles and lighting conditions.

Starting in the development phase and continuing throughout the mission, the mission planners who support operations use the requirements and resulting orbits to create high-level activity timelines, balancing the spacecraft's science objectives with its engineering and operational needs. These planners strategically lay out the timeline of when and how various science and engineering activities should be scheduled. They consider things such as periodic maintenance and calibration, Mars flyby activities, optical navigation and approach imaging, and time allocations for observing and downlinking data while orbiting the asteroid.

For the periods designated for science, a science planner then works to add the next level of detail to the plans by creating conflict-free observing timelines that align with the broader mission operations activity plan. These science planners must ensure that resources such as data volume, pointing attitude, and telemetry rates are carefully considered and allocated, so that the spacecraft can safely collect and transmit the necessary data within available capabilities. Depending on the science orbit phase, the science team may prioritize imaging of Psyche's surface or the collection of data for elemental mapping, magnetometry, or gravity science. With relatively infrequent commanding, the Magnetometer, NS, and anticoincidence shield of the GRS will operate continuously throughout the mission, with Orbit

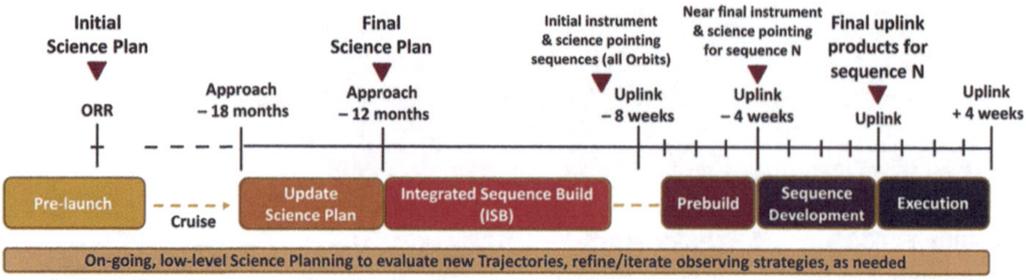

Fig. 33 Timeline for the uplink process from initial science plans through sequence development and execution. Progressive refinement of science activity plans leads to straightforward sequence implementation. Detailed science plans are developed pre-launch and then refined during cruise as additional information becomes available about the asteroid, the trajectory and arrival date, and spacecraft and instrument capabilities. The uplink planning process is designed to be strategic and iterative

D also dedicated to full gamma-ray science utilizing the instrument's high-purity germanium detector. The Imagers, which are crucial for detailed visual observations of Psyche's surface, require the most intensive observation planning. The Imager must be actively managed, and science planners must determine the optimal spacecraft pointing attitudes, ground track placements, and image footprint strategies. This includes decisions about the cadence at which individual images will be captured to achieve maximum surface coverage, along with collecting any ancillary engineering and instrument housekeeping data to support future data analysis.

Science planners work closely with instrument teams during the entire planning process to develop the detailed science observing and data collection strategies. These strategies are then captured in various ways, such as in imaging templates, and then integrated into official science activity plans. The science activity plans and supporting files then guide the instrument operations teams to add the final level of detail by crafting the specific command sequences that will be uplinked to the spacecraft from the DSN to execute the desired observations and other instrument activities. It is this top-down approach outlined in Fig. 32 that puts the Psyche mission's science goals and objectives at the heart of every decision made throughout the design and planning lifecycle and therefore ensures that the mission will provide the necessary valuable data about the target Psyche.

Throughout the mission, the high-level and detailed science operations planning processes will go through several cycles. The general timeline for process steps and refinements can be seen in Fig. 33. A couple years before approach to the asteroid, the entire science plan will be reviewed and updated as needed in response to improved information about the navigation timeline and for any changes in resources or spacecraft and instrument capabilities. At this stage in the mission, knowledge will have improved regarding arrival time, lighting conditions, timeline margins, and operational strategies for aspects like managing spacecraft momentum. The goal of this update period is to adjust and solidify the activity timeline and details enough to feed into an approximately year-long process called the Integrated Sequence Build.

During this early sequence implementation process, a full set of flight sequences will be developed for the orbital operations period, which will then serve as the initial command set to tweak and improve before actual execution once in orbit. The flight operations team will go through the process to plan, build, review, test, and document complete sequences based on the science and engineering activity plans and notional DSN tracking schedule. Besides providing a first build and validation of sequences during approach to the asteroid, science orbits, and orbit transfers between phases, this exercise also helps train and engage all flight operations team members and identify tool limitations. The output of the process will be command sequences ready for flight. Ideally only small adjustments will be needed to align with the actual DSN schedule, arrival time to Psyche, and new knowledge about orbit characteristics from data gathered while closer to the asteroid. Liens are written to document corrections and lessons learned needed for the official sequence development.

The final steps before command files are uplinked to execute on the spacecraft occur in the Science Prebuild and Sequence Development processes. For each defined sequence execution period, the Prebuild process for science is a 3-week cycle during which the science operations team takes the previously-generated instrument activity plans and sequences, current DSN track allocations, latest ephemeris and trajectory files, and any known engineering activity inputs and creates final activity plans and science-related flight sequences. During this timeframe, cross-coordination with other teams is ensured by holding a group "kick-off" meeting to review activity plans, assess the sequence development schedule, assign responsibilities, and disposition liens and changes from the Integrated Sequence Build process. A design "tabletop" meeting is also held where all teams discuss science and engineering activities in more detail, review the intent of the planned activities and changes from previous plans, and outline details of instrument activities such as off-nadir angles and data production rates. This allows overall contextual awareness and sharing of information across the broader flight operations team. Science planners generate initial command files and interface with other teams as needed for input and validation while reviewing all sequences. The output of the process is a set of pre-validated science pointing and instrument sequences that are conflict-free and constraint-checked, which become inputs to the final 4-week Sequence Development process.

A key element of the Psyche uplink process that simplifies the transition from the Integrated Sequence Build process to the Prebuild process is scheduling most spacecraft activities relative to geometric epochs related to the spacecraft's orbit around the asteroid. One common example is the epoch when the spacecraft crosses from the dark side to the lit side of each orbit. Psyche's irregular shape results in each lit-side traverse having a slightly different duration and timing relative to the subsequent terminator crossings. Without the ability to easily update the timing, a sequence built based on an early plan would become misaligned with the illumination conditions if it started one or two orbits later than the original plan. Connecting commanding for each imaging swath to the dark-to-lit terminator crossings allows for the entire plan to shift and synchronize with the current conditions by simply updating a file of epoch times generated from the latest spacecraft ephemeris. Timing uncertainties are a standard characteristic of low thrust missions and extensive use of epoch-relative timing enables the team to manage those timing uncertainties with low risk to the uplink process. The epoch update process not only streamlines the transition from the Integrated Sequence Build process to the Prebuild process but also enables final adjustments during the later Sequence Development process. Timing adjustments arise from uncertainties in arrival date, orbit transfer durations, updates to science orbit characteristics, and/or interruption of the plan by spacecraft anomalies.

The main purpose of the Sequence Development process is to generate, integrate, verify, and validate commands that respond to or implement the activity plans generated in the Prebuild process. At this stage, all operations teams submit their complete set of command sequences to be modeled as an integrated set for checking constraints, timing conflicts, flight rules, and command syntax. Sequences may also be tested on testbeds that are running the latest flight software and are connected to instrument engineering models or simulations. The Sequence Development process includes two "passes" through the review and update cycles and provides opportunities for final time epoch, image exposure, and spacecraft ephemeris updates. The last step in the process once sequence reviews and approvals are obtained from all teams is to hold a command approval meeting for all participants to give their official approval for the set of files to uplink and execute onboard the spacecraft.

5.2.2 Instrument Data Collection and Calibration During Cruise

The cruise phase provides a valuable opportunity to operate the science instrument suite and collect valuable calibration data. Early in development, the project recognized the importance of exercising both the instruments and the instrument operations teams during cruise and therefore planned for continuous operation of the Magnetometer and GRNS as a key preparation for orbital operations. Both instruments require minimal commanding unless powered off for special engineering activities or anomalies so do not add significant burden to the team during routine data collection. The Imagers are operated less frequently in cruise because they require dedicated spacecraft pointing to collect calibration data, but a robust calibration campaign was developed pre-launch.

Throughout cruise the Magnetometer monitors the interplanetary magnetic field embedded in the solar wind to provide a zero-level calibration of each sensor. The properties of the solar wind allow the zero levels of all three Magnetometer axes for each sensor unit to be determined, based on the dominant Alfvenic nature of the fluctuations in the solar wind, with fluctuations in direction but not in field strength (Leinweber et al. 2008). No special commanding is necessary although the data used for these calibrations are restricted to periods when the Hall thrusters are not operating. In addition to the interplanetary magnetic field, the Magnetometer is also sensitive to fields generated by the spacecraft subsystems (de Soria-Santacruz et al. 2020). Correlating changes in the measured field to activities on the spacecraft during cruise allows the team to remove these noise sources. For most of cruise the Magnetometer operates in its nominal 50 Hz data sampling mode although there are periods of low downlink capability where the instrument is commanded to its 10 Hz mode.

Portions of the GRNS are also powered on throughout cruise and will monitor energetic particle effects throughout the spacecraft's traverse from Earth to Psyche. Although the GRS is in reduced science mode and the HPGe sensor not cooled to cryogenic temperatures, the GRS anticoincidence shield provides energetic particle and fast neutron measurements (Lawrence et al. 2025). In addition to providing a long-term time-dependent baseline monitor of the galactic cosmic ray background, the anticoincidence shield and NS sensors observe solar energetic particle events, solar X-ray flare events, and gamma-ray burst events. Quick look data from the gamma-ray bursts are shared with the Gamma-Ray Burst Network and used along with other space assets to triangulate the location of these events in space (Hurley et al. 2010).

Every six months during cruise there are periods of higher instrument activity known as Periodic Maintenance and Calibration. The template for these instrument and engineering activities as executed onboard the spacecraft is shown in Fig. 34. This set of activities is designed to calibrate the instruments, maintain their overall health, test various behaviors, and improve the team's overall understanding of their capabilities. Each calibration period is typically four days in duration with daily two-way communications with the DSN. While most activities do not require real-time DSN coverage, these daily passes provide additional downlink opportunities and allow the calibrations to be monitored more frequently.

During a typical calibration period, both Imagers perform photometric, radiometric, and – occasionally – geometric calibrations. The first two calibrations allow the team to determine each Imager's photometric and radiometric accuracy and parameters. The third calibration characterizes their geometric distortion through the creation of a 4x3 mosaic centered on a large star cluster. While the celestial object used for each calibration is subject to

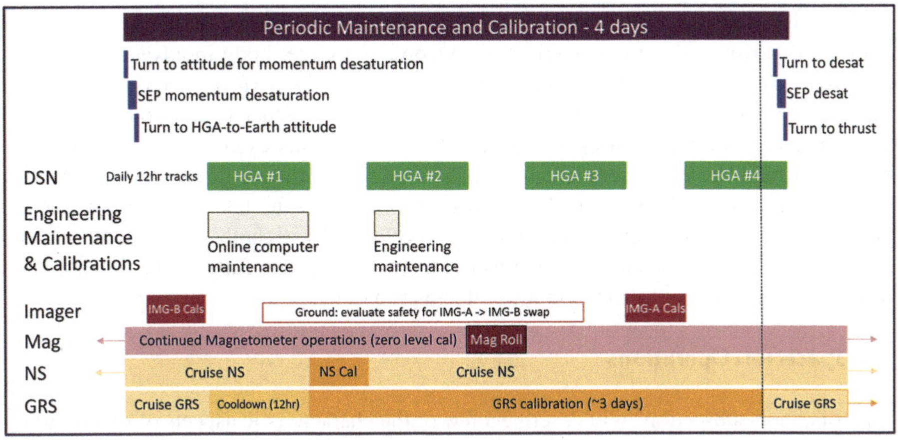

Fig. 34 Layout of the activities for a typical periodic maintenance and calibration windows. The need to thrust with the SEP system for most of cruise leaves limited time to calibrate the science instruments or perform maintenance on the spacecraft. All activities must be efficiently integrated within the 4-day timeline

change, the format of these calibrations is fairly homogeneous. First, multispectral images are collected for a bright, standard star using multiple filters and exposures. Next, the spacecraft is turned towards an object that can serve as a reflected solar light "standard" (such as Earth, Mars, Jupiter, or Saturn) and additional images are collected. If a geometric calibration is included, then the activity completes with a mosaic of a star cluster. Otherwise, a third star may be chosen for additional imaging.

The GRNS operates in its full science mode during cruise periodic maintenance and calibration windows. Early in the calibration window, the high-purity germanium (HPGe) crystal within the GRS is cooled to 85 K and a high voltage bias applied, initiating calibration data collection that lasts all four days. These measurements are designed to check the HPGe's energy resolution and track its depletion voltage over time. Meanwhile, both spectrometers are configured to collect additional raw and diagnostic data to support trend analysis of all sensors. Other parameter tuning activities and experiments utilizing the GRNS may also be included as time permits.

Finally, the Magnetometer operates continuously throughout each calibration period. During this timeframe, the spacecraft is not thrusting so the magnetic field measured is free of thruster noise and invaluable for calibrating the sensors. In addition, the spacecraft bus is commanded to roll around the Y-axis eight or nine times while the solar arrays remain fixed on the sun to measure the magnetic field created by the solar arrays. By rotating the spacecraft through this full range, the magnetic signature of the arrays as function of the array angle may be determined. The impact of the arrays on the magnetic field may then be removed from future measurements through the ground data processing pipeline.

The most unique phase during cruise is the long optimal coasting period before the Mars gravity assist. Long duration coasting provides an opportunity to perform the equivalent of a periodic maintenance and calibration activity (Magnetometer roll, GRNS calibrations, Imager geometric/photometric calibrations) without the typical timing constraints intended to limit interruptions to mission thrusting. The Magnetometer has an extended quiet period in which to perform zero level and gradiometer calibrations, and additionally the Imager is operated over a period of a few weeks to characterize the instrument's susceptibility to single event upsets caused by the space environment.

In addition, the Mars flyby provides an opportunity to rehearse aspects of science operations and perform instrument calibrations involving a target body in advance of arriving at Psyche. The GRS HPGe crystal is annealed to restore the performance (i.e. energy resolution) of the detector, after which it performs a calibration/characterization of at least 10 days including Mars closest approach. The Imager performs extended target calibrations with Mars ranging in apparent size from $> 25\%$ of to completely filling the fields of view of both cameras, and images of Mars at different exposures are taken to assess smear and characterize the in-field stray light. Imaging is performed with both Imagers (although not simultaneously, to limit risk to the hardware) to ensure they are fully calibrated and to determine alignment relative to the spacecraft and each other.

5.3 Spacecraft Operations

The following subsections will describe a few of the characteristic aspects of spacecraft operations of note for the Psyche mission. The spacecraft design is described in detail by Oh et al. (2025a). Spacecraft operations typically entail everything that needs to be done by the flight operations team on the ground as well as the functions that execute on the vehicle to accomplish the mission requirements. The ground activities include monitoring the health and performance of the system, trending the performance, responding to anomalies, planning, and building command sequences for future activities. Those ground activities have been designed in conjunction with the capabilities of the spacecraft. There are many functions that are necessary for mission success but aren't unique to the Psyche mission, such as thermal control of the spacecraft and battery charging/discharging. The subsections below are included as they are notable because the Psyche mission uses SEP to travel to a small body in deep space. NASA has flown two previous deep-space SEP missions, Deep Space 1 and Dawn. During these, JPL has developed the expertise needed to properly accommodate SEP's needs and to take best advantage of its capabilities in a deep-space mission, including systems engineering principles in development and operations, mission operations methods, trajectory design, and navigation (Rayman and Williams 2002; Rayman et al. 2007; Rayman 2020).

5.3.1 Solar Electric Propulsion in Operations

The Psyche mission utilizes SEP to provide the necessary thrust to maneuver the spacecraft after launch until arrival at Psyche and to maneuver in orbit around Psyche. The SEP system consists of several components: solar arrays, power electronics, power processing units, SPT-140 Hall effect thrusters, xenon storage and distribution, and dual axis positioning mechanisms (DAPMs). The solar arrays generate electricity from the sun's energy and are gimbaled in one axis so that they can remain pointed towards the sun. At 1 AU from the sun, the arrays can produce $\sim$21 kW, which goes down to $\sim$2.7 kW at Psyche distances from the sun ($\sim$3.1 AU). The power electronics process electricity from the arrays to regulate the voltage on the power bus, charge the battery, or discharge the battery when the arrays are not producing enough electricity. The power processing units use electricity from the power bus to operate the Hall effect thrusters, which have an input power from about 1 kW to over 4.5 kW. This input power results in about 250 mN of thrust at full power. The SPT-140 thrusters use the electricity to ionize xenon gas and accelerate the ions with a magnetic field which generates the thrust. The xenon gas is stored in seven tanks for a total of 1085 kg at launch and distributed to the thrusters via tubing, valves, and pressure regulators. Finally, the thrusters are mounted to the spacecraft body via the DAPMs with two thrusters per DAPM: one DAPM on the +X side of the spacecraft and one on the −X axis side. The DAPMs are gimbaled in two axes so that the thrusters can be positioned appropriately. One thruster is operated at a time on the spacecraft.

In order to propel the spacecraft, the thruster in use is positioned such that the thrust produced is aligned with the center of mass of the spacecraft and the flight software adjusts the gimbal position to minimize the torque generated by the thruster. A phenomenon that must be addressed for these thrusters is swirl torque: when the ions are accelerated, they produce a small, but continuous torque about the thruster exhaust direction (Oh et al. 2014, 2025a,b). This torque can't be countered by gimbaling the thruster while in use. The torque is absorbed by the reaction wheels by torquing the wheels in the opposite direction, accelerating the wheels. Since the wheels have a maximum speed, another source of torque must be applied periodically to reverse the torque buildup in the wheels. This is achieved on the Psyche spacecraft by pausing mission thrusting, turning the spacecraft approximately 35° to 40° around the Y-axis of the spacecraft to position a thruster on the opposite side to generate a torque to reverse the reaction wheel speed build up. The swirl torque magnitude is related to the thrust level – it is higher at full thrust and lower at lower thrust levels.

As described above, the thrust produced by the SPT-140 thruster is very small compared to chemical thrusters used for primary propulsion. This is countered by their much greater efficiency. Thruster efficiency, measured as specific impulse, varies with input power, which determines the thrust, and ranges from 1283 s at 58.2 mN (0.91 kW) to 1794 s at 279.3 mN (4.57 kW) (Snyder et al. 2020). Since the thrust is small, the thrusters are fired for thousands of hours (one at a time) to accomplish the interplanetary cruise to the asteroid and the maneuvers in orbit. The trajectory design accounts for all activities incompatible with optimal thrusting, primarily turning the spacecraft HGA to Earth once per week to transmit recorded engineering and instrument cruise data and to receive commands. There are also shorter, more frequent interruptions to unload the accumulated angular momentum in the reaction wheels (primarily from electric propulsion thruster swirl torque). These thrust interruptions result in a thrusting duty cycle of $\sim$85%. This percentage varies depending on the thrust level (higher thrusting levels need more frequent swirl unloads) and other spacecraft activities. The cruise phase is made up of Cruise 1 which is prior to MGA and Cruise 2 which comes after MGA. Cruise 1 requires approximately 15 months of thrusting out of the 31-month period between launch and MGA. Cruise 2 requires approximately 33 months of thrusting out of the 40-month period between MGA and arrival at Psyche. These are approximate numbers and will change as performance models are updated during cruise.

The Mission Design and Navigation team utilizes performance models to design the trajectory—the path the spacecraft will take through the solar system to meet up with Psyche—as well as the required thrusting levels, durations, and direction to achieve that trajectory. The performance models include power available from the solar array, thrust and xenon mass flow rate as a function of Power Processing Unit input power, swirl torque as a function of thrust, other spacecraft electrical loads, and available duty cycle. All these models provide a prediction of the performance for the duration of cruise, either by solar distance or time. An interesting aspect of SEP, or low thrust mission design in general, is that the maneuver or thrusting done now or in the immediate future depends on what the predicted performance will be next week, next month, next year, all the way to the destination. Therefore, a key aspect of these models is ensuring that there is appropriate margin so that it is unlikely to fall short of predicted performance in the future. Additionally, if thrusting is planned to be performed for some portion of time and that does not occur, it is termed 'missed thrust.' The Mission Design and Navigation team designs the trajectory to accommodate a minimum amount of missed thrust and still arrive on time. For the Psyche mission, the Cruise 1 period is very robust to missed thrust and the spacecraft could miss more than the minimum without impacting arrival. In fact, an anomaly on April 1, 2025, resulted in 77 days of missed thrust in Cruise 1 but had minimal impact on MGA timing and arrival at Psyche. However, Cruise 2 is generally more sensitive to missed thrust with a specific portion the most sensitive, but it is still meeting the requirements and objectives. The trajectory is being designed to build in more robustness, like planned arrival earlier than required, so that if there is another significant missed thrust event, it minimizes the impact to the science mission.

SEP missions are the most efficient, from a propulsion perspective, if they are thrusting at the 'maximum sustainable thrust.' Since the spacecraft loads vary as heaters cycle on and off, or the reaction wheels are torqued, the total load is variable. The spacecraft load on average, though, is relatively constant. To thrust at the maximum sustainable thrust level, the total average electrical load would be below, but near, the maximum power produced by the solar arrays such that when heaters cycle on, the increased load is carried by the battery, and when they cycle off, the excess power available is used to recharge the battery. Since there are inefficiencies in the system and battery charging/discharging, the average load needs to be below the maximum by some amount. If all the predictions were perfect, then each segment of thrusting would be designed at this point with no issue. However, there are uncertainties in the models and spacecraft performance (remember that each thrust segment design is unique in the spacecraft attitude relative to the sun, thrust level, and distance to the sun) that result in the thrust design not being at this exact maximum sustainable thrust. To mitigate the risk of discharging the battery too much by designing the thrust level to be closer to the peak power point intentionally, or inadvertently, there is a function in the spacecraft flight software that monitors the battery voltage so that if it goes below a threshold, the electric propulsion thrust level is autonomously reduced to allow more power to be available to recharge the battery. This functionality will continue to reduce the thrust level until the battery gets back above the threshold. Without this functionality, there is a risk that the battery would discharge to the point where fault protection would step in and terminate thrusting. While reducing thrust introduces an error from the designed trajectory, it is not nearly as impactful as missing thrust.

5.3.2 Mission Design Challenges

The Psyche mission requires an exceptionally high post-launch propulsive ΔV, 6.5 km/s, achieved entirely through a low-thrust SEP system. For comparison, NASA's only other deep space SEP missions, Deep Space 1 and Dawn, accomplished 4.3 km/s and 11.5 km/s, respectively. Of the 6.5 km/s, 5.8 km/s is needed to get the spacecraft to Psyche capture (electric propulsion ΔV). Another 0.2 km/s is necessary for the orbital transfers bringing the mission electric propulsion deterministic ΔV requirement to 6.0 km/s. Additional ΔV of about 0.5 km/s is allocated to accommodate statistical maneuvers, angular momentum management, and missed thrust. The Mars gravity assist provides an additional 0.4 km/s to the spacecraft's speed; however, the mission design is also using it for a heliocentric orbit plane change, which is very costly if done via a propulsive maneuver. The net effect of the Mars gravity assist is equivalent to a 2.0 km/s heliocentric velocity change.

While the Δv provides an interesting comparison with the far lower values for planetary missions that use chemical propulsion, it is generally considered to be a less significant metric for SEP missions. Optimizing the trajectory design is not only very complex but also very different from that for conventional missions. As indicated in Sect. 5.3.1, electric propulsion thrusting usually occurs for months or years at a time with only brief interruptions, so as to achieve the proper changes in the trajectory, subject to constraints of the mission plan. Some periods during cruise are more valuable for controlling the trajectory and some are less. Indeed, there are periods when it is optimal not to thrust. The thrust and Isp depend on available electrical power, which depends on heliocentric range (see Sect. 5.3.1). As a result, changes in when thrusting can or does occur (whether because of nominal or off-nominal changes in the mission plan) can lead to significant changes in the ΔV for the same mission objectives.

Low-thrust trajectory design is inherently more complex than traditional chemical propulsion. For instance, chemical propulsion maneuvers typically rely on linear mapping using the State Transition Matrix. In contrast, Psyche's transfers between science orbits required a sequence of maneuvers to guide the spacecraft to the next orbit's entry point—spanning multiple revolutions around the asteroid and requiring time-varying thrust vectors. Designing these low-thrust trajectories demands different capabilities from both the software and operational processes. The Psyche flight team uses Mystic (Whiffen 2006) and Veil (Parcher and Whiffen 2011) for trajectory design and maneuver planning. The continuous operation of SEP adds to the computational complexity, requiring careful optimization throughout the process (Whiffen et al. 2016). Additionally, low-thrust trajectories are highly sensitive to electrical power modeling, as the ability to sustain thrust years into the future influences present-day thrust direction decisions. Accurate power modeling is crucial to ensuring sufficient control authority throughout the mission. Errors in power predictions can compromise the spacecraft's ability to achieve its objectives.

Traditional interplanetary missions using chemical propulsion measure design margins primarily in mass or ΔV. However, low-thrust missions introduce additional constraints, including power availability, and thrusting time. Over its five-year SEP operation, Psyche must carefully manage thrusting time margins. The thrust duty cycle—the ratio of SEP thrusting time to calendar time—is determined by mission planning and must account for forced coasts, including Earth-pointing maneuvers, data downlinks, instrument activities, reaction wheel unloads, and periodic maintenance. Because SEP thrusting must be prioritized whenever possible, low-thrust missions require more integrated planning than chemical propulsion missions. Coordination between mission planners and trajectory designers ensures a robust mission plan with adequate margins for reaching Psyche.

Managing missed-thrust events is critical to mission success, so trajectory design must also account for missed thrust. The Psyche mission design is required to tolerate up to 14 days per year of unplanned coast or a single 12-day interruption during the cruise and approach phases. Pre-launch trajectory validation included margins for power availability, thrust duty cycle, and thruster performance. Throughout the mission, these margins are carefully managed and incrementally released during monthly trajectory redesigns.

The Mission Design and Navigation team routinely analyzes missed-thrust sensitivity, identifying critical mission phases where heightened readiness is required. During these periods, additional ground tracking passes are scheduled to facilitate early anomaly detection.

The Psyche mission employs two types of Thrust Verification passes to monitor SEP performance:

1. **Nominal Thrust Verification Passes:** During these 4-hour sessions, the onboard transmitter communicates via an LGA while the SEP operates at reduced power to accommodate transmitter needs. These passes provide essential spacecraft health data and tracking information for Orbit Determination.
2. **No-Downlink Thrust Verification Passes:** Lasting approximately 2 hours or less, these passes operate on the principle that if the spacecraft is thrusting nominally, no downlink

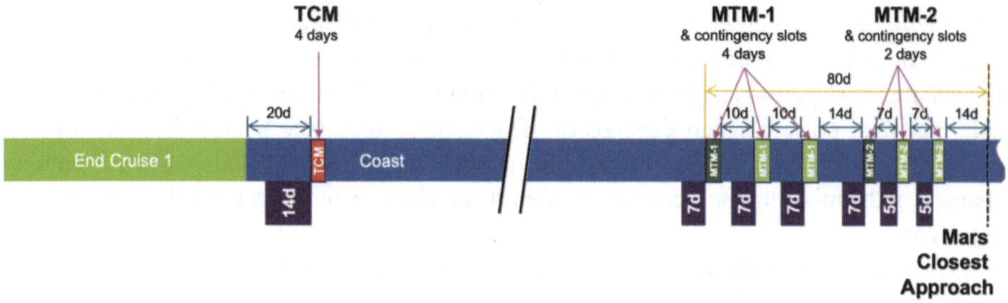

Fig. 35 Thrusting architecture for Mars gravity assist with the Trajectory Correction Maneuver (TCM) and windows for Mars Trim Maneuvers (MTMs). MTM-1 is deterministic and MTM-2 is a statistical correction. Both maneuvers include contingency cases shown in light green. This architecture was designed to achieve the most accurate delivery at Mars closest approach without overloading the flight team's ability to generate the associated spacecraft command files

signal is received. If an anomaly forces the spacecraft into safe mode, it transmits diagnostic telemetry, which DSN operators report to the flight team. Rapid recovery from safe mode is essential to minimizing missed-thrust durations, a consideration reflected in the team's flight operations procedures.

By continuously refining operational strategies and maintaining robust trajectory margins, the flight ensures resilience against anomalies while maximizing SEP efficiency to achieve mission success.

5.3.3 Mars Gravity Assist Navigation and Engineering

The Psyche mission requires a gravity assist from Mars to gain ΔV and adjust its heliocentric orbit plane. Monte Carlo simulations, run in the software package Veil, were conducted to develop a safe mission plan, incorporating both deterministic and statistical thrusting strategies for navigating the Mars flyby. In the baseline architecture (Fig. 35), two correction maneuvers follow the end of Cruise 1 to ensure a safe and optimal Mars gravity assist. A Trajectory Correction Maneuver is scheduled 20 days after Cruise 1 thrusting completes, while a Mars Trim Maneuver—part of the MGA architecture—is planned for 80 days before the spacecraft's closest approach to Mars.

Thrusting during Cruise 1, followed by the Trajectory Correction Maneuver, will target a biased aim point away from Mars to meet NASA's planetary protection requirements. The Mars Trim Maneuvers will then adjust the aim point to maximize the Mars gravity assist, optimizing the Cruise 2 trajectory for Psyche's arrival.

5.3.4 Orbit Transfers

An exciting aspect of the Psyche mission is the opportunity to explore a previously unvisited solar system body. However, this also presents challenges, such as the lack of an accurate or detailed gravity model before arrival. To address this, the mission is designed so that knowledge gained in one phase informs the next during orbit operations.

Psyche's rotation period and axis will be estimated through optical measurements during approach, while GM and low-order gravity field terms will be derived from radiometric tracking data. This information will be used to refine the Orbit A reference orbit and operations plan. Subsequently, updated gravity field and rotation data from Orbit A will guide the design and operations of Orbit B. This iterative process will continue throughout the orbit operations phase.

5.3.5 Time Correlation

The spacecraft avionics keep time by counting seconds, where zero seconds corresponds to 12:00 pm on January 1, 2000 (J2000). This time is referred to as SCLK (spacecraft clock) and is the reference for on board commanding and downlink data time tagging. The time is based on an oscillator in the electronics that may drift as result of temperature changes and age. Therefore, the team correlates the SCLK time to SCET (spacecraft event time, effectively UTC) by processing time correlation data packets from the spacecraft. The correlation activity produces a SCLKSCET file that is used to translate a given SCLK time to SCET. The SCLKSCET file contains one or more rows for each correlation, so it is a continually growing file for the life of the mission. The data in the file reflects a reconstructed time for times in the past, while the last entry is a prediction of the on board SCLK rate into the future.

Because the spacecraft uses SCLK to dispatch commands sequenced to execute at a specific time, the uplink process uses a SCLKSCET file to convert the sequence input in SCET to SCLK with the latest available correlation. The team uses the same SCLKSCET for all sequenced commanding in a given period of time so that conflicts aren't introduced by using different correlations for commands issued near the same SCET time.

Again, since the spacecraft operates in SCLK, all data is downlinked with SCLK time tags. The Ground Data System (GDS) automatically reads the latest SCLKSCET file to add the SCET time tag to the data item. Since the time correlation packets used to correlate the clock are transmitted in the same tracking pass as the other data, this latest data will not have the latest correlation applied. Therefore, to achieve the highest accuracy, the downlink data will need to be reprocessed after the SCLKSCET file is updated.

5.3.6 Data Management

Data collected by the spacecraft, engineering and instrument engineering and science data, is stored in files called Data Products (DPs). Some DPs are periodic that are continually being filled until they reach their 'full' criteria, which could be size or time open. The maximum file size is 4 MB. The other type of DP is a commanded DP, like an image or collection of specific, finite subsystem data. When a data product is created, it is given a priority value based on the criticality or urgency in getting the data to the ground. For example, critical engineering data has the highest priority so that the operations team can get insight into the state of the spacecraft in the event of an anomaly. Additionally, each DP has a tag that indicates whether it has been sent to the ground and is set to 'UNSENT' at the time of generation. During an HGA pass, the transmission of DPs is enabled by sequenced command and the flight software transmits the DPs in priority order, then oldest to newest based on generation time. As the DPs are sent, the tag is updated to 'SENT' and at the end of the pass a command is issued to dump the DP catalog DP. The catalog is used to compare what the spacecraft sent with what the GDS received. If there is any discrepancy, commands are generated to mark any DPs not received as 'UNSENT' and they'll be transmitted again when enabled. The DPs are not autonomously deleted so sequenced commands are generated to delete the DP that are already confirmed received on the ground.

The Psyche spacecraft supports data rates of 10, 40, 100, 250, 1k, 2k, 20k, 60k, 90k, 120k, 150k, 180k, 210k, 240k, 300k, and 360k bits per second (bps). The lower rates are typically used with the LGAs and the higher rates with the HGA. Which rate is used depends on the spacecraft antenna (LGA or HGA), the DSN antenna (34 m or 70 m), DSN antenna elevation, and distance to Earth. The lowest nominal rate with the HGA is 120k bps. During cruise, a single rate is used for any specific HGA pass that is supportable on a 34-m antenna, even if a 70-m antenna is scheduled so late changes in the DSN schedule would not result in sequence changes. At the asteroid, the data rate needs to be maximized to return the large amount of data collected. Therefore, the ground tools are designed to adjust the data rate during each pass as the DSN antenna elevation changes. For example, the data rate needs to be lower when the antenna is at low elevation because there is more atmospheric attenuation of the signal, so if the pass started at the lowest elevation, the sequence starts with a lower data rate and then progressively increases the rate as the elevation rises up to the maximum elevation, and then back down again as the antenna moves toward the opposite horizon, a technique referred to as a "wedding cake".

The spacecraft has three LGAs: one on the plus X face of the spacecraft, one on the minus X face, and one on the minus Z face. At the asteroid, these will be used when not in an HGA pass to collect gravity science data through navigation radiometrics. Since the spacecraft will primarily be pointed with −X close to nadir for data acquisition, different LGAs will have a view of Earth at different points in the orbit. The system can be configured to transmit engineering real time data at 10 bps if the signal strength supports it. Otherwise, it will be configured for what is called 'carrier only' which does not have any data modulated on the signal. This configuration increases the signal strength received at Earth by a small amount. In both configurations, the DSN will transmit an uplink signal that is received by the spacecraft and used to adjust the downlink frequency to account for the motion of the spacecraft relative to Earth. This is called 2-way Doppler mode and provides the navigation team with data to determine the orbital path around the body. From that, the gravity field knowledge can be refined.

5.3.7 Momentum Management in Orbit Operations

While in orbit around Psyche, the spacecraft will be subject to gravity gradient torque, a torque present when orbiting a non-symmetrical body, which the reaction wheels need to overcome. When the spacecraft is pointing the −X axis towards nadir, the gravity gradient torque is less than when the spacecraft is pointing the HGA to Earth in a fixed orientation. The spacecraft has a cold gas system consisting of thrusters that use pressurized nitrogen gas as a propellant that will be used to unload the reaction wheel momentum at the asteroid. The orbit trajectory is subject to perturbations when the thrusters fire, so these unload events must be accounted for in the trajectory design. Finalizing the operations concept for orbital operations was deferred to post launch and is still in development at the time of this writing.

5.3.8 Autonomous Fault Protection

Fault management is the system designed into the flight software and hardware to make the spacecraft robust to faults which can range from hardware failures to data corruption from solar particles to degraded or unexpected hardware performance. While the team strives to eliminate errors in the command sequences, sometimes they do happen for a range of reasons and the fault management system provides mitigations to those as well. The system consists of system level fault protection with monitors and responses to faults, as well as local fault protection. System level fault protection handles faults where normal operation cannot continue and the response is designed to resolve the fault and then reconfigure the spacecraft into a known, safe state, known as safe mode. There are three different safe modes: 'Standby-RWA,' 'Safe-RWA,' and 'Safe-CGS.' One of the differences between safe modes is whether they configure to use the reaction wheels (RWA) or the cold gas thrusters

Table 14 System Mode configurations for fault protection

System mode	Standby-RWA	Safe-RWA	Safe-CGS
Attitude Knowledge	3-axis inertial	2-axis sun reference	2-axis sun reference
Attitude Actuators	RWA	RWA	CGS
Communication Antenna	HGA	+X LGA	+X LGA
Uplink Rate	1 kbps	7.8125 bps	7.8125 bps
Downlink Rate	120 kbps	10 bps or carrier only	10 bps or carrier only
NS, Magnetometer state	Off	Off	Off
GRS state	DPU on, cryocooler on, HV ramped down	Off	Off

(CGS) for attitude control. The local fault protection operates at the lower level and can resolve the fault without interrupting other operations. Examples of system level monitor and response are the monitors on the electric propulsion system to ensure the thruster parameters are in the appropriate range. If they are not, the electric propulsion system is shut down and the system enters the 'Standby-RWA' mode which is the lowest impact safe mode. An example of local fault protection is the safety relay in the thermal subsystem. If a zone gets too hot, the safety relay is simply opened, cutting off heater power to that zone and all other operations continue.

The system modes invoked by fault protection, from least impact to greatest impact, are 'Standby-RWA', 'Safe-RWA', and 'Safe-CGS'. The high-level spacecraft configuration is shown in Table 14. In all cases, the autonomous behaviors and sequences are terminated, electric propulsion thrusting is terminated, the Magnetometer, NS, and DSOC are powered off, and GRS high voltage (HV) is ramped down. The GRS DPU and cryocooler, if they were on, are left powered. 'Standby-RWA' maintains full pointing control and points the HGA to Earth and configures for 120k bps. This is the preferred response as the high data rate and pointing control enable a faster recovery. The next level of safe mode is 'Safe-RWA' which also turns off the star tracker(s) and the GRS DPU and cryocooler, points +X to the sun using the coarse sun sensor, and begins a 1 rev/hr rotation around the +X axis. The spacecraft is configured to use the LGA which requires a 70-m DSN antenna to receive 10 bps for some parts of the mission. This mode is used when the fault indicates that it may not be possible to safely operate in the 'Standby-RWA' configuration without ground intervention. The final safe mode is 'Safe-CGS' which does everything in 'Safe-RWA', plus powers off the reaction wheels and uses the cold gas thrusters for attitude control. The pointing is the same, with +X to the sun. This mode is the most drastic and used for faults where a reaction wheel failed or a deep power fault wherein as many devices as necessary are powered off.

It should be noted that the intent of system level fault protection is to configure the spacecraft into a known, safe configuration with the minimum hardware and software enabled for that state. An exception to that design is for the GRS cryocooler. Since the GRS sensor is less susceptible to radiation degradation when cold, if the cooler is turned off in a science collection phase, it will take time to determine if an anneal is required and then many days to anneal the sensor, reducing the available time to collect the science data. This resulted in the decision to leave the GRS DPU and cryocooler on in 'Standby-RWA' as the risk to that mode of leaving these on is small.

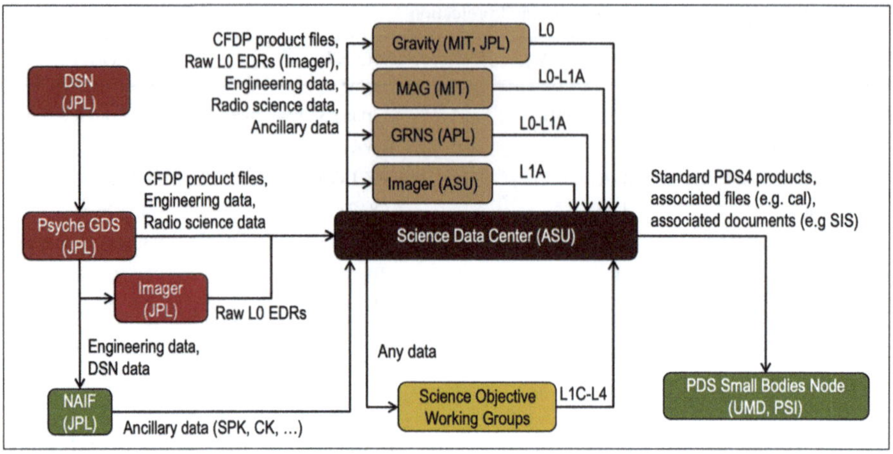

Fig. 36 Psyche science data flows from the GDS at JPL to the SDC, which coordinates all traffic of science data products between the GDS, the Psyche science team, and the PDS

6 Archiving and the Science Data Center

There are many designs that enable the analysis and processing of science data by a mission's science team. Any such design must include:

- Provision of raw data from a mission's ground data system to the science team
- Processing of raw data into standard science data products, usually by instrument-specific science teams
- Review mission-internal data products before archiving
- Archiving data products in the Planetary Data System (PDS), for use by the general science community.

In the case of the Psyche mission, we have chosen to enable this using a Science Data Center (SDC) co-located with Psyche Principal Investigator, at ASU.

The overall flow of these data and products is illustrated in Fig. 36. (In this figure, L0 refers to PDS4 Raw products, L1A refers to PDS4 Calibrated products, and L1B, L1C, L2, L3, and L4 refer to PDS4 Derived products (Planetary Data System Standards Reference 2024).)

The Psyche spacecraft downlinks science data in the form of CCSDS File Delivery Protocol (CFDP) files. While this protocol encapsulates the science data into one or more protocol data units, what we receive at JPL (and the SDC) has been unwrapped by the GDS AMPCS software into science data files (as generated by their respective instrument or the spacecraft telemetry feed, and its associated extended meta-data file that contains pertinent meta-data regarding the DSN receipt of the science file). Those, along with radio science data (originally from the DSN) and engineering data, are transferred to the SDC for pickup by the science team. The Imager team receives the science files (data file and extended meta-data file set) for each observation, in addition to an Object Description Language formatted file containing updated SCLK values, generated by MIPL. In addition, spacecraft engineering data and DSN data, especially trajectory and spacecraft attitude information, are converted to appropriate SPICE kernels (Navigation and Ancillary Information Facility 2024) and transferred to the SDC from the Navigation and Ancillary Information Facility node of the PDS.

The instrument science teams retrieve all necessary input data from the SDC to generate their standard data products at their home institutions. The results are transferred back to the SDC as PDS4 Raw and Calibrated products. Psyche's science team, often as members of Science Objective Working Groups (Sect. 5.1.1) access these data from the SDC for further processing (to PDS4 Derived products) and internal review before archiving. The SDC collects all standard science data products and packages them into the appropriate form for transfer to and archiving in the PDS. The SDC is also responsible for responding to liens identified by the PDS in external review.

6.1 Interface Between Psyche GDS and SDC

The ASU Science Data Center receives products from the Psyche GDS using the JPL developed File Exchange Interface (FEI) service. The GDS will publish instrument specific data products and requested telemetry to instrument specific file types. When those files are received at the SDC, they are ingested, cataloged, and then made available via the SDC webportal.

The ingest and catalog process updates a PostgreSQL database, running at the SDC, that contains a record for every file ingested into the SDC. This allows the SDC to track specific information about each file. Other tables aggregate the PDS4 product and its XML label, which facilitates reporting and archive generation.

The webportal allows team members to search for data products by a variety of attributes, and who can then download files of interest through the webportal (either through a web browser or the webportal ReSTful interface).

6.2 Interface Between SDC and Instrument Teams

6.2.1 Data Flow and Data Types from SDC to Instrument Teams

Instrument science teams receive their input data (as shown in Fig. 36) from the SDC. These include CFDP files of PDS4 Telemetry data, instrument and spacecraft engineering data, DSN-derived data for the Gravity Science team, and ancillary attitude and trajectory data (SPICE kernels). Each instrument operations team receives data via FEI and can also downlink the same data from the SDC. Once processed into PDS products, these are uploaded to the SDC either using the SDC web interface or a RESTful interface that allows large scale uploads (or downloads). For this paper, different levels of processed data are described in Table 15.

6.2.2 Return Flow of Data Products from Instrument Teams to SDC

The instrument science teams process their data into the products described in Table 16. These products are then transferred back to the SDC. The SDC then packages them into the "collections" and "bundles" described in Table 16 (Planetary Data System Standards Reference 2024). Note that the PDS4 Derived products are often generated by Science Objective Working Groups, using instrument data products picked up from the SDC and delivered back to there.

Table 15 Definitions of processing levels for science data, as used in this paper

NASA	PDS4	Description
Packet data	Telemetry	An encoded byte stream used to transfer data from one or more instruments to temporary storage where the raw instrument data will be extracted. PDS does not archive telemetry data. (These are supplied as CFDP files in the case of the Psyche mission.)
Level 0	Raw	Original data from an instrument. If compression, reformatting, packetization, or other translation has been applied to facilitate data transmission or storage, those processes will be reversed so that the archived data are in a PDS-approved archive format.
Level 1A	Partially Processed	Data that have been processed beyond the raw stage but which have not yet reached calibrated status.
Level 1B	Calibrated	Data converted to physical units, which makes values independent of the instrument.
Level 2 Level 3	Derived	Results that have been distilled from one or more calibrated data products (for example, maps, gravity or magnetic fields). Supplementary data used to interpret observational data, such as calibration tables or tables of viewing geometry, should also be classified as derived data if not easily matched to one of the other categories.

6.3 Role of SDC in Psyche Team Review and Analysis of Psyche Data

6.3.1 Use of SDC to Access Multiple Data Sets for Combined Higher-Level Processing

Since the SDC is the clearinghouse for Psyche mission data (from instruments and several of the working groups), the SDC will be able to facilitate aggregating and delivering information for higher-level processing. The SDC will provide web mapping services (WMS) capabilities, using OpenSource software, to deliver instrument data (observation footprints, individual map-projected imaging observations, and higher level derived products from the Magnetometer, GRNS, and Imager teams). Use of the WMS capabilities would allow the team creating geologic maps to focus on their mapping task, rather than dealing with data processing and management of individual images or mosaic products generated from the Multispectral Imager. The WMS put to good use by the Lunar Reconnaissance Orbiter Camera team with their large dataset. When regional or global derived products are generated from Magnetometer and GRNS data, these can also be offered via the WMS, allowing tools that can utilize WMS resources, such as ArcPro or QGIS, to overlay multiple datasets and answer questions.

The project will also be investigating the use of tools, such as JMARS or the Small Bodies Mapping Tool (SMBT), in conjunction with the above-mentioned WMS resource, to facilitate Psyche team members' access to mission data.

6.3.2 Pre-Delivery Internal Review of Psyche Standard Data Products

Teams that generate PDS4 products from Psyche data deliver those products to the SDC. At specified intervals (weekly, monthly and prior to deliveries), reports are generated from the SDC product holdings, which can then be sent to the generating team for review. The reports have a basic listing of products (and their attributes: filename, file size, checksum, product type) as well as information as to whether that product has passed PDS4 product validation.

Table 16 The Psyche project's four science investigations archive 11 bundles containing 60 collections

Bundle	Collections	Data volume
Magnetometer	Raw (L0) Time-stamped, raw orthogonal field components	2470.44 GB
	Partially-processed (L1A) Time series of calibrated magnetic field	1897.2 GB
	Calibrated (L1B) Time series of magnetic field (s/c signature removed)	753.3 GB
	Derived (L2) Magnetic field in J2000, etc. coordinates	44.64 GB
	Housekeeping (selected spacecraft housekeeping) data	TBD
	Calibration files	TBD
	Documents	TBD
	Browse	TBD
Magnetometer - derived	Derived (L3) Spherical harmonic coefficients (including dipole moment)	10 kB
	Derived (L3) Magnetic field map	1 GB
	Documents	TBD
	Browse image(s)	TBD
Imager	Raw (L0) Raw binary images (all images for all filters)	1198.8 GB
	Calibrated (L1B) radiance & IoF images (all images for all filters)	2397.5 GB
	Housekeeping (instrument & selected spacecraft) data	3.6 GB
	Calibration files	34.0 GB
	Documents	12 MB
	Browse images	300 GB
Imager - derived	Derived (L2) Global clear filter map at ≤ 20 m/pixel	20.4 GB
	Derived (L2) Global multispectral image maps at ≤ 200 m/pixel (for filters used in sulfide mineral detection) or ≤ 500 m/pixel (for filters used for iron-bearing silicate mineral detection)	163.2 GB
	Derived (L3) Global spectral parameter maps (3 parameters)	5.9 GB
	Documents	12 MB
	Browse images	TBD
Cartography	Derived (L3) Global topographic map (DTM), derived by SPG	2.3 GB
	Derived (L4) Global geologic unit map	3 GB
	Documents	
	Browse	TBD
GRS	Raw (L0) Time series of raw GRS counts	21.4 GB
	Calibrated (L1B) Time series of corrected calibrated gamma-ray spectra	21.4 GB
	Instrument Housekeeping data	1.5 GB
	Calibration files and geometry correction factors	<5 GB
	Documents	<10 MB
	Browse	<10 MB
GRS - derived	Derived (L3) elemental abundance map	<10 MB
	Derived (L3) Fe, Ni, Si, K, S, Al, Ca, Th, U average abundances	
	Documents	TBD
	Browse image(s)	<10 MB

Table 16 (*Continued*)

Bundle	Collections	Data volume
NS	Raw (L0) Time series of raw NS counts	7.4 GB
	Calibrated (L1B) Time series of corrected calibrated neutron spectra	7.4 GB
	Instrument Housekeeping data	1.1 GB
	Calibration files and geometry correction factors	<500 MB
	Documents	TBD
NS - derived	Derived (L3) elemental abundance map	<10 MB
	Derived (L3) neutron count rates	TBD
	Documents	TBD
	Browse	<10 MB
Radio Science	Raw (L0) Raw Radio Metric Tracking Data (TRK-2-34)	43 GB
	Raw (L0) Media Calibration File (TRK-2-23)	6.7 MB
	Raw (L0) Weather Data File (TRK-2-24)	TBD
	Raw (L0) Spacecraft Mass History File	<1 MB
	Raw (L0) Spacecraft Small-Forces File	67 MB
	Housekeeping (selected spacecraft) data	TBD
	Calibration files	TBD
	Documents	TBD
	Browse	TBD
Radio Science - derived	Derived (L3) Gravity field spherical harmonic coefficients	<1 MB
	Derived (L3) Spacecraft ephemeris (SPK file)	<1 GB
	Derived (L3) Tessellated plates shape model, derived by SPC (DSK)	<1 GB
	Documents	TBD
	Browse	TBD

Teams can review and cross-correlate with their holdings, to ensure the SDC has a complete set of products for a defined time period. This process is especially important prior to PDS4 deliveries, to make sure no issues relating to products are found during bundle/collection validation.

6.4 Creating Collections and Bundles at the SDC

The generation of archive bundles begins several months prior to the actual PDS publish date, to allow ample time for review of products, generation of the archive, and validation of the bundles. As stated in the previous section, the process begins with a review of data holdings, based on the report generated for the upcoming release.

Once the reports have been reviewed, and any issues identified and resolved, the SDC team can begin the process of generating an archive bundle and its collections. This task is performed using an automated process that queries the SDC database, creates the directory structure, and then populates the directories with the appropriate files (new files for delivery). Once the directory has been created and populated, dynamically created files such as manifest, bundle inventory, and collection inventory are generated from the newly created bundle. Once a bundle is successfully created, validation of that bundle is performed using the PDS tool VALIDATE, with the appropriate parameters and inputs. The report that is generated from the invocation is captured, as a record of the successful generation and validation of the bundle. Validation reports are ingested into the SDC, and available to the Psyche team for review. After validation, the bundle can then be transferred to the Small Bodies Node, for review by the SBN team and eventual integration into any existing released data bundles.

6.5 Advantages of Data Processing and Archiving Design

Data processing is carried out by each instrument team for their respective instrument, in order to convert science data and telemetry into PDS4 data products. As these science data products and PDS4 products are all stored at the SDC, the SDC team has coordinated with the instrument teams, the Data Archive working group, and PDS representatives on data product formats, PDS4 products, and archive bundle structures. The centralization of the PDS4 archiving allows teams to focus on their operations and processing and not be distracted by developing PDS4 expertise to support archiving functionality. ASU has a long and proven history of delivering data products to the PDS as exemplified by three instances: the Data Node for the Lunar Reconnaissance Orbiter Camera archive (Imaging Node), the THEMIS instrument archive (Imaging Node), and the TES instrument archive (Geosciences Node). The SDC expanded upon ASU's foundation and provides a unified interface for the Psyche team to access data products, and a single interface for the SBN to deal with for deliveries. The SDC team works with each of the teams creating products to advise on labels, and collection and bundle structures. The SDC team is responsible for the mission bundle and mission dictionary which the other Psyche archives reference.

7 Summary

Since 2011, the Psyche mission has grown from an idea formulated in a paper to a spacecraft with four science investigations and a successful technology demonstration on its way to explore a metallic world. The primary science objective, to determine whether Psyche is the remnant core of a planetesimal, is accomplished by observations from a series of circular mapping orbits.

The use of SEP enables the spacecraft to traverse interplanetary space and multiple different orbits about a massive, irregularly shaped, previously unexplored solar system object. Although commonly used for Earth-orbiting satellites, SEP is still relatively uncommon for deep space applications. The low-thrust mission design developed by the Dawn team and inherited by Psyche could be replicated to investigate other novel targets in the solar system.

The project developed a strong set of goals, objectives, and requirements during the proposal phase and thereafter avoided adding scope or changing the high-level approach to mapping the asteroid. Despite two changes in launch date and a significant change in arrival conditions, the science implementation plan outlined in the Concept Study Report is remarkably similar to the one detailed in this paper. The modular nature of the science orbits provides the flexibility needed to adjust the science orbit timeline to arrival conditions without changing the science observation strategies. Future surprises from the asteroid, the spacecraft, or the instruments could force changes to these plans, but the architecture provides a stable baseline from which to adjust to new challenges.

Acknowledgements The authors thank the entire Psyche team, composed of dedicated individuals whose work in many disciplines over many years has brought this project from proposal to deep space operations. We would like to specifically thank Klaus-Dieter Matz at DLR for CKVIEW results, Bruce Bills for an earlier version of the Psyche seasons graphic, David Seal for an earlier version of the mission overview graphic, and Laura Ratliff, Ellen Lamel, and Alison Johnson for help with structurally organizing and copyediting the manuscript. The flight and ground preparation for the DSOC technology demonstration owes a depth of gratitude to NASA HQ STMD/TDM, SOMD/SCaN and SMD/Discovery Program Management for guidance and support.

Author Contribution All authors wrote parts of the paper and/or commented on previous versions of the manuscript.

Funding Information The research was carried out at the Jet Propulsion Laboratory, California Institute of Technology, under a contract with the National Aeronautics and Space Administration (80NM0018D0004) and supported by NASA contract NNM16AA09, "Psyche: Journey to a Metal World." Work with CKVIEW was supported by DLR.

Declarations

Competing Interests The authors have no competing interests to declare. CAP, LTET, and JFB are guest editors of this collection, but were not involved in the peer review of this article.

References

Badolati E (2007) Gasparis, Annibale de. In: Hockey T, et al (eds) The biographical encyclopedia of astronomers. Springer, New York, p 408. https://doi.org/10.1007/978-0-387-30400-7_502

Bell JF III, Ravine MA, Caplinger MA, Schaffner JA, Brylow SM, Clark MJ, Peckham DA, Otjens PT, Price GJ, Rowell T, Ravine JW, Laramee JD, Juergens RC, Morgan W, Parker AG, Williams DA, Winhold A, Dibb S, Cisneros E, Walworth M, Zigo H, Warner N, Bates-Tarasewicz H, Amiri N, Polanskey C, Mastrodemos N, Park RS, Alonge NK, Jaumann R, Binzel RP, McCoy TJ, Martin MG, Arthur PA (2025) The Psyche multispectral imager investigation: characterizing the geology, topography, and multispectral properties of a metal-rich world Space Sci Rev, 221:47. https://doi.org/10.1007/s11214-025-01169-3

Bell JF, Davis DR, Hartmann WK, Gaffey MJ (1989) Asteroids: the big picture. In: Binzel RP, Gehrels T, Matthews MS (eds) Asteroids II. Univ. Arizona Press, Tucson, pp 921–943

Belton MJS, Chapman CR, Klaasen KP, Harch AP, Thomas PC, Veverka J, McEwen AS, Pappalardo RT (1996) Galileo's encounter with 243 Ida: an overview of the imaging experiment. Icarus 120:1. https://doi.org/10.1006/icar.1996.0032

Binzel RP, Bus SJ, Xu S, Sunshine J, Burbine TH, Neely AW, Brown RW (1995) Rotationally resolved spectra of asteroid 16 Psyche. Icarus 117:443–445

Binzel RP, Rivkin AS, Bus SJ, Sunshine JM, Burbine TH (2001) MUSES-C target asteroid 1998 SF36: a reddened ordinary chondrite. Meteorit Planet Sci Suppl 36:A20-21

Biswas A, Srinivasan M, Andrews K, Velasco A, Alerstam E, Allmaras J, Wollman E, Meenehan S, Wright M, Rogalin R (2024) Overview of the Deep Space Optical Communications (DSOC) technology demonstration. Proc SPIE 13355:133550J

Boroson DM, Robinson BS, Murphy DV, Burianek DA (2014) Overview and results of the Laser Lunar Communication Demonstration. In: Proceedings of the SPIE. 105240U-1-11

Bowman CDD, Elkins-Tanton LT, Talamante A, Bell JF III, Cisneros E, Cook A, Frieman JD, Gainor D, Hunziker J, Khan S, Lawler CR, Maschino J, McCoy TJ, Nessi K, Oran R, Seal D, Simon A, Singh R, Tolbert CM, Valentine K, Weiss B, Wenkert DD, Williams DA (2023) Mission to Psyche: including undergraduates and the public on the journey to a metal world. Space Sci Rev 219:25. https://doi.org/10.1007/s11214-023-00967-x

Buck BR, Allen GD, Duerr EK, Mcintosh KA, Moynihan ST, Shukla VN, Wang JD (2020) Photon counting camera for the NASA deep space optical communication demonstration on the Psyche mission. In: Proceedings of the SPIE. 1097809-1-10

Burks MT, Drury OB, Goldsten JO, Hagler LB, Heffern LE, Hines NR, Kim G-B, Lawrence DJ, Nelson KE, Peplowski PN, Yokley ZW (2020) GeMini: a high-resolution, low-resource, gamma-ray spectrometer for planetary science applications. Space Sci Rev 216:115

Carporzen L, Weiss BP, Elkins-Tanton LT, Shuster DL, Ebel D, Gattacceca (2011) Magnetic evidence for a partially differentiated carbonaceous chondrite parent body. Proc Natl Acad Sci USA 108(16):6386–6389. www.pnas.org/lookup/suppl/doi:10.1073/pnas.1017165108/-/DCSupplemental

CCSDS Blue Book (2019) Optical Communications Coding and Synchronization, 142.0-B-1; and Optical Communications Physical Layer 141.0-B-1

Chapman CR (1976) Asteroids as meteorite parent-bodies: the astronomical perspective. Geochim Cosmochim Acta 40:701–719

Chapman CR, Salisbury JW (1973) Comparisons of meteorite and asteroid spectral reflectivities. Icarus 19:507–522

Chapman CR, Morrison D, Zellner B (1975) Surface properties of asteroids: a synthesis of polarimetry, radiometry, and spectrophotometry. Icarus 25:104–130

Dailey JM, Dinu M, Prego R, Kolchmayer J, Berry M, Monte A, Engelberth J, Malaikal A, Piccirilli A, Le Grange J, Ahrens R, Sochor T, Jacques J, Wright MW (2019) High output power laser transmitter for high-efficiency deep-space optical communications. In: Proceedings of the SPIE. 109100M-1-12

de Gasparis A (1852) Discovery of a new planet. Mon Not R Astron Soc 12:125

de Soria-Santacruz M, Soriano M, Quintero O, Wong F, Hart S, Kokorowski M, Bone B, Solish B, Trofimov D, Bradford E, Raymond C, Narvaez P, Keys C, Lord P, Ream J, Oran R, Weiss BP, Russell C, Ascrizzi K, Elkins-Tanton L (2020) An approach to magnetic cleanliness for the Psyche mission. In: 2020 IEEE Aerospace Conference, Big Sky, MT. https://doi.org/10.1109/aero47225.2020.9172801

Deutsch L (2014) DSN future: a user perspective. In: DSN 50th anniversary symposium. https://descanso.jpl.nasa.gov/dsn50th/DSN_Symposium_2-20-14/DSN_50th_symposium_full.pdf

Dibb SD, Asphaug E, Bell JF, Binzel RP, Bottke WF, Cambioni S, Christoph JM, Elkins-Tanton LT, Jaumann R, Lawrence DJ, Oran R, O'Rourke JG, Polanskey C, Weiss BP, Wieczorek M, Williams DA (2024) A post-launch summary of the of the science of NASA's Psyche mission. AGU Adv 5:e2023AV001077. https://doi.org/10.1029/2023AV001077

Dollfus A, Mandeville J-C (1977) The M-type asteroids and the origin of iron meteorites. Meteoritics 12:206–207

Driscoll D, Zellers B, Schomacker J, Carro J, Roberts WT, Macdonald D, Guergian J, Klipstein W (2020) Deep Space Optical Communications (DSOC) beam expander design and engineering. In: Proceedings of the SPIE. 112720H-1-10

Elkins-Tanton LT, Weiss BP (eds) (2017) Planetesimals: early differentiation and consequences for planets. Cambridge University Press, Cambridge

Elkins-Tanton LT, Weiss BP, Zuber MT (2011) Chondrites as samples of differentiated planetesimals. Earth Planet Sci Lett 305:1–10

Elkins-Tanton LT, Asphaug E, Bell JF III, Bercovici H, Bills B, Binzel RP, Bottke WF, Dibb S, Lawrence DJ, Marchi S, McCoy TJ, Oran R, Park RS, Peplowski PN, Polanskey CA, Prettyman TH, Russell CT, Schaefer L, Weiss BP, Wieczorek MA, Williams DA, Zuber MT (2020) Observations, meteorites, and models: a pre-flight assessment of the composition and formation of (16) Psyche. J Geophys Res 125. https://doi.org/10.1029/2019JE006296

Elkins-Tanton LT, Asphaug E, Bell JF III, Bills B, Bottke W, Jun I, Lawrence D, Marchi S, McCoy T, Merayo J, Oran R, Park R, Peplowski P, Prettyman T, Raymond C, Weiss B, Wieczorek M, Zuber M (2022) Distinguishing the origin of asteroid (16) Psyche. Space Sci Rev 218:17. https://doi.org/10.1007/s11214-022-00880-9

Evans S, Taber W, Drain T, Smith J, Wu H-C, Guevara M, Sunseri R, Evans J (2018) MONTE: the next generation of mission design and navigation software. CEAS Space J 10:79–86. https://doi.org/10.1007/s12567-017-0171-7

Farnocchia D, Fuentes-Munoz O, Park RS, Baer J, Chesley SR (2024) Mass, density, and radius of asteroid (16) Psyche from high-precision astrometry. Astron J 168:21. https://doi.org/10.3847/1538-3881/ad50ca

Ferrari F, Tanga P (2022) Interior of top-shaped asteroids with cohesionless surface. Icarus 378:114914. https://doi.org/10.1016/j.icarus.2022.114914. ISSN 0019-1035

Gehrels T (1970) Photometry of asteroids. In: Dollfus A (ed) Surfaces and interiors of planets and satellites. Academic Press, San Diego, pp 317–375

Goldsten JO, Rhodes EA, Boynton WV, Feldman WC, Lawrence DJ, Trombka JI, Smith DM, Evans LG, White J, Madden NW, Berg PC, Murphy GA, Gurnee RS, Strohbehn K, Williams BD, Schaefer ED,

Monaco CA, Cork CP, Del Eckels J, Miller WO, Burks MT, Hagler LB, Deteresa SJ, Witte MC (2007) The MESSENGER gamma-ray and neutron spectrometer. Space Sci Rev 131:339–391. https://doi.org/10.1007/s11214-007-9262-7
Goossens S, Smith DE (2023) Gravity degree-depth relationship using point mass spherical harmonics. Geophys J Int 23:1878. https://doi.org/10.1093/gji/ggad036
Goossens SJ, Smith DE, Zuber MT, Head JW (2025) Shallow crustal density in the lunar south polar region. Planet Sci J 6:191. https://doi.org/10.3847/PSJ/adf05f
Hanuš J, Viikinkoski M, Marchis F, Ďurech J, Kaasalainen M, Delbo M, Herald D, Frappa E, Hayamizu T, Kerr S, Preston S, Timerson B, Dunham D, Talbot J (2017) Volumes and bulk densities of forty asteroids from ADAM shape modeling. Astron Astrophys 601:A114. https://doi.org/10.1051/0004-6361/201629956
Hardersen PS, Gaffey MJ, Abell PA (2005) Near-IR spectral evidence for the presence of iron-poor orthopyroxenes on the surfaces of six M-type asteroids. Icarus 175:141
Hurley K, et al (2010) The Third Interplanetary Network. Paper presented at Conference of the Gamma Ray Bursts, Annapolis, MD, Nov 01-04
Jaumann R, Bell JF III, Polanskey CA, Raymond CA, Aspaugh E, Bercovici D, Bills BR, Binzel R, Bottke W, Christoph JM, Marchi S, Neesemann A, Otto K, Park RS, Preusker F, Roatsch T, Williams DA, Wieczorek MA, Zuber MT (2022) The Psyche topography and geomorphology investigation. Space Sci Rev 218:7. https://doi.org/10.1007/s11214-022-00874-7
Kaasalainen M, Torppa J, Piironen J (2002) Models of twenty asteroids from photometric data. Icarus 159:369–395
Konopliv AS, Asmar SW, Park RS, Bills BG, Centinello F, Chamberlin AB, Ermakov A, Gaskell RW, Rambaux N, Raymond CA, Russell CT, Smith DE, Tricarico P, Zuber MT (2014) The Vesta gravity field, spin pole and rotation period, landmark positions, and ephemeris from the Dawn tracking and optical data. Icarus 240:103–117. https://doi.org/10.1016/j.icarus.2013.09.005
Lawrence DJ, Goldsten JO, Peplowski PN, Prettyman TH, Yokley ZW, Elkins-Tanton LT (2020) Preparing to measure (16) Psyche's elemental composition: interpretation of Psyche neutron measurements. In: 51st Lunar and Planetary Science Conference. Abstract #1973
Lawrence DJ, Goldsten JO, Peplowski PN, Yokley ZW, Beck AW, Burks M, Elkins-Tanton LT, Jun I, McCoy T, Prettyman TH (2021) Planetary Neutron Spectroscopy for Metal-rich Compositions: Development of Analysis Framework for Measurements at the Asteroid (16) Psyche. ESS Open Archive. https://doi.org/10.1002/essoar.10506738.1
Lawrence DJ, Goldsten JO, Peplowski PN, Burks MT, Cheng S, Cully MJ, Effron JM, Elkins-Tanton LT, Espiritu RC, Fix SG, Graziano MB, Hoffer EM, Jun I, Kim G-B, Hines NR, LeBlanc MT, Livingstone EM, Marcotte KM, McCoy TJ, Polanskey CA, Sreekantamurthy M, Yokley ZW (2025) The Psyche gamma-ray and neutron spectrometer. Space Sci Rev 221:78. https://doi.org/10.1007/s11214-025-01201-6
Leinweber HK, Russell CT, Torkar K, Zhang TL, Angelopoulos V (2008) An advanced approach to finding magnetometer zero levels in the interplanetary magnetic field. Meas Sci Technol 19(5)
Magri C, Nolan MC, Ostro SJ, Giorgini JD (2007) A radar survey of main-belt asteroids: Arecibo observations of 55 objects during 1999-2003. Icarus 186:126–151
Marchi S, Asphaug E, Bell JF III, Bottke WF, Jaumann R, Park RS, Polanskey CA, Prettyman TH, Williams DA, Binzel R, Oran R, Weiss B, Russell CT (2022) Determining the relative cratering ages of regions of Psyche's surface. Space Sci Rev 218:24. https://doi.org/10.1007/s11214-022-00891-6
Matz K-D, Roatsch T (2018) CKVIEW – a multi-mission planning tool. In: SpaceOps conference, p AIAA 2018-2392. https://doi.org/10.2514/6.2018-2392
McCord TB, Gaffey MJ (1974) Asteroids: surface composition from reflection spectroscopy. Science 186:352–355
McCoy TJ, Dibb SD, Peplowski PN, Maurel C, Bercovici HL, Corrigan CM, Bell JF III, Weiss BP, Lawrence DJ, Wenkert DD, Prettyman TH, Elkins-Tanton LT (2022) Deciphering redox state for a metal-rich world. Space Sci Rev 218:6. https://doi.org/10.1007/s11214-022-00872-9
McFadden LA, Skillman DR, Memarsadeghi N, Li J-Y, Joy SP, Polanskey CA, Rayman MD, Sykes MV, Tricarico P, Palmer E, O'Brien DP, Mottola S, Carsenty U, Mutchler M, McLean B, Schröder SE, Mastrodemos N, Schiff C, Keller HU, Nathues A, Gutiérrez-Marques P, Raymond CA, Russell CT (2015) Vesta's missing moons: comprehensive search for natural satellites of Vesta by the Dawn spacecraft. Icarus 257:207–216
McFadden L, Skillman DR, Memarsadeghi N, Carsenty U, Schröder SE, Li J-Y, Mottola S, Mutchler M, McLean B, Joy SP, Polanskey CA, Rayman MD, Fieseler PD, Sykes MV, Nathues A, Gutiérrez-Marques P, Keller HU, Raymond CA, Russell CT (2018) Dawn mission's search for satellites of Ceres: intact protoplanets don't have satellites. Icarus 316:191–204
Merayo JMG, Jørgensen JL, Friis-Christensen E, Brauer P, Primdahl F, Jørgensen PS, Allin TH, Denver T (2008) The swarm magnetometry package. In: Sandau R, Röser H-P, Valenzuela A (eds) Small satellites for Earth observation. Springer, Berlin, pp 143–151. https://doi.org/10.1007/978-1-4020-6943-7_13

Moision B, Hamkins J (2003) Deep-space, optical communications downlink budget: modulation and coding. In: IPN progress report 42-154. https://ipnpr.jpl.nasa.gov/progress_report/42-154/154K.pdf
Morrison D (1974) Radiometric diameters and albedos of 40 asteroids. Astrophys J 194:203–212
Nakamura T, Noguchi T, Tanaka M, Zolensky ME, Kimura M, Tsuchiyama A, Nakato A, Ogami T, Ishida H, Uesugi M, Yada T, Shirai K, Fujimura A, Okazaki R, Sandford SA, Ishibashi Y, Abe M, Okada T, Ueno M, Mukai T, Yoshikawa M, Kawaguchi J (2011) Itokawa dust particles: a direct link between S-type asteroids and ordinary chondrites. Science 333(6046):1113–1116. https://doi.org/10.1126/science.1207758. PMID: 21868667
National Academies of Sciences, Engineering, and Medicine (2011) Vision and voyages for planetary science in the decade 2013-2022. The National Academies Press, Washington, 398 pp. https://doi.org/10.17226/13117
National Academies of Sciences, Engineering, and Medicine (2022) Origins, worlds, and life: a decadal strategy for planetary science and astrobiology 2023-2032. The National Academies Press, Washington, 736 pp. https://doi.org/10.17226/26522
Navigation and Ancillary Information Facility (2024). https://naif.jpl.nasa.gov/naif/sitemap.html
Ockert-Bell ME, Clark BE, Shepard MK, Isaacs RA, Cloutis EA, Fornasier S, Bus SJ (2010) The composition of M-type asteroids: synthesis of spectroscopic and radar observations. Icarus 210:674–692
Oh D, Collins S, Randolph T, Vanelli C, Tilley S (2014) Feasibility of All-Electric Three Axis Momentum Management for Deep Space Small Body Rendezvous. AIAA Paper 2014-3906, July 2014. https://doi.org/10.2514/6.2014-3906
Oh DY, Collins S, Drain T, Hart W, Imken T, Larson K, Marsh D, Muthulingham D, Snyder J, Trofimov D, Elkins-Tanton L, Johnson I, Lord P, Pirkl Z (2019) Development of the Psyche mission for NASA's discovery program. In: 36th international electric propulsion conference, Vienna, Austria. IEPC-2019-192
Oh D, Collins S, Goebel D, Hart W, Imken T, Larson K, Maxwell J, Martin M, Polanskey C, Snyder J, Weise T, Elkins-Tanton L, Lord P (2022) Preparation for launch of the Psyche spacecraft for NASA's discovery program. In: 37th international electric propulsion conference, Boston, MA. IEPC-2022-452
Oh D, Alonge N, Bairstow S, Biswas A, Chesko C, Goebel D, Imken T, Na-Nakornpanom A, Nicholas A, Snyder JS, Johnson I, Linggi P, Lord P, Malone S, Pirkl Z, Ream J, Weiss B, Lawrence D, Peplowski P, Goldsten J, Burks M, Bell JF (2025b) Launch and Initial Checkout of the Psyche Spacecraft. AIAA J Spacecr Rockets. In press
Oh DY, Collins S, Drain T, Goebel D, Hart W, Imken T, Larson K, Maxwell J, Marsh D, Martin M, Muthulingam D, Snyder JS, Trofimov D, Weise T (2025a) Design of the Psyche Spacecraft for NASA's Discovery Program. Space Sci Rev 221
Ortega CA, Polanskey CA, Llopis M, Lawler CR, Rosemurgy PC, Alonge EK, Elkins-Tanton LT, Dailis ML (2021) Psyche science planning with the science opportunity analyzer. In: 16th international conference on space operations. Virtual Edition, 3 - 5 May 2021
Ostro SJ, Campbell DB, Shapiro II (1985) Mainbelt asteroids: dual-polarization radar observations. Science 229:442–446
Paper Elemental Blog (2021) Astropost: Astronomical Objects, Their Symbols, and Their Namesakes. http://paperelemental.blogspot.com/2021/07/astrosymbols.html
Parcher DW, Whiffen GJ (2011) Dawn statistical maneuver design for Vesta operations. In: AIAA/AAS astrodynamics specialist conference, New Orleans, Louisiana. AAS 2011-180
Park RS, Konopliv AS, Bills BG, Rambaux N, Castillo-Rogez JC, Raymond CA, Vaughan AT, Ermakov AI, Zuber MT, Fu RR, Toplis MJ, Russell CT, Nathues A, Preusker F (2016) A partially differentiated interior for (1) Ceres deduced from its gravity field and shape. Nature 537:515–517. https://doi.org/10.1038/nature18955
Park RS, Vaughan AT, Konopliv AS, Ermakov AI, Mastrodemos N, Castillo-Rogez JC, Joy SP, Nathues A, Polanskey CA, Rayman MD, Riedel JE, Raymond CA, Russell CT, Zuber MT (2019) High-resolution shape model of Ceres from stereophotoclinometry using Dawn imaging data. Icarus 319:812–827. https://doi.org/10.1016/j.icarus.2018.10.024
Park RS, Konopliv AS, Ermakov AI, Castillo-Rogez JC, Fu RR, Hughson KHG, Prettyman TH, Raymond CA, Scully JEC, Sizemore HG, Sori MM, Vaughan AT, Mitri G, Schmidt BE, Russell CT (2020) Evidence of non-uniform crust of Ceres from Dawn's high-resolution gravity data. Nat Astron 4:748–755. https://doi.org/10.1038/s41550-020-1019-1
Pieters CM, Noble SK (2016) Space weathering on airless bodies. J Geophys Res 121:1865–1884. https://doi.org/10.1002/2016JE005128
Planetary Data System Standards Reference (2024). https://pds.nasa.gov/datastandards/documents/sr/
Polanskey CA, Joy SP, Raymond CA (2011) Dawn science planning, operations and archiving. In: Russell C, Raymond C (eds) The Dawn mission to minor planets 4 Vesta and 1 Ceres. Springer, New York. https://doi.org/10.1007/978-1-4614-4903-4_17

Polanskey CA, Joy SP, Raymond CA (2012) Efficacy of the Dawn Vesta science plan. In: AIAA 2012-1275915. SpaceOps 2012 Conference. June 2012. https://doi.org/10.2514/6.2012-1275915
Polanskey CA, Joy SP, Raymond CA, Rayman MD (2014) Architecting the Ceres science plan. In: AIAA 2014-1720. SpaceOps 2014 Conference. May 2014. https://doi.org/10.2514/6.2014-1720
Polanskey CA, Joy SP, Raymond CA, Rayman MD (2016) Dawn Ceres mission: science operations performance. In: AIAA 2016-2442. SpaceOps 2016 Conference. https://doi.org/10.2514/6.2016-2442
Polanskey CA, Marsh DM, Moore R, Park RS, de Soria-Santacruz Pich M, Wenkert D, Elkins-Tanton LT, Williams DA, Jaumann R, Lawrence DJ, Russell CT (2018) Psyche science operations concept: maximize reuse to minimize risk. In: AIAA 2018-2703. 2018 SpaceOps Conference. May 2018. https://doi.org/10.2514/6.2018-2703
Prettyman TH, Mittlefehldt DW, Yamashita N, Lawrence DJ, Beck AW, Feldman WC, McCoy TJ, McSween HY, Toplis MJ, Titus TN, Tricarico P, Reedy RC, Hendricks JS, Forni O, Le Corre L, Li J-Y, Mizzon H, Reddy V, Raymond CA, Russell CT (2012) Elemental mapping by Dawn reveals exogenic H in Vesta's regolith. Science 338:242–246. https://doi.org/10.1126/science.1225354
Prettyman TH, Yamashita N, Toplis MJ, McSween HY, Schorghofer N, Marchi S, Feldman WC, Castillo-Rogez J, Forni O, Lawrence DJ, Ammannito E, Ehlmann BL, Sizemore HG, Joy SP, Polanskey CA, Rayman MD, Raymond CA, Russell CT (2017) Extensive water ice within Ceres' aqueously altered regolith: evidence from nuclear spectroscopy. Science 355:55–59. https://doi.org/10.1126/science.aah6765
Prettyman TH, Yamashita N, Ammannito E, Ehlmann BL, McSween HY, Mittlefehldt DW, Marchi S, Schörghofer N, Toplis MJ, Li J-Y, Pieters CM, Castillo-Rogez JC, Raymond CA, Russell CT (2019) Elemental composition and mineralogy of Vesta and Ceres: distribution and origins of hydrogen-bearing species. Icarus 318:42–55. https://doi.org/10.1016/j.icarus.2018.04.032
Prettyman TH, Mittlefehldt DW, Asphaug EI, Binzel RP, Courville SW, Jun I, Lawrence DJ, Marchi S, Merayo JMG, McCoy TJ, Weiss BP, Elkins-Tanton LT (2025) The Psyche Light Elements Investigation. Space Sci Rev 221
Primdahl F (1979) The fluxgate magnetometer. J Phys E, Sci Instrum 12:241–253
Pugh M, Kuperman I, Aguirre F, Mojaradi H, Spurgers C, Kobayashi M, Sartorius E, Jedrey T (2017) The universal space transponder: a next generation software defined radio. In: IEEE Aerospace Conference. 978-1-5090-1613-6/17
Rayman MD (2020) Lessons to the Dawn mission to Ceres and Vesta. Acta Astronaut 176:233–237. https://doi.org/10.1016/j.actaastro.2020.06.023
Rayman MD, Williams SN (2002) Design of the first interplanetary electric propulsion mission. J Spacecr Rockets 39:4. https://doi.org/10.2514/2.3848
Rayman MD, Fraschetti TC, Raymond CA, Russell CT (2007) Coupling of system resource margins through the use of electric propulsion: implications in preparing for the Dawn mission to Ceres and Vesta. Acta Astronaut 60:930–938. https://doi.org/10.1016/j.actaastro.2006.11.012
Raymond CA, Jaumann R, Nathues A, Sierks H, Roatsch T, Preusker F, Scholten F, Gaskell RW, Jorda L, Keller HU, Zuber MT, Smith DE, Mastrodemos N, Mottola S (2011) The Dawn topography investigation. Space Sci Rev 163:487–510. https://doi.org/10.1007/s11214-011-9863-z
Ream JB, Weiss BP, Oran R, Raymond CA, Polanskey CA, Wenkert D, Elkins-Tanton LT, Hart RA, Russell CT, Merayo JMG (2022) Magnetic gradiometry using frequency-domain filtering. Meas Sci Technol 33:1
Rogalin R, Grigorian E, Cho D, Tsao P, Richard N, Cheng M, Eckternach J, Okino C, Tkacenko A, Towfic Z, Xie H, Srinivasan M (2021) A high dynamic-range photon-counting receiver for deep space optical communication. In: Proceedings of the SPIE. 117680J-1-16
Russel CT, Raymond CA, Coradini A, McSween HY, Zuber MT, Nathues A, De Sanctis MC, Jaumann Konopliv RAS, Preusker F, Asmar SW, Park RS, Gaskell R, Keller HU, Mottola S, Roatsch T, Scully JEC, Smith DE, Tricarico P, Toplis MJ, Christensen UR, Feldman WC, Lawrence DJ, McCoy TJ, Prettyman TH, Reedy RC, Sykes ME, Titus TN (2012) Dawn at Vesta: testing the protoplanetary paradigm. Science 336:684
Russell CT, Raymond CA, Ammannito E, Buczkowski DL, De Sanctis MC, Hiesinger H, Jaumann R, Konopliv AS, McSween HY, Nathues A, Park RS, Pieters CM, Prettyman TH, McCord TB, McFadden LA, Mottola S, Zuber MT, Joy SP, Polanskey C, Rayman MD, Castillo-Rogez JC, Chi PJ, Combe JP, Ermakov A, Fu RR, Hoffmann M, Jia YD, King SD, Lawrence DJ, Li J-Y, Marchi S, Preusker F, Roatsch T, Ruesch O, Schenk P, Villarreal MN, Yamashita N (2016) Dawn arrives at Ceres: exploration of a small, volatile-rich world. Science 353(6303):1008–1010
Sanchez JA, Reddy V, Shepard MK, Cristina T, Cloutis EA, Takir D, Conrad A, Kiddell C, Applin D (2017) Detection of rotational spectral variation on the M-type asteroid (16) Psyche. Astron J 153:29
Shepard MK, Clark BE, Nolan MC, Howell ES, Magri C, Giorgini JD, Benner LAM, Ostro SJ, Harris AW, Warner B, Pray D, Pravec P, Fauerbach M, Bennett T, Klotz A, Behrend R, Correia H, Coloma J, Casulli S, Rivkin A (2008) A radar survey of M- and X-class asteroids. Icarus 195:184

Shepard MK, Richardson J, Taylor PA, Rodriguez-Ford LA, Conrad A, de Pater I, Adamkovics M, de Kleer K, Males JR, Morzinski KM, Close LM, Kaasalainen M, Viikinkoski M, Timerson B, Reddy V, Magri C, Nolan MC, Howell ES, Benner LAM, Giorgini JD, Warner BD, Harris AW (2017) Radar observations and shape model of asteroid 16 Psyche. Icarus 281:388–403
Shepard MK, de Kleer K, Cambioni S, Taylor PA, Virkki AK, Rívera-Valentin EG, Rodriguez Sanchez-Vahamonde C, Fernanda Zambrano-Marin L, Magri C, Dunham D, Moore J, Camarca M (2021) Asteroid 16 Psyche: shape, features, and global map. Planet Sci J 2:125
Sherwood B, McCleese D (2013) JPL innovation foundry. Acta Astronaut 89:236–247. https://doi.org/10.1016/j.actaastro.2013.04.020
Snyder JS, Lenhuito G, Frieman JD, Haag TW, Mackey JA (2020) Effects of background pressure on SPT-140 Hall thruster performance. J Propuls Power 36:5. https://doi.org/10.2514/1.B37702. http://arc.aiaa.org
Srinivasan M, Rogalin R, Lay N, Shaw M, Tkacenko A (2017) Downlink receiver algorithms for deep space optical communications. In: Proceedings of the SPIE. 100960A-1-11
Takir D, Reddy V, Sanchez J, Shepard M, Emery J (2017) Detection of water and/or hydroxyl on asteroid (16) Psyche. Astron J 153:31
Tholen DJ (1984) Asteroid Taxonomy from Cluster Analysis of Photometry. Ph.D. dissertation, Univ. Arizona, Tucson, 150 pp.
Veverka J, Belton M, Klaasen K, Chapman C (1994) Galileo's encounter with 951 Gaspra: overview. Icarus 107(1):2–17. https://doi.org/10.1006/icar.1994.1002
Veverka J, Robinson M, Thomas P, Murchie S, Bell JF III, Izenberg N, Chapman C, Harch A, Bell M, Yeomans DK, et al (2000) NEAR at Eros: imaging and spectral results. Science 289:2088–2097. https://doi.org/10.1126/science.289.5487.2088
Viikinkoski M, Vernazza P, Hanuš J, Le Corroler H, Tazhenova K, Carry B, Marsset M, Drouard A, Marchis F, Fetic R, Fusco T, Ďurech J, Birlan M, Berthier J, Bartczak P, Dumas C, Castillo-Rogez J, Cipriani F, Colas F, Ferrais M, Grice J, Jehin E, Jorda L, Kaasalainen M, Kryszczynska A, Lamy P, Marciniak A, Michalowski T, Michel P, Pajuelo M, Podlewska-Gac E, Santana-Ros T, Tanga P, Vachier F, Vigan A, Warner B, Witasse O, Yang B (2018) (16) Psyche: a mesosiderite-like asteroid? Astron Astrophys 619:L3
Walsh KJ, et al (2022) Near-zero cohesion and loose packing of Bennu's near subsurface revealed by spacecraft contact. Sci Adv 8:eabm6229. https://doi.org/10.1126/sciadv.abm6229
Weiss BP, Elkins-Tanton LT (2013) Differentiated planetesimals and the parent bodies of chondrites. Annu Rev Earth Planet Sci 41:529–560. https://doi.org/10.1146/annurev-earth-040610-133520
Weiss BP, Merayo JMG, Ream JB, Oran R, Brauer P, Cochrane CJ, Cloutier K, Elkins-Tanton LT, Jørgensen JL, Maurel C, Park RS, Polanskey CA, de Soria Santacruz-Pich M, Raymond CA, Russell CT, Wenkert D, Wieczorek MA, Zuber MT (2023) The Psyche magnetometry investigation. Space Sci Rev 219:22. https://doi.org/10.1007/s11214-023-00965-z
Whiffen GJ (2006) Mystic: implementation of the static dynamic optimal control algorithm for high-fidelity, low-thrust trajectory design. In: AIAA/AAS astrodynamics specialist conference, Keystone, Colorado. AIAA 2006-6741
Whiffen GJ, et al (2016) Dawn's mission at Ceres. In: AIAA/AAS astrodynamics specialist conference, Long Beach, California. AIAA 2016-5375
Winhold A, Bell JF III, Caplinger MA, Brylow S, Ravine M, Kubacki T, Dahl-Hansen R, Walworth M, Dibb S, Zhao A, Rodea Y, Cisneros E, Williams D, Warner N, Bates-Tarasewicz H, Monacelli B, Mastrodemos N (2025) Pre-flight and in-flight calibration of the Psyche Multispectral Imagers. Earth Space Sci. To be submitted
Wollman EE, Allmaras JP, Beyer AD, Korzh B, Frasca S, Runyan MC, Lautaro N, Farr WH, Marsili F, Briggs RM, Miles GJ, Shaw MD (2024) SNSPD-based detector system for NASA's deep space optical communications project. Opt Express 32(27):48185–48198
Yokley ZW, Peplowski PN, Goldsten JO, Lawrence DJ, Jun I, Prettyman TH, Elkins-Tanton LT (2019) Neutron spectroscopy on a metal world: tailoring the Psyche neutron spectrometer. In: 50th lunar and planetary science conf. Abstract #1295
Zellner B, Gradie J (1976) Polarimetric evidence for the albedos and compositions of 94 asteroids. Astron J 81:262–280
Zuber MT, Smith DE, Lehman DH, Hoffman TL, Asmar SW, Watkins MM (2013) Gravity Recovery and Interior Laboratory (GRAIL): mapping the lunar interior from crust to core. Space Sci Rev 178:3–24. https://doi.org/10.1007/s11214-012-9952-7
Zuber MT, Park RS, Elkins-Tanton LT, Bell JF III, Bruvold KN, Bercovici D, Bills BR, Binzel RP, Jaumann R, Marchi S, Polanskey CA, Raymond CA, Roatsch T, Wang CC, Weiss BP, Wenkert D, Wieczorek MA (2022) The Psyche gravity investigation. Space Sci Rev 218:57. https://doi.org/10.1007/s11214-0

Publisher's Note Springer Nature remains neutral with regard to jurisdictional claims in published maps and institutional affiliations.

Authors and Affiliations

Carol A. Polanskey[1] · Linda T. Elkins-Tanton[2,3] · James F. Bell III[2] · Eleanor K. Alonge[1] · Sarah H. Bairstow[1] · Richard P. Binzel[4] · Abhijit Biswas[1] · Luke Bury[1,5] · Ernest Cisneros[2] · Dongsuk Han[1] · Insoo Jun[1] · William M. Klipstein[1] · David J. Lawrence[6] · Timothy J. McCoy[7] · Nickolaos Mastrodemos[1] · José M.G. Merayo[8] · Sarah K. Noble[9] · David Y. Oh[1] · Rona Oran[4] · Carolyn A. Ortega[1] · Ryan S. Park[1] · Patrick N. Peplowski[6] · Thomas H. Prettyman[10] · Marc D. Rayman[1] · Jodie B. Ream[4] · Thomas Roatsch[11] · Timothy M. Weise[1] · Daniel D. Wenkert[1] · Benjamin P. Weiss[4] · Maria T. Zuber[4]

✉ C.A. Polanskey
Carol.A.Polanskey@jpl.nasa.gov

1 Jet Propulsion Laboratory/California Institute of Technology, Pasadena, CA, USA

2 Arizona State Univ., School of Earth & Space Exploration, Tempe, AZ, USA

3 Present address: Space Sciences Laboratory, University of California, Berkeley, Berkeley, CA, USA

4 Massachusetts Institute of Technology, Cambridge, MA, USA

5 Present address: Advanced Space, LLC, Westminster, CO, USA

6 Johns Hopkins Applied Physics Laboratory, Laurel, MD, USA

7 Smithsonian Institution, Washington, DC, USA

8 DTU Space, Technical University of Denmark, Lyngby, Denmark

9 NASA Headquarters, Washington, DC, USA

10 Planetary Science Institute, Tucson, AZ, USA

11 German Aerospace Center (DLR), Berlin, Germany

Space Science Reviews (2022) 218:17
https://doi.org/10.1007/s11214-022-00880-9

Distinguishing the Origin of Asteroid (16) Psyche

Linda T. Elkins-Tanton[1] · Erik Asphaug[2] · James F. Bell III[1] · Carver J. Bierson[1] · Bruce G. Bills[3] · William F. Bottke[4] · Samuel W. Courville[1] · Steven D. Dibb[1] · Insoo Jun[3] · David J. Lawrence[5] · Simone Marchi[4] · Timothy J. McCoy[6] · Jose M.G. Merayo[7] · Rona Oran[8] · Joseph G. O'Rourke[1] · Ryan S. Park[3] · Patrick N. Peplowski[5] · Thomas H. Prettyman[9] · Carol A. Raymond[3] · Benjamin P. Weiss[8] · Mark A. Wieczorek[10] · Maria T. Zuber[8]

Received: 6 June 2021 / Accepted: 16 March 2022 / Published online: 12 April 2022

Abstract
The asteroid (16) Psyche may be the metal-rich remnant of a differentiated planetesimal, or it may be a highly reduced, metal-rich asteroidal material that never differentiated. The NASA Psyche mission aims to determine Psyche's provenance. Here we describe the possible solar system regions of origin for Psyche, prior to its likely implantation into the asteroid belt, the physical and chemical processes that can enrich metal in an asteroid, and possible meteoritic analogs. The spacecraft payload is designed to be able to discriminate among possible formation theories. The project will determine Psyche's origin and formation by measuring any strong remanent magnetic fields, which would imply it was the core of a differentiated body; the scale of metal to silicate mixing will be determined by both the neutron spectrometers and the filtered images; the degree of disruption between metal and rock may be determined by the correlation of gravity with composition; some mineralogy (e.g., modeled silicate/metal ratio, and inferred existence of low-calcium pyroxene or olivine, for example) will be detected using filtered images; and the nickel content of Psyche's metal phase will be measured using the GRNS.

Keywords Asteroid · Psyche · Space exploration · Planetesimal · Meteorite

1 Introduction to Science Objective 1, Is Psyche a Core?

The first, and central, science objective for the NASA Psyche mission is to determine whether the asteroid (16) Psyche (henceforth referred to as "the asteroid Psyche", to distinguish it from the mission itself, which is also named Psyche) is part or the whole of a planetesimal core, or, if it is not, what is its most likely provenance.

In our summary paper, Elkins-Tanton et al. (2020), we considered all data on the physical and chemical properties of Psyche. The asteroid's bulk density appears to be between 3,400 and $4,100\ \mathrm{kg\,m^{-3}}$. Given the known densities of meteoritic metal ($\sim$ 7000–8000 $\mathrm{kg/m^{-3}}$), and assuming not more than 20% porosity (see argument in Elkins-Tanton et al. 2020), the body could contain up to 60 vol.% metal. If porosities higher than expected on a body of

The NASA Psyche Mission: Science Instruments and Investigations
Edited by James F. Bell III, Carol Polanskey and Lindy Elkins-Tanton

Extended author information available on the last page of the article

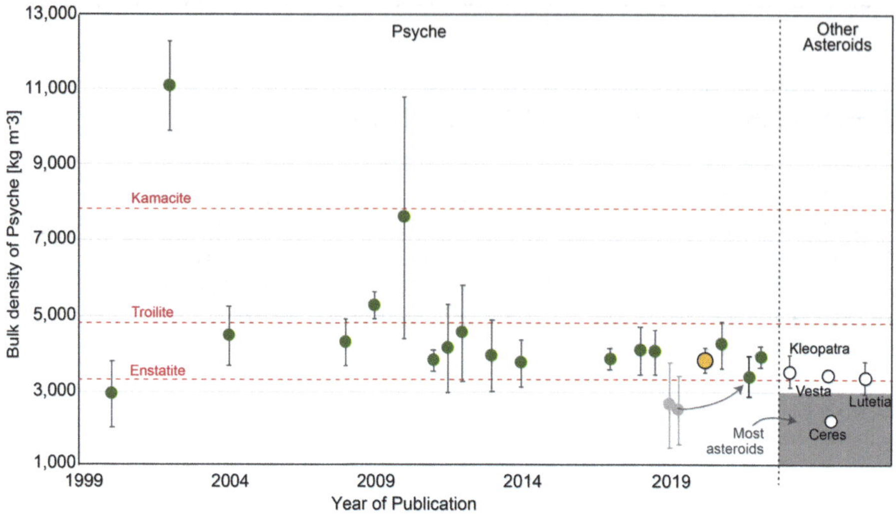

Fig. 1 The average bulk density of Psyche assumed by this project is $3,780 \pm 340\ \mathrm{kg\,m^{-3}}$ (large yellow disk; Elkins-Tanton et al. (2020)). The lowest estimates, from Siltala and Granvik (pale grey disks) were revised upward in 2021 (gray arrow to new estimate) (Siltala and Granvik 2021). Additional data from Ferrais et al. (2020) and Shepard et al. (2021) added to the data compiled in Elkins-Tanton et al. (2020). For comparison, the densities of several minerals pertinent to formation hypotheses of Psyche are included: $7,870\ \mathrm{kg\,m^{-3}}$ for kamacite, $4,840\ \mathrm{kg\,m^{-3}}$ for troilite, and $3,270\ \mathrm{kg\,m^{-3}}$ for enstatite (Mg-rich pyroxene) (Bass 1995; Smyth and McCormick 1995) (pink dashed lines). For comparison, and not placed in the year of their publication asteroid densities are: Kleopatra ($3,380 \pm 500\ \mathrm{kg\,m^{-3}}$; Marchis et al. 2021), Vesta ($3,454\ \mathrm{kg\,m^{-3}}$; Russell et al. 2012), Lutetia ($3,400 \pm 300\ \mathrm{kg\,m^{-3}}$; Vernazza et al. 2011), and Ceres ($2,161\ \mathrm{kg\,m^{-3}}$; Park et al. 2019) (empty circles). The field of most asteroids' densities from Fienga et al. (2020). Psyche remains one of the densest asteroids known

Psyche's size are enabled by the material properties of metals at relevant space temperatures, then Psyche's metal fraction could be far higher than this. In particular, if Psyche consists of exclusively metal and pore space, then the volume ratio of these components must be around 50:50.

The asteroid is therefore likely a mixture of metal and an additional lower-density component or components. Based upon modeling of available telescopic and laboratory spectral characteristics (Dibb 2021), the additional components are likely to be very low in FeO or Fe_2O_3. Likely candidates for a non-metal fraction of the asteroid include low-iron pyroxene (e.g., Hardersen et al. 2005), sulfides (e.g., Clark et al. 2004), or perhaps carbon compounds.

More recent observations and calculations have not changed this scenario. For example, an anomalously low density estimate of the asteroid Psyche by Siltala and Granvik (2019) was updated (Siltala and Granvik 2021), providing a value of $3,880\ \mathrm{kg\,m^{-3}}$. Now all recent density estimates, including Viikinkoski et al. (2018), Drummond et al. (2018), Baer and Chesley (2017), and others, falls within the range given in Elkins-Tanton et al. (2020) of 3,400 and $4,100\ \mathrm{kg\,m^{-3}}$ (Fig. 1).

Furthermore, Becker et al. (2020) obtained UV spectra of the asteroid Psyche using the Hubble Space Telescope, showing that the surface could be interpreted as pure metal. They also cautioned that when modeling the observed spectra using intimate mixture modeling, the model spectra are highly sensitive to the assumed grain sizes of the iron and silicate particles: spectra similar to the observations could be obtained with iron volume fractions as low as 10%.

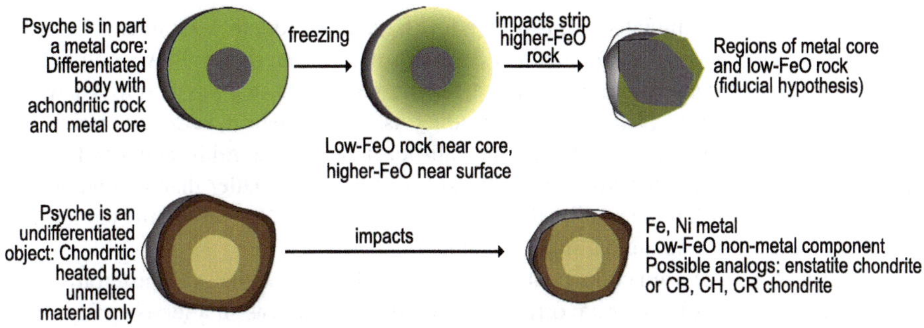

Fig. 2 Formation hypotheses that are consistent with Psyche's density and relatively featureless (low-FeO) spectra include a differentiated parent body (the fiducial hypothesis) and certain chondritic, undifferentiated materials (enstatite or CB-chondritic or similar low-FeO composition)

These density and compositional data provide constraints on asteroid Psyche's possible formation. Our favored fiducial model for the asteroid's formation is as the remains of a differentiated body containing core metal (Fig. 2). Psyche may contain a large fraction of a metal core and some silicate mantle retained on the outside, or the silicate fraction could be mixed into the metal fraction. The silicate fraction must, however, contain strikingly little iron, evidenced by the lack of strong absorption bands (for example, Hardersen et al. 2005).

Formation of such a low iron-oxide bulk rock composition seems, within the current state of planet formation theory, to require one of two processes to have occurred: (1) Silicate differentiation that produces low-iron-oxide silicates near the core and high-iron-oxides silicates nearer the surface, with subsequent removal of the outermost high-iron-oxide fraction by stripping impacts; or (2) Formation of the Psyche parent body from highly reduced primitive material in which the iron has been reduced to its metallic form and incorporated into the core, leaving the silicates naturally low in iron and high in magnesium.

Alternatively, the asteroid could be primarily composed of primitive (unmelted) high-metal material, perhaps a chondrite with silicates already low in iron-oxide, maybe even from a parent body already sampled in the terrestrial meteorite collection. We have hypothesized that Psyche could be formed from highly reduced material that formed near the Sun (Weidenschilling 1978), but we argue below that an origin beyond Jupiter might, in fact, be more likely.

When the Psyche orbiter reaches the asteroid, all its science instruments will contribute to answering this science objective. The total iron content (in metal and silicates) will be determined from element data acquired by the Gamma Ray and Neutron Spectrometer (GRNS; Lawrence et al. this journal) and from filter data from the Psyche Multispectral Imager (Bell et al. this journal). The gravity investigation (Zuber et al. this journal) will provide Psyche's bulk density, an important additional constraint on the total iron content, and combined with images and topography will inform the search for correlations between compositional and structural information. The magnetometer (Weiss et al. this journal) will seek the signature of an original core dynamo, and may be able to constrain the bulk electrical conductivity.

2 The Early Solar System Location of Psyche and M-type Asteroids from Dynamical Models

In this section, we discuss different scenarios for the formation location and early dynamical evolution of Psyche. We start by briefly describing constraints on Psyche and M-class asteroids that can help us evaluate dynamical scenarios.

The Main Belt is dominated in mass by a few large asteroids; for example, 1 Ceres has a third of the total mass (e.g., Vernazza et al. 2021). Thereafter, they follow a complicated size-frequency distribution that is often treated as a segmented or "wavy" power-law, that is, a cumulative number N larger than D going as D^{-n}, where n varies with size. The size-distribution is relatively shallow for the half-dozen largest asteroids, and then becomes steeper (larger n, more asteroids of a given size) for asteroids smaller than around 300 km diameter. Then, for asteroids smaller than 100 km, smaller than Psyche's diameter, the size-distribution transitions again to a shallower slope.

Bottke et al. (2005) used collisional evolution models to argue that most asteroids larger than about 100 km in diameter are original planetesimals, whereas smaller asteroids are increasingly the remnants of catastrophically disruptive collisions (e.g., Farinella et al. 1982). This fits with recent modeling work on planetesimal formation, where pebbles collect via streaming instabilities into spatially dense swarms that gravitationally collapse into planetesimals (e.g., Klahr and Schreiber 2021). The implication is that Psyche, being larger than 100 km in diameter, may represent a body that is potentially original and has possibly avoided catastrophic disruption. At face value this suggests that Psyche might not have originated from the disruption of a larger parent body; however see Sect. 3.

Psyche is classified as an M-type asteroid, an enigmatic taxonomic class with a red-sloped average disk-integrated spectrum that lacks evidence for significant absorption bands in visible to near-IR wavelengths (see review in Elkins-Tanton et al. 2020). In the DeMeo-Bus taxonomic system, these bodies are known as Xk (DeMeo et al. 2009). M-types have moderate albedos of 0.1 to 0.2, and this characteristic allows them to be distinguished from other classes of bodies that also have reddish slopes without absorption bands (e.g., dormant comets, which generally have albedos < 0.1).

Finally, Psyche's estimated bulk density of $3{,}780 \pm 340\ \mathrm{kg\,m^{-3}}$ (Elkins-Tanton et al. 2020) is comparable to three other M-types larger than 100 km: (21) Lutetia, (22) Kalliope, and possibly (216) Kleopatra. Vernazza et al. (2021) used high-angular-resolution imaging from VLT/SPHERE/ZIMPOL to estimate the 3D shapes for these bodies and calculated the densities of Psyche, Lutetia, Kalliope, and Kleopatra as $3{,}890 \pm 530$, $3{,}450 \pm 210$, $4{,}360 \pm 500$, and $3{,}450 \pm 410\ \mathrm{kg\,m^{-3}}$, respectively). Given these similarities, our working assumption here is that Psyche is a representative member of its taxonomic class.

To more closely examine M-type asteroids and compare them to other asteroids, we compiled the diameters and albedos of all asteroids in the AKARI and WISE datasets (Masiero et al. 2011; Alí-Lagoa et al. 2018). These data were then linked to (i) their taxonomic types as defined by DeMeo et al. (2009), Neese (2010) and DeMeo and Carry (2014) (with updates provided by F. DeMeo, personal communication), and (ii) their proper semimajor axis, proper eccentricity, and proper inclination values using the synthetic proper elements from the Asteroid Dynamics website AstDys (https://newton.spacedys.com/astdys/). With this combination, we identified all asteroids with diameter $D \geq 40$ km in the main belt; these bodies have not had their orbits meaningfully influenced by Yarkovsky thermal drift forces (e.g., Bottke et al. 2006a), this size range is observationally complete, and most of these bodies are likely to have formed in the first millions of years of the solar system, so we can use their present-day orbits to glean insights into their origins and early dynamical evolution.

From here, we combined large taxonomic classes with similar properties together and analyzed their dynamical distribution across the main belt, following the work of DeMeo et al. (2009). Our expectation is that dynamical excitation produced by early giant planet migration spread their eccentricities and inclinations values (except for a possible case we will discuss below) (Deienno et al. 2017, 2018), so here we concentrate on their semimajor axis distribution. In this manner, our work is similar to that performed by DeMeo and Carry (2014) (see also Gradie and Tedesco 1982), with diameter used as the primary discriminant rather than asteroid mass. The breakdown of our groups are as follows.

The two largest classes of S-complex bodies among our bodies are S- and L-type asteroids. The former are objects that have been linked to ordinary chondrites (e.g., Nakamura et al. 2011). The L-types have modestly different characteristics than S-types; a steep red spectrum shortward of 0.75 μm, a flat slope for wavelengths longer the 0.75 μm, and a moderate albedo (DeMeo et al. 2009, 2015). The S- and L-types have a similar semimajor axis distribution to one another, so for simplicity we put both sets together. Other S-complex asteroids have limited numbers of $D \geq 40$ km bodies, so they were excluded.

We divided asteroids resembling carbonaceous chondrites into several groups. The first group was D- and P-type asteroids, bodies whose spectroscopic signature often resembles dormant comets (e.g., DeMeo et al. 2009, 2015; Vokrouhlický et al. 2016). The second group was comprised of B- and C-type asteroids, bodies with spectra comparable to asteroids like Bennu, Ryugu, and Ceres (e.g., DeMeo et al. 2015; Hamilton et al. 2019; Kitazato et al. 2019). These taxonomic types make up the vast majority of all low albedo asteroids in the main belt (i.e., albedo $< 12\%$).

Curiously, we found that about 90 main belt bodies with $D \geq 40$ km that have no taxonomic designation. These were predominately asteroids in the outer main belt with semimajor axis $a > 2.8$ au and albedos $< 8\%$. Rather than exclude these bodies, we created albedo probability distributions for the of D/P-types and C/B-types in the outer main belt and then assigned the unclassified bodies to each group according to their albedos. Specifically, we first placed the albedos of the known D/P-types and C/B-types from the outer main belt into bins with a range between 0 and 0.08, with the bin sizes being 0.01. Next, we added the number of bodies within a bin that were D/P type or C/B-types and then divided by the total, giving us a fraction for each within the bin. From there, we took the albedos of the unassigned bodies, found the bin they were in, and used random deviates to choose a taxonomic type. This method is approximate but reasonable, given that there should be no bias between the spectral identification of a D/P type vs a C/B-type in the outer main belt.

The third group was a small unusual class called K-types, whose spectra appear to be a match with CV and CO-type carbonaceous chondrites (e.g., Burbine et al. 2002a). An example K-type asteroid would be (221) Eos, the largest member of the Eos family, while an example CV chondrite would be the calcium-aluminum inclusion-rich (CAIs,the first solids of our solar system) meteorite Allende. K-types have moderate albedos and spectra that almost seem to be a combination of C- and S-types; for this reason, they were named for the letter halfway between C and S in the alphabet. Note that many K-type $D \geq 40$ km bodies are part of the Eos family, so all in the family but (221) Eos itself were excluded to avoid bias in our calculation.

Finally, using the taxonomic catalog of M-types from our various catalogs, we found there were 41 M-types with $D \geq 40$ km in the main belt (Fig. 3). These bodies make up $\sim 5\%$ of the $D \geq 40$ km population (out of 767 such objects between 2 and 3.3 au across all taxonomic groups). Many large M-types have semimajor axes a between ~ 2.6 and ~ 2.9 au (Scott et al. 2015). The remainder are located near ~ 2.4 au and ~ 3.1–3.2 au. In general, M-types show a wide range of eccentricities (e) and inclinations (i), though in Scott et al. (2015), it was shown that for the eight M-types with $i < 5°$ and $D \geq 50$ km, three are at $a = 2.42$ au and four more are at $a = 2.67$ au. We will return to this issue below.

To make approximate comparisons between M-types and other taxonomic groups, we have divided the asteroid belt into inner, central, and outer zones. We placed our divisions between these zones at 2.5 and 2.8 au, locations defined by the 3:1 and 5:2 mean motion resonances with Jupiter, respectively. Our results for $D \geq 40$ km asteroids are shown in Fig. 4, with the error bars calculated from the square root of the number of bodies in a bin.

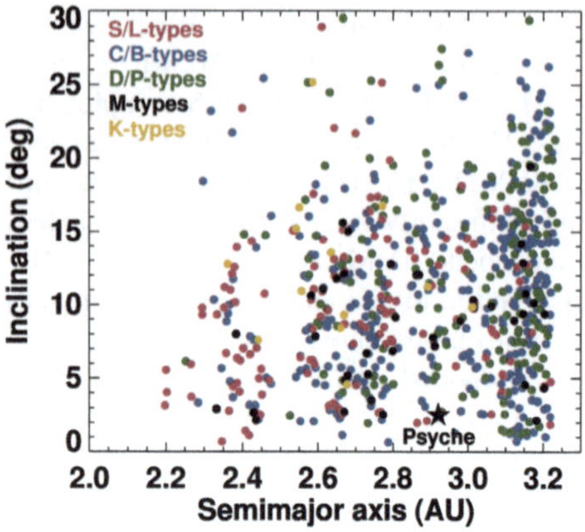

Fig. 3 Orbital distribution of $D > 40$ km asteroids in the main belt between 2.0–3.3 au that fall within the listed taxonomic groups: S/L-types (red dots), C/B-types (blue dots), D/P-types (green dots), M-types (black dots), and K-types (gold dots). All populations are dynamically excited, and there is no obvious evidence for major differences in their eccentricity and inclination distributions. The semimajor axis distribution of the classes are shown in Fig. 4

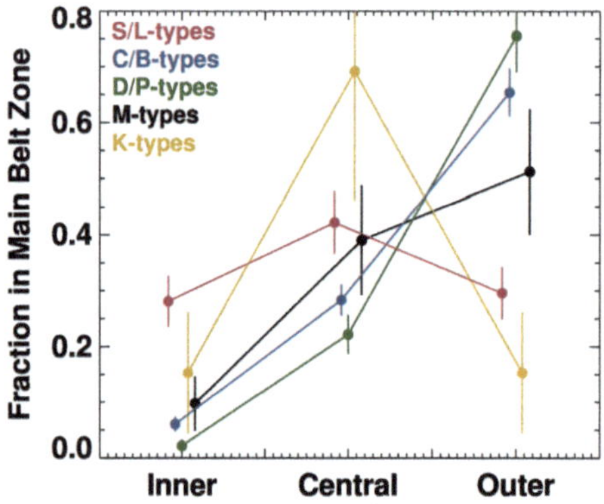

Fig. 4 The fraction of $D > 40$ km bodies in the main belt between 2.0–3.3 au with different semimajor axes and taxonomic classes. The listed taxonomic groups have the same colors as Fig. 3. We divided the inner, central, and outer main belt zones with divisions at 2.5 and 2.8 au, locations defined by the 3:1 and 5:2 mean motion resonances with Jupiter, respectively. For a given taxonomic group, error bars are the square root of the number of bodies within a zone divided by all objects in the group. The S/L-types have the largest fraction of inner main belt bodies, consistent with an origin that favors the inner solar system. The C/B and D/P have the largest fractions of bodies in the outer main belt, favoring an origin in the outer solar system. The M-types most closely resemble the C/B- and D/P-types distributions, suggesting that they may have originated in the outer solar system

Overall, we find that S/L-types have a relatively flat semimajor axis distribution across the three main belt zones, though they also have by far the largest fraction in the inner main belt. Conversely, the C/B-types and D/P-types have the strongest concentrations in the outer main belt. The M-types are closest to the distribution of the C/B-types, but there are differences as well, with the M-types having a greater fraction in the central main belt and a lower fraction in the outer main belt. Finally, the K-types seem relatively unique, with most of their bodies residing in the central main belt.

We caution that we have not renormalized the value for the fraction of the main belt depleted by the actions of giant planet migration (Nesvorny et al. 2017). This would increase the fractions in the inner main belt, which has experienced the most dynamical depletion, and would cause us to renormalize all of values. Tests suggest that the taxonomic distributions of different groups maintain their differences across the Inner, Central, and Outer main belt. For that reason, we decided not to modify the distributions using numerical modeling results, though it could be argued doing so provides a better picture of the main belt immediately after the end of giant planet migration.

To interpret Figs. 3 and 4, we need to consider what dynamical insights come from planet formation models. This subject does not make for easy distillation, so here we focus our attention on the results of numerical simulations that can reproduce many of the existing constraints (e.g., the masses and orbits of many solar system objects, including the terrestrial planets, asteroids in the main belt, and the Kuiper belt population).

Early in planet formation, when the solar nebula was present, non-carbonaceous planetesimals were growing in the inner solar system and carbonaceous planetesimals, with a range of rock and ice fractions, were forming in the outer solar system (e.g., Alexander et al. 2012; Kruijer et al. 2017; Kleine et al. 2020). The formation of Jupiter, or perhaps a pressure bump near where Jupiter formed, may have opened a gap in the solar nebula, preventing materials entrained in the gas from mixing between the inner and outer solar system (Kruijer et al. 2017; Brasser and Mojzsis 2020). This action is thought to explain the differences of isotopes in non-carbonaceous and carbonaceous meteorites (e.g., those of molybdenum Mo, ruthenium Ru, titanium Ti, and chromium Cr). It is now thought that non-carbonaceous chondrites formed in the inner solar system, while carbonaceous formed in the outer solar system, and that iron meteorites are associated with both groups. This implies that the metal-rich M-types could have presumably formed in either the inner or outer solar system.

Protoplanets and planetary embryos were also growing in these zones, though perhaps not in the asteroid belt region, via pebble accretion (where gas drag slows small objects enough that they are readily accreted by early-forming planetesimals) (e.g., Levison et al. 2015). The earliest bodies to accrete were most likely to differentiate, thanks to the heat given off by the decay of ^{26}Al. Measurements suggest that the majority of our iron and stony-iron meteorite parent bodies formed before 300,000 years after CAI formation (Kruijer et al. 2014; Scherstén et al. 2006; see review in Scott 2020), and Vesta likely formed within 1.5 Ma of CAIs (Neumann et al. 2014). Psyche and other M-types are likely to have originated at this time, but their formation location is unknown.

In the terrestrial planet region, protoplanets grew and began to dynamically excite the planetesimal populations around them. This led to mergers, collisional evolution, and possibly hit-and-run events, where the crust and mantle of some large differentiated planetesimals would be stripped by impacts with similarly sized or somewhat larger bodies (e.g. Asphaug et al. 2006; Nesvorny et al. 2021). Given its expected bulk density of $\sim 4{,}000\ \mathrm{kg\,m^{-3}}$, a single hit-and-run collision during the formation epoch could produce Psyche when a differentiated Vesta-sized progenitor collides with a still more massive (e.g., Ceres-sized) target. In this case, the Vesta-sized body escapes accretion (impacting too fast, or at too grazing an angle), losing much of its mantle in the process. Depending on the initial population that formed the planets, according to Asphaug (2010) the unaccreted population might include survivors of multiple hit-and-run collisions. This is also confirmed by several numerical studies of the late stage of terrestrial planet formation in the solar system (e.g., Chambers 2013; Emsenhhuber et al. 2020). Some of these survivors might consist primarily of mantle materials, and others may contain significant metallic core material. This scenario could apply to Psyche if it turns out to be primarily metallic (discussed in more detail below).

In the giant planet zone, similar events likely took place, and some planetary embryos were able to reach the size of several Earth masses via pebble accretion and embryo mergers (e.g., Kretke and Levison 2014). This allowed them to accrete a gaseous envelope from the solar nebula, and this core accretion scenario defines a leading hypothesis for how our system of giant planets was formed (e.g., Ormel and Klahr 2010; Lambrechts and Johansen 2012; Kretke and Levison 2014). If true, hit and run events in the outer solar system may have been common, as implied by simulations that show protoplanets and planetesimals frequently collide with one another in the same zones where the cores of the giant planets are being constructed (e.g., Levison et al. 2015; Raymond and Izidoro 2017a,b). It is highly likely that some bodies that experienced early collisional evolution in the giant planet zone made their way to the main belt.

Near the end time of the solar nebula, it is possible that a particular planet formation process took place that affected the orbits of many planetesimals. Specifically, proto-Jupiter may have grown large enough that its gravitational interactions with the solar nebula allowed it to migrate inward across the main belt zone. This event has been hypothesized to be the opening maneuver to what is now known as the "Grand Tack" (Walsh et al. 2011). Reaching ~ 1.5 au, Jupiter may have then become trapped in a mutual 2:3 resonance with an inward-migrating proto-Saturn. At that point, the collective gravitational behavior of Jupiter, Saturn, and the solar nebula may have caused Jupiter and Saturn to move away from the Sun. The outward migration ended when sufficient gas dissipated from the solar nebula that Jupiter was no longer able to move via gravitational torques. Modeling work and certain constraints indicate that this could have occurred when Jupiter reached a semimajor axis of ~ 5.5–5.6 au.

The Grand Tack and other dynamical models of planet formation and migration predict the gravitational scattering of small bodies in both the inner and outer solar system (e.g., Raymond and Izidoro 2017b). In these scenarios, S/L-type asteroids and other inner solar system bodies with characteristics representative of planetesimals from the Venus-Earth-Mars-Asteroid Belt zone were excited and then injected into the main asteroid belt. These bodies represent non-carbonaceous materials. Beyond Jupiter, C/B-type asteroids that were once denizens of the Jupiter-Saturn-Uranus-Neptune zone and beyond were captured into the main belt region. One of the strengths of the Grand Tack model is that it can plausibly reproduce the taxonomic distribution of large S/L- and C/B-type bodies found in the asteroid belt today (Walsh et al. 2011; see also Morbidelli et al. 2015).

The Grand Tack model was proposed as a means to explain the small size of Mars relative to Venus and Earth and the dynamical characteristics of the asteroid belt (Walsh et al. 2011), but it is not without its criticisms. The conditions required for, and plausibility of, the outward migration of Jupiter are still being explored (Raymond and Morbidelli 2014; Kanagawa et al. 2018). Additionally, the rapid formation of the terrestrial planets implied by the Grand Tack requires very efficient chemical equilibration during giant impacts, possibly inconsistent with fluid modeling studies (Zube et al. 2019).

Alternative models to the Grand Tack also exist. Of note, recent work has suggested that the terrestrial planets may have formed via pebble accretion (Levison et al. 2015; Johansen et al. 2021). The plausibility of this mechanism is in part controlled by an outstanding question of the ability of solids to migrate into the inner solar system after Jupiter has formed (Izidoro et al. 2021). Giant planet zone planetesimals may have been delivered to the main asteroid belt by a combination gas drag processes and scattering events produced by a growing Jupiter (e.g., Raymond and Izidoro 2017a,b; Kretke et al. 2017). There are also several other mechanisms that can potentially reproduce the small size of Mars (e.g., Levison et al. 2015; Clement et al. 2018; Johansen et al. 2021; Nesvorny et al. 2021). More work is still needed to fully explore the geochemical and dynamical implications of all of these planet formation models.

The other, independent, hypothesized dynamical event would have taken place after the solar nebula dispersed. At this time, the giant planets are thought to have been in a different configuration than the one seen today (Tsiganis et al. 2005). Gas accretion left the giant planets on nearly circular coplanar orbits between 5.5 and ~ 20 au, with most or all of these worlds locked in mutual mean motion resonances with one another. The primordial Kuiper belt existed just beyond the original orbit of Neptune, with an estimated comet population of ~ 20 Earth masses that stretched beyond 20 au. This configuration, however, is hypothesized to have eventually become unstable. As the primordial Kuiper belt lost planetesimals to the giant planet zone, it caused some giant planets to migrate far enough to either enter into or exit dynamical resonances with one another, and likely involved 5–6 giant planets, with 3 or 4 Neptune-sized bodies (e.g., Tsiganis et al. 2005; Nesvorný 2011; Nesvorný and Morbidelli 2012; Batygin et al. 2012; Vokrouhlický et al. 2016; see review in Nesvorny 2018). The culmination of these events led to the outermost giant planets migrating outward through the primordial comet disk in a violent exchange of orbital energy and angular momentum. This is thought to have led to a reorganization of the giant planets in what is now called a "giant planet instability". This instability could have occurred with or without a Grand Tack earlier on (Deienno et al. 2018).

The most important factors for the origin of Psyche in the planetesimal evolution and giant planet migration scenarios discussed above are as follows. First, giant planet encounters with one another allowed a small fraction of bodies from the primordial Kuiper belt to be captured in the central and outer main asteroid belt, the Jupiter and Neptune Trojan populations, and the irregular satellite populations (e.g., Levison et al. 2009; Nesvorný et al. 2013; Vokrouhlický et al. 2016; see review by Nesvorny 2018). This mechanism explains why primitive D- and P-type asteroids are found in those locations. Second, asteroids in the main belt at the start of the giant planet instability were dynamically excited and depleted (Nesvorny et al. 2017; Deienno et al. 2018). Third, the giant planet instability may have occurred early, possibly as soon as the solar nebula dissipated (e.g., Nesvorny et al. 2018, 2021; de Sousa et al. 2020). Accordingly, migrating giant planets may play a key role in exciting the planetesimals and protoplanets located in the terrestrial planet region, even if the giant planets never migrated that far (Agnor and Lin 2012; Kaib and Chambers 2016; Clement et al. 2018; Nesvorny et al. 2021). Note that if the instability occurs too late, the induced excitation of the terrestrial planets would likely produce orbits for them unlike those seen today.

Thus, the population of the main asteroid belt, thought to contain bodies that originated from the inner and outer solar systems, could be explained as the net result of events that took place when the nebula existed (Grand Tack and/or gas drag processes) combined with events that took place after the solar nebula was gone (giant planet instability and migration). With our discussion of early dynamical processes in place, we can now consider how to use M-type orbital constraints to glean insights into the formation location and dynamical evolution of the M-types and Psyche.

2.1 M-types from the Terrestrial Planet Region

One possible starting point for Psyche and the M-types is the terrestrial planet region. Dynamically, the primary ways such objects can get into the main belt zone early in solar system history is by scattering events from planetary embryos or gravitational interactions with Jupiter via the hypothesized Grand Tack (Bottke et al. 2006b; Walsh et al. 2011; Raymond and Izidoro 2017a; Kretke et al. 2017). Regardless of which mechanism is used here, these interactions predominantly place terrestrial planet zone planetesimals into the inner and central main belt regions, the orbital location where most large S/L-type asteroids are found today. Our expectation is that if the M-types formed in the same locale as the S/L-types, they should have been delivered to the main belt by the same mechanisms and therefore should have a similar distribution to the S/L-types. The problem is that few M-types exist in the inner main belt, a mismatch with these model predictions (Figs. 3 and 4).

A possible way out was suggested by Bottke and Asphaug (2013) (see also Scott et al. 2015), who showed that some large M-types have semimajor axis locations consistent with "fossil resonances" of Jupiter that existed when Jupiter was located at ~ 5.55 au, its putative location at the end of the Grand Tack. Prior to the giant planets taking on their present-day orbits, Jupiter was likely on a circular orbit, and many of its resonances within the main belt would have produced trapping rather than diffusive behavior. Numerical simulations showed that low eccentricity objects encountering such a resonance could enter the main belt zone. If Jupiter later migrated to its current orbit of a ~ 5.2 au via a giant planet instability, these objects would become permanently trapped in the main belt.

This scenario could potentially explain the capture of low inclination (i.e., $< 7°$) M-types at the 7:2 and 3:1 mean motion resonances with Jupiter. For Jupiter at $a \sim 5.55$ au, these resonances would be at 2.4 and 2.67 au. The problem, however, is the timescale. Simulations of this behavior indicate that putative M-types could take many tens to hundreds of millions of years to enter fossil resonances (Bottke and Asphaug 2013; Raymond and Izidoro 2017a). This is permissible if the giant planet instability occurs tens to hundreds of millions of years after the solar nebula is gone. If it occurs millions of years after the dispersion of the nebula, however, there is not enough time to get a substantial population of M-types into the main belt zone through these "dynamical doors".

For these reasons, we infer that the M-types are unlikely to be inner solar system bodies, in which case Psyche is unlikely to be made of non-carbonaceous materials.

2.2 M-types from the Asteroid Belt

It is possible that the M-types were not implanted but instead are indigenous to the central main belt. If true, the putative population could not have been dynamically affected in a profound way by the putative Grand Tack, which would have scattered M-types throughout the main belt. Any model of M-type formation and evolution would also need to explain why M-types are distributed in semimajor axis in a manner similar to C/B-types and D/P-types (Figs. 3 and 4).

Another potential problem with forming M-types in the main belt would be to explain how many of them ended up in the fossil resonances of Jupiter (Scott et al. 2015). We suspect that a population that originated in the main belt would have difficulty matching the dynamical constraints of the low inclination M-types, but this would need to be considered within a model of main belt evolution that can meet other dynamical constraints. On the other hand, it is also possible that the fossil resonance argument is weaker than suggested.

At this time, we find no strong dynamical evidence in support of the idea that M-types formed in the main belt zone. More planetesimal formation and dynamical evolution work is needed to determine whether a plausible solution exists for this hypothesis.

2.3 M-types from the Primordial Kuiper Belt

It is also possible that the M-types originated beyond Jupiter and are in fact carbonaceous in nature. In this circumstance, we need to examine dynamical processes that can capture putative M-types in the central and outer main belt but also explain their limited numbers in the inner main belt relative to other populations.

One possibility is that the M-types were originally part of the primordial Kuiper belt, a population that was once thought to have totaled ~ 20 Earth masses. Simulations indicate that approximately 99.9% of the population was scattered onto giant planet-crossing orbits when the giant planet instability caused Uranus and Neptune to migrate through the disk.

A tiny fraction of this material would have been captured in stable small body reservoirs in the inner solar system, such as the central and outer main belt and the Hilda and Jupiter Trojan populations (e.g., Vokrouhlický et al. 2016). P-types are of particular interest because their spectral signatures in the visible to short-wave near-IR are similar to that of M-types in that neither has strong absorption bands (see review by DeMeo et al. 2015). The main discriminant between the two taxonomic types is not spectra but albedo, with P-types generally having albedos much less than 10% percent and M-types having albedos that are 10–20%.

This leads to the prediction that if substantial numbers of M-types were coming from the primordial Kuiper belt, we would expect some M-types to have also been captured in the Hilda and Trojan asteroid populations, given the substantial number of large M-types found in the asteroid belt. Accordingly, the question is whether M-types could remain relatively "hidden" among a large P-type population such as the Trojans. At least so far, it appears the answer to this question is no. Observational evidence in the form of extensive Trojan observations indicate they have uniformly low albedos of around 0.05 (e.g., Emery et al. 2015).

We conclude that M-types are unlikely to be from the primordial Kuiper belt, unless their true nature can be easily hidden and that same camouflaging process works much less efficiently in the main belt, where we see M-types, than in the Hilda and Trojan populations.

2.4 M-types from the Giant Planet Zone

The evolution of planetesimals formed in the giant planet zone may provide a more plausible means to explain the M-types by implantation. An examination of the dynamical results from Walsh et al. (2011) shows that asteroids originally located between the giant planets can be dynamically captured within the main belt. This model can explain the orbital distribution of the C/B-types, and we find it interesting that the M-types have a similar semimajor axis distribution to this class (Figs. 3 and 4). That could suggest a common origin location. If there is no Grand Tack, the implantation would instead take place by gas drag processes combined with scattering events from a growing Jupiter (e.g., Raymond and Izidoro 2017a, 2017b; Kretke et al. 2017).

The capture location of giant planet zone bodies depends on the starting location within the giant planet zone. Numerical simulations of the Grand Tack or giant planet formation combined with gas drag processes suggest that giant planet zone planetesimals that are closer to Jupiter show a greater likelihood of being captured in the inner and central main belt than those that are further away from Jupiter (Walsh et al. 2011; Raymond and Izidoro 2017a, 2017b; Kretke et al. 2017). Based on this, we postulate that the M-types and Psyche were came from a zone that was closer to Jupiter than in a zone that produced most C/B-types. It has already been argued that certain carbonaceous meteorite types may have formed near Jupiter, such as the water-poor but CAI-rich CV chondrites (e.g., Desch et al. 2018). Our hypothesized scenario would potentially explain why M-types have a greater fraction of central main belt bodies and a lower fraction of outer main belt bodies than C/B-types (Figs. 3 and 4). We note that this idea is broadly consistent with existing numerical simulations but has yet to be tested by any specific simulation.

Another issue to investigate is whether some fraction of the M-types from the giant planet zone could find their way into the main belt via Jupiter's fossil resonances. For example, in the Grand Tack, M-types would get into the main belt as Jupiter migrates outward, but the capture events would happen near the migration endgame. Here the solar nebula would be nearly dissipated, and Jupiter would be grinding to a halt near its final Grand Tack location of ~ 5.55 au. This would leave M-types adjacent to the main belt to find those dynamical doors that existed at that time, namely the fossil resonances. Perhaps a comparable mechanism would take place with gras drag implantation. Our cautionary note is that there may not have been much time for them to get into the main belt if the giant planet instability occurred shortly after the end of the solar nebula.

Finally, we find it intriguing that the K-types are concentrated in the central main belt. K-types are thought to produce CV and CO chondrites (Burbine et al. 2002a), unusual carbonaceous chondrites that have a lower abundance of volatiles compared to CM or CI chondrites. Their CAI abundances are also quite high relative to other classes. It can be argued that CAIs originally formed near the Sun and were spread out through the solar nebula prior to the formation of Jupiter's core (e.g., Desch et al. 2018). Once Jupiter became large enough to cut off nebula communication between the inner and outer solar system, CAIs in the outer solar system would tend to evolve inward toward the Sun via gas drag. The net result would be higher concentrations of CAIs in a zone to the anti-sunward Jupiter than in other zones, and planetesimals forming there would be rich in CAIs (Desch et al. 2018). At that point, dynamical mechanisms akin to those discussed above would place a small fraction of these planetesimals into the main belt.

At this point, we engage in further speculation based on the limited information that exists, though the reader should be aware that the threads of evidence are thin. The results in Figs. 3 and 4 show that K-types have a different taxonomic distribution than any other group that is shown, while the M-types appear to have a distribution closer to that of the C/B group. If these trends are not dominated by small number statistics, the distribution differences could be telling us about their source regions. If we further assume that K-types are related to CV/CO chondrites, and they formed near the Jupiter zone, that would potentially rule out the possibility that the M-types formed in that zone, unless they formed at a different time (and that influenced how they get into the main belt). Finally, if we assume every asteroid group had to form in a distinct zone within the giant planet region, it forces us to put the M-types in a different place than the C/B-types and the D/P-types. We argued above that the D/P-types probably came from the primordial Kuiper belt, which leaves us with the M-types and C/B-types. If we were forced to choose an order for them, we would look to the semimajor axis distribution of the M-types, which potentially split the difference between the K-types and the C/B-types (Fig. 4). If true, this would indicate that M-types originally resided in the outer solar system between the K-types, which formed fairly close to Jupiter, and the C/B-types, which formed further away from Jupiter.

To melt and form a core as a planetesimal, Psyche would have had to accrete within ~ 1.7 Myr after CAIs if it contained the canonical concentration of ^{26}Al. If Psyche is part of a differentiated planetesimal, and it originated in the outer solar system, it would have had to either accrete earlier than the 3 to 4 Myr after CAIs assumed for most carbonaceous chondrite parent bodies, or it had to have originated with more ^{26}Al. This scenario only holds if Psyche is a part of the core of a planetesimal. If it was part of a larger body, or a droplet from a much larger collision, then the timing constraints are removed.

To match existing Psyche observations, a possible origin in the volatile- and carbon-rich outer solar system would have to be followed by crustal and mantle stripping, to accord with the relatively high bulk density of Psyche. This would include loss of exterior volatiles and any ice-shell, even in low-velocity (shock-free) mantle stripping that does not cause much heating, simply by virtue of being in the exterior and lowest gravitational potential. Further, the heating of a differentiating planetesimal liberates fluids long before metal-silicate separation, and these fluids are likely to be lost to space (Young 2001; Young et al. 2003; Fu and Elkins-Tanton 2014).

However, depending on the size of the parent body and its own thermal evolution, hydration may exist throughout the body. For example, Titan is proposed to be dominated by a mantle composed of hydrated silicates (e.g., Fortes 2012) and may even possess a hydrated core. In such a case water would be difficult or impossible to remove even in an event that removed nearly all the mantle, which is not likely in any case. In this case, these hydrated phases, stable in the original hydrostatic parent body, might not be stable near the mantle-stripped Psyche's surface, so volatiles might be expected to be present, or evidence for volatiles driven off to space with evidence for sublimation degradation and associated mass wasting.

There is much to investigate here, but we speculate that it may be possible to make a case that the M-types formed in the giant planet zone and were implanted in the central and outer main belt by the Grand Tack or by gas drag processes related to the formation of the giant planets. Their evolution in the giant planet zone could also open the door for hit and run-type encounters to be involved with their evolution, given that the formation of the cores of the giant planets led to many such events (e.g., Levison et al. 2015).

3 The Fiducial Origin Scenario: A Differentiated Parent Body for Psyche

Psyche's apparent composition of metal plus lower-density material, and the inferred paucity of FeO in that lower-density material, place constraints on its possible formation. These formation processes may also in some cases be aligned, or misaligned, with the possibility that Psyche originated in the outer solar system, as described above.

The first and simplest way to accumulate a high fraction of metal is through melting and differentiation of a body into a metal core and a silicate mantle.

3.1 The Process of Core Formation in Planetesimals

Core formation in planetesimals likely begins at the onset of partial melting under high-temperature, low-pressure and relatively dry conditions. Melting in planetesimals must be almost exclusively driven by the heat of radiogenic ^{26}Al (Fish et al. 1960; Huss et al. 2006; LaTourrette and Wasserburg 1998; Lee et al. 1976; Urey 1955). This is in sharp contrast to well-known adiabatic melting regimes on Earth, which occur at depth in the upper mantle at pressures up to 24 GPa (240 kbars) or above, or under wet crustal conditions where melting can begin at $\sim$ 800 °C. Even the centers of the largest differentiated asteroids (e.g., Vesta) reach only pressures of $\sim$ 2 kbars. The first silicate partial melts on asteroids (generally picritic, with SiO_2 below 52 wt%, and MgO up to 30 wt%) first occur when temperatures reach $\sim$ 1050 °C. However, the very first melts of a chondritic precursor are not basaltic, but Fe,Ni-FeS melts that form at $\sim$ 950 °C and contain $\sim$ 85 wt.% FeS (Kullerud 1963).

Numerous experimental studies (e.g., Gaetani and Grove 1999; Miller et al. 2016) demonstrate that because of their surface tensions these melts will not readily migrate under static conditions. Yet, melt migration did occur at low degrees of partial melting in asteroids. The preponderance of FeS (troilite) within these early formed melts and its relatively low abundance in chondritic meteorites ($\sim$ 5 wt.%) makes tracing S and/or troilite abundances a particularly powerful tool for understanding early partial melting in meteorites collectively dubbed primitive achondrites.

This group, including acapulcoites-lodranites, winonaites/IAB irons, brachinites, and ureilites, exhibit broadly chondritic compositions and mineralogies with evidence of partial melting (McCoy et al. 1996, 1997a,b; Mittlefehldt et al. 1996). Although micrometer- to centimeter-sized metal-sulfide veins are present in acapulcoites, significant sulfur depletions are only common in the lodranites, where silicate melting opened pathways for melt migration. Evidence for silicate melting and melt migration is found in plagioclase abundances. Basaltic partial melt contains ~ 55 wt.% plagioclase stiochiometrically (Morse 1980) and chondritic meteorites contain ~ 10 wt.% plagioclase. Consequently, plagioclase (or its constituent elements, most notably Al, Ca, Na, and K) is rapidly depleted in residues during partial melting. Plagioclase content is therefore a tracer for early silicate melting.

Melt migration at low degrees of partial melting may not have been a static process, but a dynamic process driven by overpressure resulting from melt expansion and/or release of volatiles, notably CO and S (McCoy et al. 1997b). The fate of these Fe,Ni-FeS and basaltic partial melts is uncertain, but two tantalizing possibilities exist in the absence of higher degrees of partial melting. Keil and Wilson (1993) suggest that on bodies only a few tens of kilometers in size, Fe,Ni-FeS melts could be buoyant in the presence of released volatiles, driven to the surface, and erupted into space at velocities exceeding the escape velocity of the body, possibly explaining the S-depletion in some iron meteorite parent bodies. On larger bodies like Psyche, these Fe,Ni-FeS melts could migrate to the center of the body, forming essentially a proto-core much smaller and richer in S than formed during later migration of metal during high degrees of partial melting more typically associated with core formation. These small S-rich proto cores may be sampled by meteorites like some of the silicate-bearing IAB irons.

Low-degree melting, however, may not have been a dominant process in planetesimal formation. Isotopic evidence from iron meteorites shows that their parent bodies accreted within 100,000 to 300,000 years after formation of calcium-aluminum inclusions (Kruijer et al. 2014; Scherstén et al. 2006; see review in Scott 2020). Results of pebble accretion models show that, given planetesimals of tens to hundreds of kilometers radius and the continued existence of a gas disk, gas giants the size of Jupiter can accrete in the outer solar system, in only an additional 100,000 years (Lambrechts and Johansen 2012; Ormel and Klahr 2010; Pollack et al. 1996). Thermal and compositional models of planetesimals show that if they formed within ~ 1.7 Ma after calcium-aluminum inclusions the heat of ^{26}Al would be sufficient to melt them completely (Elkins-Tanton et al. 2011; Sahijpal et al. 2007; Šrámek et al. 2012).

Once melting has progressed well past the eutectic, the relative densities of the materials determine whether differentiation occurs. In rapidly-accreted planetesimals that melt in the first ~ 1.7 Ma of the solar system, formation of metal cores was rapid and efficient. Thus a generation of "Ur-cores," original, first-generation cores, would form in planetesimals, and then be mixed and reformed in all subsequent accretionary collisions. These subsequent collisions occurred at a rapid pace in the first few million years of the solar system, and then more slowly as populations waned and the oligarchic stage of planetary formation commenced (Chambers 2008). The gas disk had likely locally dispersed between > 1.22 and < 3.94 Ma after CAI-formation in the noncarbonaceous reservoir, and between > 2.51 and < 4.89 Ma in the carbonaceous reservoir (95% confidence limits) (Weiss and Bottke 2021).

Especially in the outer solar system, heating is reduced not only by the possibly longer time scale of accretion and diminution of ^{26}Al, but also by the water content of the accreting materials. Castillo-Rogez and Young (2016), however, demonstrated that even with 50% water ice, if the planetesimal formed prior to 1 Myr after CAIs it could still reach the silicate solidus and differentiate into a metal core and silicate mantle. Thus, differentiation in planetesimals likely occurred in both the inner and outer solar systems, and this fiducial model for Psyche can be considered in either case.

3.2 Planetesimal Processes that can Produce a Low-FeO Non-metallic Phase in Psyche

The bulk compositions of the non-metal fractions of almost all meteorites are too FeO-rich to satisfy Psyche's spectral constraints (Dibb 2021). Thus, in this section, we present processes that can produce a low-FeO, low-density second phase that will satisfy both the spectral and density constraints of Psyche.

The melting of the parent planetesimal to allow metal core differentiation is a starting place. Fractional solidification in the molten silicate portion of a hot young planetesimal produces first high-MgO mafic minerals, gradually transitioning to more and more iron-rich mineral compositions as liquids evolve. In larger bodies, settling timescales can be significantly less than solidification timescales (Suckale et al. 2012), and thus a radial structure with high MgO at the bottom and high FeO at the top is produced. These cumulate piles are gravitationally unstable, with both thermal and compositional parameters increasing density at the top of the pile, and will overturn in the solid state (e.g. Ringwood and Kesson 1976; Spera 1992; Solomatov 2000).

In larger bodies, if the outer mantle can be removed before overturn, the body will be left with an iron-nickel core, and only the lower-mantle high-MgO, low-FeO silicates of the initial solidification remaining. Mechanisms for removing that upper mantle are complex and not well understood. Removal without mixing is a challenge, as is understanding the effects of depressurization and oxidation.

In the process of fractional solidification at the low pressures of planetesimal mantles, almost all chondritic compositions would first crystallize olivine (summary in Elkins-Tanton 2017 and see Okada et al. 1988), with the enticing exception of enstatite chondrites, which would first crystallize orthopyroxene (McCoy et al. 1999). This observation is enticing because Hardersen et al. (2005) suggested based on spectra that the non-metal phase on Psyche is orthopyroxene.

As soon as crystallization extends throughout the molten mantle, separation of crystals through floating or settling becomes difficult due to crystal interference and high viscosity (Elkins-Tanton 2017). Unlike the high gravity and often slower cooling conditions of larger planets, the solidus and liquidus of a planetesimal's silicate mantle have very little change in temperature between the core-mantle boundary and the surface. Therefore only a small amount of additional cooling past the point where the adiabat crosses the liquidus will move the adiabat entirely between the solidus and liquidus, and crystallization will commence throughout the vertical column of the planetesimal mantle. Over the ~ 0.5 kbar mantle pressure range of a planetesimal ~ 200 km in radius, the solidus will change by only about 10 °C and the adiabat by only ~ 2 °C (Elkins-Tanton 2017). A body this size would likely be the smallest possible parent body for Psyche, if Psyche is mainly a differentiated core.

Settling of crystals during solidification on planetesimal or planetary scales remains poorly understood, but geochemical and mineralogical evidence for settling (Morse 1982) and floating in the lunar magma ocean is widely accepted (e.g., Smith et al. 1970; Warren 1985; Wood et al. 1970). The difficult relevant question here is over what solidification interval settling can occur successfully on bodies with low gravity, that is, how thick a layer of low-iron early-forming cumulates can coat the core of Psyche's parent, ready to be incorporated into the Psyche of today while excluding the higher-iron, later-forming cumulates?

Solomatov and Stevenson (1993) derive an expression for the size of particles that will settle during solidification (Eq. (12)):

$$D = \left(\frac{10}{\Delta\rho g}\right)\left(\frac{\eta\alpha g F}{C_p}\right)^{0.5},$$

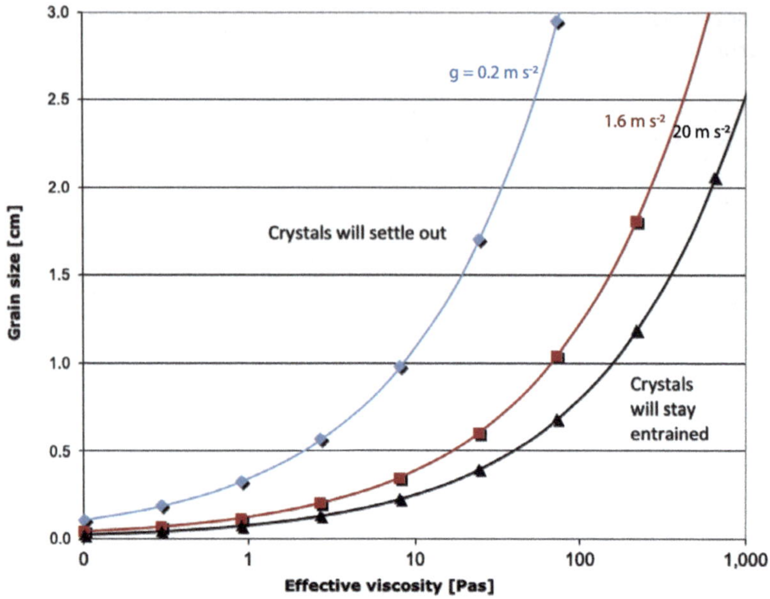

Fig. 5 Only the largest early crystals forming in a planetesimal magma ocean will settle, and only until the effective viscosity rises past a certain point. In the low gravity of a planetesimal, crystals are likely to settle only while the magma ocean remains at viscosities of picritic melts, and before a significant crystal fraction grows. Thus only the earliest-forming mineral (almost always olivine, but for enstatite chondrites, orthopyroxene) will settle to the core-mantle boundary

where D is grain size, $\Delta\rho$ is the density difference between the liquid and crystal, here taken as 100 $\mathrm{kg\,m^{-3}}$; g is the gravitational acceleration of the body, here values for Vesta, the Moon, and Mars are used; η is effective viscosity; F is heat flux from the cooling body, here taken as $10^4\ \mathrm{W\,m^{-2}}$; α is the thermal expansivity, here $3 \times 10^{-5}\ \mathrm{K^{-1}}$; and C_p is heat capacity, taken as $1{,}256.1\ \mathrm{J\,K^{-1}\,kg^{-1}}$ (Fig. 5). To settle on a small body like Vesta, crystals must grow rapidly to several millimeters while the effective viscosity of the liquid remains below 1 Pa s or so, which is consistent with an ultramafic or mafic largely crystal-free magma (Reese and Solomatov 2006). On small bodies, a critical crystal fraction forms rapidly because of the similar slopes of the adiabat and solidus, and settling of an individual phase would be impossible because of interference and networking among the solids.

Only the earliest crystals to form, therefore, have the ability to settle to the core-mantle boundary. These would be olivine in most cases, and pyroxene in the case of an enstatite chondrite bulk composition. If Psyche's low-density phase is in fact low-iron enstatite (Hardersen et al. 2005) but its bulk composition was more oxidized than enstatite chondrites, its first-crystallizing olivine must have been reduced to pyroxene, possibly through extended contact with metal.

Several processes could have produced Psyche's metal + low-density phase composition. The main goal of the three scenarios listed below is to produce a low-FeO, low-density second phase (or phase assemblage), satisfying the spectral observations of Psyche, that can be mixed with iron and nickel metal to produce the observed density of Psyche.

1. A Vesta-sized parent body is the smallest body consistent with Psyche's size and density. In this scenario, a fully molten Vesta-sized parent body settles low-iron olivine or

orthopyroxene in the earliest period of magma ocean solidification. Subsequently, the overlying cumulate mantle is removed by a series of impacts. A challenge to this scenario is the likely thinness of the olivine- or orthopyroxene-only layer at the bottom of the mantle due to the barriers to crystal settling. Only a specific series of collisions could strip only the outer mantle and leave the low-FeO silicates on the resulting Psyche.

2. A similar but possibly more plausible scenario is that Psyche's parent body was a larger body, perhaps even Moon-sized, and then suffered one or more hypervelocity collisions. This larger body allows a thicker layer of iron-poor silicates to settle to its core-mantle boundary during solidification, and subsequent impacts need only retain a portion of the core and lowest mantle to become Psyche.
3. Psyche's parent body went through the differentiation process and stripping processes described above, but its bulk composition was so reduced to begin with that the non-metallic portion was already low in FeO. Then, stripping via impacts is only required to remove sufficient silicates to match the compositional requirements. Though a simpler process, we have no examples of a meteoritic bulk composition with a silicate fraction so low in FeO.

3.3 Impact Scenarios that Could form Psyche by Stripping Mantle Material

A fully differentiated planetesimal consisting of a silicate mantle and a metallic core, with the bulk composition of almost any chondritic meteorite, would have too low a bulk density to match Psyche (with addition of expected porosity, all chondrites are less dense than the asteroid appears to be with current data), and thus would need both the removal of silicate and the exposure of metal on the surface to match Psyche observables. Thus, we consider three scenarios: The gradual erosion of the parent body by relatively small impacts, the catastrophic disruption of the parent body by one giant impact, and the stripping of the body by hit-and-run impacts.

Each scenario has testable implications. Erosion by impacts, for example, would probably start with a solidified planetesimal. Imagine a solidified $\sim$ 500-kilometer diameter differentiated planetesimal being eroded over time by hundreds of impacts comparable in magnitude to the impact that formed the large Rheasilvia Basin on Vesta (e.g., Turrini 2014). The eroded body would always retain some remnant of the mantle, but would become increasingly dominated by a well-mixed megaregolith with grains consisting of ultramafic silicates, iron-silicate breccias, and metallic iron. In this scenario the exterior of Psyche would be this highly-processed mixture of material, including parts of the impactors. For this to happen to Psyche and not other major Main Belt asteroids like Vesta and Ceres, would perhaps require it to happen before its implantation in the Main Belt.

For larger bodies, the thermodynamics becomes significantly more complex, because the parent body may well be molten, given its much larger size, with pressures of $\sim$ 10–100 kilobars or greater. Apart from any other reactions, unloading makes available considerable enthalpy (Asphaug et al. 2011). Also, giant impact and planetary differentiation models suggest that unloading may chemically reduce deep-mantle materials (Cambioni et al. 2021).

One prediction would be the efficient degassing of any escaping deep-mantle clump ejected by a disruptive impact to form Psyche; pressure release makes available significant enthalpy. The available enthalpy has implications not only for super-catastrophic giant impacts, but for smaller disruptions as well, where a change in pressure can lead to significant disequilibrium and the expansion and release of dissolved volatiles as well as droplet production.

In general, if Psyche is a largely-metallic core of a parent body, and is itself mostly iron, then by whatever mechanism it lost its mantle, it began as an approximately Vesta-sized body or larger. But to consider mantle removal from a Vesta-sized parent body by only one or two impacts rather than by erosion, one really has to think of two parent bodies. At the ~ 5 km/s velocities typical in the modern Main Belt, the projectile itself would have to be as large as Psyche (Asphaug 2010). The prediction in that case would be a globally-shocked asteroid leading to the exposure of deep mantle rock and metal.

If Psyche's progenitor was larger, e.g. a Moon-sized body, then its mantle removal by a hypervelocity collision would have to be proportionally even more intense; the projectile itself would have to be Moon- to Mars-sized, and Psyche would be a small remnant of one of the colossal "late stage" events of planet formation.

In the hit-and-run scenario (Asphaug 2017), Psyche is the unaccreted projectile (the "runner") or remnant thereof, with little compositional connection, other than coincident location, to the target. They runner and target are bodies in an accreting swarm. Because hit-and-run impacts happen during the process of accretion, the progenitor may have been molten, within the million-year timescale of ^{26}Al decay. The target that the proto-Psyche ran into (assuming proto-Psyche was originally Vesta-sized) would have been a larger planetesimal, perhaps twice as big, feeding on the more numerous smaller ones. Quite unlike ballistic disruption, mantle stripping by hit-and-run is at low velocity. During planetesimal accretion the impact velocities would be within the range of a few times the escape velocity of the larger bodies, less than 1 km/s for a Ceres-sized population, lower than the sound speed in geologic materials. The removal of a $>$ 100-km of mantle rock over the course of an hour (the gravitational encounter timescale) would have metamorphic effects on petrology, especially if the excavated material was volatile-rich or hydrated.

We think that a hit and run event is the most likely process for creating Psyche, but only if planetesimals grew hierarchically from smaller planetesimals. If they formed effectively instantaneously via pebble accretion, all the mechanisms are probable.

One question in understanding Psyche's progenitor is the absence of a collisional family. In scenarios where Psyche's parent body was disrupted or stripped before Psyche was implanted into the asteroid belt, its collisional family would be long ago lost by dynamical disruption, integration into growing planets, or loss into the Sun. Thus, the lack of a dynamical family is unsurprising, but argues against scenarios where it is disrupted in situ.

Finally, a note on processes that might happen after stripping. During solidification of metal, no matter the size of the parent, a segregated immiscible FeS liquid may be able to erupt either to the surface of the core (Abrahams and Nimmo 2019), or possibly through some remnant mantle and onto a silicate surface (Johnson et al. 2019). The Psyche spacecraft has been designed since 2012 to be able to measure sulfur on the surface of Psyche, in the hopes of detecting this possible ferromagmatism. Ferromagmatism would be a strong indicator that Psyche is the remnants of a core, that is, a mass of Fe-Ni-S metal that was once molten. But the ferromagmatism hypothesis itself does not form a complete model to explain Psyche's spectral and density characteristics. In the Johnson et al. (2019) model, the mantle still must have a low-FeO characteristic and thus the body must have followed one of the stripping, erosion, or hit-and-run scenarios described above. And in the stripped core model of Abrahams and Nimmo (2019), Psyche would need extreme porosity or a second low-density phase to explain its bulk density, thus also requiring one of the above scenarios.

4 Alternative Origin Scenarios

4.1 Metal-Rich Chondritic Meteorites: A Possible Analog if Psyche Is not a Core

We consider whether some metal-rich chondrites might serve as appropriate analogs. In particular, we focus on groups of carbonaceous chondrites that share common properties, including the CR, CH, CB, and G chondrites, along with several ungrouped or intermediate meteorites. Weisberg et al. (1995) suggested grouping CR, CH and CB chondrites as the CR clan, indicating petrogenetic kinship. We hesitate to use this nomenclature both because of the inclusion in this discussion of the isotopically-distinct G chondrites and confusion that clan designation could imply a common parent body.

The most distinctive common features of these meteorites are abundant metal, and silicates that are poor in FeO. In this respect, these meteorites are tantalizing as possibilities to match the density and inferred silicate to metal ratio and the presence of low-FeO silicates (Dibb 2021; Hardersen et al. 2005) of Psyche. Metal abundances in these meteorites range from ~ 7 vol.% in CR chondrites to more than 60 vol.% in CB chondrites (Weisberg et al. 1995). At the lower end of this metal range, the CR chondrites contain no more metal than the enstatite chondrites (Keil 1968). The average density of fragments of Bencubbin, the meteorite after which CB chondrites were named, is 5,300 $kg\,m^{-3}$ (Simpson and Murray 1932). Olivine and pyroxene compositions range widely in some of these groups, with CR chondrite olivine ranging up to Fa50 (where Fa = molar(Fe/(Fe + Mg)), but all show strong modes in the range of Fa1 to Fa4, consistent with the low-FeO content of mafic silicates inferred for Psyche from a weak ~ 0.9 micrometer absorption feature. The Psyche mission payload should be able to constrain the silicate-metal ratio from bulk density and composition, and the mafic silicate FeO relative areal abundance from multi-spectral imaging (Bell et al. this journal).

While these meteorites share a number of common properties, they differ in important ways indicative of formation mechanisms and locations. Notable among these differences are the size, texture, isotopic composition and ages of the silicate spheres (Weisberg et al. 1995). While none of these features can be determined from orbit, they are indicative of a complex set of origins. At the extremes, CR silicate spheres are mm-sized chondrules formed in the nebula, whereas cm-sized CBa silicate spheres formed ~ 1–2 Ma after nebular dissipation and appear to be the products of rapid cooling of condensates from an impact plume formed during planetesimal collision (Bollard et al. 2015; Krot et al. 2005). CH chondrites show a mixture of these two formation mechanisms (Krot et al. 2010) and the intermediate Isheyevo meteorite illustrates remarkable sedimentary textures interpreted as deposition from that plume (Garvie et al. 2017). The scale of impact plume deposition, and the possibility of resulting metal enrichment remains an open question.

The inclusion of the G chondrites (Weisberg et al. 2015), which are petrologically similar to the CBa chondrites, suggests that the process that formed these meteorites was widespread in the early solar system. The CR, CH and CB chondrites have isotopic compositions similar to other carbonaceous chondrites, presumably having formed in a common isotopic reservoir in the outer solar system (Warren 2011). In contrast, the eponymous G chondrite GRO 95551 has oxygen and chromium isotopic compositions suggestive of formation with the non-carbonaceous chondrites in the inner Solar System. Thus, impacts that produced meteorites with the cm-sized, quickly cooled silicate clasts and abundant metal typical of these meteorites occurred throughout the Solar System.

Chondrules in CB chondrites formed at ~ 3.8 ± 1 Ma and 4.81 ± 0.26 Ma after the nominal CAI-formation time (Amelin et al. 2010; Connelly et al. 2012). Large collisions between protoplanets would perhaps be expected within the giant planet zone at late nebula times when large bodies were more common. We speculate that these putative collisions could explain the existence of CH and CB chondrules, which presumably would form into M-types by planetesimal formation processes prior to injection in the asteroid belt. It could also explain the M-types as the byproduct of large and/or possibly hit-and-run collisions that could produce stony iron and iron meteorites.

If CR chondrites are primordial and formed beyond the snow line (Warren 2011), where CB chondrites formed by condensation of a high-temperature impact plume (Krot et al. 2005), it is not surprising that they also show differences in degree of aqueous alteration (Weisberg et al. 1995). CR chondrites contain ubiquitous aqueous alteration of the matrix (they could have up to 5 wt% water-equivalent H, per Schrader et al. 2014 and Alexander 2019, for example), while CH and CB chondrites contain only rare aqueously altered clasts, possibly remnants of the pre-impact assemblage (Greshake et al. 2002). This distinction, indicative of the history of the groups, should manifest in differences in hydrogen abundance determined with GRNS. If metal-rich CB chondrites formed by localized impact plume production on an otherwise aqueously altered body poorer in metal, orbital chemical measurements should detect the difference between these lithologies. This scenario is particularly intriguing given the detection of a weak ~ 3 μm spectral feature indicative of water of hydration and rotational heterogeneity (Takir et al. 2016), though the 3 μm feature would most likely result from contamination by later impacts (Avdellidou et al. 2018).

If Psyche were composed of one of these groups, it might not be the first time a spacecraft has visited such a body. The Rosetta spacecraft flew by asteroid 21 Lutetia, determining a bulk density of $3{,}400 \pm 300\ \mathrm{kg\,m^{-3}}$ and observed a largely featureless spectra lacking absorption features indicative of either FeO-bearing silicates or hydrated phases (Coradini et al. 2011; Sierks et al. 2011). This high density and lack of obvious absorption features may indicate either an enstatite chondrite composition (Vernazza et al. 2011) or a composition similar to CR, CH, CB and G chondrites (Moyano-Cambero et al. 2016), and in either case Lutetia may be partially differentiated, with a small metal core adding to density (Weiss et al. 2012). Lutetia also has lower radar albedo than Psyche has (Shepard et al. 2015, 2021). In contrast to the limited data available from the Lutetia flyby, the Psyche mission instrument package will provide much greater detail needed to identify meteorite analogs.

4.2 Highly Reduced Material from the Inner Solar System

Irresistibly, with a body like Psyche that appears to have all its iron in the metal phase, the possibility of extremely reduced materials arises. Could Psyche consist of primitive material so reduced that the iron was all driven into the metal phase? Enstatite chondrites might sample such material and, as noted above, have been previously suggested as an analog for (21) Lutetia (Vernazza et al. 2011). The bulk density of 60% of enstatite chondrites measured by Macke (2010) falls within the range of preferred densities for Psyche ($3{,}780 \pm 340\ \mathrm{kg\,m^{-3}}$). A strength of linking enstatite chondrites and Psyche is the essentially FeO-free nature of the silicates in enstatite chondrites. We note, however, that ubiquitous high-FeO silicates are observed in minimally metamorphosed type 3 enstatite chondrites (Lusby et al. 1987). Given the uncertainties in FeO concentration derived from ground based spectral analyses, enstatite chondrites should be considered as a possible analog (though Psyche's high radar albedo may be an argument against its being an enstatite chondrite). Weidenschilling (1978) proposed that Mercury might have formed similarly, explaining its large metal fraction, and Kruss and Wurm (2018, 2020) suggest that a concentration of Fe metal-rich particles in the inner solar system may explain the gradation in core sizes of the terrestrial planets. Perhaps

Psyche is an analog to that material. As outlined in McCoy et al. (this journal), determination of the redox conditions of Psyche could further confirm or refute such a link.

Kruijer et al. (2017) have shown, however, that iron meteorites are associated with both carbonaceous and non-carbonaceous materials. The Kruss and Wurm (2018, 2020) hypothesized process may not be able to produce CH and CB chondrites beyond Jupiter, where they appear to have originated.

5 How We Will Use Our Data to Determine Psyche's Origin

Discriminating whether Psyche is core material from a differentiated planetesimal, or an unusually metal-rich and FeO-poor chondritic body, will require the combination of measurements and interpretations from all the instruments. Few individual measurements would be definitive on their own; one definitive measurement might be a strong remanent magnetic field, indicating the presence in the deep past of a core dynamo and requiring a differentiated parent.

5.1 Magnetic Field

Magnetometer data may enable discrimination between the competing hypotheses that Psyche formed as the metallic core of a mantle-stripped planetesimal or as an unmelted, iron-rich chondritic aggregate. If Psyche has a strong magnetic moment, this would strongly favor its origin as a mantle-stripped planetesimal. This is because a molten core formed after mantle stripping could generate a long-lived magnetic field by the dynamo process (Bryson et al. 2017; Maurel et al. 2020). The essence of a dynamo is that large-scale advective motions of a conductive fluid can inductively generate currents that amplify an arbitrarily weak initial magnetic field (Moffatt and Dormy 2019). This magnetic field could then be recorded as remanent magnetization by ferromagnetic minerals (e.g., the iron-nickel minerals kamacite and tetrataenite) as they crystallize or cool through their magnetic ordering temperatures on the outside of the cooling core. Because such magnetization can in principle persist for longer than the age of the solar system (Nagy et al. 2019), this remanent magnetization could then be detected by a sufficiently sensitive spacecraft magnetometry investigation. On the other hand, the lack of detection of a remanent field would not exclude origin as a core for several reasons. First, the existence of a molten, advecting, conductive fluid is a necessary but not sufficient requirement for dynamo action: only some flow velocities and geometry will lead to field amplification (Moffatt and Dormy 2019). Second, the iron minerals may not have formed or cooled sufficiently at the time of dynamo generation to record the field.

Alternatively, if Psyche is an iron-rich undifferentiated chondritic body, it would never have melted and therefore generated a dynamo field that could magnetize the body. A potential complication is that the early solar nebula may have generated magnetic fields of similar intensity as dynamo surface fields on iron meteorite parent bodies (Fu et al. 2014; Weiss et al. 2021; Borlina et al. 2021). Recent magnetohydrodynamic simulations indicate this field could have been temporally steady for $> 1{,}000$ y (Weiss et al. 2021). If a chondritic body were thermally metamorphosed and cooled and/or formed secondary magnetic minerals via aqueous alteration while the nebular field was present, it could have been magnetized by this field. However, this requires that these events occurred on the parent body before the nebular field dissipated. This is challenging because paleomagnetic measurements indicate that the nebular field had already dissipated by 3.9 Ma after solar system formation (Wang et al. 2017; Weiss et al. 2021).

With respect to the nebular magnetization formed as a result of metamorphism, the estimate nebular lifetime is well before the estimated time when ordinary chondrites cooled through kamacite's 800 C Curie temperature ($\sim$ 5–60 Ma after solar system formation) (Blackburn et al. 2017). However, it is possible that other metamorphosed chondritic bodies cooled more quickly. With respect to nebular magnetization acquired by secondary alternation minerals, aqueous alteration of many carbonaceous chondrite parent bodies occurred sometime between $\sim$ 2–11 Ma after solar formation, which encompasses the lifetime of the nebula (Carporzen et al. 2011; Doyle et al. 2015; Jogo et al. 2017). For the secondary magnetic minerals to retain magnetization until the present day, the body must not subsequently heat beyond their unblocking temperatures after the nebular field dissipates. For example, chondritic bodies that form between 3.0 and 3.75 Ma may heat just enough to undergo aqueous alteration before the nebular field dissipates, but not enough to subsequently erase the magnetization (Courville et al. 2022). The solar nebular field is possibly the source of the coherent magnetization observed in the CM and CV chondrites (Cournede et al. 2015; Fu et al. 2021), although a core dynamo is also possible (Carporzen et al. 2011; Gattacceca et al. 2016). In these meteorites, aqueous alteration produced the ferromagnetic iron-oxide and iron sulfide minerals magnetite and pyrrhotite which may retain a record of the magnetic field. Therefore, if Psyche is observed to contain secondary ferromagnetic oxides or sulfides, magnetization by the nebular field should be considered as a possible explanation for some remanent magnetization (Courville et al. 2022). However, this requires a large volume fraction of Psyche (tens of percent or more) to be composed of ferromagnetic oxides and sulfides for the resultant moment of Psyche to be detectable with the magnetometer. In general, the possibility of remanent magnetization from nebular fields on Psyche merits further investigation.

Should magnetic fields be detected by the magnetometer, observations from the GRNS and Imager could potentially distinguish between a chondritic and achondritic surface. Given the relatively young cooling ages of most achondrites and the possibility that their parent bodies formed metallic cores, this will be an important indicator of whether Psyche could have been magnetized by an external source. We can further distinguish between different magnetizing fields from examining the three-dimensional field topology around the body to infer the magnetization pattern. A dynamo-generating core would produce non-uniform magnetization due to the curved geometry of the field (Runcorn 1975) and its likely temporal variation during the gradual solidification of the body (Neufeld et al. 2019). In contrast, a nebular field may impart more uniform magnetization, as the field source is external to the body (Runcorn 1975). These two types of magnetization configurations could be distinguished using a combination of data analysis and modeling tools developed by the Magnetometry Investigation (Weiss et al. this journal).

Magnetometer data can also be used to constrain the bulk electrical conductivity of Psyche by detecting induction fields due to variations of the solar wind that Psyche is exposed to. If the body has very low bulk conductivity the induced fields outside the body would be weak (a Moon-like interaction); if highly conducting, there would be a region of enhanced magnetic field in the upwind side of the body (a Venus-like interaction). The value of resistivity producing each type or reaction depends on the solar wind speed and the body size (Oran et al. 2018). A body the size of Psyche and a solar wind speed of 400 $km\,s^{-1}$ would behave as a perfect resistor for resistivity $\eta \gg 4 \times 10^4$ Ωm and as a perfect conductor for $\eta \ll 4 \times 10^4$ Ωm. Chondrites have a resistivity of 10^5 Ωm (Duba and Boland 1984), rocky achondrites materials like the lunar crust can have resistivities near 10^8 Ωm (Dyal et al. 1977) while an iron metal-rich body would have a resistivity of iron of 10^7 Ωm (Secco and Schloessin 1989). However, if the body contains a thick outer layer with high resistivity, a more conducting interior would be masked (Oran et al. 2018). Further, if the body is porous this would reduce its bulk resistivity.

5.2 Scale of Silicate/Metal Mixing

If Psyche is the stripped remnant of a differentiated planetesimal, its metal and non-metal fractions should exist in meter- to kilometer-scale or larger regions. If, alternatively, its bulk composition is chondritic, its metal and non-metal will be intimately mixed on a millimetric or centimetric scale.

For meteoritic materials, the metal-to-silicate fraction as determined by laboratory methods like XRF spectroscopy is typically defined as the total amount of metal (e.g. metallic iron, nickel) versus total silicates (elements bound as oxides). In that scenario, iron in silicates (e.g. FeO) is "book kept" with silicates. In contrast, gamma-ray and neutron spectroscopy measures elements without regard to their mineralogical state. Since the largest concentration of "metal" elements being considered for Psyche consist of Fe and Ni, for the purpose of GRNS measurements we define the "metal" content as the total concentration of Fe and Ni. "Silicates" are then defined as other major rock-forming elements (Si, O, Al, Mg, Ca, etc.).

The highest sensitivity GRNS metal-to-silicate measurements are accomplished using neutron data from the Neutron Spectrometer (NS). The NS measures neutrons in three energy bands; thermal, low-energy epithermal, and high-energy epithermal neutrons. The Gamma-Ray Spectrometer (GRS) anti-coincidence (AC) shield provides a measure of fast neutrons (Lawrence et al. 2019, 2016). Thermal neutrons and fast neutrons are independently sensitive to the concentration of Fe and Ni via their thermal-neutron absorption properties (e.g., Feldman et al. 1998; Peplowski et al. 2016) and average atomic mass (e.g., Gasnault et al. 2001; Lawrence et al. 2017), respectively. High metal fractions result in relatively low thermal neutrons and high fast neutrons; the converse is true for high silicate fractions. On Psyche, it is expected there will be variations in H concentrations that affect the neutron measurements (Reddy et al. 2018). To account for such effects, Lawrence et al. (this journal) developed an analysis framework based on principal component analysis that enables the four Psyche neutron measurements to uniquely discriminate the three compositional parameters of metal-to-silicate fraction, H concentration, and relative Ni variations.

The spatial scale of the measured metal-to-silicate fraction is determined by the altitude of the spacecraft of Psyche's surface, and the full-width at half-maximum of the footprint can be approximated as $\sim$ 1 to 1.5 times the altitude (Lawrence et al. 2003). For the lowest orbit of the mission, altitude is required to be less than one Psyche radius (R_P) and is targeting an altitude of 0.8 R_P (Polanskey this journal). This altitude corresponds to an effective spatial footprint of $\sim$ 100 to 150 km, which is a spatial scale of a large "province" on Psyche; this footprint may be reduced using spatial deconvolution techniques (Wilson et al. 2018). In a practical sense, the spatial resolution of the metal-to-silicate measurement will be similar to the gamma-ray and neutron measurements achieved with the Dawn mission at the asteroids Vesta and Ceres (Prettyman et al. 2017, 2013). Although the presence and spatial scale of any geochemically-distinct regions on Psyche are unknown, there is evidence for hemispherical-scale variations in optical albedo (Takir et al. 2016) and large-scale radar albedo (Shepard et al. 2021) that would be spatially resolvable by GRNS if they correspond to chemical differences.

The scenario is somewhat different, however, for the metal-to-silicate fraction inferred from visible to short-wave near-IR ($\sim$ 400 to 1,100 nm) multispectral imaging or spectroscopy. Specifically, the presence of an optically-significant metallic component can be inferred from an assessment of the slope of the reflectance spectrum across this full wavelength range, with the reddest slopes being consistent with the presence of a significant metallic iron and/or iron-nickel component, based on laboratory studies of iron meteorites. In contrast, specific relatively broad solid-state absorption features within this wavelength range are used to infer the presence of certain iron-bearing silicates known to be important components of chondritic meteorites and certain common classes of asteroids (like the S-types, which can exhibit diagnostic absorption features indicating the presence of olivines and/or pyroxenes).

Unlike GRNS or laboratory elemental techniques, however, multispectral observations in the visible to short-wave near-IR cannot typically be used to infer absolute elemental or mineralogic abundances. This is because of a variety of fundamental ambiguities in the interpretation and modeling of such data caused by the competing effects of particle size, particle shape, degree of compaction, viewing geometry, and other factors. Thus, multispectral observations like those to be made with the Psyche Multispectral Imager can be used to *detect* metallic iron- and nickel-bearing materials as well as to *detect* the presence of some common rock-forming iron-bearing silicates, but abundances of these phases can usually only be inferred or modeled in a relative sense. Still, the limitations on deriving absolute abundances can be offset to some degree by the ability to map relative variations in the inferred metal-to-silicate ratio at high spatial resolution across a planetary surface. As well, correlations between the imaging-based detections of such surface compositional/mineralogic components and other measured surface properties (geologic context, topography, magnetic field strength, elemental composition, etc.) can provide important additional constraints to factor in to multispectral data interpretation. Thus, each of the Psyche mission's different elemental/mineralogic measurement techniques provides a different and complementary perspective on the composition and mineralogy of the asteroid.

5.3 Correlation of Gravity and Composition

If Psyche's current internal structure predominantly reflects solidification from a liquid state, its interior would likely have a radially-distributed density structure overlain by a layer of surface materials presumably pulverized by impacts and containing exogenic material. In contrast, if Psyche was significantly disrupted and/or was never molten, it would likely have an azimuthally-varying or heterogeneous density structure at depths greater than its lid. The correlation between Psyche's gravity field with its shape could be indicative of structure and degree of spherical symmetry or breakage into rubble (Konopliv et al. 2014; RS Park et al. 2014, 2016, 2020). Furthermore, the relative amplitudes of gravity and topography over a range of wavelengths (as quantified by the admittance) will be sensitive to the density of surface materials and how density varies with depth below the surface. For example, the correlation of Vesta's gravity field with an assumed three-layer model is approximately 1 for all measurable degrees (Konopliv et al. 2014; Park et al. 2014), indicating no substantial density variations in Vesta's crust, nor a strong contrast in density between the assumed crustal layer and the mantle. Such a high correlation could also be expected for impact-related homogenization (Zuber et al. 2013). The Psyche gravity science investigation would recover a degree 10 gravity field (Zuber et al. this journal), which would allow discriminating density heterogeneity at the scale of $\sim 1/10$ of Psyche.

5.4 Silicate Compositions

GRNS measurements of the elemental composition of silicates are primarily measured using gamma-ray data, and supplemented using neutron data. For GRNS data it is important to note that the gamma-rays and neutrons are sensitive to Psyche's bulk composition and therefore do not and cannot distinguish between the surface's mineralogic form or mixture. The primary elements used to characterize Psyche's silicate compositions are Fe, Si, Al, K, and Ca. As described in Elkins-Tanton et al. (2020), various combinations of these elements can discriminate the type of material (chondritic versus achondritic) that is present on Psyche. The specific detection sensitivities for these elements are given in Lawrence et al. (this journal).

The Imager's study of silicates will be primarily based on the detection, characterization, and mapping of the so-called "1-micron" absorption band in iron-bearing silicates like pyroxene and olivine (see, e.g., reviews in Burns 1993; Gaffey et al. 1993; Bishop 2020). Specifically, the required signal to noise ratio in the Imager's four longest-wavelength near-IR filters (at 725, 850, 948, and 1042 nm; Bell et al. this journal) will provide the ability to detect weak ($\geq$ around 1% band depth) absorption in the 1-micron region potentially indicative of the presence of specific iron bearing silicates. For example, a weak feature near 900 nm has been reported by some observers in whole-disk telescopic spectra of Psyche and has been interpreted as evidence for a silicate component – potentially low-Fe pyroxene (e.g., Clark et al. 2004; Hardersen et al. 2011, 2005). Additionally, laboratory studies of metal/silicate mixtures show that the relative abundances of the kinds of iron-bearing silicate components that could reasonably be hypothesized on the surface of Psyche based on telescopic data and meteoritic analogs could indeed yield very weak absorption features. For example, Cloutis et al. (2009) showed that 90:10 mixtures of metal and olivine produce an absorption feature in the 1-micron region that is only $\sim$ 1% deep. The Imager will search for such weak absorptions by calculating (and mapping) ratios of observed surface reflectances (e.g., between 725 nm and the three longest wavelength filters), band depth maps (using 725 nm and 1042 nm images as "continuum" points), and correlations between such derived ratios and band depths with other parameters, such as 1-micron band area and/or overall albedo (e.g., Dibb 2021).

5.5 High S, K on Surface

High concentrations of sulfur and possibly potassium on the surface of Psyche would be indicative of ferrovolcanism. The immiscible, sulfur-rich liquid produced by fractional solidification of a metal core is expected to contain approximately 50 mol% sulfur (it is effectively liquid troilite; Jones and Drake 1983), and other incompatible elements like potassium may also be present. This liquid is a candidate for expulsion onto the surface of a solidifying differentiated planetesimal.

The GRNS is sensitive to K and S with gamma-ray data. K is naturally radioactive and produces a relatively high flux of gamma rays, allowing highly sensitive measurements of K abundances down to 0.02 wt.% (or 200 ppm). Numerous examples exist of K measurements at other planetary bodies using orbital gamma-ray data (e.g., Lawrence et al. 1998; Peplowski et al. 2011; Prettyman et al. 2015). The GRNS will provide a K map (to the sensitively limit of 0.02 wt%) with a spatial resolution similar to the neutron data. Thus, locations of enhanced K abundances should be detectable by GRS data. S concentrations greater than 3 wt.% can also be detected with GRS, and this capability has been demonstrated with other high-purity germanium sensors at Mars and Mercury (King and McLennon 2010; Evans et al. 2012).

The Imager provides sensitivity to S only via the search for and potential detection of specific sulfides that exhibit a diagnostic absorption feature in the visible wavelength region, such as oldhamite ((Ca,Mg,Fe)S). Specifically, the Imager will use narrowband filters

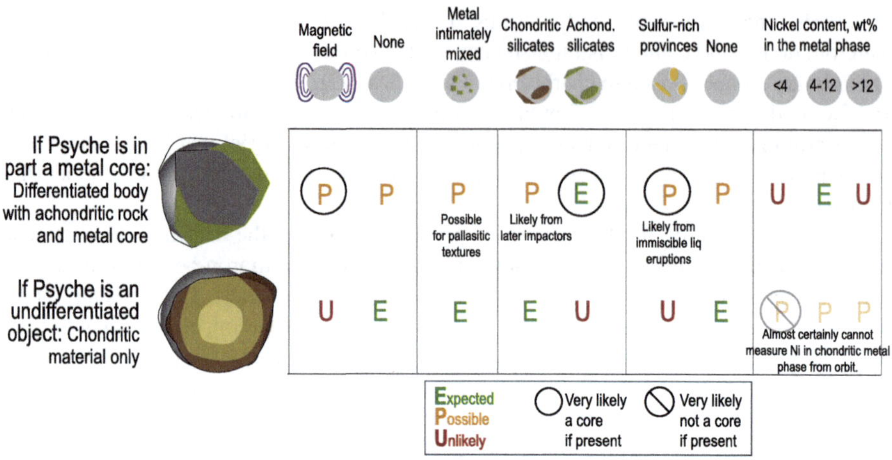

Fig. 6 This matrix of instrument measurement interpretations along the top shows the requirement for multiple measurements to support either of the origin hypotheses at left. Only a few measurements are probably definitive on their own (strong magnetic field, achondritic silicate provinces, and sulfur-rich volcanism each likely indicate Psyche is a core). Expected = Likeliest outcome according to current models; Possible = Outcome predicted in some scenarios; Unlikely = Not predicted in current models. See text for detailed explanation, and for further explanation of how the instruments can discriminate between achondritic and chondritic silicates, see Elkins-Tanton et al. (2020)

at 437, 495, and 550 nm to detect and map a weak cubic monosulfide feature in sulfides like oldhamite, which has been identified in laboratory and telescopic data at these wavelengths (e.g., Burbine 2014; Burbine et al. 2002a,b; Clark et al. 2004; Gaffey et al. 1993). Ratios among these filters and band depth maps at 495 nm using 437 nm and 550 nm as the continuum, will be the primary derived products used in the imaging search for signatures on Psyche of S-bearing phases with local mixture abundances greater that around 10%.

5.6 Nickel Content

The nickel content of the metal phase is indicative of the crystallization sequence of the core, if such it is, and of the oxidation or reduction state of the material, in the case that it is not a core. The earliest solids fractionally solidifying from a liquid core are the lowest in nickel content, and the last solidifying metal crystals contain the highest nickel content. Ni abundances are primarily measured with the 1.454 MeV gamma-ray line, however Ni produces additional line signatures that can be used to characterize Ni abundances (albeit with lower sensitivity than the 1.454 MeV line) (Peplowski et al. 2019). GRS data will measure Ni where present at concentration of 4 wt.% or higher (Lawrence et al. this journal), averaged over the spatial resolution of the GRNS measurements. In addition to the gamma-ray measurements, combined neutron data will be sensitive to relative Ni variations across Psyche's surface should they be present (Lawrence et al. 2021). Thus, GRNS measurements provide a robust means for determining Psyche's Ni concentration, which is a key measurement for understanding Psyche's origin.

5.7 Combining Instrument Measurements to Reach Conclusions

The requirement to combine many instrument observations to answer the question of Psyche's origin is shown in Fig. 6. If there is a strong magnetic field, we can conclude that

Psyche is a part of a differentiated body. The scale of metal and silicate mixing is a second key observation. If Psyche's materials contain intimately mixed metal and silicate (on a millimetric or centimetric scale, as in chondritic meteorites), then the most likely interpretation, based upon the meteorite collection, is that the material is primordial and unmelted. However, intimate mixing of silicates and metals is also characteristic of the less common pallasitic meteorites, and so intimate metal-silicate mixing is not alone a definitive measurement.

Silicate and metal provinces may also exist on a macroscopic scales, even up to tens to hundreds of kilometers. Regions of chondritic material lying on top of metal are almost certainly the result of later impactors, that is, the chondritic material is exogenous to the Psyche parent body. Achondritic impacts are far less common, judging from both asteroid spectra and meteorite incidences. Non-chondritic silicate patches would thus likely be patches of the silicate fraction of Psyche's parent body and strongly indicate that Psyche is a part of a differentiated parent body.

The fractional solidification of metal cores almost inevitably produces an immiscible liquid (Ulff-Muller 1998; Chabot 2004; Goldstein et al. 2009) with the approximate stoichiometric composition of a troilite-like monosulfide: $[(Fe,Ni)_{1-x}S]$ (Jones and Drake 1983). This immiscible liquid is less dense than its coexisting more Fe-rich liquid core, and so it is expected to rise to the surface or potentially to erupt through the contracting solid lid of a core freezing from the outside inward (Abrahams and Nimmo 2019; Elkins-Tanton et al. 2020). Large provinces of sulfur-rich material on Psyche's surface, therefore, would indicate that Psyche's parent body was differentiated; there is no obvious mechanism by which to make strongly sulfur-enriched regions on a chondritic parent body.

When metal cores fractionally solidify, the earliest-forming solids have a low nickel content, around 4 wt%. Most iron meteorites have nickel contents between 4 and 12 wt% (Buchwald 1975; Wasson et al. 2007 and references therein; Goldstein et al. 2009; Scott 2020). Above 12 wt% nickel in the metal fraction probably indicates that the parent body may have been highly oxidized (driving iron into the mantle and effectively raising the nickel fraction in the core), though IVB iron meteorites have more than 12 wt% nickel (e.g. Rasmussen et al. 1984; Goldstein et al. 2009) and they are believed to have formed in a core. (Some iron meteorites, suspected to have come from highly oxidized parent bodies, contain over 30 wt% and as much as 60 wt% nickel; these include Tishomingo and Dermbach [Buchwald 1975], and the IAB-ungrouped Oktibbeha County [Wasson and Kallemeyn 2002]).

In summary, if Psyche has a strong magnetic field, then it is almost certainly from a differentiated parent body. Similarly, large achondritic rock provinces on its surface are most likely to be remnants of its original mantle, and thus Psyche is a mixture of core material and mantle. Sulfur-rich (S) volcanism is only expected for a differentiated body. Nickel content in the metal phase below 4 wt%, however, indicates a highly oxidizing environment and is lower than iron meteorite analogs, and thus Psyche is not likely to be a core, in the absence of contradictory evidence. All other measurements need to be considered together in creating conclusions about Psyche's provenance.

6 Conclusions

The bulk density of asteroid (16) Psyche appears to be between 3,400 and 4,100 $kg\,m^{-3}$. Given the known densities of meteoritic metal, and assuming not more than 20% porosity, the body is thus predicted to contain up to 60 vol.% metal. If porosities higher than expected on a body of Psyche's size are enabled by the material properties of metals at relevant space temperatures, then Psyche's metal fraction can be far higher than 60 vol%. Indeed, in the most extreme case, the bulk density is consistent with pure iron-metal with $\sim 50\%$ porosity.

The asteroid is therefore likely a mixture of metal and an additional lower-density component or components. Based upon modeling of available telescopic and laboratory spectral characteristics, the additional components are likely to be very low in FeO or Fe_2O_3 (Dibb 2021). Likely candidates for a non-metal fraction of the asteroid include low-iron pyroxene (e.g., Hardersen et al. 2005), sulfides (e.g., Clark et al. 2004), or perhaps carbon compounds. By constraining the composition and density of the non-metal component, it will be possible to place tighter bounds on the metal volume fraction of Psyche.

Our currently favored fiducial model for the asteroid's formation is that it is the remains of a differentiated body containing core metal. Psyche may contain a large fraction of a metal core and some silicate mantle retained on the outside, or the silicate fraction could be mixed into the metal fraction. The silicate fraction must, however, be strikingly low in iron oxide.

Formation of such a low iron-oxide bulk rock composition seems, within the current state of planet formation theory, to require one of two processes to have occurred: (1) Silicate differentiation that produces low-iron-oxide silicates near the core and high-iron-oxides silicates nearer the surface, with subsequent removal of the outermost high-iron-oxide fraction by stripping impacts; or (2) Formation of the Psyche parent body from highly reduced primitive material in which the iron has been reduced to its metallic form and incorporated into the core, leaving the silicates naturally low in iron and high in magnesium.

Differentiation of a parent body followed by mantle stripping may be the simplest model to call upon to form a metal-enriched Psyche. Stripping the parent body to reduce the low-density phase and reveal metal from the core can be done via erosion by multiple impacts, by removing a portion of the deep mantle of a larger body via one catastrophic impact, or by stripping the mantle off the impactor during a hit-and-run collision. Erosion by multiple impacts may be impossible to reconcile with the preservation of the basaltic crust of Vesta. Blasting apart in a single catastrophic impact is also challenging because the parent body, being larger than Psyche, might be too large to destroy after the Main Belt was emplaced. A hit-and-run disruption seems most likely, although this would have to happen during planetesimal accretion, in the first few million years.

The parent body could be a differentiated inner-solar-system or outer-solar-system planetesimal, with different implications for Psyche's composition. The semimajor axis distribution of M-types appears different from planetesimals expected to have come from the inner solar system, and more consistent with an outer solar system origin. If so, then disruption by a hit-and-run event would have happened when interactions among similar-sized planetesimals were more common, before or during scattering and implantation in the Main Belt.

The deep interior of an outer-solar-system parent body may be volatile rich, depending on its early thermal evolution, with hydrated phases in the mantle and possibly volatiles sequestered in its core. Removal of the mantle (by whatever means) would expose these phases to low pressure, in which case a subsequent early epoch of sublimation-degradation or volatile eruption may have occurred. This might not be the case for an inner-solar-system parent body, whose deep interior may be dunitic rock grading towards metallic core-like compositions, that would be stable at lower pressures.

As an alternative to these disruption scenarios, Psyche might be an undifferentiated body consisting of a highly reduced, low-FeO material. Meteorite analog candidates include enstatite chondrite or CR, CH, or CB chondrites, some example of each of which have adequately low FeO and high density.

The Psyche mission project will determine Psyche's origin and formation of the asteroid by measuring any strong remanent magnetic field, which would imply it was the core of a differentiated body; the scale of metal to silicate mixing will be determined by both the neutron spectrometers and the filtered images; the degree of disruption between metal and rock may be determined by the correlation of gravity with composition; some mineralogy will be detected using filtered images; and the nickel content of Psyche's metal phase will be measured using the GRNS.

Acknowledgements A deep thank-you to two reviewers who spent the time and thought to truly help us improve the paper.

Author Contribution All authors wrote parts of the paper and contributed to editing.

Funding This work is supported by NASA contract NNM16AA09, "Psyche: Journey to a Metal World." The contribution of MAW was supported by CNES.

Data Availability All data and material can be found in cited material; no original data was generated for this paper.

Code Availability Not applicable.

Declarations

Conflict of Interest Not applicable.

Ethics approval Not applicable.

Consent to participate Not applicable.

Consent for publication Not applicable.

References

J.N.H. Abrahams, F. Nimmo, Ferrovolcanism: iron volcanism on metallic asteroids. Geophys. Res. Lett. **46**, 5055–5064 (2019)

C.B. Agnor, D.N.C. Lin, On the migration of Jupiter and Saturn: constraints from linear models of secular resonant coupling with the terrestrial planets. Astrophys. J. **745**, 143 (2012)

C.M.D. Alexander, Quantitative models for the elemental and isotopic fractionations in chondrites: the carbonaceous chondrites. Geochim. Cosmochim. Acta **254**, 277–309 (2019)

C.O.D. Alexander, R. Bowden, M.L. Fogel, K.T. Howard, C.D.K. Herd, L.R. Nittler, The provenances of asteroids, and their contributions to the volatile inventories of the terrestrial planets. Science **337**, 721–723 (2012)

V. Alí-Lagoa, T.G. Müller, F. Usui, S. Hasegawa, The AKARI IRC asteroid flux catalogue: updated diameters and albedos. Astron. Astrophys. **612**, A85 (2018). https://doi.org/10.1051/0004-6361/201731806

Y. Amelin, A. Kaltenbach, T. Iizuka, C.H. Stirling, T.R. Ireland, M. Petaev, S.B. Jacobsen, U–Pb chronology of the Solar System's oldest solids with variable $^{238}U/^{235}U$. Earth Planet. Sci. Lett. **300**(3–4), 343–350 (2010)

E. Asphaug, Similar-sized collisions and the diversity of planets. Geochemistry **70**(3), 199–219 (2010)

E. Asphaug, Signatures of hit-and-run collisions, in *Planetesimals: Early Differentiation and Consequences for Planets*, ed. by L.T. Elkins-Tanton, B.P. Weiss (Cambridge University Press, Cambridge, 2017), pp. 7–37

E. Asphaug, C.B. Agnor, Q. Williams, Hit-and-run planetary collisions. Nature **439**, 155–160 (2006)

E. Asphaug, M. Jutzi, N. Movshovitz, Chondrule formation during planetesimal accretion. Earth Planet. Sci. Lett. **308**, 369–379 (2011)

C. Avdellidou, M. Delbo', A. Fienga, Exogenous origin of hydration on asteroid (16) Psyche. Mon. Not. R. Astron. Soc. **475**(3), 3419–3428 (2018)

J. Baer, S.R. Chesley, Simultaneous mass determination for gravitationally coupled asteroids. Astron. J. **154**(2), 76 (2017)

J.D. Bass, Elasticity of minerals, glasses, and melts, in *Mineral Physics and Crystallography: A Handbook of Physical Constants*, vol. 2 (AGU Reference Shelf, Washington, 1995), pp. 45–63

K. Batygin, M.E. Brown, H. Betts, Instability-driven dynamical evolution model of a primordially five-planet outer solar system. Astrophys. J. **744**, L3 (2012)

T.M. Becker, N. Cunningham, P. Molyneux, L. Feaga, K.D. Retherford, R. Lotenz, Z.A. Landsman, E. Peavler, L.T. Elkins-Tanton, J.-E. Walhund, HST UV observations of asteroid (16) Psyche. Planet. Sci. J. **1**(3), 53 (2020)

D. Bell et al., this journal

J. Bishop, Visible and near-infrared reflectance spectroscopy: laboratory spectra of geologic materials, in *Remote Compositional Analysis: Techniques for Understanding Spectroscopy, Mineralogy, and Geochemistry of Planetary Surfaces*, ed. by J. Bishop, J. Moersch, J. Bell (Cambridge University Press, Cambridge, 2020), pp. 68–101

T. Blackburn, C.M.D. Alexander, R. Carlson, L.T. Elkins-Tanton, The accretion and impact history of the ordinary chondrite parent bodies. Geochim. Cosmochim. Acta **200**, 201–217 (2017)

J. Bollard, J.N. Connelly, M. Bizzarro, Pb-Pb dating of individual chondrules from the CBa chondrite Gujba: assessment of the impact plume formation model. Meteorit. Planet. Sci. **50**, 1197–1216 (2015)

C.S. Borlina, B.P. Weiss, J.F.J. Bryson, X.-N. Bai, E.A. Lima, N. Chatterjee, E.N. Mansbach, Paleomagnetic evidence for a disk substructure in the early solar system. Sci. Adv. **7**, eabj6928 (2021)

W.F. Bottke Jr., D.D. Durda, D. Nesvorný, R. Jedicke, A. Morbidelli, D. Vokrouhlický, H. Levison, The fossilized size distribution of the main asteroid belt. Icarus **174**, 11–140 (2005)

W.F. Bottke, E. Asphaug, On the origin and evolution of differentiated planetesimals, in *Lunar and Planetary Science Conference* (Lunar and Planetary Institute, Houston, 2013)

W.F. Bottke Jr., D. Vokrouhlický, D.P. Rubincam, D. Nesvorný, The Yarkovsky and YORP effects: implications for asteroid dynamics. Annu. Rev. Earth Planet. Sci. **34**, 157–191 (2006a)

W.F. Bottke, D. Nesvorný, R.E. Grimm, A. Morbidelli, D.P. O'Brien, Iron meteorites as remnants of planetesimals formed in the terrestrial planet region. Nature **439**(7078), 821–824 (2006b). https://doi.org/10.1038/nature04536

R. Brasser, S.J. Mojzsis, The partitioning of the inner and outer Solar System by a structured protoplanetary disk. Nat. Astron. **4**, 492–499 (2020)

J.F.J. Bryson, J. Harrison, A. Neufeld, F. Nimmo, J. Herrero-Albillos, F. Kronast, B.P. Weiss, Paleomagnetic evidence for dynamo activity driven by inward crystallisation of a metallic core. Earth Planet. Sci. Lett. **472**, 152–163 (2017)

V.F. Buchwald, *Handbook of Iron Meteorites*, vol. 1 (1975)

T. Burbine, Asteroids, in *Treatise on Geochemistry*, ed. by H. Holland, K. Turekian (Elsevier, Oxford, 2014), pp. 365–415

T.H. Burbine, T.J. McCoy, L.R. Nittler, G.K. Benedix, E.A. Cloutis, T.L. Dickinson, Spectra of extremely reduced assemblages: implications for Mercury. Meteorit. Planet. Sci. **37**(9), 1233–1244 (2002a)

T.H. Burbine, T.J. McCoy, A. Meibom, B. Gladman, K. Keil, Meteoritic parent bodies: their number and identification, in *Asteroids III* (2002b), pp. 653–667

R. Burns, Origin of electronic spectra of minerals in the visible and near-infrared region, in *Remote Geochemical Analysis* (1993), pp. 3–29

S. Cambioni, S.A. Jacobson, A. Emsenhuber, E. Asphaug, D.C. Rubie, T.S.J. Gabriel, S.R. Schwartz, R. Furfaro, The effect of inefficient accretion on planetary differentiation. Planet. Sci. J. **2**, 93 (2021)

L. Carporzen, B.P. Weiss, L.T. Elkins-Tanton, D.L. Shuster, D.S.Ebel.J. Gattacceca, Magnetic evidence for a partially differentiated carbonaceous chondrite parent body. Proc. Natl. Acad. Sci. USA **108**, 6386–6389 (2011). https://doi.org/10.1073/pnas.1017165108

J.C. Castillo-Rogez, E.D. Young, Origin and evolution of volatile-rich planetesimals, in *Planetesimal Differentiation*, ed. by L.T. Elkins Tanton, B.P. Weiss (Cambridge University Press, Cambridge, 2016)

N.L. Chabot, Sulfur contents of the parental metallic cores of magmatic iron meteorites. Geochim. Cosmochim. Acta **68**, 3607–3618 (2004). https://doi.org/10.1016/j.gca.2004.03.023

J.E. Chambers, Oligarchic growth with migration and fragmentation. Icarus **198**(1), 256–273 (2008)

J.E. Chambers, Late-stage planetary accretion including hit-and-run collisions and fragmentation. Icarus **224**, 43–56 (2013)

B.E. Clark, S.J. Bus, A.S. Rivkin, T. McConnochie, J. Sanders, S. Shah, T. Hiroi, M.K. Shepard, E-type asteroid spectroscopy and compositional modeling. J. Geophys. Res. **109**, E02001 (2004). https://doi.org/10.1029/2003JE002200

M.S. Clement, N.A. Kaib, S.N. Raymond, K.J. Walsh, Mars' growth stunted by an early giant planet instability. Icarus **311**, 340–356 (2018)

E.A. Cloutis, P.S. Hardersen, V. Reddy, M.J. Gaffey, D. Bailey, M.A. Craig, Metal-orthopyroxene and metal-olivine mixtures: spectral reflectance properties and implications for asteroid spectroscopy, in *Lunar and Planetary Science Conference* (2009), p. 40

J.N. Connelly, M. Bizzarro, A.N. Krot, Å. Nordlund, D. Wielandt, M.A. Ivanova, The absolute chronology and thermal processing of solids in the solar protoplanetary disk. Science **338**(6107), 651–655 (2012)

A. Coradini, F. Capaccioni, S. Erard, G. Arnold, M. De Sanctis, G. Filacchione, F. Tosi, M. Barucci, M. Capria, E. Ammannito, The surface composition and temperature of asteroid 21 Lutetia as observed by Rosetta/VIRTIS. Science **334**(6055), 492–494 (2011)

C. Cournede et al., An early solar system magnetic field recorded in CM chondrites. Earth Planet. Sci. Lett. **410**, 62–74 (2015). https://doi.org/10.1016/j.epsl.2014.11.019

S. Courville, J. O'Rourke, J. Castillo-Rogez et al., Magnetization of carbonaceous asteroids by nebular fields and the origin of CM chondrites, 12 January 2022, Preprint (Version 1) available at Research Square https://doi.org/10.21203/rs.3.rs-1196868/v1

R.R. de Sousa, A. Morbidelli, S.N. Raymond, A. Izidoro, R. Gomes, E. Vieira Neto, Dynamical evidence for an early giant planet instability. Icarus **339**, 113605 (2020)

R. Deienno, A. Morbidelli, R.S. Gomes, D. Nesvorny, Constraining the giant planets- initial configuration from their evolution: implications for the timing of the planetary instability. Astron. J. **153**, 153 (2017). https://doi.org/10.3847/1538-3881/aa5eaa

R. Deienno, A. Izidoro, A. Morbidelli, R.S. Gomes, D. Nesvorny, S.N. Raymond, Excitation of a primordial cold asteroid belt as an outcome of planetary instability. Astrophys. J. **864**, 50 (2018). https://doi.org/10.3847/1538-4357/aad55d

F.E. DeMeo, B. Carry, Solar System evolution from compositional mapping of the asteroid belt. Nature **505**, 629–634 (2014). https://doi.org/10.1038/nature12908

F.E. DeMeo, R.P. Binzel, S.M. Slivan, S.J. Bus, An extension of the Bus asteroid taxonomy into the near-infrared. Icarus **202**, 160–180 (2009)

F.E. DeMeo, C.M.O. Alexander, K.J. Walsh, C.R. Chapman, R.P. Binzel, The compositional structure of the asteroid belt, in *Asteroids IV*, ed. by P. Michel, F. DeMeo, W.F. Bottke (U. Arizona Press, Tucson, 2015), pp. 13–41

S.J. Desch, A. Kalyaan, C.M. O'D. Alexander, The effect of Jupiter's formation on the distribution of refractory elements and inclusions in meteorites. Astrophys. J. Suppl. Ser. **238**, 11 (2018). https://doi.org/10.3847/1538-4365/aad95f

S. Dibb, Exploration of spectrally featureless asteroids: Reflectance measurements of meteorites and characterization of a commercial camera system for the Psyche mission. ProQuest Dissertations Publishing (2021)

P.M. Doyle, K. Jogo, K. Nagashima, A.N. Krot, S. Wakita, F.J. Ciesla, I.D. Hutcheon, Early aqueous activity on the ordinary and carbonaceous chondrite parent bodies recorded by fayalite. Nat. Commun. **6**(1), 1–10 (2015)

J.D. Drummond, W.J. Merline, B. Carry, A. Conrad, V. Reddy, P. Tamblyn et al., The triaxial ellipsoid size, density, and rotational pole of asteroid (16) Psyche from Keck and Gemini AO observations 2004–2015. Icarus **305**, 174–185 (2018)

A.G. Duba, J.N. Boland, High temperature electrical conductivity of the carbonaceous chondrites Allende and Murchison. Proc. Lunar Planet. Sci. Conf. **15**, 232–233 (1984)

P. Dyal, C.W. Parkin, W.D. Daily, Lunar electrical conductivity, permeability and temperature from Apollo magnetometer experiments, in *NASA Special Publication*, vol. 370 (1977)

L.T. Elkins-Tanton, The taxonomy of planetesimals: consequences for planets, in *Planetesimals: Early Differentiation and Consequences for Planets*, ed. by L.T. Elkins-Tanton, B.P. Weiss (Cambridge University Press, Cambridge, 2017), pp. 365–375

L.T. Elkins-Tanton, B.P. Weiss, M.T. Zuber, Chondrites as samples of differentiated planetesimals. Earth Planet. Sci. Lett. **305**, 1–10 (2011). https://doi.org/10.1016/j.epsl.2011.03.010

L.T. Elkins-Tanton et al., Observations, meteorites, and models: a pre-flight assessment of the composition and formation of (16) Psyche. J. Geophys. Res. **125**, e2019JE006296 (2020). https://doi.org/10.1029/2019JE006296

J.P. Emery, F. Marzari, A. Morbidelli, L.M. French, T. Grav, The complex history of trojan asteroids, in *Asteroids IV*, ed. by P. Michel, F. DeMeo, W.F. Bottke (U. Arizona Press, Tucson, 2015), pp. 203–220

A. Emsenhhuber, S. Cambioni, E. Asphaug, T.S. Gabriel, S.R. Schwartz, R. Furfaro, Realistic on-the-fly outcomes of planetary collisions II. Bringing machine learning to N-body simulations. Astrophys. J. **891**, 6 (2020)

L.G. Evans, P.N. Peplowski, E.A. Rhodes, D.J. Lawrence, T.J. McCoy, L.R. Nittler, S.C. Solomon, A.L. Sprague, K.R. Stockstill-Cahill, R.D. Starr, S.Z. Weider, W.V. Boynton, D.K. Hamara, Major-element abundances on the surface of Mercury: results from the MESSENGER Gamma-Ray Spectrometer. J. Geophys. Res., Planets **117**, E00L07 (2012). https://doi.org/10.1029/2012JE004178

P. Farinella, P. Paolicchi, V. Zappalá, The asteroids as outcomes of catastrophic collisions. Icarus **52**, 409–433 (1982)

W. Feldman, B. Barraclough, S. Maurice, R. Elphic, D. Lawrence, D. Thomsen, A. Binder, Major compositional units of the Moon: lunar prospector thermal and fast neutrons. Science **281**(5382), 1489–1493 (1998)

M. Ferrais, P. Vernazza, L. Jorda, N. Rambaux, J. Hanuš et al., Asteroid (16) Psyche's primordial shape: a possible Jacobi ellipsoid. Astron. Astrophys. **638**, L15 (2020). 16 pp.

A. Fienga, C. Avdellidou, J. Hanuš, Asteroid masses obtained with INPOP planetary ephemerides. Mon. Not. R. Astron. Soc. **492**(1), 589–602 (2020)

R.A. Fish, G.G. Goles, A. Edward, The record in meteorites III. On the development of meteorites in asteroidal bodies. Astrophys. J. **132**, 243–258 (1960)

A.D. Fortes, Titan's internal structure and the evolutionary consequences. Planet. Space Sci. **60**(1), 10–17 (2012)

R.R. Fu, L.T. Elkins-Tanton, The fate of magmas in planetesimals and the retention of primitive chondritic crusts. Earth Planet. Sci. Lett. **390**, 127–138 (2014)

R.R. Fu, B.P. Weiss, E.A. Lima, R.J. Harrison, X.-N. Bai, S.J. Desch, D.S. Ebel, C. Suavet, H. Wang, D. Glenn, Solar nebula magnetic fields recorded in the Semarkona meteorite. Science **346**(6213), 1089–1092 (2014)

R.R. Fu, M.W.R. Volk, D. Bilardello, G. Libourel, G.R.J. Lesur, O. Ben Dor, The fine-scale magnetic history of the Allende meteorite: implications for the structure of the solar nebula. AGU Adv. **2**, e2021AV000486 (2021). https://doi.org/10.1029/2021AV000486

G.A. Gaetani, T.L. Grove, Wetting of mantle olivine by sulfide melt: implications for Re/Os ratios in mantle peridotite and late-stage core formation. Earth Planet. Sci. Lett. **169**(1–2), 147–163 (1999)

S. Gaffey, L. McFadden, D. Nash, C. Pieters, Ultraviolet, visible, and near-infrared reflectance spectroscopy: laboratory spectra of geologic materials, in *Remote Geochemical Analysis: Elemental and Mineralogical Composition*, vol. 43 (1993), p. 77

L.A. Garvie, L.P. Knauth, M.A. Morris, Sedimentary laminations in the Isheyevo (CH/CBb) carbonaceous chondrite formed by gentle impact-plume sweep-up. Icarus **292**, 36–47 (2017)

O. Gasnault, W.C. Feldman, S. Maurice, I. Genetay, C. d'Uston, T.H. Prettyman, K.R. Moore, Composition from fast neutrons: application to the Moon. Geophys. Res. Lett. **28**(19), 3797–3800 (2001)

J. Gattacceca, B.P. Weiss, M. Gounelle, New constraints on the magnetic history of the CV parent body and the solar nebula from the Kaba meteorite. Earth Planet. Sci. Lett. **455**, 166–175 (2016)

J.I. Goldstein, E.R.D. Scott, N.L. Chabot, Iron meteorites: crystallization, thermal history, parent bodies, and origin. Chem. Erde **69**, 293–325 (2009). https://doi.org/10.1016/j.chemer.2009.01.002

J. Gradie, E. Tedesco, Compositional structure of the asteroid belt. Science **216**, 1405–1407 (1982)

A. Greshake, A.N. Krot, A. Meibom, M.K. Weisberg, M.E. Zolensky, K. Keil, Heavily-hydrated lithic clasts in CH chondrites and the related, metal-rich chondrites Queen Alexandra Range 94411 and Hammadah al Hamra 237. Meteorit. Planet. Sci. **37**(2), 281–293 (2002)

V.E. Hamilton et al., Evidence for widespread hydrated minerals on asteroid (101955) Bennu. Nat. Astron. (2019). https://doi.org/10.1038/s41550-019-0722-2

P.S. Hardersen, M.J. Gaffey, P.A. Abell, Near-IR spectral evidence for the presence of iron-poor orthopyroxenes on the surfaces of six M-type asteroids. Icarus **175**, 141–158 (2005)

P.S. Hardersen, E.A. Cloutis, V. Reddy, T. Mothé-Diniz, J.P. Emery, The M-/X-asteroid menagerie: results of an NIR spectral survey of 45 main-belt asteroids. Meteorit. Planet. Sci. **46**, 1910–1938 (2011)

G.R. Huss, A.E. Rubin, J.N. Grossman, Thermal metamorphism in chondrites, in *Meteorites and the Early Solar System II*, vol. 943 (2006), pp. 567–586

A. Izidoro, B. Bitsch, R. Dasgupta, The effect of a strong pressure bump in the Sun's natal disk: terrestrial planet formation via planetesimal accretion rather than pebble accretion. Astrophys. J. **915**, 62 (2021)

K. Jogo, T. Nakamura, M. Ito, S. Wakita, M.Y. Zolotov, S.R. Messenger, Mn–Cr ages and formation conditions of fayalite in CV3 carbonaceous chondrites: constraints on the accretion ages of chondritic asteroids. Geochim. Cosmochim. Acta **199**, 58–74 (2017)

A. Johansen, T. Ronnet, M. Bizzarro, M. Schiller, M. Lambrechts, Å. Nordlund, H. Lammer, A pebble accretion model for the formation of the terrestrial planets in the Solar System. Sci. Adv. **7**(8), eabc0444 (2021)

B.C. Johnson, M.M. Sori, A.J. Evans, Ferrovolcanism on metal worlds and the origin of pallasites. Nat. Astron. (2019). https://doi.org/10.1038/s41550-019-0885-x

J.H. Jones, M.J. Drake, Experimental investigations of trace element fractionation in iron meteorites. III: elemental partitioning in the system Fe-Ni-S-P. Geochim. Cosmochim. Acta **47**, 1199–1209 (1983). https://doi.org/10.1016/0016-7037(86)90405-9

N.A. Kaib, J.E. Chambers, The fragility of the terrestrial planets during a giant-planet instability. Mon. Not. R. Astron. Soc. **455**, 3561–3569 (2016)

K.D. Kanagawa, H. Tanaka, E. Szuszkiewicz, Radial migration of gap-opening planets in protoplanetary disks. I. The case of a single planet. Astrophys. J. **861**(2), 140 (2018)

K. Keil, Mineralogical and chemical relationships among enstatite chondrites. J. Geophys. Res. **22**, 6945–6976 (1968)

K. Keil, L. Wilson, Explosive volcanism and the compositions of cores of differentiated asteroids. Earth Planet. Sci. Lett. **117**(1–2), 111–124 (1993)

P.L. King, S.M. McLennon, Sulfur on Mars. Elements **6**, 107–112 (2010). https://doi.org/10.2113/gselements.6.2.107

K. Kitazato et al., The surface composition of asteroid 162173 Ryugu from Hayabusa2 near-infrared spectroscopy. Science **364**, 272–275 (2019). https://doi.org/10.1126/science.aav7432

H. Klahr, A. Schreiber, Testing the Jeans, Toomre, and Bonnor-Ebert concepts for planetesimal formation: 3D streaming-instability simulations of diffusion-regulated formation of planetesimals. Astrophys. J. **911**, 9 (2021)

T. Kleine, G. Buddle, C. Burkhardt, T.S. Kruijer, The non-carbonaceous-carbonaceous meteorite dichotomy. Space Sci. Rev. **216**, 55 (2020). https://doi.org/10.1007/s11214-020-00675-w

A. Konopliv, S. Asmar, R. Park, B. Bills, F. Centinello, A. Chamberlin, A. Ermakov, R. Gaskell, N. Rambaux, C. Raymond, The Vesta gravity field, spin pole and rotation period, landmark positions, and ephemeris from the Dawn tracking and optical data. Icarus **240**, 103–117 (2014)

K. Kretke, H. Levison, Challenges in forming the solar system's giant planet cores via pebble accretion. Astron. J. **148**(6), 109 (2014)

K.A. Kretke, W.F. Bottke, H.F. Levison, D.A. Kring, Mixing of the asteroid belt due to the formation of the giant planets, in *Accretion: Building New Worlds Conference 2043* (2017), p. 2027

A. Krot, Y. Amelin, P. Cassen, A. Meibom, Young chondrules in CB chondrites from a giant impact in the early Solar System. Nature **436**(7053), 989–992 (2005)

A.N. Krot, K. Nagashima, M. Yoshitake, H. Yurimoto, Oxygen isotopic compositions of chondrules from the metal-rich chondrites Isheyevo (CH/CBb), MAC 02675 (CBb) and QUE 94627 (CBb). Geochim. Cosmochim. Acta **74**, 2190–2211 (2010)

T.S. Kruijer, M. Touboul, M. Fischer-Gödde, K.R. Bermingham, R.J. Walker, T. Kleine, Protracted core formation and rapid accretion of protoplanets. Science **344**, 1150–1154 (2014)

T.S. Kruijer, C. Burkhardt, G. Budde, T. Kleine, Age of Jupiter inferred from the distinct genetics and formation times of meteorites. Proc. Natl. Acad. Sci. USA **114**, 6712–6716 (2017)

M. Kruss, G. Wurm, Seeding the formation of Mercurys: an iron-sensitive bouncing barrier in disk magnetic fields. Astrophys. J. **869**(1), 45 (2018)

M. Kruss, G. Wurm, Composition and size dependent sorting in preplanetary growth: seeding the formation of Mercury-like planets. Planet. Sci. J. **1**(1), 23 (2020)

G. Kullerud, The Fe-Ni-S system, in *Annual Reviews of the Geophysical Laboratory* (The Carnegie Institution of Washington, Washington, 1963), pp. 175–189

M. Lambrechts, A. Johansen, Rapid growth of gas-giant cores by pebble accretion. Astron. Astrophys. **544**, A32 (2012)

T. LaTourrette, G.J. Wasserburg, Mg diffusion in anorthite: implications for the formation of early solar system planetesimals. Earth Planet. Sci. Lett. **158**, 91–108 (1998)

D. Lawrence et al., this journal

D. Lawrence, W. Feldman, B. Barraclough, A. Binder, R. Elphic, S. Maurice, D. Thomsen, Global elemental maps of the Moon: the lunar prospector gamma-ray spectrometer. Science **281**(5382), 1484–1489 (1998)

D.J. Lawrence, R.C. Elphic, W.C. Feldman, T.H. Prettyman, O. Gasnault, S. Maurice, Small area thorium features on the lunar surface. J. Geophys. Res., Planets **108**, 5102 (2003). https://doi.org/10.1029/2003JE002050

D. Lawrence, P. Peplowski, J. Goldsten, M. Burks, A. Beck, L. Elkins-Tanton, I. Jun, T. McCoy, C. Polanskey, T. Prettyman, The Psyche gamma-ray and neutron spectrometer: characterizing the composition of a metal-rich body using nuclear spectroscopy, in *Lunar and Planetary Science Conference* (2016)
D.J. Lawrence, P.N. Peplowski, A.W. Beck, W.C. Feldman, E.A. Frank, T.J. McCoy, L.R. Nittler, S.C. Solomon, Compositional terranes on Mercury: information from fast neutrons. Icarus **281**, 32–45 (2017)
D. Lawrence, M. Burks, M. Cully, L. Elkins-Tanton, J. Goldsten, I. Jun, T. McCoy, P. Peplowski, C. Polanskey, T. Prettyman, The psyche gamma-ray and neutron spectrometer: update on instrument design and measurement capabilities, in *Lunar and Planetary Science Conference* (2019)
D.L. Lawrence, J.O. Goldsten, P.N. Peplowski, Z.W. Yokley, A.W. Beck, M.T. Burks, L.T. Elkins-Tanton, I. Jun, T.J. McCoy, T.H. Prettyman, Planetary neutron spectroscopy for metal-rich compositions: development of analysis framework for measurements at the asteroid (16) Psyche. Earth Space Sci. (2021, in review)
T. Lee, D.A. Papanastassiou, G.J. Wasserburg, Demonstration of ^{25}Mg excess in Allende and evidence for ^{26}Al. Geophys. Res. Lett. **3**, 41–44 (1976)
H.F. Levison, W.F. Bottke, M. Gounelle, A. Morbidelli, D. Nesvorný, K. Tsiganis, Contamination of the asteroid belt by primordial trans-Neptunian objects. Nature **460**, 364–366 (2009)
H.F. Levison, K.A. Kretke, K.J. Walsh, W.F. Bottke, Growing the terrestrial planets from the gradual accumulation of sub-meter sized objects. Proc. Natl. Acad. Sci. **112**, 14180–14185 (2015)
D. Lusby, E.R. Scott, K. Keil, Ubiquitous high-FeO silicates in enstatite chondrites. J. Geophys. Res., Solid Earth **92**(B4), E679–E695 (1987)
R.J. Macke, Survey of meteorite physical properties: density, porosity and magnetic susceptibility, University of Central Florida (2010)
F. Marchis, L. Jorda, P. Vernazza, M. Brož, J. Hanuš, M. Ferrais, F. Vachier, N. Rambaux, M. Marsset, M. Viikinkoski, (216) Kleopatra, a low density critically rotating M-type asteroid. Astron. Astrophys. **653**, A57 (2021)
J.R. Masiero, A.K. Mainzer, T. Grav, J.M. Bauer, R.M. Cutri et al., Main belt asteroids with WISE/NEOWISE. I. Preliminary albedos and diameters. Astrophys. J. **741**, 68 (2011). https://doi.org/10.1088/0004-637X/741/2/68
C. Maurel, J.F.J. Bryson, R.J. Lyons, M.R. Ball, R.V. Chopdekar, A. Scholl, F.J. Ciesla, W.F. Bottke, B.P. Weiss, Meteorite evidence for partial differentiation and protracted accretion of planetesimals. Sci. Adv. **6**, eaba1303 (2020)
T.J. McCoy et al., this journal. https://doi.org/10.1007/s11214-022-00872-9
T.J. McCoy, K. Keil, R.N. Clayton, T.K. Mayeda, D.D. Bogard, D.H. Garrison, G.R. Huss, I.D. Hutcheon, R. Wieler, A petrologic, chemical, and isotopic study of Monument Draw and comparison with other acapulcoites: evidence for formation by incipient partial melting. Geochim. Cosmochim. Acta **60**(14), 2681–2708 (1996)
T.J. McCoy, K. Keil, R.N. Clayton, T.K. Mayeda, D.D. Bogard, D.H. Garrison, R. Wieler, A petrologic and isotopic study of lodranites: evidence for early formation as partial melt residues from heterogeneous precursors. Geochim. Cosmochim. Acta **61**(3), 623–637 (1997a)
T.J. McCoy, K. Keil, D.W. Muenow, L. Wilson, Partial melting and melt migration in the acapulcoite-lodranite parent body. Geochim. Cosmochim. Acta **61**, 639–650 (1997b)
T.J. McCoy, T.L. Dickinson, G.E. Lofgren, Partial melting of the Indarch (EH4) meteorite: a textural, chemical and phase relations view of melting and melt migration. Meteorit. Planet. Sci. **34**, 735–746 (1999)
K.J. Miller, W.-I. Zhu, L.G.J. Montési, G.A. Gaetani, V. Le Roux, X. Xiao, Experimental evidence for melt partitioning between olivine and orthopyroxene in partially molten harzburgite. J. Geophys. Res., Solid Earth **121**, 5776–5793 (2016). https://doi.org/10.1002/2016JB013122
D.W. Mittlefehldt, M.M. Lindstrom, D.D. Bogard, D.H. Garrison, S.W. Field, Acapulco-and Lodran-like achondrites: petrology, geochemistry, chronology, and origin. Geochim. Cosmochim. Acta **60**(5), 867–882 (1996)
K. Moffatt, E. Dormy, *Self-Exciting Fluid Dynamos* (Cambridge University Press, Cambridge, 2019)
A. Morbidelli, K.J. Walsh, D.P. O'Brien, D.A. Minton, W.F. Bottke, The dynamical evolution of the asteroid belt, in *Asteroids IV*, ed. by P. Michel, F. DeMeo, W.F. Bottke (U. Arizona Press, Tucson, 2015), pp. 493–508
S.A. Morse, *Basalts and Phase Diagrams: An Introduction to the Quantitative Use of Phase Diagrams in Igneous Petrology* (Springer, Berlin, 1980)
S.A. Morse, Adcumulus growth of anorthosite at the base of the lunar crust. Proc. Lunar Planet. Sci. Conf. **13**, A10–A18 (1982)
C.E. Moyano-Cambero, J.M. Trigo-Rodríguez, J. Llorca, S. Fornasier, M.A. Barucci, A. Rimola, A plausible link between the asteroid 21 Lutetia and CH carbonaceous chondrites. Meteorit. Planet. Sci. **51**(10), 1795–1812 (2016)

L. Nagy, W. Williams, L. Tauxe, A.R. Muxworthy, I. Ferreira, Thermomagnetic recording fidelity of nanometer-sized iron and implications for planetary magnetism. Proc. Natl. Acad. Sci. **116**(6), 1984–1991 (2019)

T. Nakamura et al., Itokawa dust particles: a direct link between S-type asteroids and ordinary chondrites. Science **333**, 1113 (2011). https://doi.org/10.1126/science.1207758

C. Neese, Asteroid Taxonomy V6.0. *NASA Planetary Data System*, EAR-A (2010)

D. Nesvorný, Young solar system's fifth giant planet? Astrophys. J. **742**, L22 (2011)

D. Nesvorny, Dynamical evolution of the early solar system. Annu. Rev. Astron. Astrophys. **56**, 137–174 (2018). https://doi.org/10.1146/annurev-astro-081817-052028

D. Nesvorný, A. Morbidelli, Statistical study of the early solar system's instability with four, five, and six giant planets. Astron. J. **14**, 117 (2012)

D. Nesvorný, D. Vokrouhlický, A. Morbidelli, Capture of trojans by jumping Jupiter. Astrophys. J. **768**, 45 (2013). 8 pp.

D. Nesvorny, F. Roig, W.F. Bottke, Modeling the historical flux of planetary impactors. Astron. J. **153**, 103 (2017)

D. Nesvorny, D. Vokrouhlicky, W.F. Bottke, H.F. Levison, Evidence for very early migration of the Solar System planets from the Patroclus-Menoetius binary Jupiter Trojan. Nat. Astron. **2**, 878–882 (2018)

D. Nesvorny, F.V. Roig, R. Deienno, The role of early giant-planet instability in terrestrial planet formation. Astron. J. **161**, 50 (2021). https://doi.org/10.3847/1538-3881/abc8ef

J.A. Neufeld, J.F.J. Bryson, F. Nimmo, The top–down solidification of iron as-teroids driving dynamo evolution. J. Geophys. Res., Planets (2019). https://doi.org/10.1029/2018JE005900

W. Neumann, D. Breuer, T. Spohn, Differentiation of Vesta: implications for a shallow magma ocean. Earth Planet. Sci. Lett. **395**, 267–280 (2014)

A. Okada, K. Keil, G.J. Taylor, H. Newsom, Igneous history of the aubrite parent asteroid: evidence from the Norton County enstatite achondrite. Meteorit. Planet. Sci. **23**, 59–74 (1988)

R. Oran, B.P. Weiss, O. Cohen, Were chondrites magnetized by the early solar wind? Earth Planet. Sci. Lett. **492**, 222–231 (2018)

C.W. Ormel, H.H. Klahr, The effect of gas drag on the growth of protoplanets: analytical expressions of the accretion of small bodies in laminar disks. Astron. Astrophys. **520**, A43 (2010). https://doi.org/10.1051/0004-6361/201014903

R. Park, A.S. Konopliv, S.W. Asmar, B. Bills, R.W. Gaskell, C.A. Raymond, D.E. Smith, M.T. Zuber, Gravity field expansion in ellipsoidal harmonic and polyhedral internal representations applied to Vesta. Icarus **240**, 118–132 (2014). https://doi.org/10.1016/j.icarus.2013.12.005

R. Park et al., A partially differentiated interior for Ceres deduced from its gravity field and shape. Nature **537**, 515–517 (2016)

R.S. Park, A.T. Vaughan, A.S. Konopliv, A.I. Ermakov, N. Mastrodemos, J.C. Castillo-Rogez et al., High-resolution shape model of Ceres from stereophotoclinometry using Dawn Imaging Data. Icarus **319**, 812–827 (2019)

R.S. Park, A. Konopliv, A. Ermakov, J. Castillo-Rogez, R. Fu, K. Hughson, T. Prettyman, C. Raymond, J. Scully, H. Sizemore, Evidence of non-uniform crust of Ceres from Dawn's high-resolution gravity data. Nat. Astron. **4**(8), 748–755 (2020)

P.N. Peplowski et al., Radioactive elements on Mercury's surface from MESSENGER: implications for the planet's formation and evolution. Science **333**, 1850–1852 (2011)

P.N. Peplowski, A.W. Beck, D.J. Lawrence, Geochemistry of the lunar highlands as revealed by measurements of thermal neutrons. J. Geophys. Res., Planets **121**, 388–401 (2016)

P.N. Peplowski, T.J. McCoy, M. Liebel, A.W. Beck, M.T. Burks, L.T. Elkins-Tanton, D.J. Lawrence, T.H. Prettyman, Gamma-ray spectroscopy of metal-rich meteorites: preparations for the Psyche gamma-ray spectroscopy investigation, in *50th Lunar and Planetary Science Conference, Abstract #1731* (2019)

C. Polanskey, this journal

J.B. Pollack, O. Hubickyj, P. Bodenheimer, J.J. Lissauer, M. Podolak, Y. Greenzweig, Formation of the giant planets by concurrent accretion of solids and gas. Icarus **124**(1), 62–85 (1996)

T.H. Prettyman, D.W. Mittlefehldt, N. Yamashita, A.W. Beck, W.C. Feldman, J.S. Hendricks, D.J. Lawrence, T.J. McCoy, H.Y. McSween, P.N. Peplowski, Neutron absorption constraints on the composition of 4 Vesta. Meteorit. Planet. Sci. **48**(11), 2211–2236 (2013)

T. Prettyman, N. Yamashita, R. Reedy, H. McSween Jr., D. Mittlefehldt, J. Hendricks, M. Toplis, Concentrations of potassium and thorium within Vesta's regolith. Icarus **259**, 39–52 (2015)

T. Prettyman, N. Yamashita, M. Toplis, H. McSween, N. Schörghofer, S. Marchi, W. Feldman, J. Castillo-Rogez, O. Forni, D. Lawrence, Extensive water ice within Ceres' aqueously altered regolith: evidence from nuclear spectroscopy. Science **355**(6320), 55–59 (2017)

K.L. Rasmussen, D.J. Malvin, V.F. Buchwald, J.T. Wasson, Compositional trends and cooling rates of group IVB iron meteorites. Geochim. Cosmochim. Acta **48**, 805–813 (1984)

S.N. Raymond, A. Izidoro, The empty primordial asteroid belt. Sci. Adv. **3**, e1701138 (2017a). https://doi.org/10.1126/sciadv.1701138

S.N. Raymond, A. Izidoro, Origin of water in the inner Solar System: planetesimals scattered inward during Jupiter and Saturn's rapid gas accretion. Icarus **297**, 134–148 (2017b)

S. Raymond, A. Morbidelli, The Grand Tack model: a critical review. Proc. Int. Astron. Union **9**(S310), 194–203 (2014). https://doi.org/10.1017/S1743921314008254

V. Reddy, D. Lawrence, L. Elkins-Tanton, D. Takir, Constraining hydrogen abundance on asteroid (16) Psyche, in *Lunar and Planetary Science Conference* (2018)

C.C. Reese, V.S. Solomatov, Fluid dynamics of local Martian magma oceans. Icarus **184**, 102–120 (2006). https://doi.org/10.1016/j.icarus.2006.04.008

A.E. Ringwood, S.E. Kesson, A dynamic model for mare basalt petrogenesis. Proc. Lunar Sci. Conf. **7**, 1697–1722 (1976)

S.K. Runcorn, On the interpretation of lunar magnetism. Phys. Earth Planet. Inter. **10**, 327–335 (1975)

C.T. Russell, C.A. Raymond, A. Coradini, H.Y. McSween, M.T. Zuber, A. Nathues et al., Dawn at Vesta: testing the protoplanetary paradigm. Science **336**, 684–686 (2012)

S. Sahijpal, P. Soni, G. Gupta, Numerical simulations of the differentiation of accreting planetesimals with ^{26}Al and ^{60}Fe as the heat sources. Meteorit. Planet. Sci. **42**, 1529–1548 (2007)

A. Scherstén, T. Elliott, C. Hawkesworth, S. Russell, J. Masarik, Hf-W evidence for rapid differentiation of iron meteorite parent bodies. Earth Planet. Sci. Lett. **241**(3–4), 530–542 (2006)

D.L. Schrader, J. Davidson, R.C. Greenwood, I.A. Franchi, J.M. Gibson, A water–ice rich minor body from the early Solar System: the CR chondrite parent asteroid. Earth Planet. Sci. Lett. **407**, 48–60 (2014)

E.R.D. Scott, *Iron Meteorites: Composition, Age, and Origin*. Oxford Research Encyclopedia of Planetary Science (2020). https://doi.org/10.1093/acrefore/9780190647926.013.206

E.R. Scott, K. Keil, J.I. Goldstein, E. Asphaug, W.F. Bottke, N.A. Moskovitz, Early impact history and dynamical origin of differentiated meteorites and asteroids, in *Asteroids IV* (2015), pp. 573–595

R.A. Secco, H.H. Schloessin, The electrical resistivity of solid and liquid Fe at pressures up to 7 GPa. J. Geophys. Res. **94**, 5887–5894 (1989)

M.K. Shepard et al., A radar survey of M-and X-class asteroids. III. Insights into their composition, hydration state, & structure. Icarus **245**, 38–55 (2015). https://doi.org/10.1016/j.icarus.2014.09.016

M.K. Shepard, K. de Kleer, S. Cambioni, P.A. Taylor, A.K. Virkki, E.G. Rívera-Valentin, C. Rodriguez Sanchez-Vahamonde, L. Fernanda Zambrano-Zambrano-Marin, C. Magri, D. Dunham, J. Moore, M. Camarca, Asteroid 16 Psyche: shape, features, and global map. Planet. Sci. J. **2**, 1–16 (2021)

H. Sierks, P. Lamy, C. Barbieri, D. Koschny, H. Rickman, R. Rodrigo, M.F. A'Hearn, F. Angrilli, M.A. Barucci, J.-L. Bertaux, Images of asteroid 21 Lutetia: a remnant planetesimal from the early solar system. Science **334**(6055), 487–490 (2011)

L. Siltala, M. Granvik, Asteroid mass estimation with the robust adaptive Metropolis algorithm. Astron. Astrophys. **633**, A46 (2019)

L. Siltala, M. Granvik, Mass and density of asteroid (16) Psyche. Astrophys. J. (2021). https://doi.org/10.3847/2041-8213/abe948

E.S. Simpson, D.G. Murray, A new siderolite from Bencubbin, Western Australia. Mineral. Mag. **23**, 33–37 (1932)

J.V. Smith, A.T. Anderson, R.C. Newton, E.J. Olsen, P.J. Wyllie, A.V. Crewe, M.S. Isaacson, D. Johnson, Petrologic history of the moon inferred from petrography, mineralogy and petrogenesis of Apollo 11 rocks, in *Proceedings of the Apollo 11 Lunar Science Conference*, vol. 1 (1970), pp. 897–925

J.R. Smyth, T.C. McCormick, Crystallographic data for minerals, in *Mineral Physics & Crystallography: A Handbook of Physical Constants*, vol. 2 (AGU Reference Shelf, Washington, 1995), pp. 1–17

V.S. Solomatov, Fluid dynamics of a terrestrial magma ocean, in *Origin of the Earth and Moon*, ed. by R.M. Canup, K. Righeter (University of Arizona Press, Phoenix, 2000), p. 555

V.S. Solomatov, D.J. Stevenson, Suspension in convective layers and style of differentiation of a terrestrial magma ocean. J. Geophys. Res. **98**, 5375–5390 (1993)

F.J. Spera, Lunar magma transport phenomena. Geochim. Cosmochim. Acta **56**, 2253–2265 (1992)

O. Šrâmek, L. Milelli, Y. Ricard, S. Labrosse, Thermal evolution and differentiation of planetesimals and planetary embryos. Icarus **217**, 339–354 (2012)

J. Suckale, L.T. Elkins Tanton, J.A. Sethian, Crystals stirred up: 1. Direct numerical simulations of crystal settling in nondilute magmatic suspensions. J. Geophys. Res. **117**, E08005 (2012). https://doi.org/10.1029/2012JE004067

D. Takir, V. Reddy, J. Sanchez, M.K. Shepard, J. Emery, Detection of water and/or hydroxyl on asteroid (16) Psyche. Astron. J. **153**, 31–37 (2016). https://doi.org/10.3847/1538-3881/153/1/31

K. Tsiganis, R. Gomes, A. Morbidelli, H.F. Levison, Origin of the orbital architecture of the giant planets of the Solar System. Nature **435**(7041), 459–461 (2005)

D. Turrini, The primordial collisional history of Vesta: crater saturation, surface evolution and survival of the basaltic crust. Planet. Space Sci. **103**, 82–95 (2014)
F. Ulff-Muller, Effects of liquid immiscibility on trace element fractionation in magmatic iron meteorites: a case study of group IIIAB. Meteorit. Planet. Sci. **33**, 207–220 (1998). https://doi.org/10.1111/j.1945-5100.1998.tb01626.x
H.C. Urey, The cosmic abundances of potassium, uranium, and thorium and the heat balance of the Earth, the Moon, and Mars. Proc. Natl. Acad. Sci. **41**, 127–144 (1955)
P. Vernazza, P. Lamy, O. Groussin, T. Hiroi, L. Jorda, P.L. King, M.R.M. Izawa, F. Marchis, M. Birlan, R. Brunetto, Asteroid (21) Lutetia as a remnant of Earth's precursor planetesimals. Icarus **216**, 650–659 (2011). https://doi.org/10.1016/j.icarus.2011.09.032
P. Vernazza et al., VLT/SPHERE imaging survey of the largest main-belt asteroids: final results and synthesis. Astron. Astrophys. **654**, A56 (2021)
M. Viikinkoski, P. Vernazza, J. Hanuš, H. Le Corroler, K. Tazhenova, B. Carry et al., (16) Psyche: a mesosiderite-like asteroid? Astron. Astrophys. **619**, L3 (2018)
D. Vokrouhlický, W.F. Bottke, D. Nesvorný, Capture of trans-Neptunian planetesimals in the main asteroid belt. Astron. J. **152**(2), 39 (2016)
K.J. Walsh, A. Morbidelli, S.N. Raymond, D.P. O'Brien, A.M. Mandell, A low mass for Mars from Jupiter's early gas-driven migration. Nature **475**, 206–209 (2011)
H. Wang, B.P. Weiss, X.-N. Bai, B.G. Downey, J. Wang, J. Wang, C. Suavet, R.R. Fu, M.E. Zucolotto, Lifetime of the solar nebula constrained by meteorite paleomagnetism. Science **355**(6325), 623–627 (2017)
P.H. Warren, The magma ocean concept and lunar evolution. Annu. Rev. Earth Planet. Sci. **13**, 201–240 (1985)
P. Warren, Stable-isotope anomalies and the accretionary assemblage of the Earth and Mars: a subordinate role for carbonaceous chondrites. Earth Planet. Sci. Lett. **331**, 93–100 (2011). https://doi.org/10.1016/j.epsl.2011.08.047
J.T. Wasson, G.W. Kallemeyn, The IAB iron-meteorite complex: a group, five subgroups, numerous grouplets, closely related, mainly formed by crystal segregation in rapidly cooling melts. Geochim. Cosmochim. Acta **66**, 2445–2473 (2002). https://doi.org/10.1016/S0016-7037(02)00848-7
J.T. Wasson, H. Huber, D.J. Malvin, Formation of IIAB iron meteorites. Geochim. Cosmochim. Acta **71**, 760–781 (2007). https://doi.org/10.1016/j.gca.2006.09.032
S.J. Weidenschilling, Iron/silicate fractionation and the origin of Mercury. Icarus **35**, 99–111 (1978)
M.K. Weisberg, M. Prinz, R.N. Clayton, T.K. Mayeda, M.M. Grady, C.T. Pillinger, The CR chondrite clan, in *NIPR Symp. Antarctic Meteorites*, National Institute of Polar Research (1995), pp. 11–32.
M.K. Weisberg, D.S. Ebel, D. Nakashima, N.T. Kita, M. Humayun, Petrology and geochemistry of chondrules and metal in NWA 5492 and GRO 95551: a new type of metal-rich chondrite. Geochim. Cosmochim. Acta **167**, 269–285 (2015)
B.J. Weiss et al., this journal
B.P. Weiss, W.F. Bottke, What do meteorites tell us about the formation of Jupiter? AGU Adv. **2**(2), e00376 (2021). https://doi.org/10.1029/2020AV000376
B.J. Weiss et al., Possible evidence for partial differentiation of asteroid Lutetia from Rosetta. Planet. Space Sci. **66**, 137–146 (2012)
B.P. Weiss, X.-N. Bai, R.R. Fu, History of the solar nebula from meteorite paleomagnetism. Sci. Adv. **7**(1), eaba5967 (2021)
J.T. Wilson, D.J. Lawrence, P.N. Peplowski, J.T.S. Cahill, V.R. Eke, R.J. Massey, L.F.A. Teodoro, Image reconstruction techniques in neutron and gamma-ray spectroscopy: improving lunar prospector data. J. Geophys. Res., Planets **123**, 1804 (2018). https://doi.org/10.1029/2018JE005589
J.A. Wood, J.S.J. Dickey, U.B. Marvin, B.N. Powell, Lunar anorthosites and a geophysical model of the Moon, in *Proceedings of the Apollo 11 Lunar Science Conference* (1970), pp. 965–988
E.D. Young, The hydrology of carbonaceous chondrite parent bodies and the evolution of planet progenitors. Philos. Trans. R. Soc. Lond. Ser. A, Math. Phys. Eng. Sci. **359**(1787), 2095–2110 (2001)
E.D. Young, K.K. Zhang, G. Schubert, Conditions for pore water convection within carbonaceous chondrite parent bodies–implications for planetesimal size and heat production. Earth Planet. Sci. Lett. **213**(3–4), 249–259 (2003)
N.G. Zube, F. Nimmo, R.A. Fischer, S.A. Jacobson, Constraints on terrestrial planet formation timescales and equilibration processes in the Grand Tack scenario from Hf-W isotopic evolution. Earth Planet. Sci. Lett. **522**, 210–218 (2019)
M.T. Zuber et al., this journal
M.T. Zuber, D.E. Smith, M.M. Watkins, S.W. Asmar, A.S. Konopliv, F.G. Lemoine, H.J. Melosh, G.A. Neumann, R.J. Phillips, S.C. Solomon, Gravity field of the Moon from the Gravity Recovery and Interior Laboratory (GRAIL) mission. Science **339**(6120), 668–671 (2013)

Authors and Affiliations

Linda T. Elkins-Tanton[1] · Erik Asphaug[2] · James F. Bell III[1] · Carver J. Bierson[1] · Bruce G. Bills[3] · William F. Bottke[4] · Samuel W. Courville[1] · Steven D. Dibb[1] · Insoo Jun[3] · David J. Lawrence[5] · Simone Marchi[4] · Timothy J. McCoy[6] · Jose M.G. Merayo[7] · Rona Oran[8] · Joseph G. O'Rourke[1] · Ryan S. Park[3] · Patrick N. Peplowski[5] · Thomas H. Prettyman[9] · Carol A. Raymond[3] · Benjamin P. Weiss[8] · Mark A. Wieczorek[10] · Maria T. Zuber[8]

✉ L.T. Elkins-Tanton
ltelkins@asu.edu

1 School of Earth and Space Exploration, Arizona State University, Tempe, AZ 86387-2001, USA

2 Lunar and Planetary Laboratory, University of Arizona, Tucson, AZ 85721-0092, USA

3 Jet Propulsion Laboratory, Pasadena, CA 91109, USA

4 Southwest Research Institute, Boulder, CO 80302, USA

5 Johns Hopkins University Applied Physics Laboratory, Laurel, MD 20723, USA

6 Smithsonian National Museum of Natural History, Washington, DC 20013, USA

7 National Space Institute, Danish Technical University, Lyngby, Denmark

8 Department of Earth, Atmospheric and Planetary Sciences, Massachusetts Institute of Technology, Cambridge, MA 02139-4307, USA

9 Planetary Science Institute, Tucson, AZ 85719, USA

10 Université Côte d'Azur, Observatoire de la Côte d'Azur, CNRS, Laboratoire Lagrange, Nice, France

Space Science Reviews (2022) 218:24
https://doi.org/10.1007/s11214-022-00891-6

Determining the Relative Cratering Ages of Regions of Psyche's Surface

S. Marchi[1] · E. Asphaug[2] · J.F. Bell III[3] · W.F. Bottke[1] · R. Jaumann[4] · R.S. Park[5] · C.A. Polanskey[5] · T.H. Prettyman[6] · D.A. Williams[3] · R. Binzel[7] · R. Oran[7] · B. Weiss[7] · C.T. Russell[8]

Received: 26 May 2021 / Accepted: 12 April 2022 / Published online: 2 May 2022

Abstract
The study of the cratering history of asteroid (16) Psyche is one of the investigations to be performed by the NASA *Psyche* mission. A dedicated Relative Ages Working Group will carry on these investigations using primarily imaging and topographic data, and complement the interpretation of these data with theoretical models (hydrocodes to simulate impacts) as well as laboratory experiments (impact experiments on relevant target materials). The *Psyche* Science Team will also rely on experience and lessons learned from prior space missions, such as NASA *Dawn* and ESA *Rosetta*.

The main goals of the cratering investigations are to map craters and characterize their morphology across Psyche's surface over a range of spatial resolutions. These data will then be used to constrain relative and absolute ages of Psyche's terrains, and impact-related processes will inform other investigations, such as geological mapping, surface composition, and internal structure. Psyche's cratering data will also be used to perform comparative analyses with similar data from other rocky asteroids. The present chapter provides a pre-launch view of the planned activities and methodologies of the Relative Ages Working Group.

The NASA Psyche Mission: Science Instruments and Investigations
Edited by James F. Bell III, Carol Polanskey and Lindy Elkins-Tanton

✉ S. Marchi
marchi@boulder.swri.edu

1 Southwest Research Institute, Boulder, CO, USA

2 Lunar and Planetary Laboratory, University of Arizona, Tucson, AZ, USA

3 School of Earth and Space Exploration, Arizona State University, Tempe, AZ, USA

4 Institute of geological Sciences, Free University of Berlin, Berlin, Germany

5 Jet Propulsion Laboratory, California Institute of Technology, Pasadena, CA, USA

6 Planetary Science Institute, Tucson, AZ, USA

7 Department of Earth, Atmospheric and Planetary Sciences, Massachusetts Institute of Technology, Cambridge, MA, USA

8 University of California Los Angeles, Los Angeles, CA, USA

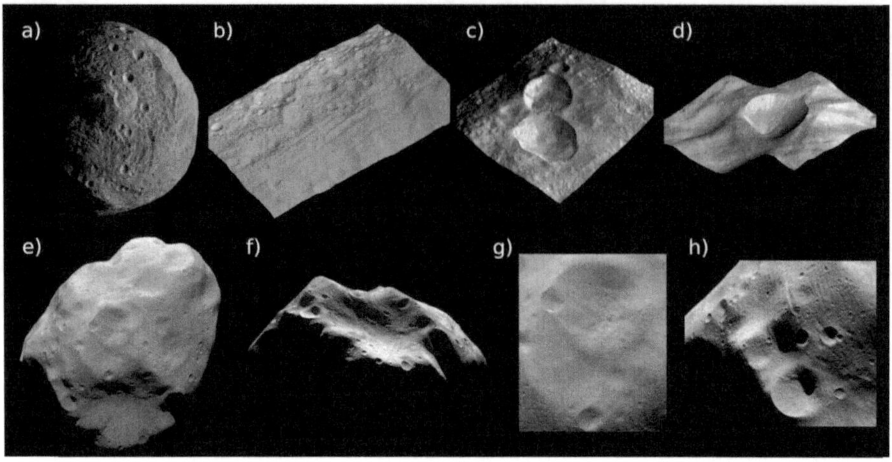

Fig. 1 A summary figure of notable craters on Vesta (**a**-**d**) and Lutetia (**e**-**h**). These asteroids have mean diameters of about 525 and 100 km, respectively, thus bracketing Psyche's size (mean diameter 222 km). **a**) Vesta's southern hemisphere highlighting the Rheasilvia basin, and its central mound; **b**) 3D rendering of Divalia Fossae; **c**) Superposition of craters (the 70-km diameter Marcia crater is the lowest, largest crater in the image); **d**) 17-km diameter Antonia crater, an irregular impact structure located on a steep slope; **e**) Lutetia's irregular shape; **f**) 55-km diameter Massalia crater; **g**) Lutetia's Baetica region sculpted by possibly two or three superposed impacts; **h**) Lutetia's Narbonensis region with throughs and lineaments. Credit: **a**)-**d**) NASA/JPL-Caltech/UCLA/MPS/DLR/IDA; **e**)-**h**) MPS/UPD/LAM/IAA/RSSD/INTA/UPM/DASP/IDA

1 Introduction

Impact cratering is one of the primary processes producing surface modifications on asteroids observed by spacecraft (Marchi et al. 2015, for a review). The largest impacts sculpt the overall shape of asteroids, while smaller impacts excavate and redistribute target material across the surface leading to complex landscapes. The surface spatial density of craters is key to both identify process-specific geologic units as part of geologic mapping, and reconstruct the asteroid's past evolution. In this chapter, we present a brief summary of cratering on notable rocky asteroids, including Vesta and Ceres as revealed by the NASA *Dawn* mission and Lutetia visited by the ESA *Rosetta* mission. The choice to focus on these asteroids is motivated by their bulk properties that are closest to what is currently known about Psyche: Vesta's and Lutetia's sizes bracket that of Psyche and their high densities are similar to Psyche's estimated density (see below). Furthermore, these asteroids share a similar collisional environment, thus facilitating comparative investigations that may help understand target-specific cratering outcomes, for instance due to different material properties. Finally, we note that the *Dawn* and *Psyche* missions have similar observational architectures, therefore the lessons learned from *Dawn* will inform the planning and execution of cratering investigations at Psyche. Additional observations from smaller asteroids (e.g., Eros) will be discussed as relevant.

Vesta's overall geology is the result of impact cratering and its subsequent effects. Two giant impacts (Rheasilvia and Veneneia basins, about 500 and 460 km in diameter, respectively; Marchi et al. 2012a; Schenk et al. 2012) at its south pole altered the original asteroid's ellipsoidal shape (Fig. 1). These impacts excavated crustal and upper mantle material that ended up covering much of the surface, and also was ejected from the asteroid and became the source of the Vesta's asteroid family (Asphaug 1997). Vesta's asteroid family is the probable source of the howardite, eucrite and diogenite (HED) meteorites reaching Earth (McCord et al. 1970; Russell et al. 2012; McSween et al. 2013; Schenk et al. 2021b). The excavation and deposition of Rheasilvia and Veneneia accounts for much of the topographic structure of Vesta, with a relief of about 42 km (Jaumann et al. 2012; Jutzi et al. 2013). In addition, the energy of these impacts transformed Vesta's surface via impact-induced seismic activity to form tectonic ridge-and-trough systems along the equator and in the northern hemisphere (Divalia and Saturnalia Fossae, respectively; Jaumann et al. 2012; Bowling 2015). Smaller impacts have excavated subsurface materials at multiple locations, including possibly buried volatile-rich materials as a source of localized pitted terrains (as in the 70 km across Marcia crater; Denevi et al. 2012), and igneous intrusions (e.g., Brumalia Tholus, a putative laccolith exposed by Teia crater; Buczkowski et al. 2014). Many localized color variations on Vesta's surface (e.g., Oppia, Octavia craters) appear to represent ejecta from impacts of compositionally distinct material, different from the typical vestan surface (Garry et al. 2014; Williams et al. 2014a). Vesta's bulk density is 3.46±0.01 g/cm^3 (Russell et al. 2012), thus within the range of estimated density for Psyche (Elkins-Tanton et al. 2020, 2021).

Ceres's surface is also largely controlled by cratering (Park et al. 2019, 2020), but there are several notable differences compared to its smaller sibling Vesta. Overall, Ceres' topography has only about 10 km of vertical relief, and it appears to be modulated horizontally by a few large-scale depressions (e.g., Vendimia Planitia, 800 km across) and raised terrains (e.g., Hanami Planum, 560 km across). These depressions are compatible with ancient impact structures, but unlike the Rheasilvia basin on Vesta their origin remains unclear. The largest well-defined impact crater on Ceres is Kerwan (284 km across, Williams et al. 2018), thus highlighting significant depletion of large craters compared to Vesta. Because Vesta and Ceres experienced comparable collision evolution histories over the last 4.5 Gyr or so, this suggests that many large craters on Ceres were not preserved (Marchi et al. 2016), possibly due to its rheologically weak crust that would allow for the gradual removal of complex topography (Fu et al. 2017; Bland et al. 2016; Otto et al. 2019). Ceres' low bulk density (~2.16 g/cm^3; Russell et al. 2016) indicates a higher volatile content than Vesta, Lutetia and Psyche, and as such Ceres' cratering may not be directly relevant to Psyche. Nevertheless, Ceres is one of the Main Belt asteroids with high resolution coverage and cratering data, therefore it will be useful in comparative analyses.

Important cratering data has also been gathered by other missions. For instance, ESA's *Rosetta* mission flew-by the 100-km asteroid Lutetia, revealing a highly irregular shape sculpted by large craters, such as the 55-km diameter crater Massalia (see Fig. 1). Lutetia's bulk density is 3.4±0.3 g/cm^3 (Sierks et al. 2011). Post-cratering downslope movement of material has also been shown to be an important surface modification process on Lutetia, despite the low surface gravity. For example, Lutetia's Baetica regions show signs of regolith downslope motion on a flank of a possible impact crater, along with a concentration of boulders at the base of the slope (Massironi et al. 2012; Kueppers et al. 2012). Similarly, downslope movement of regolith, and dust transport from impact and solar effects, has created slope streaks, boulder tracks, and flat sedimentary "pond" deposits on the floors of some impact craters on smaller asteroids such as Eros (e.g., Robinson et al. 2001). Eros also revealed a number of tectonic features (Prockter et al. 2002), and a prominent ridge (18-km long Hinks Dorsum) has been interpreted as impact-induced compression due to the formation of Shoemaker crater (Watters et al. 2011). A detailed review of asteroid cratering is beyond the scope of this chapter, and the interested reader is referred to Marchi et al. (2015), and Bottke et al. (2020).

In addition to its other novelties (Elkins-Tanton et al. 2020), Psyche is intermediate in size between Lutetia and Vesta. It will be the first asteroid ever visited in the size range 100-500 km diameter, which may be the size range of the largest asteroids to have been catastrophically disrupted during Solar System history (Bottke et al. 2005), so we can for now only speculate on how global scale (Asphaug et al. 2015) and sub-catastrophic to catastrophic impacts (Holsapple and Housen 2019) will have altered its geology, and on the related question of the absence of a Psyche-derived asteroid family (e.g., Davis et al. 1999).

In addition to space mission data, advanced adaptive optics imaging techniques applied to the largest asteroids are widening the scope of comparative investigations. For example, for the second largest asteroid Pallas (528 km mean diameter), imaging by Marsset et al. (2020) reveal a highly saturated "golf ball" like surface at a scale of about 4 km/pixel. Comparatively, the same technique applied to Psyche by Viikinkoski et al. (2018) achieves a similar spatial resolution but Psyche's reconstructed surface image does not appear crater saturated (see below). It is worth noting Pallas and Psyche are in different collisional environments (e.g., higher average impact velocities for Pallas) and successful recognition of small craters by adaptive optics imaging may be subject to subtle variations in lighting or the crater depth-to-diameter ratios. For example, adaptive optics imaging of Ceres (Vernazza et al. 2020) shows few craters in contrast to findings by the *Dawn* mission, likely due to shallow craters on Ceres compared to Vesta's craters (Schenk et al. 2021a).

This brief summary of the complex nature of large asteroid cratering indicates that the outcome of impact processes is controlled by the asteroid's size, composition, and material properties. Ground-based occultation and high angular imaging resolved Psyche's ellipsoidal shape to have principal axes of 277x238x168 km^3 (222±4 km mean diameter; Ferrais et al. 2020). Uncertainties in volume and mass estimates, result in a density estimate with a most likely range from 3.4 to 4.1 g/cm^3 (Elkins-Tanton et al. 2020). Because of the density and other surface properties, it is thought that Psyche may have a metal content up to 60% (Elkins-Tanton et al. 2020, 2021), and thus cratering may behave in radically different ways compared to rocky asteroids (e.g., Marchi et al. 2020).

It is expected that the gamut of cratering processes described above may play an important role in interpreting Psyche's crater populations. Impact crater abundances are expected to be different on geologic surfaces produced at different times, and thus could be a key referent in determining the relative stratigraphy of Psyche's surface. Craters may also expose materials from the interior of Psyche, or host exogenic materials delivered by comets or asteroids (e.g., water ice in permanently-shadowed regions of worlds like the Moon; e.g., Luchsinger et al. 2021). In addition, the cumulative effects of impacts can add exogenous materials to Psyche's surface over billions of years (e.g., Marchi et al. 2019), and in case of Psyche's high metal content, they could result in complex metal-silicate mixing. Thus, it is anticipated that Psyche's cratering may reveal peculiar traits when compared to previously visited asteroids. Cratering may also lead to the acquisition or modification of remnant magnetization through impact heating or shocks (e.g., Gattacceca et al. 2010). Magnetization of small bodies by impact plasmas has also been proposed (e.g., Muxworthy et al. 2017). Furthermore, determining the relative ages of different regions on the surface may help trace the history of any asteroid-generated dynamo field that may have operated on the body and can now be revealed from the existence of remnant magnetization.

The following sections lay out the *Psyche* Science Team plan to characterize crater populations on Psyche, as informed by pre-launch mission planning and theoretical modeling.

2 Mapping Craters on Psyche

Specific science objectives related to impact cratering have been identified as important goals for the *Psyche* mission to be carried out by the Relative Ages Working Group. In addi-

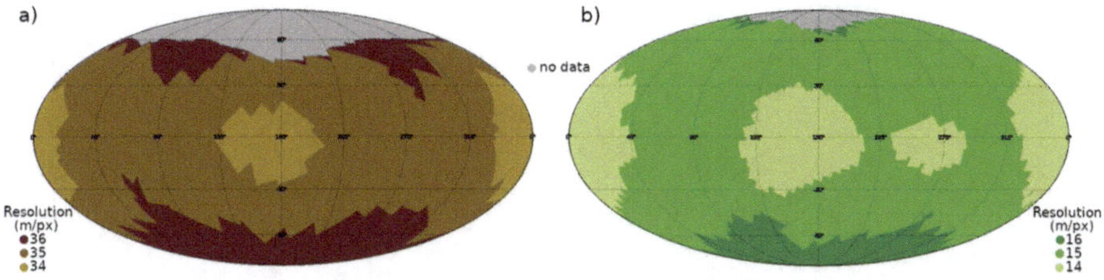

Fig. 2 Mollweide projection of resolution maps for planned observations at Psyche for Orbit A (**a**) and Orbit B (**b**), with an altitude of 700 and 290 km, respectively. Colors indicate the best spatial resolution available and surface coverage. The cumulative area coverage at resolutions higher than 36 m/px (Orbit A) and 16 m/px (Orbit B) is 93% and 98%, respectively. Additional higher resolution imaging of localized areas will be acquired in Orbit C and D (altitude 170 and 85 km, respectively), but with much less surface coverage. Note the gray region near the north pole is not imaged in Orbit A and B, but it will be imaged by Orbit C and D

tion to simply characterizing the overall geology and topography of Psyche at spatial scales of $\leq$ 200 m/pix horizontally and $\leq$ 50 m/pix vertically (Jaumann et al. 2021), these goals also include determining the relative ages of surface regions by counting craters with diameters larger than 1 km over at least 50% of the asteroid. Crater identification and mapping relies on a combination of imaging resolution, coverage and illumination and the mission's cratering objectives are expected to be met using data from the *Psyche* Multispectral Imager investigation (Bell et al. 2021) acquired during a series of increasingly higher-resolution imaging campaigns from lower orbital altitudes over the course of the mission (Polanskey et al. 2021). Comparison with the *Dawn* mission indicates that crater identification can be reliably conducted with 5 pixels per diameter, and possibly down to 3 pixels per diameter under ideal illumination conditions and benign local topography (e.g., smooth terrains as in Marcia crater on Vesta; Marchi et al. 2014). The planned resolution maps are presented in Fig. 2, for the two primary mission phases devoted to imaging (called Orbit A and B). For example, Orbit A (altitude of 700 km) will achieve a resolution of 35 m/px on about 80% of the surface; while in Orbit B (altitude 290 km) will achieve a resolution of 15 m/px on about 90% of the surface, demonstrating that the mission cratering goal of mapping craters larger than 1 km will likely be exceeded (see later for expected crater statistics).

Initial photogeologic analysis of images of Psyche begins with the identification of geologic features, including impact craters. Crater size, shape, morphology (simple, complex, multi-ring structures, etc.) and distribution will be noted, along with identification and mapping the extents of continuous and discontinuous ejecta blankets from impact craters. Identifying contacts between craters and other features and geologic units are essential to determine the stratigraphic history of the asteroid's surface. Cratering data will be obtained primarily using georeferenced images from the Multispectral Imager and derived products, such as mosaics. We will also analyze the topographic data for evidence of large depressions that could represent ancient, degraded impact structures. Here we conveniently discuss specific issues associated with the identification of large and small craters.

Large craters. Psyche's shape is significantly elongated, raising the question of how much of its shape is a byproduct of large impacts. Here we define as "large", those craters with diameter larger than about 1/5th of the asteroid's diameter. Early asteroid investigations showed that the shapes of relatively small asteroids (31-km Ida, 17-km Eros, etc) were primarily controlled by large impacts (e.g., Thomas 1999; Thomas et al. 2002). This conclusion holds true for larger asteroids, and observations of Vesta and Lutetia have shown that large craters are also important to control their overall shape (Fig. 1).

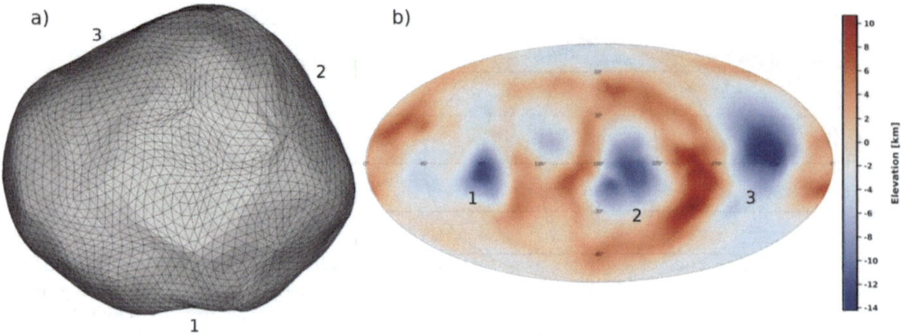

Fig. 3 Psyche shape model and topography. **a**) A polar view of Psyche shape model revealing an equatorial silhouette that largely departs from a regular ellipsoid. Three putative large craters are indicated with 1, 2, 3. **b**) A Mollweide projection of the elevation with respect to the best fit ellipsoid (277x238x168 km^3). The equatorial depressions are indicated. The range of elevation is $[-14.2, 10.70]$ km. Adapted from Ferrais et al. (2020)

On this basis, we expect that large collisions are responsible for some irregularities in Psyche's appearance. A radar shape model (Shepard et al. 2017) shows two distinct depressions, of approximately ~ 67 and ~ 53 km in diameter, which could be impact structures. Viikinkoski et al. (2018) identified two additional depressions ~ 90 km in diameter, while Ferrais et al. (2020) identified three large crater-like depressions larger ~ 44, 60, 80 km in diameter fairly close to Psyche's equator (Fig. 3). Assuming these features are truly impact craters, we might expect they significantly altered Psyche's overall shape and surface properties. The ultimate morphological consequences of these events is a strong function of Psyche's bulk mechanical properties, which remain poorly constrained for the time being. We anticipate, however, that these putative large collisions could have resulted in significant crustal porosity as well as the redeposition of subsurface materials across the Psyche's surface. This may have led to the generation of fine particulates on Psyche's surface. This conclusion holds true also in case of a high metal content for Psyche, as laboratory experiments reveal that cratering in Fe-Ni meteorites and ingots can produce small metal fragments similar to rocky regolith (Marchi et al. 2020; Christoph et al. 2021). In addition, large impacts may also generate widespread tectonic features (ridges, graben, etc.), as proposed for Eros, Ida and Vesta (Buczkowski et al. 2008; Asphaug et al. 1996; Bowling 2015). For Eros, it has been shown that the consecutive formation of the observed three largest craters (Himeros, Psyche, and Shoemaker) could generate up to 20% bulk porosity (Tonge et al. 2016).

While current data provide tantalizing evidence of large craters on Psyche, the extent to which even larger impacts may have been responsible for the Psyche's shape remains to be understood and it is one of the goals of the *Psyche* mission.

Small craters. High resolution imaging of the small crater distribution on Psyche will open a new window in comparative planetology for cratering processes on a wide variety of planetary surfaces. When examined at high resolution in small local regions, Vesta and Lutetia are heavily cratered down to the imaging resolution limit (Fig. 1). Surfaces such as Achaia on Lutetia and Vesta's northern hemisphere are indicative of heavily cratered terrains resulting from the cumulative effect of billion years of collisions (Marchi et al. 2012a,b). These craters are responsible for the gardening of the top layers and therefore may reveal small scale compositional variations resulting from the dredging up of subsurface materials, which may be compositionally distinct from the surface and/or less weathered.

Such processes have been observed on Vesta (Reddy et al. 2012; Pieters et al. 2012), but not on Lutetia, probably because Lutetia appears to be compositionally uniform (Coradini et al. 2011; Masoumzadeh and Boehnhardt 2019). Small craters will also be used for comparative analysis with other smaller asteroids to better constrain cratering processes on bodies with different bulk properties (e.g., Bottke et al. 2020).

3 Synergies Between Cratering and Other Psyche Mission Investigations

It is expected that the Psyche's surface will be heavily cratered at crater sizes smaller than a few km in diameter. The exposure of subsurface materials in recent small impacts may allow for a better characterization of compositional variability across the surface and reveal the presence of crustal stratification (Bell et al. 2021). An interesting and possibly peculiar aspect about Psyche is the interaction of impact-deposited materials (mostly silicates from rocky asteroids) with Psyche's endogenic material which could have high metal content. This situation could lead to metal-silicate mixing resulting in complex surface morphologies and variable composition. Thus, the coupling of cratering and compositional investigations by the Imager and Gamma Ray and Neutron Spectrometer (GRNS) have the potential to disentangle pristine and accreted materials.

As an example, hydrated minerals were found on Vesta's otherwise anhydrous, basaltic surface (De Sanctis et al. 2012; Prettyman et al. 2012; McCord et al. 2012; Reddy et al. 2012). Hydrated species, thought to result from the infall of carbonaceous chondrite impactors, were most abundant in more heavily cratered regions containing basaltic rock types representative of Vesta's ancient crust (e.g., McSween et al. 2013; Ammannito et al. 2013; Prettyman et al. 2013). Moreover, a portion of the dark material may have been emplaced as ejecta from the impact that formed the Veneneia basin, which is overlain by the younger Rheasilvia basin in Vesta's southern hemisphere (e.g., Reddy et al. 2012; Turrini et al. 2014). As such, the distribution of hydrated minerals provides information on Vesta's chronostratigraphy. Based on telescopic observations in the near-IR, Psyche too is thought to be contaminated with water/OH-bearing minerals (Takir et al. 2016). Consequently, measurements of hydrogen by Psyche's GRNS (Lawrence et al. 2021) may provide additional information on the timing, flux and makeup of the impactors.

In addition, Psyche's reflectance spectra have been interpreted as a metal powder mixture with up to 10 wt.% orthopyroxenes (e.g., Ockert-Bell et al. 2010; Fornasier et al. 2010; Hardersen et al. 2011), although this interpretation is not unique. Interestingly, spatial heterogeneities in surface properties have been proposed based on optical, radar and ALMA data (Shepard et al. 2021; de Kleer et al. 2021). Lacking detailed information about the spatial distribution of the silicate component (e.g., intimate vs areal mixing) it is not clear if these silicates are the result of impact contamination. Spatially resolved color data by Psyche's imager could provide further insight on the origin of the silicate component (Bell et al. 2021).

One of the primary objectives of the Psyche mission is determining whether the asteroid carries remnant magnetization that could be a record of a past dynamo (Elkins-Tanton et al. 2021). There are three possible synergies between the study of cratering processes and the magnetometry investigation. These all relate to the central goal of the magnetometry investigation to characterize the magnetization in Psyche to understand the asteroid's thermal evolution and the history of past magnetic fields (Weiss et al. 2021). First, combining information about the ages of different regions with their inferred magnetization could constrain

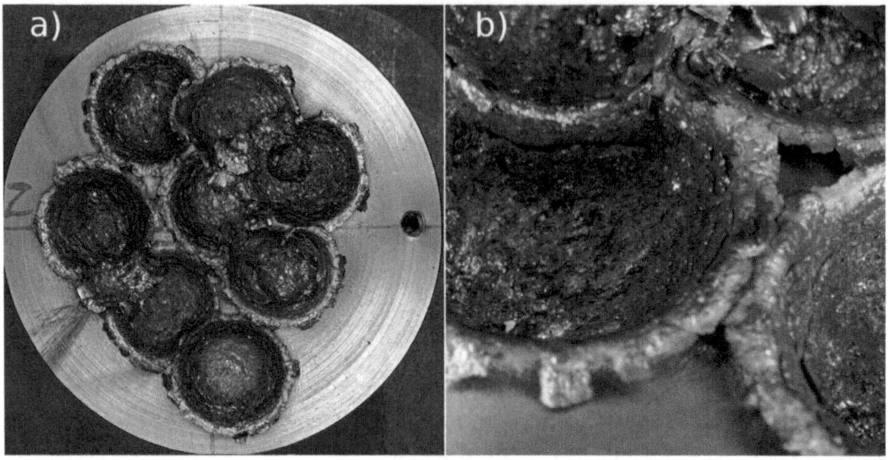

Fig. 4 Consecutive impacts on a Fe-Ni ingot with a diameter of 7 cm (**a**), and close up views of interesting features (**b**). The petal structure of adjacent craters interact to form a complex intersecting pattern, while the superposition of craters produces a large topographic excursion. The 13 superposed craters were produced by silica impactors with diameters of 3.175 and 6.35 mm, and impact velocities ranging from 4.5–5.5 km/s. Adapted from Marchi et al. (2020)

temporal changes in the core dynamo field. Second, because impact processes may heat, shock, and/or fracture, they can affect the magnetization in the body. If a field is present during an impact, this can lead to the acquisition of thermal or shock remnant magnetization of basin and ejecta materials. Alternatively, if no field is present, impact heating and shock can lead to demagnetization. Lastly, impact in the presence of a dynamo could generate plasmas that modify the dynamo field (Hood and Artemieva 2008; Oran et al. 2020). However, studies of how impacts may modify an ambient field and whether such a modified field could magnetize crustal rocks have so far been limited largely to the Moon and other rocky bodies (Chen et al. 1995; Hood and Artemieva 2008; Weiss et al. 2010; Oran et al. 2020). Revealing the magnetic history of Psyche may therefore benefit from crater chronology while also potentially being complicated by the contemporaneous processes of dynamo evolution, impact cratering, and thermal evolution.

Another aspect of interest is the interaction between topography and crater morphology. As discussed above, cratering is sensitive to material properties. The possible metal-rich nature of Psyche may result in interesting crater morphologies. Laboratory impact experiments in iron meteorites and Fe-Ni ingots have shown that craters exhibit sharp rims that retain a competent structure, as opposed to cratering in rocks (Libourel et al. 2019; Marchi et al. 2020). Also, crater floors are characterized by the presence of concentric segments, resembling somewhat the structure of rose petals. These craters have a typical depth-to-diameter ratio of ~ 0.4, compared to 0.15-0.25 for rocky targets (Marchi et al. 2020, and references therein). Processes such as cracking and spallation are also less frequent owing to the high strength of metals, implying that a metallic surface can retain more craters per unit area (Fig. 4). We stress that these peculiar crater morphologies observed in metal targets are dominated by the material strength, and their applicability to larger craters is unclear (see below).

Furthermore, available topography data reveal Psyche's complex topography with a global vertical excursion in excess of 20 km (Fig. 3b). This could generate locally steep slopes, thus resulting in asymmetric craters such as those observed on Vesta and Lutetia

(Massironi et al. 2012; Marchi et al. 2012a; Jaumann et al. 2012; Krohn et al. 2014; Fig. 1). The morphology of craters may also inform us about subsurface inhomogeneities, such as void spaces resulting from fracturing or a rubble pile structure. Interestingly, the presence of significant macroporosity could account for Psyche's reduced density compared to iron meteorites (Elkins-Tanton et al. 2020; Marchi et al. 2020). Under these circumstances, it is expected that craters with sizes comparable with average block size or fracturing spacing would have irregular, polygonal shapes. Similar explanations have been proposed for irregular craters on Vesta and Ceres (e.g., Schenk et al. 2021a). See Jaumann et al. (2021), for further details.

4 Crater Chronologies for Main Belt Asteroids

As outlined in the previous section, cratered landscapes provide valuable insights in the overall geological evolution and physical properties of asteroids. In addition, crater statistics is a powerful tool to interpret the temporal evolution of the surface. In general, higher crater spatial densities imply older terrains. This simple observation constitutes the basis of chronostratigraphic analyses (see next section). Here we will discuss the basic principles of crater formation and interpretation of crater spatial densities.

The crater size is a complex function of impactor size, mass, velocity, and the mechanical properties of the target. Simplified analytical expressions have been derived (e.g., Holsapple and Housen 2007) to better understand how to turn these parameters into craters of a given size. The so-called Pi-scaling law allows computation of the transient crater diameter (D_t) as a function of impactor size (d), impact conditions and material properties, and reads:

$$D_t = kd\left[\frac{gd}{2v_\perp^2}\left(\frac{\rho}{\delta}\right)^{2\nu/\mu} + \left(\frac{Y}{\rho v_\perp^2}\right)^{(2+\mu)/2}\left(\frac{\rho}{\delta}\right)^{\nu(2+\mu)/\mu}\right]^{-\mu/(2+\mu)} \quad (1)$$

where g is the target gravitational acceleration, $v_\perp$ is the perpendicular component of the impactor velocity, δ is the projectile density, ρ and Y are the density and "cratering strength" of the target, k and μ depend on the cohesion of the target material and ν on its porosity. Therefore, the nature of the terrain affects the crater efficiency and the functional dependence of the crater size with respect to the input parameters (e.g., impactor size and velocity).

A critical parameter is target cratering strength. This is an empirical quantity associated with but not equivalent to material tensile-shear strength, and it is particularly important for small craters, which form in the so-called strength regime. The effects of material strength are negligible in larger collisions, which are solely controlled by the target's gravity. This is supported by numerical modeling of the largest craters on Vesta (Jutzi and Asphaug 2015) show an increased influence of material friction at these sizes, further complicating the analysis, but indicating that the largest craters are roughly as predicted by gravity scaling.

Although it is not expected to be abrupt, a transition from strength to gravity dominance is where the left and right terms inside the brackets of Eq. (1) are equal, i.e. impactor diameter d_{sg}:

$$d_{sg} = 2\frac{v_\perp^2}{g}\left(\frac{Y}{\rho v_\perp^2}\right)^{(2+\mu)/2}\left(\frac{\rho}{\delta}\right)^{\nu} \quad (2)$$

Equation (1) provides the so-called transient crater size resulting from the direct excavation and removal of target material. Large craters on planetary surfaces are expected to

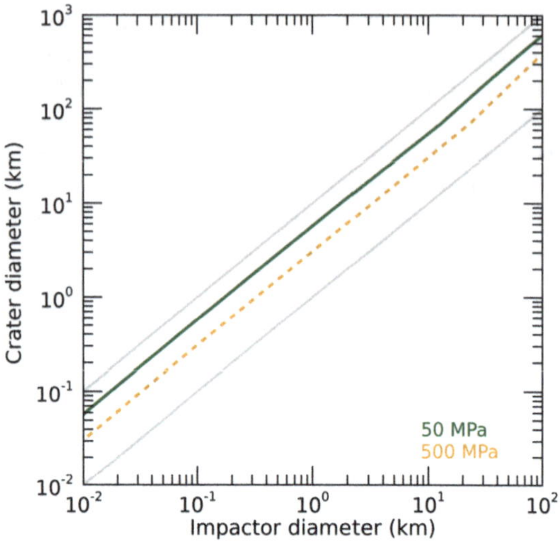

Fig. 5 Final crater size vs impactor size as a function of target strength (as indicated). The chosen strength values are representative values for metal-rich targets (see text). For reference, the gray lines correspond to $D_f = d$ and $D_f = 10d$

undergo a phase of crater modification in which highly fractured materials flow back toward the cavity. The resulting final crater (D_f) is typically from 20% to 50% larger than the transient crater in rocky targets. The modification stage is expected to suppress some of the peculiar crater morphologies observed in laboratory impacts on metallic targets. Figure 5 shows the computed crater size vs impactor size for average impact velocity at Psyche. It is interesting to use Eq. (2) to calculate the impactor diameter corresponding to the transition from strength to gravity regime. Using $Y = 50$ MPa (see below for more details) and a computed average impact velocity for Psyche of 3.7 km/s (so the vertical component is 2.6 km/s), $d_{sg} \sim 40$ km, and $D_f > 200$ km. This indicates that cratering on Psyche may be largely dominated by the strength regime. If correct, this prediction would have important consequences as it indicates that crater morphology is controlled by target strength making the morphologies observed in small-scale laboratory experiments more relevant.

If the asteroid has uniform properties across the surface, the spatial density of small craters superposed on different terrains can be used to infer relative age. In reality, however, this may not be the case. Consider for instance a scenario in which Psyche's surface has metal-rich and silicate-rich terrains. If the two terrains have the same age, we expect a similar total number of impacts, yet craters in the metal-rich terrain would be systematically smaller than in the silicate-rich terrain. Therefore, at a given crater size, it would appear that the metal-rich terrain has less craters. This may erroneously suggest these terrains are considerably younger than the silicate-rich terrains.

Craters are also subject to erasure processes. This is obviously the case when consecutive impacts occur in the same location. Older craters can be totally or partially removed by a younger impact, depending on their relative sizes. Craters can also be eroded by ballistic emplacement of ejecta from distant impacts, or slope failure due to seismic shaking (O'Brien et al. 2006; Richardson 2009; Minton et al. 2019). These processes have been modeled for certain rocky asteroids (e.g., Marchi et al. 2015 for a review) and their applicability to Psyche remains to be investigated.

In order to assess the crater production rate on a given asteroid surface, we need several quantities: the collision probabilities and impact velocities between the target asteroid and the Main Belt population, a crater scaling law that can turn projectiles into craters (or vice

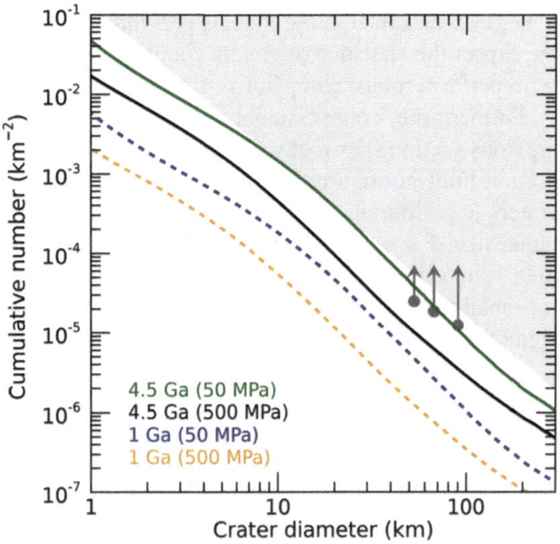

Fig. 6 Psyche's cratering models. The curves indicate the model production functions for the anticipated cumulative number of craters for different strength values and terrain ages, as indicated. The model neglects crater erasing. The data points indicate the four independent large-scale cavities (putative impact craters) inferred from ground-based data (see text). The shaded area indicates crater saturation (the lower bound is 10% of the geometric saturation, as commonly used for rocky terrains)

versa), and the physical properties of the target body. We demonstrate this here by creating a test model of crater production on Psyche.

We use Eq. (1) in its general formulation (including the strength to gravity regime transition) as our crater scaling law for this exercise ($\mu = 0.55$; $\nu = 0.4$). We adopt a Psyche bulk density of 4.1 g/cm^3, a mean diameter of 222 km (Ferrais et al. 2020) and an impactor density of 2.6 g/cm^3. This latter value is an approximation, given that Main Belt projectiles have a variety of compositions and bulk densities. We computed the asteroid impact velocity distribution for Psyche by considering asteroids larger than 50 km in diameter crossing Psyche's orbit (see Marchi et al. 2012b, 2016), obtaining an average value of 3.7 km/s and an intrinsic probability of collision of 4.13×10^{-18} km^{-2} yr^{-1}. We also adopted an impactor size-frequency distribution from the Main Belt evolution model of Bottke et al. (2020). We considered two cratering strength values for illustrative purposes: 500 MPa (for solid metal; Marchi et al. 2020); and a reduced strength of 50 MPa more appropriate if Psyche is heavily fractured and/or contains a significant fraction of silicates, perhaps as indicated by its inferred bulk density (Elkins-Tanton et al. 2020). For reference, intact basalt has a strength of 20 MPa (e.g., Cotto-Figueroa et al. 2016). Finally, we implemented a correction for transient-to-final crater $D_f = 1.3 D_t$, as typically done for rocky targets, but we note that the validity of this correction is questionable for a metallic target.

The resulting model production functions (MPFs), that is the computed number of craters per unit surface for 4.5 and 1 Ga, and for two values of strength, are reported in Fig. 6. The data points indicate the inferred crater cumulative numbers corresponding to the four large cavities larger than 50 km in diameter. This is likely a lower limit of the real number of large craters (as indicated by the vertical arrows). This analysis indicates that if Psyche's strength is close to 500 MPa, then the putative large craters would be very old. Conversely, a greater number of large craters on Psyche would imply that the strength has to be significantly lower than ~500 MPa. Therefore, the number of large craters on Psyche is expected to provide an important constraint on the bulk strength, and its putative metallic nature.

The MPFs are a tool to assess the number of small craters as a function of time. For instance, for a Psyche surface of 1 Ga, it is expected to have between 5 to 30 craters larger than 10 km, or between 500 and 3,000 craters larger than 1 km, depending on the target strength. The number of craters could be significantly higher if the target strength were lower than assumed in the models presented in Fig. 6. Even in the case of higher strength, we expect the statistics of craters coupled with surface coverage (see Sect. 2) would allow us to perform robust comparative statistical analysis on units across the surface.

Furthermore, craters cannot accumulate indefinitely due to erasure, as discussed in the previous section. For rocky objects such as the Moon and Vesta, the theoretical crater saturation limit is often referred to as 10% of the geometrical saturation for closely packed craters (e.g., Marchi et al. 2012b; Fig. 6). This would translate in a total number of craters larger than 1 km over Psyche's surface of $\sim 10^4$. Thus, the actual number of craters larger than 1 km observed on Psyche could provide an indirect constraint on the terrain strength: if a small number of 1 km craters (close to 500, as noted above) is observed and no obvious sign of recent erasure is found, then the strength of the terrains is metal-like.

5 Toward a Chronostratigraphy for Psyche

Mapping impact crater abundances and distributions are an essential part of geologic mapping of planetary surfaces. Identifying contacts between craters and other features and geologic units are essential to determine the stratigraphic history of the asteroid's surface. In particular, we will be looking for impact crater distributions on homogeneous units, i.e., surfaces that have morphological, compositional, or topographic attributes indicative of being formed by a temporally-distinct process or set of geologic processes. In general, for a given set of geologic units, the greater the number of craters on the unit, the older the unit. Older units also have a great abundance of larger diameter craters compared to younger units. All impact craters down to a given diameter limit (based on image resolution) in a given unit are mapped and plotted in a size-frequency distribution graph, and comparison of these graphs aids in determination of the relative ages of the surface units.

Additionally, by applying established impact crater chronology functions to these size-frequency distributions (e.g., the MPFs presented in the previous section), it is possible to derive absolute model ages for each geologic unit or area in which craters have been counted. These absolute model ages, typically reported in million or billions of years old, provide a numerical referent for comparison of times of geologic activity on Psyche, that can be compared with other objects in the Solar System. Application of crater statistical methods to age date planetary surfaces has been well established (Arvidson et al. 1978). For example, during the geologic mapping of Vesta, identification of the largest impact events, and recognition of geologic units produced by those events, were essential to develop a chronostratigraphy and geologic time scale for that asteroid. Specifically, the two largest south polar impact craters, Veneneia and Rheasilvia, were used to demarcate the bulk of Vesta's geologic history: Pre-Veneneian Period, the oldest crustal material, ~2-4.6 Ga; Veneneian Period, the materials produced between the Veneneia and Rheasilvia impact events, ~1-2 Ga; and Rheasilvian Period, the materials produced by the Rheasilvia impact, $\lesssim 1$ Ga. A final period, Marcian, was added for one of the youngest large craters, ~200-390 Ma to present (Fig. 7; Williams et al. 2014a,b).

A similar approach will be developed for asteroid Psyche. Global mosaics and topography maps will be used to produce a global geologic map. This map, in which variations in albedo, morphology, roughness, topography, color (composition), and crater abundances will be used to define and characterize geologic units indicative of the geologic processes that formed them, will enable identification of the major, global-affecting processes that will define major units suitable for a stratigraphy system. Lithostratigraphic units will be used to determine chronostratigraphic units, and the corresponding chronology units that will define Psyche's geologic time scale (Williams et al. 2014a).

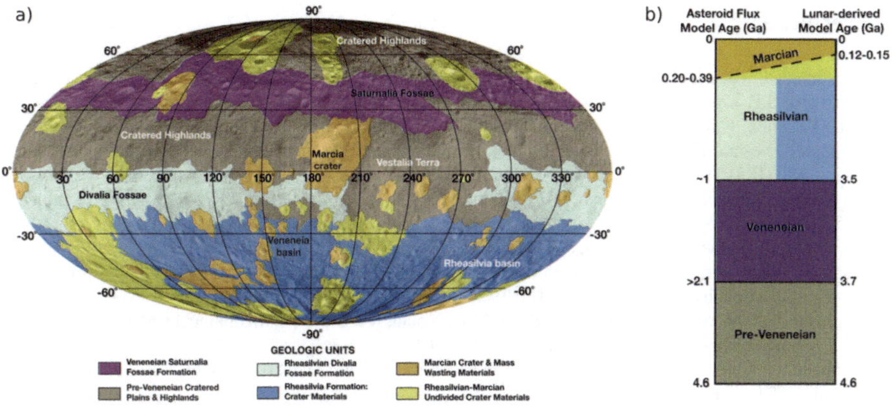

Fig. 7 Mollweide projection of a simplified global geologic map of asteroid Vesta (**a**) consistent with global chronostratigraphy and (**b**) geologic time scale. Adapted from Williams et al. (2014a)

6 Conclusions

In this chapter we presented the planned activities of the *Psyche* mission Relative Ages Working Group and associated cratering studies. Craters are expected to be a primary evolutionary process for Psyche's surface, as indicated by prior studies of asteroids. However, the outcome of impacts on Psyche could radically differ from rocky siblings. If Psyche has a high metal abundance, for instance, craters could have different morphologies than on other rocky bodies. Processes associated with impacts, such as fracturing and generation of fine particulates, could also be different from rocky asteroids, as revealed by laboratory impact experiments on iron meteorites. The presence and number of large craters coupled with Main Belt collisional models will provide important constraints on the bulk properties of Psyche. Smaller craters may be used to derive relative ages of different geological units, and provide absolute cratering age. Relative and absolute ages will be used to derive a chronostratigraphy of Psyche surface in conjunction with geological mapping.

The activities of the Relative Ages Working Group will be carried on synergistically with those of the other working groups as outlined in several chapters in this book. In particular, it is expected that impact-related processes affect topography (Jaumann et al. 2021), internal structure (Zuber et al. 2021), surface composition (Lawrence et al. 2021; Bell et al. 2021), and magnetization history (Weiss et al. 2021). Thus, the products derived by the Relative Ages Working Group (e.g., crater catalogs) will be distributed and utilized by the whole *Psyche's* Science Team to carry on the mission scientific objectives.

References

E. Ammannito, M.C. De Sanctis, F. Capaccioni, M. Teresa Capria, F. Carraro, J.-P. Combe, S. Fonte, A. Frigeri, S.P. Joy, A. Longobardo, G. Magni, S. Marchi, T.B. McCord, L.A. McFadden, H.Y. McSween, E. Palomba, C.M. Pieters, C.A. Polanskey, C.A. Raymond, J.M. Sunshine, F. Tosi, F. Zambon, C.T. Russell, Vestan lithologies mapped by the visual and infrared spectrometer on Dawn. Meteorit. Planet. Sci. **48**, 2185–2198 (2013). https://doi.org/10.1111/maps.12192

R. Arvidson, J. Boyce, C. Chapman, M. Cintala, M. Fulchignoni, H. Moore, G. Neukum, P. Schultz, L. Soderblom, R. Strom, A. Woronov, R. Young, Standard techniques for presentation and analysis of crater size-frequency data. Icarus **37**, 467–474 (1978)

E. Asphaug, Impact origin of the Vesta family. Meteorit. Planet. Sci. **32**, 965–980 (1997). https://onlinelibrary.wiley.com/doi/abs/10.1111/j.1945-5100.1997.tb01584.x

E. Asphaug, J.M. Moore, D. Morrison, W. Benz, M.C. Nolan, R.J. Sullivan, Mechanical and geological effects of impact cratering on Ida. Icarus **120**, 158–184 (1996)

E. Asphaug, G. Collins, M. Jutzi, Global scale impacts, in *Asteroids IV* (2015), pp. 661–678

Bell et al. (2021). This book

M.T. Bland, C.A. Raymond, P.M. Schenk, R.R. Fu, T. Kneissl, J.H. Pasckert, H. Hiesinger, F. Preusker, R.S. Park, S. Marchi, S.D. King, J.C. Castillo-Rogez, C.T. Russell, Composition and structure of the shallow subsurface of Ceres revealed by crater morphology. Nat. Geosci. **9**, 538–542 (2016). https://doi.org/10.1038/ngeo2743

W.F. Bottke Jr., D.D. Durda, D. Nesvorný, R. Jedicke, A. Morbidelli, D. Vokrouhlický, H.F. Levison, Linking the collisional history of the main asteroid belt to its dynamical excitation and depletion. Icarus **179**(1), 63–94 (2005)

W.F. Bottke, D. Vokrouhlický, R.-L. Ballouz, O.S. Barnouin, H.C. Connolly, C. Elder, S. Marchi, T.J. McCoy, P. Michel, M.C. Nolan, B. Rizk, D.J. Scheeres, S.R. Schwartz, K.J. Walsh, D.S. Lauretta, Interpreting the cratering histories of Bennu, Ryugu, and other spacecraft-explored asteroids. Astron. J. **160**, 14 (2020). https://doi.org/10.3847/1538-3881/ab88d3

T.J. Bowling, Global surface modification of Asteroid 4 Vesta following the Rheasilvia impact. Thesis (Ph.D.)–Purdue University (2015); Publication Number: AAT 3736753; ISBN: 9781339256887; Source: Dissertation Abstracts International, Volume: 77-04(E), Section: B.; 174 p.

D.L. Buczkowski, O.S. Barnouin-Jha, L.M. Prockter, 433 Eros lineaments: global mapping and analysis. Icarus **193**, 39–52 (2008). https://doi.org/10.1016/j.icarus.2007.06.028

D.L. Buczkowski, D.Y. Wyrick, M. Toplis, R.A. Yingst, D.A. Williams, W.B. Garry, S. Mest, T. Kneissl, J.E.C. Scully, A. Nathues, M.C. De Sanctis, L. LeCorre, V. Reddy, M. Hoffmann, E. Ammannito, A. Frigeri, F. Tosi, F. Preusker, T. Roatsch, C.A. Raymond, R. Jaumann, C.M. Pieters, C.T. Russell, The unique geomorphology and physical properties of the Vestalia Terra plateau. Icarus **244**, 89–103 (2014). https://doi.org/10.1016/j.icarus.2014.03.035

G. Chen, T.J. Ahrens, R. Hide, Hypervelocity impacts and magnetization of small bodies in the Solar System. Icarus **115**(1), 86–96 (1995)

J.M. Christoph, T. Sharp, S. Marchi, L.T. Elkins-Tanton, Characterizing ejecta fragments from impact experiments into meteoric iron using Scanning Electron Microscopy (SEM), in *52nd Lunar and Planetary Science Conference*, held virtually, 15–19 March, 2021, LPI Contribution No. 2548, id. 2730

A. Coradini, F. Capaccioni, S. Erard et al., The surface composition and temperature of asteroid 21 Lutetia as observed by Rosetta/VIRTIS. Science **334**(6055), 492 (2011)

D. Cotto-Figueroa, E. Asphaug, L.A.J. Garvie, A. Rai, J. Johnston, L. Borkowski et al., Scale-dependent measurements of meteorite strength: implications for asteroid fragmentation. Icarus **277**, 73–77 (2016)

D.R. Davis, P. Farinella, F. Marzari, The missing Psyche family: collisionally eroded or never formed? Icarus **137**(1), 140–151 (1999)

K. de Kleer, S. Cambioni, M. Shepard, The surface of (16) Psyche from thermal emission and polarization mapping. Planet. Sci. J. **2**, 149 (2021). https://doi.org/10.3847/PSJ/ac01ec

M.C. De Sanctis, J.P. Combe, E. Ammannito, E. Palomba, A. Longobardo, T.B. McCord, S. Marchi, F. Capaccioni, M.T. Capria, D.W. Mittlefehldt, C.M. Pieters, J. Sunshine, F. Tosi, F. Zambon, F. Carraro, S. Fonte, A. Frigeri, G. Magni, C.A. Raymond, C.T. Russell, D. Turrini, Detection of widespread hydrated materials on Vesta by the VIR imaging spectrometer on board the Dawn mission. Astrophys. J. **758**, L36 (2012). https://doi.org/10.1088/2041-8205/758/2/l36

B.W. Denevi, D.T. Blewett, D.L. Buczkowski, F. Capaccioni, M.T. Capria, M.C. De Sanctis, W.B. Garry, R.W. Gaskell, L. Le Corre, J.-Y. Li, S. Marchi, T.J. McCoy, A. Nathues, D.P. O'Brien, N.E. Petro, C.M. Pieters, F. Preusker, C.A. Raymond, V. Reddy, C.T. Russell, P. Schenk, J.E.C. Scully, J.M. Sunshine, F. Tosi, D.A. Williams, D. Wyrick, Pitted terrain on Vesta and implications for the presence of volatiles. Science **338**, 246–249 (2012). https://doi.org/10.1126/science.1225374

Elkins-Tanton et al. (2021). This book

L.T. Elkins-Tanton, E. Asphaug, J.F. Bell, H. Bercovici, B. Bills, R. Binzel, W.F. Bottke, S. Dibb, D.J. Lawrence, S. Marchi, T.J. McCoy, R. Oran, R.S. Park, P.N. Peplowski, C.A. Polanskey, T.H. Prettyman, C.T. Russell, L. Schaefer, B.P. Weiss, M.A. Wieczorek, D.A. Williams, M.T. Zuber, Observations, meteorites, and models: a preflight assessment of the composition and formation of (16) Psyche. J. Geophys. Res., Planets **125**, e2019JE006296 (2020). https://doi.org/10.1029/2019JE006296

M. Ferrais, P. Vernazza, L. Jorda, N. Rambaux, J. Hanuš, B. Carry, F. Marchis, M. Marsset, M. Viikinkoski, M. Brož, R. Fetick, A. Drouard, T. Fusco, M. Birlan, E. Podlewska-Gaca, E. Jehin, P. Bartczak, J. Berthier, J. Castillo-Rogez, F. Cipriani, F. Colas, G. Dudziński, C. Dumas, J. Ďurech, M. Kaasalainen, A. Kryszczynska, P. Lamy, H. Le Coroller, A. Marciniak, T. Michalowski, P. Michel, T. Santana-Ros, P. Tanga, F. Vachier, A. Vigan, O. Witasse, B. Yang, Asteroid (16) Psyche's primordial shape: a possible Jacobi ellipsoid. Astron. Astrophys. **638**, L15 (2020). https://doi.org/10.1051/0004-6361/202038100

S. Fornasier, B.E. Clark, E. Dotto, A. Migliorini, M. Ockert-Bell, M.A. Barucci, Spectroscopic survey of M-type asteroids. Icarus **210**, 655–673 (2010)

R.R. Fu, A.I. Ermakov, S. Marchi, J.C. Castillo-Rogez, C.A. Raymond, B.H. Hager et al., The interior structure of Ceres as revealed by surface topography. Earth Planet. Sci. Lett. **476**, 153–164 (2017). https://doi.org/10.1016/j.epsl.2017.07.053

W.B. Garry, D.A. Williams, R.A. Yingst, S.C. Mest, D.L. Buczkowski, F. Tosi, M. Schäfer, L. Le Corre, V. Reddy, R. Jaumann, C.M. Pieters, C.T. Russell, C.A. Raymond, Geologic mapping of ejecta deposits in Oppia Quadrangle, Asteroid (4) Vesta. Icarus **244**, 104–119 (2014)

J. Gattacceca, M. Boustie, L. Hood, J.-P. Cuq-Lelandais, M. Fuller, N.S. Bezaeva, T. de Resseguier, L. Berthe, Can the lunar crust be magnetized by shock: experimental groundtruth. Earth Planet. Sci. Lett. **299**, 42–53 (2010). https://doi.org/10.1016/j.epsl.2010.08.011

P.S. Hardersen, E.A. Cloutis, V. Reddy, T. Mothé-Diniz, J.P. Emery, The M-/X-asteroid menagerie: results of an NIR spectral survey of 45 main-belt asteroids. Meteorit. Planet. Sci. **46**, 1910–1938 (2011)

K.A. Holsapple, K.R. Housen, A crater and its ejecta: an interpretation of deep impact. Icarus **191**(2), 586–597 (2007). https://doi.org/10.1016/j.icarus.2006.08.029

K.A. Holsapple, K.R. Housen, The catastrophic disruptions of asteroids: history, features, new constraints and interpretations. Planet. Space Sci. **179**, 104724 (2019)

L.L. Hood, N.A. Artemieva, Antipodal effects of lunar basin-forming impacts: initial 3D simulations and comparisons with observations. Icarus **193**(2), 485–502 (2008)

Jaumann et al. (2021). This book

R. Jaumann, D.A. Williams, D.L. Buczkowski, R.A. Yingst, F. Preusker, H. Hiesinger, N. Schmedemann, T. Kneissl, J.B. Vincent, D.T. Blewett, B.J. Buratti, U. Carsenty, B.W. Denevi, M.C. De Sanctis, W.B. Garry, H.U. Keller, E. Kersten, K. Krohn, J.-Y. Li, S. Marchi, K.D. Matz, T.B. McCord, H.Y. McSween, S.C. Mest, D.W. Mittlefehldt, S. Mottola, A. Nathues, G. Neukum, D.P. O'Brien, C.M. Pieters, T.H. Prettyman, C.A. Raymond, T. Roatsch, C.T. Russell, P. Schenk, B.E. Schmidt, F. Scholten, K. Stephan, M.V. Sykes, P. Tricarico, R. Wagner, M.T. Zuber, H. Sierks, Vesta's shape and morphology. Science **336**, 687–690 (2012). https://doi.org/10.1126/science.1219122

M. Jutzi, E. Asphaug, The shape and structure of cometary nuclei as a result of low-velocity accretion. Science **348**(6241), 1355–1358 (2015)

M. Jutzi, E. Asphaug, P. Gillet et al., The structure of the asteroid 4 Vesta as revealed by models of planet-scale collisions. Nature **494**, 207–210 (2013). https://doi.org/10.1038/nature11892

K. Krohn, R. Jaumann, D. Elbeshausen, T. Kneissl, N. Schmedemann, R. Wagner, J. Voigt, K. Otto, K.D. Matz, F. Preusker, T. Roatsch, K. Stephan, C.A. Raymond, C.T. Russell, Asymmetric craters on Vesta: impact on sloping surfaces. Planet. Space Sci. **103**, 36–56 (2014)

M. Kueppers, R. Moissl, J.-B. Vincent, S. Besse, S.F. Hviid, B. Carry, B. Grieger, H. Sierks, H.U. Keller, S. Marchi (the OSIRIS Team), Boulders on Lutetia. Planet. Space Sci. **66**, 71–78 (2012)

Lawrence et al. (2021). This book

G. Libourel, A.M. Nakamura, P. Beck, S. Potin, C. Ganino, S. Jacomet et al., Hypervelocity impacts as a source of deceiving surface signatures on iron-rich asteroids. Sci. Adv. **5**, eaav3971 (2019)

K.M. Luchsinger, N.J. Chanover, P.D. Strycker, Water within a permanently shadowed lunar crater: further LCROSS modeling and analysis. Icarus **354**, 114089 (2021)

S. Marchi, M. Massironi, J.-B. Vincent, A. Morbidelli, S. Mottola, F. Marzari, M. Küppers, S. Besse, N. Thomas, C. Barbieri, G. Naletto, H. Sierks, The cratering history of asteroid (21) Lutetia. Planet. Space Sci. **66**, 87–95 (2012a). https://doi.org/10.1016/j.pss.2011.10.010

S. Marchi, H.Y. McSween, D.P. O'Brien, P. Schenk, M.C. De Sanctis, R. Gaskell, R. Jaumann, S. Mottola, F. Preusker, C.A. Raymond, T. Roatsch, C.T. Russell, The violent collisional history of asteroid 4 Vesta. Science **336**, 690–694 (2012b). https://doi.org/10.1126/science.1218757

S. Marchi, W.F. Bottke, D.P. O'Brien, P. Schenk, S. Mottola, M.C. De Sanctis, D.A. Kring, D.A. Williams, C.A. Raymond, C.T. Russell, Small crater populations on Vesta. Planet. Space Sci. **103**, 96–103 (2014). https://doi.org/10.1016/j.pss.2013.05.005

S. Marchi, C.R. Chapman, O.S. Barnouin, J.E. Richardson, J.-B. Vincent, Cratering on asteroids, in *Asteroids IV*, ed. by P. Michel, F.E. DeMeo, W.F. Bottke (University of Arizona Press, Tucson, 2015), pp. 725–744. 895 pp. ISBN 978-0-816-53213-1

S. Marchi, A.I. Ermakov, C.A. Raymond, R.R. Fu, D.P. O'Brien, M.T. Bland, E. Ammannito, M.C. de Sanctis, T. Bowling, P. Schenk, J.E.C. Scully, D.L. Buczkowski, D.A. Williams, H. Hiesinger, C.T. Russell, The missing large impact craters on Ceres. Nat. Commun. **7**, 12257 (2016)

S. Marchi, A. Raponi, T.H. Prettyman, M.C. De Sanctis, J. Castillo-Rogez, C.A. Raymond, E. Ammannito, T. Bowling, M. Ciarniello, H. Kaplan, E. Palomba, C.T. Russell, V. Vinogradoff, N. Yamashita, An aqueously altered carbon-rich Ceres. Nat. Astron. **3**, 140–145 (2019). https://doi.org/10.1038/s41550-018-0656-0

S. Marchi, D.D. Durda, C.A. Polanskey, E. Asphaug, W.F. Bottke, L.T. Elkins-Tanton, L.A.J. Garvie, S. Ray, S. Chocron, D.A. Williams, Hypervelocity impact experiments in iron-nickel ingots and iron meteorites: implications for the NASA Psyche mission. J. Geophys. Res., Planets **125**(2), e05927 (2020)

M. Marsset, M. Brož, P. Vernazza et al., The violent collisional history of aqueously evolved (2) Pallas. Nat. Astron. **4**, 569–576 (2020). https://doi.org/10.1038/s41550-019-1007-5

N. Masoumzadeh, H. Boehnhardt, Global spectrophotometric properties of Asteroid (21) Lutetia using Rosetta-OSIRIS images. Icarus **326**, 1–9 (2019)

M. Massironi, S. Marchi, M. Pajola et al., Geological map and stratigraphy of asteroid (21) Lutetia. Planet. Space Sci. **66**, 125–136 (2012)

T.B. McCord, J.B. Adams, T.V. Johnson, Asteroid Vesta: spectral reflectivity and compositional implications. Science **168**, 1445–1447 (1970). https://doi.org/10.1126/science.168.3938.1445

T.B. McCord, J.Y. Li, J.P. Combe, H.Y. McSween, R. Jaumann, V. Reddy, F. Tosi, D.A. Williams, D.T. Blewett, D. Turrini, E. Palomba, C.M. Pieters, M.C. De Sanctis, E. Ammannito, M.T. Capria, L. Le Corre, A. Longobardo, A. Nathues, D.W. Mittlefehldt, S.E. Schröder, H. Hiesinger, A.W. Beck, F. Capaccioni, U. Carsenty, H.U. Keller, B.W. Denevi, J.M. Sunshine, C.A. Raymond, C.T. Russell, Dark material on Vesta from the infall of carbonaceous volatile-rich material. Nature **491**, 83 (2012). https://doi.org/10.1038/nature11561

H.Y. McSween, R.P. Binzel, M.C. De Sanctis, E. Ammannito, T.H. Prettyman, A.W. Beck, V. Reddy, L. Le Corre, M.J. Gaffey, T.B. McCord, C.A. Raymond, C.T. Russell, Dawn; the Vesta-HED connection; and the geologic context for eucrites, diogenites, and howardites. Meteorit. Planet. Sci. **48**, 2090–2104 (2013). https://doi.org/10.1111/maps.12108

D.A. Minton, C.I. Fassett, M. Hirabayashi, B.A. Howl, J.E. Richardson, The equilibrium size-frequency distribution of small craters reveals the effects of distal ejecta on lunar landscape morphology. Icarus **326**, 63–87 (2019). https://doi.org/10.1016/j.icarus.2019.02.021

A.R. Muxworthy, P.A. Bland, T.M. Davison, J. Moore, G.S. Collins, F.J. Ciesla, Evidence for an impact-induced magnetic fabric in Allende, and exogenous alternatives to the core dynamo theory for Allende magnetization. Meteorit. Planet. Sci. **52**, 2132–2146 (2017)

D. O'Brien, R. Greenberg, J. Richardson, Craters on asteroids: reconciling diverse impact records with a common impacting population. Icarus **183**, 79–92 (2006). https://doi.org/10.1016/j.icarus.2006.02.008

M.E. Ockert-Bell, B.E. Clark, M.K. Shepard, R.A. Isaacs, E.A. Cloutis, S. Fornasier, S.J. Bus, The composition of M-type asteroids: synthesis of spectroscopic and radar observations. Icarus **210**, 674–692 (2010)

R. Oran, B.P. Weiss, Y. Shprits, K. Miljković, G. Tóth, Was the moon magnetized by impact plasmas? Sci. Adv. **6**(40), eabb1475 (2020)

K.A. Otto, S. Marchi, A. Trowbridge, H.J. Melosh, H.G. Sizemore, Ceres crater degradation inferred from concentric fracturing. J. Geophys. Res., Planets **124**, 1188–1203 (2019). https://doi.org/10.1029/2018JE005660

R.S. Park et al., High-resolution shape model of Ceres from stereophotoclinometry using Dawn Imaging Data. Icarus **319**, 812–827 (2019). https://doi.org/10.1016/j.icarus.2018.10.024

R.S. Park et al., Evidence of non-uniform crust of Ceres from Dawn's high-resolution gravity data. Nat. Astron. **4**, 748–755 (2020). https://doi.org/10.1038/s41550-020-1019-1

C.M. Pieters, E. Ammannito, D.T. Blewett, B.W. Denevi, M.C. De Sanctis, M.J. Gaffey, L. Le Corre, J.-Y. Li, S. Marchi, T.B. McCord, L.A. McFadden, D.W. Mittlefehldt, A. Nathues, E. Palmer, V. Reddy, C.A. Raymond, C.T. Russell, Distinctive space weathering on Vesta from regolith mixing processes. Nature **491**, 79–82 (2012). https://doi.org/10.1038/nature11534

Polanskey et al. (2021). This book

T.H. Prettyman, D.W. Mittlefehldt, N. Yamashita, D.J. Lawrence, A.W. Beck, W.C. Feldman, T.J. McCoy, H.Y. McSween, M.J. Toplis, T.N. Titus, P. Tricarico, R.C. Reedy, J.S. Hendricks, O. Forni, L. Le Corre, J.Y. Li, H. Mizzon, V. Reddy, C.A. Raymond, C.T. Russell, Elemental mapping by Dawn reveals exogenic H in Vesta's regolith. Science **338**, 242–246 (2012). https://doi.org/10.1126/science.1225354

T.H. Prettyman, D.W. Mittlefehldt, N. Yamashita, A.W. Beck, W.C. Feldman, J.S. Hendricks, D.J. Lawrence, T.J. McCoy, H.Y. McSween, P.N. Peplowski, R.C. Reedy, M.J. Toplis, L. Le Corre, H. Mizzon, V. Reddy, T.N. Titus, C.A. Raymond, C.T. Russell, Neutron absorption constraints on the composition of 4 Vesta. Meteorit. Planet. Sci. **48**, 2211–2236 (2013). https://doi.org/10.1111/maps.12244

L. Prockter, P. Thomas, M. Robinson, J. Joseph, A. Milne, B. Bussey, J. Veverka, A. Cheng, Surface expressions of structural features on Eros. Icarus **155**, 75–93 (2002). https://doi.org/10.1006/icar.2001.6770

V. Reddy, L. Le Corre, D.P. O'Brien, A. Nathues, E.A. Cloutis, D.D. Durda, W.F. Bottke, M.U. Bhatt, D. Nesvorny, D. Buczkowski, J.E.C. Scully, E.M. Palmer, H. Sierks, P.J. Mann, K.J. Becker, A.W. Beck, D. Mittlefehldt, J.-Y. Li, R. Gaskell, C.T. Russell, M.J. Gaffey, H.Y. McSween, T.B. McCord, J.-P. Combe, D. Blewett, Delivery of dark material to Vesta via carbonaceous chondritic impacts. Icarus **221**, 544–559 (2012). https://doi.org/10.1016/j.icarus.2012.08.011

J.E. Richardson, Cratering saturation and equilibrium: a new model looks at an old problem. Icarus **204**, 697–715 (2009)

M. Robinson, P. Thomas, J. Veverka et al., The nature of ponded deposits on Eros. Nature **413**, 396–400 (2001). https://doi.org/10.1038/35096518

C.T. Russell, C.A. Raymond, A. Coradini, H.Y. McSween, M.T. Zuber, A. Nathues, M.C. De Sanctis, R. Jaumann, A.S. Konopliv, F. Preusker, S.W. Asmar, R.S. Park, R. Gaskell, H.U. Keller, S. Mottola, T. Roatsch, J.E.C. Scully, D.E. Smith, P. Tricarico, M.J. Toplis, U.R. Christensen, W.C. Feldman, D.J. Lawrence, T.J. McCoy, T.H. Prettyman, R.C. Reedy, M.E. Sykes, T.N. Titus, Dawn at Vesta: testing the protoplanetary paradigm. Science **336**, 684–686 (2012). https://doi.org/10.1126/science.1219381

C.T. Russell, C.A. Raymond, E. Ammannito, D.L. Buczkowski, M.C. De Sanctis, H. Hiesinger, R. Jaumann, A.S. Konopliv, H.Y. McSween, A. Nathues, R.S. Park, C.M. Pieters, T.H. Prettyman, T.B. McCord, L.A. McFadden, S. Mottola, M.T. Zuber, S.P. Joy, C. Polanskey, M.D. Rayman, J.C. Castillo-Rogez, P.J. Chi, J.P. Combe, A. Ermakov, R.R. Fu, M. Hoffmann, Y.D. Jia, S.D. King, D.J. Lawrence, J.-Y. Li, S. Marchi, F. Preusker, T. Roatsch, O. Ruesch, P. Schenk, M.N. Villarreal, N. Yamashita, Dawn arrives at Ceres: exploration of a small, volatile-rich world. Science **353**, 1008–1010 (2016). https://doi.org/10.1126/science.aaf4219

P. Schenk, D.P. O'Brien, S. Marchi, R. Gaskell, F. Preusker, T. Roatsch, R. Jaumann, D. Buczkowski, T. McCord, H.Y. McSween, D. Williams, A. Yingst, C. Raymond, C. Russell, The geologically recent giant impact basins at Vesta's South pole. Science **336**, 694–697 (2012). https://doi.org/10.1126/science.1223272

P. Schenk, J. Castillo-Rogez, K.A. Otto, S. Marchi, D. O'Brien, M. Bland, K. Hughson, B. Schmidt, J. Scully, D. Buczkowski, K. Krohn, T. Hoogenboom, G. Kramer, V. Bray, A. Neesemann, H. Hiesinger, T. Platz, M.C. De Sanctis, S. Schroeder, L. Le Corre, L. McFadden, M. Sykes, C. Raymond, C.T. Russell, Compositional control on impact crater formation on mid-sized planetary bodies: dawn at Ceres and Vesta, Cassini at Saturn. Icarus **359**, 114343 (2021a). https://doi.org/10.1016/j.icarus.2021.114343

P.M. Schenk, A. Neesemann, S. Marchi, K. Otto, T. Hoogenboom, D.P. O'Brien, J. Castillo-Rogez, C.A. Raymond, C.T. Russell, G. Osinski, A young age of formation of Rheasilvia basin on Vesta from floor deformation patterns and crater counts. Meteorit. Planet. Sci. **57**, 22–47 (2021b). https://doi.org/10.1111/maps.13772

M.K. Shepard et al., Radar observations and shape model of asteroid 16 Psyche. Icarus **281**, 388–403 (2017)

M.K. Shepard, K. de Kleer, S. Cambioni, P.A. Taylor, A.K. Virkki, E.G. Rívera-Valentin, C. Rodriguez Sanchez-Vahamonde, L. Fernanda Zambrano-Marin, C. Magri, D. Dunham, J. Moore, M. Camarca, Asteroid 16 Psyche: shape, features, and global map. Planet. Sci. J. **2**, 125 (2021). https://doi.org/10.3847/PSJ/abfdba

H. Sierks, P. Lamy, C. Barbieri, D. Koschny, H. Rickman, R. Rodrigo, M.F. A'Hearn, F. Angrilli, M.A. Barucci, J.-L. Bertaux, I. Bertini, S. Besse, B. Carry, G. Cremonese, V. Da Deppo, B. Davidsson, S. Debei, M. De Cecco, J. De Leon, F. Ferri, S. Fornasier, M. Fulle, S.F. Hviid, R.W. Gaskell, O. Groussin, P. Gutierrez, W. Ip, L. Jorda, M. Kaasalainen, H.U. Keller, J. Knollenberg, R. Kramm, E. Kuhrt, M. Kuppers, L. Lara, M. Lazzarin, C. Leyrat, J.J.L. Moreno, S. Magrin, S. Marchi, F. Marzari, M. Massironi, H. Michalik, R. Moissl, G. Naletto, F. Preusker, L. Sabau, W. Sabolo, F. Scholten, C. Snodgrass, N. Thomas, C. Tubiana, P. Vernazza, J.-B. Vincent, K.-P. Wenzel, T. Andert, M. Patzold, B.P. Weiss, Images of asteroid 21 Lutetia: a remnant planetesimal from the early solar system. Science **334**, 487–490 (2011). https://doi.org/10.1126/science.1207325

D. Takir, V. Reddy, J.A. Sanchez, M.K. Shepard, J.P. Emery, Detection of water and/or hydroxyl on asteroid (16) Psyche. Astron. J. **153**, 31 (2016). https://doi.org/10.3847/1538-3881/153/1/31

P. Thomas, Large craters on small objects: occurrence, morphology, and effects. Icarus **142**, 89–96 (1999). https://doi.org/10.1006/icar.1999.6211

P.C. Thomas, L. Prockter, M. Robinson, J. Joseph, J. Veverka, Global structure of asteroid 433 Eros. Geophys. Res. Lett. **29**, 46-1–46-4 (2002). https://doi.org/10.1029/2001GL014599

A.L. Tonge, K.T. Ramesh, O. Barnouin, A model for impact-induced lineament formation and porosity growth on Eros. Icarus **266**, 76–87 (2016). https://doi.org/10.1016/j.icarus.2015.11.018

D. Turrini, J.P. Combe, T.B. McCord, N. Oklay, J.B. Vincent, T.H. Prettyman, H.Y. McSween, G.J. Consolmagno, M.C. De Sanctis, L. Le Corre, A. Longobardo, E. Palomba, C.T. Russell, The contamination of the surface of Vesta by impacts and the delivery of the dark material. Icarus **240**, 86–102 (2014). https://doi.org/10.1016/j.icarus.2014.02.021

P. Vernazza, L. Jorda, B. Carry, J. Hanuš, M. Marsset, M. Viikinkoski, F. Marchis, M. Brož, A. Drouard, T. Fusco, R. Fétick, M. Ferrais (HARISSA team), SPHERE unveils the true face of the largest main belt asteroids. Messenger **179**, 13–16 (2020). https://doi.org/10.18727/0722-6691/5187

M. Viikinkoski, P. Vernazza, J. Hanus, H.L. Coroller, K. Tazhenova, B. Carry, M. Marsset, A. Drouard, F. Marchis, R. Fetick, T. Fusco, J. Durech, M. Birlan, J. Berthier, P. Bartczak, C. Dumas, J. Castillo-Rogez, F. Cipriani, F. Colas, M. Ferrais, J. Grice, E. Jehin, L. Jorda, M. Kaasalainen, A. Kryszczynska, P. Lamy, A. Marciniak, T. Michalowski, P. Michel, M. Pajuelo, E. Podlewska-Gaca, T. Santana-Ros, P. Tanga, F. Vachier, A. Vigan, B. Warner, O. Witasse, B. Yang, (16) Psyche: a mesosiderite-like asteroid? Astron. Astrophys. **619**, L3 (2018). https://doi.org/10.1051/0004-6361/201834091

T.R. Watters, P.C. Thomas, M.S. Robinson, Thrust faults and the near-surface strength of asteroid 433 Eros. Geophys. Res. Lett. **38** (2011). https://doi.org/10.1029/2010GL045302

Weiss et al. (2021). This book

B.P. Weiss, S. Pedersen, I. Garrick-Bethell, S.T. Stewart, K.L. Louzada, A.C. Maloof, N.L. Swanson-Hysell, Paleomagnetism of impact spherules from Lonar crater, India and a test for impact-generated fields. Earth Planet. Sci. Lett. **298**(1–2), 66–76 (2010)

D.A. Williams, B.W. Denevi, D.W. Mittlefehldt, S.C. Mest, P.M. Schenk, R.A. Yingst, D.L. Buczkowski, J.E.C. Scully, W.B. Garry, T.B. McCord, J.-P. Combe, R. Jaumann, C.M. Pieters, A. Nathues, L. Le Corre, M. Hoffmann, V. Reddy, M. Schäfer, T. Roatsch, F. Preusker, S. Marchi, T. Kneissl, N. Schmedemann, G. Neukum, H. Hiesinger, M.C. De Sanctis, E. Ammannito, A. Frigeri, T.H. Prettyman, C.T. Russell, C.A. Raymond, The geology of the Marcia quadrangle of asteroid Vesta: assessing the effects of large, young craters. Icarus **244**, 74–88 (2014a). Special Issue: The Geology of Vesta. https://doi.org/10.1016/j.icarus.2014.01.033

D.A. Williams, R. Jaumann, H.Y. McSween Jr., C.A. Raymond, S. Marchi, N. Schmedemann, C.T. Russell, The chronostratigraphy of protoplanet Vesta. Icarus **244**, 158–165 (2014b). https://doi.org/10.1016/j.icarus.2014.06.027

D.A. Williams, T. Kneissl, A. Neesemann, S.C. Mest, E. Palomba, T. Platz, A. Nathues, A. Longobardo, J.E.C. Scully, A. Ermakov, R. Jaumann, D.L. Buczkowski, M. Schäfer, G. Thangjam, C.M. Pieters, T. Roatsch, F. Preusker, S. Marchi, N. Schmedemann, H. Hiesinger, A. Frigeri, C.A. Raymond, C.T. Russell, The geology of the Kerwan quadrangle of dwarf planet Ceres: investigating Ceres' oldest, largest impact basin. Icarus **316**, 99–113 (2018). The geologic mapping of Ceres. https://doi.org/10.1016/j.icarus.2017.08.015

Zuber et al. (2021). This book

Space Science Reviews (2025) 221:110
https://doi.org/10.1007/s11214-025-01240-z

The Psyche Light Elements Investigation

Thomas H. Prettyman[1] · David W. Mittlefehldt[2] · Erik I. Asphaug[3] · Richard P. Binzel[4] · Samuel W. Courville[5] · Linda T. Elkins-Tanton[5,6] · David J. Lawrence[7] · Simone Marchi[8] · José M.G. Merayo[9] · Timothy J. McCoy[10] · Benjamin P. Weiss[4]

Received: 20 March 2025 / Accepted: 27 October 2025 / Published online: 11 November 2025

Abstract
Light elements, such as C, S, Si, O, C, and H, are thought to be present in Earth's liquid-Fe outer core. These elements lower melting temperatures, thereby allowing the core to remain in liquid state at high pressure and influencing magnetic and geodynamic processes. However, the identity and abundance of the light elements in the cores of terrestrial planets and how they were delivered to these cores is not well known. The NASA Psyche mission will travel to and explore (16) Psyche, which may be the metal-rich core of a differentiated planetesimal exposed by collisional stripping. If so, the Psyche mission could provide a direct assessment of the light element content of an asteroidal core, allowing comparisons to the inferred composition of planetary cores and the parent bodies of the magmatic iron group meteorites. In particular, Earth's high-pressure core formed gradually (over ∼100 Myr), in a multistage process, under increasingly oxidizing conditions, whereas the cores of planetesimals formed quickly (within 10 Myr) at low pressure, likely in chemical equilibrium with their mantles. The trace element systematics and mineral composition of magmatic iron meteorites indicate the presence of C, P, and S in planetesimal cores prior to solidification. Such elements would have played a role in core dynamics, including dynamo generation. Their low solubility combined with the immiscibility of their mineral precipitates would have resulted in their separation from Fe upon crystallization and their eruption onto the surface of a stripped core (via ferrovolcanism). The Psyche spacecraft will detect their elemental, mineral, and magnetic signatures with the payload instruments, which include a Gamma Ray and Neutron Spectrometer, a Multispectral Imager, and a Magnetometer. Additional constraints on interior composition and processes influenced by light elements will be provided by Psyche's gravity and geomorphology investigations. We provide a brief introduction to the topic of light elements along with prospects for (16) Psyche. While we emphasize core formation processes, we also consider other possibilities for the origin and evolution of this metal-rich body.

Keywords Planetary cores · Asteroid (16) Psyche · NASA Psyche mission · Planetary cores · Light elements · Iron meteorites · Ferrovolcanism · Metal-rich chondrite meteorites

1 Introduction

The NASA Psyche spacecraft is on its way to explore the main-belt asteroid (16) Psyche (Elkins-Tanton et al. 2020, 2022; Dibb et al. 2024). (16) Psyche is thought to be metal-

Extended author information available on the last page of the article

rich, perhaps the stripped core of a differentiated protoplanet, a layered asteroid on which the surface was enriched in metal by some process, or an undifferentiated body consisting of primitive, metal-rich materials similar in composition to CB chondrites (Bierson et al. 2025). The Psyche spacecraft will acquire remote sensing data from close-proximity orbits. The observations will unravel many aspects of (16) Psyche's origins and evolution, and – if (16) Psyche is a core – provide a deeper understanding of processes that occurred in the cores of terrestrial planets and their building blocks.

The Psyche mission's gravity investigation will determine the asteroid's density and interior structure (Zuber et al. 2022). The Multispectral Imager will determine (16) Psyche's shape and morphology and place constraints on the relative surface age and mineralogy (Jaumann et al. 2022; Marchi et al. 2022; Dibb et al. 2023; Bell et al. 2025). The Gamma-Ray and Neutron Spectrometer (GRNS) will determine the elemental composition of (16) Psyche's regolith (Lawrence et al. 2025). And, the Magnetometer will constrain core formation processes by searching for a remanent magnetic field from an ancient dynamo (Weiss et al. 2023).

The Light Elements Working Group directly supports one of the primary mission objectives, Psyche Mission Objective C:

> "*Determine whether small metal bodies incorporate the same light elements as are expected in the Earth's high-pressure core.*"

The objectives of the Working Group overlap with and complement other aspects of the Psyche mission's investigation, including the "Is (16) Psyche a Core?" and "Oxidation-Reduction" Working Groups (Elkins-Tanton et al. 2022; McCoy et al. 2022). The Light Elements Working Group will provide a forum for the Psyche team and members of the planetary and Earth sciences community to discuss these topics. Goals of the working group include anticipation of light element measurement scenarios based on plausible hypotheses for origins and formation prior to the arrival of the spacecraft at (16) Psyche, and the discussion, analysis, and interpretation of light element data acquired during orbital operations.

The possibility that (16) Psyche is the residual metallic core of a differentiated planetesimal may lead to new insights into the interiors of the terrestrial planets. As Earth's deep interior is not accessible, our understanding of the evolution and current state of Earth's core is inferred from seismology, the geochemistry of the crust and mantle, cosmochemical constraints, knowledge of geologic processes supported by high pressure experiments (Dziewonski and Anderson 1981; Stevenson 1981; Fischer and McDonough 2025), as well as the study of the geomagnetic field over large time scales based on paleomagnetism (Smirnov et al. 2011; Zhou et al. 2022).

Earth's core is thought to consist of a liquid outer layer that is gradually freezing on top of a solid Fe-Ni-rich inner core. The latent heat released by this process drives mass convection in the outer core, which in turn sustains Earth's dynamo magnetic field. The outer core is a liquid iron alloy containing nickel and light elements, such as H, C, O, Si, P, S, and K (e.g., Hirose et al. 2021). The presence of light elements influences the thermodynamic properties of the fluid, lowering the melting temperature and allowing the outer core to remain in liquid state at high pressure (Stevenson 1981).

The protracted, multi-stage process of the formation of Earth's high-pressure core (tens to 100 Myr) contrasts with the likely path of smaller planetesimal cores, many of which would have formed more quickly (within a few Myr) and at low pressure (e.g., Bizzarro et al. 2005; Wood et al. 2006). The formation of Earth's core occurred continuously with accretion, with a portion of the metal derived from the infall of differentiated planetesimals (Anderson and Hanks 1972; Poirier 1994; Wood et al. 2006). Accreted metal equilibrated with a silicate-rich magma ocean on Earth, delivering siderophile elements and light elements to the core (Wade and Wood 2005; Wood et al. 2006). In the earliest stages, conditions in Earth's core would have resembled the low-pressure conditions expected during the growth of planetesimals. As interior pressures increased, crystallization of perovskite at the base of the magma ocean led to increased oxygen fugacity for metal-silicate equilibrium between the core and the mantle (Wade and Wood 2005). This resulted in oxidation of accreted iron, a transition to the addition of iron sulfide to the core, and ultimately the cessation of core growth (e.g., Newsom and Sims 1991; Wade and Wood 2005). Delivery of light elements to the Earth's core occurred in a multistage process with oxygen fugacity, pressure and temperature all increasing with time.

The high-pressure conditions during Earth's core formation contrast with planetesimal formation, which occurred at low pressure. Core-mantle partition coefficients derived from low pressure experiments, relevant to the formation of planetesimals, are much higher for many siderophile elements than inferred for Earth, which has an excess of siderophile elements in its mantle (e.g., Wood et al. 2006). Wood et al. (2006) argue that the excess indicates that the Earth's core could not have formed under conditions of chemical equilibrium, with constant pressure, temperature and oxygen fugacity, whereas core formation on protoplanets, like (4) Vesta, may well have occurred under those conditions (e.g., Toplis et al. 2013). Consequently, the distribution of light elements in planetesimal cores is expected to differ from that of Earth's core and other terrestrial planets (Fig. 1).

The Psyche mission provides an opportunity to directly examine the relict core of a planetesimal stripped of its outer silicate layers by collisions. If (16) Psyche is a core, then the planned observations will enable inferences of the role light elements played in the evolution of this small, metal-rich body. Observations of light elements and their associated minerals within metallic phases on the surface will provide constraints on the light element content of (16) Psyche's ancient, molten core. Additional constraints on present bulk composition will be provided by gravimetry. The detection of remanent magnetism from an ancient dynamo would provide additional clues to the role of light elements in core dynamics. Geomorphology will provide information on processes like volcanism that are controlled by light element content. The inferred pattern of light elements within (16) Psyche's ancient, molten core can be compared with that inferred for the cores of terrestrial planets, including Earth, as well as the parent bodies of the magmatic iron meteorites.

If (16) Psyche is not a stripped core, then the distribution of light elements would likely be different. For example, if (16) Psyche is a layered, differentiated asteroid, then the metal enrichment might have occurred via ferrovolcanism (Johnson et al. 2020; Courville et al. 2025). The metal-rich magmas could have been Fe-Ni-S cotectic liquids and could have a predictable S content (e.g., Kitakaze et al. 2016). Should (16) Psyche prove to be a metal-rich primitive body, the measurement of light elements could constrain the sulfide and graphite contents on the surface.

This paper provides a brief introduction to light elements in Earth's core, including a description of metal-silicate partitioning at high pressure and constraints on core composition (Sect. 2). Core formation on protoplanets differentiated by igneous processes is supported by meteorite studies (Sect. 3) and inferences about (4) Vesta by Dawn (Sect. 4). These sections support the Psyche mission's top-level objective to determine if small metal bodies include the same light elements as Earth's core, as well as providing background needed to understand possible paths for Psyche's origins and evolution and insights into what can be learned about small bodies from spacecraft remote sensing data. Prospects for observations at (16) Psyche and their interpretation are presented in Sect. 5. This work supplements several prior publications that assess possibilities for (16) Psyche's origins and evolution.

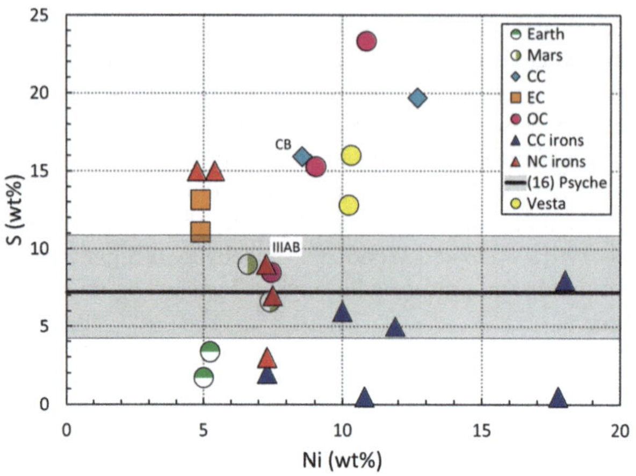

Fig. 1 Comparison of sulfur and Ni contents in planetary cores. Estimates include: the liquid outer core of Earth by Hirose et al. (2021); Earth's bulk core (Fischer and McDonough 2025); two recent estimates of the martian core (Yoshizaki and McDonough 2020; Khan et al. 2022); the molten cores of the parent bodies of magmatic iron meteorites (IIIAB iron core labeled) (Chabot and Zhang 2022; Zhang et al. 2024); (4) Vesta's core based on geochemical models, constraints from HED meteorites, and data acquired by Dawn (Toplis et al. 2013); a rough estimate for (16) Psyche's core from (this study, see the Appendix and Fig. 4); and bulk metal+sulfide of different chondrite types (CB chondrites labeled) (Jarosewich 1990, 2006). (The chondrites are effectively plotted where cores on their bodies would plot had they differentiated; they are thus directly comparable to the iron meteorite and planetary core compositions shown, but different than where bulk chondrites would plot.) The grey band on the (16) Psyche estimate is the uncertainty derived from the modeling done here. The magmatic iron meteorites are thought to have formed as molten materials cooled within their parent bodies, allowing equilibration with their silicate mantles. This process likely occurred on (4) Vesta; however, (16) Psyche's remnant core may have cooled in the absence of a mantle following collisional stripping of the mantle (e.g., Elkins-Tanton et al. 2020, 2022), possibly influencing the light element composition of the liquid. Differences between the light element content of Earth, which is thought to include S, Si, O, C, and H, and the cores of magmatic planetesimals, known to include S, O, C, and P from the iron meteorites, are a result of many factors, including accreted materials, the complexity of core-mantle evolution, and the pressure, temperature, and redox conditions during core growth. Abbreviations: CC – carbonaceous chondrites; EC – enstatite chondrites; OC – ordinary chondrites; CC irons – magmatic iron meteorites of the carbonaceous isotopic cluster; NC irons – magmatic irons of the noncarbonaceous isotopic cluster. For the chondrites, values are wt% of the metal+sulphide system, not of the bulk chondrite

2 Light Elements in Earth's Core

That Earth's interior is denser than its crust was inferred by gravimetry over 250 years ago (e.g., Mills 2022). The modern view of Earth's interior was informed primarily by the measurement of seismic waves (e.g., Lehmann 1936; Brush 1980; cf., Donini et al. 2019). These observations revealed the layered structure of the Earth, including the presence of a solid iron-rich inner core surrounded by a liquid outer core containing iron alloyed with nickel and light elements. Geochemical evidence for the core, and constraints on the partitioning of light elements between the mantle and the core, comes from the relative abundances of many siderophile and chalcophile elements in bulk silicate Earth, which are suppressed relative to lithophile elements when controlled for volatility (McDonough 2003). A combination of geological and cosmochemical constraints is needed to constrain the composition of the core (Fischer and McDonough 2025) (Fig. 1).

The core makes up 32% of Earth's mass and 16% of its volume (e.g., Hirose et al. 2021; Fischer and McDonough 2025). The outer core makes up approximately 96 vol.% of the total core (Hirose et al. 2021). Extreme conditions are found within the core, with estimated temperatures and pressures of 4300 K and 136 GPa, respectively at the core-mantle boundary and 6300 K and 229 GPa at the boundary between the inner and outer core (Dehant et al. 2022).

Birch (1952) noted that the density of the Earth's liquid outer core calculated from seismic data was slightly lower than expected for pure Fe at the temperatures and pressures of the deep interior; he posited that this was caused by light elements being contained within the molten Fe. Current estimates are that the liquid outer core density is reduced by 6–8%, and the solid inner core density is reduced by 4–5%, compared to those of pure Fe (e.g., Hirose et al. 2021). The question of which light elements are in the core continues to be an immensely active area of research, especially in the experimental determination of metal-silicate partitioning at high temperature and pressure. The literature on the topic is too vast to summarize here; we will focus on recent results and how they can inform us of the Earth's core composition. Commonly considered light element candidates are H, C, O, Si and S (e.g., Poirier 1994), to which we will add P.

Experimental studies have shown that molten metal-molten silicate partitioning occurs via two mechanisms depending on the element (M) (e.g., Chidester et al. 2022):

$$\text{exchange reactions:} \qquad MO_{n/2} + \tfrac{n}{2}\mathrm{Fe} \rightleftharpoons M + \tfrac{n}{2}\mathrm{FeO}; \text{ and} \tag{1}$$

$$\text{dissociation reactions:} \qquad MO_{n/2} \rightleftharpoons M + \tfrac{n}{2}\mathrm{O}. \tag{2}$$

Exchange reactions result in a decrease in the mass and density of the core because the light element is replacing Fe, while dissociation reactions increase core mass while decreasing density by adding light elements to the core. Wüstite dissociation is a special case because it adds both Fe and O to the core:

$$\mathrm{Fe}_{1-x}\mathrm{O} \rightleftharpoons (1-x)\,\mathrm{Fe} + \mathrm{O}. \tag{3}$$

Light element partitioning between molten silicates and molten metal can be compositionally dependent; for example, dissociation reactions (Eq. 2) that add O to the molten metal will depend in part on the cation M.

Beyond the simple compositional dependency inherent in Eq. 2, parameterizations of partitioning results show a dependency on the bulk composition of the molten silicate and molten metal. The most comprehensive treatment of this for O, S and Si is that of Chidester et al. (2022), who combined their results with the existing body of experimental partitioning data to parameterize partition coefficients for these light elements, the lithophile elements Mg, Al, and Ca, and the radioactive lithophile elements K, U and Th. An important feature of their parameterization is that they model the molten silicate using a multicomponent regular solution model. Other studies typically model the molten silicate using the commonly used non-bridging oxygens per tetrahedral cation (nbo/t) formalism (e.g., Ikuta et al. 2024). Chidester et al. (2022) noted that at mid-mantle pressures where metal-silicate equilibration might have occurred, network forming cations (Si, Al) transition from tetrahedral to octahedral structures in the melt which is not accounted for using the nbo/t parameter. Chidester et al. (2022) found that their parameterizations for partition coefficients gave good fits to the body of experimental data, and that excluding O, simple exchange reactions (Eq. 1) were for the most part sufficient to model element partitioning.

Element partitioning can have a pressure dependency that can reverse partitioning trends established at lower pressures. This has been demonstrated for P in which increasing siderophile behavior with increasing temperature at modest pressures reverses at pressures greater than ~15 GPa (Ikuta et al. 2024); these authors found the coefficient for pressure dependency at <15 GPa was strongly positive, while above that pressure it was strongly negative. This demonstrates the complexity of light element partitioning and hints at the uncertainties that may arise when experimental data are used to estimate the composition of planetary cores.

Hydrogen becomes significantly siderophile at higher temperatures and pressures. For example, Tagawa et al. (2021) found that for experiments run between 30 and 60 GPa pressure and 3100 and 4600 K, H partitioning can be modeled as an exchange reaction (Eq. 1) yielding $D_{\mathrm{H}}^{\mathrm{metal/silicate}} \geq 29$. These authors parameterized their results with respect to pressure and temperature but not specifically with respect to composition or melt structures. Using a range of literature estimates for fixed P and T of core formation and bulk silicate Earth water content, they estimated a range of H contents of the core of 0.32 to 0.61 wt% (Tagawa et al. 2021).

Considerable experimental work has been done on C partitioning between molten silicate and metal, but recent work has suggested that much of the earlier studies might not be germane to the core forming process. Grewal et al. (2021) found that the partitioning of C between molten silicate and metal is dependent on the C concentration of the system. Many of the previous experimental studies were done using graphite capsules and therefore the system was graphite-saturated; graphite-saturation is not thought to have pertained during core formation from a terrestrial magma ocean (Grewal et al. 2021). The experiments were done at modest pressure (3 GPa), a temperature of 1973 K and reducing conditions (2.86 to 6.67 log units below the iron-wüstite buffer). Under these conditions, C is strongly siderophile, with $D_{\mathrm{C}}^{\mathrm{metal/silicate}}$ of 840-8040 (Grewal et al. 2021). Thus, one can expect that C was quantitatively partitioned into the molten metallic phase during core formation.

The difficulty in using partition coefficients to model the light element inventory of the core lies in the little-constrained nature of the physical models. It is likely that core formation was an ongoing process during accretion of the Earth. At some point during accretion the proto-Earth's interior became hot enough for significant melting of the silicate and metallic systems and core formation began. As additional planetesimals were accreted, they too separated metal from silicate, growing the core. As the Earth grew, the temperatures and pressures at which planetesimals segregated generally increased; the light element contents of metallic material added to the core would have varied with time even if the accreted planetesimals were identical in composition.

Chidester et al. (2022) modeled core compositions for several accretion scenarios with the baseline case assuming the molten metal could be modeled as having equilibrated with molten silicate at a fixed temperature and pressure; this is often assumed to be equivalent to the average temperature and pressure of metal-silicate equilibration. Continuous growth models of both constant and varying compositions and assuming complete and partial equilibration were tested. Their modeling shows that wide ranges in light element contents can result, depending on the model assumptions. For example, they find a range of Si in the core of ~1-8 wt% for all the models tested, but this range can be narrowed to ~2-4 wt% by requiring the model to also reproduce the estimated Si content of the mantle. Similarly, their models that match the Fe content of the mantle and the mass of the core result in ≤1 wt% O in the core. None of the models of Chidester et al. (2022) result in significant S in the core; they found ≤0.2 wt%. However, their formalism treats S in the silicate melt as S^0, whereas it can have valences between S^{2-} and S^{6+} depending on the f_{O2}; additional partitioning experiments for S are required to determine the robustness of the S partitioning model (Chidester et al. 2022).

3 Cores of Differentiated Protoplanets

While no meteorites have been identified as originating from (16) Psyche, iron meteorites provide key clues to understanding the role of light elements in core formation in planetesimals in the early solar system. Iron meteorites present abundant evidence for light elements having been components of asteroidal cores; this evidence is in the form of minor and accessory minerals included within them (Buchwald et al. 1977). Troilite (FeS) nodules are a ubiquitous and obvious feature of irons. Typically less abundant but still ubiquitous is schreibersite ($(Fe,Ni)_3P$). Carbon phases – graphite, cohenite ($(Fe,Ni)_3C$), and haxonite ($(Fe,Ni)_{23}C_8$) – are not as common and are present in only a few iron meteorite groups. Chromite ($FeCr_2O_3$) is an accessory phase in many iron meteorites and daubréelite ($FeCr_2S_4$) is common in some groups. Silicon is present in some highly-reduced iron meteorites, sometimes forming perryite ($(Ni,Fe)_8(Si,P)_3$). Together, these attest to S, P, C, O, and Si having been components of the metallic melts which formed the iron meteorites.

The classification of iron meteorites is based on their chemical composition, originally using their Ni, Ga and Ge contents as developed through a series of papers by the Wasson group at UCLA (e.g., Wasson 1967), but now including other elements. The current classification system recognizes 13 distinct groups (Goldstein et al. 2009, with the exception that IIICD irons are not included within the IAB complex), where a group is defined based on there being five or more meteorites having compositional characteristics that allow them to be distinguished from other irons. However, compositional and isotopic evidence indicates the IIAB and IIG irons are likely from the same planetesimal core (Wasson and Choe 2009; Anand et al. 2024). Not all iron meteorites can be grouped and of the 1279 iron meteorites with known compositions, roughly 12.5% cannot be associated with defined groups; they are classified as ungrouped irons. In addition, the compositions of the metal phase of stony-iron meteorites can be compared to the iron meteorites and inferences can be drawn regarding their origin. There are three such groups: main-group pallasites; Eagle Station pallasites; and mesosiderites.

Most iron meteorite groups display negative linear trends on log(E) vs. log(Ni) or log(Au) plots where E is a refractory siderophile element such as Re, Os or Ir; the IIIAB irons are an example (Fig. 2). For those groups with many members, variations in refractory siderophile element contents are large, a factor of 1600 for IIIAB irons and 5000 for IIAB irons. These trends are consistent with fractional crystallization of metallic cores in which iron meteorites with high Ir and low Au contents represent early formed solids and those with low Ir and high Au contents late-forming solids (Scott 1972); there are 11 of these groups (counting IIAB and IIG separately) which are referred to as magmatic irons. The compositional characteristics of the metal in the two pallasite groups and the mesosiderites indicate that they also were likely derived from magmatic cores. Main-group pallasites and mesosiderites have metal compositions similar to those of IIIAB irons, but there is no conclusive evidence that they were formed in a single core (e.g., Hassanzadeh et al. 1990; Wasson and Choi 2003; Chabot and Zhang 2022). Similarly, there are compositional similarities between IIIF irons and Eagle Station pallasites, but it seems unlikely that they were formed on a single parent object (Hilton et al. 2020).

Iron meteorite group IIE shows a limited range in refractory siderophile element contents and there is no anti-correlation between these elements and Ni or Au (Fig. 2). The IAB irons do have a wide range in refractory siderophile element contents (Ir by a factor of 1000) but their correlations with Ni or Au are more diffuse than those of the IIIAB irons. The IAB and IIE irons contain abundant silicate inclusions (e.g., Goldstein et al. 2009). These two groups are not thought to have originated in magmatic cores. They are sometimes called "non-magmatic," although they clearly formed from metallic melts (Maurel et al. 2020).

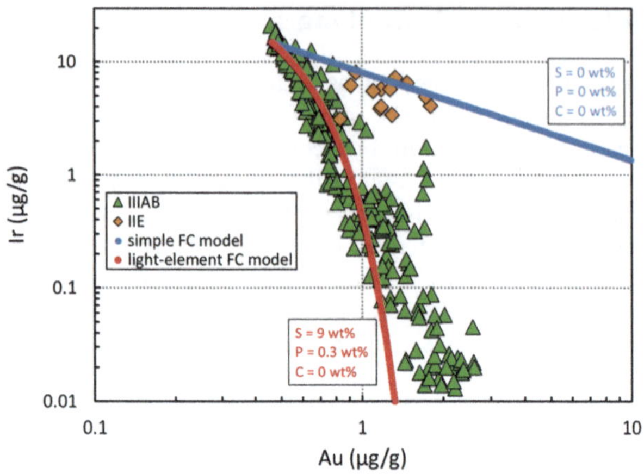

Fig. 2 Comparison of trace element systematics for exemplar magmatic (IIIAB) and non-magmatic (IIE) iron meteorite groups. A simple fractional crystallization model based on experimentally determined partition coefficients for the pure Fe-Ni system (Chabot et al. 2017) cannot reproduce the IIIAB iron trend. When partition coefficients that consider the effects of S and P are used in the model, the IIIAB data can be moderately well fit. IIIAB data and model are from Chabot and Zhang (2022); IIE data are from Malvin et al. (1984), Wasson and Wang (1986) and Wasson et al. (1989)

The ungrouped irons and pallasites present a problem when attempting to established how they were formed. Scott (1979) suggested that ungrouped irons represent probably at least 50 distinct parent objects; including the grouped irons and stony irons, the total would be >65. However, Scott (1979) knew of only 69 ungrouped irons, while currently there are 160; the number of distinct parent objects represented by ungrouped irons is undoubtedly much higher than 50, on the order of 115. Scott (1979) also noted that compositional trends among all ungrouped irons mimicked trends observed within individual magmatic iron meteorite groups and suggested that the ungrouped irons most likely had similar evolutionary histories; they were formed in fractionally crystallized cores. However, it is difficult to be confident of the petrogenetic history of one-off meteorites with such simple mineralogy. We can be confident that we have samples of 14 planetesimal cores, and an unknown, but probably large, number more are plausibly represented by the ungrouped irons and pallasites (e.g., Goldstein et al. 2009).

As noted in relation to Fig. 2, the light element contents of molten Fe,Ni affect the partitioning of trace siderophile elements between solid and liquid Fe,Ni (e.g., Jones and Drake 1983). This was ascribed to the light element as interacting with other elements in metallic systems (Jones and Malvin 1990). The latter study assumed that the light element either repelled or attracted other elements in the metallic liquid, the result being that those other elements either partitioned more strongly into solid metal or had greater solubility in the liquid metal.

Carbon and P also affect siderophile element partitioning in metallic systems (e.g., Chabot et al. 2017). In many cases, the effects of these elements are in the same direction as S (see Chabot et al. 2017, Fig. 3) and modeling assuming S only will slightly overestimate the core S content. Chabot and Zhang (2022) and Zhang et al. (2024) have modeled trace siderophile element trends for all magmatic iron groups to derive core P and S contents. They assumed that C was not significant, which is a safe assumption given that C < 0.2 wt% in iron meteorites (see Grewal and Asimow 2023) and calculated core compositions are similarly low (Buchwald et al. 1977; Hirschmann et al. 2021). The model results show that S contents in planetesimal cores varied from ∼0 to 15 wt% (Fig. 1).

Warren (2011) summarized the data on isotopic anomalies in the Earth, Moon, Mars, chondrites, achondrites and the lithic components of stony irons and demonstrated that these materials fall into one of two clusters, the carbonaceous (CC, so-called because it contains all carbonaceous chondrite groups) and the noncarbonaceous (NC). These two clusters were posited to represent separate reservoirs in the early Solar System, which were suggested to be the outer Solar System for the carbonaceous cluster, and the inner Solar System for the noncarbonaceous cluster. Warren (2011) further showed that main-group pallasites and mesosiderites are in the noncarbonaceous cluster, while the Eagle-Station pallasites are in the carbonaceous cluster. Kruijer et al. (2017) used Mo isotopic systematics to show that seven iron meteorite groups (including both non-magmatic groups) are members of the noncarbonaceous cluster; and five are members of the carbonaceous cluster. Kruijer et al. (2017) did not analyze samples of IIG irons, but Anand et al. (2024) have shown that these are members of the noncarbonaceous cluster. Thus, differentiation and core formation occurred in planetesimals of the carbonaceous and noncarbonaceous clusters, and if Warren's hypothesis is correct, in the inner and outer Solar System. The models of iron meteorite cores show that those from the carbonaceous cluster tend to have lower S contents than do those of the noncarbonaceous cluster (Fig. 1).

4 (4) Vesta – a Case Study

The NASA Psyche mission derives much of its scientific heritage from the NASA Dawn mission, which explored the two largest bodies in the main asteroid belt, (4) Vesta and (1) Ceres (Russell and Raymond 2011). These protoplanets have survived largely intact, providing a window into the earliest epoch of the Solar System. (4) Vesta is a differentiated, igneous body (520 km diameter), with a basaltic crust, ultramafic mantle, and core. Prior to Dawn, (4) Vesta was thought to be the source of the howardite, eucrite, and diogenite (HED) meteorites. Data acquired by Dawn at (4) Vesta strengthened the connection between (4) Vesta and these achondrites (McSween et al. 2013). Examining interpretations of (4) Vesta derived from Dawn mission data can give us a perspective on what might be learned from the Psyche mission, including through evaluation of the light element data.

The composition of (4) Vesta's surface as revealed by Dawn is like that of howardite, a brecciated mixture of basaltic eucrites, cumulate eucrites, and diogenites, representative of crustal materials. A portion of howardites – the so called "regolithic howardites" – contain H-bearing, carbonaceous chondrite inclusions (Cartwright et al. 2013), likely delivered from the infall of carbonaceous impactors to (4) Vesta's basaltic surface (McCord et al. 2012).

The detection of OH-bearing phases and H by Dawn's Visible to Infrared mapping spectrometer (VIR) and the Gamma Ray and Neutron Detector (GRaND), in association with low albedo regions on (4) Vesta, supports this hypothesis (De Sanctis et al. 2012; Prettyman et al. 2012). Exogenic carbonaceous chondrite materials are a minor component of (4) Vesta's regolith, less than about 5 wt.% based on the analysis of GRaND data (Mittlefehldt and Prettyman 2025). Much of the observed carbonaceous material may have been delivered by a small number of impacts (Reddy et al. 2012).

This assessment is supported by meteorite studies. The most common exogenous component found in HED polymict breccias is CM chondrite clasts (Zolensky et al. 1996), and similarly, excess siderophile and volatile element contents of howardites are best matched by CM chondrites (Chou et al. 1976). Regolithic howardites are thought to represent ancient, well-mixed regolith (Warren et al. 2009). Using the Ni, Ir and Au contents from a literature database (Mittlefehldt 2015) as a basis, regolithic howardites (Warren et al. 2009; Mittlefehldt et al. 2013) average ~4 wt% CM material. *If* HEDs are indeed from (4) Vesta, then this is an estimate for the vestan regolith.

Although large impact basins like Rheasilvia and Veneneia are present on (4) Vesta, evidence for endogenous olivine is lacking (Le Corre et al. 2015). This suggests a crust-mantle boundary deeper than the floors of these craters, with diogenitic plutonic rocks dominating the lower crust sampled by impacts (e.g., Barrat et al. 2010; Toplis et al. 2013; Vaci et al. 2021). A pure olivine mantle may not be present at depths expected if (4) Vesta formed a magma ocean (Ermakov et al. 2014). However, laboratory spectroscopic measurements of olivine-bearing and olivine-free diogenites indicate that 70:30 orthopyroxene:olivine mixtures might be difficult to resolve by the Dawn VIR instrument (Beck et al. 2013).

Dawn's gravity investigation revealed a central mass concentration, consistent with an Fe,Ni core, with a radius between 107 and 113 km (Russell et al. 2012), making up about 8% of (4) Vesta's volume and 17% of the mass (Ermakov et al. 2014). The presence of a core along with geochemical constraints from meteorite studies and remote sensing data by Dawn, confirms that (4) Vesta underwent igneous differentiation from a chondritic source. However, recent modeling of Dawn geophysical data indicates that the core is much smaller, ~50 km in radius, and has a lower density and smaller density contrast with the mantle than previously modeled; the core might be a metal-sulfide-olivine mixture (Park et al. 2025).

Efficient melting of accreted chondritic material on small bodies resulted from the presence of short-lived radioisotopes, like ^{26}Al and ^{60}Fe, present in the early solar system (McCoy et al. 2006; Zuber et al. 2011; Formisano et al. 2013). The details of the differentiation process, including the silicate melt fraction required for segregation of metal and the style of differentiation (e.g., formation of a magma ocean vs. limited partial melting and fractional crystallization), have been explored through the lens of the HED meteorites and geochemical modeling (e.g., Stolper 1977; Righter and Drake 1996, 1997; Ruzicka et al. 1997; Warren 1997; Mandler and Elkins-Tanton 2013).

Toplis et al. (2013) used mass balance constraints and thermodynamic modeling combined with data from HED meteorites and Dawn to constrain the bulk composition of (4) Vesta. Their analysis considered 12 chondritic compositions and found that (4) Vesta's bulk composition was best matched by a Na-depleted H-chondrite with the possible addition of a CM-like component. The crustal composition was taken to be that of the primitive, main group eucrite Juvinas, which was assumed to be in equilibrium with the olivine and pyroxene rich mantle (Fig. 3). Core and mantle composition was determined using an iterative mass balance procedure, resulting in estimates of the mass of metal (Fe,Ni) and sulfide (Fe,Ni-S) consistent with gravity constraints from Dawn (Fig. 1). A 60:40 mixture of olivine to pyroxene was estimated for the mantle, with 9% porosity required to match density constraints from gravimetry, possibly resulting from fracturing of the interior by the Rheasilvia impact (Park et al. 2014).

Fu et al. (2012) reported evidence that (4) Vesta formed an advecting metallic core with a magnetic dynamo, dissipating within about 6 Myr following the formation of the solar system. Remanent magnetism in Allan Hills A81001, a eucrite meteorite with an ^{40}Ar-^{39}Ar plateau age (cooling age) of 3.69 Ga, suggests a surface magnetic field of $<2\ \mu T$ was present at that time. The field strength is sufficient to withstand irradiation by solar wind, perhaps explaining limited weathering by solar wind ions. While the meteorite could not have been magnetized directly by the dynamo, the study provides evidence that a dynamo was once active in (4) Vesta's early history. An ancient dynamo on (4) Vesta is supported by thermophysical modeling by Formisano et al. (2016), who argued that an active core dynamo

Fig. 3 (4) Vesta from the outside in. On the left, the surface morphology observed by Dawn's Framing Camera is shown. Superimposed in blue is the distribution of H determined by Dawn's Gamma Ray and Neutron Detector (Prettyman et al. 2012), which is similar to the distribution of hydrated minerals found by Dawn's Visible to Infrared Mapping Spectrometer (De Sanctis et al. 2012) and low albedo regions on (4) Vesta's surface (McCord et al. 2012; Jaumann et al. 2014). The average concentration of H was 200 μg/g, which implies accreted carbonaceous chondrite materials comprise less than about 5% of (4) Vesta's crust (Mittlefehldt and Prettyman 2025). Otherwise, the elemental composition and mineralogy of (4) Vesta's surface is consistent with that of howardite, which consists of igneous minerals, including Fe-rich pyroxene and Na-poor plagioclase (e.g., Ammannito et al. 2013). On the right is the interior structure and composition inferred from gravimetry and thermodynamic modeling (after Toplis et al. 2013). (4) Vesta's original basaltic crust was assumed to be derived from melts with the same composition as the primitive, main group eucrite, Juvinas. The current crust is a complex mix of basalts and cumulative lithologies. The mantle consists of olivine and pyroxene, and the core is assumed to consist of Fe,Ni metal surrounded by a metal-sulfide cotectic mix

field with intensity up to 2600 μT could have persisted for 150-500 Myr with surface fields similar to that predicted by Fu et al. (2012). Dynamos may have been active on other small differentiated protoplanets, including the residual core of asteroid (16) Psyche following collisional stripping.

5 Prospects for Psyche

A strong case can be made that (4) Vesta is the parent body of most meteorites in the HED clan (McSween et al. 2013). This argument is based on telescopic observations of (4) Vesta, analyses of Dawn mission data, and meteorite studies. To make more definitive statements about the identity of the HED parent body, acquiring representative samples from Vesta may be required. The likely valid HED-Vesta connection supports many of the detailed and very specific geochemical inferences described in the foregoing section. On the other hand, no such connection exists for asteroid (16) Psyche. However, it's metallic character and other properties inferred from telescopic observations provide some hints as to what (16) Psyche might be (Elkins-Tanton et al. 2022; Bierson et al. 2025).

In this section, we briefly review the available observations of (16) Psyche and provide a partial list of hypotheses for (16) Psyche's origins and evolution. Observations planned for the Psyche spacecraft are described, focusing on light element signatures and how they would constrain aspects of (16) Psyche's magmatic evolution, if indeed (16) Psyche is a stripped core. If large portions of Psyche's surface are metallic, then minerals bearing light elements that precipitated from the melt may be present and could be detected by instruments on board the Psyche spacecraft. We note that an important outcome of the Psyche mission would be the identification of meteorites that best represent (16) Psyche's composition, such that the detailed information provided by meteorite studies can be brought to bear on the analysis of Psyche mission data.

5.1 Summary of Current Observations

Asteroid (16) Psyche is in the main asteroid belt (2.92 AU semi-major axis, 0.134 eccentricity, and 3.097° inclination, retrieved from the IAU Minor Planet Center on 15-Feb-2025). (16) Psyche's irregular shape is roughly an oblate spheroid, with ellipsoidal dimensions of $278 \times 238 \times 171$ km, with uncertainties of a few km (Shepard et al. 2021). The asteroid has a rotational period of about 4.2 hours with the spin axis tipped towards the plane of the ecliptic. Such high obliquity suggests (16) Psyche experienced at least one giant impact. It is difficult if not impossible to remove the mantle of a differentiated protoplanet to produce a body the size of (16) Psyche in a single collision (Bottke et al. 2005). Asphaug et al. (2006) argue that hit and run collisions during the pairwise accretion of planetesimals may have been relatively common. These collisions would have stripped protoplanets of their outer silicate layers, resulting in the exposure of some of their cores while asteroids were still accreting (Yang et al. 2007).

Earth based remote sensing observations of (16) Psyche reveal a body that is likely metal rich, with a density that is higher than silicate minerals, but much lower than iron (e.g., Elkins-Tanton et al. 2020; Cantillo et al. 2021; Elkins-Tanton et al. 2022). The asteroid's high radar albedo suggests the presence of Fe,Ni metal on or near the surface (e.g., Ostro et al. 1985; Shepard et al. 2017). Reflectance spectra bear similarities to iron meteorites; and analyses of infrared reflectance spectra suggest high metal content (e.g., 82.5% metal per Cantillo et al. 2021). Data from the Atacama Large Millimeter Array (ALMA) show that (16) Psyche's surface is heterogeneous, with some regions rich in metal and others rich in silicates (Cambioni et al. 2022). Rotationally-resolved spectra in the mid-infrared lack spectral features and is consistent with that expected for a metal-rich body (Arredondo et al. 2024). A metal-rich surface is also supported by thermal infrared interferometric observations (Matter et al. 2013).

The most recent published measurement of (16) Psyche's density is 4172 ± 145 kg/m^3 from high-precision astronomy (Farnocchia et al. 2024). This can be compared to the bulk density of (4) Vesta of 3459 kg/m^3 reported by Konopliv et al. (2014), which is understood to be a largely intact differentiated protoplanet. Like (4) Vesta, the surface of (16) Psyche is thought to include exogenic contamination based on observations of the 3 μm spectral region, which suggest the presence of OH- and water-bearing mineral phases (Takir et al. 2016). Jarmak et al. (2024) used James Webb Space Telescope data spanning 1.1-6.63 μm to estimate the hydrogen abundance, which was found to be in the 250-400 μg/g range, similar to that found on (4) Vesta (Prettyman et al. 2012).

5.2 Possible Origins and Evolution of (16) Psyche

Several hypotheses have been suggested to explain the foregoing observations. (16) Psyche may be the stripped core of a differentiated protoplanet like (4) Vesta (e.g., Burbine et al. 2024), as proposed by the Psyche mission (Elkins-Tanton et al. 2020); this is the homogeneous high-metal, high-porosity asteroid model of Bierson et al. (2025). This explanation has fallen in and out of favor, given improvements in density estimates and analyses of remote sensing data (e.g., Viikinkoski et al. 2018; Shepard et al. 2021). However, if (16) Psyche were a residual core, then materials enriched in sulfur may have been erupted onto the surface by ferrovolcanism (Johnson et al. 2020; Courville et al. 2025). This scenario involves outside-in cooling of the core which is expected for asteroid-sized bodies (Williams 2009), which may have been covered by a thin mantle. The formation of iron-nickel dendrites as the core cooled (Haack and Scott 1992) would produce pockets of melt, increasingly enriched in sulfur. It is hypothesized that excess pressures would result in the intrusion of the dense, sulfur-rich metallic magma through dykes into the mantle and/or eruption onto the surface (Johnson et al. 2020).

The fluid likely to be intruding or erupting would be an immiscible liquid that forms after fractional solidification begins in an Fe-Ni-S system. This immiscible liquid is stoichiometrically nearly identical to the mineral troilite (Jones and Drake 1983), though it would carry other light elements that are incompatible with iron-nickel metallic liquid (e.g., Corgne et al. 2008). The most likely ferrovolcanism, therefore, would not actually be metallic but rather sulfide-rich (Bercovici et al. 2022).

Curiously, this cooling scenario could also apply to a largely intact, differentiated silicate body that accreted primitive materials with high metal content, with a bulk composition like that of the enstatite or CB/CH chondrite meteorites; this is the internally layered asteroid model of Bierson et al. (2025). As suggested by Shepard et al. (2021), radar bright regions would correspond to local ferrovolcanic eruptions/intrusions through the silicate mantle and crust. As with Mercury, the silicate layers of such a planetesimal would be relatively thin compared to the core.

Alternatively, Viikinkoski et al. (2018) suggested (16) Psyche may be a mesosiderite-like asteroid, with a pyroxene rich outer layer. This hypothesis is genetically related to the possibility of stripped core in that the proposed collisional stripping could have removed the outer layers of a differentiated protoplanet, followed by the accretion of core and silicate materials. Kim and Hirabayashi (2022) modeled (16) Psyche as differentiated body with an outer silicate layer and core, providing constraints on minimum core size (up to 40% of the asteroid's volume). They argue that ferrovolcanic surface eruptions could be a source of metal-rich surface materials. However, the lack of FeO spectral features for (16) Psyche make a mesosiderite-like material on its surface unlikely (Elkins-Tanton et al. 2020).

The observed, coarse heterogeneity of the surface does not seem to fit the hypothesis that (16) Psyche is an undifferentiated body that accreted primitive, metal rich material (e.g., Elkins-Tanton et al. 2022); this is the homogeneous silicate-rich asteroid model of Bierson et al. (2025). If this model pertains, the bulk density could be consistent with that of the CB chondrites (average grain density 5650 kg m^{-3}), with added porosity (e.g., Macke et al. 2011; Dibb et al. 2023). The CB chondrite meteorites have higher, but variable, concentrations of metal compared to all other chondrite groups (25-80 vol.%; Weisberg et al. 2001; Krot et al. 2002, 2024) and the existence of these primitive meteorites suggests metal-rich planetesimals formed in the early solar system, possibly in close proximity to the Sun, with implications for the formation of the inner planets (Weisberg et al. 2001). Whether the parent body of the CB chondrite meteorites formed by direct condensation from the solar nebula or from the ejecta plume of a giant impact is unknown (Krot et al. 2005; Weyrauch et al. 2019), but recent work favors the latter (e.g., Krot et al. 2024).

5.3 Inferring Psyche's Sulfur Content Based on Current Observations

Assuming (16) Psyche is a residual core, it is possible to predict the range of bulk composition and porosity of the body (Elkins-Tanton et al. 2020, 2022). In this study, we inferred bulk properties, including the concentration of sulfur, using additional information from reflectance spectroscopy and meteorite studies.

We fit a model of core-mantle composition to the density of (16) Psyche reported by Farnocchia et al. (2024). The asteroid's bulk density was modeled as a mixture of void space (porosity) and solid consisting of Fe,Ni metal, with variable Ni content, troilite (FeS), and olivine, with variable amounts of forsterite and fayalite (Mg- and Fe-rich end members). Because the model has multiple parameters and there is only one observational constraint (density), the analysis can only describe the range of model parameters in the absence of other information.

However, some additional information is available from analyses of spectral reflectance observations in the visible to near infrared (0.35 to 2.5 μm) (Cantillo et al. 2021), which provide constraints on (16) Psyche's regolith composition. They found 82.5 wt.% metal, 7 wt.% low Fe pyroxene, and 10.5 wt.% carbonaceous chondrite by mass best fit the reflectance data. Their best fit to the density data resulted in a macroporosity of 35%, assuming (16) Psyche's bulk composition was identical to the surface.

Our analysis takes a slightly different approach. We assume that the carbonaceous chondrites contaminate the near surface and are not a significant bulk component, analogous to (4) Vesta (e.g., De Sanctis et al. 2012; McCord et al. 2012; Prettyman et al. 2012). However, the reflectance data along with meteorite studies can aid in coarsely constraining the four volume fractions (porosity, metal, troilite, and silicate) in our core-mantle model.

In our analysis, we assume that the ratio of silicates to metal determined by reflectance spectroscopy is the same as the bulk and that the density range for spanned by olivine end members forsterite and fayalite is representative of the bulk silicate composition. Our ingoing assumption for the concentration of sulfur in the solid bulk material is about 6 wt.%, with the range of that inferred for the magmatic irons (Fig. 1). This would imply that much of the sulfur in (16) Psyche's liquid core was retained. Finally, we assumed initially that the macroporosity is on the order of 10% to represent a core that was not substantially broken apart by collisional processes. The foregoing assumptions were used to construct a prior probability distribution for the model parameters.

We used Bayesian inference with Markov Chain Monte Carlo (MCMC) to sample model parameters that are consistent with (16) Psyche's density and uncertainty. The inferred parameters are shown in Fig. 4 and details of the inversion can be found in the Appendix. The results depend on the model assumptions (core-mantle composition) as well as the foregoing assumptions about (16) Psyche's bulk composition; however, the process illustrates the potential range of compositional parameters and their correlations, given what is currently known about (16) Psyche.

The density inversion results in a porosity (void volume fraction) of $0.31^{+0.05}_{-0.06}$, consistent with the value reported by (Cantillo et al. 2021), but larger than the upper bound of 0.2 discussed by Elkins-Tanton et al. (2020, 2022). Porosity is characteristic of the internal structure of small bodies, with coherent bodies having low porosity and bodies with high porosity consisting of rubble/aggregate held together loosely by gravity (e.g., Britt et al. 2002). Asteroids with greater than 30% porosity are likely rubble piles. Moreover, cold compaction in metal-rich rubble piles is thought to be minimal, leading to relatively high porosities (Zhang et al. 2022). As such, (16) Psyche may not be an intact core but instead may be re-accreted collisional debris. The abundance of Fe,Ni metal in the solid portion of the asteroid ranges from 55 wt.% to 76 wt.% (16th to 84th percentile), with a median of 66 wt.%. Such high metal fractions are consistent with core material, but are also consistent with CB chondrites once the 25-80 vol.% metal is converted to a wt.% basis.

The values of sulfur in the form of troilite in the solid asteroid material sampled by MCMC ranged from 4.2 wt.% to 11 wt.% (16th to 84th percentiles) with the median of 7.2 wt.% taken to be a representative value (Figs. 1 and 7). This is consistent with our ingoing

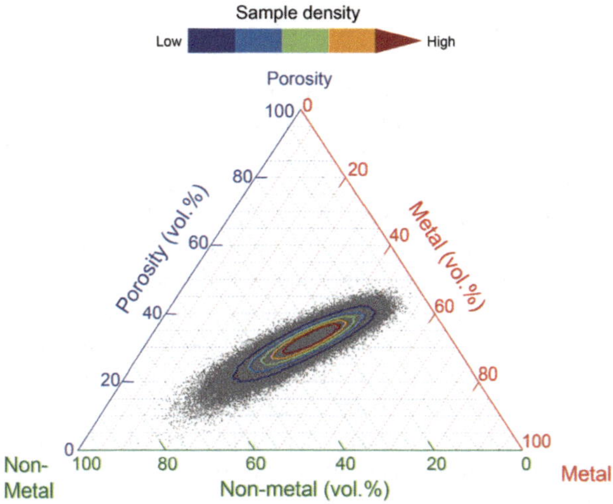

Fig. 4 The ternary plot shows the posterior distribution of parameters used to model Psyche's density. Bayesian inversion with Markov Chain Monte Carlo (MCMC) was used to fit the density of (16) Psyche given prior information from spectral reflectance observations and meteorite studies. The MCMC samples are shown as points, with sample density indicated by contours. The density of the asteroid was modeled as mixture of core material (Fe,Ni metal and troilite), lower mantle material (olivine), and void (porosity). The Ni fraction is allowed to vary, but was limited to <30 wt.%, consistent with the range for iron meteorites. The non-metal component is a mixture of troilite, the only S-bearing mineral in our model, and silicates. Silicates are modeled as a mixture of forsterite (Mg-olivine) and fayalite (Fe-olivine). The sampled points yield densities consistent with the most recently reported density of asteroid (16) Psyche using high-precision astrometry of 4172 ± 145 kg/m^3 (Farnocchia et al. 2024). The sampled distribution is consistent with the schema presented by Elkins-Tanton et al. (2020). Marginal distributions for each parameter are shown in the Appendix and the 16th, 50th, and 84th percentiles are used to estimate parameters and uncertainties. This approach gives, 0.35 ± 0.06, $0.32^{+0.11}_{-0.10}$, and $0.31^{+0.05}_{-0.06}$, respectively, for the volume fraction of metal, nonmetal, and void. The estimated sulfur content of the asteroid is $7.2^{+3.7}_{-3.0}$ wt.%. Pure troilite was not sampled (36 wt.% S)

assumptions (prior) and shows that (16) Psyche's density is consistent with models with sulfur in the range expected for the parent cores of the magmatic iron meteorites. Few analyses of S or modal calculations of troilite are available for CB chondrites and the data plotted in Fig. 1 are based on duplicate analyses of a single meteorite which have different metal and troilite contents (Jarosewich 1990, 2006). Limited data indicate that CB chondrites are depleted in volatile elements (see for example Krot et al. 2002),and likely contain much less S than we calculate for (16) Psyche. The data shown in Fig. 1 likely are higher in S than a truly representative CB chondrite would have. On a primitive, undifferentiated body, we would expect the S to be effectively uniformly distributed across the surface.

It would be difficult to distinguish the residual core scenario from that of an undifferentiated, metal-rich body based on density alone. The CB chondrite meteorites have 25-80 vol.% metal and minor amounts of troilite (Weisberg et al. 2001; Krot et al. 2002, 2024). Silicates are often in the form of coarse olivine-rich chondrules. The same mineral components were used in our mixing model. To fit the density data (Fig. 4 trend), an undifferentiated, CB-chondrite-like body would also require significant macroporosity.

5.4 Spacecraft Observations

Spacecraft observations will resolve much of the uncertainty in Earth-based remote sensing data. The Gravity Investigation will accurately determine the mass of (16) Psyche and provide constraints on the interior density structure (Zuber et al. 2022). The Multispectral Imager will reveal the detailed shape and geomorphology of the surface (Jaumann et al. 2022). Crater morphology will be a key observable of the Psyche mission (Marchi et al. 2022), as bulk composition and porosity will influence the size, shape, and distribution of ejecta. For instance, the presence of light elements may alter the mechanical properties of the target, thus cratering efficiency and morphology. Crater counting will be used to determine the relative chronology of surface units. Geomorphological evidence for magmatic volcanism is abundant on all of the terrestrial planets; although (4) Vesta's volcanic past is primarily informed by geochemical signatures. Perhaps (16) Psyche will reveal distinct morphologies associated with ferrovolcanism, such as vents, floodplains, domes, or dykes. If so, this would provide evidence that light elements, likely sulfur, played a role in (16) Psyche's evolution.

Psyche's Multispectral Imager includes seven narrow band filters spanning the visible to near infrared (0.4 to 1.1 μm) (Dibb et al. 2023; Bell et al. 2025). The filters provide information on the slope/shape of the reflectance spectrum (e.g., sensitive to the presence of Fe,Ni metal) and absorption features characteristic of specific minerals, including Fe-bearing silicates and sulfides. Laboratory studies of meteorites show that diagnostic sulfide minerals, like troilite (FeS), which is expected if (16) Psyche is a core, and oldhamite (CaS), which is not commonly found in iron meteorites, are detectable by the imager. Detection limits for sulfur-bearing minerals are expected to be on order of 10 wt.% (Elkins-Tanton et al. 2022). Enstatite chondrites contain oldhamite (Keil 1968), which may be present if (16) Psyche accreted primitive materials formed in a highly reducing environment (McCoy et al. 2022). Studies of the mineralogy of iron meteorites show that sulfur and phosphorus were present in the molten cores of protoplanets, with sulfur expected to be the primary light element. As such, phosphides may be present on (16) Psyche's surface; however, their abundance is expected to be much lower than that of the sulfides. Additional work is needed to determine if iron-bearing phosphides like schreibersite (Fe_3P) found in meteorites could be detected by the imager.

The Psyche mission's Gamma-Ray and Neutron Spectrometer (GRNS) includes a high resolution gamma-ray spectrometer (GRS) capable of resolving characteristic gamma rays produced by the decay of radioactive elements and excitation of nuclei by cosmic ray interactions (Lawrence et al. 2025). The GRNS will measure and map the elemental composition of (16) Psyche's surface on coarse spatial scales (on the order of 10s of km, compared to about 10 m expected for the Multispectral Imager). The instrument is sensitive to composition in the outermost few decimeters of the surface, including the abundance of radioelements, major elements in silicate minerals, Fe, Ni, and compositional macroparameters such as the thermal neutron absorption cross section and effective atomic mass of the materials that make up the regolith. The GRS is similar to the GRS flown on the MESSENGER mission to Mercury (Goldsten et al. 2007).

Sulfur was found to be abundant in surface materials of Mercury (up to 4 wt.%) as determined by X-Ray Spectroscopy (Nittler et al. 2011). The MESSENGER GRS detected gamma rays produced by neutron inelastic scattering and radiative neutron capture with ^{32}S (Evans et al. 2012). The former produced a doppler broadened peak at 2230 keV in a complex spectral region with gamma rays from reactions with H, Al, and the decay of ^{214}Bi (^{238}U decay chain) and ^{24}Na (multiple reaction channels). At Mercury, spectral fitting was required to determine the intensity of the 2230 keV peak separately from the other gamma rays in the region. The analysis of a low altitude spectrum acquired at an effective altitude of 0.4 Mercury body radii (1.88 steradians) and accumulated for 11.5 days resulted in about 10% relative uncertainty (precision only) in the intensity of the 2230 keV peak. This demonstrates the feasibility of the detection of S using high resolution gamma ray spectroscopy.

Using MESSENGER GRNS-like backgrounds and energy resolution sensitivity (e.g., Evans et al. 2017), along with conservative estimates of the cosmic-ray environment and a measurement altitude of 1-Psyche radii, it was determined that the GRS would quantify Psyche's S abundance at a relative precision of 20% if it were present with an abundance of 3 wt.% (Lawrence et al. 2025). Given that the Psyche GRS energy resolution performance is over a factor of two better than MESSENGER (Lawrence et al. 2024, 2025) and the expected range of orbit altitudes is ~0.5 to ~0.8 Psyche radii, we expect to achieve a S abundance sensitivity of less than 3 wt.%. Thus, quantification of S if present at the 4 wt.% level (about 10 wt.% troilite) is expected and some level of mapping may be possible, assuming that the current GRS performance remains consistent and the goals for proximity and accumulation time for the Psyche spacecraft's low altitude orbit are met. This orbit is currently planned to have an average altitude of about 0.65 body radii, close to that achieved by MESSENGER at Mercury and much lower than the altitudes sampled by Dawn at (4) Vesta (Evans et al. 2012; Prettyman et al. 2019; Polanskey et al. 2025). This orbit would enable mapping on coarse spatial scales (up to 200 km^2 areas) (Lawrence et al. 2025).

Lawrence et al. (2025) reported sensitivities for other light elements using gamma ray spectroscopy, including K and Si (with 200 mg/g and 2.5 wt.% detection limits when mapped on 200 km^2 areas). The average concentrations of carbon and phosphorus may be measurable over the entire surface sample by the orbit if they were present in concentrations greater than 1 wt.%.

The detection of sulfur in abundances higher than that found in chondritic meteorites accompanied by other observations indicating (16) Psyche is a core (Elkins-Tanton et al. 2020, 2022), would provide strong indication that sulfur was the primary light element in (16) Psyche's core, similar to that expected for the parent bodies of the magmatic iron meteorites. High sulfur concentrations may also be observed if (16) Psyche were accreted from primitive materials formed in a reducing environment, like the enstatite chondrites with 2-5.6 wt.% S (Mason 1966; Lodders and Fegley 1998). The droplets from which metal grains found in CB chondrites formed may have contained up to 3 wt.% S (Srinivasan et al. 2017).

The Psyche mission's Magnetometer investigation will determine if (16) Psyche has a remanent magnetic field produced by an ancient dynamo (Weiss et al. 2023). Attributes of the magnetic field as measured from orbit, including the magnetic moment, provide clues to the strength and duration of a dynamo, which in turn depend on the concentration of light elements present in the core. Detection of a magnetic field of about 6 nT in the lowest altitude mapping orbit would be consistent with this scenario. On the other hand, the lack of a remanent field would be indeterminate with respect to whether Psyche formed from a core since there are many reasons a core might not have generated a strong and long dynamo field. The interpretation of orbital data is made further ambiguous if (16) Psyche was substantially disrupted by collisions that could have reset magnetization due to shock heating. Finally, ancient ferrovolcanism may have left a magnetic signature that could be detected by Psyche's Magnetometer (Courville et al. 2025).

5.5 Light Elements in Metal-Rich Regions

Science requirements for the mission include determining the abundance of light elements in regions of Psyche's surface that appear in the metal phase and comparing the results to

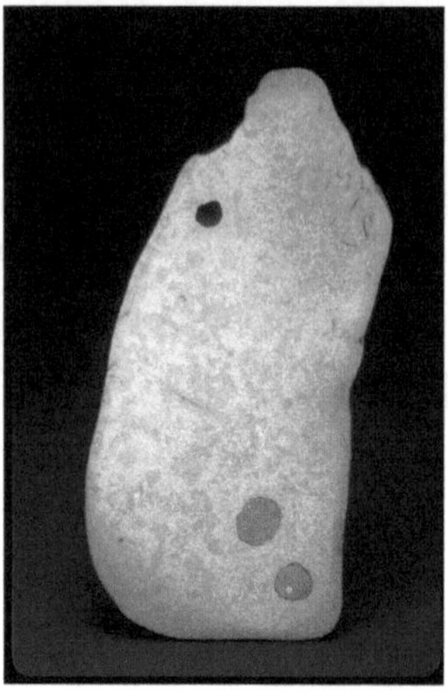

Fig. 5 Section of the IIC iron meteorite Ballinoo (image courtesy Smithsonian National Museum of Natural History). The section includes large, troilite nodules (medium yellowish grey inclusions, lower right) that precipitated as the molten Fe,Ni alloy crystallized. The dark circular feature at upper left is a hole in the sample. The dimensions of the section are 19 × 9 cm as listed in Buchwald (1975)

their inferred abundance in Earth's core. If metallic regions on the scale of the areas mappable by GRNS, about 200 km^2 are exposed on the surface of (16) Psyche, then the GRNS and Multispectral Imager could determine the composition and mineralogy of light-element precipitates in the metal phase. Regions larger than a few 10s of meters would be accessible to the Multispectral Imager. These observations would constrain the light element content of the ancient, molten core. In meteorites, such precipitates are in the form of nodules, most commonly of troilite (Fig. 5), but in some meteorite groups, troilite-graphite or graphite (Buchwald et al. 1977). Metallic regions smaller than the GRNS footprint could be analyzed using spatial unmixing guided by higher resolution mineralogy data acquired by the imager. This approach is model based and adds uncertainty but has been demonstrated on other missions (e.g., Prettyman et al. 2004; Hagerty et al. 2006).

During crystallization of metallic cores, the S and P contents of the melt phase will build up as Fe,Ni crystallizes. If the initial P content was low, the system will reach the Fe,Ni-FeS cotectic and metal plus troilite will crystallize. However, with sufficient P in the melt, the melt could enter a two-phase melt field in which immiscible S-rich, P-poor and P-rich, S-poor liquids are present (Jones and Drake 1983). Liquid immiscibility has been suggested to have occurred during crystallization of the IIAB (Wasson et al. 2007) and IIIAB (Ulff-Møller 1998) iron meteorite cores based on trace siderophile element distributions. However, in the case of the IIIAB irons, more recent geochemical modeling indicates that liquid immiscibility might not be necessary to explain the trace siderophile element distributions (Chabot and Zhang 2022; see Fig. 2, this study). Experimental studies show that multiple immiscible melts should be expected in the low pressure environments of asteroidal cores that contain S, C and O; one will be C-rich but low in S and O and the other will be S-rich, have high O but low C (Corgne et al. 2008).

If (16) Psyche is a largely intact core, the near surface of the asteroid could expose metallic regions containing crystalline products of one or more of the immiscible melts. Jones and Drake (1983) ran one experiment in the Fe-Ni-S-P system to estimate the effects of Ni on the boundaries of the two-liquid immiscibility field. The S-rich liquid contained 28.9 wt% S but only 0.16 wt% P, while the P-rich liquid contained 3.0 wt% S and 5.7 wt% P. It is not possible to make firm predictions for compositions of possible crystalline immiscible liquids on (16) Psyche, but using the experimental results as a guide, the S content might be in the 10-20 wt% range while the P content could be a few wt%. It is difficult to predict what might be found on the surface of the asteroid. If crystallization occurred at the core-mantle boundary or if the liquids erupted onto the surface (Bercovici et al. 2019), then high concentrations of S and P might be distributed on scales that can be sensed by the instruments. Absent experiments in more complex systems, predictions as to the number of immiscible melts that would form if C, O, P and S were all present in significant amounts cannot be made.

6 Summary

Light elements play an important role in the dynamics of planetary cores, including Earth's core and the ancient cores of differentiated protoplanets. The addition of light elements to molten Fe,Ni metal, changes the properties of the fluid, lowering melting temperatures. This contributes to the formation and persistence of magnetic dynamos. Despite advances in seismometry and high-pressure experimental methods, including diamond anvil and shock wave experiments, the light element content of Earth's liquid outer core is still uncertain. What is known is that light elements account for roughly 10 wt.% of the outer core, with S, Si, and O as likely primary light elements with the possibility of smaller amounts of C and H.

The protracted, and complex evolution of Earth's active core contrasts with that of protoplanets. (4) Vesta and the parent bodies of the magmatic iron meteorites formed quickly (within about 10 Myr), with light elements partitioned into the core in chemical equilibrium with their mantles. The mineralogy of magmatic iron meteorites shows that C, O, S, and P were present in the cores of protoplanets, with S as the dominant light element. Analyses of trace element systematics show that the molten cores represented by these meteorites would have contained a wide range of S (approaching 20 wt.% S in some cases). As these cores cooled, light elements would have been enriched in an increasingly buoyant melt, potentially resulting in ferrovolcanism, causing the intrusion of S-rich magmas into the mantles of intact protoplanets or the eruption of these materials onto the surface of residual molten cores stripped of their outer silicate layers by collisions.

Constraints from the NASA Dawn mission and studies of meteorites originating from (4) Vesta, show that the smallest terrestrial planet is a mostly intact, fully differentiated world with a Fe,Ni core containing S, an ultramafic mantle, and basaltic crust. Moreover, studies of remanent magnetism in meteorites suggest (4) Vesta sustained an ancient dynamo. Our knowledge of asteroid (16) Psyche is limited. Earth based observations constrain the asteroid's surface composition, mass, shape, and volume. These observations suggest an asteroid rich in metal, but with a lower bulk density than expected for a planetary core. This leaves open many possible pathways for (16) Psyche's origins and development. The presence of metal implies formation in a reducing environment. (16) Psyche may be a homogeneous body accreted from primitive, metal-rich materials, a residual core exposed by collisions, or a fully intact, differentiated body with a large core. As Dawn did with (4) Vesta, the Psyche mission will enable a detailed investigation of the geology of (16) Psyche.

The Psyche spacecraft's instruments are capable of characterizing sulfide mineralogy, determining the concentration of sulfur on (16) Psyche's surface, and measuring remanent magnetic fields if a dynamo was active. (16) Psyche's gravity investigation will provide constraints on interior structure and composition, placing limits on the thickness of silicate/sulfide layers and porosity. High resolution imagery will enable investigation of surface morphology, which may reveal evidence for ferrovolcanism.

The data acquired by the Psyche mission will enable further studies of the partitioning and segregation of light elements into and from the metal phase during accretion, core formation, and subsequent cooling. If (16) Psyche formed a core, then the Psyche mission will enable a first, direct investigation of core formation processes on a protoplanet, which will have implications for understanding the evolution of Earth's early core. The Psyche mission's Light Elements investigation will provide context for studies of iron meteorites and insights into the inner workings of the terrestrial planets.

Appendix

We used Bayesian inference with Markov Chain Monte Carlo (MCMC) to sample from the posterior distribution of the parameters of a density model of (16) Psyche. This approach samples parameters that are consistent with (16) Psyche's density given the observational uncertainties and prior information from reflectance spectroscopy, meteorite studies, and astrometric observations. The model was fitted to a recently reported value for (16) Psyche's density 4172 ± 145 kg/m^3 determined by high-precision astrometry (Farnocchia et al. 2024).

The density model considered a mixture of void (porosity), metal, troilite (FeS), and silicates. The metal fraction was represented by an Fe,Ni alloy free of impurities, with Ni allowed to vary up to 30 wt.%, consistent with the range for iron meteorites. The silicate component was modeled as olivine, with a linear mixture of forsterite (Mg) and fayalite (Fe) end members. This gives 6 parameters, including 4 for the metal, non-metal, and void fractions, and an additional 2 for the weight fraction of Ni in metal and the weight fraction of forsterite in olivine. Component densities were assumed to be 7870, 8900, 4600, 3220, and 4390 kg/m^3 for iron metal, nickel metal, troilite, forsterite, and fayalite, respectively. Because porosity, metal, and non-metal fractions must sum to 1, the model effectively has 5 parameters, which were compared to a single observation. Since this system is underdetermined, a precise set of parameters that matches (16) Psyche's density cannot be found. However, prior information is available to guide estimates of (16) Psyche's bulk composition.

Cantillo et al. (2021) constrained (16) Psyche's regolith composition using visible to near infrared spectra (0.35-2.5 μm), finding a best fit for 82.5% metal, 7% low Fe pyroxene, and 10.5% carbonaceous chondrite by mass. Their density inversion estimated a porosity of 35%, assuming the asteroid's bulk composition was the same as the surface. In this study, we assumed that carbonaceous materials on (16) Psyche are accreted asteroid debris that contaminate the near surface and do not make up an appreciable portion of the bulk. Our ad hoc model includes only core material and silicates modeled as olivine, a mixture of forsterite and fayalite end members. Our initial guess at the silicate to metal fraction was taken to be the same as pyroxene to metal ratio found by Cantillo et al. (2021). Baer et al. (2011) estimated the porosity of 26 asteroids based on astrometric observations. They found that asteroids with radii below 300 km have a wide range of bulk porosities, ranging from about 5% to 60%, with a reported value for (16) Psyche of 10%. For our inversion, we started with the assumption that (16) Psyche's porosity was 10% to represent a core that was exposed while it was still partially molten vs. a reaccreted rubble pile formed by the disruption of a solid body, which would be expected to have high porosity. We assumed the volume fraction of troilite was 0.2. Given the assumed porosity and silicate to metal ratio, this corresponds to roughly 6 wt.% S, within the range found in magmatic meteorites shown in Fig. 1.

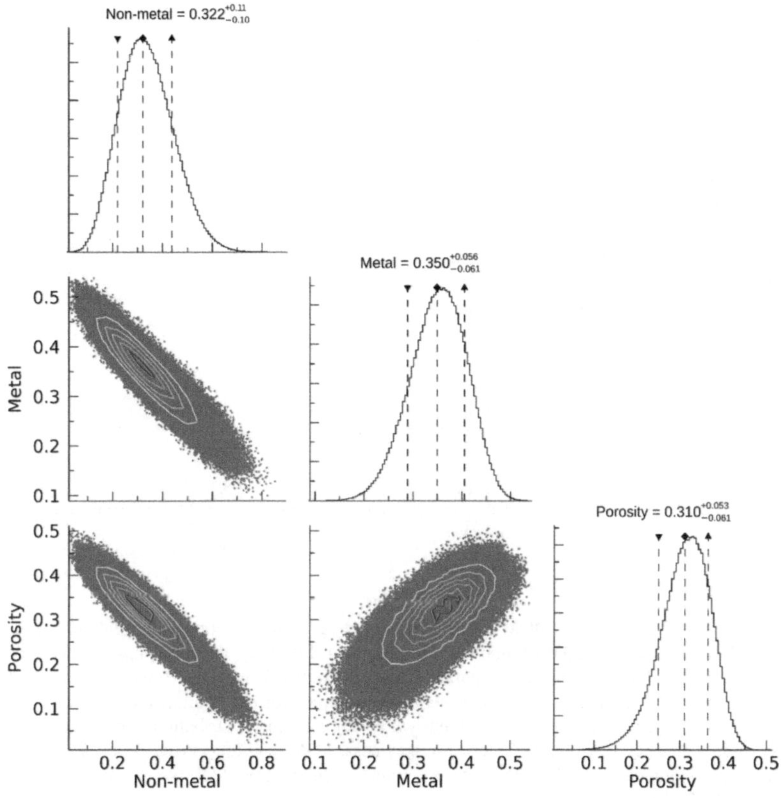

Fig. 6 Corner plot summarizing MCMC samples of the non-metal, metal, and porosity fractions and their correlations. The MCMC samples are the individual points in the correlation charts below the diagonal. The diagonal contains the marginal distributions for the parameters, which we use to estimate their values. Each marginal distribution is marked with the 16th (triangle down), 50th (diamond), and 84th (triangle up) percentiles. The range between the 16th and 84th percentiles provides a measure of uncertainty, with the median taken to be a representative value. Parameter estimates, which appear above each marginal distribution, use the standard notation for asymmetric errors, with the upper uncertainty given by the difference between the 84th percentile and the median, and the lower uncertainty given by the difference between the median and the 16th percentile

Given the assumed porosity, silicate-to-metal mass fraction, and troilite volume fraction, we constructed a Dirichlet prior (e.g., Gelman et al. 2013) for the volume fractions of core and mantle materials. We set the concentration parameters to $2 \times [1.0, 5.8, 2.0, 1.2]$, corresponding to the porosity and volume fractions of metal, troilite, and olivine, respectively. These parameters establish the expected proportions of the components, while their magnitudes control the strength of the prior. The parameters selected resulted in a broad prior distribution skewed towards high metal fractions.

The results of the inversion are shown in Figs. 6, 7, and Table 1. Figure 6 is a corner plot summarizing the distribution of metal and non-metal fractions and porosity. For example, to fit (16) Psyche's density, an increase in metal fraction may require an increase in porosity as indicated by the sample correlation (metal vs. porosity). The marginal distributions shown on the diagonal of the corner plot in Fig. 6 and in Fig. 7 are used to determine parameter confidence intervals (16th to 84th percentiles) and representative values (median). These give

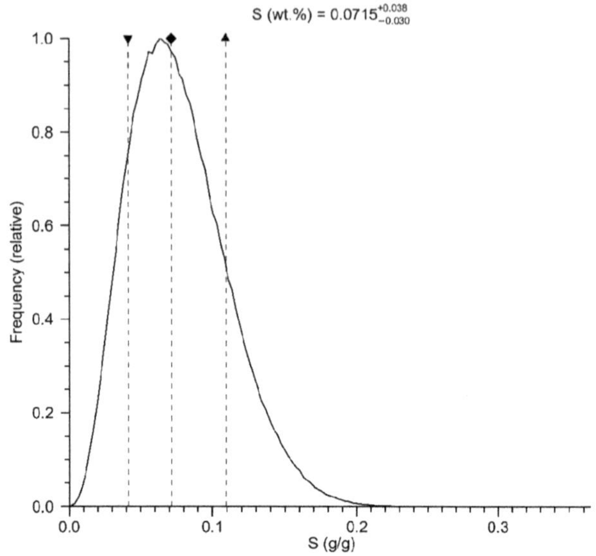

Fig. 7 Marginal distribution for the mass fraction of S (from FeS) in the non-porous asteroid material. The marginal distribution of S is marked with the 16th (triangle down), 50th (diamond), and 84th percentiles (triangle up). The range between the 16th and 84th percentiles, which we use as an estimate of uncertainty, is 4.2 wt.% to 11 wt.% with the median of 7.2 wt.% taken to be a representative value

Table 1 Summary of results of the MCMC inversion, including volume fractions and solid phase parameters. For individual MCMC samples, the volume fractions sum to 1. Similarly, the solid phase components, Fe,Ni metal, olivine, and troilite, are treated as separate phases, such that their weight fractions sum to 1 for each sample. Percentiles for the marginal distributions of the samples for each parameter are shown. The 16th, 50th (median), and 84th percentiles are provided, which for a normal distribution correspond to the -1σ, mean, and $+1\sigma$ values (V). A figure of merit for skewness, FOM $= (V_{84} - V_{50})/(V_{50} - V_{16})$, is provided as an indicator of symmetry (FOM $=1$ indicates a symmetric distribution; FOM>1 skewed to the right of the median; and FOM<1 skewed left)

	Percentiles			Skewness
	16th	50th	84th	FOM
Volume fractions				
Porosity	0.250	0.310	0.363	0.87
Fe,Ni metal	0.289	0.350	0.405	0.92
Troilite	0.105	0.180	0.276	1.26
Olivine	0.053	0.118	0.213	1.47
Density (kg/m^3)	4088	4231	4374	1.00
Solid phase				
Fe,Ni metal (g/g)	0.55	0.66	0.76	0.90
Olivine (g/g)	0.047	0.10	0.19	1.45
Troilite (g/g)	0.11	0.20	0.30	1.25
S (g/g)	0.042	0.072	0.11	1.25

broad ranges for the distribution of model parameters. The marginal distribution of sulfur in troilite (FeS) is shown in Fig. 7. Confidence intervals for the volume mixing fractions and solid phase weight fractions are shown in Table 1. The median volume fractions for MCMC samples of the posterior distribution were proportional to [2.1, 2.3, 1.6, 1]. In order to fit

Psyche's density, the inversion favored higher porosity and lower metal volume fractions than assumed in formulating the prior.

MCMC methods The differential evolution Markov Chain Monte Carlo (DE-MC) algorithm of Nelson et al. (2014) was used to draw samples given the log likelihood of the posterior distribution (Ter Braak 2006). The DE-MC algorithm operates on an ensemble of walkers. Jumps are determined by fixed multiples of differences between randomly selected parameter vectors, and the acceptance probability is determined using the Metropolis ratio, ensuring detailed balance with respect to the posterior distribution. The posterior distribution is proportional to the product of the likelihood and the probability density function for the prior. A Dirichlet distribution was used as the prior for the volume mixing fractions, from which the initial population of walkers was selected. The uncertainty in the density observation was modeled as a normal distribution with the quoted 1-sigma uncertainty. Solutions with parameter values outside their allowable range [0,1] were excluded, and the sum of the core/mantle material volume fractions was constrained to 1. The Markov Chain included 1000 walkers and sampled 20,000 steps, with an acceptance fraction of 25%. Convergence of the Markov Chain was assessed using batch means, with the autocorrelation time of 37(10), allowing most of the 20 million samples to be used following burn in. The reconstructed density was $4231 \pm 143\ kg/m^3$, consistent with the observational density constraint.

Acknowledgements The authors thank the reviewers for their thorough and constructive comments, which significantly improved the manuscript. The Light Elements investigation would not be possible without the dedication and perseverance of the entire Psyche team.

Funding Information This work is supported by NASA contract NNM16AA09, "Psyche: Journey to a Metal World." DWM's participation was supported as a Heritage Fellow at the Lunar and Planetary Institute. The LPI is operated by the Universities Space Research Association. This is LPI Contribution number 3154.

Declarations

Competing Interests The authors declare no competing interests. LTET, a guest editor of this collection, was not involved in the review of this manuscript.

References

Ammannito E, De Sanctis MC, Capaccioni F, Teresa Capria M, Carraro F, Combe J-P, Fonte S, Frigeri A, Joy SP, Longobardo A, Magni G, Marchi S, McCord TB, McFadden LA, McSween HY, Palomba E, Pieters CM, Polanskey CA, Raymond CA, Sunshine JM, Tosi F, Zambon F, Russell CT (2013) Vestan lithologies mapped by the visual and infrared spectrometer on Dawn. Meteorit Planet Sci 48(11):2185–2198. https://doi.org/10.1111/maps.12192

Anand A, Spitzer F, Hopp T, Windmill R, Kruttasch P, Burkhardt C, Dauphas N, Greenwood R, Hofmann B, Mezger K, Kleine T (2024) Isotopic evidence for a common parent body of IIG and IIAB iron meteorites. Geochim Cosmochim Acta 382:118–127. https://doi.org/10.1016/j.gca.2024.07.025

Anderson DL, Hanks TC (1972) Formation of the Earth's core. Nature 237(5355):387–388. https://doi.org/10.1038/237387a0

Arredondo A, McAdam MM, Becker TM, Elkins-Tanton L, Landsman Z, Müller T (2024) Rotationally resolved mid-infrared spectroscopy of (16) Psyche. Planet Sci J 5(2):33. https://doi.org/10.3847/PSJ/ad16ec

Asphaug E, Agnor CB, Williams Q (2006) Hit-and-run planetary collisions. Nature 439(7073):155–160. https://doi.org/10.1038/nature04311

Baer J, Chesley SR, Matson RD (2011) Astrometric masses of 26 asteroids and observations on asteroid porosity. Astron J 141(5):143. https://doi.org/10.1088/0004-6256/141/5/143

Barrat J-A, Yamaguchi A, Zanda B, Bollinger C, Bohn M (2010) Relative chronology of crust formation on asteroid Vesta: insights from the geochemistry of diogenites. Geochim Cosmochim Acta 74(21):6218–6231. https://doi.org/10.1016/j.gca.2010.07.028

Beck AW, McCoy TJ, Sunshine JM, Viviano CE, Corrigan CM, Hiroi T, Mayne RG (2013) Challenges in detecting olivine on the surface of 4 Vesta. Meteorit Planet Sci 48(11):2155–2165. https://doi.org/10.1111/maps.12160

Bell JF, Ravine MA, Caplinger MA, Schaffner JA, Brylow SM, Clark MJ, Peckham DA, Otjens PT, Price GJ, Rowell T, Ravine JW, Laramee JD, Juergens RC, Morgan W, Parker AG, Williams DA, Winhold A, Dibb S, Cisneros E, Walworth M, Zigo H, Auchterlonie L, Warner N, Bates-Tarasewicz H, Amiri N, Polanskey C, Mastrodemos N, Park RS, Alonge NK, Jaumann R, Binzel RP, McCoy TJ, Martin MG, Arthur PA (2025) The Psyche multispectral imager investigation: characterizing the geology, topography, and multispectral properties of a metal-rich world. Space Sci Rev 221(4):47. https://doi.org/10.1007/s11214-025-01169-3

Bercovici H, Elkins-Tanton L, Schaefer L (2019) The effect of bulk composition on the behavior of sulfur during core formation. In: 50th Lunar and Planetary Science Conference. Abstract #1366

Bercovici HL, Elkins-Tanton LT, O'Rourke JG, Schaefer L (2022) The effects of bulk composition on planetesimal core sulfur content and size. Icarus 380:114976. https://doi.org/10.1016/j.icarus.2022.114976

Bierson CJ, Courville SW, Ermakov A, Elkins-Tanton LT, Wieczorek M, Park RS, Baijal N (2025) (16) Psyche's different possible formation scenarios and internal structures from current constraints. J Geophys Res Planets 130(4):e2024JE008640. https://doi.org/10.1029/2024JE008640

Birch F (1952) Elasticity and constitution of the Earth's interior. J Geophys Res 57(2):227–286. https://doi.org/10.1029/SP026p0031

Bizzarro M, Baker JA, Haack H, Lundgaard KL (2005) Rapid timescales for accretion and melting of differentiated planetesimals inferred from 26 Al- 26 Mg chronometry. Astrophys J 632(1):L41–L44. https://doi.org/10.1086/497638

Bottke WF, Durda DD, Nesvorný D, Jedicke R, Morbidelli A, Vokrouhlický D, Levison H (2005) The fossilized size distribution of the main asteroid belt. Icarus 175(1):111–140. https://doi.org/10.1016/j.icarus.2004.10.026

Britt DT, Yeomans D, Housen K, Consolmagno G (2002) Asteroid density, porosity, and structure. In: Bottke WF, Cellino A, Paolicchi P, Binzel R (eds) Asteroids III. University of Arizona Press, Tucson, pp 485–500

Brush SG (1980) Discovery of the Earth's core. Am J Phys 48(9):705–724. https://doi.org/10.1119/1.12026

Buchwald VF (1975) Handbook of iron meteorites. Their history, distribution, composition and structure. Center for Meteorite Studies, Arizona State University, University of California Press. https://hdl.handle.net/10524/33750

Buchwald VF, Horne JET, Dunham KC (1977) The mineralogy of iron meteorites. Philos Trans R Soc Lond Ser A, Math Phys Sci 286(1336):453–491. https://doi.org/10.1098/rsta.1977.0127

Burbine TH, Greenwood RC, Zhang B, Buchanan PC (2024) How many Vesta-like bodies existed in the asteroid belt? Meteorit Planet Sci 59(4):878–894. https://doi.org/10.1111/maps.14134

Cambioni S, de Kleer K, Shepard M (2022) The heterogeneous surface of asteroid (16) Psyche. J Geophys Res Planets 127(6):e2021JE007091. https://doi.org/10.1029/2021JE007091

Cantillo DC, Reddy V, Sharkey BNL, Pearson NA, Sanchez JA, Izawa MRM, Kareta T, Campbell TS, Chabra O (2021) Constraining the regolith composition of asteroid (16) Psyche via laboratory visible near-infrared spectroscopy. Planet Sci J 2(3):95. https://doi.org/10.3847/PSJ/abf63b

Cartwright JA, Ott U, Mittlefehldt DW, Herrin JS, Herrmann S, Mertzman SA, Mertzman KR, Peng ZX, Quinn JE (2013) The quest for regolithic howardites. Part 1: two trends uncovered using noble gases. Geochim Cosmochim Acta 105:395–421. https://doi.org/10.1016/j.gca.2012.11.047

Chabot NL, Zhang B (2022) A revised trapped melt model for iron meteorites applied to the IIIAB group. Meteorit Planet Sci 57(2):200–227. https://doi.org/10.1111/maps.13740

Chabot NL, Wollack EA, McDonough WF, Ash RD, Saslow SA (2017) Experimental determination of partitioning in the Fe-Ni system for applications to modeling meteoritic metals. Meteorit Planet Sci 52(6):1133–1145. https://doi.org/10.1111/maps.12864

Chidester BA, Lock SJ, Swadba KE, Rahman Z, Righter K, Campbell AJ (2022) The lithophile element budget of Earth's core. Geochem Geophys Geosyst 23(2):e2021GC009986. https://doi.org/10.1029/2021GC009986

Chou C-L, Boynton WV, Bild RW, Kimberlin J, Wasson JT (1976) Trace element evidence regarding a chondritic component in howardite meteorites. In: Proceedings of the lunar science conference, 7th, pp 3501–3518

Corgne A, Keshav S, Wood BJ, McDonough WF, Fei Y (2008) Metal–silicate partitioning and constraints on core composition and oxygen fugacity during Earth accretion. Geochim Cosmochim Acta 72(2):574–589. https://doi.org/10.1016/j.gca.2007.10.006

Courville SW, Sanderson HR, Bierson CJ, Elkins-Tanton LT, Oran R, O'Rourke JG, Russell CT, Weiss BP, Williams DA (2025) Ferromagmatic intrusions on asteroid (16) Psyche may be magnetized. J Geophys Res Planets 130(8):e2025JE009031. https://doi.org/10.1029/2025JE009031

De Sanctis MC, Combe JP, Ammannito E, Palomba E, Longobardo A, McCord TB, Marchi S, Capaccioni F, Capria MT, Mittlefehldt DW, Pieters CM, Sunshine J, Tosi F, Zambon F, Carraro F, Fonte S, Frigeri A, Magni G, Raymond CA, Russell CT, Turrini D (2012) Detection of widespread hydrated materials on Vesta by the VIR imaging spectrometer on board the Dawn mission. Astrophys J 758(2):L36. https://doi.org/10.1088/2041-8205/758/2/l36

Dehant V, Campuzano SA, De Santis A, van Westrenen W (2022) Structure, materials and processes in the Earth's core and mantle. Surv Geophys, 43(1):263–302. https://doi.org/10.1007/s10712-021-09684-y

Dibb SD, Bell JF III, Elkins-Tanton LT, Williams DA (2023) Visible to near-infrared reflectance spectroscopy of asteroid (16) Psyche: implications for the Psyche mission's science investigations. Earth Space Sci 10(1):e2022EA002694. https://doi.org/10.1029/2022EA002694

Dibb SD, Asphaug E, Bell JF, Binzel RP, Bottke WF, Cambioni S, Christoph JM, Elkins-Tanton LT, Jaumann R, Lawrence DJ, Oran R, O'Rourke JG, Polansky C, Weiss BP, Wieczorek M, Williams DA (2024) A post-launch summary of the science of NASA's Psyche mission. AGU Adv 5(2):e2023AV001077. https://doi.org/10.1029/2023AV001077

Donini A, Palomares-Ruiz S, Salvado J (2019) Neutrino tomography of Earth. Nat Phys 15(1):37–40. https://doi.org/10.1038/s41567-018-0319-1

Dziewonski AM, Anderson DL (1981) Preliminary reference Earth model. Phys Earth Planet Inter 25(4):297–356. https://doi.org/10.1016/0031-9201(81)90046-7

Elkins-Tanton LT, Asphaug E, Bell JF III, Bercovici H, Bills B, Binzel R, Bottke WF, Dibb S, Lawrence DJ, Marchi S, McCoy TJ, Oran R, Park RS, Peplowski PN, Polanskey CA, Prettyman TH, Russell CT, Schaefer L, Weiss BP, Wieczorek MA, Williams DA, Zuber MT (2020) Observations, meteorites, and models: a preflight assessment of the composition and formation of (16) Psyche. J Geophys Res Planets 125(3):e2019JE006296. https://doi.org/10.1029/2019JE006296

Elkins-Tanton LT, Asphaug E, Bell JF, Bierson CJ, Bills BG, Bottke WF, Courville SW, Dibb SD, Jun I, Lawrence DJ, Marchi S, McCoy TJ, Merayo JMG, Oran R, O'Rourke JG, Park RS, Peplowski PN, Prettyman TH, Raymond CA, Weiss BP, Wieczorek MA, Zuber MT (2022) Distinguishing the origin of asteroid (16) Psyche. Space Sci Rev 218(3):17. https://doi.org/10.1007/s11214-022-00880-9

Ermakov AI, Zuber MT, Smith DE, Raymond CA, Balmino G, Fu RR, Ivanov BA (2014) Constraints on Vesta's interior structure using gravity and shape models from the Dawn mission. Icarus 240:146–160. https://doi.org/10.1016/j.icarus.2014.05.015

Evans LG, Peplowski PN, Rhodes EA, Lawrence DJ, McCoy TJ, Nittler LR, Solomon SC, Sprague AL, Stockstill-Cahill KR, Starr RD, Weider SZ, Boynton WV, Hamara DK, Goldsten JO (2012) Major-element abundances on the surface of Mercury: Results from the MESSENGER Gamma-Ray Spectrometer. J Geophys Res Planets 117(E12). https://doi.org/10.1029/2012JE004178

Evans LG, Peplowski PN, Rhodes EA, Goldsten JO, Starr RD, Solomon SC (2017) The MESSENGER gamma-ray spectrometer: calibration and operations. Icarus 288:186–200. https://doi.org/10.1016/j.icarus.2017.01.022

Farnocchia D, Fuentes-Muñoz O, Park RS, Baer J, Chesley SR (2024) Mass, density, and radius of asteroid (16) Psyche from high-precision astrometry. Astron J 168(1):21. https://doi.org/10.3847/1538-3881/ad50ca

Fischer RA, McDonough WF (2025) Earth's core composition and core formation. In: Anbar A, Weis D (eds) Treatise on geochemistry, 3rd edn. Elsevier, Oxford, pp 17–71. https://doi.org/10.1016/B978-0-323-99762-1.00116-9

Formisano M, Federico C, Turrini D, Coradini A, Capaccioni F, De Sanctis MC, Pauselli C (2013) The heating history of Vesta and the onset of differentiation. Meteorit Planet Sci 48(11):2316–2332. https://doi.org/10.1111/maps.12134

Formisano M, Federico C, De Angelis S, De Sanctis MC, Magni G (2016) A core dynamo in Vesta? Mon Not R Astron Soc 458(1):695–707. https://doi.org/10.1093/mnras/stw337

Fu RR, Weiss BP, Shuster DL, Gattacceca J, Grove TL, Suavet C, Lima EA, Li L, Kuan AT (2012) An ancient core dynamo in asteroid Vesta. Science 338(6104):238. https://doi.org/10.1126/science.1225648

Gelman A, Carlin JB, Stern HS, Dunson DB, Vehtari A, Rubin DB (2013) Bayesian data analysis, 3rd edn. Chapman & Hall/CRC texts in statistical science, vol 106. Chapman and Hall/CRC, Boca Raton

Goldstein JI, Scott ERD, Chabot NL (2009) Iron meteorites: crystallization, thermal history, parent bodies, and origin. Geochemistry 69(4):293–325. https://doi.org/10.1016/j.chemer.2009.01.002

Goldsten JO, Rhodes EA, Boynton WV, Feldman WC, Lawrence DJ, Trombka JI, Smith DM, Evans LG, White J, Madden NW, Berg PC, Murphy GA, Gurnee RS, Strohbehn K, Williams BD, Schaefer ED, Monaco CA, Cork CP, Del Eckels J, Miller WO, Burks MT, Hagler LB, DeTeresa SJ, Witte MC (2007) The MESSENGER gamma-ray and neutron spectrometer. Space Sci Rev 131(1):339–391. https://doi.org/10.1007/s11214-007-9262-7

Grewal DS, Asimow PD (2023) Origin of the superchondritic carbon/nitrogen ratio of the bulk silicate Earth – an outlook from iron meteorites. Geochim Cosmochim Acta 344:146–159. https://doi.org/10.1016/j.gca.2023.01.012

Grewal DS, Dasgupta R, Aithala S (2021) The effect of carbon concentration on its core-mantle partitioning behavior in inner Solar System rocky bodies. Earth Planet Sci Lett 571:117090. https://doi.org/10.1016/j.epsl.2021.117090

Haack H, Scott ERD (1992) Asteroid core crystallization by inward dendritic growth. J Geophys Res Planets 97(E9):14727–14734. https://doi.org/10.1029/92JE01497

Hagerty JJ, Lawrence DJ, Hawke BR, Vaniman DT, Elphic RC, Feldman WC (2006) Refined thorium abundances for lunar red spots: Implications for evolved, nonmare volcanism on the Moon. J Geophys Res Planets 111(E6). https://doi.org/10.1029/2005JE002592

Hassanzadeh J, Rubin AE, Wasson JT (1990) Compositions of large metal nodules in mesosiderites: links to iron meteorite group IIIAB and the origin of mesosiderite subgroups. Geochim Cosmochim Acta 54(11):3197–3208. https://doi.org/10.1016/0016-7037(90)90134-7

Hilton CD, Ash RD, Walker RJ (2020) Crystallization histories of the group IIF iron meteorites and Eagle Station pallasites. Meteorit Planet Sci 55(12):2570–2586. https://doi.org/10.1111/maps.13602

Hirose K, Wood B, Vočadlo L (2021) Light elements in the Earth's core. Nat Rev, Earth Environ 2(9):645–658. https://doi.org/10.1038/s43017-021-00203-6

Hirschmann MM, Bergin EA, Blake GA, Ciesla FJ, Li J (2021) Early volatile depletion on planetesimals inferred from C–S systematics of iron meteorite parent bodies. Proc Natl Acad Sci USA 118(13):e2026779118. https://doi.org/10.1073/pnas.2026779118

Ikuta N, Sakamoto N, Tagawa S, Hirose K, Tsutsumi Y, Yokoo S, Yurimoto H (2024) Pressure dependence of metal–silicate partitioning explains the mantle phosphorus abundance. Sci Rep 14(1):1194. https://doi.org/10.1038/s41598-024-51662-y

Jarmak SG, Becker TM, Woodward CE, Honniball CI, Rivkin AS, McAdam MM, Landsman ZA, Cambioni S, Müller TG, Takir D, Retherford KD, Arredondo A, Elkins-Tanton LT (2024) Estimate of water and hydroxyl abundance on asteroid (16) Psyche from JWST data. Planet Sci J 5(8):183. https://doi.org/10.3847/PSJ/ad66b9

Jarosewich E (1990) Chemical analyses of meteorites: a compilation of stony and iron meteorite analyses. Meteoritics 25(4):323–337. https://doi.org/10.1111/j.1945-5100.1990.tb00717.x

Jarosewich E (2006) Chemical analyses of meteorites at the Smithsonian institution: an update. Meteorit Planet Sci 41(9):1381–1382. https://doi.org/10.1111/j.1945-5100.2006.tb00528.x

Jaumann R, Nass A, Otto K, Krohn K, Stephan K, McCord TB, Williams DA, Raymond CA, Blewett DT, Hiesinger H, Yingst RA, De Sanctis MC, Palomba E, Roatsch T, Matz KD, Preusker F, Scholten F, Russell CT (2014) The geological nature of dark material on Vesta and implications for the subsurface structure. Icarus 240:3–19. https://doi.org/10.1016/j.icarus.2014.04.035

Jaumann R, Bell JF, Polanskey CA, Raymond CA, Aspaugh E, Bercovici D, Bills BR, Binzel R, Bottke W, Christoph JM, Marchi S, Neesemann A, Otto K, Park RS, Preusker F, Roatsch T, Williams DA, Wieczorek MA, Zuber MT (2022) The Psyche topography and geomorphology investigation. Space Sci Rev 218(2):7. https://doi.org/10.1007/s11214-022-00874-7

Johnson BC, Sori MM, Evans AJ (2020) Ferrovolcanism on metal worlds and the origin of pallasites. Nat Astron 4(1):41–44. https://doi.org/10.1038/s41550-019-0885-x

Jones JH, Drake MJ (1983) Experimental investigations of trace element fractionation in iron meteorites, II: the influence of sulfur. Geochim Cosmochim Acta 47(7):1199–1209. https://doi.org/10.1016/0016-7037(83)90062-5

Jones JH, Malvin DJ (1990) A nonmetal interaction model for the segregation of trace metals during solidification of Fe-Ni-S, Fe-Ni-P, and Fe-Ni-S-P alloys. Metall Mater Trans B 21(4):697–706. https://doi.org/10.1007/BF02654248

Keil K (1968) Mineralogical and chemical relationships among enstatite chondrites. J Geophys Res 73(22):6945–6976. https://doi.org/10.1029/JB073i022p06945

Khan A, Sossi PA, Liebske C, Rivoldini A, Giardini D (2022) Geophysical and cosmochemical evidence for a volatile-rich Mars. Earth Planet Sci Lett 578:117330. https://doi.org/10.1016/j.epsl.2021.117330
Kim Y, Hirabayashi M (2022) A numerical approach using a finite element model to constrain the possible interior layout of (16) Psyche. Planet Sci J 3(5):122. https://doi.org/10.3847/PSJ/ac6b39
Kitakaze A, Machida T, Komatsu R (2016) Phase relations in the Fe–ni–s system from 875 to 650 °C. Can Mineral 54(5):1175–1186. https://doi.org/10.3749/canmin.1500087
Konopliv AS, Asmar SW, Park RS, Bills BG, Centinello F, Chamberlin AB, Ermakov A, Gaskell RW, Rambaux N, Raymond CA, Russell CT, Smith DE, Tricarico P, Zuber MT (2014) The Vesta gravity field, spin pole and rotation period, landmark positions, and ephemeris from the Dawn tracking and optical data. Icarus 240:103–117. https://doi.org/10.1016/j.icarus.2013.09.005
Krot AN, Meibom A, Weisberg MK, Keil K (2002) The CR chondrite clan: implications for early Solar System processes. Meteorit Planet Sci 37(11):1451–1490. https://doi.org/10.1111/j.1945-5100.2002.tb00805.x
Krot AN, Amelin Y, Cassen P, Meibom A (2005) Young chondrules in CB chondrites from a giant impact in the early Solar System. Nature 436(7053):989–992. https://doi.org/10.1038/nature03830
Krot AN, Nagashima K, Ivanova MA, Lauretta D, Libourel G, Johnson BC, Brenker FE, Hoffman V, Bizzarro M (2024) Mineralogy, petrology, and oxygen isotopic compositions of chondritic and achondritic lithologies in the anomalous CB carbonaceous chondrites Sierra Gorda 013 and Fountain Hills. Meteorit Planet Sci 59(4):754–777. https://doi.org/10.1111/maps.14072
Kruijer TS, Burkhardt C, Budde G, Kleine T (2017) Age of Jupiter inferred from the distinct genetics and formation times of meteorites. Proc Natl Acad Sci USA 114(26):6712–6716. https://doi.org/10.1073/pnas.1704461114
Lawrence D, Peplowski P, Goldsten J, Burks M, Espiritu R, Yokley Z, Alonge N, Bairstow S, De Soria-Santacruz Pich M, Polanskey C (2024) The Psyche gamma-ray and neutron spectrometer: "first light" measurements. In: 55th Lunar and Planetary Science Conference. Abstract #1992
Lawrence DJ, Goldsten JO, Peplowski PN et al (2025) The Psyche Gamma-Ray and Neutron Spectrometer. Space Sci Rev 221:78. https://doi.org/10.1007/s11214-025-01201-6
Le Corre L, Reddy V, Sanchez JA, Dunn T, Cloutis EA, Izawa MRM, Mann P, Nathues A (2015) Exploring exogenic sources for the olivine on asteroid (4) Vesta. Icarus 258:483–499. https://doi.org/10.1016/j.icarus.2015.01.018
Lehmann I (1936) P'. Publications du Bureau Central Seismologique International, Série A, Travaux Scientifique
Lodders K, Fegley B Jr (1998) The planetary scientist's companion. Oxford University Press, New York
Macke RJ, Consolmagno GJ, Britt DT (2011) Density, porosity, and magnetic susceptibility of carbonaceous chondrites. Meteorit Planet Sci 46(12):1842–1862. https://doi.org/10.1111/j.1945-5100.2011.01298.x
Malvin DJ, Wang D, Wasson JT (1984) Chemical classification of iron meteorites—X. Multielement studies of 43 irons, resolution of group IIIE from IIIAB, and evaluation of Cu as a taxonomic parameter. Geochim Cosmochim Acta 48(4):785–804. https://doi.org/10.1016/0016-7037(84)90101-7
Mandler BE, Elkins-Tanton LT (2013) The origin of eucrites, diogenites, and olivine diogenites: magma ocean crystallization and shallow magma chamber processes on Vesta. Meteorit Planet Sci 48(11):2333–2349. https://doi.org/10.1111/maps.12135
Marchi S, Asphaug E, Bell JF, Bottke WF, Jaumann R, Park RS, Polanskey CA, Prettyman TH, Williams DA, Binzel R, Oran R, Weiss B, Russell CT (2022) Determining the relative cratering ages of regions of Psyche's surface. Space Sci Rev 218(4):24. https://doi.org/10.1007/s11214-022-00891-6
Mason B (1966) The enstatite chondrites. Geochim Cosmochim Acta 30(1):23–39. https://doi.org/10.1016/0016-7037(66)90089-5
Matter A, Delbo M, Carry B, Ligori S (2013) Evidence of a metal-rich surface for the asteroid (16) Psyche from interferometric observations in the thermal infrared. Icarus 226(1):419–427. https://doi.org/10.1016/j.icarus.2013.06.004
Maurel C, Bryson JFJ, Lyons RJ, Ball MR, Chopdekar RV, Scholl A, Ciesla FJ, Bottke WF, Weiss BP (2020) Meteorite evidence for partial differentiation and protracted accretion of planetesimals. Sci Adv 6(30):eaba1303. https://doi.org/10.1126/sciadv.aba1303
McCord TB, Li JY, Combe JP, McSween HY, Jaumann R, Reddy V, Tosi F, Williams DA, Blewett DT, Turrini D, Palomba E, Pieters CM, De Sanctis MC, Ammannito E, Capria MT, Le Corre L, Longobardo A, Nathues A, Mittlefehldt DW, Schröder SE, Hiesinger H, Beck AW, Capaccioni F, Carsenty U, Keller HU, Denevi BW, Sunshine JM, Raymond CA, Russell CT (2012 Dark material on Vesta from the infall of carbonaceous volatile-rich material. Nature 491:83. https://doi.org/10.1038/nature11561
McCoy TJ, Mittlefehldt DW, Wilson L (2006) Asteroid differentiation. Meteorites and the early Solar System II. University of Arizona Press, Tucson. https://muse.jhu.edu/book/85537
McCoy TJ, Dibb SD, Peplowski PN, Maurel C, Bercovici HL, Corrigan CM, Bell JF, Weiss BP, Lawrence DJ, Wenkert DD, Prettyman TH, Elkins-Tanton LT (2022) Deciphering redox state for a metal-rich world. Space Sci Rev 218(2):6. https://doi.org/10.1007/s11214-022-00872-9

McDonough W (2003) Compositional model for the Earth's core. Treatise on geochemistry. Elsevier, Inc.
McSween HY, Binzel RP, De Sanctis MC, Ammannito E, Prettyman TH, Beck AW, Reddy V, Le Corre L, Gaffey MJ, McCord TB, Raymond CA, Russell CT (2013) Dawn; the Vesta-HED connection; and the geologic context for eucrites, diogenites, and howardites. Meteorit Planet Sci 48(11):2090–2104. https://doi.org/10.1111/maps.12108
Mills V (2022) A Song of Schiehallion. Royal Society Blog. https://royalsociety.org/blog/2022/08/song-of-schiehallion/
Mittlefehldt DW (2015) Asteroid (4) Vesta: I. The howardite-eucrite-diogenite (HED) clan of meteorites. Chem Erde 75(2):155–183. https://doi.org/10.1016/j.chemer.2014.08.002
Mittlefehldt DW, Prettyman TH (2025) Modeling Dawn mission GRaND data to evaluate 4 Vesta as the mesosiderite parent asteroid. In: 56th Lunar and Planetary Science Conference. Abstract #2472
Mittlefehldt DW, Herrin JS, Quinn JE, Mertzman SA, Cartwright JA, Mertzman KR, Peng ZX (2013) Composition and petrology of HED polymict breccias: the regolith of (4) Vesta. Meteorit Planet Sci 48(11):2105–2134. https://doi.org/10.1111/maps.12182
Nelson B, Ford EB, Payne MJ (2014) Run DMC: an efficient, parallel code for analyzing radial velocity observations using N-body integrations and differential evolution Markov Chain Monte Carlo. Astrophys J Suppl Ser 210(1):11. https://doi.org/10.1088/0067-0049/210/1/11
Newsom HE, Sims KWW (1991) Core formation during early accretion of the Earth. Science 252(5008):926–933. https://doi.org/10.1126/science.252.5008.926
Nittler LR, Starr RD, Weider SZ, McCoy TJ, Boynton WV, Ebel DS, Ernst CM, Evans LG, Goldsten JO, Hamara DK, Lawrence DJ, McNutt RL, Schlemm CE, Solomon SC, Sprague AL (2011) The major-element composition of Mercury's surface from MESSENGER X-ray spectrometry. Science 333(6051):1847–1850. https://doi.org/10.1126/science.1211567
Ostro SJ, Campbell DB, Shapiro II (1985) Mainbelt asteroids: dual-polarization radar observations. Science 229(4712):442–446. https://doi.org/10.1126/science.229.4712.442
Park RS, Konopliv AS, Asmar SW, Bills BG, Gaskell RW, Raymond CA, Smith DE, Toplis MJ, Zuber MT (2014) Gravity field expansion in ellipsoidal harmonic and polyhedral internal representations applied to Vesta. Icarus 240:118–132. https://doi.org/10.1016/j.icarus.2013.12.005
Park RS, Ermakov AI, Konopliv AS, Vaughan AT, Rambaux N, Bills BG, Castillo-Rogez JC, Fu RR, Jacobson SA, Stewart ST, Toplis MJ (2025) A small core in Vesta inferred from Dawn's observations. Nat Astron 9(6):824–834. https://doi.org/10.1038/s41550-025-02533-7
Poirier J-P (1994) Light elements in the Earth's outer core: a critical review. Phys Earth Planet Inter 85(3):319–337. https://doi.org/10.1016/0031-9201(94)90120-1
Polanskey CA, Elkins-Tanton LT, Bell JF et al (2025) Psyche Mission description and design rationale. Space Sci Rev 221:95. https://doi.org/10.1007/s11214-025-01218-x
Prettyman TH, Feldman W, Mellon M, McKinney G, Boynton W, Karunatillake S, Lawrence D, Maurice S, Metzger A, Murphy J (2004) Composition and structure of the Martian surface at high southern latitudes from neutron spectroscopy. J Geophys Res 109(E5). https://doi.org/10.1029/2003je002139
Prettyman TH, Mittlefehldt DW, Yamashita N, Lawrence DJ, Beck AW, Feldman WC, McCoy TJ, McSween HY, Toplis MJ, Titus TN, Tricarico P, Reedy RC, Hendricks JS, Forni O, Le Corre L, Li JY, Mizzon H, Reddy V, Raymond CA, Russell CT (2012) Elemental mapping by Dawn reveals exogenic H in Vesta's regolith. Science 338(6104):242–246. https://doi.org/10.1126/science.1225354
Prettyman TH, Yamashita N, Ammannito E, Ehlmann BL, McSween HY, Mittlefehldt DW, Marchi S, Schörghofer N, Toplis MJ, Li JY, Pieters CM, Castillo-Rogez JC, Raymond CA, Russell CT (2019) Elemental composition and mineralogy of Vesta and Ceres: distribution and origins of hydrogen-bearing species. Icarus 318:42–55. https://doi.org/10.1016/j.icarus.2018.04.032
Reddy V, Le Corre L, O'Brien DP, Nathues A, Cloutis EA, Durda DD, Bottke WF, Bhatt MU, Nesvorny D, Buczkowski D, Scully JEC, Palmer EM, Sierks H, Mann PJ, Becker KJ, Beck AW, Mittlefehldt D, Li J-Y, Gaskell R, Russell CT, Gaffey MJ, McSween HY, McCord TB, Combe J-P, Blewett D (2012) Delivery of dark material to Vesta via carbonaceous chondritic impacts. Icarus 221(2):544–559. https://doi.org/10.1016/j.icarus.2012.08.011
Righter K, Drake MJ (1996) Core formation in Earth's Moon, Mars, and Vesta. Icarus 124(2):513–529. https://doi.org/10.1006/icar.1996.0227
Righter K, Drake MJ (1997) A magma ocean on Vesta: core formation and petrogenesis of eucrites and diogenites. Meteorit Planet Sci 32(6):929–944. https://doi.org/10.1111/j.1945-5100.1997.tb01582.x
Russell CT, Raymond CA (2011) The Dawn mission to Vesta and Ceres. Space Sci Rev 163(1–4):3–23. https://doi.org/10.1007/s11214-011-9836-2
Russell CT, Raymond CA, Coradini A, McSween HY, Zuber MT, Nathues A, De Sanctis MC, Jaumann R, Konopliv AS, Preusker F, Asmar SW, Park RS, Gaskell R, Keller HU, Mottola S, Roatsch T, Scully JEC, Smith DE, Tricarico P, Toplis MJ, Christensen UR, Feldman WC, Lawrence DJ, McCoy TJ, Prettyman TH, Reedy RC, Sykes ME, Titus TN (2012) Dawn at Vesta: testing the protoplanetary paradigm. Science 336(6082):684–686. https://doi.org/10.1126/science.1219381

Ruzicka A, Snyder GA, Taylor LA (1997) Vesta as the howardite, eucrite and diogenite parent body: implications for the size of a core and for large-scale differentiation. Meteorit Planet Sci 32(6):825–840. https://doi.org/10.1111/j.1945-5100.1997.tb01573.x

Scott ERD (1972) Chemical fractionation in iron meteorites and its interpretation. Geochim Cosmochim Acta 36(11):1205–1236. https://doi.org/10.1016/0016-7037(72)90046-4

Scott ERD (1979) Origin of anomalous iron meteorites. Mineral Mag 43(327):415–421. https://doi.org/10.1180/minmag.1979.043.327.15

Shepard MK, Richardson J, Taylor PA, Rodriguez-Ford LA, Conrad A, de Pater I, Adamkovics M, de Kleer K, Males JR, Morzinski KM, Close LM, Kaasalainen M, Viikinkoski M, Timerson B, Reddy V, Magri C, Nolan MC, Howell ES, Benner LAM, Giorgini JD, Warner BD, Harris AW (2017) Radar observations and shape model of asteroid 16 Psyche. Icarus 281:388–403. https://doi.org/10.1016/j.icarus.2016.08.011

Shepard MK, de Kleer K, Cambioni S, Taylor PA, Virkki AK, Rívera-Valentin EG, Rodriguez Sanchez-Vahamonde C, Fernanda Zambrano-Marin L, Magri C, Dunham D, Moore J, Camarca M (2021) Asteroid 16 Psyche: shape, features, and global map. Planet Sci J 2(4):125. https://doi.org/10.3847/PSJ/abfdba

Smirnov AV, Tarduno JA, Evans DAD (2011) Evolving core conditions ca. 2 billion years ago detected by paleosecular variation. Phys Earth Planet Inter 187(3):225–231. https://doi.org/10.1016/j.pepi.2011.05.003

Srinivasan P, Jones RH, Brearley AJ (2017) Compositions and microstructures of CB sulfides: implications for the thermal history of the CB chondrite parent body. Meteorit Planet Sci 52(10):2193–2219. https://doi.org/10.1111/maps.12921

Stevenson DJ (1981) Models of the Earth's core. Science 214(4521):611. https://doi.org/10.1126/science.214.4521.611

Stolper E (1977) Experimental petrology of eucritic meteorites. Geochim Cosmochim Acta 41(5):587–611. https://doi.org/10.1016/0016-7037(77)90300-3

Tagawa S, Sakamoto N, Hirose K, Yokoo S, Hernlund J, Ohishi Y, Yurimoto H (2021) Experimental evidence for hydrogen incorporation into Earth's core. Nat Commun 12(1):2588. https://doi.org/10.1038/s41467-021-22035-0

Takir D, Reddy V, Sanchez JA, Shepard MK, Emery JP (2016) Detection of water and/or hydroxyl on asteroid (16) Psyche. Astron J 153(1):31. https://doi.org/10.3847/1538-3881/153/1/31

Ter Braak CJF (2006) A Markov Chain Monte Carlo version of the genetic algorithm differential evolution: easy Bayesian computing for real parameter spaces. Stat Comput 16(3):239–249. https://doi.org/10.1007/s11222-006-8769-1

Toplis MJ, Mizzon H, Monnereau M, Forni O, McSween HY, Mittlefehldt DW, McCoy TJ, Prettyman TH, Sanctis MC, Raymond CA, Russell CT (2013) Chondritic models of 4 Vesta: implications for geochemical and geophysical properties. Meteorit Planet Sci 48(11):2300–2315. https://doi.org/10.1111/maps.12195

Ulff-Møller F (1998) Effects of liquid immiscibility on trace element fractionation in magmatic iron meteorites: a case study of group IIIAB. Meteorit Planet Sci 33(2):207–220. https://doi.org/10.1111/j.1945-5100.1998.tb01626.x

Vaci Z, Day JMD, Paquet M, Ziegler K, Yin Q-Z, Dey S, Miller A, Agee C, Bartoschewitz R, Pack A (2021) Olivine-rich achondrites from Vesta and the missing mantle problem. Nat Commun 12(1):5443. https://doi.org/10.1038/s41467-021-25808-9

Viikinkoski M, Vernazza P, Hanuš J, Le Coroller H, Tazhenova K, Carry B, Marsset M, Drouard A, Marchis F, Fetick R, Fusco T, Ďurech J, Birlan M, Berthier J, Bartczak P, Dumas C, Castillo-Rogez J, Cipriani F, Colas F, Ferrais M, Grice J, Jehin E, Jorda L, Kaasalainen M, Kryszczynska A, Lamy P, Marciniak A, Michalowski T, Michel P, Pajuelo M, Podlewska-Gaca E, Santana-Ros T, Tanga P, Vachier F, Vigan A, Warner B, Witasse O, Yang B (2018) (16) Psyche: a mesosiderite-like asteroid? Astron Astrophys 619:L3. https://doi.org/10.1051/0004-6361/201834091

Wade J, Wood BJ (2005) Core formation and the oxidation state of the Earth. Earth Planet Sci Lett 236(1):78–95. https://doi.org/10.1016/j.epsl.2005.05.017

Warren PH (1997) Magnesium oxide-iron oxide mass balance constraints and a more detailed model for the relationship between eucrites and diogenites. Meteorit Planet Sci 32(6):945–963. https://doi.org/10.1111/j.1945-5100.1997.tb01583.x

Warren PH (2011) Stable-isotopic anomalies and the accretionary assemblage of the Earth and Mars: a subordinate role for carbonaceous chondrites. Earth Planet Sci Lett 311(1):93–100. https://doi.org/10.1016/j.epsl.2011.08.047

Warren PH, Kallemeyn GW, Huber H, Ulff-Møller F, Choe W (2009) Siderophile and other geochemical constraints on mixing relationships among HED-meteoritic breccias. Geochim Cosmochim Acta 73(19):5918–5943. https://doi.org/10.1016/j.gca.2009.06.030

Wasson JT (1967) The chemical classification of iron meteorites: I. A study of iron meteorites with low concentrations of gallium and germanium. Geochim Cosmochim Acta 31(2):161–180. https://doi.org/10.1016/S0016-7037(67)80043-7

Wasson JT, Choe W-H (2009) The IIG iron meteorites: probable formation in the IIAB core. Geochim Cosmochim Acta 73(16):4879–4890. https://doi.org/10.1016/j.gca.2009.05.062

Wasson JT, Choi B-G (2003) Main-group pallasites: chemical composition, relationship to IIIAB irons, and origin. Geochim Cosmochim Acta 67(16):3079–3096. https://doi.org/10.1016/S0016-7037(03)00306-5

Wasson JT, Wang J (1986) A nonmagmatic origin of group-IIE iron meteorites. Geochim Cosmochim Acta 50(5):725–732. https://doi.org/10.1016/0016-7037(86)90348-0

Wasson JT, Ouyang X, Wang J, Eric J (1989) Chemical classification of iron meteorites: XI. Multi-element studies of 38 new irons and the high abundance of ungrouped irons from Antarctica. Geochim Cosmochim Acta 53(3):735–744. https://doi.org/10.1016/0016-7037(89)90016-1

Wasson JT, Huber H, Malvin DJ (2007) Formation of IIAB iron meteorites. Geochim Cosmochim Acta 71(3):760–781. https://doi.org/10.1016/j.gca.2006.09.032

Weisberg MK, Prinz M, Clayton RN, Mayeda TK, Sugiura N, Zashu S, Ebihara M (2001) A new metal-rich chondrite grouplet. Meteorit Planet Sci 36(3):401–418. https://doi.org/10.1111/j.1945-5100.2001.tb01882.x

Weiss BP, Merayo JMG, Ream JB, Oran R, Brauer P, Cochrane CJ, Cloutier K, Elkins-Tanton LT, Jørgensen JL, Maurel C, Park RS, Polanskey CA, de Soria Santacruz-Pich M, Raymond CA, Russell CT, Wenkert D, Wieczorek MA, Zuber MT (2023) The Psyche magnetometry investigation. Space Sci Rev 219(3):22. https://doi.org/10.1007/s11214-023-00965-z

Weyrauch M, Zipfel J, Weyer S (2019) Origin of metal from CB chondrites in an impact plume – a combined study of Fe and Ni isotope composition and trace element abundances. Geochim Cosmochim Acta 246:123–137. https://doi.org/10.1016/j.gca.2018.11.022

Williams Q (2009) Bottom-up versus top-down solidification of the cores of small Solar System bodies: constraints on paradoxical cores. Earth Planet Sci Lett 284(3):564–569. https://doi.org/10.1016/j.epsl.2009.05.019

Wood BJ, Walter MJ, Wade J (2006) Accretion of the Earth and segregation of its core. Nature 441(7095):825–833. https://doi.org/10.1038/nature04763

Yang J, Goldstein JI, Scott ERD (2007) Iron meteorite evidence for early formation and catastrophic disruption of protoplanets. Nature 446(7138):888–891. https://doi.org/10.1038/nature05735

Yoshizaki T, McDonough WF (2020) The composition of Mars. Geochim Cosmochim Acta 273:137–162. https://doi.org/10.1016/j.gca.2020.01.011

Zhang Z, Bercovici D, Elkins-Tanton L (2022) Cold compaction and macro-porosity removal in rubble-pile asteroids: 1. Theory. J Geophys Res Planets 127(10):e2022JE007342. https://doi.org/10.1029/2022JE007342

Zhang B, Chabot NL, Rubin AE (2024) Compositions of iron-meteorite parent bodies constrain the structure of the protoplanetary disk. Proc Natl Acad Sci USA 121(23):e2306995121. https://doi.org/10.1073/pnas.2306995121

Zhou T, Tarduno JA, Nimmo F, Cottrell RD, Bono RK, Ibanez-Mejia M, Huang W, Hamilton M, Kodama K, Smirnov AV, Crummins B, Padgett F (2022) Early Cambrian renewal of the geodynamo and the origin of inner core structure. Nat Commun 13(1):4161. https://doi.org/10.1038/s41467-022-31677-7

Zolensky ME, Weisberg MK, Buchanan PC, Mittlefehdt DW (1996) Mineralogy of carbonaceous chondrite clasts in HED achondrites and the Moon. Meteorit Planet Sci 31:518–537. https://doi.org/10.1111/j.1945-5100.1996.tb02093.x

Zuber MT, McSween HY, Binzel RP, Elkins-Tanton LT, Konopliv AS, Pieters CM, Smith DE (2011) Origin, internal structure and evolution of 4 Vesta. Space Sci Rev 163(1–4):77–93. https://doi.org/10.1007/s11214-011-9806-8

Zuber MT, Park RS, Elkins-Tanton LT, Bell JF, Bruvold KN, Bercovici D, Bills BG, Binzel RP, Jaumann R, Marchi S, Raymond CA, Roatsch T, Wang CC, Weiss BP, Wenkert D, Wieczorek MA (2022) The Psyche gravity investigation. Space Sci Rev 218(8):57. https://doi.org/10.1007/s11214-022-00905-3

Authors and Affiliations

Thomas H. Prettyman[1] · David W. Mittlefehldt[2] · Erik I. Asphaug[3] · Richard P. Binzel[4] · Samuel W. Courville[5] · Linda T. Elkins-Tanton[5,6] · David J. Lawrence[7] · Simone Marchi[8] · José M.G. Merayo[9] · Timothy J. McCoy[10] · Benjamin P. Weiss[4]

✉ T.H. Prettyman
prettyman@psi.edu

[1] Planetary Science Institute, Tucson, AZ, USA

[2] Lunar and Planetary Institute, Universities Space Research Association, Houston, TX, USA

[3] Lunar and Planetary Laboratory, University of Arizona, Tucson, AZ, USA

[4] Department of Earth, Atmospheric, and Planetary Sciences, Massachusetts Institute of Technology, Cambridge, MA, USA

[5] School of Earth and Space Exploration, Arizona State University, Tempe, AZ, USA

[6] Present address: Space Sciences Laboratory, University of California, Berkeley, Berkeley, CA, USA

[7] Johns Hopkins University Applied Physics Laboratory, Laurel, MD, USA

[8] Southwest Research Institute, Boulder, CO, USA

[9] DTU Space, Technical University of Denmark, Lyngby, Denmark

[10] Smithsonian National Museum of Natural History, Washington, DC, USA

Space Science Reviews (2022) 218:6
https://doi.org/10.1007/s11214-022-00872-9

Deciphering Redox State for a Metal-Rich World

Timothy J. McCoy[1] · Steven D. Dibb[2] · Patrick N. Peplowski[3] · Clara Maurel[4] · Hannah L. Bercovici[2] · Catherine M. Corrigan[1] · James F. Bell III[2] · Benjamin P. Weiss[4] · David J. Lawrence[3] · Daniel D. Wenkert[5] · Thomas H. Prettyman[6] · Lindy T. Elkins-Tanton[2]

Received: 24 May 2021 / Accepted: 3 February 2022 / Published online: 1 March 2022

Abstract
The Psyche mission's Oxidation-Reduction Working Group is focused on understanding, determining, and applying the redox state of (16) Psyche to understand the origin of a metal-rich world. The oxidation-reduction state of an asteroid, along with its temperature, parent body size, and composition, is a key parameter in determining the history of an asteroid. Determining the redox state from spacecraft data is most easily done by examining potential metal-oxide buffer pairs. The occurrence of Ni, Fe, C, Cr, P and Si, in that order, in the metal or sulfide phase of an asteroidal body indicates increasingly reduced conditions. Key observations by the Imager and Gamma-Ray and Neutron Spectrometer (GRNS) of Psyche can bracket the redox state using metal-oxide buffers. The presence of Fe,Ni metal can be confirmed by the ratios of Fe/O or Fe/Si and the concentration of Ni variability in metal across the asteroid can be determined by GRNS. The FeO concentration of silicates is complementary to the Ni concentration of metal and can be constrained using filters on the Imager. The presence of FeO in silicates from ground-based observations is one of the few measurements we already have of redox state, although available data permit a wide range of silicate compositions and mineralogies. The presence of C, P or Si concentrated in the metallic, Fe-rich portion of the asteroid, as measured by GRNS, or Ca-sulfide, determined by imaging, would indicate increasingly reducing conditions. Linkage to known types of meteorites, whether metal-rich chondrites, stony-irons or irons, expands the mineralogical, chemical and isotopic data not available from remote observations alone. Redox also controls both silicate and metal mineralogy, influencing differentiation, solidification, and subsolidus cooling, including the relative abundance of sulfur in the core and possible magnetic signatures. The redox state of Psyche, if a fully-differentiated metallic core, might constrain the location and timing of both the formation of Psyche and any oxidation it might have experienced.

Keywords Psyche · Oxidation-reduction · Iron meteorites · Chemical composition · Spectroscopy

---

The NASA Psyche Mission: Science Instruments and Investigations
Edited by James F. Bell III, Carol Polanskey and Lindy Elkins-Tanton

---

Extended author information available on the last page of the article

1 Introduction

The instrument suite of the Psyche spacecraft will measure the chemical composition, provide constraints on the mineralogy, measure the magnetic signature, and determine the mass and size of the asteroid. These observations are key to understanding the history of (16) Psyche, ranging from processes occurring today such as space weathering, to those that occurred 4.6 billion years ago including the formation of the chondritic protolith and, possibly, the differentiation of the asteroid. One goal of the mission is to relate the history of Psyche to that of the terrestrial planets to understand shared features and histories.

As early as Prior (1916), oxidation-reduction (redox) reactions were recognized as a fundamental control on the mineralogy and mineral compositions of meteorites. Prior's rule states that when less metal is present in meteorites, the existing metal is richer in nickel. In addition, the associated silicates are richer in FeO. In essence, Prior (1916) envisioned meteorites, particularly the undifferentiated chondritic meteorites, as isochemical, with redox as the major control. This view was widely accepted until the work of Urey and Craig (1953), who established that different meteorite groups were, in fact, not isochemical. Nonetheless, chondritic meteorites, considered the precursor to differentiated planetary objects, exhibit a wide range of oxidation states (Righter et al. 2016). Likewise, the terrestrial planets range widely in oxidation state from the highly reduced Mercury to the relatively oxidized Mars. The difference in the redox state of elements – particularly for iron – is found in bodies ranging from undifferentiated chondrites to fully differentiated planets. Thus, the difference is a property likely established in the precursor chondrite and inherited during differentiation. Redox state variation within meteorites and terrestrial planets served as the impetus for our effort to constrain the oxidation state of Psyche.

In this chapter, we explore the potential role of redox on the formation of Psyche. For clarity, the use of "Psyche" refers to the asteroid, whereas the terms "spacecraft", "mission" and "project" are used for those aspects of the mission. We begin with an exploration of redox – what is it, how do we quantify it, and how does it control element partitioning between minerals (e.g., silicates, metal)? In the latter part of the chapter, we discuss how the mission team, specifically the Oxidation-Reduction Working Group, will use data from the spacecraft's instruments to constrain the oxidation state of the asteroid. Finally, we address how redox-controlled mineralogy influences the history of a body and may help us constrain the origin of Psyche.

2 Oxidation-Reduction: A Primer for Psyche

From a chemical perspective, oxidation is the loss of electrons by an element and, conversely, reduction is the gain of electrons by an element. A useful example is the combustion of carbon in the presence of oxygen:

$$C + O_2 = CO_2 \qquad \text{Reaction 1}$$

In this reaction, carbon is being oxidized from 0 to 4+, whereas oxygen is being reduced from 0 to -2. A useful way to think of this is that carbon bonds with oxygen during combustion, transforming from the reduced elemental state to the oxygen-bearing state. As with all redox reactions, the reduction of one element is balanced by the oxidation of another. This reaction – and its variant to produce carbon monoxide – are both societally relevant with the mass burning of fossil fuels contributing to the increase in atmospheric CO_2 and to understanding Psyche. The reverse of this reaction – transformation of CO_2 to carbon – occurs during photosynthesis through a complex set of reactions occurring during the Calvin cycle which incorporate CO_2 from the atmosphere into glucose. As carbon comprises about 50% of the dry mass of trees, about half the mass of the largest tree you have ever seen materialized out of thin air during photosynthesis!

To quantify the redox state, geologists typically use oxygen fugacity, which is an equivalent of the partial pressure of oxygen in a particular environment, most commonly rocks, corrected for the nonideal behavior of the gas. An instructive example is the oxygen fugacity of air. At 21% O_2, the oxygen fugacity of air is $\sim 10^{-0.7}$. Oxygen fugacities are typically plotted on diagrams of log (fO_2) vs. 1/T (K). Thus, air plots at about -0.7. One of the non-intuitive parts of oxygen fugacity is that it measures the partial pressure of gaseous oxygen, rather than the concentration of oxygen that might be bound in a rock. Thus, rocks, which typically have 40–50 wt.% oxygen plot many orders of magnitude lower in oxygen fugacity than air, which contains 21 wt.% oxygen.

To circumvent the difficulty of comparing temperature dependent oxygen fugacity values for various geological equilibria, pairs of metal and oxide for the same element are commonly used. At constant pressure and temperature, the *co-existence* of metal and oxide fixes the fO_2 at a specific value, such that an increase or decrease in the *amount* of oxide or metal will not change the fO_2. Hence the use of the word "buffer" or "oxygen buffer" for these metal-oxide pairs. Oxygen fugacities for these metal-oxide pairs are often determined by electrochemical or calorimetric means (e.g., Sato 1971). The slope of these buffer curves in $\log fO_2 - 1/T$ space is negative (the fO_2 decreases with decreasing temperature), because of the relationship between fO_2 and free energy (G). As most buffers are metal and oxide, their free energy and thus their slopes are similar, making their use convenient for a temperature-independent comparison (Ellingham 1944). In addition to removing the strong dependence of fO_2 on temperature, they have the decided advantage when working with spacecraft data that the occurrence of an element as a metal – either in a metallic phase or a sulfide – can constrain the oxygen fugacity relative to the metal-oxide buffers. Here we discuss several metal-oxide buffers relevant to our investigation of Psyche, including iron, nickel, carbon, phosphorus, chromium and silicon.

For the benefit of those unfamiliar with meteorite nomenclature used here (Weisberg et al. 2006), individual meteorites are given names based on the site where they fell or were found (e.g., Canyon Diablo, Arizona). Undifferentiated meteorites, called chondrites, are divided into classes (e.g., C for carbonaceous, E for enstatite) and groups and given letter designations (e.g., H for high-iron ordinary chondrites, CB for carbonaceous chondrites similar to the type specimen Bencubbin). Differentiated meteorite groups (e.g., eucrites, brachinites) either have historical names (e.g., eucrite from the Greek referring to the easily distinguished mineral grains) or are named after the type specimen (e.g., brachinites from Brachina, Australia). Iron meteorites are subdivided on the basis of chemical composition – originally Ga, Ge and Ni – using a Roman numeral (I-IV) and letter (A-G). In some cases, groups thought to be distinct (e.g., IIA and IIB) were found with analyses of additional meteorites to be end members of a common group (e.g., IIAB). Meteorites within a type (e.g., H, CM, IIAB) are thought to have originated on a common parent body.

2.1 Iron

With valence states of 0, 2+ and 3+, iron is one of the most widely used elements in metal-oxide buffers, as it covers a wide range of oxidation states applicable from the interior (core, mantle) to the surface (crust) of the Earth. Under highly oxidizing conditions, a buffer exists

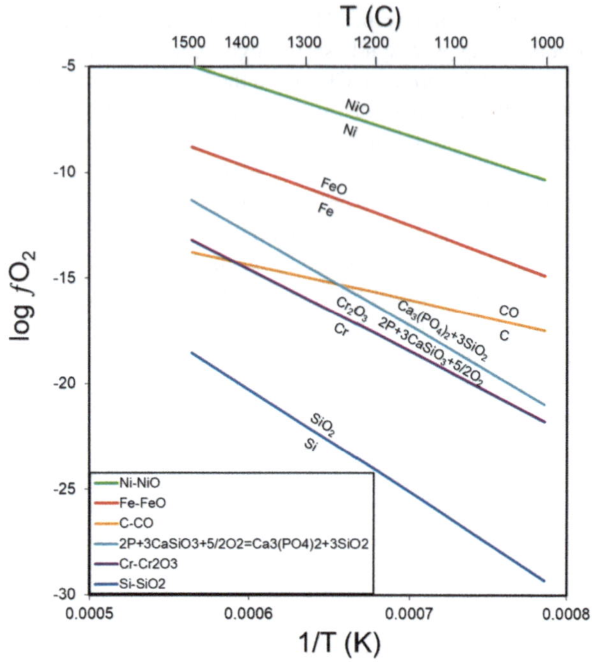

Fig. 1 Log fO_2 vs 1/T (K) for metal-oxide or mineral buffers in the range of 1000–1500 °C for Ni, Fe, C, P, Cr and Si using data from O'Neill and Pownceby (1993) (Fe, Ni), Holzheid and O'Neill (1995) (Cr), French and Eugster (1965) (C), and Robie et al. (1978) (P, Si). Among these elements, Si is the most readily and Ni the least readily oxidized

between magnetite ($Fe^{2+}Fe_2^{3+}O_4$) and hematite ($Fe_2^{3+}O_3$). This buffer is seldom applied in meteorites as most hematite likely forms by alteration of metal in Earth's oxygen-rich atmosphere. At conditions slightly more reducing is a buffer between FeO and $Fe^{2+}Fe_2^{3+}O_4$. FeO is commonly called wüstite after the mineral of that formula ($Fe_{(1-x)}O$). The wüstite-magnetite buffer is likewise seldom invoked, as geologists tend to prefer a mineral buffer reaction between quartz, fayalite and magnetite, commonly termed FMQ:

$$3Fe_2SiO_4 \text{ (fayalite)} + O_2 = 2Fe_3O_4 \text{ (magnetite)} + 3SiO_2 \text{ (quartz)} \qquad \text{Reaction 2}$$

Magnetite is present in some chondritic meteorites (e.g., CK, CI, CM), typically because of aqueous alteration of metal on the parent asteroid early in the history of the Solar System, not as a result of formation in an oxidizing region of the nebula.

Most relevant to Psyche is the iron-wüstite (IW) buffer (Fig. 1), in large part because iron is certainly the most abundant element in any asteroid core. The position of the buffer was calculated using electrochemical data from O'Neill and Pownceby (1993) over the range of temperatures (1000–1500 °C) at which melting, differentiation and core formation occur on asteroids. At this buffer, iron metal (Fe) coexists with iron oxide (FeO), typically in silicates (olivine, $(Mg,Fe)_2SiO_4$; pyroxene $(Mg,Fe)SiO_3$). For an oxygen fugacity above this buffer, iron metal is unstable, but below the buffer, FeO in silicates decreases with increasing reduction but remains present until extremely reducing conditions, a topic we discuss later. Oxygen fugacity is typically reported relative to this buffer, so IW+1 is one log unit above the IW buffer and IW-2 is two log units below the IW buffer. The identification of indigenous metallic iron, even in the presence of FeO, on a planetary body constrains the oxygen fugacity to at or below the IW buffer.

The redox relationship between Fe in metal and sulfides, collectively grouped as reduced iron, and FeO in silicates and oxides, grouped as oxidized iron, in meteorites was the impetus behind one part of Prior's rule – the less metal, the more FeO in the associated silicates.

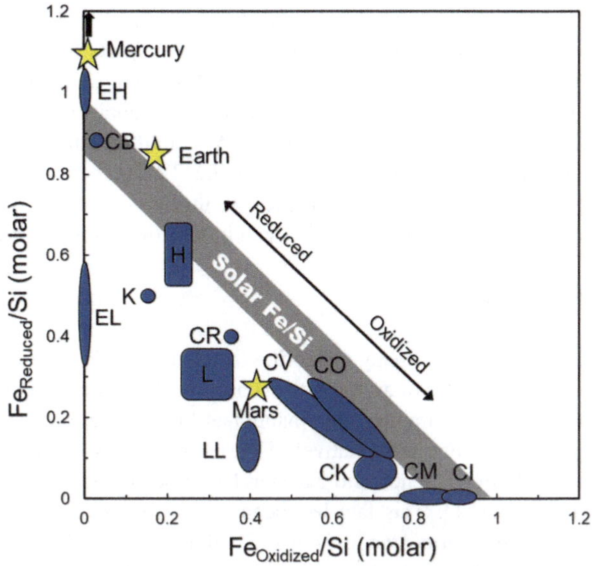

Fig. 2 Modified Urey-Craig plot of $Fe_{oxidized}$/Si vs $Fe_{reduced}$/Si (molar), after Yoshizaki et al. (2021) with additional data for the CB chondrite Bencubbin (Jarosewich 1990), Mars (Yoshizaki and McDonough 2020) and Mercury (Nittler et al. 2018). The $Fe_{reduced}$/Si ratio of CB chondrites likely is from a sample depleted in metal relative to the bulk meteorite. Note that data for CH and G chondrites is not available, but they likely plot close to CB chondrites. Redox control can vary a composition along a tie line of total Fe/Si, such as the Solar Fe/Si ratio, whereas differences in total Fe/Si, such as between H, L and LL ordinary chondrites, cannot be redox controlled

Urey and Craig (1953) would use superior wet chemical analyses of ordinary chondrites which measured reduced iron in metal and sulfides and oxidized iron in silicates and oxides, as well as total iron, to demonstrate that two groups of ordinary chondrites, called H and L chondrites, differed in total iron and must sample distinct geochemical reservoirs. The Urey-Craig diagram (Fig. 2) serves to illustrate the differences in redox between different chondritic meteorites and planetary bodies and provides a useful conceptual framework for observations of the relative abundance of reduced and oxidized iron on Psyche. Meteorite groups plotting at high reduced iron and low oxidized iron values, including the EH and EL enstatite chondrites and the metal-rich CB chondrites (and, likely, the closely related CH and G chondrites, for which data is unavailable), have iron largely in the reduced state, with low FeO in silicates. In contrast, meteorite groups plotting at high oxidized iron and low reduced iron, including CI and CM chondrites, have little iron in the reduced state with FeO-rich silicates and, typically, magnetite. Oxidation-reduction moves compositions along a constant total Fe/Si ratio, such as shown by Solar Fe/Si. Thus, from the perspective of total Fe/Si, H and CO chondrites, neither of which experienced significant aqueous alteration, are close to isochemical but differ in their proportions of oxidized and reduced iron. Different total Fe/Si ratios cannot be related simply through redox control, as redox does not change the total abundance of iron relative to silicon, but result from other processes including accretion and, in the case of asteroids, later impact. Compositions of planets are less constrained than meteorite groups with considerably larger uncertainties, but illustrate increasing oxidation (and, interestingly, decreasing total Fe/Si) and decreasing core size from Mercury ($\sim 70\%$ of total radius) to Earth (32%) to Mars (21%) (e.g., Righter et al. 2006). Interestingly, the biggest uncertainty in planetary compositions is often the size and composition of the core, whereas if Psyche is a core, it may offer the inverse problem of a chemically well-characterized core possibly with remnants of the less well constrained mantle or crust.

2.2 Nickel

As the second most abundant element in meteoritic metal, with concentrations ranging from 5–60 wt.% in iron meteorites, nickel exhibits significant variation both within and between iron meteorite groups. The Ni-NiO buffer from 1000–1500 °C calculated using electrochemical data from O'Neill and Pownceby (1993) plots from 4.6 to 3.8 log units above Fe-FeO (IW) (Fig. 1). Thus, Fe in the metallic form will oxidize much more readily than Ni in the metallic form. Therefore, oxidation, even at conditions below the IW buffer, will enrich the residual metal in Ni relative to Fe. At conditions more oxidizing than IW, Ni is incorporated into sulfide phases like pentlandite. This is the case for the CK chondrites, which lack metallic iron. Later in this chapter, we discuss the role of redox reactions in establishing the differences in composition between groups of iron meteorites.

2.3 Carbon

Carbon occurs in many meteorites, often as soluble or insoluble organic matter in carbonaceous chondrites or as graphite in enstatite chondrites, some groups of achondrites (ureilites, acapulcoites-lodranites, winonaites) and iron meteorites (notably the IAB irons). Graphite typically occurs with relatively FeO-poor silicates in meteorites. Not surprisingly then, the C-CO buffer (French and Eugster 1965) occurs at conditions more reducing than the IW buffer (Fig. 1). This fact has been exploited for centuries when coke (coal heated in an oxygen-poor atmosphere) is mixed with iron oxides during the smelting of iron metal. Thus, a finding of indigenous graphite in an asteroidal core would indicate reducing conditions. A more interesting feature of the C-CO buffer is its slope on a plot of $\log fO_2$ vs 1/T (K) (Fig. 1). Most metal-oxide buffer pairs are subparallel, as observed with Fe-FeO and Ni-NiO. In contrast, C-CO differs dramatically in slope from other metal-oxide buffers. At 1000 °C, C-CO is IW-2.6, but is IW-5.0 by 1500 °C. This slope results from the fact the reaction contains both a solid (C) and a gas (CO), whereas other buffers are solid-solid buffers (e.g., Fe-FeO). Thus, while graphite can serve as a reductant for iron at the entire relevant range of temperatures, it can also serve as a reductant for elements like Cr and P at higher temperatures. Modern industrial processes exploit this by mixing carbon and silica sand at extremely high temperatures to produce silicon metal as an input product for the manufacturing of silicon wafers for computers. Finally, the solid-gas nature of the C-CO buffer results in it being particularly pressure sensitive, shifting by ~ 4 log units to more oxidizing conditions (closer to IW) with a pressure increase from 1–100 bars (Brett and Sato 1984; Walker and Grove 1993; Benedix et al. 2005). This has interesting consequences in a C-buffered system if, for example, the mantle of a partially molten asteroid was stripped resulting in a lower pressure in the now exposed core.

2.4 Chromium and Phosphorus

These two buffers are similar enough – and serve to illustrate the same geochemical behavior – that we treat them together in this section. The metal-oxide buffer for $Cr\text{-}Cr_2O_3$ can be calculated using electrochemical data from Holzheid and O'Neill (1995) and ranges from IW-6.9 at 1000 °C to IW-4.4 at 1500 °C (Fig. 1). A buffer for phosphorus and phosphorus oxide is not well determined, but a buffer for the reaction:

$$2P + 3CaSiO_3 + 5/2O_2 = Ca_3(PO_4)_2 + 3SiO_2 \qquad \text{Reaction 3}$$

can be calculated from the thermodynamic data of Robie et al. (1978). This buffer ranges from IW-6.1 at 1000 °C to IW-2.5 at 1500 °C (Fig. 1), ~ 1–2 log units higher than the $Cr\text{-}Cr_2O_3$ buffer. At these redox conditions, a fundamental shift in geochemical behavior occurs, with elements that normally occur with oxygen to form minerals in the crust of the

Earth instead forming metal or sulfides that would potentially concentrate in the core of a planet. In the early part of the 20th century, Goldschmidt separated elements into those that occurred in the crust (lithophile), metal (siderophile) or sulfides (chalcophile) (e.g., Mason 1992). This classification is applicable to the redox state at the surface of the Earth. In contrast, the buffers for Cr and P suggest that at ~ 5 log units more reducing than IW, elements will change their behavior. This is observed in meteorites (e.g., IAB and IIAB irons; Buchwald 1975) with the formation of the phosphide schreibersite, which can form either during crystallization of a reduced metallic magma or during later solid-state diffusive processes. In more oxidizing conditions, phosphorus typically occurs in phosphates. Similarly, chromium is typically observed under oxidizing conditions as chromite ($FeCr_2O_4$), but under reducing conditions becomes chalcophile forming daubréelite ($FeCr_2S_4$), which is a common exsolution product from troilite (FeS). These phases are observed in meteorites that formed under reducing conditions, such as enstatite meteorites, but are also found in somewhat more oxidized meteorites, like IAB irons where they coexist with silicates with modest FeO concentrations (e.g., Fa_{10}, where Fa is the molar $(Fe/(Fe + Mg)) * 100$) in olivine and the corresponding value in pyroxene (Fs) is the molar $(Fe/(Fe + Mg + Ca)) * 100$. If Psyche is a core remnant rich in metal and/or sulfide, the identification of P and/or Cr as a significant component of that metal (as indicated by, e.g., a high iron concentration) would provide evidence of reducing conditions well below that of the IW buffer.

2.5 Silicon

The Si-SiO_2 buffer, calculated using thermodynamic data from Robie et al. (1978), marks the onset of what might be termed ultra-reducing conditions, which range from IW-14.4 at 1000 °C to IW-9.7 at 1500 °C (Fig. 1). Metallic silicon occurs in the Fe,Ni metal of enstatite meteorites, often comprising several weight percent. In these meteorites, silicon-bearing metal coexists with essentially FeO-free silicates. In addition to silicon becoming siderophile within enstatite meteorites, a wide range of normally lithophile elements become chalcophile, including Mg, Mn, Ti, Ca, Na, and K, forming an array of exotic sulfides including niningerite (MgS), alabandite (MnS), wassonite (TiS), oldhamite (CaS), and djerfisherite ($K_6Na(Fe,Cu)_{24}S_{26}Cl$) (Keil 2010; Rubin and Ma 2017). Spectral identification of any of these phases within sulfide regions and/or identification of silicon alloyed in large metallic regions of Psyche would indicate ultra-reducing conditions.

3 Key Observations

With an understanding of redox processes in general, we can address key observations to be made by the spacecraft payload to constrain the origin of Psyche. As outlined in the chapter by Elkins-Tanton et al. (this journal), revised density estimates of Psyche permit a broad range of interpretations, from a largely metallic core with significant macroporosity to a silicate-dominated assemblage similar to enstatite or CB chondrites. In this section, we first discuss the case of a largely metallic core but address how our observations are applicable to silicate-dominated cases.

3.1 Presence of Fe,Ni Metal

The first and perhaps most robust indicator of redox state is the presence of Fe,Ni metal. Although expected, as nearly all asteroidal meteorites contain this phase, the presence of Fe,Ni metal indicates formation at a redox conditions at or below the IW buffer. The presence of Fe,Ni metal in a largely metallic asteroid could be indicated by any number of morphologic (e.g., crater rims) or spectral (e.g., reddening) features. Compositionally, iron dominates the composition of Fe,Ni metal and determination of a high percentage of iron (e.g., > 30 wt.% Fe, based on meteorite analyses in Jarosewich 1990) on the surface of Psyche would be a strong indicator of a large metallic component. That observation might be coupled with low concentrations of elements typically bound with iron in non-metallic phases, including Mg, Si, or O. These measurements are enabled by the orbital nature of the mission, substantially improving constraints on origin compared to the pioneering flyby mission of M-class asteroid Lutetia by Rosetta (Coradini et al. 2011). Even if Fe were less than 30 wt.% and a substantial silicate component was indicated, it may still be possible to confidently determine the presence of Fe,Ni metal. McCoy et al. (2001) used elemental ratios (e.g., Fe/Si, Fe/O) constrained for individual phases and compared them to the value determined for the bulk to show that a significant fraction of Fe must be in a metallic phase on asteroid 433 Eros.

3.2 Ni Concentration of Metal

If Fe,Ni metal comprises a significant portion of the asteroid, the Ni concentration of that metal can be a strong indicator of formation processes, including oxidation. Fractional crystallization is probably the best-known process for producing intragroup Ni variations. However, the solid metal/liquid metal partition coefficient for Ni is close to 1 (~ 0.9–0.95; Campbell and Humayun 2005), so the composition of the crystallizing solid only differs slightly from the composition of the liquid. This leads to a relatively restricted range of Ni concentration within any group of iron meteorites and, by extension, the common core from which they crystallized. As an example, IIIAB irons, which are the most abundant and well-studied group, range from 7.1–10.6 wt.% Ni (Wasson 1999). Condensation in the solar nebula can also produce a range of Ni concentrations in metal that is incorporated into chondritic meteorites and, later, in the cores of differentiating asteroids. In a cooling nebular gas, the first metal to condense has a Ni concentration of 19 ± 3 wt.%, which decreases with decreasing temperature (Kelly and Larimer 1977). Thus, differences in condensation temperature of the precursor chondritic meteorites that formed iron cores could, in theory, explain the range of Ni concentrations between the major groups of iron meteorites, which range from 5.3–6.5 wt.% in IIAB irons up to 16–18 wt.% in IVB irons (Goldstein et al. 2009).

While condensation could produce the range of intergroup variations observed in iron meteorites, significant evidence suggests that oxidation was the major factor in determining the Ni concentration in some groups. This would not be surprising given the relative positions of the Fe-FeO and Ni-NiO buffers discussed above. As iron is much more readily oxidized than nickel, oxidizing conditions would oxidize Fe to FeO, which would be incorporated into silicates, leaving Ni in the metallic form and enriching the metal in Ni. Consequently, an asteroid that has experienced oxidation should have a mantle and crust enriched in FeO and a core with a higher concentration of Ni. The evidence that this indeed happened comes from the study of siderophile trace elements in iron meteorites. During condensation, elements of similar volatility should be incorporated in similar concentrations, relative to the composition of the condensing gas, in metal. In contrast, during oxidation, elements that are more readily oxidized (e.g., have metal-oxide buffer curves at higher oxygen fugacity) can be fractionated relative to elements of similar volatility. This latter behavior is exactly what is observed for readily oxidized elements like W, Mo, P and Fe in some high-Ni iron meteorites (e.g., IVB irons, Campbell and Humayun 2005; Milton-South Byron trio irons,

McCoy et al. 2019). Thus, a core composition enriched in Ni likely formed through oxidation. Although Fe,Ni metal indicates formation at or below the IW buffer, high Ni irons likely formed at oxidation conditions close to IW, while lower-Ni irons formed at more reducing conditions. We return later to this topic to explore in what environment and when this oxidation may have occurred.

The spacecraft's Gamma-Ray and Neutron Spectrometer (GRNS) will provide measurements that will be used to determine the concentration and variability of elemental Ni on Psyche. The GRNS instrument design and observation plan were developed to enable detection of ≥ 4 wt% (global average) Ni content on Psyche with a one-standard-deviation uncertainty of 20% relative (Lawrence et al. this journal). This is accomplished via gamma-ray measurements of the characteristic Ni gamma rays at 1454 keV. Mapping of the Ni content on Psyche using the gamma-ray measurement is possible using summed spectra, albeit with higher uncertainty relative to the global value. Our expected statistical precision supports reporting five separate Ni concentrations, which correspond to $\sim$ 140-km-diameter pixels along the equatorial ground track of the spacecraft in the lowest altitude orbit (see Chap. 1 for details). This spatial resolution is comparable to the intrinsic spatial resolution of our individual (instantaneous) gamma-ray measurements, which have a full-width at half maximum spatial resolution of 90–120 km. GRNS neutron measurements can also be used to infer the Ni content, particularly when combined with the gamma-ray measurements (Lawrence et al. 2020). These neutron data are best suited toward providing higher spatial resolution measurements of the variability of Ni across Psyche's surface. A spatial resolution of $\sim$ 75 km is possible, which will facilitate correlating Ni concentrations with large geologic units and craters. At present, at least two craters with $\sim$ 100-km diameters are known to be present on Psyche (Viikinkoski et al. 2018).

3.3 FeO Concentration in Silicates

If iron is oxidized from metal prior to or during core formation, it will be incorporated as FeO into the silicate magma that crystallizes to form the mantle and crust of the asteroid. Thus, low-Ni irons, where most of the iron remained in the core, should coexist with low FeO silicates. Conversely, high-Ni irons, where much of the iron was oxidized into the silicate shell of the asteroid, should coexist with high FeO silicates. To first order, observations of meteorites support this general relationship. As an example, the olivine-rich brachinites contain both Ni-rich metal and FeO-rich olivine (29.7 wt.% Ni, $Fa_{32.3}$ in ALH 84025; Day et al. 2012), whereas pyroxene-rich aubrites (enstatite achondrites) contain Ni-poor metal and pyroxene that is essentially FeO-free enstatite (5.8 wt.% Ni, $Fs_{0.02}$ in Aubres; Watters and Prinz 1979). In practice, many iron meteorites do not contain coexisting silicates and few, if any, groups of irons can be confidently linked to an associated group of silicate-rich stony or stony-iron meteorites. An additional complication is that while Ni fractionates little between the parent metallic magma and the crystallizing solid metal (Campbell and Humayun 2005), the same is not true for iron within the silicates. Fractionation between silicate magmas and crystallizing silicates (e.g., olivine) are well-documented, as are fractionations between different silicate minerals (e.g., olivine and pyroxene) (e.g., Righter and Drake 1997; Ruzicka et al. 1997; McCoy et al. 2006; Mandler and Elkins-Tanton 2013). Thus, silicates that do coexist with iron meteorites may not be representative of the bulk silicate magma from which they formed. As most of these silicates are likely early-crystallizing phases at the core-mantle boundary, they are likely less iron rich than the bulk of silicates crystallized from the same body. For this reason, care must be taken to constrain both the FeO concentration of the silicates and, if possible, the mineral phases present.

Mineralogical constraints on the surface compositions of asteroids can be made using remote sensing spectroscopy at UV to near-IR wavelengths (~ 0.1–3.0 μm). These measurements reveal diagnostic spectral behavior based on comparisons to spectra of meteorites, silicates, synthetic minerals, and their mixtures (Bell et al. 2021 and references therein). For example, laboratory studies demonstrate that the band centers of ~ 1.0 μm absorption features in reflectance spectra of Fe-bearing silicates (e.g., olivine and pyroxene) shift to longer wavelengths with increasing Fe content (e.g., Adams 1975; Cloutis and Gaffey 1991). Visible to near-IR telescopic reflectance spectra of Psyche are relatively featureless and red-sloped (reflectance increasing from $\sim 10\%$ to $\sim 25\%$ from 0.3 to 2.5 μm), consistent with spectra of many iron meteorites (e.g., Bus and Binzel 2002; Clark et al. 2004; DeMeo et al. 2009; Fornasier et al. 2010; Sanchez et al. 2017). Some published spectra of Psyche, though, also exhibit a broad and subtle absorption feature centered near 0.92 to 0.95 μm and varying from $\sim 0\%$ to $\sim 3\%$ band depth (Fornasier et al. 2010; Ockert-Bell et al. 2010; Sanchez et al. 2017). This weak feature is consistent with the presence on the asteroid of Fe-bearing pyroxenes, which are also a common constituent of metal-rich meteorites (e.g., Mittlefehldt et al. 1998). The weak band depth and relatively short-wavelength band center of this feature has been interpreted as evidence that pyroxene on Psyche occurs primarily as low-Fe, low-Ca enstatite (e.g., Fig. 3; Hardersen et al. 2011).

Laboratory spectra of both natural and synthetic pyroxenes indicate that pyroxene on Psyche could be much more Fe-rich, however. For example, Cloutis and Gaffey (1991) showed that pyroxenes with band centers between ~ 0.91 and 0.95 μm could have between ~ 0–65 mol.% ferrosillite (Fs = molar $Fe^{2+}/(Ca + Mg + Fe^{2+})$) and up to 30 mol.% wollastonite (Wo = molar $Ca/(Ca + Mg + Fe^{2+})$). Furthermore, Klima et al. (2007) showed that orthopyroxenes with 25 to 100 mol.% Fs also exhibited band centers between 0.913 and 0.94 μm (Fig. 3). It is thus possible that silicates on Psyche are not homogenously mixed with metal and may even be of regionally varying composition. This interpretation is consistent with whole-disk rotational spectral variations of Psyche that showed band centers ranging from 0.92 μm to 0.94 μm (Sanchez et al. 2017) and consistent with pyroxene chemistries of $Fs_{26.7}Wo_{3.5}$ to $Fs_{48.2}Wo_{11.8}$ (Fig. 3) based on analyses of HED meteorites (Burbine et al. 2007a).

High-spatial resolution mapping of Psyche at diagnostic wavelengths could therefore provide important constraints on Psyche's silicate composition. Indeed, the narrowband near-IR filters on the spacecraft's Multispectral Imager were chosen specifically to attempt to discriminate between low-Ca pyroxene, high-Ca pyroxene, olivine, and Fe-bearing glass by specifically characterizing the reflectance of the surface at 725, 850, 948, and 1041 nm (e.g., Dibb and Bell 2018; Bell et al. 2016, 2021). With the multispectral imaging data, these diagnostic silicates on Psyche can be globally mapped at spatial scales down to ~ 5 meters/pixel. Although the elemental data is available on much coarser spatial scales, the optical and nuclear spectroscopy data sets can be combined to further constrain mineralogy and characterize geochemical trends (e.g., Marchi et al. 2019; Prettyman et al. 2019).

3.4 Concentration of C, P or Si in Metal

The incorporation of carbon, phosphorus, or silicon, in that order, in iron meteorites, either as included phases or dissolved in the metallic phase, indicates increasingly reducing conditions at a temperature of 1000 °C (Fig. 1).

3.4.1 Carbon

Determining the concentration of these elements within individual iron meteorites is notoriously difficult. Carbon can occur as a trace element in the metallic phase or as in-

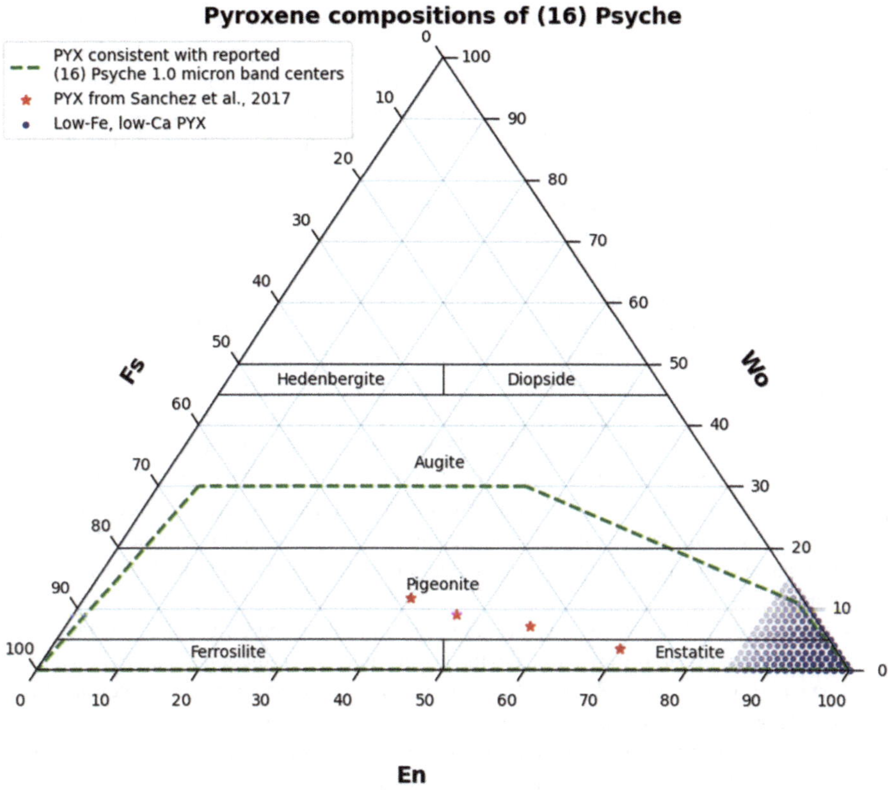

Fig. 3 Possible pyroxene compositions on Psyche as suggested by spectroscopy of meteorites and silicates. The interpretation of Psyche as dominated by low-Fe, low-Ca orthopyroxene (blue dots) is not necessarily consistent with the range of reported band centers in a wider range of laboratory spectra of pyroxenes (red stars and green dashed field)

cluded graphite, which is sometimes shock-transformed to diamond or lonsdaleite, or carbides, including cohenite ($(Fe, Ni, Co)_3C$), haxonite ($(Fe, Ni, Co)_{23}C_6$) or the rare edscottite (Fe_5C_2). Many analyses of iron meteorites, particularly those done by instrumental neutron activation analysis (INAA), do not routinely include C, P or Si. Moore et al. (1969) analyzed 10–20 gram samples of iron meteorites for Ni, Co, P, C, and S. Among the 100 distinct meteorites analyzed, the Rifle mass of the Canyon Diablo meteorite – the meteorite that formed Meteor Crater in Arizona – contained the highest carbon concentration at 0.185 wt.%. However, the Rifle specimen contains a graphite nodule of 4 cm in diameter, which was not included in the analyses of Moore et al. (1969), and Buchwald (1975) and Goldstein et al. (2017) noted the presence of graphite nodules up to 5 kg within Canyon Diablo. Many low-Ni IAB irons contain abundant mixed graphite-troilite (FeS) inclusions. Thus, Canyon Diablo and some other low-Ni group IAB irons may have C concentrations that exceeded a few weight percent (Buchwald 1975; Scott and Goldstein 2012). IIAB, IIIAB and IIIE irons contain carbides and, more rarely, graphite (Scott and Goldstein 2012), consistent with lower C concentrations (Buchwald 1975).

Obtaining true bulk compositions has typically required labor- and time-intensive planimetric analyses, in which the abundance of included phases is measured on multiple large slabs, often covering areas of more than 1000 cm^2. The mineralogy of each inclusion is determined optically and recombined into a bulk composition for the whole meteorite. One goal of the project's educational student capstone projects is to automate this process using a conveyor belt photography approach followed by implementation of machine-learning algorithms. This would enable the recognition of included phases and calculation of bulk compositions (Bowman et al. this journal), substantially increasing the meteorite database for comparison to data from Psyche.

The spacecraft GRNS is based on the MESSENGER GRNS (Goldsten et al. 2007), an instrument that detected both Si and C on Mercury. Both elements are detectable via gamma-ray measurements. Detection of C via gamma-ray measurements is possible, however it is complicated by large instrument backgrounds originating from the carbon-bearing GRNS anti-coincidence shield. Peplowski et al. (2015a) reported a C concentration on Mercury of 1.4 ± 0.9 wt% C from observations of the 4438-keV gamma ray signature. There is no requirement for or formal estimate of C sensitivity for the Psyche GRNS gamma-ray measurements, however comparable performance to the MESSENGER result is likely. The most robust C measurements from MESSENGER were made with a combination of neutron, x-ray, and spectral reflectance data (Peplowski et al. 2016). That study yielded higher-precision detection of C in $\sim$ 100-km deposits in three locations across Mercury's surface. The success of a similar investigation at Psyche will depend on the chemical makeup of Psyche's surface and the quality of elemental and mineralogical data provided by the GRNS and Imager.

A possible complication in determining the origin of carbon on Psyche's surface is the contribution of exogenous impactors. On the howardite-eucrite-diogenite parent body, likely (4) Vesta, carbon-bearing CM2 chondrites are the most common exogenic material (Zolensky et al. 1996) and contain up to a few wt.% carbon. A useful discriminator in this respect is hydrogen, as CM2 chondrites contain abundant hydrated silicates not known from carbon-bearing iron meteorites. This approach was applied to understanding exogenic contributions to the regolith of Vesta using gamma-ray and neutron spectroscopy with the GRaND instrument on Dawn (Prettyman et al. 2012). Remote-sensing observations of Psyche have reported the presence of a spectral absorption feature associated with hydrogen, and a cross correlation of observations of Vesta and Psyche with orbital H measurements at Vesta suggests Psyche may have 200–300 ppm H (Reddy et al. 2018).

The GRNS flying on the Psyche mission is expected to have comparable performance as prior GRNS instruments. There are three means of hydrogen detection with the Psyche GRNS. For concentrations higher than $\sim$ 0.5 wt.%, hydrogen can be detected via the 2223-keV gamma ray (e.g., Boynton et al. 2002). For lower concentrations, like those observed at the Moon, Vesta, and expected at Psyche, epithermal neutron measurements can identify and map hydrogen concentrations as low as 50 ppm (e.g., Feldman et al. 1998; Prettyman et al. 2012; Lawrence et al. 2015). Finally, the relative fluxes of same-element gamma rays (e.g. Fe inelastic/capture ratio) and thermal neutron absoprtion can be used to measure hydrogen concentrations of $>$ 200 ppm using a technique developed and applied to data from asteroid 433 Eros (Peplowski et al. 2015b).

3.4.2 Phosphorus

Phosphorus occurs in iron meteorites in solid solution within the metallic phase, as a wide variety of Ca-, Na-, Mg- and Fe-bearing phosphates (Buchwald 1984), and as the phosphides schreibersite ($(Fe,Ni,Co)_3P$), rarer nickelphosphide ($(Ni,Fe,Co)_3P$) and the exceptionally rare allabogdanite ($(Fe,Ni)_2P$) and unnamed ($(Fe,Ni)_4P$). Most iron meteorites contain less than 1 wt.% phosphorus (Esbensen and Buchwald 1979; Buchwald 1984), which is the limit for detection with the spacecraft GRNS. It is worth noting, however, that phosphorus is an incompatible element during core crystallization, and is therefore concentrated in later-forming, higher-Ni iron meteorites. Scott (1972), Buchwald (1975) and Yang and Goldstein (2005) showed strong positive correlations between P and Ni in IIAB, IIIAB, IVA and IVB irons. In the case of IIIAB irons, the later-crystallizing irons contained up to 1 wt.% P. While a general trend of faster cooling with lower Ni in IIIAB irons, indicative of early-formed, low-Ni irons occurring closer to the surface of the core, suggests inwards concentric or dendritic crystallization (Yang and Goldstein 2006), a core that experienced concentric outwards crystallization could, in theory, have a surface sufficiently enriched in P to be detectable via gamma-ray measurement, as outlined below.

An interesting phenomenon – liquid immiscibility – may occur during crystallization of a reduced, P- and S-rich magma to produce core regions with unusually high P concentrations. In this respect, IIAB and IIG irons are particularly interesting. Esbensen and Buchwald (1979) and Buchwald (1984) reported P concentrations in IIAB irons up to 0.9 wt.%. An individual analysis for the Ponca Creek mass of Ainsworth by Moore et al. (1969) contained 1.13 wt.% P, although this value is likely unrepresentative (Buchwald 1975). Chabot and Drake (2000) noted that fractional crystallization could encounter a field of liquid immiscibility in the Fe-Ni-S-P system, producing immiscible liquids with one highly enriched in S and the other moderately enriched in P. Wasson et al. (2007) argued that IIAB irons experienced liquid immiscibility with some IIAB irons containing a high fraction of trapped P-rich melt. These authors suggested that IIAB irons sample slightly less than 50% of the fractional crystallization sequence and that late-stage immiscible liquids may have separated into distinct, density separated layers, with evolution of these melts dominating the latter half of crystallization of the core. A S-rich upper layer would have crystallized greater than 90% FeS, although no meteorites apparently sampled this layer. Wasson et al. (2007) argue that these materials are more readily destroyed by erosional processes in space. Wasson and Choe (2009) argued that the six known IIG irons sample the lower P-rich immiscible liquid from crystallization of the IIAB core, an idea broadly supported by Chabot et al. (2020). A bulk P concentration of 1.7–2.1 wt.% (Buchwald 1975; Wasson and Choe 2009), if in a laterally or vertically extensive unit, might be measurable by GRNS. Further, observations of Psyche could determine the existence or fate of a hypothesized S-rich layer in such a scenario.

Phosphorous detection is possible via gamma-ray measurements. Based on a comparison of the relative gamma-ray emission probabilities and measurement efficiencies for the Si 1778-keV and P 1266-keV gamma-ray peaks, we expect that the GRNS will be sensitive to concentrations of > 1 wt% P by the end of the primary mission (75+ days of observing time). All element detection limits, P included, can be improved through lower-altitude observations, or added observation time, which may be possible in a potential Psyche extended mission.

3.4.3 Silicon

At the most reducing conditions calculated for meteorites, silicon is incorporated into the metallic phase in enstatite chondrites. Silicon-bearing metal is inherited by related achondrites, stony-iron and iron meteorites during differentiation. Unfortunately, relatively few silicon-bearing iron meteorites exist. Wai and Wasson (1969) observed silicon concentrations below detection limits (~ 25 ppm) in all but two of the iron meteorites they studied sampling the nine groups known at the time, suggesting that more than 99% of iron meteorites contained less than 30 ppm Si in the metal. Silicon is found in metal particles within the highly-reduced aubrites, with average concentrations for individual aubrites ranging from 0.12 to 2.44 wt.% Si (Watters and Prinz 1979) and individual grains in aubrites exhibiting a similar range (Casanova et al. 1993). Mt. Egerton is the only well-described enstatite stony-iron meteorite and contains 4.0 wt.% Si in its metallic phase (Watters and Prinz 1979). Like a few aubrites and at least one silicon-bearing iron, Mt. Egerton also contains the nickel silicide perryite ($(Ni, Fe)_5(Si, P)_2$) as an exsolved phase from the metal (Casanova et al. 1993). Silicon-bearing irons include the impact-melted and quenched iron Nedagolla (0.14 wt.% Si) and Tucson (0.8 wt.% Si), which also contains FeO-free silicates (Buchwald 1975), as well as LEW 85369. Of special note is the Horse Creek iron meteorite, with a bulk Si concentration of 2.5 wt.%, including ~ 3 vol.% perryite (Wai and Wasson 1969). Although few in number, collectively these samples point to the fact that the cores of ultra-reduced asteroids could contain Si concentrations measurable by remote gamma-ray spectroscopy as follows.

We expect to be able to detect ≥ 2.5 wt.% (global average) of Si concentration on Psyche with a one-standard-deviation uncertainty of 20% relative (Lawrence et al. this journal). This is accomplished via gamma-ray measurements of the characteristic Si gamma rays at 1778 keV. Mapping of the Si content will have higher uncertainty and a very coarse spatial resolution of ~ 5 pixels over the entire asteroid. A possible complication is the occurrence of exogenous or indigenous silicates admixed with metal. The most likely source of exogenous silicates is the previously mentioned CM2 chondritic impactors (Zolensky et al. 1996). Hydrogen abundances could distinguish that contribution. Silicon dissolved in metal might be distinguished from silica in silicates by examining other elements that typically co-occur in silicates, including aluminum (feldspar), magnesium (olivine, pyroxene) and oxygen (all silicates). Al and Ca are detectable when present at concentrations greater than ~ 1 wt.%, and the S detection limit is 3 wt.%. Mg detection is not a formal requirement of the mission, however based on comparison to unpublished MESSENGER GRS analyses, we expect to be sensitive to at least ~ 5 wt.% Mg. Oxygen should be detectable at approximately twice the concentration of silicon, suggesting a detection limit of ~ 5 wt.%. Fortuitously, the most common mafic silicates – olivine ($(Mg,Fe)_2SiO_4$) and pyroxene ($(Mg,Fe)SiO_3$) – have O/Si ratios of 1.7–2.3 by weight, suggesting the minimum detectable ~ 2.5 wt.% silicon should be accompanied by ~ 5 wt.% oxygen if bound in silicates. An O/Si ratio lower than ~ 1.7 might indicate silicon dissolved in metal, particularly if coupled with the presence of sulfide minerals that include Mg, Mn, Ti, Ca, Na, and K. Because K is a naturally radioactive element, GRNS is sensitive to concentrations as low as 200 ppm, making this element a particularly sensitive diagnostic measurement.

3.5 Sulfide Mineralogy

Sulfur is a ubiquitous component of iron meteorites. Highly incompatible in crystallizing metal, sulfur forms sulfides that sometimes occur as trapped or included melt within iron meteorites or, more often, is excluded during crystallization. The fate of this excluded sulfur-rich melt is highly uncertain, but sulfide-rich (< 10 wt.% sulfide) meteorites are exceedingly rare in the world's collection and large, sulfide meteorites are unknown. Estimates of the initial S concentration in the cores of iron meteorite groups range from 1 wt.% in IVB irons to 17 wt.% in IIAB irons (Chabot 2004) and up to 27 wt.% in low-Ni IAB irons (Goldstein et al. 2009). Within the iron meteorites themselves, Buchwald (1975) noted his planimetric analyses showed that S concentrations of 0.2–1 wt.% are quite common and above 1 wt.% not infrequent. Troilite (FeS) is the most common sulfide in iron meteorites, so much so that Buchwald (1975) specifically noted the only two sulfide-bearing irons that did not contain troilite. Daubréelite ($FeCr_2S_4$) occurs as an exsolution product from troilite, most abundantly in the low-Ni members of groups IAB, IIAB and IIIAB typically formed under reducing conditions. Interestingly, daubréelite occurs, sometimes in association with chromite ($FeCr_2O_4$), even in high-Ni IVB irons that experienced oxidation at high-temperature, leading Campbell and Humayun (2005) to argue that the redox state of the body must have changed with temperature from oxidizing at igneous temperatures to reducing at the temperature of sulfide exsolution.

As noted earlier, a wide range of normally lithophile elements become chalcophile at ultra-reducing oxygen fugacities where silicon is incorporated into metal. These include Mg, Mn, Ti, Ca, Na, and K, with an array of exotic sulfides including, most frequently, niningerite (MgS), alabandite (MnS), and oldhamite (CaS). The abundances of these phases likely to occur within a crystallizing sulfide melt are essentially unknown, as they have never been sampled. However, partial melting experiments on enstatite chondrites (McCoy et al. 1999) produced sulfide melts with 2–6 wt.% each of Mg, Ca and Cr and more than 10 wt.% Mn, suggesting these may form significant fractions during crystallization. Further, the densities of oldhamite (2.6 g/cm^3) and niningerite (3.2 g/cm^3) are significantly less than that of troilite (4.6 g/cm^3). Thus, these phases might separate by buoyancy within a crystallizing sulfide melt. Although rare, oldhamite-rich clasts, with concentrations of oldhamite up to 30 vol.%, are known from aubrites (Wheelock et al. 1994; McCoy 1998), perhaps indicative of this process.

In contrast to the metal compositions, where GRNS will likely provide the main indication of composition, sulfide mineralogy is likely best characterized by multispectral imaging. The relationship between composition and spectroscopic features exhibited by sulfides has not been characterized as well as that of the iron-bearing silicates. Most endmember diatomic sulfides (e.g., troilite – FeS; oldhamite – CaS; niningerite – MgS; and alabandite – MnS) exhibit either an absorption feature or a change in spectral slope at visible (0.3–0.7 μm) wavelengths (e.g., Britt et al. 1992; Cloutis and Gaffey 1994; Burbine et al. 2002; Helbert et al. 2013), caused by either crystal-field effects or semiconductor band-gap electronic transitions (Vaughan and Craig 1978; Cloutis and Gaffey 1994; Clark 1999).

Spectra from natural and synthetic sulfides can have markedly different absorption features, however, presumably because of differing degrees of crystallinity at microscales. For example, meteoritic troilite is characterized by a dark (less than $\sim 10\%$ reflectance), red-sloped spectrum from 0.35 to 2.5 μm (Britt et al. 1992; Cloutis and Gaffey 1994), with a break to a less-red slope beyond 0.6–0.7 μm. Synthetic troilite measured by Cloutis and Gaffey (1994) has much stronger absorption features from 0.3–0.6 μm, as well as a strong ~ 1.0 μm feature attributed to Fe^{2+} crystal field effects. On the other hand, spectra of oldhamite from the Norton County aubrite exhibit strong absorption features centered near ~ 0.5 μm and ~ 0.95 μm (Burbine et al. 2002). Spectra of synthetic oldhamite measured by Helbert et al. (2013) only exhibit weak features near 0.55 and 0.75 μm. These distinctive absorption features of oldhamite, if present in spectra from Psyche, might provide the best evidence of formation at ultra-reducing conditions.

Several of the Multispectral Imager's narrowband filters have been chosen to specifically attempt to discriminate between different potential sulfides on the asteroid. Specifically, measurements of the reflectance at 0.437, 0.495, and 0.55 μm could allow the detection of troilite or oldhamite on the surface, if they are present in significant quantities. A study of spectral properties of iron meteorites and their common accessory phases by Cloutis et al. (1990) indicates that the reflectance of a 90/10 wt.% mixture of iron meteorite metal and troilite powders is several percent darker than a spectrum of pure iron meteorite powder. Further work on metal-troilite mixtures by Dibb et al. (2021) shows the development of the visible wavelength absorption feature in spectra of troilite when the metal-to-troilite ratio is 60/40 wt.%. Additional evidence for oldhamite could come from detection of the narrow ~ 0.95 μm oldhamite feature (like that in Norton County) using the Imager's near-IR filters (e.g., Dibb and Bell 2018; Bell et al. 2016, 2021). Just like for the iron-bearing silicates, using Imager and GRNS data together could significantly help to constrain the identification of sulfides on Psyche.

4 Redox Constraints on the Genesis of Psyche

The discussion above demonstrates how redox controls the mineralogy and phase compositions of an asteroidal body and, inversely, how determining the mineralogy and phase compositions can reveal the redox state of Psyche. In this section, we discuss how determining the redox state of Psyche can help us reveal the genesis of this metal-rich world.

4.1 Meteorite Analogs as Clues to Origin

Throughout the history of asteroid spacecraft missions, links to meteorite analogs – whether argued to originate on the target asteroid or on that class of asteroids – have extended interpretations not available from spacecraft data alone. Examples include data from the Near Earth Asteroid Rendezvous (NEAR) mission suggesting links between (253) Mathilde and heated CM chondrites (Rivkin et al. 1997) and (433) Eros and ordinary chondrites (Trombka et al. 2000), data from the Dawn mission that supported Vesta as the parent body of the howardite-eucrite-diogenite meteorites (McSween et al. 2013), and orbital data from the OSIRIS-REx mission linking (101955) Bennu to hydrated CI or CM chondrites (Hamilton et al. 2019). The links strengthen meteorite-asteroid associations that can be applied to hundreds of thousands of asteroids that will never be visited by spacecraft. In addition, such links provide the distinct advantage for the mission of potentially expanding the toolset available for interpretation to mineralogical, chemical, and isotopic (including formation, impact, and exposure ages) data not attainable by spacecraft-based instruments. We will discuss several potential analogs, focusing on the how the redox state of the meteorite might further inform the redox state of Psyche.

4.1.1 Iron Meteorites

Early density estimates for Psyche (see Elkins-Tanton et al. this journal, and references therein) pointed to a metal-rich asteroid akin to the parent bodies of iron meteorites. These meteorites, composed largely of Fe,Ni metal, are commonly thought of as reduced and must have formed at or below the IW buffer. While true, iron meteorites sample a remarkable range of redox conditions. The IVB irons, enriched in Ni by oxidation and depleted in elements incorporated in metal under reducing conditions (C, P and Si), formed at an oxygen fugacity of $\sim$ IW-1 (Campbell and Humayun 2005), among the most oxidized of the iron meteorites. The Ni-poor, C-rich IAB irons formed at $\sim$ IW-2.5 to IW-3 (Benedix et al. 2005) likely similar to the P-rich IIAB irons (Fig. 1), intermediate among iron meteorites in redox state. This is consistent with their low Ni concentrations and incorporation of the moderately reduced elements C and P. Finally, rare Si-bearing irons, the most reduced of the iron meteorites, formed in the range of IW-5 to IW-7 if similar to Si-bearing metal in aubrites (Righter et al. 2016) to as reducing as IW-9 to IW-10 if on the Si-SiO_2 buffer (Fig. 1), bookending a 4–9 log unit difference in redox among irons.

4.1.2 Stony-Iron Meteorites

Recent estimates of the density of Psyche between 3.4 and 4.1 g/cm^3, corresponding to 30–60 vol.% metal (Elkins-Tanton et al. 2020 and references therein), expand the range of meteorite analogs. Stony-iron pallasites and mesosiderites are permissible meteorite analogs based on their density (Macke 2010).

Pallasites range from subequal mixtures of metal and silicate to metal-dominated meteorites, all with magnesian olivine of Fa_{11-18} (Mittlefehldt et al. 1998). Pallasites have long been viewed as sampling the core-mantle boundary of a differentiated asteroid. Yang et al. (2010) suggested that non-uniform cooling rates require a mixed metal-silicate body formed by collision, a view which has gained increasing support.

Mesosiderites are mixtures of metal and silicates and, with only a few exceptions, silicate-dominated. There is overall agreement that impact played a role in forming mesosiderites, with their mixture of core metal and crustal silicates (Mittlefehldt et al. 1998). It remains an open question whether collisional silicate-metal mixing could produce a body of the size of Psyche with intimate mixing of metal and silicate at the cm-scale throughout the body or whether mixing would have occurred on a more localized or smaller body scale. Silicates are relatively FeO-rich (pyroxene up to Fs_{40}) basaltic to orthopyroxenitic clasts (Mittlefehldt et al. 1998). Viikinkoski et al. (2018) suggested mesosiderites as a meteoritic analog for Psyche, but, unlike Psyche, mesosiderite spectra exhibit deep absorption features (Burbine et al. 2007b) consistent with their FeO-rich silicates (Mittlefehldt et al. 1998).

While pallasites and mesosiderites differ in important aspects, the silicate portions of these bodies appear to have formed at similar oxygen fugacities at or near the IW buffer (Hewins and Ulmer 1984; Righter et al. 1990). The differences therefore do not result from redox state, but rather from fractionation that occurs during differentiation of the silicate mantle and crust (e.g., Righter and Drake 1997; Ruzicka et al. 1997; McCoy et al. 2006; Mandler and Elkins-Tanton 2013), a topic discussed earlier.

4.1.3 Chondritic Meteorites

Enstatite chondrites (Keil 1969) and CB, CH, and G chondrites (Weisberg et al. 1995) are permissible chondritic analogs for Psyche based on density alone (Macke 2010). From a redox perspective, these plot at a high ratio of reduced to oxidized iron on a modified Urey-Craig diagram (Fig. 2), suggesting they formed in a reducing environment.

Enstatite chondrites have essentially FeO-free silicates and up to 28 wt.% of Ni-poor (3.3–8.2 wt.%), Si-bearing metal (1.1–3.6 wt.%) (Keil 1969), indicative of formation at or below the Si-SiO_2 buffer (Fig. 1). Enstatite chondrites would seem to be a poor match to Psyche. The FeO-free nature of their silicates is inconsistent with a weak 1 µm absorption feature and their densities marginally overlap with the low range of currently accepted densities, although we reiterate that considerable uncertainty exists in these derived values for Psyche.

CB, CH and G chondrites have slightly more iron-rich mafic silicates (olivine of Fa_3) and up to 60 vol.% metal, with a Ni concentration of 6–12 wt.% (e.g., Weisberg et al. 1995). These meteorites illustrate the interesting point that although the compositions of their metal and silicates point to a common redox value, the metal to silicate ratios are an unreliable indicator of redox state. CB/CH/G chondrites have higher metal to silicate than enstatite chondrites but are less reduced. The high metal to silicate ratio in the former likely results from their formation as impact products, as argued by Krot et al. (2005) based on young silicate ages. It is unclear whether this high metal to silicate ratio reflects the composition of an entire asteroid or a localized region on a larger asteroid.

4.2 The Influence of Redox-Controlled Mineralogy on Psyche

Redox is an important controlling factor in determining the mineralogy of both silicate and metal-sulfide fractions of any asteroidal body. As a result, the differentiation, solidification, and subsolidus cooling histories of these bodies can differ significantly.

4.2.1 Silicate Mineralogy

As discussed earlier, oxidation-reduction of iron between silicates and metal should result in low-Ni metal coexisting with low FeO silicates, whereas high-Ni metal should coexist with high FeO silicates. We noted that observations of meteorites support this general relationship, citing the olivine-rich brachinites (Ni-rich metal and FeO-rich olivine (29.7 wt.% Ni, $Fa_{32.3}$ in ALH 84025; Day et al. 2012)) and aubrites (Ni-poor metal and pyroxene that is essentially FeO-free enstatite (5.8 wt.% Ni, $Fs_{0.02}$ in Aubres; Watters and Prinz 1979)). A corollary to this, demonstrated by Mueller and Olsen (1967) for ordinary chondrites, is that the silicate mineralogy also changes with redox. Olivine is chemically equivalent to pyroxene with metallic iron and oxygen:

$$FeMgSiO_4 = MgSiO_3 + Fe^0 + 1/2O_2 \qquad \text{Reaction 5}$$

As a result, the chondritic precursors of asteroids will tend to be olivine-rich under oxidizing conditions and pyroxene-rich under more reducing conditions. Again, this is consistent with, brachinite formation from R chondrite-like, olivine-rich precursors (Gardner-Vandy et al. 2013), and aubrite formation from pyroxene-rich, enstatite chondrite-like precursors (Keil 2010). Of relevance to the Psyche mission is that redox controlled mineralogy should also be reflected in the ratio of, for example, silicon to oxygen, with an olivine-dominated Psyche having a lower Si/O ratio and a pyroxene-dominated Psyche a higher Si/O ratio.

4.2.2 Thermal History

Redox control of mineralogy and mineral compositions can influence thermal histories during differentiation. It is widely accepted that short-lived ^{26}Al was the primary heat source for asteroid differentiation (McSween et al. 2002). The absence of reduced Al, even at the ultra-reducing conditions of enstatite meteorites, suggest that the Al-Al_2O_3 buffer occurs at more reducing conditions than the Si-SiO_2 buffer and Al was not incorporated into planetary cores. Unlike aluminum, iron is redox sensitive. Reduced to its metallic form, eat-producing ^{60}Fe, if incorporated into the core at the upper limit of allowable initial Solar System concentrations, could extend core solidification or slow subsolidus cooling, as suggested by Moskovitz and Walker (2011) for IVA irons. Among the possible influences of this slower cooling, formation of tetrataenite during prolonged cooling could alter the magnetic signature of Psyche, as discussed below.

4.2.3 Core and Asteroid Size

As discussed above, redox can serve as a control on the Ni concentration of the core in response to the oxidation or reduction of iron. Concomitant to this difference in Ni are differences in the size, S fraction, and crystallization temperature of the core. Bercovici (2020) calculated core radii and compositions for ordinary and carbonaceous chondrites.

If we consider only those that did not experience aqueous alteration (H, L, LL, CO), core radii range from 38 to 49% of the body. Toplis et al. (2013) considered a larger range of chondrites (EH, EL, H, L, CH, CO), calculated core radii vary from 38% (L) to 55% (EH) of the body. If Psyche, with a radius of 112 km, samples a core of a differentiated body, similar core radii percentages would yield an intact parent body of 205 to 296 km in radius. These bodies would differ in volume by a factor of three and the largest of these calculated bodies would be larger than Vesta.

4.2.4 Core Sulfur Concentration

As the concentration of iron in the core changes because of oxidation, sulfur concentration should change as well. Bercovici et al. (2019) and Bercovici (2020) calculated that S varies from 9.4–21.5 wt.% of the core among H, L, LL and CO chondrites, as fO_2 varies from IW-1.4 in H chondrites to IW-0.54 in CO chondrites. Thus, an oxidized Psyche might be expected to have a higher sulfur concentration in its core than a reduced Psyche with a larger core and metallic fraction. Toplis et al. (2013) calculated S concentrations of 1 wt.% (CH) to 25 wt.% (L, EL) for non-hydrated compositions. Both Bercovici (2020) and Toplis et al. (2013) emphasized that bulk composition, as much as redox, is perhaps the most significant control on S concentration in the core, as evidenced by large, S-rich cores for the CI chondrites (Bercovici et al. 2019; Bercovici 2020) and the EL chondrites (Toplis et al. 2013). Thus, while S concentration varies with oxygen fugacity, it is perhaps not the dominant control.

4.2.5 Core Crystallization

Sulfur concentration can also have a pronounced effect on the crystallization temperature of the core. In the Fe-Ni-S system (Kullerud 1963), the liquidus temperature ranges from 1539 °C in the S-free case to 988 °C at the Fe-FeS cotectic at 31 wt.% S. Thus, even among the non-aqueously altered chondritic compositions considered by Toplis et al. (2013) and Bercovici (2020), the liquidus temperature could vary from more than 1500 °C for CH chondrites to $\sim$ 1350 °C for H chondrites to $\sim$ 1200 °C for L, LL, CO and EL chondrites. These differences might influence the onset of core formation and the duration of a molten core during asteroid cooling and crystallization. The metal-sulfide component of S-rich chondritic meteorites might be completely molten at lower degrees of silicate melting and remain molten longer during crystallization and cooling of the silicate shell. One possible outcome of this extended thermal history could be the migration of an immiscible, S-rich melt to the surface of a core stripped of its silicate shell prior to complete differentiation (Bercovici et al. 2019; Bercovici 2020), with possible detection of such S-rich materials on the surface of Psyche if it was stripped of its silicate shell while the core was still partially molten.

4.2.6 Subsolidus History

At subsolidus temperatures, redox is a major control on the Ni and P concentrations of the core and, as a result, the metallic mineralogy established during solid state exsolution. Above 912 °C, Fe,Ni metal, irrespective of Ni concentration, occurs as the disordered γ phase, termed taenite in meteorites. Below this temperature, iron meteorites, with Ni concentrations of 5–60 wt.%, exsolve in the solid state to a mixture of higher-Ni, γ-Fe,Ni (taenite) and low-Ni, α-Fe,Ni (kamacite). This transition is commonly envisioned as $\gamma \rightarrow \gamma + \alpha$ with exsolution temperatures ranging from more than 800 °C at 5 wt.% Ni to $\sim$ 600 °C at 20 wt.% Ni (e.g., Romig and Goldstein 1980 and references therein). Yang and Goldstein (2005) showed that this reaction does not occur in the formation of iron meteorites. Instead, formation of kamacite is controlled by the Fe-Ni-P system with intermediate steps forming γ + phosphide, $\alpha + \gamma$ + phosphide, or α_2 (martensite). Irrespective of the mechanism, the relative abundance of taenite (γ-Fe,Ni) and kamacite (α-Fe,Ni) will depend on the bulk concentration of Ni in the metallic phase, with higher-Ni irons (and the cores from which they come) having greater abundances of taenite. In addition, high-Ni irons can, if cooled slowly, form a metastable, ordered γ'' phase called tetrataenite (Yang et al. 1996). The relative abundances of these phases could potentially influence impact dynamics and, particularly, magnetics of Psyche. Microhardness measurements of taenite yield values approximately twice that of kamacite (Buchwald 1975). Thus, a greater abundance of taenite in an oxidized, high-Ni core might be more resistant to impact, although a variety of other parameters (e.g., grain boundaries per unit volume) likely also influence the response to impact (Petrovic 2001).

4.2.7 Magnetic Signature

Tetrataenite, disordered taenite and kamacite are the dominant ferromagnetic minerals in iron meteorites, which control the magnetic remanence. Tetrataenite and disordered taenite are found in µm-scale and larger rims, as nm- to µm-size grains intergrown with kamacite (plessite) and as < 200-nm grains embedded in a Ni-poor matrix (cloudy zone). Combined with GNRS measurements of Ni abundances, measurements of Psyche's remanent magnetic field and/or induced magnetic field by the Magnetometry Investigation could help constrain the kamacite to taenite ratio.

The variation in grain size and abundance of taenite/tetrataenite in different iron meteorites typically leads to a > 2 order of magnitude difference in saturation remanent magnetization (i.e., the magnetization produced after switching on and off a strong external field) between Ni-poor and Ni-rich meteorites (Dos Santos et al. 2015). Magnetic field measurements at Psyche will not provide an estimate of the saturation remanent magnetization, but will provide an estimate of the natural remanent magnetization (NRM; magnetic moment per unit mass) that these minerals may have acquired if crystallizing or cooling in a magnetic field. Both saturation and natural remanent magnetization are proportional (e.g., Wieczorek et al. 2012), which implies that taenite/tetrataenite-rich (i.e., Ni-rich) iron-meteorite-like material might be expected to carry a higher NRM than kamacite-rich (i.e., Ni-poor) material. Unfortunately, many iron meteorite samples are subject to various magnetic contaminations on Earth (hand magnets, weathering, and viscous effects from the geomagnetic field), such that this trend has not been observed in laboratory experiments (Terho et al. 1993). However, Psyche will not have been affected by such remagnetization, making magnetic measurements an additional tool to possibly constrain the Ni abundances.

The Magnetometry Investigation will also conduct electromagnetic sounding to measure the asteroid's inductive response to the solar wind magnetic field. This method probes the internal structure of electrically conducting planetary bodies. Different induced field geometries will arise depending on the composition of the surface (and interior) of the asteroid (Elkins-Tanton et al. 2020). For example, if the oxidation state of the surface is such that iron oxides formed above the IW buffer dominate the composition, the asteroid will behave as a poor electrical conductor and have a weak induced field. This signal would likely be distinguishable from the NRM possibly carried by the iron oxides. On the other hand, a more reduced body formed at or below IW dominated by Fe-Ni will be highly conducting, such that it may be challenging to discriminate between a weak remanent field and induced field.

4.3 Redox Constraint on Where and When Psyche Formed

4.3.1 Inner and Outer Solar System Isotopic Reservoirs

Isotopic anomalies thought to have a nucleosynthetic origin have been known from stony meteorites since the pioneering work of Clayton et al. (1973) on oxygen isotopes. Warren (2011) summarized a variety of nucleosynthetic isotopic data for chromium, titanium, nickel, and oxygen in chondrites, achondrites, stony-irons, and silicate-bearing IIIAB irons, as well as Earth, the Moon and Mars. Samples divide into two distinct populations with carbonaceous chondrites, including the metal-rich CB chondrites, forming one cluster and other meteorites (ordinary and enstatite chondrites, R chondrites, achondrites, most stony-irons, Earth, Moon and Mars) forming a second cluster. Warren (2011) argued that this bimodality may reflect episodic accretion or, alternatively, a division of material that accreted in the inner (non-carbonaceous) and outer (carbonaceous) Solar System. If the latter is true, association with a meteorite analog might provide insight on whether Psyche formed in the inner or outer Solar System.

In the case of an undifferentiated, metal-rich Psyche with low, but detectable, FeO in silicates, the inclusion of CB chondrites in the carbonaceous group might point to an outer Solar System origin. We caution, however, that the spacecraft's instrument suite cannot measure isotopic abundances for Cr, Ti, Ni and O. Indeed, Cr and Ti concentrations will likely not be measurable remotely. Further, although the CB chondrites fall into the carbonaceous group, the petrologically-similar G chondrites are in the non-carbonaceous cluster, inferred to have an inner Solar System origin. Thus, whatever process produced these metal-rich undifferentiated meteorites operated in both isotopic reservoirs and, by extension, throughout the Solar System.

Kruijer et al. (2017) extended this dichotomy of meteorites by measuring Mo and W isotopes in iron meteorites. Like stony and stony-iron meteorites, irons also divided into two groups, termed NC and CC irons. NC irons included groups IC, IIAB, IIIAB, IIIE and IVA, whereas CC irons included IIC, IID, IIF, IIIF and IVB irons as well as the ungrouped iron Wiley. Although there were relatively few meteorite groups reported by both Warren (2011) and Kruijer et al. (2017), IIIAB irons were included in the non-carbonaceous group in both studies. Interestingly, Kruijer et al. (2017) reported that NC and CC irons differed in their two-stage Hf-W model ages, with NC iron core formation occurring ~ 0.3–1.8 Myr after CAI formation, whereas CC iron core formation occurred ~ 2.2–2.8 Myr after CAI formation. These authors argued that these two isotopic reservoirs were separated because of the formation of Jupiter ~ 1–4 Ma after CAI formation.

4.3.2 A Proxy for Psyche

The instrument payload of the mission will be unable to measure Mo or W isotopes or the concentrations of these elements. Yet, the concentration of Ni may serve as a reasonable proxy for these groupings, as noted by Rubin (2018). Except for the IIIF irons, all of the groups within the CC cluster and the ungrouped Wiley have average Ni concentrations in excess of 10 wt.% (Mittlefehldt et al. 1998, and references therein). In contrast, none of the irons within the NC cluster have average Ni concentrations greater than 9 wt.% (Fig. 4). Although with somewhat greater overlap, the same holds for ungrouped irons (Spitzer et al. 2019). Thus, a Ni concentration equal to or in excess of 10 wt.% for a metal-dominated, differentiated Psyche would strongly favor an origin in the CC reservoir, presumably the outer Solar System beyond Jupiter. In contrast, a Ni concentration less than 10 wt.% would strongly, but perhaps not uniquely, favor an inner Solar System origin.

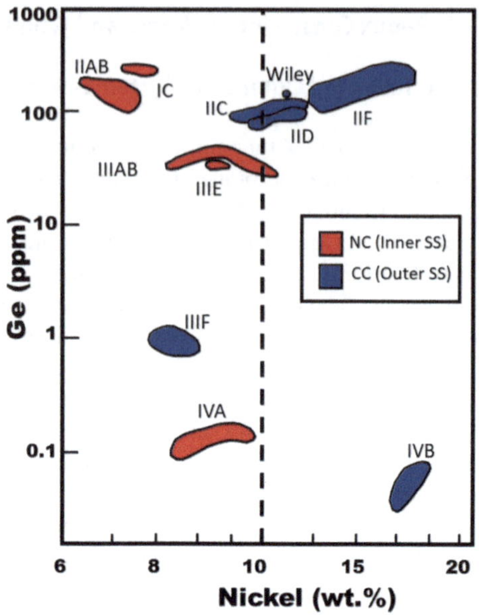

Fig. 4 Ni (wt.%) vs Ge (ppm) for groups of iron meteorites and the ungrouped iron Wiley determined to have formed in the NC (inner Solar System) or CC (outer Solar System) isotopic reservoirs by Kruijer et al. (2017). Ni and Ge values from J.T. Wasson, as summarized in Goldstein et al. (2009) and Buchwald (1975). Excepting group IIIF irons, CC iron groups have greater than 10 wt.% Ni, while NC irons have less than 10 wt.% Ni. Thus, Ni, which is controlled in part by oxidation, may serve as a useful proxy to identify the formation location of Psyche

The difference in Ni concentrations between the NC and CC clusters, particularly given that these are average compositions for the various groups, almost certainly reflects oxidation of iron. The work of Kruijer et al. (2017) might provide some insights on when and where that oxidation occurred. Oxidation of Fe,Ni metal in a parent body setting would likely require interaction between metal and water, which was incorporated into the parent body as ice. The higher Ni concentrations in CC irons is, from this perspective, consistent with their origin as the cores of ice-rich asteroids accreted in the outer Solar System. However, the ages of Ni-rich CC irons present an interesting conundrum. Kruijer et al. (2017) calculated accretion times for CC iron groups, assuming chondritic Al abundances, of ~ 0.7–1.2 Ma. These accretion times equate to 1–2 half lives of ^{26}Al (720,000 years), implying a decrease in thermal energy compared to NC bodies accreted earlier. Grimm and McSween (1993) were among the earliest to suggest that accretion of ice, coupled with later accretion times in the outer Solar System, might preclude silicate melting. Castillo and Young (2016) required accretion within 1 Myr of CAIs to reach even the silicate solidus. Together, these observations might suggest that relatively late accretion of asteroids in the CC region would have prevented differentiation of ice-rich asteroids required for parent body oxidation of Fe,Ni metal to FeO. Instead, the alterative scenario is that oxidation of Fe,Ni metal occurred while still unaggregated particles in the solar nebula prior. Thus, the metal present in the CC region during late accretion of Ni-rich, CC iron meteorite parent bodies was already oxidized.

5 Conclusions

The mission's Oxidation-Reduction Working Group is focused on understanding, constraining, and applying the redox state of Psyche to decipher the origin of a metal-rich world. Specifically:

- Oxidation-reduction state of an asteroid is a key parameter in determining its history.

- The occurrence of Ni, Fe, C, Cr, P and Si, in that order, in the metal or sulfide phase of an asteroidal body indicates increasingly reduced conditions.
- Key observations made by the Imager and Gamma-ray and Neutron Spectrometer of Psyche to constrain the redox state include: the presence of Fe,Ni metal, the Ni concentration of that metal, the FeO concentration and mineralogy of silicates, the presence of C, P or Si in the metallic phase, and the mineralogy of sulfides.
- Constraining the redox state of Psyche can help constrain its origin, as well as the location and timing of its formation and possible oxidation event(s). Combined meteorite analog studies will provide a better understanding of the controls that redox has on silicate and metal mineralogy, associated differences during melting, crystallization and subsolidus cooling, and magnetic signatures.

Acknowledgements The authors thank the entire Psyche team of engineers and scientists for supporting the work of the Oxidation-Reduction Working Group and enabling the observations discussed here. We thank Kevin Righter for an informative and helpful pre-submission review of the first sections of this paper, which clarified some of the content on oxygen buffers. Helpful reviews and comments by Edward Scott, Guest Editor Carol Polanskey, and an anonymous reviewer significantly improved and clarified some aspects of the paper.

References

J.B. Adams, Interpretation of visible and near-infrared diffuse reflectance spectra of pyroxenes and other rock-forming minerals, in *Infrared and Raman Spectroscopy of Lunar and Terrestrial Minerals*, ed. by C.J. Karr (Academic Press, New York, 1975), pp. 91–116

J.F. Bell III, M.A. Ravine, M.A. Caplinger, J.A. Schaffner, S.M. Brylow, M.J. Clark, D.A. Peckham, P.T. Otjens, J.W. Ravine, J.D. Laramee, R.C. Juergens, W. Morgan, A.G. Parker, D. Williams, A. Winhold, S. Dibb, E. Cisneros, M. Walworth, N. Amiri, C. Polanskey, N. Mastrodemos, R. Park, N.K. Alonge, R. Jaumann, R.P. Binzel, T. McCoy, M.G. Martin, P.A. Arthur, The Psyche Multispectral Imager investigation: characterizing the geology, topography, and color properties of a potentially metallic world. Space Sci. Rev. (2021, this journal)

J.F. Bell, L.T. Elkins-Tanton, C. Polanskey, M.A. Ravine, M.A. Caplinger, E. Asphaug, D. Bercovici, B. Bills, R.P. Binzel, W. Bottke, R. Jaumann, S. Marchi, R.S. Park, C.A. Raymond, D. Wenkert, M. Wieczorek, B. Weiss, M.T. Zuber, The Psyche Multispectral Imager investigation: characterizing the geology, topography, and compositional properties of a metallic world, in *47th Lunar Planet. Sci. Conf., #1366* (2016)

G.K. Benedix, D.S. Lauretta, T.J. McCoy, Thermodynamic constraints on the formation conditions of winonaites and silicate-bearing IAB irons. Geochim. Cosmochim. Acta **69**, 5123–5131 (2005)

H.L. Bercovici, The effect of bulk composition on the sulfur content of cores. MS Thesis, Arizona State University (2020). 34 pp.

H.L. Bercovici, L.T. Elkins-Tanton, L. Schaefer, The effect of bulk composition on the behavior of sulfur during core formation, in *50th Lunar Planet. Sci. Conf., #1366* (2019)

W.V. Boynton, W.C. Feldman, S.W. Squyres, T.H. Prettyma, J. Brückner, L.G. Evans, R.C. Reedy, R. Starr, J.R. Arnold, D.M. Drake, P.A.J. Englert, Distribution of hydrogen in the near surface of Mars: evidence for subsurface ice deposits. Science **297**, 81–85 (2002)

R. Brett, M. Sato, Intrinsic oxygen fugacity measurements on seven chondrites, a pallasite, and a tektite and the redox state of meteorite parent bodies. Geochim. Cosmochim. Acta **48**(1), 111–120 (1984)

D.T. Britt, J.F. Bell, H. Haack, E.R.D. Scott, The reflectance spectrum of troilite and the T-type asteroids (abstract). Meteoritics **27**, 207 (1992)
V.F. Buchwald, *Handbook of Iron Meteorites* (Univ. of California Press, Berkeley, 1975)
V.F. Buchwald, Phosphate minerals in meteorites and lunar rocks, in *Phosphate Minerals*, ed. by J.O. Nriagu, P.B. Moore (Springer, Berlin, 1984). https://doi.org/10.1007/978-3-642-61736-2_5
T.H. Burbine, T.J. McCoy, L.R. Nittler, G.K. Benedix, E.A. Cloutis, T.L. Dickinson, Spectra of extremely reduced assemblages: implications for Mercury. Meteorit. Planet. Sci. **37**, 1233–1244 (2002)
T.H. Burbine, P.C. Buchanan, R.P. Binzel, Deriving formulas from HED spectra for determining the pyroxene mineralogy of Vesta and Vestoids, in *38th Lunar Planet. Sci. Conf., #2117* (2007a)
T.H. Burbine, R.C. Greenwood, P.C. Buchanan, I.A. Franchi, C.L. Smith, Reflectance spectra of mesosiderites: implications for asteroid 4 Vesta, in *38th Lunar Planet. Sci. Conf., #2119* (2007b)
S.J. Bus, R.P. Binzel, Phase II of the small main-belt asteroid spectroscopic survey: the observations. Icarus **158**, 106–145 (2002)
A.J. Campbell, M. Humayun, Compositions of group IVB iron meteorites and their parent melt. Geochim. Cosmochim. Acta **69**, 4733–4744 (2005)
I. Casanova, K. Keil, H.E. Newsom, Composition of metal in aubrites: constraints on core formation. Geochim. Cosmochim. Acta **57**, 675–682 (1993)
J.C. Castillo, E.D. Young, Origin and evolution of volatile-rich planetesimals, in *Planetesimal Differentiation*, ed. by L. Elkins-Tanton, B. Weiss (Cambridge University Press, Cambridge, 2016)
N.L. Chabot, Sulfur contents of the parental metallic cores of magmatic iron meteorites. Geochim. Cosmochim. Acta **68**, 3607–3618 (2004)
N.L. Chabot, M.J. Drake, Crystallization of magmatic irons meteorites: the effect of phosphorus and liquid immiscibility. Meteorit. Planet. Sci. **35**, 807–816 (2000)
N.L. Chabot, R.H. Cueva, A.W. Beck, R.D. Ash, Experimental partitioning of trace elements into schreibersite with applications to IIG iron meteorites. Meteorit. Planet. Sci. **55**, 726–743 (2020)
R.N. Clark, Spectroscopy of rocks and minerals, and principles of spectroscopy, in *Manual of Remote Sensing, Vol. 3, Remote Sensing for the Earth Sciences* (1999) 58 pp.
B.E. Clark, S.J. Bus, A.S. Rivkin, M.K. Shepard, S. Shah, Spectroscopy of X-type asteroids. Astron. J. **128**, 3070–3081 (2004)
R.N. Clayton, L. Grossman, T.K. Mayeda, A component of primitive nuclear composition in carbonaceous meteorites. Science **182**, 485–488 (1973)
E.A. Cloutis, M.J. Gaffey, Pyroxene spectroscopy revisited: spectral-compositional correlations and relationship to geothermometry. J. Geophys. Res., Planets **96**, 22809–22826 (1991)
E.A. Cloutis, M.J. Gaffey, An X-ray diffraction and reflectance spectroscopy study of iron sulphides (abstract), in *25th Lunar Planet. Sci. Conf.* (1994), pp. 273–274
E.A. Cloutis, M.J. Gaffey, D.G.W. Smith, R. St. J. Lambert, Reflectance spectra of "featureless" materials and the surface mineralogies of M- and E-class asteroids. J. Geophys. Res., Solid Earth **95**, 281–293 (1990)
A. Coradini, F. Capaccioni, S. Erard, G. Arnold, M.C. De Sanctis, G. Filacchione, F. Tosi, M.A. Barucci, M.T. Capria, E. Ammannito, D. Grassi, G. Piccioni, S. Giuppi, G. Bellucci, J. Benkhoff, J.P. Bibring, A. Blanco, M. Blecka, D. Bockelee-Morvan, F. Carraro, R. Carlson, U. Carsenty, P. Cerroni, L. Colangeli, M. Combes, M. Combi, J. Crovisier, P. Drossart, E.T. Encrenaz, C. Federico, The surface composition and temperature of asteroid 21 Lutetia as observed by Rosetta/VIRTIS. Science **334**, 492–494 (2011)
J.M.D. Day, R.J. Walker, R.D. Ash, Y. Liu, D. Rumble III, A.J. Irving, C.A. Goodrich, K. Tait, W.F. McDonough, L.A. Taylor, Origin of felsic achondrites Graves Nunataks 06128 and 06129, and ultramafic brachinites and brachinite-like achondrites by partial melting of volatile-rich primitive parent bodies. Geochim. Cosmochim. Acta **81**, 94–128 (2012)
F.E. DeMeo, R.P. Binzel, S.M. Slivan, S.J. Bus, An extension of the Bus asteroid taxonomy into the near-infrared. Icarus **202**, 160–180 (2009)
S.D. Dibb, J.F. Bell, Optimization of narrowband visible to near-IR filters for the Psyche Multispectral Imager, in *49th Lunar Planet. Sci. Conf., #2203* (2018)
S.D. Dibb, J.F. Bell III, L.A.J. Garvie, Reflectance spectra of metal-troilite mixtures: implications for M/X-type asteroid exploration, in *52nd Lunar Planet. Sci. Conf., #1543* (2021)
E. Dos Santos, J. Gattacceca, P. Rochette, R.B. Scorzelli, G. Fillion, Magnetic hysteresis properties and ^{57}Fe Mössbauer spectroscopy of iron and stony-iron meteorites: implications for mineralogy and thermal history. Phys. Earth Planet. Inter. **242**, 50–64 (2015)
L.T. Elkins-Tanton, E. Asphaug, J.F. Bell III, H. Bercovici, B. Bills, R. Binzel, W.F. Bottke, S. Dibb, D.J. Lawrence, S. Marchi, T.J. McCoy, R. Oran, R.S. Park, P.N. Peplowski, C.A. Polanskey, T.H. Prettyman, C.T. Russell, L. Schaefer, B.P. Weiss, M.A. Wieczorek, D.A. Williams, M.T. Zuber, Observations, meteorites, and models: a preflight assessment of the composition and formation of (16) Psyche. J. Geophys. Res., Planets **125**, e2019JE006296 (2020). https://doi.org/10.1029/2019JE006296

H.J.T. Ellingham, Reducibility of oxides and sulphides in metallurgical processes. J. Soc. Chem. Ind. **63**, 125–160 (1944)

K.H. Esbensen, V.F. Buchwald, A database of iron meteorite data – statistical analysis of the global iron meteorite population (abstract). Meteoritics **14**, 573 (1979)

W.C. Feldman, S. Maurice, A.B. Binder, B.L. Barraclough, R.C. Elphic, D.J. Lawrence, Fluxes of fast and epithermal neutrons from Lunar Prospector: evidence for water ice at the lunar poles. Science **281**, 1496–1500 (1998)

S. Fornasier, B.E. Clark, E. Dotto, A. Migliorini, M. Ockert-Bell, M.A. Barrucci, Spectroscopic survey of M-type asteroids. Icarus **210**, 655–673 (2010)

B.M. French, H.P. Eugster, Experimental control of oxygen fugacities by graphite-gas equilibriums. J. Geophys. Res. **70**, 1529–1539 (1965). https://doi.org/10.1029/JZ070i006p01529

K.G. Gardner-Vandy, D.S. Lauretta, T.J. McCoy, A petrologic, thermodynamic and experimental study of brachinites: partial melt residues of an R chondrite-like precursor. Geochim. Cosmochim. Acta **133**, 36–57 (2013)

J.I. Goldstein, E.R.D. Scott, N. Chabot, Iron meteorites: crystallization, thermal history, parent bodies and origin. Chem. Erde, Geochem. **69**, 293–325 (2009)

J.I. Goldstein, G.R. Huss, E.R.D. Scott, Ion microprobe analyses of carbon in Fe-Ni metal in iron meteorites and mesosiderites. Geochim. Cosmochim. Acta **200**, 367–407 (2017)

J.O. Goldsten, E.A. Rhodes, W.V. Boynton, W.C. Feldman, D.J. Lawrence, J.I. Trombka, D.M. Smith, L.G. Evans, J. White, N.W. Madden, P.C. Berg, G.A. Murphy, R.S. Gurnee, K. Strohbehn, B.D. Williams, E.D. Schaefer, C.A. Monaco, C.P. Cork, J. Del Eckels, W.O. Miller, M.T. Burks, L.B. Hagler, S.J. DeTeresa, M.C. Witte, The MESSENGER Gamma-ray and Neutron spectrometer. Space Sci. Rev. **131**, 339–391 (2007)

R.E. Grimm, H.Y. McSween Jr., Heliocentric zoning of the asteroid belt by aluminum-26 heating. Science **259**, 653–655 (1993)

V.E. Hamilton, A.A. Simon, P.R. Christensen, D.C. Reuter, B.E. Clark, M.A. Barucci, N.E. Bowles, W.V. Boynton, J.R. Brucato, E.A. Cloutis, H.C. Connolly Jr., K.L. Donaldson Hanna, J.P. Emery, H.L. Enos, S. Fornasier, C.W. Haberle, R.D. Hanna, E.S. Howell, H.H. Kaplan, L.P. Keller, C. Lantz, J.-Y. Li, L.F. Lim, T.J. McCoy, F. Merlin, M.C. Nolan, A. Praet, B. Rozitis, S.A. Sandford, D.L. Schrader, C.A. Thomas, X.-D. Zou, D.S. Lauretta (the OSIRIS-REx team), Evidence for widespread hydrated mineral on asteroid (101955) Bennu. Nat. Astron. (2019). https://doi.org/10.1038/s41550-019-0722-2

P.S. Hardersen, E.A. Cloutis, V. Reddy, T. Mothé-Diniz, J.P. Emery, The M-/X-asteroid menagerie: results of an NIR spectral survey of 45 main-belt asteroids. Meteorit. Planet. Sci. **46**, 1910–1938 (2011)

J. Helbert, A. Maturilli, M. D'Amore, Visible and near-infrared reflectance spectra of thermally processed synthetic sulfides as a potential analog for the hollow forming materials on Mercury. Earth Planet. Sci. Lett. **369**, 233–238 (2013)

R.H. Hewins, G.C. Ulmer, Intrinsic oxygen fugacities of diogenites and mesosiderite clasts. Geochim. Cosmochim. Acta **48**, 1555–1560 (1984)

A. Holzheid, H.S.C. O'Neill, The Cr-Cr_2O_3 oxygen buffer and the free energy of formation of Cr_2O_3 from high-temperature electrochemical measurements. Geochim. Cosmochim. Acta **59**, 475–479 (1995)

E. Jarosewich, Chemical analyses of meteorites: a compilation of stony and iron meteorite analyses. Meteoritics **25**, 323–337 (1990)

K. Keil, Mineralogical and chemical relationships among enstatite chondrite. J. Geophys. Res. **73**, 6945–6976 (1969)

K. Keil, Enstatite achondrite meteorites (aubrites) and the histories of their asteroidal parent bodies. Geochemistry **70**, 295–317 (2010)

W.R. Kelly, J.W. Larimer, Chemical fractionations in meteorites. VIII – iron meteorites and the cosmochemical history of the metal phase. Geochim. Cosmochim. Acta **41**, 93–111 (1977)

R.L. Klima, C.M. Pieters, M.D. Dyar, Spectroscopy of synthetic Mg-Fe pyroxenes I: spin-allowed and spin-forbidden crystal field bands in the visible and near-infrared. Meteorit. Planet. Sci. **42**, 235–253 (2007)

A.N. Krot, Y. Amelin, P. Cassen, A. Meibom, Young chondrules in CB chondrites from a giant impact in the early Solar System. Nature **436**, 989–992 (2005)

T.S. Kruijer, C. Burkhardt, G. Budde, T. Kleine, Age of Jupiter inferred from the distinct genetics and formation times of iron meteorites. Proc. Natl. Acad. Sci. **114**, 6712–6716 (2017)

G. Kullerud, The Fe-Ni-S system. Ann. Rep. Geophys. Lab **67**, 4055–4061 (1963)

D.J. Lawrence, P.N. Peplowski, J.B. Plescia, B.T. Greenhagen, S. Maurice, T.H. Prettyman, Bulk hydrogen abundances in the lunar highlands: measurements from orbital neutron data. Icarus **255**, 127–134 (2015)

D.J. Lawrence, J.O. Goldsten, P.N. Peplowski, T.H. Prettyman, Z.W. Yokley, L.T. Elkins-Tanton, Preparing to measure (16) Psyche's elemental composition: interpretation of Psyche neutron measurements, in *51st Lunar and Planetary Science Conference, Abstract #1973* (2020)

R.J. Macke, Survey of meteorite physical properties: density, porosity and magnetic susceptibility. PhD dissertation, Univ. of Central Florida (2010). 332 pp.

B.E. Mandler, L.T. Elkins-Tanton, The origin of eucrites, diogenites, and olivine diogenites: magma ocean crystallization and shallow magma chamber processes on Vesta. Meteorit. Planet. Sci. **48**, 2333–2349 (2013)

S. Marchi, A. Raponi, T.H. Prettyman, M.C. De Sanctis, J. Castillo-Rogez, C.A. Raymond et al., An aqueously altered carbon-rich Ceres. Nat. Astron. **3**(2), 140–145 (2019). https://doi.org/10.1038/s41550-018-0656-0

B.H. Mason, *Victor Moritz Goldschmidt: Father of Modern Geochemistry (Vol. 4)* (Geochemical Society, San Antonio, 1992)

T.J. McCoy, A pyroxene-oldhamite clast in Bustee: igneous aubritic oldhamite and a mechanism for Ti enrichment in aubritic troilite. Antarct. Meteor. Res. **11**, 32–48 (1998)

T.J. McCoy, T.L. Dickinson, G.E. Lofgren, Partial melting of the Indarch (EH4) meteorite: a textural, chemical, and phase relations view of melting and melt migration. Meteorit. Planet. Sci. **34**, 735–746 (1999)

T.J. McCoy, T.H. Burbine, L.A. McFadden, R.D. Starr, M.J. Gaffey, L.R. Nittler, L.G. Evans, N. Izenberg, P. Lucey, J.I. Trombka, J.F. Bell III, B.E. Clark, P.E. Clark, S.W. Squyres, C.R. Chapman, W.V. Boynton, J. Veverka, The composition of 433 Eros: a mineralogical-chemical synthesis. Meteorit. Planet. Sci. **36**, 1661–1672 (2001)

T.J. McCoy, D.W. Mittlefehldt, L. Wilson, Asteroid differentiation, in *Meteorites and the Early Solar System II*, ed. by D. Lauretta, H.Y. McSween Jr. (Univ. of Arizona Press, Tucson, 2006), pp. 733–745

T.J. McCoy, C.M. Corrigan, K. Nagashima, V.S. Reynolds, R.D. Ash, W.F. McDonough, J. Yang, J.I. Goldstein, C.D. Hilton, The Milton Pallasite and South Byron Trio irons: evidence for oxidation and core crystallization. Geochim. Cosmochim. Acta **259**, 358–370 (2019)

H.Y. McSween Jr., R.P. Binzel, M.C. De Sanctis, E. Ammannito, T.H. Prettyman, A.W. Beck, V. Reddy, L. Le Corre, M.J. Gaffey, T.B. McCord, C.A. Raymond, C.T. Russell (the Dawn Science Team), Dawn, the Vesta-HED connection; and the geologic context for eucrites, diogenites, and howardites. Meteorit. Planet. Sci. **48**, 2090–2104 (2013)

H. McSween, A. Ghosh, R. Grimm, L. Wilson, E. Young, Thermal evolution models of asteroids, in *Asteroids III*, ed. by W. Bottke et al. (Univ. of Arizona, Tucson, 2002), pp. 559–571

D.W. Mittlefehldt, T.J. McCoy, C.A. Goodrich, A. Kracher, Non-chondritic meteorites from asteroidal bodies, in *Planetary Materials*, ed. by J.J. Papike. Reviews in Mineralogy, vol. 36 (1998), pp. 4.1–4.195

C.B. Moore, C.F. Lewis, D. Nava, Superior analyses of iron meteorites, in *Meteorite Research* (1969), pp. 738–748

N.A. Moskovitz, R.J. Walker, Size of the group IVA iron meteorite core: constraints from the age and composition of Muonionalusta. Earth Planet. Sci. Lett. **308**, 410–416 (2011)

R.F. Mueller, E.J. Olsen, The olivine, pyroxene, and metal content of chondritic meteorites as a consequence of Prior's rule. Mineral. Mag. **36**, 311–318 (1967)

L.R. Nittler, N.L. Chabot, T.L. Grove, P.N. Peplowski, The chemical composition of Mercury, in *Mercury: The View After MESSENGER*, ed. by S.C. Solomon, L.R. Nittler, B.J. Anderson (Cambridge University Press, Cambridge, 2018), pp. 30–51

M.E. Ockert-Bell, B.E. Clark, M.K. Shepard, R.A. Isaacs, E.A. Cloutis, S. Fornaiser, S.J. Bus, The composition of M-type asteroids: synthesis of spectroscopic and radar observations. Icarus **210**, 674–692 (2010)

H.S.C. O'Neill, M.I. Pownceby, Thermodynamic data from redox reactions at high temperatures. I. An experimental and theoretical assessment of the electrochemical method using stabilized zirconia electrolytes, with revised values for the Fe-"FeO", Co-CoO, Ni-NiO and Cu-Cu_2O oxygen buffers, and new data for the W-WO_2 buffer. Contrib. Mineral. Petrol. **114**, 296–314 (1993)

P.N. Peplowski, D.J. Lawrence, L.G. Evans, R.L. Klima, D.T. Blewett, J.O. Goldsten, S.L. Murchie, T.J. McCoy, L.R. Nittler, S.C. Solomon, R.D. Starr, Constraints on the abundance of carbon in near-surface materials on Mercury: results from the MESSENGER Gamma-Ray Spectrometer. Planet. Space Sci. **108**, 98–107 (2015a)

P.N. Peplowski, D. Bazell, L.G. Evans, J.O. Goldsten, D.J. Lawrence, L.R. Nittler, Hydrogen and major element concentrations on 433 Eros: evidence for an L- or LL-chondrite-like surface composition. Meteorit. Planet. Sci. **50**, 353–367 (2015b)

P.N. Peplowski, R.L. Klima, D.J. Lawrence, C.M. Ernst, B.W. Denevi, E.A. Frank, J.O. Goldsten, S.L. Murchie, L.R. Nittler, S.C. Solomon, Remote sensing evidence for an ancient carbon-bearing crust on Mercury. Nat. Geosci. **9**, 273–276 (2016)

J.J. Petrovic, Mechanical properties of meteorites and their constituents. J. Mater. Sci. **36**, 1579–1583 (2001)

T.H. Prettyman, D.W. Mittlefehldt, N. Yamashita, D.J. Lawrence, A.W. Beck, W.C. Feldman, T.J. McCoy, H.Y. McSween, M.J. Toplis, T.N. Titus, P. Tricarico, R.C. Reedy, J.S. Hendricks, O. Forni, L. Le Corre, J.-Y. Li, H. Mizzon, V. Reddy, C.A. Raymond, C.T. Russell, Elemental mapping by Dawn reveals exogenic H in Vesta's regolith. Science **338**, 242–246 (2012). https://doi.org/10.1126/science.1225354

T.H. Prettyman, N. Yamashita, E. Ammannito, B.L. Ehlmann, H.Y. McSween, D.W. Mittlefehldt et al., Elemental composition and mineralogy of Vesta and Ceres: distribution and origins of hydrogen-bearing species. Icarus **318**, 42–55 (2019). https://doi.org/10.1016/j.icarus.2018.04.032

G.T. Prior, On the genetic relationship and classification of meteorites. Mineral. Mag. **18**, 26–44 (1916)

V. Reddy, D.J. Lawrence, L. Elkins-Tanton, D. Takir, Constraining hydrogen abundance on asteroid (16) Psyche, in *Lunar Planet. Sci. Conf., #1344* (2018)

K. Righter, M.J. Drake, A magma ocean on Vesta: core formation and petrogenesis of eucrites and diogenites. Meteorit. Planet. Sci. **32**, 929–944 (1997)

K. Righter, R.J. Arculus, J.W. Delano, C. Paslick, Electrochemical measurements and thermodynamic calculations of redox equilibria in pallasite meteorites: implications for the eucrite parent body. Geochim. Cosmochim. Acta **54**, 1803–1815 (1990)

K. Righter, M.J. Drake, E.R.D. Scott, Compositional relationships between meteorites and terrestrial planets, in *Meteorites and the Early Solar System II* (2006), pp. 803–828

K. Righter, S.R. Sutton, L. Danielson, K. Pando, M. Newville, Redox variations in the inner solar system with new constraints from vanadium XANES in spinels. Am. Mineral. **101**, 1928–1942 (2016)

A.S. Rivkin, B.E. Clark, D.T. Britt, L.A. Lebofsky, Infrared spectrophotometry of the NEAR flyby target 253 Mathilde. Icarus **127**, 255–257 (1997)

R.A. Robie, B.S. Hemingway, J.R. Fisher, Thermodynamic properties of minerals and related substances at 298.15 K and 1 Bar (10^5 Pascals) pressure and at higher temperatures, in *USGS Bulletin 1452* (1978). 456 pp.

A.D. Romig Jr., J.I. Goldstein, Determination of the Fe-Ni and Fe-Ni-P phase diagrams at low temperatures (700 to 300 °C). Metall. Trans. A **11A**, 1151–1159 (1980)

A.E. Rubin, Carbonaceous and noncarbonaceous iron meteorites: differences in chemical, physical, and collective properties. Meteorit. Planet. Sci. **53**, 2357–2371 (2018)

A.E. Rubin, C. Ma, Meteoritic minerals and their origins. Geochemistry **77**, 325–385 (2017)

A. Ruzicka, G.A. Snyder, L.A. Taylor, Vesta as the howardite, eucrite and diogenite parent body: implications for the size of a core and for large-scale differentiation. Meteorit. Planet. Sci. **32**, 825–840 (1997)

J.A. Sanchez, V. Reddy, M.K. Shepard, C. Thomas, E.A. Cloutis, D. Takir, A. Conrad, C. Kiddell, D. Applin, Detection of rotational spectral variation on the M-type asteroid (16) Psyche. Astron. J. **153**, 29 (2017)

M. Sato, Electrochemical measurements and control of oxygen fugacity and other gaseous fugacities with solid electrolyte sensors, in *Research Techniques for High Pressure and High Temperature*, ed. by G.C. Ulmer (Springer, Berlin, 1971), pp. 43–99

E.R.D. Scott, Chemical fractionation in iron meteorites and its interpretation. Geochim. Cosmochim. Acta **36**, 1205–1236 (1972)

E.R.D. Scott, J.I. Goldstein, Occurrence of carbides and graphite in iron meteorites and origin of C-rich irons (abstract), in *43rd Lunar Planet. Sci. Conf., #2671* (2012)

F. Spitzer, C. Burkhardt, G. Budde, T.S. Kruijer, T. Kleine, Genetic heritage and chronology of ungrouped iron meteorites, in *50th Lunar Planet. Sci. Conf., #2592* (2019)

M. Terho, L.J. Pesonen, I.T. Kukkonen, The petrophysical classification of meteorites: new results. Geol. S. Finl. Q **29.1/91/1**, 1–68 (1993)

M.J. Toplis, H. Mizzon, M. Monnereau, O. Forni, H.Y. McSween, D.W. Mittlefehldt, T.J. McCoy, T.H. Prettyman, M.C. Sanctis, C.A. Raymond, C.T. Russell, Chondritic models of 4 Vesta: implications for geochemical and geophysical properties. Meteorit. Planet. Sci. **48**, 2300–2315 (2013)

J.I. Trombka, S.W. Squyres, J. Brückner, W.V. Boynton, R.C. Reedy, T.J. McCoy, P. Gorenstein, L.G. Evans, J.R. Arnold, R.D. Starr, L.R. Nittler, M.E. Murphy, I. Mikheeva, R.L. McNutt Jr., T.P. McClanahan, E. McCartney, J.O. Goldsten, R.E. Gold, S.R. Floyd, P.E. Clark, T.H. Burbine, J.S. Bhangoo, S.H. Bailey, M. Petaev, The elemental composition of asteroid 433 Eros: results of the NEAR-Shoemaker X-ray spectrometer. Science **289**, 2101–2105 (2000)

H.C. Urey, H. Craig, The composition of the stone meteorites and the origin of the meteorites. Geochim. Cosmochim. Acta **4**, 36–82 (1953)

D.J. Vaughan, J.R. Craig, *Mineral Chemistry of Metal Sulfides* (Cambridge University Press, Cambridge, 1978). 493 pp.

M. Viikinkoski, P. Vernazza, J. Hanuš, H. Le Coroller, K. Tazhenova, B. Carry, M. Marsset, A. Drouard, F. Marchis, R. Fetick, T. Fusco, (16) Psyche: a mesosiderite-like asteroid? Astron. Astrophys. **619**, L3 (2018)

C.M. Wai, J.T. Wasson, Silicon concentrations in the metal of iron meteorites. Geochim. Cosmochim. Acta **33**, 1465–1468 (1969)

D. Walker, T. Grove, Ureilite smelting. Meteoritics **28**, 629–636 (1993)

P.H. Warren, Stable-isotopic anomalies and the accretionary assemblage of the Earth and Mars: a subordinate role for carbonaceous chondrites. Earth Planet. Sci. Lett. **311**, 93–100 (2011)

J.T. Wasson, Trapped melt in IIIAB irons; solid/liquid elemental partitioning during the fractionation of the IIIAB magma. Geochim. Cosmochim. Acta **63**, 2875–2889 (1999)
J.T. Wasson, W.-H. Choe, The IIG iron meteorites: probable formation in the IIAB core. Geochim. Cosmochim. Acta **73**, 4879–4890 (2009)
J.T. Wasson, H. Huber, D.J. Malvin, Formation of IIAB iron meteorites. Geochim. Cosmochim. Acta **71**, 760–781 (2007)
T.R. Watters, M. Prinz, Aubrites – their origin and relationship to enstatite chondrites, in *Proc. 10th Lunar Planet. Sci. Conf.* (1979), pp. 1073–1093
M.K. Weisberg, M. Prinz, R.N. Clayton, T.K. Mayeda, M.M. Grady, C.T. Pillinger, The CR chondrite clan. Proc. NIPR Symp. Antarct. Meteor. **8**, 11–32 (1995)
M.K. Weisberg, T.J. McCoy, A.N. Krot, Systematics and evaluation of meteorite classification, in *Meteorites and the Early Solar System II*, ed. by D. Lauretta, H.Y. McSween Jr. (Univ. of Arizona Press, Tucson, 2006), pp. 19–52
M.M. Wheelock, K. Keil, C. Floss, G.J. Taylor, G. Crozaz, REE geochemistry of oldhamite-dominated clasts from the Norton County aubrite: igneous origin of oldhamite. Geochim. Cosmochim. Acta **58**, 449–458 (1994)
M.A. Wieczorek, B.P. Weiss, S.T. Stewart, An impactor origin for lunar magnetic anomalies. Science **335**, 1212–1215 (2012)
J. Yang, J.I. Goldstein, The formation of the Widmanstätten structure in meteorites. Meteorit. Planet. Sci. **40**, 239–253 (2005)
J. Yang, J.I. Goldstein, Metallographic cooling rates of the IIIAB iron meteorites. Geochim. Cosmochim. Acta **70**, 3197–3215 (2006)
C.-W. Yang, D.B. Williams, J.I. Goldstein, A revision of the Fe-Ni phase diagram at low temperatures ($< 400\ ^\circ$C). J. Phase Equilib. **17**, 522–531 (1996)
J. Yang, J.I. Goldstein, E.R.D. Scott, Main-group pallasites: thermal history, relationship to IIIAB irons, and origin. Geochim. Cosmochim. Acta **74**, 4471–4492 (2010)
T. Yoshizaki, W.F. McDonough, The composition of Mars. Geochim. Cosmochim. Acta **273**, 137–162 (2020). https://doi.org/10.1016/j.gca.2020.01.011
T. Yoshizaki, R.D. Ash, M.D. Lipella, T. Yokoyama, W.F. McDonough, Variable refractory lithophile element compositions of planetary building blocks: insights from components of enstatite chondrites. Geochim. Cosmochim. Acta (2021, in press)
M.E. Zolensky, M.K. Weisberg, P.C. Buchanan, D.W. Mittlefehldt, Mineralogy of carbonaceous chondrite clasts in HED achondrites and the Moon. Meteorit. Planet. Sci. **31**, 518–537 (1996)

Authors and Affiliations

Timothy J. McCoy[1] · Steven D. Dibb[2] · Patrick N. Peplowski[3] · Clara Maurel[4] · Hannah L. Bercovici[2] · Catherine M. Corrigan[1] · James F. Bell III[2] · Benjamin P. Weiss[4] · David J. Lawrence[3] · Daniel D. Wenkert[5] · Thomas H. Prettyman[6] · Lindy T. Elkins-Tanton[2]

✉ T.J. McCoy
mccoyt@si.edu

1 Dept. of Mineral Sciences, National Museum of Natural History, Smithsonian Institution, Washington, DC 20560-0119, USA

2 School of Earth and Space Exploration, Arizona State University, Tempe, AZ 85287, USA

3 Johns Hopkins Applied Physics Laboratory, Laurel, MD 20723, USA

4 Dept. of Earth, Atmospheric and Planetary Sciences, Massachusetts Institute of Technology, Cambridge, MA 02139, USA

5 Mission Systems and Operations Division, Jet Propulsion Laboratory, California Institute of Technology, Pasadena, CA 91109, USA

6 Planetary Science Institute, Tucson, AZ 85719, USA

Space Science Reviews (2022) 218:7
https://doi.org/10.1007/s11214-022-00874-7

The Psyche Topography and Geomorphology Investigation

Ralf Jaumann[1] · James F. Bell III[2] · Carol A. Polanskey[3] · Carol A. Raymond[3] · Erik Aspaugh[4] · David Bercovici[5] · Bruce R. Bills[3] · Richard Binzel[6] · William Bottke[7] · John M. Christoph[2] · Simone Marchi[7] · Alicia Neesemann[1] · Katharina Otto[8] · Ryan S. Park[3] · Frank Preusker[8] · Thomas Roatsch[8] · David A. Williams[2] · Mark A. Wieczorek[9] · Maria T. Zuber[6]

Received: 7 June 2021 / Accepted: 15 February 2022 / Published online: 8 March 2022

Abstract
Detailed mapping of topography is crucial for the understanding of processes shaping the surfaces of planetary bodies. In particular, stereoscopic imagery makes a major contribution to topographic mapping and especially supports the geologic characterization of planetary surfaces. Image data provide the basis for extensive studies of the surface structure and morphology on local, regional and global scales using photogeologic information from images, the topographic information from stereo-derived digital terrain models and co-registered spectral terrain information from color images. The objective of the Psyche topography and geomorphology investigation is to derive the detailed shape of (16) Psyche to generate orthorectified image mosaics, which are needed to study the asteroids' landforms, interior structure, and the processes that have modified the surface over geologic time. In this paper we describe our approaches for producing shape models, and our plans for acquiring requested image data to quantify the expected accuracy of the results. Multi-angle images obtained by Psyche's camera will be used to create topographic models with about 15 m/pixel horizontal resolution and better than 10 m height accuracy on a global scale. This is slightly better as global imaging obtained during the Dawn mission, however, both missions yield resolutions of a few m/pixel locally. Two different techniques, stereophotogrammetry and stereophotoclinometry, are used to model the shape; these models will be merged with the gravity fields obtained by the Psyche spacecraft to produce geodetically controlled topographic models. The resulting digital topography models, together with the gravity data, will reveal the tectonic, volcanic, impact, and gradational history of Psyche, and enable co-registration of data sets to determine Psyche's geologic history.

Keywords Psyche · Asteroide topograph · Asteroide geology · Mapping strategy

1 Introduction/Background

The objectives of the Psyche mission are to investigate the processes and conditions that formed the asteroid Psyche that also affected its subsequent geologic evolution. The shape

The NASA Psyche Mission: Science Instruments and Investigations
Edited by James F. Bell III, Carol Polanskey, and Lindy Elkins-Tanton

Extended author information available on the last page of the article

and detailed topography of asteroid Psyche may have recorded much of its history, including differentiation, possible magmatism, convective motions and bombardment history, as well as the nature of surface alteration by gradational (weathering, erosion and depositional) processes.

Asteroid Psyche's high density of 3.88 ± 0.25 g/cm^3 (Siltala and Granvik 2021) suggests an unexplored surface mineralogy partly composed of metallic compounds with specific reaction to geological processes (for a review, see Elkins-Tanton et al. 2020). The reconstruction of the shape, along with the gravity fields, will allow inferences on the internal structure, while basins, crater shapes, lineaments, ridges, graben, and volcanic constructs will provide a framework for interpreting the images and spectral data to determine the geologic history of Psyche. By exploring the formation and evolution of Psyche, we glean insights into the nature of other objects in the asteroid belt and possibly the interiors of differentiated objects that coalesced to form the terrestrial planets.

Image data provide the basis for extensive studies of the surface structure and morphology on local, regional, and global scales. This is accomplished by extracting geomorphologic information from images, topographic information as derived from digital terrain models (DTMs), and spectral information from color images (i.e. Jaumann et al. 2012). Knowledge of the magnitudes as well as the directions of slopes is needed to understand such quantities as the depths of craters and depressions, the viscous response of materials to stress, the heights of mountains and thus the volumes of geological features. Direct topographical measurements are also needed to determine whether more than one phase of deformation (Fossen et al. 2019) is present and the degree of erosion of the surface from meteoritic and thermal processes (i.e. Delbo et al. 2014). Additionally, for asteroids like Psyche with a surface gravity less than 1% that of the Earth, and much lower interior stresses, these topographic expressions are likely to be more global in scale and can constrain the asteroid's internal structure, formation and geology. Deeply excavating craters can thus probe the global properties (e.g. Jutzi et al. 2013), as can global faults (Buczkowski et al. 2012) and antipodal units.

Determining physical surface properties also provides geologic context. Identification of geologic units in terms of morphology, structure, age and compositional heterogeneities depends on local relief and the direction of slopes. Relationships of different geologic units define stratigraphic sequences, which are used to model the processes that formed specific surface features. Topographic measurements are thus crucial for these investigations and are also used to refine the geodetic control network needed for gravimetric investigations (Raymond et al. 2011; Park et al. 2016). A comprehensive evaluation of the data from the Psyche Multispectral Imagers, based on the image-derived topographic image base maps, will enable combined structural/compositional investigations using camera and spectrometers.

Like Dawn, Psyche is principally a systematic mapping mission to a previously unexplored object. Thus, there are no requirements for targeted observations; nor can the shape and gravity field be sufficiently characterized until arrival. As demonstrated by Dawn close-orbiting experience (Rayman and Mase 2014; Parcher and Whiffen 2011; and Kennedy et al. 2013) a large, unmapped asteroid requires a recursive approach in which science observations in one orbit enable refinement of the gravity field and shape model so as to plan the transfer to the next lower orbit. Psyche will execute a series of four progressively lower circular orbits and spend 20 months in orbit observing Psyche simultaneously with all science instruments. Topographic maps are produced using the stereo photogrammetry (SPG) and stereo photoclinometry (SPC) techniques mainly developed by the Dawn mission (Raymond et al. 2011; Polanskey et al. 2012; Jaumann et al. 2012; Preusker et al. 2016).

2 Current Geomorphological Knowledge

2.1 Landforms

Asteroid Psyche is suggested to have at least one large impact crater near its south pole, similar to that found on (4) Vesta (Shepard et al. 2017; Viikinkoski et al. 2018). In addition, the shape of Psyche as modelled by Ferrais et al. (2020) indicate deviations from an ellipsoid, interpreted as three large depressions along its equator. According to the model of Ferrais et al. (2020) the flatness and density of Psyche are compatible with a formation at hydrostatic equilibrium with a shorter rotation period that may have later slowed down by impact events. With its low gravity, the interior stresses inside of Psyche are much less than 1 kbar, comparable the stress conditions found a few 100 meters below the surface of the Earth; a solid Psyche can maintain a substantial disequilibrium shape attributed to its formation.

Mostly due to the two large south polar impacts Veneneia and Rheasilvia, the surface of Vesta exhibits extended undulated landforms (Thomas et al. 1997; Jaumann et al. 2012; Schenk et al. 2012). The ratio of the amplitude of surface relief to the radius of Vesta is $\sim 15\%$, meaning that topography locally adds or subtracts 1/5th of the body's average radius! Compared to a $\sim 1\%$ ratio for the Moon and Mars steep slopes $> 10°$ up to $30°$ are much more common on Vesta. And like Vesta, Psyche has a rapid rotation (4.1 hr), which means that mega-crater formation happens in the presence of Coriolis forces that could dramatically influence the final topography (Jutzi and Asphaug 2011; Otto et al. 2016). Also Pallas records a violent collisional history, with numerous craters larger than 30 km in diameter populating its surface and two large impact basins which might consider Pallas a preview for Psyche although the surface strength of Psyche is most likely much higher than Pallas (Marsset et al. 2020).

Considering the expected large impact craters on Psyche, the ratio of the amplitude of surface relief to radius is likely to be at least 10% also resulting in steep slopes, depending upon how much silicate regolith is contained on the surface. Steep slopes will also encourage gravitationally driven mass wasting, which is a significant landscape forming and erosive processes on Vesta (Otto et al. 2013; Krohn et al. 2014). Next to granular landslides and impact ejecta related flow-like features, Vesta also exhibits pervasive slumping of more competent material along the crater rim of the large impact basin Rheasilvia. Given the potentially high strength of Psyche's crust and the steep slopes, large scale mass-wasting including slumping can therefore be expected. Therefore, Psyche is expected to exhibit Vesta like asymmetric craters formed by impacts on steep slopes, which results in a sharp upslope crater rim, and a muted downslope crater rim because of preferential accumulation of ejecta downslope (Jaumann et al. 2012; Krohn et al. 2014). However, the shapes of craters on Psyche may differ significantly from what we know from Vesta due to a high content of metal and differing impact velocities (Libourel et al. 2019; Marchi et al. 2019). The dependence of crater collapse on target strength has already been shown for silicate Vesta and ice-rich Ceres (Schenk et al. 2021). Larger craters on Psyche are also expected to undergo gravity-driven collapse at the end of the excavation phase but on high strength metallic bodies, collapse might be suppressed (Elkins-Tanton et al. 2020). In addition, the distribution of silicates on Psyche will therefore result in different stratifications of the near-surface that are exposed by impact cratering (Elkins-Tanton et al. 2020). Extensional features caused by the pressure response of large impacts could be produced, such as the Divalia and Saturnalia Fossae observed on Vesta (Jaumann et al. 2012; Buczkowski et al. 2012). In this context steep hills and elongated depressions would provide an environment for mass wasting and collapse. Psyche's higher density compared to Vesta, and thus its metal fraction, could yield crater morphologies with steeper slopes. More competent material could result in ejecta dominated by large blocks and less sorting of regolith particles. From more details about putative impact structure on Psyche see Marchi et al. (2022) this issue.

Although impact craters are expected to dominate Psyche's morphology, a large impact basin and/or large impact crater could cause global features such as ancient compressional blocks resulting in mountainous complexes like Vestalia Terra on Vesta, possibly including dike-like features (Buczkowski et al. 2014). Johnson et al. (2020) describes a process where core material intrudes into the overlying rocky mantle or possibly even erupts onto the surface as ferrovolcanism eruptions that may produce volcanic landforms on Psyche.

2.2 Surface Alteration

The evidence of a thick regolith layer on Vesta (Jaumann et al. 2012; Denevi et al. 2016), the Moon, and all planetary airless bodies with a solid surface prove the effectiveness of weathering by solar wind, meteoroid, asteroid, and cometary bombardment. Primary planetary crusts are covered by globally and regionally distributed ejecta. Craters are eroded by bombardment and covered by subsequent ejecta of larger impacts resulting in impact-induced resurfacing by the formation of multiple regolith layers. Steep slopes and failure support mass wasting and scarp formation yielding additional resurfacing by landslides.

However, regolith on Psyche will be subject to a possible metallic component mixed into silicate particles. In addition to the metallic component originating from Psyche's bulk composition, metal-bearing silicate, sulfide, and oxide minerals are chemically reduced by space weathering processes including solar wind ion irradiation (Pieters et al. 2000) and micrometeoroid impact (Sasaki et al. 2001), forming metallic iron particles at a variety of length scales up to ~ 10 microns (Noguchi et al. 2014; Matsumoto et al. 2020). Thus, components of different density, size and sorting are predicted to dominate the regolith resulting in complex block-dominated mass wasting features (Kleinhans et al. 2011). The angle of repose of freely moving grains is higher on smaller bodies: lesser gravity allows grains to pile at higher angles before slipping downslope. Ermakov et al. (2019) finds that the angle of repose on Vesta and Ceres is around 34°, while on the Moon and Mars it is closer to 30°, and on the Earth lower still. This means that where regolith is formed on asteroids, we can predict it to pile into steeper slopes assuming the same regolith particle shapes and sizes. Thus, with Psyche's even lower gravity, its angle of repose may be higher than that of Vesta.

Importantly, it is not clear if metal forms a regolith at all. Regolith on Psyche may be mainly rocky material. If a metallic regolith exists, grains might be less regular, and therefore more interlocking capable of forming even steeper slopes? Or would it be mainly smooth tektites formed by impacts into metal, and therefore slipperier and forming shallower slopes? If metal and silicate particles are mixed the regolith become more complex with respect to its physical properties. Haack et al. (2020) mixed spherical and angular particles and measured the tensile strength of different mixing ratios (proxy for the steepness of the slopes). They found that the strength of spherical and angular particle mixtures is dominated by the stronger component (angular particles) and that it varies up to two orders of magnitude between purely spherical and purely angular particles. It is also possible that larger agglutinates will form by impact gluing together silicate particles by molten metal increasing the grain size of the regolith. The grain sizes of regolith on asteroids can be predicted including thermal, polarimetric, photometric, and spectroscopic data (e.g. Hicks et al. 2014; Hiroi et al. 1994). Capria et al. (2014) used VIR thermal data to estimate the thermal inertia of Vesta's surface, and thereby estimate the surface particle size range. They conclude that Vesta has a regolith with grain sizes and properties similar to that of the Moon, with a size range from microscopic to multi-meter blocks. Thermal inertia estimates of Psyche are 133 or 114 $J\,m^{-2}\,s^{-0.5}\,K^{-1}$ (Matter et al. 2013), which is among the highest thermal inertia values measured for an asteroid of this size, supporting evidence of a metal-rich surface for Psyche. However, a study by Landsman et al. (2018) reported thermal inertia values that were representative of a silicate regolith.

Considering a metallic substrate on Psyche will additionally cause fracture from impact shock resulting in large-scale displaced bedrock units and metallic blocks in the meters to tens-of-meters range. However, it is unclear whether blocks in this size range remain angular and intact or the heat transfer will partially melt the metals forming breccias? It is also unclear what the smallest possible block originating from impact shock without melting may be in metal, whether meter-scale boulders, tens-of-cm-scale cobbles, cm-scale gravel, or smaller. Investigations into shock darkening e.g. Moreau et al. (2019) have found that even mm-scale differences in shock impedance, as in a regolith with mixed metallic and non-metallic particles, are enough to drive significant local heterogeneity in shock pressure and temperature, resulting in partial melting of the metallic fraction. Whether similar effects occur at cm-scales or larger is unclear.

Impact experiments on metal yield a variety of metal ejecta fragments (Marchi et al. 2019), with preliminary characterization indicating a wide range of alteration textures from fully-melted microtektites to shock-fractured shards preserving pre-impact microstructure (Christoph et al. 2021). Therefore mm- and cm-sized metal impact melted spherules and impact shards could interact with silicate particles. Shaking from impacts and sliding may sort this material. Impact experiments on metal yield a variety of metal microtektites (Marchi et al. 2019). Therefore mm- and cm-sized metal tektite spheres may interact with silicate particles. Shaking from impacts and sliding may sort this material. The interactions of metal and silicate particles, thus, are possibly complex and difficult to predict. Variations in surface roughness, mixtures of minerals or meteorite types, and optical properties like phase angle and polarization of collected asteroid spectral data (i.e., Britt and Pieters 1988; Cloutis et al. 1990a,b; Lupishko and Belskaya 1989; De Sanctis et al. 2012, 2013; Stephan et al. 2014; Figeri et al. 2015) also expect Psyche having a surface that is dominated by relatively fine-grained (20–50 μm in size) iron-nickel metal, but with a significant component of low-Fe, low-Ca orthopyroxene. The brittle-ductile transition for metal will also affect the outcome of cratering on Psyche because the strength of the metal increases for lower temperatures (Marchi et al. 2020, 2022 this issue).

3 Topography of Asteroids

Asteroids can be highly irregular planetary bodies, with some showing an order of magnitude higher topography to radius ratio than planets or moons. Detailed mapping of topography is central to understanding of processes shaping surfaces. Topography is fundamental for studying their surfaces and interiors, yielding information on landforms, slopes and local surface properties such as roughness as well as internal density variations. The shape of Psyche is essential to properly invert gravity data of asteroids and derive a robust modeled interior structure. Detailed impact crater shapes, slopes, as well as the form of any surface feature provides fundamental morphometric constraints on geological processes that have operated on surfaces (Jaumann et al. 2012).

On asteroids, the shape of craters, possible surface deflection by mass loads, and mass wasting and regolith movement and front shapes, provide constraints on rheology over time and depth. Topography is obtained by a combined solution using camera views at various angles and multiple views of each surface element. Camera-derived surface heights enable the development of regional grids for quantitative studies of surface processes. Stereo information, combined with precise gravity field models, also provides a geodetic control grid at sufficient scale, in which to co-locate other data types at absolute positions on the bodies. Both the imaging and the gravity datasets can provide information needed to resolve internal heterogeneity. At shorter scales and at lowest altitudes the thickness of the crust can be deduced. Geomorphologic studies require slope information at a few percent at the scale of the smallest resolved craters. Optical observation contributes to the topography objective via stereo imaging, from which digital terrain models at high spatial resolution (on the order of the camera pixel size) are derived (i.e. Jaumann et al. 2007; Raymond et al. 2011; Preusker et al. 2012; Gwinner et al. 2016). These digital terrain models provide fundamental insight into geological processes and help constrain studies of the formation and evolution of the crust.

The combination of gravity, topography, and surface composition data is key to modeling the entire structure and composition of a body. Image data provide the basis for extensive studies of the surface structure and morphology on local, regional, and global scales using the photogeological information, the topographic information from the derived digital terrain model, spectral information from color images, and spectral reflectance data. Knowledge of the magnitudes as well as the directions of slopes is needed to understand such quantities as the viscosities of flow features and lavas, and hence their possible composition, the volumes of constructs and flows, the depths of craters/depressions, and the heights of mountains.

Direct topographical measurements are also needed to determine whether polyphase deformation (i.e. Fossen et al. 2019) is present and the degree of erosion of the surface. Identification of geologic units in terms of structure, age and compositional heterogeneities depends on local relief and the direction of slopes. Relationships of different geologic units define stratigraphic sequences, which are used to model the processes that formed specific surfaces. Topographic measurements are crucial for these investigations and are also used to refine the geodetic control network needed for gravimetric investigations. Topography of small bodies improved the science output of the orbiting missions Dawn, Rosetta, Hayabusa 2 and Osiris-REx yielding detailed insight into the shape and structure of Vesta, Ceres, Ryugu and Bennu (i.e. Jaumann et al. 2012; Preusker et al. 2012, 2016; Fu et al. 2017; Watanabe et al. 2019; Sugita et al. 2019; Lauretta et al. 2019) and supported landing and sampling (i.e. Preusker et al. 2019; Scholten et al. 2019; Park et al. 2019).

4 Methodology

Imaging data have been used to generate topography data set in planetary exploration. Two approaches will be used to quantify the topography of the surface of Psyche; these are stereo-photogrammetry (SPG) and stereo-photoclinometry (SPC), which also uses a shape-from-shading approach. Both techniques are applied to multiple optically observed images of the same region (Jaumann et al. 2007; Gaskell et al. 2008; Raymond et al. 2011). *Photogrammetry*, the geometrical analysis of stereo pairs of optical images, has provided much of our knowledge about the quantitative topography of Earth and other planets. *Photoclinometry* is an alternative approach that can be applied to unpaired images and can achieve single-pixel resolution, because it is based on interpreting brightness (shading) variations in slopes rather than measuring geometrical parallaxes. It is more model-dependent and requires additional data for scaling and absolute referencing. The Psyche team will also use a refined photoclinometric technique (photoclinometry by deformation, or PCD) that can iteratively solve for both the slopes and albedo variations, thus reducing the model-dependence in the photoclinometric solution (Raymond et al. 2011).

4.1 Stereophotogrammetric Mapping (SPG)

Photogrammetric processing will derive three-dimensional models of Psyche's surfaces and ortho-rectified images (orthoimages). The raw camera data received from the spacecraft are sorted, adjusted by acquisition times, decompressed and radiometrically calibrated. Camera ancillary data (geometric distortion model), as well as other external geometric information such as orbit and attitude data. A variety of panchromatic and color three-dimensional products can be achieved by means of stereo-photogrammetric processing, i.e., Digital Terrain Models (DTMs), multi-spectral orthoimages, image mosaics, and finally various types of 3D-visualizations and cartographic products. DTMs are the main high-level products that serve as a precise description of Psyche's topography for science and for ortho-rectification of image data at high spatial resolution. Considerable experience related to the photogrammetric processing of planetary data has been gained over more than 15 years from various planetary missions (e.g., Jaumann et al. 2007; Scholten et al. 2009; Raymond et al. 2011; Jaumann et al. 2012; Preusker et al. 2015, 2016; Gwinner et al. 2016). The stereo-photogrammetric derivation of a DTM relies on area-based multi-image matching of two or more images. From the combination of the resulting tie point coordinates, combined with the interior camera geometry, and information of the orbit position and pointing, a 3-dimensional forward ray intersection yields an irregular grid of object points, which is used to interpolate a final raster DTM. Generally, the quality of DTMs and orthoimages strongly depends on the accuracy of the orientation data such as the well calibrated interior camera geometry, the reconstructed position and attitude spacecraft pointing data derived from radio tracking and optical navigation and combined with results from photogrammetric block adjustment information.

The photogrammetric DTM generation process chain (Fig. 1) is embedded in an iterative procedure. A coarse-to-fine iteration includes photogrammetric block adjustment in order to improve the initial spacecraft/position, pointing direction, and the alignment of the camera to the spacecraft. Subsequently, intermediate DTMs are calculated and serve as input for pre-rectification. Once iterations of all global stereo image dataset have yielded full-resolution tie point coordinates the final orientation is derived by an overall photogrammetric adjustment of the entire global block yielding the final stereo DTM interpolation.

Stereo data are pre-rectified to a constant pixel scale in common geometry based on preliminary orbit and pointing information. With remaining parallaxes, the deviation of the *a priori* topography assumption to the true shape of the body can be estimated to enable linking the pre-rectification geometry to the raw image geometry. Applying a multi-image matching technique to pre-rectified image allow to derive conjugate points in each of the stereo observations. The algorithm uses area-based correlations in image pyramids to derive approximate values for the image coordinates, refine sub-pixel accuracy by least-squares matching (Gwinner et al. 2009). After multi-image matching, the derived image coordinates of all stereo observations are transformed back to raw radiometrically corrected data geometry, using the history files generated during the pre-rectification. The accuracy of this back-transformation is better than one tenth of a pixel (Scholten et al. 2005). Lines of sight as defined by the image coordinate pair for each stereo observation, the geometric calibration, and the orientation data, with least-squares adjustment are used for forward ray intersection,

Fig. 1 Processing line for DTM generation from stereo data (Raymond et al. 2011)

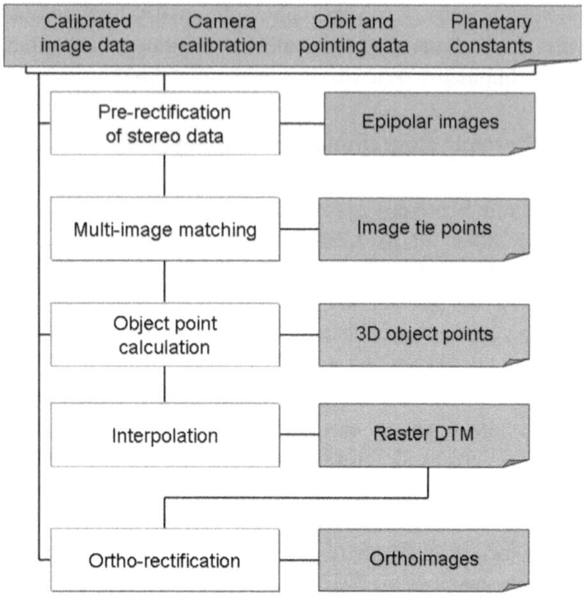

resulting in object points defined in body-fixed Cartesian coordinates. Multi-stereo capability enables minimization of the influence of occasional gross matching errors, which cannot be avoided by simple two-image matching.

Object points are transformed from body-fixed Cartesian coordinates to geographic latitude/longitude/height values based on a three-dimensional reference system. The geographic latitude/longitude coordinates of all object points from all stereo image sets are transformed to standard map projections (equidistant projection for equatorial regions up to $\pm 85°$, stereographic projection for polar regions). Object points located within a DTM pixel are averaged and can optionally be combined by different filtering techniques with those within a chosen neighborhood (Gwinner et al. 2009). For regions that lack any object point information, a final gap-filling algorithm using DTM pyramids with reduced resolution can be applied.

Ortho-rectification uses ray-tracing of the regular pixel positions of the input radiometrically-corrected pixel matrix with the DTM based on interior camera calibration and exterior orientation data (Scholten et al. 2005). A 3-D intersection point with the DTM is derived for each grid point and transformed to the map projection of the output orthoimage with the same standard projections as for the DTM. For each pixel within a mesh of these transformed grid points, an indirect resampling based on different selectable interpolation methods is performed within the radiometrically corrected input matrix in order to derive the final orthoimage grey value. Finally, ortho-rectified color and panchromatic channels, if rectified to the same map projection and pixel scale, fit to each other with the accuracy provided by the DTM and the applied orientation data and can be superimposed as co-registered multi-spectral/multi-phase data sets (Scholten et al. 2005). Ortho-rectified images of different orbits are combined to one common image mosaic using image information in overlapping areas for producing seamless transition from image to image by weighted averaging of overlapping image parts. The geometric relationship between the images to be mosaicked is considered to be fixed and is defined by each image's label information comprising nominal map projection parameters, scale, and the offset of its upper left pixel with respect to the projection center. Statistics of the density of object points (Gwinner et al. 2009), which might

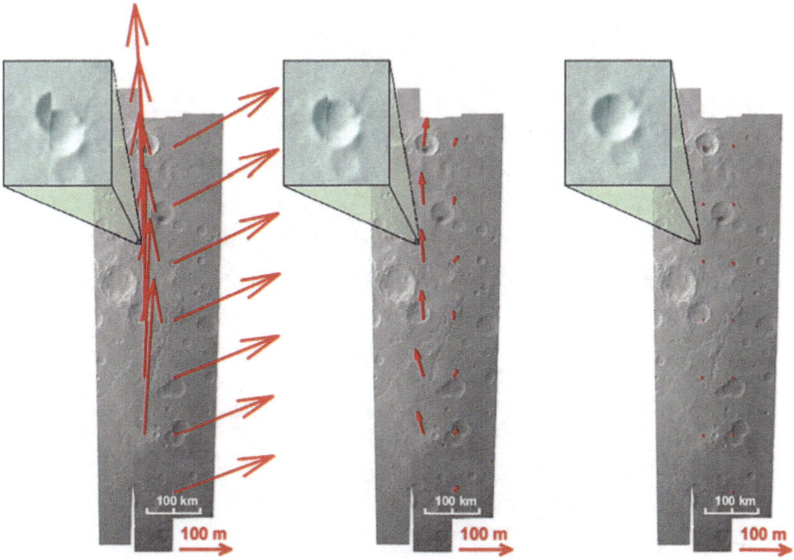

Fig. 2 Example of displacements (red arrows). Identical points are identified in overlapping blocks of Mars Express HRSC images and corrected by bundle block adjustment (Gwinner et al. 2016)

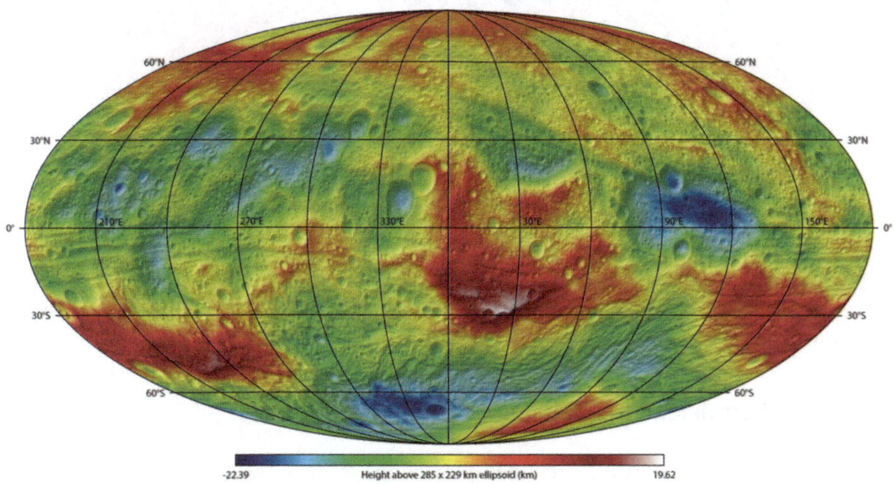

Fig. 3 HAMO-based SPG-derived global topography of Vesta (e.g. Jaumann et al. 2012; Preusker et al. 2012, 2016) superposed on a hill shade model derived thereof. The map is in equal area Mollweide projection in the Claudia Double Prime reference system for Vesta (Li 2012)

be affected by compression effects, variable local texture and varying illumination conditions, will allow a quality assessment of the stereo data processing accuracy. Displacements identified in overlapping images can be corrected by bundle block adjustment (Gwinner et al. 2016) (Fig. 2).

Figures 3 and 4 show final DTM products resulting from stereo data of the Dawn and Mars Express missions.

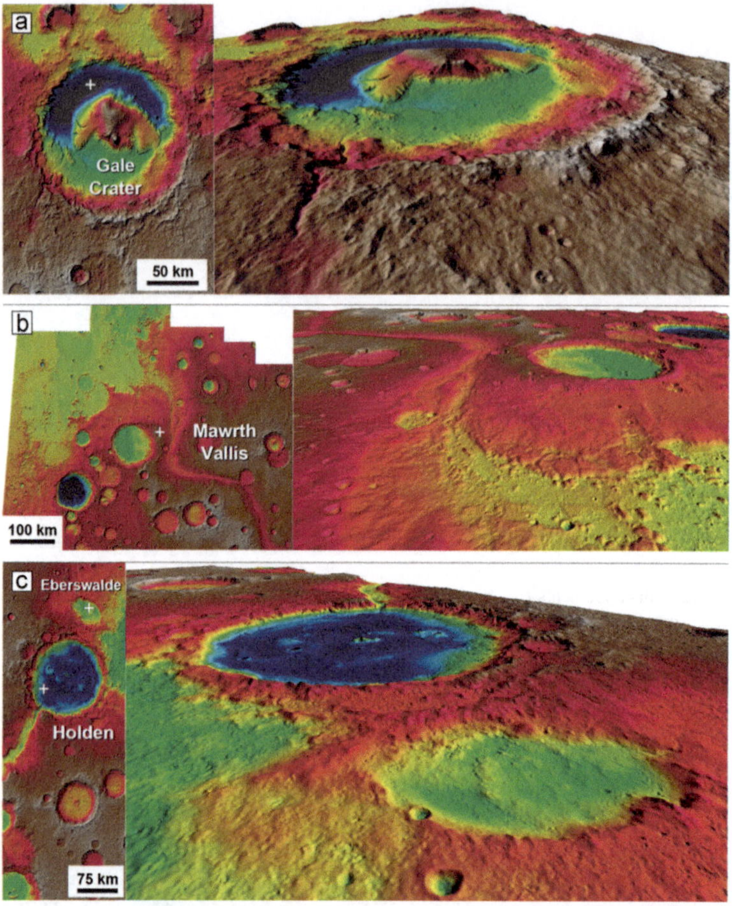

Fig. 4 Examples of final DTM products. Mars Express HRSC multi-orbit DTMs of the final four candidate landing sites of the Curiosity (white crosses centers of proposed ellipses). On the left plan views (norths up); right-side perspective views; color-code elevations (red = high, blue = low). (**a**) Gale crater; (**b**) MawrthVallis; (**c**) Holden and Eberswalde craters (Gwinner et al. 2016)

4.2 Stereophotoclinometric Mapping (SPC)

Stereophotoclinometry uses orthorectified projections of many images of a surface patch, observed under differing illumination conditions, to construct a landmark template called a maplet (Raymond et al. 2011; Park et al. 2019). The slope and albedo at each pixel of a maplet is determined by least square minimizing the brightness residuals between the images and illuminated maplet via least square estimation. The slopes are then integrated to produce a digital terrain map (DTM). Figure 5 shows the topo/albedo maplet constructed from 62 NEAR images of Eros, re-illuminated and compared with projected image data. The maplets display the major features of the images and reproduce the brightness variations well.

A maplet identified in a large number of images allows determination of its surface location very precisely, by averaging down spacecraft position errors and camera pointing uncertainties, while a large number of maplets in a single image allow refining spacecraft

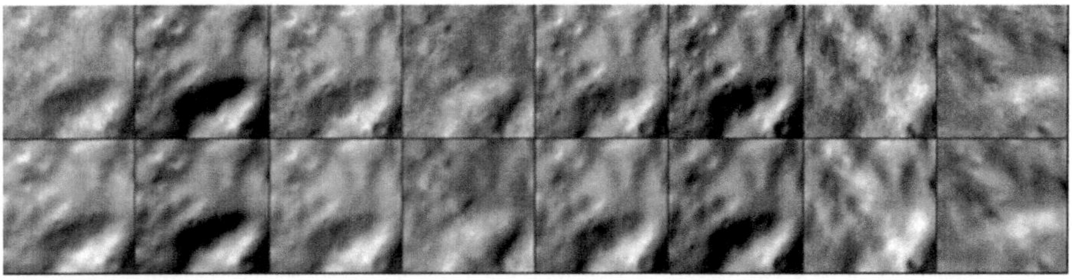

Fig. 5 Illuminated maplet (bottom row) and projected image data (top row) of Dawn stereoscopic data (from Raymond et al. 2011)

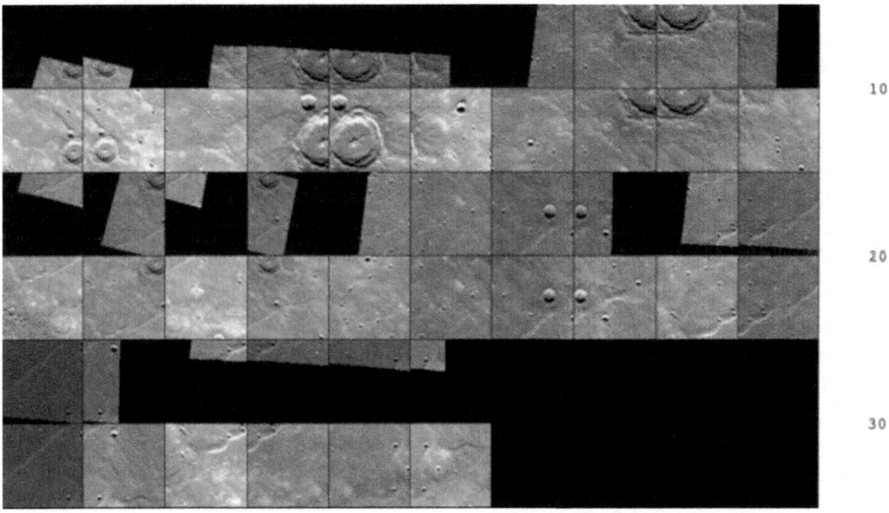

Fig. 6 Auto registration of projected MDIS (MESSENGER) data. Notice that the maplet surface does not have to be completely within an image to be usable (from Raymond et al. 2011)

position and camera pointing estimates. Figure 5 shows an example from an auto registration that solves for the spacecraft state of the MESSENGER mission with the ensemble of existing maplets.

A maplet ensemble is more powerful than a collection of single maplets. Overlapping maplets provide a three-dimensional constraint between the body-fixed locations of their centers. Finally, if the ensemble tiles the entire body, then they can be combined to produce a global topography model (GTM).

An initial set of maplets will be constructed from the first images. As Psyche rotates, a surface feature is seen at varying local sun elevations, allowing to construct the topography from the changing illumination. Correlation of re-illuminated maplet images with the original allow to determine the precise surface position of each maplet to a few tenths of a pixel (Fig. 6).

SPC processing (Fig. 7) has been used on small bodies such as asteroids Eros (Thomas et al. 2002) and Itokawa (Gaskell et al. 2006), and large ones such as Mercury (e.g. Tenthoff et al. 2020), Vesta (Gaskell 2012), and Ceres (Park et al. 2019). Psyche is comparable with the Vesta case and as maplets can be located in images at any resolution, illumination and viewing geometry, they provide excellent navigation tie points as well.

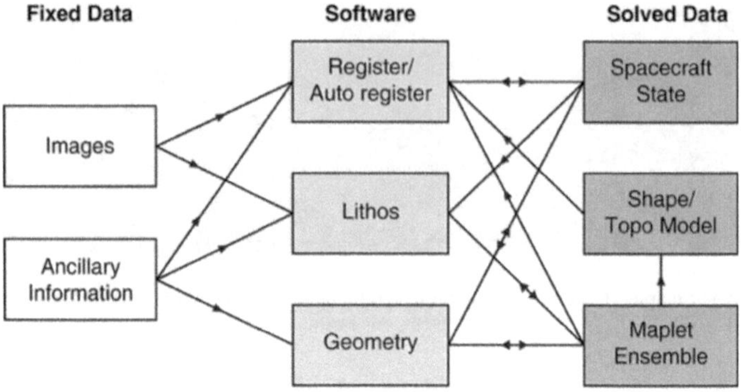

Fig. 7 SPC processing flow from inputs (left) through principal software functions (middle) to products (right). The process is iterative within convergence limits (from Raymond et al. 2011)

An extension of the SPC technique is photoclinometry by deformation (PCD) (i.e. Raymond et al. 2011). This is a technique that allows refinement of shape models or DTMs ("maplets") derived using other techniques such as stereophotogrammetry and stereophotoclinometry described above. The method is applied to a collection of DTMs, typically the SPC DTMs produced after different orbit altitudes. PCD requires as input the set of DTMs describing the surface of the body, and accurate geometric information derived by the navigation and further refined with the above-mentioned SPC and stereo techniques: the pole orientation, the S/C-object vector in the body-fixed frame and the rotation matrix from the body-fixed frame to the camera frame for each camera image. It also requires input "photometric parameters" allowing calculation of the bi-directional reflectance of the surface; an albedo map can be taken as input. The free parameters of the methods are the heights of the vertices of the DTM above the surface, with the DTM being described as a triangular mesh ("plate model").

Although it has been successfully applied to the shape model of asteroid Steins (Jorda et al. 2012) and to DTMs of the asteroid Lutetia (Capanna et al. 2012) observed by the OSIRIS cameras aboard ROSETTA, PCD is a newly developed method. It is also a CPU intensive technique. On Dawn, PCD DTMs and derived photometric parameters have been compared to the products of the stereo and SPC processing pipelines successfully (Raymond et al. 2011; Konopliv et al. 2011) and are be used as inputs to the geodetic DTM. PCD DTMs will also help refine the contributions of brightness variations caused by slope-related surface properties.

4.3 Geodetic-Referenced DTM

The derived models of Psyche's shape and gravity will be integrated to produce a single combined and internally consistent model of topography and gravity field. The various shape models will be combined using their covariance matrices and statistical estimates of their quality. In addition, radiometric tracking data of the Psyche spacecraft will be re-analyzed using the best gravity model, together with the shape models, for potential improvement of both the shape and gravity. The analysis will be conducted with the GEODYN geophysical estimation program (Rowlands et al. 1993; Pavlis et al. 2001) from imaging data, and Doppler tracking data of various kinds (Zuber et al. 2000, 2022; Neumann et al. 2001). This combination of approaches and solutions will allow parameters to be derived using complementary techniques, thereby offering a potential improvement for the complete geodetic

Table 1 Requirements for optimal and useable observing geometry for SPG and SPC (see also Polanskey et al. 2018)

Parameters	Optimal conditions		Useable	
	SPG	SPC	SPG	SPC
Illumination Variation	< 10°	20°–70°	< 15°	10°–85°
Stereo Separation	15°, 15°–65°	0°, 20°–90°	15°, 10°–65°	0°, 15°–90°
Incidence Angle	10°–60°	20°–70°	5°–70°	10°–85°
Emission Angle	< 55°	20°–50°	0°–75°	0°–70°
Phase Angle	5°–90°	0°–90°	5°–90°	0°–90°

and geophysical model of Psyche. The primary objective of this approach is to obtain the best overall topography of Psyche relative to an equipotential surface, with a corresponding gravity field for interpretation of Psyche's interior and dynamics. For more information about gravity see Zuber et al., this issue.

4.4 Observing Requirements for Image-Derived Topography

For stereo imaging the flight system pointing knowledge (i.e., the estimated orientation of the S/C reference frame) should be accurate to 1.5 mrad per axis (3 sigma, zero-to-peak), recorded at least once per second. The Spacecraft System shall provide a resolution of better than 0.005 mrad per axis in its pointing commands and shall be able to point the mounting reference of either imaging camera at a ground-prescribed inertial orientation, with an accuracy of better than 4.65 mrad per axis (3 sigma, zero-to-peak). Note: The accuracy is defined as the total difference between the commanded and actual pointing. Any imaging target should be tracked with an accuracy of better than 5 microrad/sec per axis (3 sigma, zero-to-peak). This includes errors due to tracking rate resolution with a rate resolution of better than 0.25 microrad/sec per axis for tracking rates that it provides to the Flight System. The Spacecraft System shall provide a rate resolution of better than 0.25 microrad/sec per axis in its rate commands.

The stereophotogrammetric method just requires overlapping images from different observation directions. But quality and quantity of the DTM are significantly affected by the imaging and illumination conditions involved. Psyche's stereophotogrammetric observation and illumination requirements are defined based on the experience from stereo processing of other planetary missions, mainly Dawn (Raymond et al. 2011). In a similar manner, conditions that contribute to better accuracy photoclinometrically-derived heights can also be defined (Raymond et al. 2011). Table 1 lists the optimal conditions for each technique, as well as the useable conditions that may not always produce the best height accuracy.

Requirements for SPG, in order of priority:

- the variation of the illumination direction of stereo images should be < 10°, to minimize mis-matching caused by variable illumination. Angles up to 15° are useable.
- the angle between stereo images should be > 15° and < 65°, to provide sufficient vertical and horizontal accuracy. Angles as small as 10° can be used.
- the incidence angle for each image segment should be > 10° and < 60°, to provide sufficient contrast and to minimize the amount of cast shadow areas. Incidence as low as 5° and as high as 70° can be used with care.
- the emission angle for each image segment should be < 55°, to exclude near-limb image segments. Emission as high as 70° can be used, and

- the phase angle for each image segment should be > 5°, to avoid influence from opposition effect.

Basic requirements for SPC:

- the variation of the illumination should be 20°–70° (10°–85° might be useable).
- Nadir image with the emission angle < 20°.
- Image in emission angle bins: 20°–50° (0°–70° might be usable).
- Image in incidence angle bins: 10°–75° (85° might be usable).
- Phase angle 0°–90°.
- At least one image with camera azimuth in each of the bins: 0°–120°, 120°–240°, 240°–360°, with a maximum camera azimuth separation between adjacent azimuth bins of no larger than 150°.
- At least two images separated by solar azimuth angle by no less than 150°.

Considering these requirements, it is evident that SPG and SPC techniques are complementary in exploiting different observing conditions, with some overlap. This fits the Psyche mission concept, because the observing geometry is such that both techniques can be used effectively during the mission, as described below

4.5 Imaging Strategy

At Psyche the topographic model will be constructed using data from Orbit B at an altitude of 293 km. Data from other orbits will also be used when appropriate. Psyche's imaging plans include ~ 19,720 images in Orbit B (5,916 with the clear filter and a total of 13,804 with the seven color filters), at ~ 15 m pixel scale, each of which covers an area of ~ 600 km^2. Each image in Orbit B will span a range of phase angles, including angles < 5 degrees. The main phase of multi-angle imaging occurs in the Orbit B phase; multiple complete global mappings with overlapping images are planned. For Orbit B, two nadir and seven off-nadir acquisitions are planned, with a minimum phase angle of 30 degrees.

Psyche will orbit for 20 months. Four coast-only science orbits are foreseen reaching within one body radius in altitude (Fig. 8). Results from each orbit enable final design of, and navigation into, the successive orbits – a method demonstrated by Dawn at both Vesta and Ceres. Each science orbit is designed around a particular driving requirement (for A, gravity for safe navigation to B; for B, topography; for C, gravity science; for D, composition), leaving margins for the other measurements reaching into orders of magnitude (Table 2; Figs. 8 and 9). The downlink budget is sufficient. The science team is organized into both instrument teams and multi-disciplinary working groups supporting requirements definition and data processing.

Psyche is principally a systematic mapping mission to a previously unvisited body. Thus, there are no requirements for targeted observations; nor can the shape and gravity field be characterized sufficiently until arrival. As demonstrated by Dawn (Rayman and Mase 2014; Parcher and Whiffen 2011; Kennedy et al. 2013), close orbiting of a large, unmapped asteroid requires a recursive approach in which science observations in one orbit enable refinement of the gravity field and shape model so as to plan the transfer to the next lower orbit. A key observation strategy for Psyche is the use of multiple-angle images to derive topography for gravity science, and for morphology. This technique perfected by Dawn (Raymond et al. 2011; Polanskey et al. 2012), avoids the need for a laser altimeter on the spacecraft. It does require systematic coverage by overlapping, fixed-off-nadir images, which Psyche's sequence of long-duration orbits provides in abundance. The major imaging requirements are: image pixel scale < 20 m at mean altitude < 400 km; imaging ground track with 20%

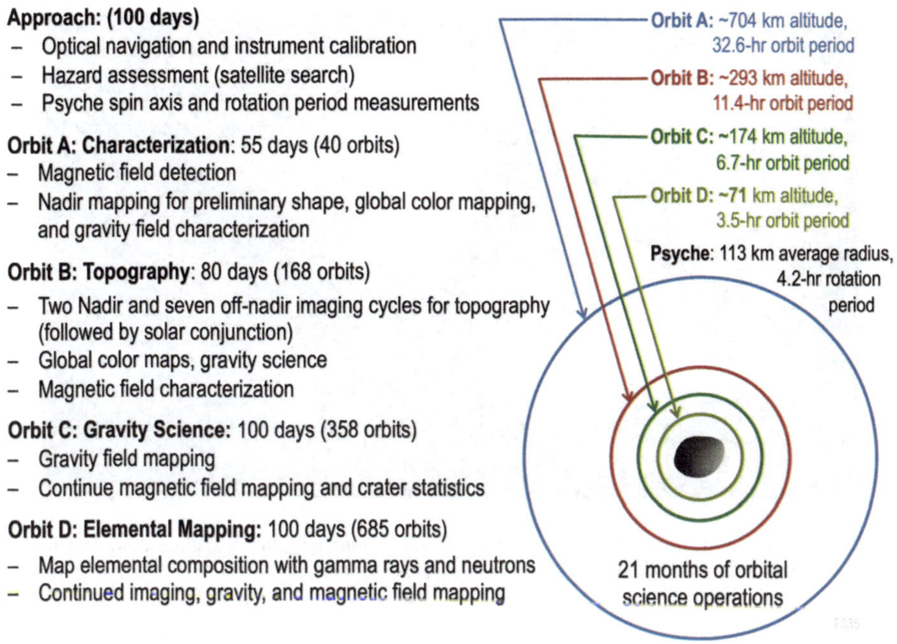

Fig. 8 Psyche science profile uses orbits that progressively improve data quality as knowledge of the gravity field allows lower altitudes (orbit-transfer durations not shown)

Table 2 Relevant orbit parameters yield coverage, resolution & range needed. R/R_P is the ratio of the Psyche spacecraft orbit radius to the Psyche body radius

Science orbit or phase	Ave. altitude (km)	R/R_P	Ave. period (hr)	Inclination	Phase duration (days)	Cycle duration (days)	Repeat cycles/phase	Orbits/repeat cycle	Ave. pixel scale (m)
Approach					100				
Orbit A	704	7.2	32.6	90	55	12.2	4.5	9	35
Transition to Orbit B					18				
Orbit B	293	3.5	11.4	90	80+19	9	9	17	15
Transition to Orbit C					23				
Orbit C	174	2.5	6.7	89	100	14.5	7	49	9
Transition to Orbit D					138				
Orbit D	71	1.7	3.5	160	100	30	3.3	206	4

overlap at equator, for full coverage in one cycle (also achieves gravity coverage requirement); beta angle 30–40°, fixed or slowly varying; Lighting conditions > 80% of the body illuminated or ability to complete coverage later, which will be fulfilled during Orbit B.

Science operations start during the approach phase with an initial assessment of Psyche and its environment using radiometric and optical-image data to estimate spin state, develop a preliminary shape model, calibrate image exposures, and search for orbiting secondaries. At an average altitude of 704 km, Orbit A allows global characterization of the body including first time gravity data; combining these data with the global imaging coverage obtained

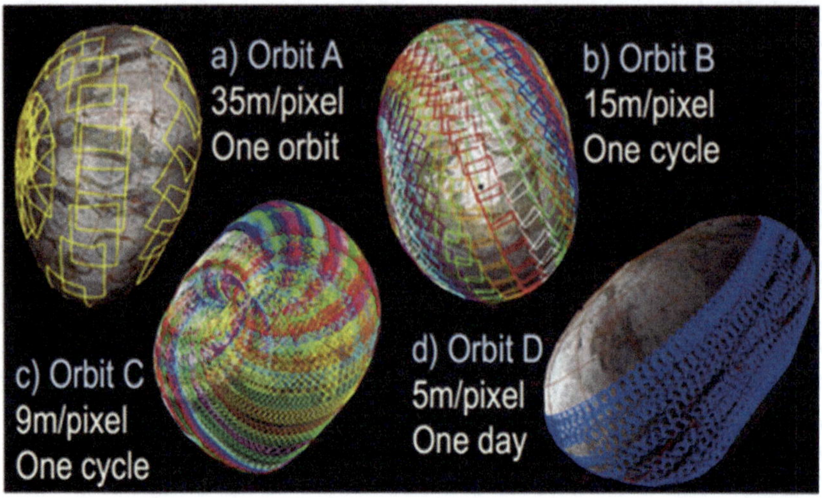

Fig. 9 Resonant Psyche orbits allow superb coverage with routine mapping operations down to Orbit C; stable retrograde Orbit D yields 70% compositional mapping coverage

in this orbit allows to derive the shape-constrained gravity model. A ground-track repeat every 4th cycle would provide complete imaging coverage in a few days. Each cycle generates a complete map of the illuminated hemisphere estimated to exceed 90% of the body based on the currently known pole model, using all eight filters, at 35 m/pixel. The resulting maps and shape model satisfy the geomorphology and color objectives. Figure 9 illustrates the imaging coverage for one mapping orbit. 61 images cover Psyche from terminator to terminator and at least 10% along-track overlap based on Dawn's high-fidelity science planning tools).

Detailed geological mapping using morphological and topographic information enables identification of the geologic context of surface features. Identification of geologic units in terms of morphology, structure, age and compositional heterogeneities is enhanced by information on local relief and the direction of slopes. Relationships of different geologic units define stratigraphic sequences, enabling the identification of relative ages and the modeling of the processes that formed specific surfaces. Topographic measurements are crucial for these investigations and are also used to refine the geodetic control network needed for gravimetric investigations. A comprehensive evaluation of the measurements from the Psyche instruments, based on the image-derived topographic base maps, will enable combined structural/ compositional investigations using the camera and spectrometers.

At Psyche the topographic model will be constructed using data from the Orbit B at a mean planetocentric radius of 399 km. Data from other orbits will also be used when appropriate. Psyche's imaging plans include $\sim$ 19,720 images in Orbit B (5,916 with the clear filter and a total of 13,804 with the seven color filters), at $\sim$ 15 m pixel scale, each of which covers an area of $\sim$ 600 km^2. Each image in Orbit B will span a range of phase angles, including angles < 5 degrees. The main phase of multi-angle imaging occurs in the Orbit B phase; multiple complete global mappings with overlapping images are planned. For Orbit B, two nadir and seven off-nadir acquisitions are planned, with a minimum phase angle of 30 degrees.

Fig. 10 Illustration of the Orbit B ground track for 17 orbits. The imager footprints for each orbit are given a different color and lay down an image swath with overlapping frames that butts against, or overlaps, the neighboring swath

The main topographic mapping data set is obtained in Orbit B. Once the shape models are developed and validated, we will assess them and merge the information with the gravity field to produce the 'best' DTM that is referenced to the Psyche center-of-mass.

Global mapping with the imager requires 17 orbits over 8.4 days to complete; the image footprints relative to the ground-track are shown in Fig. 10. The plans for Orbit B include seven off-nadir global mapping cycles and two nadir mappings; each off-nadir global mapping is done at a fixed attitude. The off-nadir angles were chosen such that both techniques will get adequate coverage. The viewing geometry is shown graphically in Fig. 11. Coverage plots for stereo, are shown in Fig. 9, which also illustrates the fact that most data that respect the stereo illumination variation conditions have a very narrow range of illumination, $< 5°$ (bottom panel), which is a natural consequence of the polar sun-synchronous orbit.

In Fig. 12 the stereo coverage plots for SPG and SPC in Orbit B are shown, illustrating the excellent coverage with many useable data. Combining both data set will result in the required stereo coverage of 80% which meet the mission requirements.

In addition, the combination of SPG and SPC improves the topographic resolution significantly (Fig. 13) as demonstrated by Dawn results of Ceres (Neesemann et al. 2022).

Topographic maps are produced using the stereo photogrammetry (SPG) and stereo photoclinometry (SPC) techniques developed by Dawn [Raymond et al. 2011; Polanskey et al. 2012]; The topography model from clear-filter images taken from multiple view angles using one imager at a time. The spacecraft is nadir-pointed for the first and fifth mapping cycles; and at a fixed off-nadir attitude for the rest. Each mapping cycle uses one fixed off-nadir angle. The set of angles for the series optimizes among multiple factors: stereo separation between image pairs ($\geq 15°$); and incidence, emission, and phase angles. Four ground-track repeat cycles, taking 33.6 d total in Orbit B, are required to meet the topography objective; five additional cycles are added for margin, driving the science portion of Orbit B to 80 d total. The spacecraft ground track provides uniform coverage with 20% overlaps between imaging swaths. The lighting conditions and data budget allow 99% of the surface to be mapped in Orbit B. In order to verify the coverage, the received images will be georectified based on the preliminary shape model and mosaicked immediately after each download to support operations planning (Fig. 14).

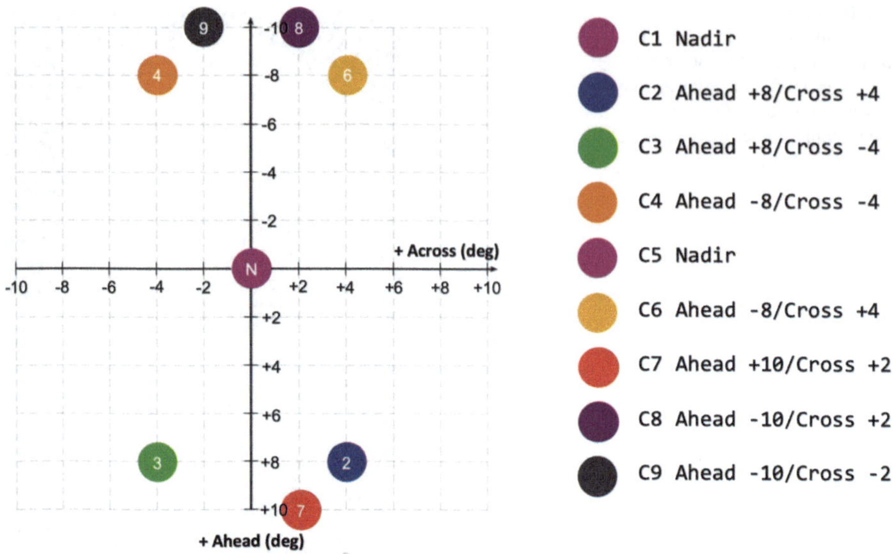

Fig. 11 Illustration of the off nadir viewing geometry as referenced to spacecraft coordinates for the nine global mapping cycles (Polanskey et al. 2022). For a polar orbit, constant illumination is in the "Ahead" direction, while maximum illumination variation is achieved in the "Cross" direction. The angles in the Cross direction are maximized within the constraints of keeping on the body and achieving proper incidence and emission angles

5 Data Products

The following high-level products will be produced:

1. local, regional, and global Digital Terrain Models (DTMs)
2. local, regional, and global clear filter mosaics
3. local, regional, and global color mosaics

Figure 15 shows the tiling scheme for quadrangle mapping used for the Dawn mission.

The Psyche coordinate system will be defined in two steps. In close cooperation with the IAU, a tiny crater near the equator will be selected to define the longitude system. Next, the rotation axis and period are determined during the photogrammetric processing of the stereo images. Together these will define the coordinate system. The east longitude system will be used and the thematic convention for naming craters and other features will be defined in close cooperation with the IAU. The Psyche topographic atlas will be produced in a scale of 1:250,000 or 1:500,000 and will consist of 15 tiles that conform to the quadrangle scheme. This tiling schema was recently used for the Saturnian satellites like Enceladus and Vesta and Ceres (Roatsch et al. 2012, 2013, 2016).

Detailed geological mapping using morphological and topographic information enables identification of the geologic context of surface features. Identification of geologic units in terms of morphology, structure, age and compositional heterogeneities is enhanced by information on local relief and the direction of slopes. Relationships of different geologic units define stratigraphic sequences, enabling the identification of relative ages and the modeling of the processes that formed specific surfaces. Topographic measurements are crucial for these investigations and are also used to refine the geodetic control network needed for

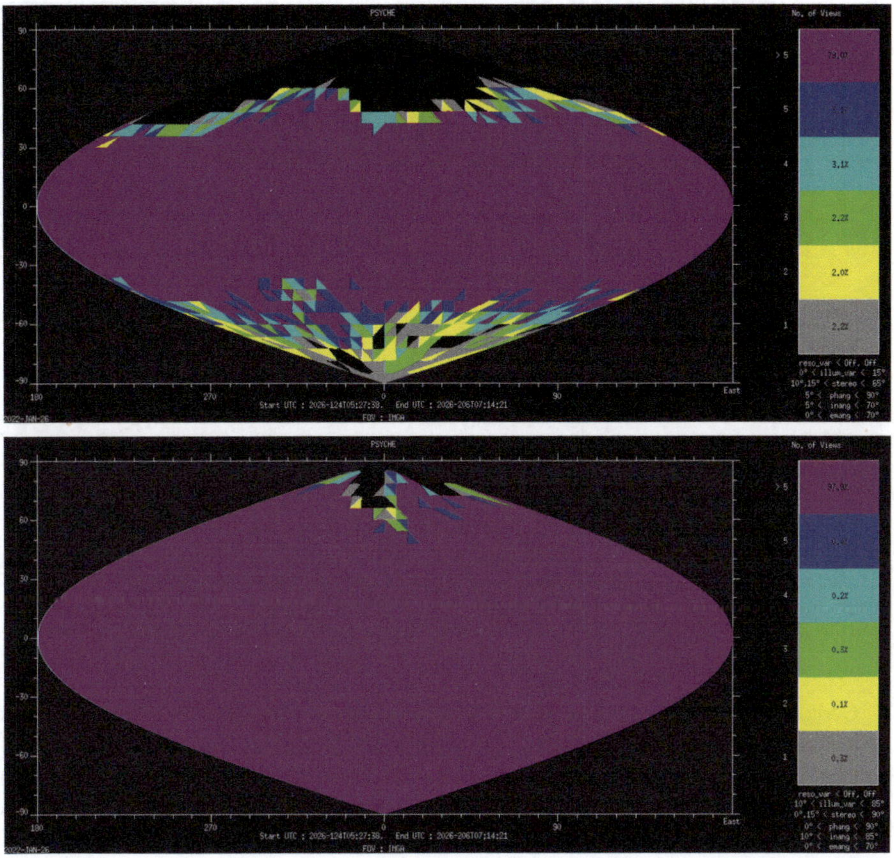

Fig. 12 In Orbit B, SPG coverage meets the useable stereo conditions for incidence angle up to 70° (top) and resolutions of 10 m/pixel – –20 m/pixel. The bottom plot shows the coverage for SPC in Orbit B with the same resolution. The colors indicate the number of independent datums in each .25° bin that collectively meet the criteria outlined in Table 1. The area covered at a given number of views is indicated by the percentage number in the legend

gravimetric investigations. A comprehensive evaluation of the measurements from the Psyche instruments, based on the image-derived topographic base maps, will enable combined structural/compositional investigations using the camera and spectrometers

A geologic mapping campaign for asteroid Psyche will be performed during the approximately two-year (21 month) nominal mission which will result in a global geologic map of Psyche at a map scale of 1:500,000. This map scale is based on the size of Psyche (geometric mean diameter of 186 km). However, given the Orbit B image spatial resolution of $\sim$ 15 m/pixel, the digital geologic mapping of Psyche can be done at a map scale as high as 1:60,000. Depending upon the variety and types of surface features discovered on Psyche, additional regional/local geologic maps as needed/justified by the geology of Psyche will be produced, in which cartographic image maps and DTMs are used as our mapping bases. As was done with Vesta and Ceres, geologic mapping will be done digitally using ArcGIS™ software and in consultation with mapping specialists from the U.S. Geological Survey, using community standards (i.e., Skinner et al. 2018).

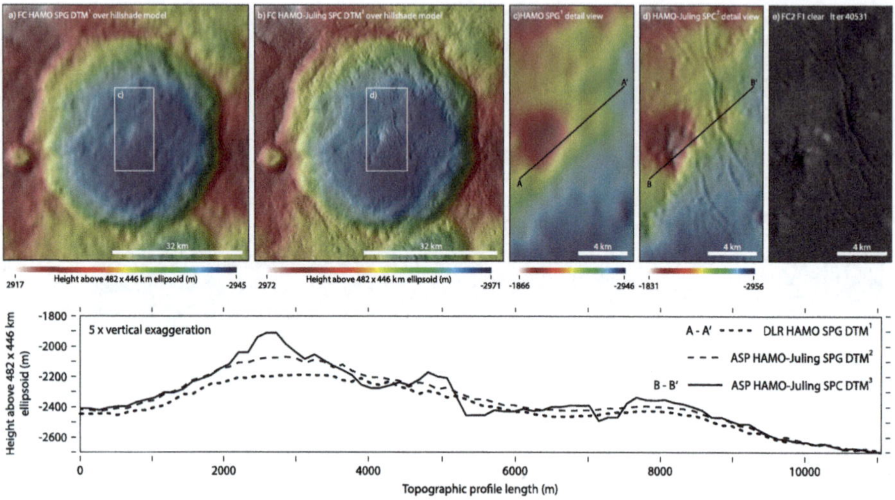

Fig. 13 Topography of Azacca impact crater on Ceres based on Dawn Framing Camera (FC) observations. (**a**) The SPG DTM as derived from HAMO data. (**b**) FC HAMO and Juling 1 orbit frames bundle adjusted and photometric corrected frames SPG processed, followed by SPC processing. This combination allows topographic resolutions as of the HAMO input frames (~ 136–140 m/pixel). (**c**, **d**, **e**) Show subsets of the central peak and the topographic cross section illustrate the improvement of the SPG and SPC combination

Geological mapping of a new planetary object with a totally different surface will arise new issues. Like that used for asteroid (4) Vesta on the Dawn mission surface features will be identified during approach and refined during the first orbital phase (Russell et al. 2015). Based on modeled processes of solidification possible for a stripped, molten metal core the following developed hypotheses for features are expected on Psyche's surface (see also Elkins-Tanton et al. 2020):

- Scarps, perhaps on the order of 10 km tall, from contraction of the body.
- Sulfur-rich lava flows from the original solidification and liquid expulsion.
- Craters of all sizes and perhaps with different morphologies depending on whether they occurred on the warm or cold side of Psyche.
- Slabs, probably on the order of 100 km scale, of achondritic original mantle material from the Psyche parent body that fell back onto the surface during stripping.
- Exogenic chondritic silicate deposits delivered by impactors.

At present it is unclear how these and other features will look on a metallic object. For example, one might expect a harder metallic surface to be more resistant to the effects of space weathering, such that the surface could have less regolith than silicate objects have. We will follow standard best geological mapping practices to define and characterize the surface into geologic units based on physical attributes (morphology, albedo, color, texture, contact relations with neighboring units and mainly topography), and then work to interpret these features in terms of all the possible geologic processes that might produce them (e.g. Jaumann et al. 2012; Buczkowski et al. 2016). As the spacecraft enters lower orbits and spatial resolution of images improve, it is likely (as we found at Vesta) that one or more hypotheses regarding formation processes for some geologic units can be eliminated (e.g. Jaumann et al. 2012; Buczkowski et al. 2016). This would improve our understanding of the geology of Psyche and enable us to derive its geologic history. Hypothetical surface

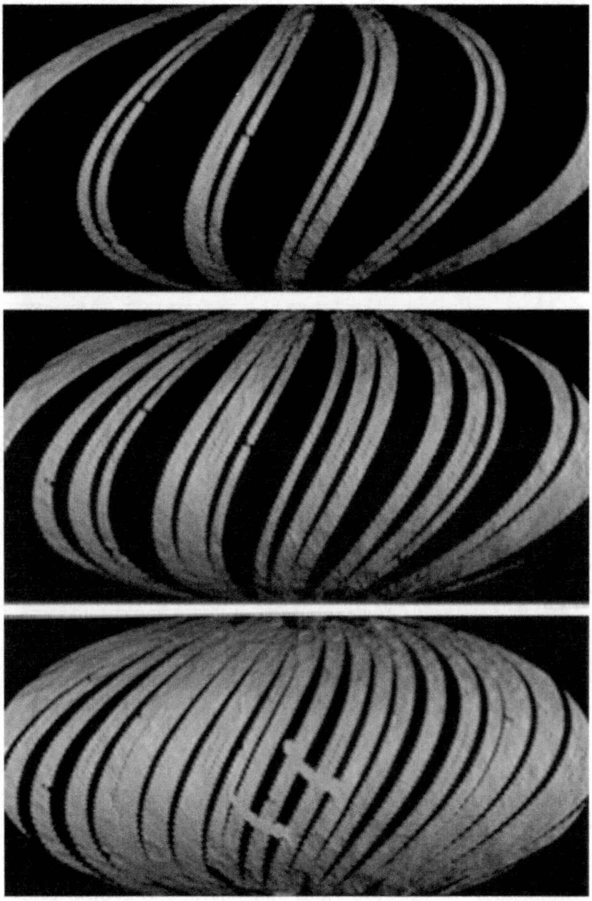

Fig. 14 Example of Dawn Ceres coverage verification

features expected on Psyche are also known on Earth and other previously mapped solar system bodies. Contraction scarps, surface expression of thrust faults, where the crust is broken along an inclined plane and pushed upward are observed on Mercury (e.g. Watters 2021) and the Moon (e.g. Banks et al. 2012) and Earth (e.g. Plescia and Golombek 1986). Sulfur-rich lava flows exist on Io (e.g. Lopes 2007), and lava flows consisting primarily of calcium occur on Earth in Africa (Bondre 2008). However, despite of the abundance of basalt on Vesta lava flows were not identified due to regolith an ejecta burial (e.g. Russell et al. 2015). The same effect might obscure possible lava flows on Psyche while spectroscopic observation might detect basaltic compositions. As exogenic silicates (e.g. Le Corre et al. 2015) delivered by impactors have also already been mapped on Vesta and the Moon, this should also be expected on Psyche (Elkins-Tanton et al. 2020).

6 Expected Results

Imagery contributes to many of Psyche's scientific objectives, often in combination with other complementary data sets. For example, the average density is calculated from the volume as measured by the camera and the mass is derived from radio science. The prime task

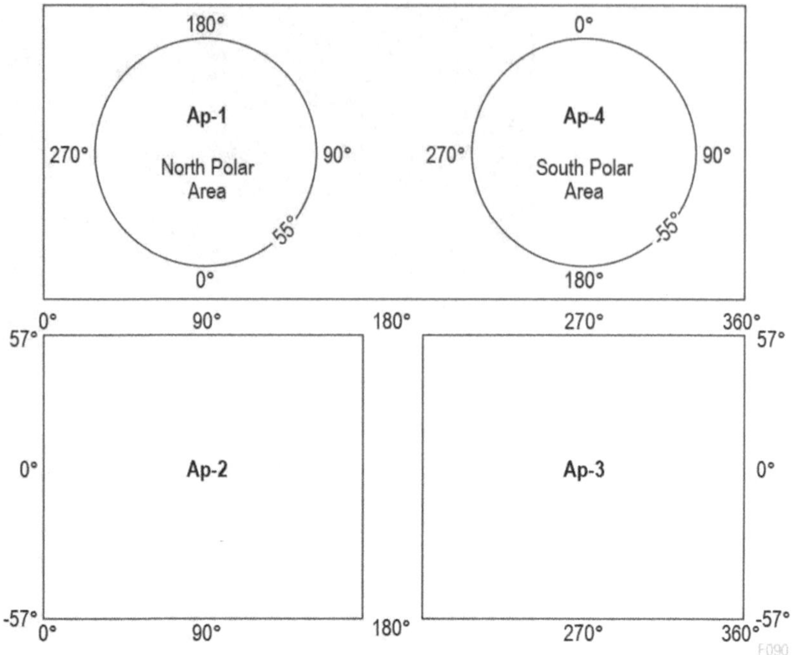

Fig. 15 4-quadrangle tilling scheme of the Psyche topographic atlas

of the camera is to image (map) the surface of Psyche in high spatial resolution, under various illumination conditions and from multiple directions, and in different spectral bands. The images are then combined into mosaics for each spectral band, thus presenting an important tool for the study and analysis of the geomorphology and decipher the geological history and the evolution of the surface. Imaging will be performed from far-field global maps to local-site characterization under various observing geometries and lighting conditions yielding surface resolution of about 15 m/pixel globally and a few m/pixel locally comparable what was achieved by the Dawn mission (e.g. Raymond et al. 2011; Russell et al. 2015). The analysis of geomorphologic features (down to the scale of a few 10's of meters in the high-resolution phase yields exclusive evidence on the different geologic processes responsible for the state of the surfaces, including impact cratering, tectonism, gradation (weathering, erosional, and depositional features like mass movements) and possible volcanism (e.g. Jaumann et al. 2012; Russell et al. 2015; Schmedemann et al. 2014; Williams et al. 2014; Buczkowski et al. 2016; Hiesinger et al. 2016). Clues to the formation age of surface units and its stratigraphy can be derived from the frequency-size distribution of craters, their morphology and interior structure, all of which requires topographic information (e.g. Schmedemann et al. 2014; Williams et al. 2014; Hiesinger et al. 2016). The vertical accuracy of the stereo data is like the spatial pixel resolution that allows a detailed 3-dimensional analysis of surface features (e.g. Raymond et al. 2011; Jaumann et al. 2012; Preusker et al. 2016). Multispectral images allow mapping compositional surface variations (e.g. De Sanctis et al. 2013, 2016; Ammannito et al. 2016), whereas multi-angle images provide hints to the textural properties of the regolith and the surface roughness (e.g. Schroeder et al. 2013).

Slopes of movable surface material are determined not just by the geologic processes that create topography, but by the nature of the material (regolith) (e.g. Pieters et al. 2012) and the gravity (e.g. Konopliv et al. 2011) of the body. Knowledge of the magnitudes as well as directions of slopes is needed to answer questions about the viscosities and strengths, and hence the possible composition of materials (e.g. Duarte et al. 2019). Information of the physical surface properties by the means of multi-phase angle observations also supports the geologic context characterization (e.g. Schroeder et al. 2013). The major result of these studies is the identification of geologic units in terms of morphology, structure, age, and compositional heterogeneities with evidence of the presence of local relief. The relationship of different geologic units will build up stratigraphic sequences, which are used to model the processes that formed the specific surface (e.g. Williams et al. 2014). Stereoscopic measurements are crucial for these investigations and are also used to refine the geodetic control network and thus support gravimetric investigations (e.g. Konopliv et al. 2011).

The shapes of craters on Psyche will be of great interest, especially if there are large areas of metal (see also Marchi et al., this issue). Each body in the solar system has yielded new examples and types of shapes of impact craters, and craters on asteroids have been particularly interesting. For example, Vesta has many 'asymmetric craters', formed by impacts on steep slopes, which results in sharp upslope crater rim, and a muted downslope crater rim (because of preferential accumulation of ejecta downslope) (Jaumann et al. 2012; Krohn et al. 2014). The angle of repose of freely moving grains is higher on smaller bodies: lesser gravity allows grains to pile at higher angles before slipping down. This means that where regolith is formed on asteroids, we can expect it to pile into steeper slopes than we are used to seeing on Earth. If a metal substrate can form a regolith is unknown and on Psyche regolith may be mainly rocky material, either native or delivered by later impacts. Slopes as estimated by topographic analysis will therefore constrain the nature of surface material. Shaking from impacts and sliding on hillslopes may sort the material.

Impact crater on asteroids like Vesta and Ceres show related fracturing of different occurrence. Global fracture systems induced by large impacts dominate Vesta (Jaumann et al. 2012; Buczkowski et al. 2012), impact related fracture systems on Ceres show concentrical and radial fractures which are interpreted due to creep of low viscosity subsurface layers or upwelling processes (Otto et al. 2019). Another impact related fracture system are concentric or radial floor-fracture craters on Ceres due to mechanisms include cooling/melting processes, degassing, and subsidence of the crater floor by up-doming of subsurface material because of absolute tensile stresses (Krohn et al. 2020). The spin of asteroids evolves due to collisions (e.g. Mao and McKinnon 2020) and thus may imply fracturing (e.g. Fu et al. 2014, 2017; Otto et al. 2016). Impact induced fracturing as observed on Vesta might also occur on Psyche while the volatile related fracturing as observed on Ceres is rather unlikely (see also Marchi et al. 2022 this issue; Zuber et al. 2022 this issue).

A comprehensive evaluation of the measurements from instruments based on the stereo derived topographic image base maps will improve the overall mission output by combined structural/compositional investigations using the camera and spectrometer. The high-resolution digital shape models are derived using various techniques and merged into a geodetically controlled topographic model will enable investigation of the geology expected on Psyche. Slumps and landslides can be identified and tied to source regions. Craters will be well resolved topographically, and their degradation states revealed. The high-resolution slope data will enable brightness variations to be confidently identified and related to spectral variations in terms of composition, grain sizes, or other factors. Psyche's high-resolution topographic models, together with the gravity data will enable co-registration of data sets to determine the geologic history.

Acknowledgements The authors are grateful for the experience gained with the Dawn project. A portion of this work was performed at the Jet Propulsion Laboratory, California Institute of Technology, the Freie Universitaet Berlin and the Arizona State University in Tempe.

Funding Note Open Access funding enabled and organized by Projekt DEAL.

References

E. Ammannito, M.C. DeSanctis, M. Ciarniello, A. Frigeri, F.G. Carrozzo, J.-Ph. Combe, B.L. Ehlmann, S. Marchi, H.Y. McSween, A. Raponi, M.J. Toplis, F. Tosi, J.C. Castillo-Rogez, F. Capaccioni, M.T. Capria, S. Fonte, M. Giardino, R. Jaumann, A. Longobardo, S.P. Joy, G. Magni, T.B. McCord, L.A. McFadden, E. Palomba, C.M. Pieters, C.A. Polanskey, M.D. Rayman, C.A. Raymond, P.M. Schenk, F. Zambon, C.T. Russell, Distribution of phyllosilicates on the surface of Ceres. Science **353**, 1006 (2016). https://doi.org/10.1126/science.aaf4279

M.E. Banks, T.R. Watters, M.S. Robinson, L.L. Tornabene, T. Tran, L. Ojha, N.R. Williams, Morphometric analysis of small-scale lobate scarps on the Moon using data from the Lunar Reconnaissance Orbiter. J. Geophys. Res. **117**, E00H11 (2012). https://doi.org/10.1029/2011JE003907

N. Bondre, Molten carbonate eruptions. Nat. Geosci. **1**, 498 (2008). https://doi.org/10.1038/ngeo274

D.T. Britt, C.M. Pieters, Bidirectional reflectance properties of iron-nickel meteorites, in *Lunar and Planetary Science Conference 18* (1988)

D.L. Buczkowski, D.Y. Wyrick, K.A. Iyer, E.G. Kahn, J.E.C. Scully, A. Nathues, R.W. Gaskell, T. Roatsch, F. Preusker, P.M. Schenk, L. Le Corre, V. Reddy, R.A. Yingst, S. Mest, D. Williams, W.B. Garry, O.S. Barnouin, R. Jaumann, C.A. Raymond, C.T. Russell, Large-scale troughs on Vesta: a signature of planetary tectonics. Geophys. Res. Lett. **39**(17), L18205 (2012). https://doi.org/10.1029/2012GL052959

D.L. Buczkowski, D.Y. Wyrick, M. Toplis, R.A. Yingst, D.A. Williams, W.B. Garry, S. Mest, T. Kneissl, S.E.C. Scully, A. Nathues, M.C. De Sanctis, L. LeCorre, V. Reddy, M. Hoffmann, E. Ammannito, A. Frigeri, F. Tosi, F. Preusker, T. Roatsch, C.A. Raymond, R. Jaumann, C.M. Pieters, C.T. Russell, The unique geomorphology and physical properties of the Vestalia Terra plateau. Icarus **244**, 89–103 (2014). https://doi.org/10.1016/j.icarus.2014.03.035

D.L. Buczkowski, B.E. Schmidt, D.A. Williams, S.C. Mest, J.E.C. Scully, A.I. Ermakov, F. Preusker, P. Schenk, K.A. Otto, H. Hiesinger, D. O'Brien, S. Marchi, H. Sizemore, K. Hughson, H. Chilton, M. Bland, S. Byrne, N. Schorghofer, T. Platz, R. Jaumann, T. Roatsch, M.V. Sykes, A. Nathues, M.C. De Sanctis, C.A. Raymond, C.T. Russell, The geomorphology of Ceres. Science **353**, 1004 (2016). https://doi.org/10.1126/science.aaf4332

C. Capanna, L. Jorda, G. Gesquière, P. Lamy, M. Kaasalainen, Lutetia global shape and topography reconstructed with photoclinometry by deformation. Vol. 7, EPSC2012-616 (2012)

M.T. Capria, F. Tosi, M.C. De Sanctis, F. Capaccioni, E. Ammannito, A. Frigeri, F. Zambon, S. Fonte, E. Palomba, D. Turrini, T.N. Titus, S.E. Schröder, M. Toplis, J.-Y. Li, J.-P. Combe, C.A. Raymond, C.T. Russell, Vesta surface thermal properties map. Geophys. Res. Lett. **41**, 1438–1443 (2014). https://doi.org/10.1002/2013GL059026

J.M. Christoph, T. Sharp, S. Marchi, L.T. Elkins-Tanton, Characterizing ejecta fragments from impact experiments into meteoric iron using scanning electron microscopy (SEM), in *LPSC 52 Abstract #2730* (2021)

E.A. Cloutis, M.J. Gaffey, D.G. Smith, R.S.J. Lambert, Metal silicate mixtures: spectral properties and applications to asteroid taxonomy. J. Geophys. Res. **95**, 8323–8338 (1990a)

E.A. Cloutis, M.J. Gaffey, D.G. Smith, R.S.J. Lambert, Reflectance spectra of "featureless" materials and the surface mineralogies of M- and E-class asteroids. J. Geophys. Res. **95**(B1), 281–293 (1990b)

M.C. De Sanctis, E. Ammannito, M.T. Capria, F. Tosi, F. Capaccioni, F. Zambon, F. Carraro, S. Fonte, A. Frigeri, R. Jaumann, G. Magni, S. Marchi, T.B. McCord, L.A. McFadden, H.Y. McSween, D.W. Mittlefehldt, A. Nathues, E. Palomba, C.M. Pieters, C.A. Raymond, C.T. Russell, M.J. Toplis, D. Turrini, Spectroscopic characterization of mineralogy and its diversity across Vesta. Science **336**, 697–700 (2012). https://doi.org/10.1126/science.1219270

M.C. De Sanctis, E. Ammannito, T.M. Capria, F. Capaccioni, J.-P. Combe, A. Frigeri, A. Longobardo, G. Magni, S. Marchi, T.B. McCord, E. Palomba, F. Tosi, F. Zambon, F. Carraro, S. Fonte, J.-Y. Li, L.A. McFadden, D.W. Mittlefehkdt, C.M. Pieters, R. Jaumann, K. Stephan, C.A. Raymond, C.T. Russell, Vesta's mineralogical composition as revealed by the visible and infrared spectrometer on Dawn. Meteorit. Planet. Sci. **48**(11), 2166–2184 (2013). https://doi.org/10.1111/maps.12138

M.C. De Sanctis, A. Raponi, E. Ammannito, M. Ciarniello, M.J. Toplis, H.Y. McSween, J.C. Castillo-Rogez, B.L. Ehlmann, F.G. Carrozzo, S. Marchi, F. Tosi, F. Zambon, F. Capaccioni, M.T. Capria, S. Fonte, M. Formisano, A. Frigeri, M. Giardino, A. Longobardo, G. Magni, E. Palomba, L. McFadden, C.M. Pieters, R. Jaumann, P. Schenk, R. Mugnuolo, C.A. Raymond, C.T. Russell, Bright carbonate deposits as evidence of aqueous alteration on (1) Ceres. Nature **536**, 54–57 (2016). https://doi.org/10.1038/nature18290

M. Delbo, G. Libourel, J. Wilkerson et al., Thermal fatigue as the origin of regolith on small asteroids. Nature **508**, 233–236 (2014). https://doi.org/10.1038/nature13153

B.W. Denevi, A.W. Beck, E.I. Coman, B.J. Thomson, E. Ammannito, D.T. Blewett, J.M. Sunshine, M.C. De Sanctis, J.-Y. Li, S. Marchi, D.W. Mittlefehldt, N.E. Petro, C.A. Raymond, C.T. Russell, Global variations in regolith properties on asteroid Vesta from Dawn's low-altitude mapping orbit. Meteorit. Planet. Sci. **51**, 2366–2386 (2016). https://doi.org/10.1111/maps.12729

K.D. Duarte, B. Schmidt, H. Chilton, K. Hughson, H. Sizemore, K.L. Ferrier, J. Buffo, J. Scully, A. Nathues, T. Platz, M. Landis, S. Byrne, M. Bland, C.T. Russell, C.A. Raymond, Landslides on Ceres: diversity and geologic context. J. Geophys. Res., Planets **124**, 3329–3343 (2019). https://doi.org/10.1029/2018JE005673

L.T. Elkins-Tanton, E. Asphaug, J.F. Bell III, H. Bercovici, B. Bills, R. Binzel et al., Observations, meteorites, and models: a preflight assessment of the composition and formation of (16) Psyche. J. Geophys. Res., Planets **125**, e2019JE006296 (2020). https://doi.org/10.1029/2019JE006296

A.I. Ermakov, M. Kreslavsky, J.E.C. Scully, K.H.G. Hughson, R.S. Park, Surface roughness and gravitational slope distributions of Vesta and Ceres. J. Geophys. Res., Planets **124**, 14–30 (2019). https://doi.org/10.1029/2018JE005813

M. Ferrais, P. Vernazza, L. Jorda, N. Rambaux, J. Hanuš, B. Carry, F. Marchis, M. Marsset, M. Viikinkoski, M. Brož, R. Fetick, A. Drouard, T. Fusco, M. Birlan, E. Podlewska-Gaca, E. Jehin, P. Bartczak, J. Berthier, J. Castillo-Rogez, F. Cipriani, F. Colas, G. Dudziński, C. Dumas, J. Ďurech, M. Kaasalainen, A. Kryszczynska, P. Lamy, H. Le Coroller, A. Marciniak, T. Michalowski, P. Michel, T. Santana-Ros, P. Tanga, F. Vachier, A. Vigan, O. Witasse, B. Yang, Asteroid (16) Psyche's primordial shape: a possible Jacobi ellipsoid. Astron. Astrophys. **638**, L15 (2020). https://doi.org/10.1051/0004-6361/202038100

A. Figeri, M.C. De Sanctis, E. Ammannito, F. Tosi, A. Lomgobardo, F. Zambon, T. McCord, J.P. Combe, R. Jaumann, C.A. Raymond, C.T. Russell, The spectral parameter maps of Vesta from VIR data. Icarus **259**, 10–20 (2015). https://doi.org/10.1016/j.icarus.2015.06.027

H. Fossen, G. Carolina, G. Cavalcante, R. Vizeu, L. Pinheiro, C.J. Archanjo, Deformation – progressive or multiphase? J. Struct. Geol. **125**, 82–99 (2019). https://doi.org/10.1016/j.jsg.2018.05.006

R.R. Fu, B.H. Hager, A.I. Ermakov, M.T. Zuber, Efficient early global relaxation of asteroid Vesta. Icarus **240**, 133–145 (2014)

R.R. Fu, A.I. Ermakov, S. Marchi, J.C. Castillo-Rogez, C.A. Raymond, B.H. Hager, M.T. Zuber, S.D. King, M.T. Bland, M.C. De Sanctis, F. Preusker, R.S. Park, C.T. Russell, The interior structure of Ceres as revealed by surface topography. Earth Planet. Sci. Lett. **476**, 153–164 (2017)

R. Gaskell, SPC shape and topography of Vesta from DAWN imaging data, in *AAS, DPS XLIV, 209.03* (2012)

R. Gaskell, J. Saito, M. Ishiguro, T. Kubota, T. Hashimoto, N. Hirata, S. Abe, O.S. Barnouin-Jha, D. Scheeres, Global topography of asteroid 25143 Itokawa, in *LPSC 37 Abstract #1826* (2006)

R.W. Gaskell, O.S. Barnouin-Jha, D.J. Scheeres, A.S. Konopliv, T. Mukai, S. Abe, J. Saito, M. Ishiguro, T. Kubota, T. Hashimoto, J. Kawaguchi, M. Yoshikawa, K. Shirakawa, T. Kominato, N. Hirata, H. Demura, Characterizing and navigating small bodies with imaging data. Meteorit. Planet. Sci. **43**, 1049–1062 (2008)

K. Gwinner, F. Scholten, M. Spiegel, R. Schmidt, B. Giese, J. Oberst, C. Heipke, R. Jaumann, G. Neukum, Derivation and validation of high-resolution digital terrain models from Mars Express HRSC data. Photogramm. Eng. Remote Sens. **75**(9), 1127–1142 (2009)

K. Gwinner, R. Jaumann, H.H. Hauber, C. Heipke, J. Oberst, G. Neukum, V. Ansan, J. Bostelmann, A. Dumke, S. Elgner, G. Erkeling, F. Fueten, N.M. Hoekzema, E. Kersten, D. Loizeau, D.-D. Matz, V. Mertens, G. Michael, A. Pasewaldt, P. Pinet, F. Preusker, D. Reiss, T. Roatsch, R. Schmidt, F. Scholten,

M. Spiegel, R. Stesky, D. Tirsch, S. van Gasselt, M. Wählisch, K. Willner, The high resolution stereo camera (HRSC) of Mars Express and its approach to science analysis and mapping for Mars and its satellites. Planet. Space Sci. **196**, 93–138 (2016). https://doi.org/10.1016/j.pss.2016.02.014

D. Haack, K. Otto, B. Gundlach, C. Kreuzig, D. Bischoff, E. Kührt, J. Blum, Tensile strength of dust-ice mixtures and their relevance as cometary analog material. Astron. Astrophys. **642**, A218 (2020). https://doi.org/10.1051/0004-6361/202037763

M.D. Hicks, B.J. Buratti, K.J. Lawrence, J. Hillier, J-Y. Li, V. Reddy, S. Schröder, A. Nathues, M. Hoffmann, L. Le Corre, R. Duffard, H-B. Zhao, C. Raymond, C. Russell, T. Roatsch, R. Jaumann, H. Rhoades, D. Mayes, T. Barajas, T-T. Truong, J. Foster, A. McAuley, Spectral diversity and photometric behavior of main-belt and near-Earth vestoids and (4) Vesta: a study in preparation for the Dawn encounter. Icarus **235**, 60–74 (2014). https://doi.org/10.1016/j.icarus.2013.11.011

H. Hiesinger, S. Marchi, N. Schmedemann, P. Schenk, J.H. Pasckert, A. Neesemann, D.P.O. Brien, T. Kneissl, A.I. Ermakov, R.R. Fu, M.T. Bland, A. Nathues, T. Platz, D.A. Williams, R. Jaumann, J.C. Castillo-Rogez, O. Ruesch, B. Schmidt, R.S. Park, F. Preusker, D.L. Buczkowski, C.T. Russell, C.A. Raymond, Cratering on Ceres: implications for its crust and evolution. Science **353**, 1003 (2016). https://doi.org/10.1126/science.aaf4759

T. Hiroi, C.M. Pieters, H. Takeda, Grain size of the surface regolith of asteroid 4 Vesta estimated from its reflectance spectrum in comparison with HED meteorites. Meteoritics **29**, 394–396 (1994). https://doi.org/10.1111/j.1945-5100.1994.tb00603.x

R. Jaumann, G. Neukum, T. Behnke, T.C. Duxbury, K. Eichentopf, J. Flohrer, S.v. Gasselt, B. Giese, K. Gwinner, E. Hauber, H. Hoffmann, A. Hoffmeister, U. Köhler, K.-D. Matz, T.B. McCord, V. Mertens, J. Oberst, R. Pischel, D. Reiss, E. Ress, T. Roatsch, P. Saiger, F. Scholten, G. Schwarz, K. Stephan, M. Wählisch (the HRSC Co-Investigator Team), The high-resolution stereo camera (HRSC) experiment on Mars Express: instrument aspects and experiment conduct from interplanetary cruise through the nominal mission. Planet. Space Sci. **55**, 928–952 (2007)

R. Jaumann, D.A. Williams, D.L. Buczkowski, R.A. Yingst, F. Preusker, H. Hiesinger, N. Schmedemann, T. Kneisl, J.B. Vincent, D.T. Blewett, B.J. Buratti, U. Carsenty, B.W. Denevi, M.C. De Sanctis, W.B. Garry, H.U. Keller, E. Kersten, K. Krohn, J. Li, S. Marchi, K.D. Matz, T.B. McCord, H.Y. McSween, S.C. Mest, D.W. Mittlefehldt, S. Mottola, A. Nathues, G. Neukum, D.P. O'Brien, C.M. Pieters, T.H. Prettyman, C.A. Raymond, T. Roatsch, C.T. Russell, P. Schenk, B.E. Schmidt, F. Scholten, K. Stephan, M.V. Sykes, P. Tricarico, R. Wagner, M.T. Zuber, H. Sierks, Vesta's shape and morphology. Science **336**, 687–690 (2012). https://doi.org/10.1126/science.1219122

B.C. Johnson, M.M. Sori, A.J. Evans, Ferrovolcanism on metal worlds and the origin of pallasites. Nat. Astron. **4**, 41–44 (2020). https://doi.org/10.1038/s41550-019-0885-x

L. Jorda, P.L. Lamy, R.W. Gaskell, M. Kaasalainen, O. Groussin, S. Besse, G. Faury, Asteroid (2867) Steins: shape, topography and global physical properties from OSIRIS observations. Icarus **221**, 1089–1100 (2012). https://doi.org/10.1016/j.icarus.2012.07.035

M. Jutzi, E. Asphaug, Mega-ejecta on asteroid Vesta. Geophys. Res. Lett. **38**, L01102 (2011). https://doi.org/10.1029/2010GL045517

M. Jutzi, E. Asphaug, P. Gillet, J.A. Brarrat, W. Benz, The structure of the asteroid 4 Vesta as revealed by models of planet-scale collisions. Nature **494**, 207–210 (2013). https://doi.org/10.1038/nature11892

B. Kennedy, M. Abrahamson, A. Ardito, D. Han, R. Haw, N. Mastrodemos, S. Nandi, R. Park, B. Rush, A. Vaughan, Dawn orbit determination team: trajectory and gravity prediction performance during Vesta science phases, in *Paper AAS 2013-345, 23rd AAS/AIAA Spaceflight Mechanics Meeting*, Kauai, Hawaii, Feb. 10–14, 2013 (2013)

M.G. Kleinhans, H. Markies, S.J. de Vet, A.C. In 'T Veld, F.N. Postema, Static and dynamic angles of repose in loose granular materials under reduced gravity. J. Geophys. Res. **116**, E11004 (2011). https://doi.org/10.1029/2011JE003865

A.S. Konopliv, S.W. Asmar, F.G. Bills, N. Mastrodemos, R.S. Park, C.A. Raymond, D.E. Smith, M.T. Zuber, The dawn gravity investigation at Vesta and Ceres. Space Sci. Rev. **163**, 461–486 (2011)

K. Krohn, R. Jaumann, D. Elbeshausen, T. Kneissl, N. Schmedemann, J. Voigt, K. Otto, K.-D. Matz, F. Preusker, T. Roatsch, K. Stephan, R. Wagner, C.A. Raymond, C.T. Russell, Asymmetric craters on Vesta: impact on sloping surfaces. Planet. Space Sci. **103**, 36–56 (2014). https://doi.org/10.1016/j.pss.2014.04.011

K. Krohn, I. von der Gathen, D.L. Buczkowski, R. Jaumann, K. Wickhusen, F. Schulzeck, K. Stephan, R. Wagner, J.C.E. Scully, C.A. Raymond, C.T. Russell, Fracture geometry and statistics of Ceres' floor fractures. Planet. Space Sci. **187**, 104955 (2020). https://doi.org/10.1016/j.pss.2020.104955

Z.A. Landsman, J.P. Emery, H. Campins, J. Hanuš, L.F. Lim, D.P. Cruikshank, Asteroid (16) Psyche: evidence for a silicate regolith from spitzer space telescope spectroscopy. Icarus **304**, 58–73 (2018). https://doi.org/10.1016/j.icarus.2017.11.035

D.S. Lauretta, D.N. DellaGiustina, C.A. Bennett et al., The unexpected surface of asteroid (101955) Bennu. Nature **568**, 55–60 (2019). https://doi.org/10.1038/s41586-019-1033-6

L. Le Corre, V. Reddy, J. Sanchez, T. Dunn, E. Cloutis, M. Izawa, P. Mann, A. Nathues, Exploring exogenic sources for the olivine on Asteroid (4) Vesta. Icarus **258**, 483–499 (2015). https://doi.org/10.1016/j.icarus.2015.01.018

J. Li, *Body-Fixed Coordinate Systems for Asteroid (4) Vesta, Planetary Data System* (2012)

G. Libourel, A.M. Nakamura, P. Beck, S. Potin, C. Ganino, S. Jacomet et al., Hypervelocity impacts as a source of deceiving surface signatures on iron-rich asteroids. Sci. Adv. **5**(8), eaav3971 (2019). https://doi.org/10.1126/sciadv.aav3971

R.M.C. Lopes, Io: the volcanic moon, in *Encyclopedia of the Solar System*, ed. by L.-A. McFadden, P.R. Weissman, T.V. Johnson 2nd edn. (Academic Press, San Diego, 2007), pp. 419–430. https://doi.org/10.1016/B978-012088589-3/50026-8. Chap. 22

D.F. Lupishko, I.N. Belskaya, On the surface composition of the M-type asteroids. Icarus **78**, 395–401 (1989)

X. Mao, W.B. McKinnon, Spin evolution of Ceres and Vesta due to impacts. Meteorit. Planet. Sci. **55**, 2493–2518 (2020). https://doi.org/10.1111/maps.13594

S. Marchi, D.D. Durda, C.A. Polanskey, E. Asphaug, W.F. Bottke, L.T. Elkins-Tanton et al., Hypervelocity impact experiments in iron-nickel ingots and meteorites: implications for the NASA Psyche mission. J. Geophys. Res., Planets **125**(2), e2019JE005927 (2019). https://doi.org/10.1029/2019JE005927

S. Marchi, D.D. Durda, C.A. Polanskey, E. Asphaug, W.F. Bottke, L.T. Elkins-Tanton et al., Hypervelocity impact experiments in iron-nickel ingots and iron meteorites: implications for the NASA Psyche mission. J. Geophys. Res., Planets **125**, e2019JE005927 (2020). https://doi.org/10.1029/2019JE005927

S. Marchi, E. Asphaug, J. Bell, W.F. Bottke, R. Jaumann, R.S. Park, C.A. Polanskey, T. Prettyman, D.A. Williams, W.E. Page, Determining the relative ages of regions of Psyche's surface. Space Sci. Rev. (2022, this journal)

M. Marsset, M. Brož, P. Vernazza et al., The violent collisional history of aqueously evolved (2) Pallas. Nat. Astron. **4**, 569–576 (2020). https://doi.org/10.1038/s41550-019-1007-5

T. Matsumoto, D. Harries, F. Langenhorst et al., Iron whiskers on asteroid Itokawa indicate sulfide destruction by space weathering. Nat. Commun. **11**, 1117 (2020). https://doi.org/10.1038/s41467-020-14758-3

A. Matter, M. Delbo, L.S. Carry, Evidence of a metal-rich surface for the Asteroid (16) Psyche from interferometric observations in the thermal infrared. Icarus **226**, 419–427 (2013). https://doi.org/10.1016/j.icarus.2013.06.004

J-G. Moreau, T. Kohout, K. Wünnemann, P. Halodova, J. Haloda, Shock physics mesoscale modeling of shock stage 5 and 6 in ordinary and enstatite chondrites. Icarus **332**, 50–65 (2019). https://doi.org/10.1016/j.icarus.2019.06.004

A. Neesemann, S. van Gasselt, S.H.G. Walter, R. Jaumann, J.C. Castillo-Rogez, C.A. Raymond, C.T. Russell, F. Postberg, Mapping of Ceres by Dawn: generation of higher level cartographic products based on Framing Camera Data, in *53rd Lunar and Planetary Science Conference*, March 7–11, 2022 (2022)

G.A. Neumann, D.D. Rowlands, F.G. Lemoine, D.E. Smith, M.T. Zuber, Crossover analysis of Mars Orbiter Laser Altimeter data. J. Geophys. Res. **106**, 23,753–23,768 (2001). https://doi.org/10.1029/2000JE001381

T. Noguchi, M. Kimura, T. Hashimoto, M. Konno, T. Nakamura, M.E. Zolensky, R. Okazaki, M. Tanaka, A. Tsuchiyama, A. Nakato, T. Ogami, H. Ishida, R. Sagae, S. Tsujimoto, T. Matsumoto, J. Matsuno, A. Fujimura, M. Abe, T. Yada, T. Mukai, M. Ueno, T. Okada, K. Shirai, Y. Ishibashi, Space weathered rims found on the surfaces of the Itokawa dust particles. Meteorit. Planet. Sci. **49**, 88–214 (2014)

K.A. Otto, R. Jaumann, K. Krohn, K.-M. Matz, F. Preusker, T. Roatsch, P. Schenk, F. Scholten, K. Stephan, C.A. Raymond, C.T. Russell, Mass wasting features and processes in Vesta's south polar basin Rheasilvia. J. Geophys. Res. **118**, 2263–2437 (2013). https://doi.org/10.1002/2013JE004333

K.A. Otto, R. Jaumann, K. Krohn, F. Spahn, C.A. Raymond, C.T. Russell, The Coriolis effect on mass wasting during the Rheasilvia impact on asteroid Vesta. Geophys. Res. Lett. **43**, 12340–12347 (2016). https://doi.org/10.1002/2016GL071539

K.A. Otto, S. Marchi, A. Trowbridge, H.J. Melosh, H.G. Sizemore, Ceres Crater degradation inferred from concentric fracturing. J. Geophys. Res., Planets **124**, 1188–1203 (2019). https://doi.org/10.1029/2018JE005660

D.W. Parcher, G.J. Whiffen, Dawn statistical maneuver design for Vesta operations, in *Paper AAS 2011-180, AIAA/AAS Astrodynamics Specialist Conference*, New Orleans, Louisiana, USA, Feb. 13–17, 2011 (2011)

R.S. Park, A.S. Konopliv, B.G. Bills, N. Rambaux, J.C. Castillo-Rogez, C.A. Raymond, A.T. Vaughan, A.I. Ermakov, M.T. Zuber, R.R. Fu, M.J. Toplis, C.T. Russell, A. Nathues, F. Preusker, A partially differentiated interior for Ceres deduced from its gravity field and shape. Nature **537**, 515–517 (2016). https://doi.org/10.1038/nature18955

R.S. Park, A.T. Vaughan, A.S. Konopliv, A.I. Ermakov, N. Mastrodemos, J.C. Castillo-Rogez, S.P. Joy, A. Nathues, C.A. Polanskey, M.D. Rayman, J.E. Riedel, C.A. Raymond, C.T. Russell, M.T. Zuber, High-resolution shape model of Ceres from stereophotoclinometry using Dawn Imaging Data. Icarus **319**, 812–827 (2019). https://doi.org/10.1016/j.icarus.2018.10.024

D.E. Pavlis et al., GEODYN operations manuals. Raytheon ITTS Contractor Report, Lanham, MD (2001)

C.M. Pieters, L.A. Taylor, S.K. Noble, L.P. Keller, B. Hapke, R.V. Morris, C.C. Allen, D.S. McKay, S. Wentworth, Space weathering on airless bodies: resolving a mystery with lunar samples. Meteorit. Planet. Sci. **35**, 1101–1107 (2000)

C. Pieters, E. Ammannito, D. Blewett et al., Distinctive space weathering on Vesta from regolith mixing processes. Nature **491**, 79–82 (2012). https://doi.org/10.1038/nature11534

J.B. Plescia, M.P. Golombek, Origin of planetary wrinkle ridges based on the study of terrestrial analogs. Geol. Soc. Am. Bull. **97**, 1289–1299 (1986)

C.A. Polanskey, S.P. Joy, C.A. Raymond, Efficacy of the Dawn Vesta science plan, in *12th International Conference on Space Operations 2012*, Stockholm, Sweden, June 2012 (2012)

C.A. Polanskey, L. Elkins-Tanton, R. Jaumann, D.J. Lawrence, D.M. Marsh, R.R. Moore et al., Psyche science operations concept: maximize reuse to minimize risk, in *2018 SpaceOps Conference* (2018), p. 2703

C.A. Polanskey et al., Space Sci. Rev. (2022, this journal)

F. Preusker, F. Scholten, J. Knollenberg, E. Kührt, K.-D. Matz, S. Mottola, T. Roatsch, N. Thomas, The northern hemisphere of asteroid (21) Lutetia – topography and orthoimages from Rosetta OSIRIS NAC image data. Planet. Space Sci. **66**, 54–63 (2012)

F. Preusker, F. Scholten, K.-D. Matz et al., Shape model, reference system definition, and cartographic mapping standards for comet 67P/Churyumov-Gerasimenko – stereo-photogrammetric analysis of Rosetta/OSIRIS image data. Astron. Astrophys. **583**, A33 (2015). https://doi.org/10.1051/0004-6361/201526349

F. Preusker, F. Scholten, K.D. Matz, S. Elgner, R. Jaumann, T. Roatsch, S.P. Joy, C.A. Polanskey, C.A. Raymond, C.T. Russell, Dawn at Ceres – shape model and rotational state, in *47th Lunar and Planetary Science Conference [#1954]*, March 21–25, 2016 (2016)

F. Preusker, F. Scholten, S. Elgner, K.-D. Matz, R.T. Kameda, R. Jaumann, S. Sugita, S. Honda, T. Morota, E. Tatsumi, Y. Cho, K. Yoshioka, S. Sawada, Y. Yokota, N. Sakatani, M. Hayakawa, M. Matsuoka, M. Yamada, T. Kouyama, H. Suzuki, C. Honda, K. Ogawa, The MASCOT landing area on asteroid (162173) Ryugu: stereo-photogrammetric analysis using images of the ONC onboard the Hayabusa2 spacecraft. Astron. Astrophys. **632**, L4 (2019). https://doi.org/10.1051/00046361/201936759

M.D. Rayman, R.A. Mase, Dawn's exploration of Vesta. Acta Astronaut. **94**, 159–167 (2014)

C.A. Raymond, R. Jaumann, A. Nathues, H. Sierks, T. Roatsch, F. Preusker, F. Scholten, R.W. Gaskell, L. Jorda, H-U. Keller, M.T. Zuber, D.E. Smith, N. Mastrodemos, S. Mottola, The Dawn topography investigation. Space Sci. Rev. **163**, 487–510 (2011). https://doi.org/10.1007/s11214-011-9863-z

T. Roatsch, E. Kersten, K.-M. Matz, F. Preusker, F. Scholten, R. Jaumann, C.A. Raymond, C.T. Russell, High resolution Vesta High Altitude Mapping Orbit (HAMO) Atlas derived from Dawn framing camera images. Planet. Space Sci. **73**, 283–286 (2012). https://doi.org/10.1016/j.pss.2012.08.021

T. Roatsch, E. Kersten, A. Hoffmeister, M. Wählisch, K.-D. Matz, C.C. Porco, Recent improvements of the Saturnian satellite atlases: Mimas, Enceladus, and Dione. Planet. Space Sci. **77**, 118–125 (2013). https://doi.org/10.1016/j.pss.2012.02.016

T. Roatsch, E. Kersten, K.-D. Matz, F. Preusker, F. Scholten, R. Jaumann, C.A. Raymond, C.T. Russell, Ceres Survey Atlas derived from Dawn Framing Camera images. Planet. Space Sci. **121**, 115–120 (2016). https://doi.org/10.1016/j.pss.2015.12.005

D.D. Rowlands et al., GEODYN II system description. Hughes-STX Contractor Report, Greenbelt, MD (1993)

C.T. Russell, H.Y. McSween, R. Jaumann, C.A. Raymond, The Dawn mission to Vesta and Ceres, in *Asteroids IV*, ed. by P. Michel, F.E. DeMeo, W.F. Bottke (Univ. of Arizona Press, Tucson, 2015), pp. 419–433

S. Sasaki, K. Nakamura, Y. Hamabe, E. Kurahashi, T. Hiroi, Production of iron nanoparticles by laser irradiation in a simulation of lunar-like space weathering. Nature **410**, 555–557 (2001)

P. Schenk, D.P. O'Brien, S. Marchi, R. Gaskell, F. Preusker, T. Roatsch, R. Jaumann, D. Buczkowski, T. McCord, H.Y. McSween, D. Williams, A. Yingst, R. Raymond, C. Russell, The geologically recent giant impact basins at Vesta's south pole. Science **336**, 694–697 (2012). https://www.science.org/doi/full/10.1126/science.1223272

P. Schenk, J. Castillo-Rogez, K.A. Otto, S. Marchi, D. O'Brien, H.K. Bland, B. Schmidt, J. Scully, D. Buczkowski, K. Krohn, T. Hoogenboom, G. Kramer, V. Bray, A. Neesemann, H. Hiesinger, T. Platz, M.C. De Sanctis, S. Schroeder, L. Le Corre, L. McFadden, M. Sykes, C. Raymond, C.T. Russell, Compositional control on impact crater formation on mid-sized planetary bodies: Dawn at Ceres and Vesta, Cassini at Saturn. Icarus **359**, 114343 (2021). https://doi.org/10.1016/j.icarus.2021.114343

N. Schmedemann, T. Kneissl, B.A. Ivanov, G.G. Michael, R.J. Wagner, G. Neukum, O. Ruesch, H. Hiesinger, K. Krohn, P.F. Roatsch, H. Sierks, R. Jaumann, V. Reddy, A. Natheus, S.H.G. Walter, A. Neesemann, C.A. Raymond, C.T. Russell, The cratering record, chronology, and surface ages of (4) Vesta in comparison to smaller asteroids and the ages of HED meteorites. Planet. Space Sci. **103**, 104–130 (2014). https://doi.org/10.1016/j.pss.2014.04.004

F. Scholten, K. Gwinner, T. Roatsch, K.-D. Matz, M. Wählisch, B. Giese, J. Oberst, R. Jaumann, G. Neukum (HRSC Co-Investigator Team), Mars Express HRSC data processing – methods and operational aspects. Photogramm. Eng. Remote Sens. **71**(10), 1143–1152 (2005)

F. Scholten, J. Oberst, K.-D. Matz, T. Roatsch, M. Wählisch, M.S. Robinson (the LROC Team), Towards global lunar topography using LROC WAC stereo data, in *41st Lunar and Planetary Science Conference, #2111* (2009)

F. Scholten, F. Preusker, S. Elgner, K.-D. Matz, R. Jaumann, M. Hamm, S.E. Schroeder, A. Koncz, N. Schmitz, F. Trauthan, M. Grott, J. Biele, T.-M. Ho, S. Kameda, S. Sugita, The Hayabusa2 lander MASCOT on the surface of asteroid (162173) Ryugu – stereo-photogrammetric analysis of MASCam image data. Astron. Astrophys. **632**, L5 (2019). https://doi.org/10.1051/0004-6361/201936760

S.E. Schroeder, S. Mottola, H.U. Keller, C.A. Raymond, C.T. Russell, Resolved photometry of Vesta reveals physical properties of crater regolith. Planet. Space Sci. **85**, 198–213 (2013). https://doi.org/10.1016/j.pss.2013.06.009. ISSN 0032-0633

M. Shepard, J. Richardson, P.A. Taylor, L.A. Rodrigues-Ford, A. Conrad, I. de Pater, M. Adamkovics, K. de Kleer, J.R. Males, K.M. Morzinski, L.M. Close, M. Kaasalainen, M. Viikinkoski, B. Timerson, V. Reddy, C. Magri, M.C. Nolan, E.S. Howell, L.A.M. Benner, J.D. Giorgini, B.D. Warner, A.W. Harris, Radar observations and shape model of asteroid 16 Psyche. Icarus **281**, 388–403 (2017). https://doi.org/10.1016/j.icarus.2016.08.011

L. Siltala, M. Granvik, Mass and density of asteroid (16) Psyche. ApJL (2021, in press). 2103.01707v1 [astro-ph.EP]

J.A. Skinner, A.E. Huff Jr., C.M. Fortezzo, T.A. Giather, T.M. Hare, M.A. Hunter, Planetary Geologic Mapping Protocol – 2018 (2018). Downloadable at: https://astropedia.astrogeology.usgs.gov/download/Docs/Mappers/PGM_Protocol_March_2018.pdf

K. Stephan, R. Jaumann, M.C. De Sanctis, F. Tosi, E. Ammannito, K. Krohn, F. Zambon, S. Marchi, O. Ruesch, K.D. Matz, F. Preusker, T. Roatsch, C.A. Raymond, C.T. Russell, Small fresh impact craters on asteroid 4 Vesta: a compositional and geological fingerprint. J. Geophys. Res., Planets **119**(4), 2013JE004388 (2014)

S. Sugita, R. Honda, T. Morota, S. Kameda, H. Sawada, E. Tatsumi, M. Yamada, C. Honda, Y. Yokota, T. Kouyama, N. Sakatani, K. Ogawa, H. Suzuki, T. Okada, N. Namiki, S. Tanaka, Y. Iijima, K. Yoshioka, M. Hayakawa, Y. Cho, M. Matsuoka, N. Hirata, H. Miyamoto, D. Domingue, M. Hirabayashi, T. Nakamura, T. Hiroi, T. Michikami, P. Michel, R.-L. Ballouz, O.S. Barnouin, C.M. Ernst, S.E. Schröder, H. Kikuchi, R. Hemmi, G. Komatsu, T. Fukuhara, M. Taguchi, T. Arai, H. Senshu, H. Demura, Y. Ogawa, Y. Shimaki, T. Sekiguchi, T.G. Müller, A. Hagermann, T. Mizuno, H. Noda, K. Matsumoto, R. Yamada, Y. Ishihara, H. Ikeda, H. Araki, K. Yamamoto, S. Abe, F. Yoshida, A. Higuchi, S. Sasaki, S. Oshigami, S. Tsuruta, K. Asari, S. Tazawa, M. Shizugami, J. Kimura, T. Otsubo, H. Yabuta, S. Hasegawa, M. Ishiguro, S. Tachibana, E. Palmer, R. Gaskell, L. Le Corre, R. Jaumann, K. Otto, N. Schmitz, P.A. Abell, M.A. Barucci, M.E. Zolensky, F. Vilas, F. Thuillet, C. Sugimoto, N. Takaki, Y. Suzuki, H. Kamiyoshihara, M. Okada, K. Nagata, M. Fujimoto, M. Yoshikawa, Y. Yamamoto, K. Shirai, R. Noguchi, N. Ogawa, F. Terui, S. Kikuchi, T. Yamaguchi, Y. Oki, Y. Takao, H. Takeuchi, G. Ono, Y. Mimasu, K. Yoshikawa, T. Takahashi, Y. Takei, A. Fujii, C. Hirose, S. Nakazawa, S. Hosoda, O. Mori, T. Shimada, S. Soldini, T. Iwata, M. Abe, H. Yano, R. Tsukizaki, M. Ozaki, K. Nishiyama, T. Saiki, S. Watanabe, Y. Tsuda, The geomorphology, color, and thermal properties of Ryugu: implications for parent-body processes. Science **364**, 252 (2019)

M. Tenthoff, K. Wohlfarth, C. Wöhler, High resolution digital terrain models of Mercury. Remote Sens. **12**(23), 3989 (2020). https://doi.org/10.3390/rs12233989

P.C. Thomas, R.R. Benzel, M.J. Gaffey, B.H. Zellner, A.D. Storrs, E. Wells, Vesta: spin pole, size, and shape from HST images. Icarus **128**, 88–94 (1997)

P.C. Thomas, J. Joseph, B. Carcich, J. Veverka, B.E. Clark, J.F. Bell, A.W. Byrd, R. Chomko, M. Robinson, S. Murchie, L. Prockter, A. Cheng, N. Izenberg, M. Malin, C. Chapman, L.A. McFadden, R. Kirk, M. Gaffey, P.G. Lucey, Eros: shape, topography, and slope processes. Icarus **155**(1), 18–37 (2002). https://doi.org/10.1006/icar.2001.6755

M. Viikinkoski, P. Vernazza, J. Hanuš, H. Le Corroler, K. Tazhenova et al., (16) Psyche: a mesosiderite-like asteroid? Astron. Astrophys. **619**, 619 (2018). https://doi.org/10.1051/0004-6361/201834091

S. Watanabe, M. Hirabayashi, N. Hirata, N. Hirata, R. Noguchi, Y. Shimaki, H. Ikeda, E. Tatsumi, M. Yoshikawa, S. Kikuchi, H. Yabuta, T. Nakamura, S. Tachibana, Y. Ishihara, T. Morota, K. Kitazato, N. Sakatani, K. Matsumoto, K. Wada, H. Senshu, C. Honda, T. Michikami, H. Takeuchi, T. Kouyama, R.

Honda, S. Kameda, T. Fuse, H. Miyamoto, G. Komatsu, S. Sugita, T. Okada, N. Namiki, M. Arakawa, M. Ishiguro, M. Abe, R. Gaskell, E. Palmer, O.S. Barnouin, P. Michel, A.S. French, J.W. McMahon, D.J. Scheeres, P.A. Abell, Y. Yamamoto, S. Tanaka, K. Shirai, M. Matsuoka, M. Yamada, Y. Yokota, H. Suzuki, K. Yoshioka, Y. Cho, N. Nishikawa, T. Sugiyama, H. Kikuchi, R. Hemmi, T. Yamaguchi, N. Ogawa, G. Ono, Y. Mimasu, K. Yoshikawa, T. Takahashi, Y. Takei, A. Fujii, C. Hirose, T. Iwata, M. Hayakawa, S. Hosoda, O. Mori, H. Sawada, T. Shimada, S. Soldini, H. Yano, R. Tsukizaki, M. Ozaki, Y. Iijima, K. Ogawa, M. Fujimoto, T.-M. Ho, A. Moussi, R. Jaumann, J.-P. Bibring, C. Krause, F. Terui, T. Saiki, S. Nakazawa, Y. Tsuda, Hayabusa2 arrives at the carbonaceous asteroid 162173 Ryugu – a spinning top-shaped rubble pile. Science **364**, 268–272 (2019)

T.R. Watters, A case for limited global contraction of Mercury. Commun. Earth Environ. **2**, 9 (2021). https://doi.org/10.1038/s43247-020-00076-5

D.A. Williams, R. Jaumann, H.Y. McSween Jr., C.A. Raymond, S. Marchi, N. Schmedemann, C.T. Russell, The chronostratigraphy of protoplanet Vesta. Icarus **244**, 158–165 (2014). https://doi.org/10.1016/j.icarus.2014.06.027

M.T. Zuber, D.E. Smith, A.F. Cheng, J.B. Garvin, O. Aharonson, T.D. Cole, P.J. Dunn, Y. Guo, F.G. Lemoine, G.A. Neumann, D.D. Rowlands, M.H. Torrence, The shape of 433 Eros from the NEAR-shoemaker laser rangefinder. Science **289**, 2097–2101 (2000). https://doi.org/10.1126/science.289.5487.2097

M.T. Zuber, R.S. Park, L.T. Elkins-Tanton, J. Bell, K.N. Bruvold, D. Bercovici, B.R. Bills, R.P. Binzel, R. Jaumann, S. Marchi, C.A. Polanski, C.A. Raymond, T. Roatsch, C. Wang, B.P. Weiss, M.A. Wieczorek, The Psyche gravity investigation. Space Sci. Rev. (2022, this journal)

Publisher's Note Springer Nature remains neutral with regard to jurisdictional claims in published maps and institutional affiliations.

Authors and Affiliations

Ralf Jaumann[1] · James F. Bell III[2] · Carol A. Polanskey[3] · Carol A. Raymond[3] · Erik Aspaugh[4] · David Bercovici[5] · Bruce R. Bills[3] · Richard Binzel[6] · William Bottke[7] · John M. Christoph[2] · Simone Marchi[7] · Alicia Neesemann[1] · Katharina Otto[8] · Ryan S. Park[3] · Frank Preusker[8] · Thomas Roatsch[8] · David A. Williams[2] · Mark A. Wieczorek[9] · Maria T. Zuber[6]

✉ R. Jaumann
ralf.jaumann@fu-berlin.de

1 Freie Universitaet Berlin, Institute of Geological Sciences, 12249 Berlin, Germany

2 School of Earth and Space Exploration, Arizona State University, Tempe, AZ 86387-2001, USA

3 Jet Propulsion Laboratory, Pasadena, CA 91109, USA

4 Lunar and Planetary Laboratory, Tucson, AZ, USA

5 Department of Earth and Planetary Sciences, Yale University, New Haven, CT 06520, USA

6 Department of Earth, Atmospheric, and Planetary Sciences, Massachusetts Institute of Technology, Cambridge, MA 02139, USA

7 Southwest Research Institute, Boulder, CO 80302, USA

8 Deutches Zentrum für Luft- und Raumfahrt, Institute of Planetary Research, Berlin, Germany

9 Université Côte d'Azur, Observatoire de la Côte d'Azur CNRS, Laboratoire Lagrange, Nice, France

Space Science Reviews (2023) 219:22
https://doi.org/10.1007/s11214-023-00965-z

The Psyche Magnetometry Investigation

Benjamin P. Weiss[1] · José M.G. Merayo[2] · Jodie B. Ream[1] · Rona Oran[1] · Peter Brauer[2] · Corey J. Cochrane[3] · Kyle Cloutier[3] · Linda T. Elkins-Tanton[4] · John L. Jørgensen[2] · Clara Maurel[1] · Ryan S. Park[3] · Carol A. Polanskey[3] · Maria de Soria Santacruz-Pich[3] · Carol A. Raymond[3] · Christopher T. Russell[5] · Daniel Wenkert[3] · Mark A. Wieczorek[6] · Maria T. Zuber[1]

Received: 3 December 2021 / Accepted: 1 March 2023 / Published online: 28 March 2023

Abstract
The objective of the Psyche Magnetometry Investigation is to test the hypothesis that asteroid (16) Psyche formed from the core of a differentiated planetesimal. To address this, the Psyche Magnetometer will measure the magnetic field around the asteroid to search for evidence of remanent magnetization. Paleomagnetic measurements of meteorites and dynamo theory indicate that a diversity of planetesimals once generated dynamo magnetic fields in their metallic cores. Likewise, the detection of a strong magnetic moment ($> 2 \times 10^{14}$ Am^2) at Psyche would likely indicate that the body once generated a core dynamo, implying that it formed by igneous differentiation. The Psyche Magnetometer consists of two three-axis fluxgate Sensor Units (SUs) mounted 0.7 m apart along a 2.15-m long boom and connected to two Electronics Units (EUs) located within the spacecraft bus. The Magnetometer samples at up to 50 Hz, has a range of ±80,000 nT, and an instrument noise of 39 pT axis^{-1} 3σ integrated over 0.1 to 1 Hz. The two pairs of SUs and EUs provide redundancy and enable gradiometry measurements to suppress noise from flight system magnetic fields. The Magnetometer will be powered on soon after launch and acquire data for the full duration of the mission. The ground data system processes the Magnetometer measurements to obtain an estimate of Psyche's dipole moment.

The NASA Psyche Mission: Science Instruments and Investigations
Edited by James F. Bell III, Carol Polanskey and Lindy Elkins-Tanton

✉ B.P. Weiss

1 Department of Earth, Atmospheric and Planetary Sciences, Massachusetts Institute of Technology (MIT), Cambridge, MA, USA

2 DTU Space, Technical University of Denmark (DTU), Copenhagen, Denmark

3 Jet Propulsion Laboratory (JPL), California Institute of Technology, Pasadena, CA, USA

4 School of Earth and Space Exploration, Arizona State University, Tempe, AZ, USA

5 Department of Earth and Space Sciences, University of California, Los Angeles, Los Angeles, CA, USA

6 Observatoire de la Côte d'Azur, Centre National de la Recherche Scientifique (CNRS), Laboratoire Lagrange, Université Côte d'Azur, Nice, France

Keywords Fluxgate magnetometer · Gradiometry · Paleomagnetism · Magnetosphere · Dynamo · Psyche · Planetary differentiation · Asteroids and planetesimals

1 Introduction

1.1 Psyche Mission Overview

The Psyche mission aims to characterize the structure, composition, geology and origin of the main belt asteroid (16) Psyche (Elkins-Tanton et al. 2022; Polanskey et al. 2023). With a mean radius, R_P, of $\sim$ 111-km (Shepard et al. 2021), Psyche is likely the largest iron-rich body in the solar system and will be the first small, metal-rich world to be encountered by a spacecraft. The mission has three science goals: (1) understand a previously-unexplored building block of planet formation: iron-rich planetesimals; (2) look inside terrestrial planets, including Earth, by directly examining the interior of a differentiated body, which otherwise could not be seen and (3) explore a new type of world that is rich in metal rather than in rock or ice. One of the three science instruments is the Magnetometer, which will measure the magnetic field around the asteroid. Here we describe the science objectives of the Magnetometry Investigation, their relationship to mission objectives, the context of previous magnetic studies of asteroids and meteorites, the design, performance, and accommodation of the instrument, and the expected data products and calibration pipeline.

This article is structured as follows. In the rest of Sect. 1, we describe the science objectives of the Magnetometry Investigation, how these translate into high level engineering requirements, and an overview of the Magnetometer instrument. In Sect. 2, we review our understanding of remanent magnetization in meteorites and small bodies. In Sect. 3, we describe the design, ground calibration, accommodation, operations and performance of the Magnetometer. In Sects. 4 and 5, we describe the data and in-flight calibration pipelines. We summarize the Investigation in Sect. 6.

1.2 Science Objectives

The three mission science goals (Sect. 1.1) are addressed with five science objectives (Polanskey et al. 2023). Of these, the Magnetometry Investigation was designed to address the central mission science objective: determine whether Psyche formed from a core or if it is primordial unmelted material (Elkins-Tanton et al. 2022).

Based upon the currently available data, the team has assembled models for what Psyche might be (Elkins-Tanton et al. 2020). The asteroid's bulk density is estimated to be between 3,400 and 4,100 $\mathrm{kg\,m^{-3}}$. Because meteoritic metal has a density of about 7,000 $\mathrm{kg\,m^{-3}}$, Psyche must be a mixture of metal and lower-density components including silicates and/or pore space. Given the range of likely secondary components, Psyche is probably between ~30–60 vol.% (~70–100 wt.%) metal (Elkins-Tanton et al. 2020; Zhang et al. 2022). Psyche thus appears to be more metal-rich than the Earth (~32 wt.% metal) (McDonough and Sun 1995) and ordinary (LL, L, and H) chondritic materials (~2–20 wt.% metal) (Dunn et al. 2010).

Core formation may be the simplest model for forming a metal-enriched parent body for Psyche. Our current fiducial model for the asteroid's formation is therefore that it is the remains of a differentiated body that had formed a metallic core. Alternatively, Psyche might be an undifferentiated body consisting of a highly reduced, low-FeO material. High density, low-FeO meteorite analog candidates include enstatite, CH, and CB chondrites (Dibb et al. 2022; Elkins-Tanton et al. 2020).

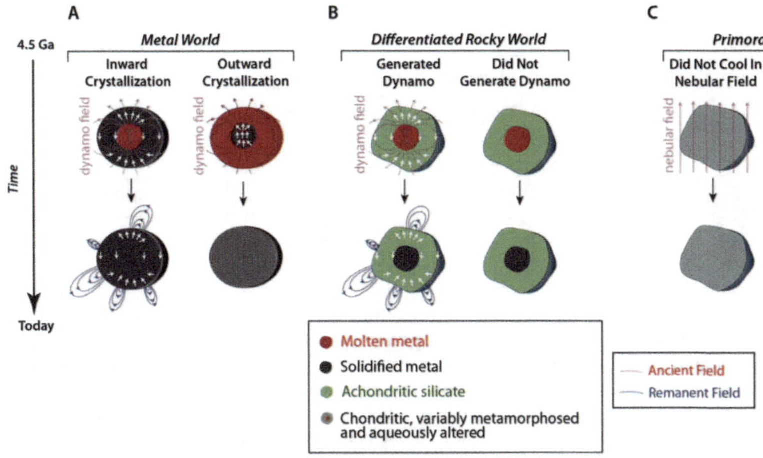

Fig. 1 Some possible magnetization states for Psyche and their relationships to possible formation and evolution models. Top and bottom rows show Psyche's structure 4.5 billion years ago (Ga) and today, respectively. Red and blue fields denote dynamo and remanent magnetic fields. Red, dark grey and green shading denote molten metal, solid metal, and melted silicates, respectively in (**A**) and (**B**). Grey and red-gray shading denote unheated and heated but unmelted primordial materials in (**C**). (**A**) Formation as a differentiated mantle-stripped metallic planetesimal with a molten core. Such a body could crystallize from the outside-in (left) or inside-out (right). Both scenarios might generate a dynamo but only the former could produce remanent magnetization. (**B**) Formation as a differentiated or partially differentiated body with a silicate crust and molten iron core. Such a body might (left) or might not (right) generate a dynamo that could magnetize the silicate mantle. Similarly, a mantle-stripped metallic world (**A**) might also not generate a dynamo (not shown here). (**C**) Formation as an unmelted primordial aggregate. Such a body might not (left) or might (right) have been heated and/or aqueously altered while a nebular magnetic field was present. Only the latter could magnetize the asteroid

Measurement of a strong remanent magnetic field would perhaps be the single most definitive determiner of Psyche's origin: if Psyche is magnetized, the long-past action of a core dynamo (Sect. 1.3) is the most likely cause (Fig. 1). Psyche would thus be determined to be a core or parts of a core. On the other hand, if Psyche is composed of metal-rich undifferentiated material, it would never have generated a dynamo. The absence of a remanent field would certainly be indeterminate: even if Psyche was a core, its materials may have not been magnetized by an existing dynamo field, or else the core may never have generated a dynamo field.

Psyche could become magnetized by the dynamo if part of it cooled and/or underwent low-temperature recrystallization while the field was still being generated in the asteroid's molten interior (Fig. 1). Acquiring magnetization would require that ferromagnetic materials on Psyche were below their characteristic critical ordering temperatures (e.g., 780 °C for the Fe mineral kamacite and as low as 320 °C for the FeNi mineral tetrataenite) while the dynamo was still active (Scheinberg et al. 2017). Therefore, such a body could have recorded a paleofield only if it crystallized from the outside-in (rather than from the inside-out) where the materials were sufficiently cool. Inward crystallization will occur if the liquidus (melting temperature) increases more rapidly with pressure than does the internal temperature profile of the body, a condition that is likely but not guaranteed for small bodies depending on particular on their sulfur-contents (Williams 2009). The outer layers could be composed of metal (if Psyche is a mantle-stripped metallic body) (Neufeld et al. 2019; Scheinberg et al. 2015), an achondritic silicate mantle (if Psyche is fully differentiated but its mantle was

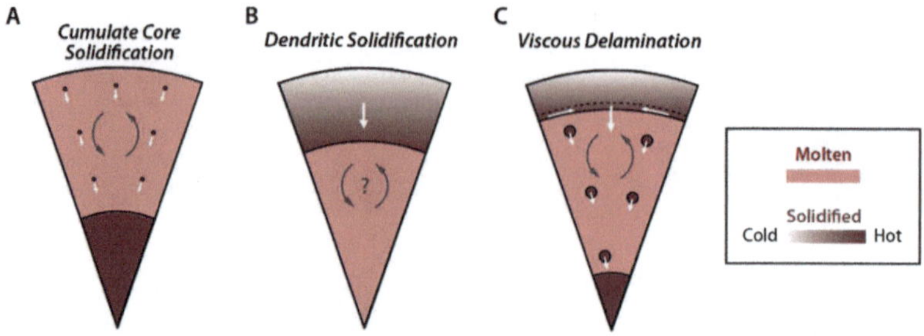

Fig. 2 Three evolutionary scenarios for an inwardly crystallizing mantle-stripped metallic planetesimal. (**A**) The crystallites, which are denser than the surrounding liquid, sink to the bottom and accumulate as a detrital core that remains above the Curie temperature during the lifetime of any convective dynamo (Scheinberg et al. 2015). Such a body would not be magnetized. (**B**) The core undergoes dendritic inward crystallization without foundering, thereby building a solid crust. However, this crust may only cool below the Curie temperature after any convection ceases, such that it would not be magnetized (Scheinberg et al. 2015). (**C**) A stable crust forms and undergoes large-scale foundering at its low viscosity, warm base. Such a body might maintain a dynamo sufficiently long for the outer part of the crust to cool below its Curie point and become magnetized (Neufeld et al. 2019). Light red color denotes molten iron and brown shading denotes solid iron with varying temperature

not fully stripped away) (Bryson et al. 2015; Weiss et al. 2008) or even an undifferentiated, primordial crust (if Psyche is partially differentiated) (Weiss and Elkins-Tanton 2013). However, even for an inwardly crystallized body, only some crystallization scenarios would lead to magnetization acquisition (Fig. 2). Alternatively, a mantle-stripped core may also have solidified from the inside outward such that there was no solid core material lying above the dynamo.

Although an undifferentiated body could not have generated a dynamo, an external field like that of the solar nebula (Weiss et al. 2021) might magnetize such a body if it was aqueously altered and/or thermally metamorphosed before the nebula dispersed (Courville et al. 2022). Such magnetization could potentially be distinguished from an internal field by the fact that it may be unidirectional throughout the body given that the nebular field is expected to be spatially uniform over scales of at least 10^{-2} AU (Wang et al. 2017). Furthermore, such a body could be identified by the fact that it is expected to compositionally and mineralogically resemble chondritic materials as indicated by Gamma-Ray and Neutron Spectrometer and Multispectral Imager data (Elkins-Tanton et al. 2022).

1.3 Detecting Psyche's Magnetic Field

Advecting planetary metallic cores can generate magnetic fields by the dynamo process in which motion of a conductive fluid amplifies a seed magnetic field by the process of induction (Moffatt and Dormy 2019). Such a dynamo field can then imprint natural remanent magnetization (NRM), which is the semi-permanent alignment of electron spins in ferromagnetic minerals (see Table 1 for a list of all acronyms and variables used in this article). As introduced in Sect. 1.2), NRM can take the form of thermoremanent magnetization when ferromagnetic minerals cool through their critical ordering temperatures or as crystallization remanent magnetization when these minerals crystallize or form as transformation products from pre-existing minerals during aqueous alteration or subsolidus cooling (Dunlop and Özdemir 1997). Although planetesimal dynamos are thought to only have lasted for at most several hundred million years (Ma) due to the efficient radiative cooling of small bodies

Table 1 Acronyms and variables used in this manuscript

Acronym or variable	Definition
MIT	Massachusetts Institute of Technology
DTU	Technical University of Denmark
JPL	Jet Propulsion Laboratory
CNRS	Centre National de la Recherche Scientifique
NRM	natural remanent magnetization
IMF	interplanetary magnetic field
SU, SU1, SU2	Sensor Unit, Sensor Unit 1, Sensor Unit 2
EU, EU1, EU2	Electronics Unit, Electronics Unit 1, Electronics Unit 2
FS	Flight system (frame)
Ma	million years
CAI	calcium aluminum-rich inclusion
XPEEM	X-ray photoemission electron microscopy
MG	main group
VFM	Vector Field Magnetometer
CHAMP	Challenging Microsatellite Project
CDC	Compact Detector Coils
CSC	Compact Spherical Coils
FPGA	field-programmable gate array
eu	engineering units
LVDS	low-voltage differential signaling
UART	universal asynchronous receiver-transmitter
SPUS	SPUS (Swarm Packet Utilization Standard)
PPS	pulse-per-second
STM	spacecraft time message
DSOC	Deep Space Optical Communication
TWTA	traveling-wave tube amplifiers
CCSDS	Consultative Committee for Space Data Systems
CFDP	CCSDS File Delivery Protocol
PBF	Psyche-body-fixed (frame)
PSO	Psyche-solar-orbit (frame)
J2E	Psyche-centered EMO J2000 (frame)
SCLK	spacecraft clock
NAIF	Navigation and Ancillary Information Facility
CDF	Common Data Format
SADA	solar array drive assembly
R_P	mean radius of Psyche
m_S	specific NRM
M_P	dipole moment of Psyche
$+X, +Z, -Z$	identities of decks of Psyche spacecraft bus
r_P	Psyche-centric distance
B_MP	equatorial remanent magnetic field
P_B	magnetic pressure
μ_0	permeability of free space

Table 1 (*Continued*)

Acronym or variable	Definition
P_{SW}	solar wind dynamic pressure
ρ_{SW}	solar wind density
v_{SW}	solar wind velocity
m_{H}	mass of hydrogen atom
n_{SW}	solar wind number density
m_{P}	magnetization of Psyche
F	upstream field enhancement
r_{MP}	radius of Psyche's magnetosphere
f_s	sampling frequency of Magnetometer
x, y, z	Cartesian coordinates for measurement location
$\vec{r}$	position with respect to spacecraft bus
$\vec{r}_1$	position of SU1
$\vec{r}_2$	position of SU2
$\vec{B}_{\mathrm{M}}$	measured magnetic field
$\Delta\vec{B}$	difference between field measured at SU2 and SU1
$\vec{B}_{\mathrm{amb}}$	ambient magnetic field
$\vec{B}_{\mathrm{FS}}$	flight system magnetic field
$\vec{B}_{\mathrm{FS,DC}}$	steady part of flight system magnetic field
$\vec{B}_{\mathrm{FS,AC}}$	time-variable part of flight system magnetic field
$\vec{B}_{\mathrm{N}}$	instrument noise
$\vec{B}_{\mathrm{N},2}$ and $\vec{B}_{\mathrm{N},1}$	instrument noise for measurements acquired with SU2 and SU1
a	index of field component measured by SU1
b	index of field component measured by SU2
$\boldsymbol{\alpha}$ and α_{ab}	coupling matrix and its entries for SU1 and SU2
$\vec{B}'_{\mathrm{amb}}$	estimated ambient magnetic field
$\vec{M}'_{\mathrm{FS}}$	estimated magnetic moment of flight system
$\vec{B}_{\mathrm{ppr},n}$	Partially Processed (ppr) data
$\vec{O}_n$	offset vector for SU*n*/EU*n*
$\boldsymbol{S_n}$	scale factor matrix for SU*n*/EU*n*
$\boldsymbol{V_n}$	orthogonalization matrix for SU*n*
$\boldsymbol{R_n}$	rotation matrix for SU*n*/EU*n*
n	index for SU1 and SU2
t_{ABS}	spacecraft time of the start of each packet
$t_{m,a}$	measurement time of each sample in a science packet
$t_{\mathrm{MAG},a}$	time for each science packet measurement relative to the start of the packet
Δ_{PPS}	offset of the time of a field measurement with respect to the arrival of the PPS
t_d	instrument latency
c	index for each of c50 measurements in a science packet
$\vec{B}_{\mathrm{raw},n}$	Raw data vector for SU*n*
$eu_{n,a}$	entry for a^{th} field component in Raw data vector for SU*n*
$T_{n,i}$	temperature reading for SU*n*/EU*n* and associated CSC, CDC, or EU

Table 1 (*Continued*)

Acronym or variable	Definition
i	index for CSC, CDC, or EU temperature sensors
$C_{n,0,i} \dots C_{n,7,i}$	coefficients for each temperature sensor from ground calibration
$VET_{n,i}$	voltage measured by each temperature sensor
$\vec{E}_n$	vector of linearized field data for SUn
$E_{n,a}$	entry for a^{th} field component in $\vec{E}_n$
$k_{n,2,j}$ and $k_{n,3,j}$	linearity parameters from ground calibration
$O_{n,a}$	entry for a^{th} field component in $\vec{O}_n$
$O_{n,0,a}$, $O_{n,\text{CDC},a}$, $O_{n,\text{EU},a}$ and $O_{n,t,a}$	coefficients for offset matrix from ground calibration
$\Delta O_{n,\text{CDC},a}(T_{n,\text{CDC}})$	look-up table for offset matrix from ground calibration
t_L	time since launch (in years)
$S_{n,a}$	entry for a^{th} field component in $\boldsymbol{S_n}$
$S_{n,0,a}$, $S_{n,\text{CSC},a}$, $S_{n,\text{EU},a}$, and $S_{n,t,a}$	coefficients for scale factor matrix from ground calibration
$u_{n,1}$, $u_{n,2}$, and $u_{n,3}$	coefficients for orthogonalization matrix from ground calibration
$\vec{B}_{\text{cal,FS},n}$	Partially Processed data rotated into FS coordinates
f	frequency of magnetic field
$\vec{B}_{\text{cal,zl},n}$	Partially Processed data rotated into FS coordinates with zero-level correction
$\vec{B}_{\text{zl},n}$	Estimate of the Magnetometer zero-level
$\vec{B}_{\text{SA},n}$	solar array magnetic field
θ_S	SADA angle
$\vec{B}_{\text{cal,zl,SA},n}$	Partially Processed data rotated into FS coordinates with zero-level and solar array correction applied
$\vec{B}_{\text{cal,zl,SA,FDF},n}$	Partially Processed data rotated into FS coordinates with zero-level and solar array correction applied and frequency domain filtering applied
f_{FDF}	frequency domain filtering function
$\vec{B}_{\text{cal,zl,SA,FDF,SC},n}$	Partially Processed data rotated into FS coordinates with zero-level and solar array correction applied, frequency domain filtering applied, and step-changes suppressed
f_{SC}	field step change suppression function
$\vec{B}_{\text{cal},n}$	Calibrated data
$\boldsymbol{R}_q$	rotation matrix from FS to J2E, PBF, or PSO frames
q	index for J2E, PBF, or PSO frames
$\vec{B}_{\text{der},q}$	Derived data
r_0	reference distance from Psyche
θ_P	Psyche-centric latitude
ϕ_P	Psyche-centric longitude
$\vec{B}_{\text{der,i}}$	part of derived field generated internally to r_0
$\vec{B}_{\text{der,e}}$	part of derived field generated externally to r_0
$\vec{r}_P$	position with respect to Psyche
$\vec{B}'_{\text{der,i}}$	estimate of Psyche's internal field
$\vec{B}'_{\text{der,e}}$	estimate of the external field at Psyche
$g_l^{m,\text{i}}$ and $h_l^{m,\text{i}}$	internal Gauss coefficients
l	degree of Gauss coefficients

Table 1 (*Continued*)

Acronym or variable	Definition
m	order of Gauss coefficients
R_l	power spectrum of field
$\vec{B}_{\mathrm{cal,SA,FDF,SC},n}$	Partially Processed data rotated into FS coordinates with solar array correction applied, frequency domain filtering applied, and step-changes suppressed
$\boldsymbol{A_n}$	covariance matrix of magnetic field data used in solar wind monitoring
$\vec{b}_n$	Right hand size of system of equations solved for zero-level estimates using solar wind monitoring
B_x, B_y, B_z	Cartesian components of $\vec{B}_{\mathrm{cal,SA,FDF,SC},n}$

Notes: The top of the table lists acronyms in order of appearance and the bottom lists variables in order of appearance. The first column lists acronym or variable and second column lists its definition

(e.g., Maurel et al. 2021 and Nichols et al. 2021). NRM can potentially persist until today (Shah et al. 2018). The Magnetometry Investigation addresses the key mission science objective by constraining the remanent magnetization of Psyche.

Gravity power spectra of planetary bodies have been empirically found to be well-described by the so-called Kaula rule, which assumes that planetary gravity fields reflect randomly distributed density anomalies of small spatial coherence distributed throughout the outer regions of the bodies and where gravitational stress is constant and limited by crustal elastic strength (Ermakov et al. 2018; Zuber et al. 2022). However, there is no equivalent a priori estimate for the intrinsic magnetic field of a magnetized body because magnetization depends on the unknown past magnetic field history experienced by the body and the unknown form of any magnetization (e.g., the various aforementioned forms of NRM as well as induced magnetization). As such, the Magnetometry Investigation was designed to distinguish between a broad range of possible magnetization states that relate to aspects of the body's, structure origin and geologic history.

The Psyche Magnetometry Investigation is designed to be capable of detecting an asteroid magnetic field that is sufficiently strong to be the product of a past dynamo. The specific NRM intensities, m_s, of known iron meteorites have been found to range from $\sim 10^{-4}$ to $\sim 10^{-2}$ Am^2 kg^{-1}, where the latter value excludes samples most likely to have been remagnetized by hand magnets (Pesonen et al. 1993) (see Sect. 2.2). In the absence of a magnetic equivalent for the Kaula rule, we assume as a limiting case that any fine-scale NRM in Psyche formed by a dynamo has an intensity at the lower end of this range. It has been found that the crustal magnetic fields for Earth, Mars, and the Moon are each well described by magnetization models consisting of randomly oriented dipoles, perhaps with some lateral spatial correlation (Voorhies 2008; Voorhies et al. 2002; Wieczorek 2018). As such, we also assume that Psyche's shell can be described by such a pattern (see also Sect. 4.6.3).

To estimate the thickness of such a magnetized layer, we consider a recent thermal evolution model of a Psyche-like, mantle-stripped iron planetesimal which found that the outer 40% of its radius could be magnetized by an internal core dynamo (Neufeld et al. 2019). If we also therefore assume Psyche's crust is randomly magnetized with a spatial scale given by $\sim$ 40% of its current radius (e.g., $\sim$ 40-km coherence scale) and with a fine-scale magnetization at the lower end of known iron meteorites, we find that Psyche would have a dipole moment $M_{\mathrm{P}} = 2 \times 10^{14}$ Am^2. The sole level 1 requirement of the Magnetometry Investigation is to be able to detect this limiting magnetic moment. Note that if we instead assume a fine-scale magnetization at the upper end of known iron meteorites with 40-km coherence scale, the moment would be two orders of magnitude larger than this value (Oran et al. 2022), while a uniformly magnetized body with this magnetization would be three orders of magnitude larger. Note that although a spherical body containing magnetization acquired solely by a steady field of internal origin will produce no external remanent field (Runcorn 1975), Psyche's nonsphericity (Shepard et al. 2021), combined with the likelihood of time-variability of the dynamo field and magnetostatic interactions between metallic materials within Psyche mean that any dynamo-generated remanent magnetization in the asteroid will almost certainly generate an external field (e.g., Aharonson et al. 2004; Stephenson 1976; Srnka 1976).

The Magnetometry Investigation constrains the asteroid's magnetic moment by sensing its remanent magnetic field. The minimum detectable magnetic moment can be roughly translated into a minimum detectable field as a function of distance from Psyche under the approximation that Psyche's magnetic field is that of a dipole in a vacuum, such that the field scales as r_P^{-3} for Psyche-centric distance, r_P. The Magnetometer will continuously acquire data after launch including during mapping orbits at increasingly lower altitudes (Polanskey et al. 2023). At the two closest mapping orbits C and D, which are currently estimated to have altitudes of 75 and 190 km (Polanskey et al. 2023), the level-1 minimum moment has a polar field of ~ 6 nT and ~ 2 nT. The Magnetometry Investigation is designed such that these remanent fields at these altitudes should be larger than estimated noise sources. The latter are expected to be dominated by fluctuations of the interplanetary magnetic field (IMF) ($\sim \pm 2$ nT per axis in the asteroid belt), Instrument Uncertainty (intrinsic instrument noise, estimated to be < 0.17 nT axis^{-1} 3σ), and Flight System Field Knowledge (uncertainty in knowledge of the flight system field after it has been suppressed during ground data processing; required to be < 0.4 nT axis^{-1} 3σ) (see Sects. 3 and 4). The Magnetometry Investigation's single level 2 requirement is that the combined Instrument Uncertainty and Flight System Field Knowledge, termed the Reconstructed Measurement Uncertainty, be less than 1.5 nT axis^{-1} 3σ (Fig. 3). We currently estimate that we meet this requirement with a margin of 0.4 nT axis^{-1} 3σ.

1.4 Overview of the Psyche Magnetometer

The Psyche Magnetometer is designed to meet this level 2 requirement with margin. It consists of two three-axis fluxgate sensor units (SUs) on a 2.15-m long boom, each connected by a harness to electronics units (EUs) within the spacecraft bus (Figs. 4, 5). Each EU has two redundant two sides (named A and B) that can connect to the same SU via a cross-strapping board that switches the analog signals. The outboard sensor, located at the outer end of the boom, is named Sensor Unit 1 (SU1) and the inboard sensor, located 0.7 m inward of the end of the boom, is named Sensor Unit 2 (SU2). The two reasons that the mission carries two sensors are to enable gradiometry measurements to reduce noise from the flight system magnetic fields and to provide redundancy. With respect to the latter, the Magnetometry Investigation is designed to meet its level 1 requirement in the event of failure of either SU1 or SU2.

MIT leads the science investigation, operations, ground data processing and trending, DTU designed, built, and calibrated the Magnetometer and leads anomaly resolution, and the Jet Propulsion Laboratory leads commanding. MIT, JPL, and the University of California, Los Angeles jointly developed the level 1 and level 2 Magnetometer science requirements. All four institutions will participate in science operations and data analysis.

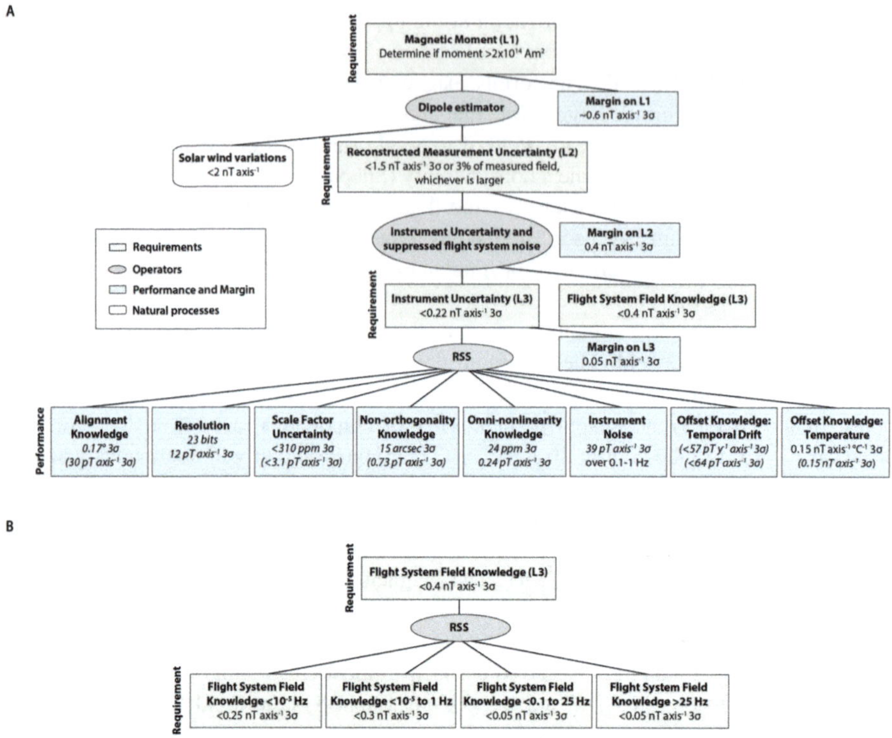

Fig. 3 Magnetometer and flight system knowledge requirements, performance, and uncertainty budget. Green boxes contain requirements, diagonal lines denote the flowdown to child requirements and performance, blue boxes contain performance and margin, white box contains uncertainty from solar wind variations and ovals denote operators (RSS = sum in quadrature). (**A**) Overall budget. The level 1 requirement is that the Magnetometry Investigation be able to detect a remanent field from (16) Psyche if the asteroid's moment exceeds 2×10^{14} Am^2. The level 2 requirement is that the Reconstructed Measurement Uncertainty, which combines Instrument Uncertainty with uncertainty in Flight System Field Knowledge, enables such a detection given typical solar wind variations in the asteroid belt. The Instrument Uncertainty is composed of uncertainties in knowledge of the orientation of each Sensor Unit (SU) relative to the Flight System (FS) frame (Alignment Knowledge), digitization (Resolution), knowledge of the scale factors, $\boldsymbol{S_n}$ (Scale Factor Uncertainty), knowledge of the departure from orthogonality for the SU axes (Non-orthogonality Knowledge), knowledge of the interactions between the three SU axes (Omni-nonlinearity Knowledge), instrument noise integrated over 0.1–1 Hz (Instrument Noise), knowledge of the long-term drift of the offsets, $\vec{O}$ (Offset Knowledge: Temporal Drift), and knowledge of the offset dependence on temperature (Offset Knowledge: Thermal). The requirement levels (L1 = level 1, L2 = level 2, and L3 = level 3) are listed in parentheses next to each requirement name. The requirements assume measurements acquired in a 10 nT ambient field (for the Alignment Knowledge, Resolution, Scale Factor Uncertainty, Non-orthogonality Knowledge, and Omni-nonlinearity Knowledge) with temperature variations at the sensor cores of ±1 °C (for Offset Knowledge: Temperature), and an assumed lapse of 402 d since the last zero-level field correction (for Offset Knowledge: Temporal Drift; Sect. 4.4). (**B**) Breakdown of total Flight System Field Knowledge requirement. This flows down to requirements on knowledge of the integrated field in four frequency ranges after use of planned mitigations (gradiometry, zero-level corrections, and corrections for solar array fields): $< 10^{-5}$ Hz, 10^{-5} to 0.1 Hz, 0.1 to 25 Hz, and 25 Hz to 100 kHz

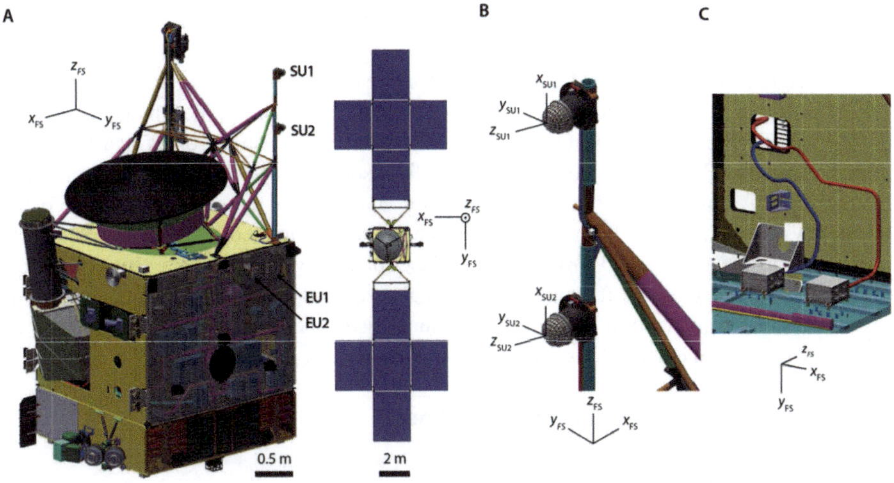

Fig. 4 Psyche spacecraft and location of Magnetometer Sensor Units (SUs) and Electronics Units (EUs). (**A**) SU1 and SU2 are mounted along a boom of length 2.15 m pointing away from the $+Z$ deck. Some of the most magnetic subsystems were placed at distal locations on the spacecraft: Hall thrusters ($-Z$ deck), Deep Space Optical Communication (DSOC) ($+X$ deck). (**B**) Close-up of SUs. (**C**) Close-up of EUs and harnesses inside spacecraft bus. Compasses show orientation of FS coordinate system

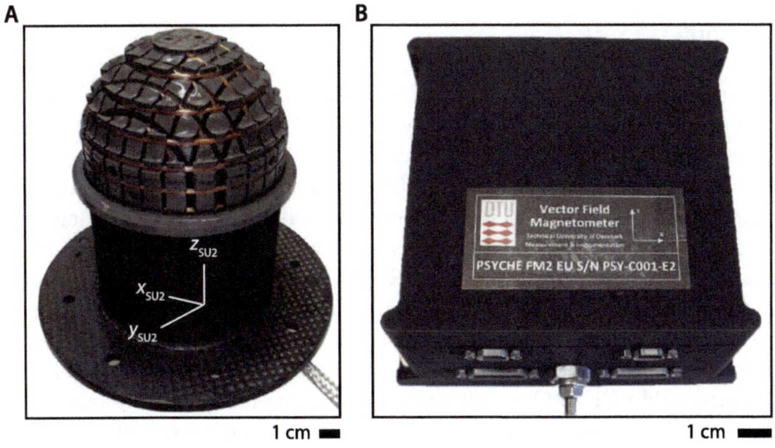

Fig. 5 Photographs of Flight Model 2. (**A**) Sensor Unit 2 (SU2). Compass shows orientation of SU2 coordinate system. (**B**) Electronics Unit 2 (EU2)

2 Planetesimal and Asteroid Magnetism

2.1 Introduction to Remanent Magnetization

Most meteorites are samples of planetesimals and many contain NRM records of past magnetic fields on these bodies. The ferromagnetic minerals carrying these records are mainly iron-nickel metals, iron oxides, and iron sulfides. The NRM intensities of meteorites depends on their mineralogy, grain size, history of exposure to ancient fields, and preservation state.

The origin of meteorite NRM was debated for decades following its identification in the 1960s. However, it is now understood that some meteorites acquired their magnetic records during or after their formation during the first several Ma to hundreds of Ma after solar system formation (Weiss et al. 2010). Two main field sources have been identified for known meteorites: a dynamo field generated by advection of a metallic core in the meteorite parent body (Bryson et al. 2019a; Weiss et al. 2008) and a field generated by the solar nebula (Fu et al. 2014; Borlina et al. 2021; Weiss et al. 2021). Such NRM can survive to the present day depending on a meteorite's ferromagnetic crystal sizes and mineralogy, which control the resistance to remagnetization by thermal metamorphism, shocks, terrestrial weathering and contact with hand magnets (Weiss et al. 2010; Gattacceca et al. 2010; Uehara et al. 2012).

2.2 Magnetic Record of Meteorite Parent Bodies

2.2.1 Introduction

NRM attributed to early solar system magnetic fields has been identified in several chondrite and achondrite groups. Here, we focus on recent advances in revealing ancient magnetic records carried by chondrites, iron meteorites and stony-iron meteorites. Paleomagnetic studies conducted on basaltic achondrites, which sample the silicate mantles of differentiated planetesimals, were more recently reviewed elsewhere (Scheinberg et al. 2017).

2.2.2 Chondrites

Bulk chondrites and their consitutent chondrules have been found to contain ancient NRM. Individual chondrules, which cooled as free-floating objects in the solar nebula, have been found to contain pre-accretional thermoremanent NRM from cooling in the presence of a nebular magnetic field (Weiss et al. 2021). This field appears to have persisted until about ~ 4 Ma after the formation of calcium aluminum-rich inclusions (CAIs) (Wang et al. 2017; Weiss et al. 2017; Borlina et al. 2022). Because chondrules accreted onto their parent body in random orientations, such pre-accretional NRM is expected to be randomly oriented within the parent body (Weiss and Elkins-Tanton 2013), as has now been observed for two chondrite groups (Borlina et al. 2021; Fu et al. 2014). Given, the measured chondrule-scale NRM intensities (which range up to $\sim 10^{-3}$ $\mathrm{Am^2\,kg^{-1}}$) (Borlina et al. 2021; Fu et al. 2014), the bulk remanent field of a 100-km radius asteroid composed of randomly oriented 0.1-mm radius magnetized chondrules is expected to be many orders of magnitude below the sensitivity of the Magnetometer even at Orbit D (Polanskey et al. 2023).

Bulk chondrite NRM was acquired after accretion of the parent body during low-temperature crystallization (Cournède et al. 2015) and/or thermal metamorphic events (Bryson et al. 2019b; Carporzen et al. 2011; Gattacceca et al. 2016). This NRM is unidirectional across meteorites up to at least the hand sample scale (up to > 10 cm). The source of this field has been mainly attributed to either a nebular field (if the meteorites were magnetized while the nebula was still present) (Fu et al. 2021) or an internal core dynamo in a partially differentiated body (Bryson et al. 2019b; Carporzen et al. 2011; Elkins-Tanton et al. 2011; Weiss et al. 2012; Weiss and Elkins-Tanton 2013; Maurel et al. 2021). Amplification of the IMF by conducting asteroid interiors (Anand et al. 2022; O'Brien et al. 2020) and by impact-generated plasmas (Muxworthy et al. 2017) have also been recently proposed as a source for carbonaceous chondrite NRM. However, the first mechanism has difficulty explaining the intensity of the observed NRM given the temporal variability of the IMF (Oran et al. 2018) and the second is challenged by the spatial uniformity of the peak laboratory demagnetization temperatures observed amongst meteorite subsamples (Gattacceca et al. 2016). Only the external field sources could be spatially uniform across a planetary body (e.g., Runcorn 1975) and so could conceivably produce uniform magnetization at the ~ 200-km scale (Courville et al. 2022).

2.2.3 Iron and Stony-Iron Meteorites

Given that many iron and stony-iron meteorites typically cooled and underwent subsolidus recrystallization several Ma to hundreds of Ma after the dissipation of the solar nebula, this would leave a core dynamo as the most likely magnetizing field. However, it had long been assumed that iron and stony-iron meteorites formed in metallic cores that were largely isothermal due to thermally insulatation by thick, overlying silicate mantles. In such a scenario, by the time they cooled through the Curie temperature, the entire core would already be solid and therefore any dynamo would have ceased. Therefore, it was expected that iron meteorites would not contain ancient NRM (Cisowski 1987). However, over the last 15 years, it has been realized that many iron meteorite parent bodies either underwent catastrophic collisions that stripped away part of their silicate mantles (Matthes et al. 2020; Yang and Goldstein 2006; Yang et al. 2008, 2010), or alternatively contained reservoirs of metal within their silicate mantles located at shallower depths than their core (Bryson et al. 2015; Maurel et al. 2020; Nichols et al. 2021, 2016). On such planetesimals, meteorites could have met the conditions required for NRM acquisition while their parent body's core dynamo was still active (Neufeld et al. 2019; Scheinberg et al. 2015).

In iron and stony-iron meteorites, reliable paleomagnetic records can be carried by microstructures called cloudy zones (Bryson et al. 2014a,b; Einsle et al. 2018; Mansbach et al. 2022; Maurel et al. 2019; Uehara et al. 2011). Cloudy zones are ensembles of < 200-nm-diameter, ferromagnetic "islands" that acquire their NRM during their crystallographic transformation into the mineral tetrataenite at its 320 °C ordering temperature. A high spatial resolution, synchrotron-based technique called X-ray photoemission electron microscopy (XPEEM) (Scholl 2002) has recently been adapted to measure cloudy zone magnetizations (Bryson et al. 2014b). XPEEM magnetic measurements rely on the preferential absorption of X-rays depending on the orientation of the sample's local magnetization to obtain an average magnetization vector of the cloudy zone. Since 2014, XPEEM has enabled the study of a diversity of iron and stony-iron meteorite parent bodies. In particular, recent studies of IVA irons, IAB irons, IIE irons and main group (MG) pallasites collectively demonstrate that like stony meteorite parent bodies, iron and stony iron meteorite parent bodies had a diversity of differentiation states and magnetic field environments. These studies illustrate the wide range of possibilities for Psyche to have experienced and retained a record of an ancient magnetic field. We review each of these in turn.

Essentially all known IVA irons, IIE irons, and MG pallasites likely cooled too slowly to record a nebular magnetic field, so any NRM would likely be a record of dynamo. Cooling rate studies of IVA irons indicated that their parent body likely lost most of its silicate mantle through impacts occurring early in its history (Moskovitz and Walker 2011; Yang et al. 2008). Combined cooling rate and major element analyses indicate that the exposed core likely crystallized from the outside-in, akin to one of the scenarios envisioned for Psyche (left side of panel A of Fig. 1). An XPEEM study of the IVA meteorite Steinbach found that its cloudy zones carry NRM attributed to a $\gg 50$-μT dynamo field powered by advection of the IVA core (Bryson et al. 2017). The fact that such a mantle-stripped core could generate

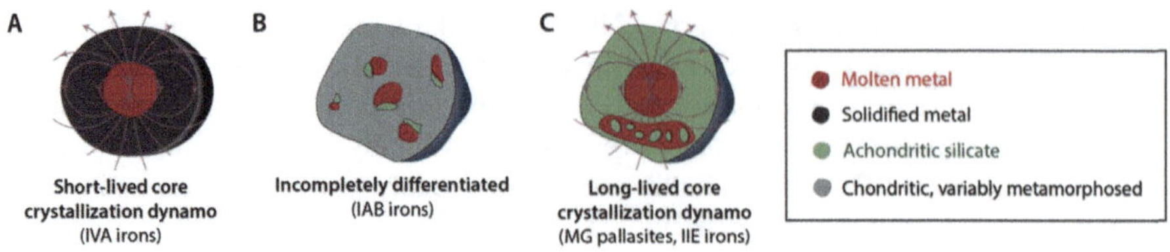

Fig. 6 Magnetic histories inferred from paleomagnetic studies of iron and stony iron meteorites. (**A**) Magnetization in main group pallasites and IIE iron meteorites suggests that their parent bodies generated a dynamo and formed a molten metallic core. Mineralogical and chemical studies separately indicate that that their parent body was differentiated. These meteorites could have cooled through their Curie points during the dynamo lifetime and record these field if they formed in metal pools trapped within the crust. (**B**) Lack of magnetization in IAB iron meteorites suggests their body may not have generated a dynamo. Mineralogical and chemical studies suggest that their parent body was partially differentiated. (**C**) Remanent magnetization in IVA irons is consistent with formation in the crust of an inwardly crystallizing, mantle stripped metallic planetesimal. Such a model was first motivated by the cooling rate and Ni content data. See text for references

a dynamo field, recorded by its outermost layers, is supported by the numerical simulations (Neufeld et al. 2019) mentioned in Sect. 1.3 (Fig. 6 panel A).

The IAB iron meteorites are thought to have orginated from at least three parent bodies (Worsham et al. 2017). Two parent bodies (sampled by the MG-sLL and sLM-sLH subgroups) are thought to either have undergone only localized differentiation (Hunt et al. 2018) or have remained undifferentiated and experienced multiple metal-pool forming impacts (Worsham et al. 2017). XPEEM analyses of one sLL and one sLH IAB iron meteorite found that neither of these meteorites carried significant NRM (Fig. 6 panel B) (Bryson et al. 2014b; Nichols et al. 2018).

Both the MG pallasites and the silicate-bearing IIE irons likely formed via impacts that mixed exogenous or endogenous metal with silicates in their parent bodies' mantles (Bryson et al. 2015; Maurel et al. 2020, 2021; Nichols et al. 2021, 2016; Tarduno et al. 2012). These events led to the meteoritic metal cooling through the ordering temperature while core dynamos were active (Fig. 6 panel C).

Among five MG pallasites analyzed with XPEEM, two reached the ordering temperature prior to 120 Ma after accretion and do not carry significant NRM (Nichols et al. 2016). In contrast, the other three later-cooled meteorites were magnetized by a field with an intensity between 2 and 200 μT (Bryson et al. 2015; Nichols et al. 2021). The MG pallasites therefore appear to have captured the onset of a dynamo. These data are consistent with a dynamo powered by core crystallization (Nimmo 2009) whose initiation appears to have been delayed until $>$ 100 Ma after parent-body accretion when the core cooled to its liquidus temperature.

Three IIE iron meteorites were analyzed with XPEEM and found to have cooled in a field between 5 and 300 μT (Maurel et al. 2020, 2021). Their NRM are dated to $\sim$ 80, $\sim$ 100 and $\sim$ 160 Ma after CAI-formation using $^{40}Ar/^{39}Ar$ geochronology (Bogard et al. 2000). These late ages and high paleointensities essentially rule out all magnetic field sources except that of a dynamo powered by core crystallization.

2.3 Asteroid Magnetospheres

The dynamo magnetic field intensities inferred from meteorite paleomagnetic studies are in some cases sufficiently high that such ancient dynamos would have formed magnetospheres, regions surrounding the bodies dominated by their internal magnetic field from which external plasmas like the solar wind are excluded. Here we examine whether the remanent field produced by the NRM in asteroids could lead to the formation of a present-day magnetosphere. A minimum condition for the formation of a magnetosphere is that the body's equatorial remanent magnetic field, B_{MP}, generates a magnetic pressure $P_B = B_{MP}^2/2\mu_0$, that is sufficiently strong to balance the dynamic pressure of the solar wind, $P_{sw} = \rho_{sw} v_{sw}^2$, at the subsolar point of the body's surface [see Gombosi 2009]. Here, v_{sw} and $\rho_{sw} = m_H n_{sw}$ are the solar wind bulk velocity and density, n_{sw} is the solar wind number density, m_H is the mass of the hydrogen atom, and μ_0 is the permeability of free space. At the magnetopause, the equatorial field of a spherical, uniformly magnetized Psyche with magnetization, m_P, is $B_{MP} = F\mu_0 m_P R_P^3/(9r_{MP}^3)$ where $F \sim 4$ is the upstream field enhancement relative to that of a dipole in a vacuum (Gombosi 2009). Setting $r_{MP} = R_P$ in the equation $P_{sw} = P_B$ gives

$$m_P > v_{sw}\sqrt{\frac{18\rho_{sw}}{\mu_0 F}} \qquad (1)$$

For a Psyche-sized body and using $v_{sw} = 400\ \mathrm{km\,s^{-1}}$ and $n_{sw} = 1\ \mathrm{cm^{-3}}$ (Gombosi 2009), this gives a minimum magnetization $m_P \sim 0.04\ \mathrm{A\,m^{-1}}$ and a minimum moment of $M_P \sim 2 \times 10^{14}\ \mathrm{Am^2}$. Therefore, coincidentally, the minimum detectable moment specified by the level 1 requirement (Sect. 1.3) would just barely produce a magnetosphere and exclude solar wind protons with mean solar wind velocities (gyroradius $\sim$ 150 km) and would easily exclude electrons (gyroradius $\sim$ 5 km). By comparison, a uniformly magnetized Psyche-like body with the maximal magnetization found in iron meteorites (e.g., moment of $2 \times 10^{17}\ \mathrm{Am^2}$; Sect. 1.3) would have $r_{MP} \sim 5\ R_P$.

Equation (1) provides a first order means to predict the size of the magnetosphere on its dayside. Three-dimensional modeling of the interaction is needed to appropriately plan how Magnetometer data would be used to determine the magnetization. The large planetary magnetospheres in the solar system are typically simulated using the magnetohydrodynamic equations, the system of equations combining the hydrodynamic Euler equations and Maxwell's equations of electromagnetism and neglecting the displacement current. This approximation allows us to treat the solar wind plasma as a fluid, greatly increasing computational efficiency.

However, because the solar wind electron gyroradius but not proton gyroradius is small compared to the size of the minimum magnetosphere, only the electrons but not the protons can be treated with a fluid prescription, making the magnetohydrodynamic equations inaccurate. Rather, the protons must be modeled such that their gyration motion is resolved using particle-in-cell methods. These methods solve the equation of motion of "macroparticles" representing a large number of particles as an approximation of the true particle distribution function (Sugiyama and Kusano 2007). However, such codes are much more computationally expensive than fluid codes: in a fluid code representing a quasi-neutral electron and proton plasma, the domain is divided into cells for which only eight variables are solved (the fluid density, pressure, three components of the velocity and three components of the magnetic field). Running full kinetic codes for three-dimensional global simulations is not feasible on any existing supercomputing platform and this will likely remain the case for the coming years (Gombosi et al. 2021).

Systems whose lengthscale falls between the electron and proton gyroradii can be simulated by a hybrid computational model in which protons are modeled with a particle-in-cell approach and the electrons are treated as a fluid (Kallio et al. 2008). Because even such hybrid codes are expensive, most numerical models of asteroid magnetic fields have used either three-dimensional magnetohydrodynamic codes (Baumgartel et al. 1997; Blanco-Cano et al. 2003; Omidi et al. 2002) or 2–2.5 dimensional hybrid codes (Baumgartel et al. 1994;

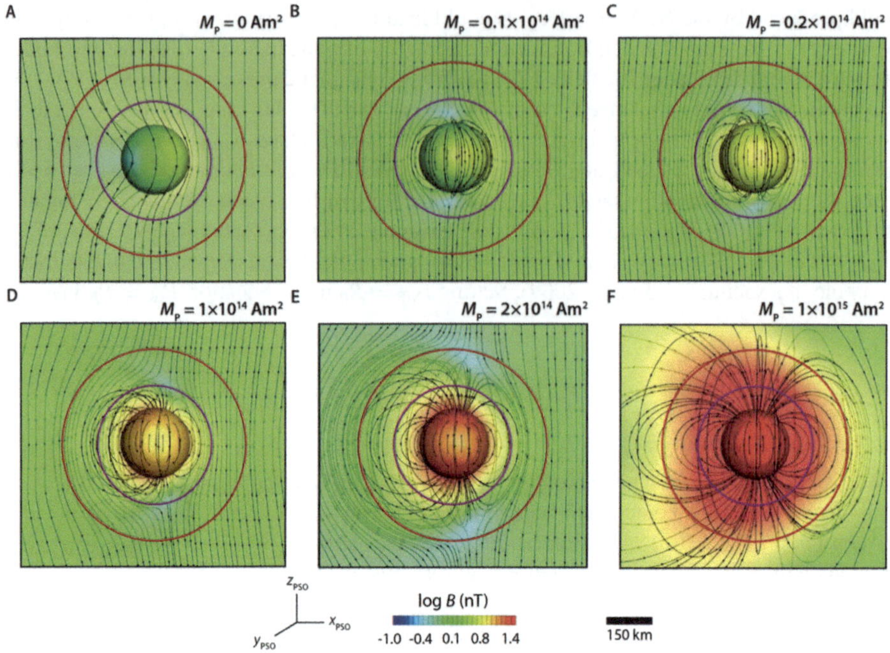

Fig. 7 Hybrid simulations showing the magnetic field of a perfectly conducting spherical body of radius 111-km with a remanent magnetic moment of (**A**) 0, (**B**) 0.05×, (**C**) 0.1×, (**D**) 0.5×, (**E**) 1× and (**F**) 5× the level 1 requirement moment ($2 \times 10^{14}\ \mathrm{Am^2}$). The simulation is performed in the PSO frame. The body's magnetic moment is aligned along the spin axis and oriented toward $+z$ direction. The solar wind is flowing toward the $-x$ direction and the IMF is aligned along the $+z$ direction. The solar wind has a density of 1 particle $\mathrm{cm^{-3}}$, a speed of 400 $\mathrm{km\,s^{-1}}$, an IMF magnitude of 1 nT, an electron temperature of 100,000 K and a proton temperature of 40,000 K. These are typical conditions for slow wind streams in the asteroid belt (Fatemi and Poppe 2018; Oran et al. 2022). The simulations were performed using the HYB code (Kallio et al. 2008). Each panel shows a spherical surface representing the surface of the body and a plane passing through the day-night plane (x–z plane), both colored by the magnetic field magnitude. The magnetic field topology is traced by black curves. The 190 and 75 km mean altitudes of the nominal Orbits C and D are shown as red and purple circles, respectively. The scale bar shows the size of the proton gyroradius in the nearby undisturbed IMF. Compasses show orientation of PSO coordinate system

Blanco-Cano et al. 2003, 2004; Omidi et al. 2002; Wang and Kivelson 1996). Despite these approximations, these simulations broadly validated condition (1) (e.g., Blanco-Cano et al. 2004 and Omidi et al. 2002). More recently, three-dimensional hybrid codes have modeled asteroids with permanent moments at or above the upper limit expected for Psyche (Fatemi and Poppe 2018; Oran et al. 2022; Simon et al. 2006).

We have conducted new three-dimensional hybrid simulations of electrically conducting Psyche-like bodies with lower remanent magnetization (following the approach described in Oran et al. 2022) (Fig. 7). These simulations show that Eq. (1), which corresponds to the moment of the simulation shown in Fig. 7E, slightly underestimates the minimum moment: a true magnetosphere that excludes protons and electrons occurs at moments $>\sim 10^{15}\ \mathrm{Am^2}$ (Fig. 7F). We also see that for moments between $\sim 10^{13}$ and $\sim 10^{14}\ \mathrm{Am^2}$, a volume with closed magnetic field lines forms around the body, but only electrons are excluded from the region (Fig. 7B-E). Finally, for unmagnetized, conducting bodies, there is only a weak perturbation to the IMF near the body and no region of closed field lines (Fig. 7A).

2.4 Past Asteroid and Comet Magnetic Field Measurements

Six previous missions acquired magnetic measurements in the vicinity of ten asteroids. All but two of these asteroid encounters were flybys. No clear indication of a remanent magnetic field was obtained for these bodies, such that these missions only provided upper limits on the asteroids' magnetic moments (Table 2). A signature of an internal field can be an observed as an increase of the ambient magnetic field compared to the IMF as the spacecraft approaches the object and/or as changes in the magnetic field and turbulence levels due to the change in plasma environment as the spacecraft enters a magnetosphere (Greenstadt 1971).

At large distances where the spacecraft remains outside the magnetopause, a signature of the magnetosphere would be manifested as draping of the IMF around the obstacle (e.g., Fig. 7). In this case, one can deduce the existence of an obstacle from comparing disturbances in the IMF far from the body with theoretical predictions of the interaction of the body with the solar wind, as was done for (951) Gaspra (Baumgartel et al. 1994; Kivelson et al. 1993). This indirect method can be used to place an upper limit on the magnetic field (see Table 2) but such draping can be either due to NRM or, if the asteroid is conductive, to induced magnetic fields. If the asteroid is indeed unmagnetized but conductive, the Magnetometry Investigation may be able to constrain its internal conductivity profile (Elkins-Tanton et al. 2020) which would constrain its composition (metallic versus stony). Finally, there are indirect methods of inferring the presence of an intrinsic magnetic field that do not rely on magnetometry data such as the presence of a bow shock, as was done in the case of the Dawn mission orbiting (1) Ceres (Russell et al. 2016).

3 The Psyche Magnetometer

3.1 Introduction

The Magnetometer was designed and built by DTU. It has high heritage from the Vector Field Magnetometer (VFM) onboard the European Space Agency's Swarm mission (2013–) (Merayo et al. 2008) with the main changes being an expanded range, additional qualification to cope with cryogenic temperatures expected during flight and shock and random vibrations during launch, as well as additional electronics radiation protection. Each field component is measured with a single dynamic range of $\pm 80{,}000$ nT with a noise of $< 30\ \mathrm{pT\,(Hz)^{-1/2}\,axis^{-1}}\ 3\sigma$ at 1 Hz ($< 39\ \mathrm{pT\,axis^{-1}}\ 3\sigma$ integrated over 0.1–1 Hz) (Figs. 3 and 8). The VFM in turn has extensive heritage and in particular is the main instrument in missions mapping the magnetic field of the Earth: Ørsted (1999–) (Nielsen et al. 1995) and the Challenging Microsatellite Project (CHAMP) (2000–2010) (Merayo et al. 2005).

DTU designed, built, and delivered engineering models and ultimately three Flight Models for the Psyche project. Flight Models 1 and 2 became the pair SU1 and EU1 and the pair SU2 and EU2, respectively. Flight Model 3 is held in reserve at JPL for anomaly resolution and as a backup flight spare.

3.2 Fluxgate Magnetometry

The Magnetometer uses the fluxgate principle for measuring magnetic fields (Primdahl 1979). The basis of this technique is that a high permeability material (ring core) is periodically driven in and out of saturation by a toroidally-wound excitation coil wrapped with

Table 2 Summary of past missions constraining asteroidal magnetic fields

Body	m_S (Am^2 kg^{-1})	Estimation method	Caveats	Encounter	CA (body radii/km)	References
(951) Gaspra	$< 10^{-2}$	IMF rotation + modeling	Signature could be solar wind feature (Blanco-Cano et al. 2003)	Galileo 10/29/1991 Flyby	230/1600	Baumgartel et al. (1994) and Kivelson et al. (1993)
(243) Ida	–	IMF rotation + modeling	Signature is a solar wind feature (Blanco-Cano et al. 2003)	Galileo 8/28/1993 flyby	169/2700	
(253) Mathilde	–	No detection		Near-Earth Asteroid Rendezvous (NEAR) 7/27/1997 flyby	46/1212	Cheng (2002)
(433) Eros	$< 1.9 \times 10^{-6}$	No conclusive detection, upper limit only		NEAR 12/23/1998 landing	0/1	Acuña et al. (2002)
(9969) Braille	$< 10^{-2}$	Magnetometer data fit to possible dipole	Signature suspected to be spacecraft noise	Deep Space 17/28/1999 flyby	36/28	Richter et al. (2001)
(2867) Šteins	$< 10^{-5}$–10^{-3}	No detection, upper limit only		Rosetta* 9/5/2008 flyby	160/800	Auster et al. (2010)
(21) Lutetia	$< 5.9 \times 10^{-7}$	No conclusive detection, upper limit only		Rosetta* flyby	62/3120	Richter et al. (2012)
(4) Vesta	$< 10^{-5}$	No magnetometer, upper limit only		Dawn Orbit	0.4/210	Villarreal (2018)
(1) Ceres	–	No magnetometer, detection of energetic particles consistent with a bow shock	Bow shock may be due to a transient atmosphere	Dawn 6/27/2015 Orbit	4.7/4400	Russell et al. (2016)
(162173) Ryugu	$< 10^{-6}$	No conclusive detection, upper limit only	Hayabusa2 Mobile Asteroid Scout 6/27/2018 (arrival) 7/11/2019 (landing)	0/1	Herčik et al. (2020)	

Notes: The first column lists the object name. The second column lists the upper limit on the specific magnetization, m_S, as deduced from the observations. The third column described the method used to constrain the magnetization in the second column. The fourth column lists any caveats to the detection method. The fifth column lists the mission, date of the encounter, and type of encounter. The sixth column lists the altitude of closest approach (CA) during the encounter (0 for missions that ended in a landing) in units of body radii and km. For elongated bodies, the equivalent radius or mean radius were used. For Ceres, this column lists the distance at which energetic particles were detected (not the mission's CA). The last column lists the reference for the information

*For the Rosetta mission, two planned flybys of (140) Siwa and (4979) Otawara were planned (Pätzold et al. 2001) but were cancelled during the mission

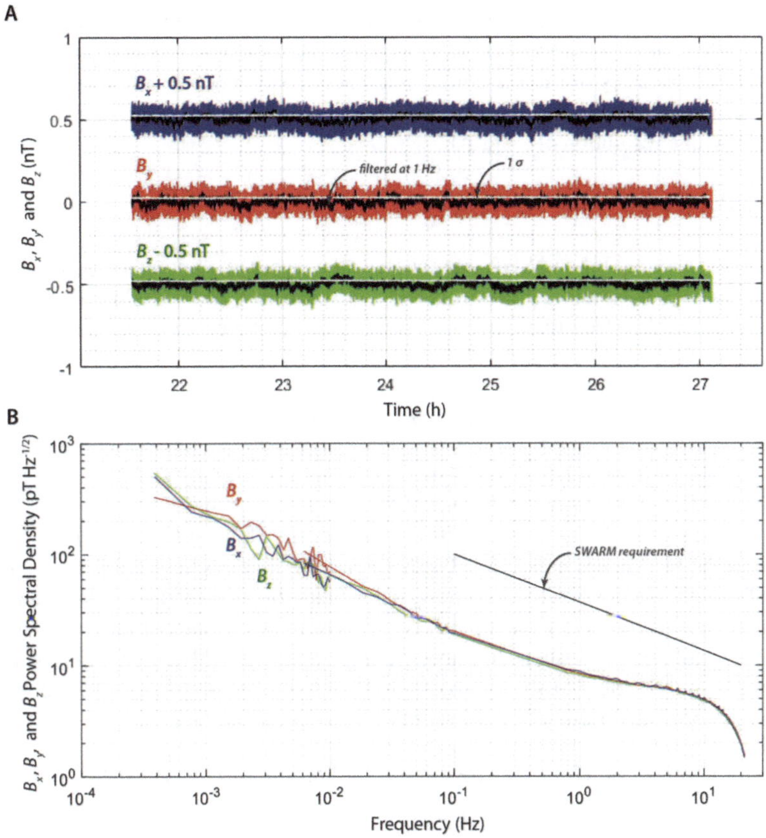

Fig. 8 End-to-end noise of Flight Model 2 (SU2 paired with EU2 side A) (**A**) Time series of three field components B_x (blue), B_y (red), and B_z (green). B_x is offset by 0.5 nT and B_z is offset by -0.5 nT for clarity. (**B**) Power spectral density (PSD) of B_x (blue), B_y (red), and B_z (green). Shown for comparison was the noise requirement for the Swarm Mission Vector Field Magnetometer, which provides heritage for the Psyche Magnetometer

pickup coils. If an ambient field is present, the part of the toroid wrapped with a pickup coil oriented parallel to the ambient field will remain saturated for a longer duration than the part of the toroid wrapped with a pickup coil oriented antiparallel to the ambient field. The flux induced in the pickup coils detects this temporal asymmetry in the saturation state which can subsequently be detected as a second harmonic in the pick-up coils. Finally, compensation coils act to maintain zero field in the pickup coils, serving as a null detector such that when the external field changes, the compensating current is accordingly adjusted to bring the sensor into balance.

The Psyche Magnetometer's core consists of 11 wraps of an amorphous magnetic metal ribbon, which is by heat treatment optimized for several magnetic properties (e.g., nearly zero magnetostriction and magnetic hysteresis) (Nielsen et al. 1997, 1990). As result of this treatment, the sensor has superior performance in applications where thermal and long-term stability are driving parameters. The excitation coil with a current at a frequency of 15.625 kHz drives the core into saturation twice per cycle. The pickup coil is short-circuited by an operation amplifier (Primdahl et al. 1991) producing the fluxgate signal at 31.25 kHz (second harmonic of the excitation frequency), which together with all even harmonics are subsequently demodulated and fed to an integrator. The output of the integrator is connected through a feedback resistor to the compensation coil, which defines the magnetic axis. The voltage across the feedback resistor is digitized by an effective 23.5-bit analog-to-digital converter with a resolution of ~ 11 pT.

3.3 Gradiometry

As discussed above, SU1 and SU2 are mounted along a boom pointing along the FS $+z$-axis at distances of 1.45 m and 2.15 m from the $+Z$ spacecraft deck (Fig. 4). This enables the differenced field between these two positions to be computed from the measured field at each sensor, $\vec{B}_{\mathrm{M}}(\vec{r}) = (B_x, B_y, B_z)$:

$$\Delta\vec{B} = \vec{B}_{\mathrm{M}}(\vec{r}_2) - \vec{B}_{\mathrm{M}}(\vec{r}_1) \tag{2}$$

where $\vec{r}$ is the position relative to the spacecraft and subscripts 1 and 2 refer to field measurements at SUn for sensor number $n = 1{:}2$.

The measured field, $\vec{B}_{\mathrm{M}} = \vec{B}_{\mathrm{amb}} + \vec{B}_{\mathrm{FS}} + \vec{B}_{\mathrm{N}}$, for the ambient field of scientific interest, $\vec{B}_{\mathrm{amb}}$, and the instrument noise, $\vec{B}_{\mathrm{N}}$. Furthermore, the flight system field, $\vec{B}_{\mathrm{FS}} = \vec{B}_{\mathrm{FS,DC}} + \vec{B}_{\mathrm{FS,AC}}$ for flight system steady and time-variable fields $\vec{B}_{\mathrm{FS,DC}}$ and $\vec{B}_{\mathrm{FS,AC}}$. Now, the ambient field (e.g., the asteroid's remanent field) is expected to be constant at the two SUs because $r_2 - r_1$ is small relative to their distances from the ambient field source (e.g., the asteroid). By comparison, the field of the nearby flight system will be stronger at SU2 compared to SU1. Therefore, the differenced field is only a function of $\vec{B}_{\mathrm{FS}}$ and $\vec{B}_{\mathrm{N}}$:

$$\begin{aligned}\Delta\vec{B} &= \vec{B}_{\mathrm{FS}}(\vec{r}_2) + \vec{B}_{\mathrm{amb}}(\vec{r}_2) + \vec{B}_{\mathrm{N},2} - \left[\vec{B}_{\mathrm{FS}}(\vec{r}_1) + \vec{B}_{\mathrm{amb}}(\vec{r}_1) + \vec{B}_{\mathrm{N},1}\right] \\ &= \vec{B}_{\mathrm{FS}}(\vec{r}_2) - \vec{B}_{\mathrm{FS}}(\vec{r}_1) + \vec{B}_{\mathrm{N},2} + \vec{B}_{\mathrm{N},1}\end{aligned}$$

where $\vec{B}_{\mathrm{N},2}$ and $\vec{B}_{\mathrm{N},1}$ are the instrument noise for measurements with SU1 and SU2. For a given flight system component, the ratio of the ath component of the flight system field at SU1 and the bth component of the flight system field at SU2 is known as the coupling matrix $\alpha_{ab} = B_{\mathrm{FS},a}(r_1)/B_{\mathrm{FS},b}(r_2)$. If is $\boldsymbol{\alpha}$ known a priori, then an estimate of the ambient field can be obtained from (Ness et al. 1971):

$$\vec{B}'_{\mathrm{amb}} = (1 - \boldsymbol{\alpha})^{-1}\left[\vec{B}_{\mathrm{M}}(\vec{r}_1) - \boldsymbol{\alpha}\vec{B}_{\mathrm{M}}(\vec{r}_2)\right] \tag{3}$$

If $\boldsymbol{\alpha}$ is not known a priori, it can be estimated by fitting the differenced field with a dipole estimate, $\vec{M}'_{\mathrm{FS}}$ (or in principle, multipole) representing the flight system field:

$$\vec{B}_{\mathrm{FS}}(\vec{r}) = \frac{\mu_0}{4\pi}\left[\frac{3\boldsymbol{r}(\vec{M}'_{\mathrm{FS}} \cdot \vec{r})}{r^5} - \frac{\vec{M}'_{\mathrm{FS}}}{r^3}\right] \tag{4}$$

where here the origin is assumed to be located at the dipole location.

The spacing of the Psyche SUs along the boom was determined to optimize the use of gradiometry to characterize and remove the spacecraft fields from the measured field. Our detailed approach is described in Cochrane et al. (2023) [see their approach for two sensors]. We conducted simulations to estimate the uncertainty in knowledge of the flight system field as a function of spacing between the sensors. Using a multiple dipole model of the flight system field (Sect. 3.6.3), we modelled the flight system DC and AC field at

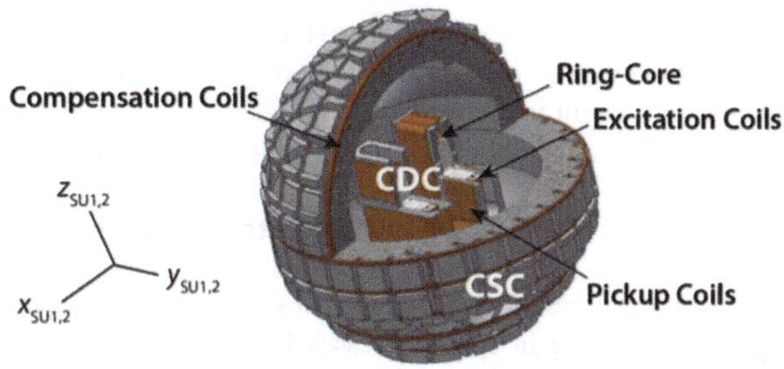

Fig. 9 Cutaway of a Sensor Unit (SU). Shown is the CSC (grey sphere wrapped with compensation coils) and interior CDC (wrapped ring core wrapped with pickup coils). Compass shows orientation of SU1 and SU2 coordinate systems

each sensor as a function of their spacing along the boom. SU1 was fixed at the end of the boom while SU2 was sequentially moved along the length of the boom. We then used the predicted fields at the two SUs to estimate the moment and position of a dipole (six total parameters) representing the flight system, $\vec{M}'_{\mathrm{FS}}$. We repeated this for 1,000 instances of the flight system field configuration in which the orientations of the dipoles in the multiple dipole flight system model were randomly varied. We found that when SU2 is too close to the spacecraft deck, the offset dipole approximation [Eq. (4)] is a poor approximation of the flight system field. At the other extreme, when SU2 is too close to SU1, the difference between their two measurements becomes comparable or less than the sensor noise [i.e., $\vec{B}_{\mathrm{FS}}(\vec{r}_2) - \vec{B}_{\mathrm{FS}}(\vec{r}_1) << +\vec{B}_{\mathrm{N},2} + \vec{B}_{\mathrm{N},1}$]. This resulted in the identification of a range of optimum positions for DC field removal between 1.35 m and 1.65 m from the spacecraft deck and for AC and step removal between 1.15 and 1.5 m from the spacecraft deck. A location near the outer edge of the optimum positions for AC removal was selected. Considerations for this choice included placing SU2 as far from the spacecraft as possible to allow for graceful degradation in case SU1 failed while still optimizing spacecraft field removal with the gradiometer.

3.4 Description of Instrument

Three ring core elements in an orthogonal configuration and wrapped with excitation and pickup coils form the Compact Detector Coils (CDC). In principle, the pickup coils could also used for compensating the external field, such that the CDC could be used as a fluxgate magnetometer with excellent performance. However, the nonspherical shape of its pickup coils limits the omnidirectional linearity (Brauer et al. 1997) (Fig. 9). In particular, a nonlinearity in the range of Earth's field is present due to the proximity of the detector coils along each of the three axes, which mutually interfere due to stray fields produced during compensation. This problem is overcome with the addition of the surrounding Compact Spherical Coils (CSC), whose spherical shapes create an internal field that is nearly zero (Primdahl and Jensen 1982) (Fig. 9). The excitation of the three excitation coils are series coupled and therefore share the excitation drive, whereas the detector and compensation coils have each their own circuitry. The sensor elements are carefully adjusted as to reduce as much as possible the feedthrough (Petersen et al. 1992) for optimal signal to noise ratio. As a result the uncertainty associated with interactions between the axes (omni-nonlinearity knowledge) for each SU is only 24 ppm 3σ, which translates to 0.24 $\mathrm{pT\,axis^{-1}}$ 3σ for a measured field of 10 nT (Fig. 3).

The Magnetometer is equipped with a field-programmable gate array (FPGA) that controls all digital functions in the instrument: it produces a highly stable excitation square wave signal with a frequency of 15.625 kHz and a detection signal of frequency 31.25 kHz that are used by the excitation drive and phase detector blocks, respectively. After power on, the instrument operates autonomously. It activates the phase detector with respect to the excitation current in the sensor using the pre-flight predetermined default parameters for the frequency, delay and width, and transmits data at up to 50 Hz.

3.5 Pre-flight Calibration and Expected Performance

The Magnetometer design incorporates components with high thermal and temporal stability against degassing in space, stresses from changing temperatures, and total ionizing dose from high energy particles. Nevertheless, its performance is ultimately limited by tolerances in the machining processes, electronic parts and assembly of the individual parts and their temperature dependencies. A calibration process has been implemented to correct for these effects. The raw analog-to-digital converter readings from the Magnetometer are transmitted through spacecraft telemetry to ground, where they are converted from engineering units (eu) to magnetic field values (nT) as discussed in Sect. 4. This requires knowledge of the offset vector, $\vec{O}_n$, the scale factor matrices, $\boldsymbol{S}_n$, the orthogonality matrices, $\boldsymbol{V}_n$, and the rotation matrices, $\boldsymbol{R}_n$, that transform the orientation of the data from the individual SU1 and SU2 frames to the FSe and other frames. Note that in addition to depending on the SU-EU number, n, $\vec{O}_n$ and $\boldsymbol{S}_n$ also depend on the choice of EU side (A or B); for simplicity, we have omitted a subscript denoting the latter dependency here and below.

Because conversion from eu to nT occurs during ground data processing, the Psyche Magnetometry uncertainty budget does not depend on the values of the coefficients in $\vec{O}_n$, $\boldsymbol{R}_n$, $\boldsymbol{S}_n$ and $\boldsymbol{V}_n$ but rather on our knowledge of them. Therefore, we conducted a pre-flight calibration campaign at DTU to measure these parameters based on the methods described in Merayo et al. (2000a,b). We also have designed an in-flight validation campaign that will yield estimates of the sum of $\vec{O}_n$ and the flight system DC field at each sensor throughout the mission (Sect. 5).

The parameters $\vec{O}_n$, $\boldsymbol{S}_n$ and $\boldsymbol{V}_n$ are slightly temperature and time-dependent. For this reason, each SU contains two temperature sensors (located in the CDC and CSC) and each EU incorporates a third temperature sensor. These have an accuracy of less than 0.27 °C 3σ in the operating temperature range following ground calibration. The Magnetometer transmits these temperatures, which are then used in the ground data processing.

Offset knowledge as a function of temperature is the dominant term in our Instrument Uncertainty and is specified by a driving requirement for the Magnetometer design (Fig. 3). Both the SU and EU contribute to the offset. These were characterized independently with pre-flight tests in which the SU and EU were scanned separately in temperature. The SU offset drift was measured such that after calibration, our knowledge of its contribution to the offset is better than 0.25 nT over the whole operational temperature range of the SUs (~ -125 to $+50$ °C) (e.g., Table 3). For the much smaller temperature ranges that the SUs are expected to experience during Orbit D (± 1 °C), the repeatability of the offset is better than 0.15 nT axis^{-1} 3σ after calibration (Fig. 3). The knowledge of the EUs' contribution after calibration is better than 0.2 nT for their whole operational temperature range (-20 to $+50$ °C) (e.g., Table 3) and just 0.01 nT over the temperature range they will experience in Orbit D.

Based on heritage from the Swarm VFM, the long-term temporal drift of the Magnetometer offset is expected to be exceptionally low (well below 58 pT axis^{-1} y^{-1} 3σ). Given our regular in-flight zero-level calibration activities (Sect. 5), it should contribute negligibly to the Instrument Uncertainty (Fig. 3). In particular, for Psyche with a moment equal to that of the level 1 requirement (Sect. 1.3), the spacecraft would pass within the asteroid's magnetosphere at altitudes below ~ 170 km (just below that of the nominal orbit C) where it would

Table 3 Magnetometer mass, volume and power

Magnetometer component	Final mass (g)	Final size (mm)	Power (W)
SU1 including harness and bracket	896	CSC diameter: 82 bracket diameter 140	
SU2 including harness and bracket	892	CSC diameter: 82 bracket diameter 140	
EU1	924	100 × 124 × 66	2.2
EU2	923	100 × 124 × 66	2.2

Note: The first column lists the Magnetometer component, the second column lists its mass, the third column lists its dimensions, and the fourth column lists its power. Power is for operation in near-zero field and 0 °C like that expected for level 1 requirement Psyche dipole moment in orbit. Operation in an 80 μT field at 70 °C would increase the power to 2.6 and 2.3 W for EU1 and EU2, respectively

be shielded from the solar wind. We currently estimate that the spacecraft will spend 402 days at or below this altitude, which sets the maximum time since the last estimate of the Magnetometer zero-level from monitoring of solar wind fluctuations (Sects. 4.4 and 5.3).

Calibration of the scale factor as a function of temperature was also conducted for both the SUs and EUs. The scale factor drift of the sensor depends uniquely on changes in the volume of the CSC shell, which has an isotropic thermal expansion rate of 30 $\text{ppm}\,°\text{C}^{-1}$. The scale factor of the EUs is on the order of 1 $\text{ppm}\,°\text{C}^{-1}$. After compensation, the uncertainty on the measurement due to the residual effect is better than 3.1 $\text{pT}\,\text{axis}^{-1}\,3\sigma$ in a 10 nT field (Fig. 3).

Due to the design of the Magnetometer, the nonorthogonality angles are constant with temperature. After ground calibration, knowledge uncertainty on these angles is better than 15 arcsec 3σ (i.e., 0.73 $\text{pT}\,\text{axis}^{-1}\,3\sigma$ for a 10 nT field) (Fig. 3), consistent with extensive data from heritage instruments (Yin and Lühr 2011). The main contribution to field orientation uncertainty is knowledge of the SU axes relative to the FS frame which totals 0.17° 3σ (i.e., 30 $\text{pT}\,\text{axis}^{-1}\,3\sigma$ for a 10 nT field).

3.6 Accommodation

3.6.1 Resources

The total masses of each of the two EUs and each of the two SUs including their brackets and harnesses are 0.92 g and 0.9 g, respectively. The length of the intra-harness between each EU and SU is selectable without degrading the performance and for Psyche was required to be 3.85 m for all Flight Models. The masses of each harness and bracket are 120 $\text{g}\,\text{m}^{-1}$ and 135 g, respectively. Finally, the mass of each CSC, which has a diameter of 82 mm, is 310 g. The size of each EU is 100 × 124 × 66 mm. The power dissipation of each Flight Model is about 2.2 W (due to a higher radiation hardened DC-DC converter). Table 3 provides details.

3.6.2 Interfaces

To mount the SU onto the spacecraft, a carbon fiber reinforced polymer bracket was developed with a finger shape, with one end glued onto the CSC so as not to deform the sensor's structure during temperature changes. The other end of the bracket has a plate with four holes used to attach the sensor to a Ti node on the boom. The EU has a standard interface with four bolts into the panel of the spacecraft. Attitude transfer from the SU axes to the

FS coordinate system was achieved by a reference cube assembly temporarily mounted on each SU. The orientations of the three normal faces of the cube were determined with respect to the SU coil axes following the triaxial coil method of Primdahl et al. (2002). During assembly onto the spacecraft, the rotation matrices relating the two reference cubes and the spacecraft were measured. After this, the reference cube assemblies were dismounted.

Upon application of power to one of the EU sides (A or B), that side will connect to the SU autonomously. Each SU was designed to be nonredundant as it does not contain any electronic components. Each EU side contains all resources to operate independently and it is not expected that both sides will be powered on at the same time during science operations. The EU chassis walls are composed of aluminum with side wall thicknesses and top and bottom wall thicknesses of 3 mm and 6 mm, respectively. This provides radiation protection to the EUs at a level that is about 10 times larger than the expected total ionizing dose (Oran et al. 2022).

Although the Magnetometer can operate with a wide range of power supply voltages, for Psyche the nominal voltage is 31 V (with a low and high interval of 27.2 and 33.6 V, respectively). Table 3 shows the power consumption of the units. To avoid current loops that could create disturbing magnetic fields, the power connector provides the nominal voltage and return, but the grounding of the unit is performed through a stud that is connected separately to the reference grounding point of the spacecraft power supply. The grounding of the SU harness is also carefully separated and routed as to reduce any interference between the two SU-EU pairs or any other equipment near the location of the Magnetometer hardware.

Using a low-voltage differential signaling (LVDS) electrical layer, the Magnetometer communication features a universal asynchronous receiver-transmitter (UART) serial interface at 115,200 bps with a structure of 1 start bit followed by 8 bits and then a stop bit. Data and commands are organized in packets following the Swarm Packet Utilization Standard (PUS) protocol (European Cooperation for Space Standardization 2003). A few seconds after power on, the Magnetometer enters the default mode (50 Hz sampling) and begins to transmit science data at a rate of approximately 1 s where each packet contains 50 vectors in addition to time information, temperature and other flags that provide the status of the instrument. To synchronize the science data, a pulse-per-second (PPS) signal is provided to the instrument, which is used to relate the time of the first sample of each packet with respect to this signal with an interval of 125 μs. The end-to-end time characterization is verified on the ground and is typically 0.2 ms. A spacecraft time message (STM) is sent to the Magnetometer every second enabling the Magnetometer to report the information needed to fully recover the absolute time of the data on the ground. Should the PPS or STM be absent, the Magnetometer will propagate the time from the last time it was synchronized to them since being powered on. Housekeeping data are transmitted once per minute. Uncompressed science data packets contain 500 bytes, but if the field magnitude does not change significantly, the packet can be compressed to about 200 bytes, which yields a data volume of less than 2 kbps assuming a realistic 90% compression rate. Telecommands can be sent to the Magnetometer, but it is not expected that this will be performed in-flight apart from the power on sequence. The telecommands can also change the sampling frequency to 10 Hz, 1 Hz, and 0 Hz (the latter would just provide housekeeping data and no science data). They can also change the excitation frequency and the delay of the detector, but these are not expected to be used in-flight.

The allowable temperatures for Magnetometer components are given in Table 4. The instrument has been qualified, tested, and calibrated to this range. Importantly, the spherical shape of the CSC, which undergoes isotropic thermal expansion and contraction, means that Magnetometer does not require active thermal control in-flight despite the very wide range of temperatures experienced.

Table 4 Allowable flight temperatures for the SUs, EUs, and harnesses

Magnetometer component	Operational minimum (°C)	Operational maximum (°C)	Non-operational minimum (°C)	Non-operational maximum (°C)
SUs	−125	50	−125	50
EUs	−20	50	−30	50
Harnesses	−125	50	−125	50

Note: The first column lists the Magnetometer component, the second and third columns list the minimum and maximum operational temperatures and the fourth and fifth columns list the minimum and maximum non-operational temperatures

3.6.3 Magnetic Cleanliness

As discussed in Sect. 1.3, the dominant contribution to the Magnetometer's Reconstructed Measurement uncertainty is the flight system's magnetic field (Fig. 3A). Even though the science data are acquired at 50 Hz, the Investigation is sensitive to fields ranging from DC up to 100 kHz, with frequencies > 50 Hz aliased into the measured signal. We use two methods to minimize this noise source. First, as discussed in this section, a magnetic cleanliness program was implemented to design and build a flight system that produces minimal DC and AC fields at the SUs. Second, as discussed in Sect. 5, the flight system field measured at the sensors is suppressed during ground data processing using gradiometry and solar wind monitoring.

The Psyche magnetic cleanliness program uses three broad approaches: prevention, characterization, and mitigation of flight system fields. For more details, we refer the reader to de Soria Santacruz-Pich et al. (2020). Our foremost approach was prevention: to design the flight system to minimize stray DC and AC fields. This led us to two sets of requirements. First, we require that our knowledge of the flight system field should be better than 0.4 nT per axis 3σ after suppression using gradiometry and other techniques while using both SU1/EU1 and SU2/EU2 (Sect. 4.4) (Fig. 3A). This flows down to requirements on knowledge uncertainty for fields in four frequency ranges: $< 10^{-5}$ Hz, 10^{-5} to 0.1 Hz, 0.1 to 25 Hz, 25 Hz to 100 kHz (Fig. 3B). Second, the flight system field system from 0.1 Hz to 100 kHz should be sufficiently low that our level 1 requirement could be met even if SU1 failed and we were only left with measurements from the inboard sensor, SU2.

These metrics then drove the development of magnetic cleanliness requirements and guidelines that were then flowed to designers of spacecraft subsystems and components early in the development process. These included maximizing the distance between magnetic field sources [e.g., Hall thrusters and cold gas thrusters, components of the Deep Space Optical Communication (DSOC) payload, latch vales, and traveling-wave tube amplifiers (TWTAs)] and the SUs, minimizing the usage of ferromagnetic and high magnetic permeability materials (e.g., by using aluminum for structural supports), reducing current loops (e.g., in the solar arrays), and self-compensation of the fields within a component to minimize magnetic poles (e.g., by using back-to-back mounting of latch vales and arranging ferromagnetic struts in the DSOC in a closure configuration). We also ensured that the spacecraft has no conductive paths along which there could be strong thermal gradients that could generate thermoelectric currents during flight (e.g., Brauer et al. 2018). Because many of these mitigations are cost-neutral or low-cost, they led to major efficiencies in cost and schedule by minimizing the need for later mitigation activities.

Our second approach was to characterize the AC and DC flight system magnetic field. This was conducted at the subsystem level prior to assembly of the spacecraft and at the

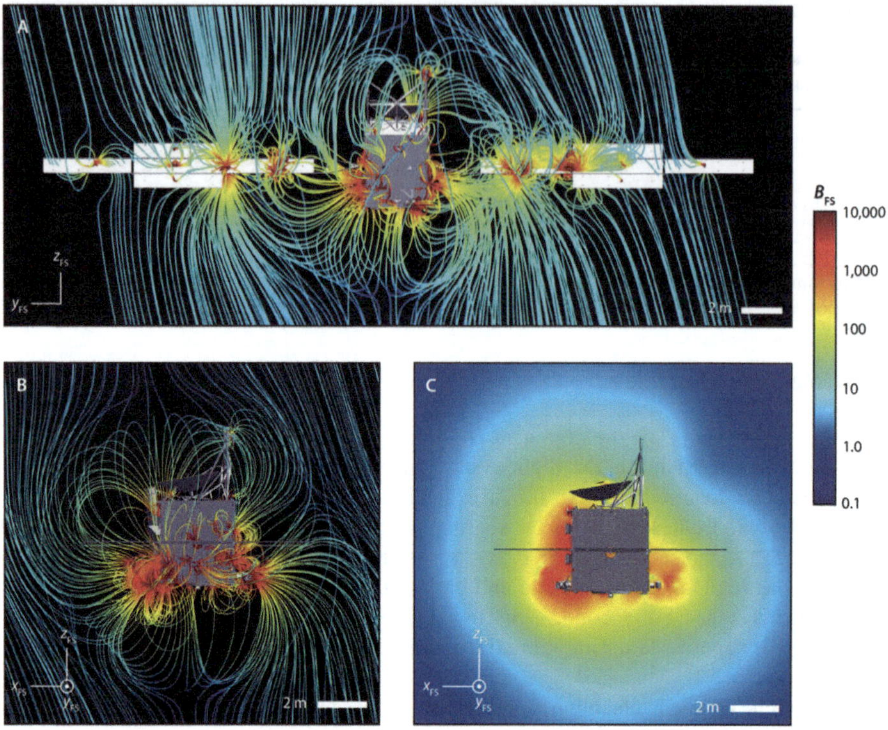

Fig. 10 Psyche flight system magnetic field. (**A**, **B**) Two views of the modeled flight system field plus a background IMF, $\vec{B}_{\text{IMF}} = (1\ \text{nT}, 1\ \text{nT}, 5\ \text{nT})$ and superposed on the spacecraft model (grey and white). Shown are magnetic field lines simulated from more than 200 individual dipole sources using a Runge–Kutta method. The magnetic field lines are color-coded according to field strength, with blue representing weaker field strengths and red representing stronger field strengths. Note that the solar array currents are modeled by multiple dipole field point sources (**C**) Magnetic field intensity in the flight system in plane with surface normal in the y direction containing the boom and two SUs. Same color intensity as in (**A**, **B**). Compasses in (**A**–**C**) show orientation of FS coordinate system

flight system-level during assembly in the JPL high-bay. With respect to pre-assembly characterization, all spacecraft subsystems were evaluated for their impact to the magnetics budget and those considered to likely produce the strongest fields at SU2 were tested and characterized early in the development phase of the mission. In particular, the magnitudes of the magnetic moments of the prioritized subsystems were measured. Results from subsystem-level tests were used to construct a spacecraft-level magnetics model that approximates the DC and AC field using dipoles positioned at the location of each subsystem (following Mehlem 1978a,b; Cochrane et al. 2023; Mehlem and Wiegand 2010; Neubauer and Schatten 1974; Zhang et al. 2007). Measured subsystems were assigned a dipole magnitude equal to the value measured with dipole orientation in a random direction, while unmeasured subsystems were assigned dipole magnitudes estimated to be upper limits from heritage measurements of analog subsystems. This multiple-dipole model was then used to estimate the flight system field at the locations of SU1 and SU2 (Fig. 10).

The flight system-level testing involved measuring the field at multiple locations around the spacecraft and then fitting for the spacecraft net dipole moment. System-level testing was conducted as part of the campaign to characterize the electromagnetic interference and elec-

Table 5 Psyche flight system magnetic field measured at JPL

Field quantity	Measured value
DC field at SU2[a]	18 nT
Ratio of DC field at two SUs[b]	1.7
AC field at SU2 integrated from 10^{-5} to 0.1 Hz[c]	< 1.4 nT axis^{-1} 3σ
Ratio of AC field at SU2 integrated from 10^{-5} to 0.1 Hz at SUs[d]	> 1.4
AC field at SU2 integrated from 0.1 to 25 Hz[e]	< 1.2 nT axis^{-1} 3σ
Ratio of AC field integrated from 0.1 to 25 Hz at SUs[f]	> 2.5
AC field at SU2 integrated from 25 Hz to 100 kHz as sensed by Magnetometer[g]	< 0.4 nT axis^{-1} 3σ

Note: The first column lists the spacecraft field quantity measured and the second column lists the value measured

[a]Total DC ($< 10^{-5}$ Hz) field at the location of the inboard sensor, $B_{\mathrm{FS,DC}}(r_2)$

[b]Ratio of the DC field at the inboard to that at the outboard sensor, $B_{\mathrm{FS,DC}}(r_2)/B_{\mathrm{FS,AC}}(r_1)$

[c]Estimated maximum 3σ per axis variation of frequency-integrated AC field for all spacecraft subsystems at inboard sensor, $3[\sum_P \int_{10^{-5}\,\mathrm{Hz}}^{0.1\,\mathrm{Hz}} \mathrm{PSD}_{p,s}(r_2, f)\mathrm{d}f]^{1/2}$. Here $\mathrm{PSD}_{p,s}$ is the power spectral density ($\mathrm{nT}^2\,\mathrm{Hz}^{-1}$) for spacecraft subsystem, p, and field axis $s = x$, y or z and f is frequency

[d]Ratio of the integrated AC field from 10^{-5} to 0.1 Hz at the inboard to that at the outboard sensor, $[\sum_P \int_{10^{-5}\,\mathrm{Hz}}^{0.1\,\mathrm{H}} \mathrm{PSD}_{p,s}(r_2, f)\mathrm{d}f]^{1/2}/[\sum_P \int_{10^{-5}\,\mathrm{Hz}}^{0.1\,\mathrm{Hz}} \mathrm{PSD}_{p,s}(r_1, f)\mathrm{d}f]^{1/2}$

[e]Estimated maximum 3σ per axis variation of frequency-integrated AC field for all spacecraft subsystems at inboard sensor, $3[\sum_P \int_{0.1\,\mathrm{Hz}}^{25\,\mathrm{Hz}} \mathrm{PSD}_{p,s}(r_2, f)\mathrm{d}f]^{1/2}$

[f]Ratio of the integrated AC field from 0.1–25 Hz at the inboard to that at the outboard sensor, $[\sum_P \int_{0.1\,\mathrm{Hz}}^{25\,\mathrm{Hz}} \mathrm{PSD}_{p,s}(r_2, f)\mathrm{d}f]^{1/2}/[\sum_P \int_{0.1\,\mathrm{Hz}}^{25\,\mathrm{Hz}} \mathrm{PSD}_{p,s}(r_1, f)\mathrm{d}f]^{1/2}$

[g]Estimated maximum 3σ per axis variation of frequency-integrated AC field for all spacecraft subsystems at inboard sensor as measured by the Magnetometer, $3[\sum_P \int_{25\,\mathrm{Hz}}^{100\,\mathrm{kHZ}} \mathrm{PSD}_{p,s}(r_2, f)R(f)^2\mathrm{d}f]^{1/2}$. Here, $R(f) \leq 1$ is the Magnetometer response function

tromagnetic compatibility of the spacecraft during assembly in the high-bay at the JPL. The results (Table 5) indicate that the spacecraft is remarkably magnetically clean and will enable the Magnetometry Investigation to meet the two performance metrics discussed above.

3.7 Operations

The nominal operations plan for the Magnetometer is simple: it will be turned on soon after launch and remain on for the remainder of the mission. All data will be sent to the ground at 50 Hz and will be processed at MIT (Sect. 4). All measurements are expected to be usable for science except those acquired during periods of thrusting by the electric propulsion system, which will occur periodically during cruise and transfer orbits, and during DSOC operations, which will occur prior to arrival at Psyche (Polanskey et al. 2023). Real-time commands (e.g., power cycling and changing fluxgate tuning parameters) are only expected to be issued to resolve anomalies. MIT will conduct science planning and JPL will develop formal commands following communications with MIT. Anomalies will be resolved through cooperation between MIT, DTU and the JPL. MIT will conduct long-term monitoring of magnetic field data, temperature measurements, and internal instrument voltages to assess instrument performance.

Table 6 Magnetometer data levels

Data level	Reference frames	Units	Description
Packet	SU1 and SU2 axes 1, 2, and 3	DN	Telemetry packets received from the instrument
Raw	SU1 and SU2 axes 1, 2, and 3	eu	Unpacked telemetry including time-tagged magnetic field and temperature measurements
Partially Processed	SU1, SU2 axes x, y, and z	nT, °C	Magnetic field vectors corrected for temperature dependent scale factors, offsets, non-linearity, and non-orthogonality using ground calibration data,
Calibrated	FS	nT	Magnetic field vectors with flight system field and residual instrument offets suppressed
Derived	PBF, PSO, J2E	nT	Calibrated, rotated magnetic field vectors
Remanent Field	PBF	Am^2, nT	Estimate of dipole moment and possibly also higher-order internal Gauss coefficients and magnetic anomaly maps

Note: The first column lists the PDS4 data level, the second column lists the reference frame (SU1/SU2 = Sensor Unit 1/2 frames, FS = flight system, PBF = Psyche-body-fixed, PSO = Psyche-Sun-orbit, and J2E = Psyche-centered EMO J2000), the third column lists the units (DN = digital numbers, eu = engineering units), and the fourth column gives a description of the data level

4 Data Pipeline

4.1 Overview of Data Levels

The data products produced by the Psyche Magnetometry Investigation are organized into six data levels (Fig. 11): packet, Raw, Partially Processed (ppr), Calibrated (cal), Derived (der) and Remanent Field (rf) (Fig. 11, Table 6). The processing for each is presented in the following sections.

Packet data are the raw telemetry sent from the instrument to the spacecraft and subsequently packaged into Consultative Committee for Space Data Systems (CCSDS) File Delivery Protocol (CFDP) data products for transmission. Science packets include three magnetic field axes and temperature measurements required for calibration and the information to accurately timestamp the measurements. The observations are nominally reported with a frequency of $f_s = 50$ Hz using a lossless compression scheme in which the change in the field is reported rather than the total field. These data are decompressed and time-tagged to generate the Raw data. Corrupted data are also flagged and removed at this data level.

Partially Processed data have been corrected for temperature-dependent scale factors, offsets, non-linearity and non-orthogonality using ground calibration data. The measurements are in physical units (nT and °C) and are given in both the individual sensor frames (SU1 and SU2) and in the FS reference frame. These corrections are based on calibration tables developed during the ground testing campaign prior to launch (Sect. 3.5). At the packet, Raw, and Partially Processed data levels, the two SU/EU pairs are treated separately so there are two files per day (one for each SU) for these data levels.

Calibrated data have been processed to suppress contributions from flight system generated-fields and residual instrument offsets and field interactions between the two SUs. Data at this level are given in nT in the FS frame. We plan to use two main methods to suppress flight system fields and residual instrument offsets. First, measurements from both SUs will enable gradiometry methods in which the flight system field can be characterized based

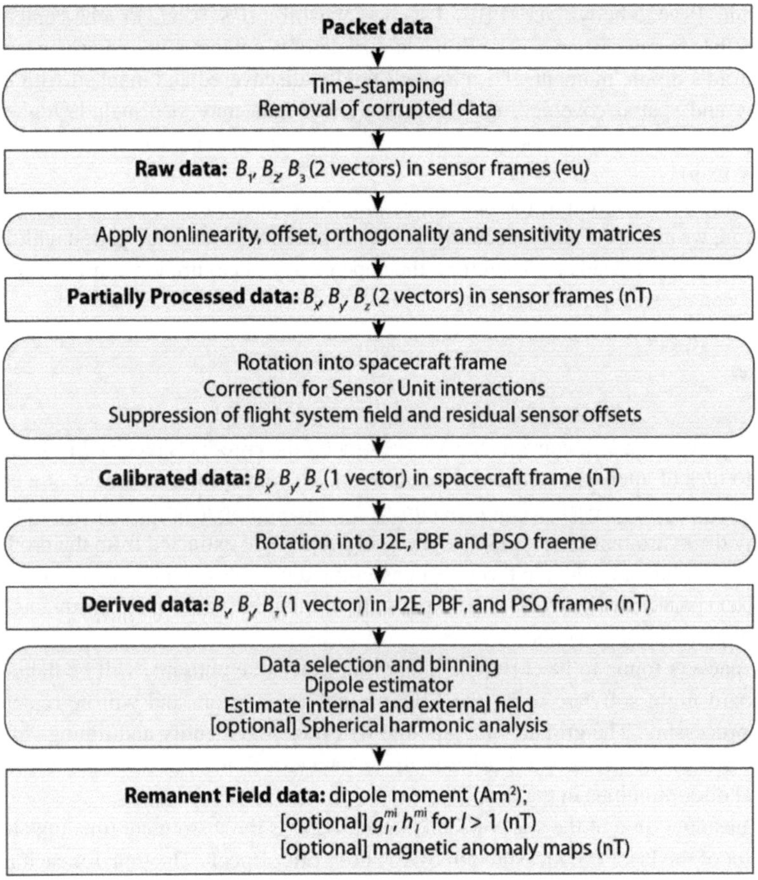

Fig. 11 Psyche Magnetometry Investigation data levels. Packet data received from the spacecraft (pink box) are time-stamped, cleaned of corrupted data, and organized into files of approximately 1 day in duration (second blue oval) to produce Raw data: a magnetic field vector for each SU in SU coordinates with units of eu [Eq. (5)] (top green box). These are then calibrated (second blue oval) to produce Partially Processed data: a magnetic field vector for each SU in SU coordinates with units of nT [Eq. (7)] (second green box). These data are then rotated into the FS frame, flight system fields are suppressed, and residual senser offsets removed (third blue oval) to produce Calibrated data (third green box): a single magnetic field vector in the spacecraft frame with units of nT. These data are then rotated into the J2E, PBF, and PSO frames (fourth blue oval) to produce Derived data in units of nT [Eq. (9)] (fourth green box). The data are then binned and down-selected and the internal field and externally generated field are estimated (blue box). These are used to produce Remanent Field data, which at a minimum are a constraint on the asteroid's dipole moment [e.g., Eq. (10)] (bottom green box). Optionally, if a remanent field is detected and mapped with sufficient coverage in PBF coordinates, a magnetic anomaly map and possibly also internal Gauss coefficients with degree $l > 1$ can be estimated [Eq. (12)]

on its property of being stronger in intensity at the inboard relative to the outboard sensor (Ream et al. 2021). Second, monitoring variations of the solar wind (Leinweber et al. 2008) during flight will be used to correct for residual instrument offsets and the flight system DC field (see Sect. 4.4 and 5.3).

The final two data levels are used for assessing the Magnetometer level 1 requirement (see Sect. 1.3). The Derived data products consist of files which contain the Calibrated data rotated into Psyche-body-fixed (PBF), Psyche-solar-orbit (PSO), and Psyche-centered EMO J2000 (J2E) reference frames. The Remanent Field data contain at minimum an estimate of the asteroid's dipole moment. If a remanent field is discovered and mapped with sufficient accuracy and spatial coverage, the Remanent Field data may also include higher-degree internal Gauss coefficients and possibly even a map of the remanent field (e.g., magnetic anomaly map).

We now discuss in detail the data processing flow for each of these data levels. In this discussion, we assume a nominal configuration in which SU1 has been paired with EU1 side A and SU2 with EU2 side A. Therefore, the dependency of calibration constants on EU side has not been noted explicitly below.

4.2 Raw

The Raw processing step converts the Magnetometer data products received from the spacecraft into files of approximately 1 day in duration. The data products consist of a collection of packets transmitted each second containing 50 three-component field measurements collected by the instruments every 0.02 s. The data packets are extracted from the product files, converted to a human readable format and assembled into files containing a time interval of 1 day. The spacecraft clock (SCLK) along with internal timing for the instruments are used to time-tag the measurements.

Any packets found to be corrupted and/or are otherwise unusable will be flagged either by onboard flight software or by the MIT ground data system and will be rejected from further processing. The ground data pipeline will track the identity and timings of rejected packets. The data pipeline will further scan the data stream to find, report, and resolve, any temporal discontinuities in the data.

The absolute time of the start of each packet, t_{ABS}, is the instrument time updated based on receipt of the last PPS, measured in SCLK ticks after Epoch. The time for each measurement within the packet relative to the start of the packet, t_{MAG}, is given by

$$t_m(c) = t_{\mathrm{ABS}} + t_{\mathrm{MAG}}(c)$$

Here

$$t_{\mathrm{MAG}}(c) = \frac{\Delta_{\mathrm{PPS}}}{8 \times 10^6} + t_d + c/f_s$$

where t_d is the instrument latency as found by the timing characterization test, Δ_{PPS} is the relative time of the sample with respect to the last PPS received by the Magnetometer, and $c = 0{:}49$ is the index of each of the 50 measurements in the packet.

Raw data are uncalibrated and in units of eu for each of the three magnetometer axes $a = 1{:}3$ in each sensor frame, SU1 and SU2:

$$\vec{B}_{\mathrm{raw},n} = \begin{bmatrix} eu_{n,1} \\ eu_{n,2} \\ eu_{n,3} \end{bmatrix}, \tag{5}$$

where $eu_{n,a} = e_{n,a}/2^{6.5}$ where $e_{n,a}$ are the raw uncompressed measurement data.

4.3 Partially Processed Data

The production of Partially Processed data involves converting the three axes from each SU into field vector components. This involves three steps: housekeeping calibration, linearization, and linear calibration. During the housekeeping calibration, the temperature readings for the CSC, CDC and EU are linearized and converted to °C using a 7th order polynomial:

$$T_{n,i} = C_{n,0,i} + C_{n,1,i} VET^1_{n,i} + C_{n,2,i} VET^2_{n,i} + C_{n,3,i} VET^3_{n,i} + C_{n,4,i} VET^4_{n,i} + C_{n,5,i} VET^5_{n,i} + C_{n,6,i} VET^6_{n,i} + C_{n,7,i} VET^7_{n,i}$$

where i denotes either the CSC, CDC, or EU temperature sensor, $C_{n,0,i}$, $C_{n,1,i}$, $C_{n,2,i}$, $C_{n,3,i}$, $C_{n,4,i}$, $C_{n,5,i}$, $C_{n,6,i}$, and $C_{n,7,i}$ are the coefficients for each temperature sensor determined by thermal tests carried out prior to integrating the instrument on the spacecraft (Sect. 3.5), and $VET_{n,i}$ is the voltage measured by each temperature sensor.

The next step is to apply a linearization algorithm to the Raw measurements:

$$\vec{E}_n = \begin{bmatrix} E_{n,1} \\ E_{n,2} \\ E_{n,3} \end{bmatrix}, \quad \text{where } E_{n,a} = \left[1 + (k_{n,2,a} + k_{n,3,a} eu_{n,a}) eu_{n,a}\right] eu_{n,a} \tag{6}$$

and where the linearity parameters $k_{n,2,a}$ and $k_{n,3,a}$ were determined by testing measurements against an in-house high-performance voltmeter.

Finally, linear calibration applies the orthogonality matrix, temperature dependent offset vector, and sensitivity matrix to determine the applied magnetic field vector in nT in the orthogonalized sensor measurement frame coordinates (x, y, z). The measurements are calibrated using

$$\vec{B}_{\text{ppr},n} = \boldsymbol{V_n} \boldsymbol{S_n} (\vec{E}_n - \vec{O}_n) \tag{7}$$

The offset (units of eu) is:

$$\vec{O}_n = \begin{bmatrix} O_{n,1} \\ O_{n,2} \\ O_{n,3} \end{bmatrix},$$

$$\text{where } O_{n,a} = O_{n,0,a} + O_{n,\text{CDC},a} T_{n,\text{CDC}} + \Delta O_{n,\text{CDC},a}(T_{n,\text{CDC}}) + O_{n,\text{EU},a} T_{n,\text{EU}} + O_{n,t,a} t_L$$

and where $T_{n,\text{CDC}}$ is the CDC temperature for SU1 or SU2, $T_{n,\text{EU}}$ is the temperature of EU1 or EU2, t_L is the time in years since the launch of the spacecraft, $O_{n,0,a}$, $O_{n,\text{CDC},a}$, $O_{n,\text{EU},a}$ and $O_{n,t,a}$ are calibration coefficients determined prior to launch (Sect. 3.5) and $\Delta O_{n,\text{CDC},a}(T_{n,\text{CDC}})$ is a look-up table for correcting minor deviations from the temperature-dependent calibration also determined prior to launch (Sect. 3.5). The scale factor matrix $\boldsymbol{S_n}$ (units of $\text{nT}\,\text{eu}^{-1}$) is:

$$\boldsymbol{S_n} = \begin{bmatrix} S_{n,1} & 0 & 0 \\ 0 & S_{n,2} & 0 \\ 0 & 0 & S_{n,3} \end{bmatrix}, \quad \text{where } S_{n,a} = S_{n,0,a} + S_{n,\text{CSC},a} T_{n,\text{CSC}} + S_{n,\text{EU},a} T_{n,\text{EU}} + S_{n,t,a} t_L$$

Table 7 Flight system field frequency ranges and expected mitigation approaches

Frequency range, f (Hz)	Mitigation technique
$< 10^{-5}$	solar wind monitoring (when not in magnetosphere) and time-domain gradiometry (when in magnetosphere)
10^{-5} to 0.1	solar array look-up tables
> 0.1	gradiometry using frequency-domain filtering
Step changes	time-domain gradiometry

Note: The first column lists the frequency range of pure tones (first three rows) and step changes in the DC field (fourth row) and the second column lists the associated mitigation technique for suppressing the flight system field

for calibration coefficients $S_{n,0,a}$, $S_{n,\mathrm{CSC},a}$, $S_{n,\mathrm{EU},a}$, and $S_{n,t,a}$ determined prior to launch (Sect. 3.5). The orthogonalization matrix

$$\boldsymbol{V}_n = \begin{bmatrix} 1 & -\sin u_{n,1} & -\sin u_{n,2} \\ 0 & \cos u_{n,1} & -\sin u_{n,3} \\ 0 & 0 & \sqrt{1 - \sin^2 u_{n,2} - \sin^2 u_{n,3}} \end{bmatrix}$$

where the angles, $u_{n,1}$, $u_{n,2}$, and $u_{n,3}$ are determined on the ground during instrument testing (Sect. 3.5). While the offsets and sensitivity matrix values are dependent on SU temperatures, EU temperatures and time, the non-orthogonality is constant over the operational temperature range and the lifetime of the mission.

4.4 Calibrated Data

The Calibrated data processing begins by rotating the Partially Processed data from each SU frame into the common FS frame:

$$\vec{B}_{\mathrm{cal,FS},n} = \boldsymbol{R}_n \vec{B}_{\mathrm{cal},n} \tag{8}$$

for rotation matrices, $\boldsymbol{R}_n$.

The rotated data can then be processed to suppress flight system fields. Several such methods are under development. We consider pure tones in the three lowest frequency bins described in Sect. 3.6.3. Additionally, we separately consider broad-spectrum flight system signals associated with step changes in the quasi-static field to a new DC value. These four categories are defined based on the methods available to identify and remove their field signatures (Table 7). As described next, the suppression of signals in each category is associated with a separate processing step. These processing steps can be conducted in any order; the following description arbitrarily follows the order starting with DC fields, to 10^{-5} to 0.1 Hz fields, to to >0.1 Hz fields, and finally to step changes.

Estimated flight system DC fields and instrument offsets remaining after the production of Partially Processed data, $\vec{B}_{\mathrm{zl},n}$, will be suppressed to removed the zero-level corrected field:

$$\vec{B}_{\mathrm{cal,zl},n} = \vec{B}_{\mathrm{cal,FS},n} - \vec{B}_{\mathrm{zl},n}$$

We estimate $\vec{B}_{\mathrm{zl},n}$ in two different ways. When the spacecraft is immersed in the solar wind, we will use estimates of the flight system DC field at each SU derived from monitoring

Alfvénic fluctuations of the IMF (termed solar wind monitoring). These DC field estimates will be periodically derived from regular calibration activities during cruise (Sect. 5.3). Second, if and when the spacecraft enters a magnetosphere around (16) Psyche, the spacecraft would then be shielded from the solar wind (e.g., Fig. 7E, F) and solar wind monitoring would then no longer be feasible. In this case, we will estimate the flight system DC field using time-domain gradiometry. The last available zero-level field level calculations for each instrument will be used as an estimate for the spacecraft field at the locations of the SUs. These values can either be removed directly from the measurements, or, as discussed in Sect. 3.3, a coupling matrix $\boldsymbol{\alpha}$ associated with this estimate can then be constructed and the ambient field estimate using Eq. (3).

The dominant noise source in the 10^{-5} to 0.1 Hz range is expected to be from slow rotation of the solar panels, which generate magnetic fields mainly due to DC photoelectric currents. Therefore, as described in Sect. 5.4, another calibration activity will be to generate tables estimating contributions to $\vec{B}_{\mathrm{FS}}$ at each SU from the solar panels as a function of their orientation, $\vec{B}_{\mathrm{SA},n}(\theta_S)$, where θ_S is the angle of the solar array normal with respect to the spacecraft bus. These will be based on data collected during spacecraft rolls. The solar array-generated field values at each SU will be subtracted from measurements at each sensor:

$$\vec{B}_{\mathrm{cal,zl,SA},n} = \vec{B}_{\mathrm{cal,zl},n} - \vec{B}_{\mathrm{SA},n}(\theta_S).$$

Estimated flight system >0.1 Hz fields will be identified and removed using frequency-domain filtering of gradiometry measurements (Ream et al. 2021). In particular, intervals containing flight system >0.1 Hz fields can be identified when a simultaneous change occurs in the rolling maximum and minimum (upper and lower envelope) of the differenced field, $\Delta\vec{B}$. For each such interval containing >0.1 Hz field noise, a power spectrum of the differenced field is calculated. This enables identification of flight system-generated spectral peaks by selecting frequencies with power in the differenced field Fast Fourier Transform (FFT) above a specified threshold (e.g., 98th percentile of amplitude values). These contributions can then be suppressed by reducing the power in the identified frequencies to the noise level (Ream et al. 2021). Because the differenced field does not contain ambient field fluctuations, this filtering will not affect the spectral peaks of the time-variable ambient fields. An inverse FFT is then used to recreate a cleaned time series for each sensor. This yields:

$$\vec{B}_{\mathrm{cal,zl,SA,FDF},n} = f_{\mathrm{FDF}}(\vec{B}_{\mathrm{cal,zl,SA},n})$$

Step changes in the flight system fields are expected when components that generate dominantly DC fields (e.g., heaters) turn on and off and when subsystems switch between different modes (e.g., the Multispectral Imager, high gain antenna, and power distribution assembly). These changes will be identified in two ways. First, pattern-recognition algorithms can identify rapid field changes in time series of the differenced field like those expected for near-instantaneous changes in the power state of the flight system. The algorithm currently implemented in the data pipeline uses an 0.1 s running average of the differenced field $\Delta\vec{B}$ and checks for changes in the field greater than twice the instrument noise. Coupling coefficients can be derived from simultaneously observed field changes at each sensor. Second, time-tagged spacecraft ancillary data can be used to independently identify step changes from known spacecraft subsystems. In this case, coupling coefficients can be derived during initial checkout calibration activities (Sect. 5.2). When a step change is identified it is removed from the applicable interval for each instrument:

$$\vec{B}_{\mathrm{cal,zl,SA,FDF,SC},n} = f_{\mathrm{SC}}(\vec{B}_{\mathrm{cal,zl,SA,FDF},n})$$

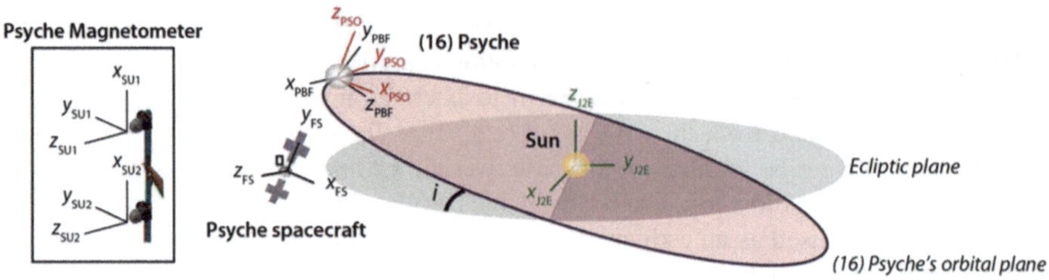

Fig. 12 Coordinate frames used for creating Derived data products. The pink ellipse signifies the orientation of (16) Psyche's orbital plane with respect to the ecliptic plane (gray shaded ellipse). The Sun (yellow sphere) is the center of the J2E frame (green arrows). The PSO and PBF frames (red and black arrows, respectively) are centered on the body (gray sphere) and are translated and rotated with respect to the inertial frame together as the body moves in its orbit. The PSO and PBF frames rotate with respect to each other due to both the body's spin and the motion of the body with respect to the Sun. The Psyche flight system (Fig. 4) defines the FS frame and the individual SUs define the SU1 and SU2 frames (Figs. 4, 5, 9)

The step removal and peak suppression algorithms can be iterated to further clean the data. After these calibration steps are complete—rotation into FS coordinates and suppression of flight system fields and residual instrument offsets—and the final differenced field is determined to be less than then the acceptable noise threshold, the final result is calibrated, cleaned, science-grade data, $\vec{B}_{\mathrm{cal}}$.

4.5 Derived Data

The Derived data are magnetic field measurements rotated from the FS frame into coordinate frames oriented relative to (16) Psyche's orbit, a frame rotating with (16) Psyche, and the Sun (Fig. 12):

$$\vec{B}_{\mathrm{der},q} = \boldsymbol{R}_q \vec{B}_{\mathrm{cal}} \tag{9}$$

where $q =$ "J2E", "PBF", or "PSO". We discuss each of these frames below.

The J2E frame is the same as the J2000 Earth mean orbit inertial frame but centered at (16) Psyche's instantaneous location:

- The x-axis is the intersection of Earth's equatorial plane and the ecliptic plane as it was oriented at the epoch J2000, at which time it pointed to the first point of Aries.
- The z-axis points in the direction perpendicular to Earth's orbital plane.
- The y-axis completes a right-handed system.

The PBF frame is a non-inertial frame centered at the center of mass of Psyche and rotating with the body:

- The z-axis is aligned with Psyche's rotation axis.
- The x-axis connects the origin with the body's prime meridian (currently defined in Shepard et al. 2021), crossing the surface at the body's equator.
- The y-axis completes a right-handed system.

The PSO frame is an inertial frame centered at the center of mass of Psyche

- The x-axis points to Sun.
- The z-axis is perpendicular to the orbital plane of the asteroid.
- The y-axis completes a right-handed system.

The spacecraft ephemerides, as well as the ephemerides of solar system bodies are provided by the Navigation and Ancillary Information Facility (NAIF) at JPL in the J2E, PBF, and PSO frames. The data pipeline uses the SPICE toolkit to extract the information in both frames for the spacecraft trajectory and performs conversions of the magnetic field from the spacecraft frame (NAIF name "PSYC") to the other frames as needed for different products.

The PBF frame will be used for mapping the magnetic field with respect to geographical location on the body's surface, while the PSO frame will be used to map the field with respect to the magnetosphere, which rotates both with the body and with the direction of the incoming solar wind flow. All conversions between the frames are time-dependent and rely on knowledge of the ephemerides of Earth, (16) Psyche, the Sun, and the spacecraft, as well as the spacecraft attitude, as provide by NAIF.

4.6 Remanent Field Data

4.6.1 Overview of Remanent Field Data

The purpose of high-level data processing is to determine whether Psyche contains remanent magnetization consistent with formation by a dynamo paleofield, and if so, to constrain the strength and pattern of this magnetization. This will yield at least one and up to three different data products. First, we will constrain the asteroid's dipole moment and, in particular, establish whether it exceeds the threshold specified by the level 1 requirement that would indicate Psyche has magnetization like that known for iron meteorites at spatial scales of 40 km or greater. If a remanent dipole moment is detected, then depending on the ratio of the remanent field to the noise and on how spatially complete are the measurements, we will optionally create a magnetic anomaly map and possibly even estimate higher-order multipole field terms.

4.6.2 Dipole Moment

Given that the flight system magnetic field is suppressed during the production of $\vec{B}_{\mathrm{cal}}$ data, the $\vec{B}_{\mathrm{der}}$ data are expected to be dominated by contributions with sources external to the spacecraft (from hereon we drop the q subscript that denotes the frame of the data). This flight system-cleaned field can be as the sum of a field internal to a radial distance $r_P = r_0$ from (16) Psyche's center-of-mass that encloses the entire asteroid, $\vec{B}_{\mathrm{der,i}}$, and a field generated external to this distance, $\vec{B}_{\mathrm{der,e}}$:

$$\vec{B}_{\mathrm{der}}(r_0, \theta, \phi) = \vec{B}_{\mathrm{der,i}} + \vec{B}_{\mathrm{der,e}}$$

for Psyche-centric colatitude, θ_P, and longitude, ϕ_P. The main contributions to $\vec{B}_{\mathrm{der,i}}$ are the asteroid's remanent field and fields due to currents induced inside the body by the changing IMF. The main contributions to $\vec{B}_{\mathrm{der,e}}$ are the IMF and fields due to current systems created by the interaction of the magnetized body with the solar wind.

The dipole moment of the asteroid will be estimated from $\vec{B}_{\mathrm{der}}$ data. Our hybrid simulations (Fig. 7) indicate that if $M_{\mathrm{P}} \gtrsim 1.5 \times 10^{14}$ Am^2 or $\lesssim 2 \times 10^{13}$ Am^2, then Orbit D should lie entirely inside and outside the asteroid's magnetosphere, respectively. For dipole moments between these two limits, we expect that the spacecraft will pass in and out of the magnetosphere during Orbit D. Equivalent but higher limiting moment values can be assigned for Orbits A, B, and C and the transfer orbits. Each of these three cases will be analyzed to obtain a moment estimate using a distinct approach.

Similarly, we can obtain a constraint on Psyche's dipole moment by recasting [Eq. (1)] in terms of the subsolar magnetopause stand-off distance, R_{MP}, at which the magnetic pressure due to the body is balanced by the solar wind pressure:

$$M_{\mathrm{P}} \approx 4\pi\, v_{\mathrm{sw}} R_{\mathrm{MP}}^{3} \sqrt{\frac{2\rho_{\mathrm{sw}}}{F\mu_0}} \tag{10}$$

This relation, as well as the aforementioned hybrid simulations, requires knowledge of the solar wind speed. Because the Psyche spacecraft does not carry a solar wind particle instrument, we can use solar wind velocity and IMF estimates from semi-empirical and physics-based solar wind propagation models that are constrained by measurements taken at 1 AU (Merkin et al. 2016; Odstrcil 2003; Oran et al. 2013).

The simplest situation will be if the spacecraft never passes through a magnetopause boundary even during Orbit D, such that $\vec{B}_{\mathrm{der}}$ is dominated by $\vec{B}_{\mathrm{der,e}}$. In this case, the analytical relationship [Eq. (10)] and our hybrid simulations (Fig. 7) indicate that $M_{\mathrm{P}} \lesssim 10^{13}\ \mathrm{Am}^2$.

Due to the variability of the solar wind and the body's rotation, we expect the magnetospheric topology to significantly change over time. While large planetary magnetospheres have extended regions with a quasi-fixed topology, a small magnetosphere of a quickly rotating body would continuously change its configuration as it reconnects to the external fields. Because of the periodicity of the rotation and the fact that similar wind condition would repeat during the mission, we will collect data from various times and bin them as a function of spacecraft locations in the PBF and PSO frames. This should enable us to construct several separate mean configurations for further study. For orbits that pass into and out of the magnetosphere, we will use the binned data to map the magnetopause surface. Magnetopause crossings can be identified from observing abrupt changes in the angle of $\vec{B}_{\mathrm{der}}$ (e.g., by tens of degrees), increases in B_{der} (e.g., by factors of several or more), and/or changes in the several Hz frequency variations of B_{der} (e.g., by up 1–2 orders of magnitude) (Winslow et al. 2013). These indicators can be collected and used to determine the magnetopause boundary. These indicators can be collected and used to determine the magnetopause boundary to estimate M_{P} from Eq. (10).

Orbits for which the spacecraft spends a large fraction of the orbit inside the magnetosphere will yield $\vec{B}_{\mathrm{der}}$ that may be dominated by $\vec{B}_{\mathrm{der,i}}$. Such measurements will enable a first order estimate of dipole moment magnitude by fitting the data to a simple dipole approximation analogous to Eq. (4):

$$\vec{B}_{\mathrm{der}}(\vec{r}_p) = \frac{\mu_0}{4\pi}\left[\frac{3\vec{r}(\vec{M}_{\mathrm{P}}\cdot\vec{r}_p)}{r_P^5} - \frac{\vec{M}_{\mathrm{P}}}{r_P^3}\right] \tag{11}$$

Although the magnetosphere is compressed by the wind and deviates from a pure dipole, this effect is most prominent at the outer regions of the magnetosphere, while closer to the body the field becomes increasingly dipolar (e.g., within 2 R_{P} for body with dipole moment $> 10^{15}\ \mathrm{Am}^2$; Fig. 7F).

We can use the above estimates of the moment to model the solar wind interaction with the asteroid using hybrid simulations (e.g., Fig. 7) to obtain a measure of how the wind distorts the field away from the dipole field in Eq. (11). Such a simulation will solve for the magnetospheric currents and induced currents self-consistently. We can then subtract the simulated external field from the data, and thus remove the contribution of the Sun-tied, time dependent external field to yield a map of Psyche's internal field,

$$\vec{B}'_{\mathrm{der,i}} \approx \vec{B}_{\mathrm{der}} - \vec{B}'_{\mathrm{der,e}}$$

This procedure would be done iteratively to yield a more accurate estimate of $\vec{M}_{\mathrm{P}}$.

4.6.3 Magnetic Anomaly Map and Higher-Order Gauss Coefficients

Although only an estimate of the dipole moment is required for the level 1 requirement, spatial maps of the remanent magnetic field of Psyche (i.e., a magnetic anomaly map) and associated higher order Gauss coefficients would provide bonus science. As discussed below, the Gauss coefficient power spectrum could constrain the depth of the magnetization and its horizontal spatial scale. Combined with anomaly maps, the magnetic power spectrum could enable a correlation between the NRM and possible geologic terranes (e.g., craters, volcanic edifices, tectonic features, and/or endogenous and exogenous surface materials with varying compositions and mineralogies). This could in turn constrain the age and source of the NRM (internal versus external fields) and in turn constrain the thermal and magnetic evolution of the asteroid. For example, the demonstration that Psyche's magnetization was unidirectional or not could be used to distinguish between external field sources like the nebular field and internal field sources like a dynamo, assuming scrambling of the magnetization by impact-induced block rotation can be excluded (Sect. 1.2).

Obtaining a magnetic anomaly map would require that Psyche's dipole moment was sufficiently large that the spacecraft spends a large fraction of time in a magnetosphere (e.g., at or below 170 km for a Psyche moment at the level 1 value). In this case, a map could be obtained of $\vec{B}'_{\mathrm{der,i}}$ as a function of radial coordinates r, θ_P and ϕ_P. This map could be downward- or upward-continued to a fixed altitude r_0 to yield a magnetic anomaly map,

$$\vec{B}_{\mathrm{rf,anom}}(r_0, \theta_P, \phi_P) = \vec{B}'_{\mathrm{der,i}}(r_0, \theta_P, \phi_P).$$

If this map has sufficiently complete coverage in θ_P and ϕ_P, we can use least squares fitting to obtain the higher order Gauss coefficients:

$$\vec{B}_{\mathrm{rf,anom}}(r_0, \theta_P, \phi_P) = -\nabla \sum_{l=1}^{N_\mathrm{i}} \sum_{m=0}^{l} r_0 \left(\frac{r_0}{r_P}\right)^{l+1} P_l^m(\cos\theta_P)\left(g_l^{m,\mathrm{i}} \cos m\phi_P + h_l^{m,\mathrm{i}} \sin m\phi_P\right) \tag{12}$$

where l and m are the degree and order (integers), $g_l^{m,\mathrm{i}}$ and $h_l^{m,\mathrm{i}}$ are the internal Gauss coefficients, N_i is an integer specifying the highest orders in the internal and external fields, P_l^m are the Schmidt semi-normalized associated Legendre polynomials (Blakely 1996). Note that the dipole moment estimate (Sect. 4.6.2) can be expressed as:

$$M_\mathrm{P} = \left(4\pi r_0^3/\mu_0\right)\sqrt{\left(g_1^{0,\mathrm{i}}\right)^2 + \left(h_1^{0,\mathrm{i}}\right)^2 + \left(h_1^{1,\mathrm{i}}\right)^2}$$

As dicussed below, the Gauss coefficients provide information about the angular spatial scale and depth of the magnetization distribution with the body.

The power of the magnetic field is defined as the mean-squared average of the magnetic-field intensity at a given radius r:

$$R_l = \left(\frac{a}{r}\right)^{2l+4} (l+1) \sum_{m=0}^{l} \sqrt{\left(g_l^{m,i}\right)^2 + \left(h_l^{m,i}\right)^2} \tag{13}$$

This quantity, and how it varies as a function of l, can be computed easily when the Gauss coefficients are known (Lowes 1966). To assess how the strength of the magnetic field might vary beyond the simple dipole term, we make use of a stochastic magnetization model (Voorhies et al. 2002; Wieczorek 2018). The model considered here consists of an ensemble of magnetized thin spherical caps that are confined to lie between two depths below the 111 km mean radius of the body. Each spherical cap is set to have the same angular radius and is assumed to possess a random magnetization intensity, a random direction of magnetization and a random location. After averaging over the random variables, a simple analytic equation can be shown to exist for the power spectrum.

The number of spherical caps and their mean-squared magnetization are part of a factor that multiplies the theoretical power spectra and that does not otherwise affect the power spectrum shape (Wieczorek 2018). Furthermore, Wieczorek (2018) demonstrated that the depth of the bottom of the magnetized zone, d_b, has only a small impact on the spectra, especially at the low spherical-harmonic degrees that will be resolved by the Psyche mission. Wieczorek (2018) also demonstrated that different assumptions about the directions of magnetization would affect the magnitude of the power spectrum by no more than a factor of about two. The two parameters that have the largest effect on the shape of the spectra are the radii of the magnetized disks, θ_0, and the depth to the top of the magnetized region, d_t.

To understand the dependence of the Gauss coefficients on magnetization depth and spatial scale, we calculated three sets of magnetic power spectra. First, we set the angular radius of the spherical caps to 1° and varied the depth to the top of the magnetized region from the surface to 60 km (Fig. 13A). We fixed the thickness of the magnetized zone to be equal to the 40-km coherence scale discussed in Sect. 1.3. For spherical-harmonic degrees near 10, the magnitudes of the spectra vary by about 6 orders of magnitude. Next, we placed the top of the magnetized zone at the surface and the bottom depth at 40 km, and calculated spectra where the angular radii of the spherical caps vary from 0.1 to 20° (corresponding to diameters of about 0.5 to 80 km at the surface) (Fig. 13B). Near degree 10, this parameter is seen to affect the magnitude of the power spectra by about 2 orders of magnitude. In particular, these models show that the maximum power occurs near degrees 3–4 and that the power for these degrees can be 4–5 times larger than the corresponding power of the dipole term. Finally, we again placed the top of the magnetized zone at the surface and varied the bottom depth from 10 to 80 km (Fig. 13C). The latter affirms the lack of sensitivity of the spectra to bottom depth previously observed by Wieczorek (2018).

These example spectra show that if the Gauss coefficients of Psyche can be determined for degrees beyond the dipole term, that it should be possible to constrain the size and depth of the magnetic sources. We emphasize that the theoretical power spectra of Wieczorek (2018) were developed for planetary bodies that could be approximated as being spherical. Although we do not expect the nonspherical shape of Psyche to affect the broad conclusions taken from Fig. 13, more complex magnetization models that consider the real shape of Psyche will need to be investigated.

4.7 Archiving

Data products will be archived as daily Common Data Format (CDF) files with Planetary Data Systems (PDS) Small Bodies Node using the PDS4 standards. Data products at the Raw, Partially Processed and Calibrated levels will be archived continuously throughout the mission as the data are received from the spacecraft and processed through the Magnetometer data pipeline. Data products up to the Calibrated data level will have a measurement cadence of $f_s = 50$ Hz. The Derived products will be archived continuously during orbital operations with a measurement cadence of 1 Hz and the Remanent Field products will be archived at the end of the primary mission.

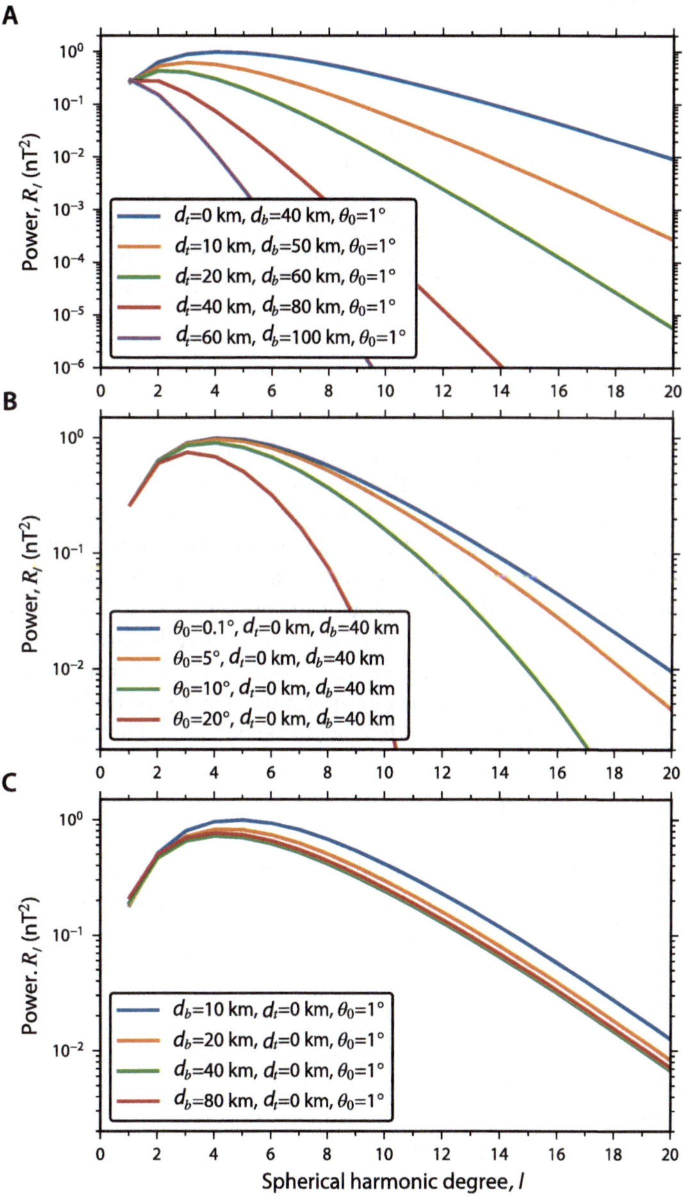

Fig. 13 Magnetic field power spectra predicted by a stochastic magnetization model. The magnetization consists of thin spherical caps that have random magnetization directions and intensities. The magnetized spherical caps are defined by their angular radius, θ_0, and reside within an annulus that is defined by the depths to the top and bottom of the magnetized zone, d_t and d_b, respectively. The power spectra are computed at a radius of 145 km and it is assumed that Psyche can be approximated by a sphere with a mean radius of 113 km. (**A**) Dependence of the power spectra on the depth to the top of the magnetized region beneath the surface, which varies from 0 to 60 km. (**B**) Dependence of the power spectra on the angular radius of the individual magnetized spherical caps, which varies from 0.1 to 20°. (**C**) Dependence of the power spectra on the depth to the bottom of the magnetized region beneath the surface, which varies from 10 to 80 km

5 In-Flight Calibration

5.1 Overview

The pre-flight calibration campaign focused on determining the scale factors, orthogonality and offsets, of the SUs and EUs (Sect. 3.5). However, because this campaign was conducted before the Magnetometer was mounted on the spacecraft did not quantify noise from the flight system. To address the latter, there will be three additional in-flight campaigns: an initial checkout campaign, solar wind monitoring, and solar panel rotation. These will give regular estimates of the offsets and flight system field during cruise and while in orbit around the asteroid.

5.2 Initial Checkout

During initial checkout activities occurring just after launch, the Magnetometer will be turned on and passively measure the flight system magnetic field as other subsystems are progressively turned on and cycled through their various modes and power states. The data collected during initial checkout will be used to build calibration tables that include both the expected field contribution from each subsystem as well as coupling coefficients based on the changes in the observed field resulting from the subsystems. These tables will later be updated using information collected throughout cruise and during a flyby of Mars. During orbital operations, the calibration table will be used along with ancillary data indicating what subsystems are being commanded to inform the removal of flight system generated fields using gradiometry and frequency filtering of the differenced field (Sect. 4.4).

5.3 Solar Wind Monitoring

In addition to the passive monitoring of the flight system components, there will be a dedicated solar wind monitoring interval during initial checkout for determination of the zero-level field, $\vec{B}_{\mathrm{zl}}$. Correcting for the zero-level field is part of the Calibrated data processing step (Sect. 4.4). Solar wind monitoring takes advantage of the Alfvénic nature of the IMF (Belcher 1973; Davis and Smith 1968) for which the ratio of the magnitude of the compressional fluctuations to the transverse fluctuations is low. This means that when there are changes in the direction of the IMF, its magnitude stays relatively constant. However, when a non-rotating spacecraft field is added to the IMF, the magnitude of the total observed field will change. The zero levels, which include both the flight system DC field and the residual instrument zero level offsets, are given by the corrections to the magnetic field components that keep the magnitude of the field fixed during the observed IMF rotations (Leinweber et al. 2008).

In particular, time sequences of the ambient field can be used to estimate the zero-level field for each SU-EU pair as those values that minimize the variance of the squared magnitude of the measured field $|\vec{B}_{\mathrm{cal,SA,FDF,SC},n}|^2$ where $\vec{B}_{\mathrm{cal,SA,FDF,SC},n}$ data have been already cleaned of 10^{-5} to 0.1 Hz, >0.1 Hz, and step changes (note that other intermediate Calibrated data products like $\vec{B}_{\mathrm{cal,FDF,SC},n}$ could be used instead here) (Leinweber et al. 2008). This is equivalent to solving the system of linear equations for $\vec{B}_{\mathrm{zl}}$:

$$\boldsymbol{A}_n \vec{B}_{\mathrm{zl},n} = \vec{b}_n \tag{14}$$

Here, the covariance matrix

$$A_n = \begin{bmatrix} \langle B_x^2\rangle - \langle B_x\rangle^2 & \langle B_x B_y\rangle - \langle B_x\rangle\langle B_y\rangle & \langle B_x B_z\rangle - \langle B_x\rangle\langle B_z\rangle \\ \langle B_y B_x\rangle - \langle B_y\rangle\langle B_x\rangle & \langle B_y^2\rangle - \langle B_y\rangle^2 & \langle B_y B_z\rangle - \langle B_y\rangle\langle B_z\rangle \\ \langle B_z B_x\rangle - \langle B_z\rangle\langle B_x\rangle & \langle B_z B_y\rangle - \langle B_z\rangle\langle B_y\rangle & \langle B_z^2\rangle - \langle B_z\rangle^2 \end{bmatrix}$$

where B_x, B_y, B_z are the Cartesian components of $\vec{B}_{\mathrm{cal,SA,FDF,SC},n}$, the angle brackets denote temporal means and

$$\vec{b} \equiv \frac{1}{2}\begin{bmatrix} \langle B_x|\vec{B}_{\mathrm{cal,SA,FDF,SC},n}|^2\rangle - \langle B_x\rangle\langle|\vec{B}_{\mathrm{cal,SA,FDF,SC},n}|^2\rangle \\ \langle B_y|\vec{B}_{\mathrm{cal,SA,FDF,SC},n}|^2\rangle - \langle B_y\rangle\langle|\vec{B}_{\mathrm{cal,SA,FDF,SC},n}|^2\rangle \\ \langle B_z|\vec{B}_{\mathrm{cal,SA,FDF,SC},n}|^2\rangle - \langle B_z\rangle\langle|\vec{B}_{\mathrm{cal,SA,FDF,SC},n}|^2\rangle \end{bmatrix}$$

and where $|\vec{B}_{\mathrm{cal,SA,FDF,SC},n}|^2 = B_x^2 + B_y^2 + B_z^2$.

5.4 Solar Panel Rotation

The solar panels are expected to be the main subsystem contributing to the flight system field in the frequency range 10^{-5}–0.1 Hz due to their rotation about the spacecraft y axis as they track the sun. This is a critical frequency range because it also contains the expected frequencies that will be observed at the asteroid as the spacecraft observes the remanent field while in orbit. An additional complication is that unlike most other flight system components with substantial magnetic fields, the solar arrays are larger in size than their minimum distance to SU2, such that their fields at SU2 will be highly nondipolar and therefore not easily removed with single dipole-fitting gradiometry techniques (Sect. 4.4). Furthermore, the solar arrays rotate about the FS y axis, such that the fields at the SUs will depend on the solar array drive assembly (SADA) angle, θ_S. Note that the solar panels are expected to always remain within 5° of the Sun and to maintain an approximately (within 10%) constant current state during the full mission, such that the main source of field variation at the Magnetometer should be changes in the relative orientation of the SADA relative to the spacecraft bus rather than changes in currents.

We will therefore characterize the solar array-generated fields at each SU as a function of θ_S. This will be accomplished by performing a series of calibration rolls where the spacecraft is rotated along the solar array axis for a duration of 8 hours over a total of 8 full 360° rotations while keeping the solar arrays facing the sun. This activity will be repeated periodically throughout cruise and approach. The measurements collected by the Magnetometer during the rolls, $\vec{B}_{\mathrm{cal,FDF,SC},n}$, will be used to compile a look-up table of the observed solar array field at each SU as a function of θ_S. The value of θ_S at the start of the rolls is also recorded since the zero-level calibration depends on this. This table will then be used to subtract the solar array field from the Magnetometer data as part of producing Calibrated data from Partially Processed data (Sect. 4.4). These calibration rolls will be repeated periodically throughout cruise, during the Mars gravity assist and during approach to Psyche.

6 Summary

The Psyche Magnetometry Investigation is designed to search for a remanent field around asteroid (16) Psyche that would provide evidence of past dynamo action. Evidence for a core dynamo would demonstrate that the asteroid was partially or fully differentiated rather than being an unmelted accretionary aggregate, thereby addressing one of the key mission science objectives. Using our knowledge of the age and intensity of NRM in meteorites and a nascent but growing understanding of dynamo generation on planetesimals, we conclude that a demonstration that Psyche has a moment of at least $2 \times 14\ \mathrm{Am}^2$, which would produce a field of up to ~ 6 nT at the lowest planned mapping altitude (75 km), would be consistent with a past core dynamo. However, given the wide range of NRM intensities in iron meteorites, the field at this altitude might range up $\sim 10{,}000$ nT, which would make Psyche one of the most magnetic objects encountered by a spacecraft. As such, a key challenge was to design a Magnetometry Investigation with sufficient Reconstructed Measurement Uncertainty for detecting the minimum dipole moment while also having a high dynamic range to avoid saturation of the most strongly magnetized conceivable possibilities. The Psyche Magnetometry Investigation achieves this using high-heritage dual three-axis fluxgate sensors derived from the VFM on the Swarm mission, gradiometry combined with a stringent magnetic cleanliness program for the flight system, a series of pre-flight and in-flight calibration activities, and a ground data processing program that applies these calibrations, suppresses flight system fields, and constructs magnetic moment estimates from modeling of magnetic field data.

Acknowledgements This work is supported by NASA contract NNM16AA09, "Psyche: Journey to a Metal World." We thank the many engineers and scientists at JPL and UCLA who helped us develop this concept over the last decade including Kathryn Rowe, Steven Joy, Ryan Caron, Ryan Seaton, and Edwin Schauble. We also thank Brenda Carbone and Kerin Willis at MIT for administrative support. The research was carried out at the Jet Propulsion Laboratory, California Institute of Technology, under a contract with the National Aeronautics and Space Administration (80NM0018D0004).

Declarations

Competing Interests The authors declare no competing interests.

References

Acuña MH, Anderson BJ, Russell CT, Wasilewski P, Kletetschka G, Zanetti L, Omidi N (2002) NEAR magnetic field observations at 433 Eros: first measurements from the surface of an asteroid. Icarus 155:220–228

Aharonson O, Zuber MT, Solomon SC (2004) Crustal remanence in an internally magnetized non-uniform shell: a possible source for Mercury's magnetic field? Earth Planet Sci Lett 218:261–268

Anand A, Carroll-Nellenback J, Blackman EG, Tarduno JA (2022) Asteroid magnetization from the early solar wind. Mon Not R Astron Soc 509:2957–2968

Auster HU, Richter I, Glassmeier K-H, Berghofer G, Carr CM, Motschmann U (2010) Magnetic field investigations during ROSETTA's 2867 Šteins flyby. Planet Space Sci 58:1124–1128

Baumgartel K, Sauer K, Bogdanov A (1994) A magnetohydrodynamic model of solar wind interaction with asteroid Gaspra. Science 263:653

Baumgartel K, Sauer K, Story TR, McKenzie JF (1997) Solar wind response to a magnetized asteroid: linear theory. Icarus 129:94–105

Belcher JW (1973) A variation of the Davis-Smith method for in-flight determination of spacecraft magnetic fields. J Geophys Res 78:6480–6490

Blakely RJ (1996) Potential theory in gravity and magnetic applications. Cambridge University Press, New York

Blanco-Cano X, Omidi N, Russell CT (2003) Hybrid simulations of solar wind interaction with magnetized asteroids: comparison with Galileo observations near Gaspra and Ida. J Geophys Res 108:1216

Blanco-Cano X, Omidi N, Russell CT (2004) How to make a magnetosphere. Astron Geophys 45:3.14–13.17

Bogard DD, Garrison DH, McCoy TJ (2000) Chronology and petrology of silicates from IIE iron meteorites: evidence for a complex parent body evolution. Geochim Cosmochim Acta 64:2133–2154

Borlina CS, Weiss BP, Bryson JFJ, Bai X-N, Lima EA, Chatterjee N, Mansbach EN (2021) Paleomagnetic evidence for a disk substructure in the early solar system. Sci Adv 7:eabj6928

Borlina CS, Weiss BP, Bryson JFJ, Armitage P Schulz R (2022) Lifetime of the outer solar system nebula from carbonaceous chondrites. J Geophys Res 127:e2021JE007139

Brauer R, Merayo JMG, Nielsen OV, Primdahl F, Petersen JR (1997) Transverse field effect in fluxgate sensors. Sens Actuators A, Phys 59:70–74

Brauer P, Merayo JMG, Thamm H, Tøfnner-Clausen L, Amann M, Hulot G, Vogel P, Vigneron P, Flobergha-gen R (2018) Magnetic perturbations from thermoelectric currents in Swarm thermal blankets. Geophys Res Abstr 20:EGU2018-14561-14561

Bryson JFJ, Church NS, Kasama T, Harrison RJ (2014a) Nanomagnetic intergrowths in Fe–Ni meteoritic metal: the potential for time-resolved records of planetesimal dynamo fields. Earth Planet Sci Lett 388:237–248

Bryson JFJ, Herrero-Albillos J, Kronast F, Ghidini M, Redfern SAT, van der Laan G, Harrison RJ (2014b) Nanopaleomagnetism of meteoritic Fe–Ni studied using X-ray photoemission electron microscopy. Earth Planet Sci Lett 396:125–133

Bryson JFJ, Nichols CIO, Herrero-Albillos J, Kronast F, Kasama T, Alimadadi H, van der Laan G, Nimmo F, Harrison RJ (2015) Long-lived magnetism from solidification-driven convection on the pallasite parent body. Nature 517:472–475

Bryson JFJ, Weiss BP, Harrison RJ, Herrero-Albillos J (2017) Paleomagnetic evidence for dynamo activity driven by inward crystallisation of a metallic asteroid. Earth Planet Sci Lett 472:152–163

Bryson JFJ, Neufeld JA, Nimmo F (2019a) Constraints on asteroid magnetic field evolution and the radii of meteorite parent bodies from thermal modelling. Earth Planet Sci Lett 521:68–78

Bryson JFJ, Weiss BP, Getzin B, Abrahams JNH, Nimmo F, Scholl A (2019b) Paleomagnetic evidence for a partially differentiated ordinary chondrite parent asteroid. J Geophys Res 124:1880–1898

Carporzen L, Weiss BP, Elkins-Tanton LT, Shuster DL, Ebel DS, Gattacceca J (2011) Magnetic evidence for a partially differentiated carbonaceous chondrite parent body. Proc Natl Acad Sci USA 108:6386–6389

Cheng AF (2002) Near Earth asteroid rendezvous: mission summary. In: Bottke WF, Cellino A, Paolicchi P, Binzel RP (eds) Asteroids III. University of Arizona Press, Tucson, pp 351–366

Cisowski SM (1987) Magnetism of meteorites. In: Jacobs JA (ed) Geomagnetism. Academic Press, London, pp 525–560

Cochrane CJ, Murphy N, Raymond CA, Biersteker JB, Dang K, Jia X, Korth H, Narvaez P, Ream JB, Weiss BP (2023) Magnetic field modeling and visualization of the Europa Clipper spacecraft. Space Sci Rev. In preparation

Cournède C, Gattacceca J, Gounelle M, Rochette P, Weiss BP, Zanda B (2015) An early solar system magnetic field recorded in CM chondrites. Earth Planet Sci Lett 410:62–74

Courville SW, O'Rourke JG, Castillo-Rogez JC, Fu RR, Oran R, Weiss BP, Elkins-Tanton LT (2022) Acquisition and preservation of remanent magnetization in carbonaceous asteroids. Nat Astron 6:1387–1397

Davis L, Smith EJ (1968) The in-flight determination of spacecraft magnetic field zeros. Eos Trans AGU 49:257

de Soria Santacruz-Pich M, Soriano M, Quintero O, Wong F, Hart S, Kokorowski M, Bone B, Solish B, Trofimov D, Bradford E, Raymond CA, Narvaez P, Keys C, Lord P, Ream J, Oran R, Weiss BP, Russell C, Ascrizzi K, Elkins-Tanton L (2020) An approach to magnetic cleanliness for the Psyche mission. In: 2020 IEEE Aerosp. Conf. Proc.

Dibb SD, Bell JF III, Garvie LAJ (2022) Spectral reflectance variations of aubrites, metal-rich meteorites, and sulfides: implications for exploration of (16) Psyche and other "spectrally featureless" asteroids. Meteorit Planet Sci 57:1570

Dunlop DJ, Özdemir Ö (1997) Rock magnetism: fundamentals and frontiers. Cambridge University Press, Cambridge. 573 pp.

Dunn TJ, Cressey G, McSween HY Jr, McCoy TJ (2010) Analysis of ordinary chondrites using powder X-ray diffraction: 1. Modal mineral abundances. Meteorit Planet Sci 45:123–134

Einsle JF, Eggeman AS, Martineau BH, Saghi Z, Collins SM, Blukis R, Bagot PAJ, Midgley PA, Harrison RJ (2018) Nanomagnetic properties of the meteorite cloudy zone. Proc Natl Acad Sci USA 115:E11436–E11445

Elkins-Tanton LT, Weiss BP, Zuber MT (2011) Chondrites as samples of differentiated planetesimals. Earth Planet Sci Lett 305:1–10

Elkins-Tanton LT, Asphaug E, Bell JF III, Bercovici H, Bills B, Binzel R, Bottke WF, Dibb S, Lawrence DJ, Marchi S, McCoy TJ, Oran R, Park RS, Peplowski PN, Prettyman TH, Russell CT, Schaefer L, Weiss BP, Wieczorek MA, Williams DA, Zuber MT (2020) Observations, meteorites, and models: a pre-flight assessment of the composition and formation of (16) Psyche. J Geophys Res 125:e2019JE006296

Elkins-Tanton LT, Asphaug E, Bell J, Bills B, Bottke W, Courville s, Jun I, Lawrence D, Marchi S, McCoy T, Merayo J, Oran R, O'Rourke J, Park R, Peplowski P, Prettyman T, Raymond C, Weiss B, Wieczorek M, Zuber M (2022) Distinguishing the origin of (16) Psyche. Space Sci Rev 218:17

Ermakov AI, Park RS, Bills BG (2018) Power laws of topography and gravity spectra of the solar system bodies. J Geophys Res 123:2038–2064

European Cooperation for Space Standardization (2003) Ground systems and operations: telemetry and telecommand packet utilization, p 228

Fatemi S, Poppe AR (2018) Solar wind interaction with asteroid 16 Psyche: implication for formation theories. Geophys Res Lett 45:39–48

Fu RR, Weiss BP, Lima EA, Harrison RJ, Bai X-N, Desch SJ, Ebel DS, Suavet C, Wang H, Glenn D, Le Sage D, Kasama T, Walsworth RL, Kuan AT (2014) Solar nebula magnetic fields recorded in the Semarkona meteorite. Science 346:1089–1092

Fu RR, Volk MWR, Bilardello D, Libourel G, Lesur GRJ, Ben Dor O (2021) The fine-scale magnetic history of the Allende meteorite: implications for the structure of the solar nebula. AGU Adv 2:e2021AV000486

Gattacceca J, Boustie M, Lima E, Weiss BP, de Resseguier T, Cuq-Lelandais JP (2010) Unraveling the simultaneous shock magnetization and demagnetization of rocks. Phys Earth Planet Inter 182:42–49

Gattacceca J, Weiss BP, Gounelle M (2016) New constraints on the magnetic history of the CV parent body and the solar nebula from the Kaba meteorite. Earth Planet Sci Lett 455:166–175

Gombosi TI (2009) Physics of the space environment. Cambridge University Press, Cambridge

Gombosi TI, Chen Y, Glocer A, Huang Z, Jia X, Liemhohn MW, Manchester WB, Pulkinnen T, Sachdeva N, Al Shidi Q, Sokolov IV, Szente J, Tenishev V, Toth G, van der Holst B, Welling DT, Zhao L, Zou S (2021) J Space Weather Space Clim 11:42

Greenstadt EW (1971) Conditions for magnetic interaction of asteroids with the solar wind. Icarus 14:374–381

Herčik D, Auster H-U, Constantinescu D, Blum J, Fornaçon K-H, Fujimoto M, Gebauer K, Grundmann J-T, Güttler C, Hillenmaier O, Ho T-M, Hördt A, Krause C, Kürht E, Matsuoka A, Motschmann U, Moussei-Soffys A, Richter I, Sasaki K, Stoll B, Weiss BP, Wolf F, Glassmeier K-H (2020) Magnetic properties of asteroid (162173) Ryugu. J Geophys Res 125:e2019JE006035

Hunt AC, Cook DL, Lichtenberg T, Reger PM, Ek M, Golaberk GJ, Schönbachler M (2018) Late metal–silicate separation on the IAB parent asteroid: constraints from combined W and Pt isotopes and thermal modelling. Earth Planet Sci Lett 482:490–500

Kallio E, Zhang TL, Barabash S, Jarvinen R, Sillanpää I, Janhunen P, Fedorov A, Sauvaud J-A, Mazelle C, Thocaven J-J, Gunell H, Andersson H, Griogoriev A, Brinkfeldt K, Futaana Y, Holmström M, Lundin R, Yamauchi M, Asamura K, Baumjohann W, Lammer H, Coates AJ, Linder DR, Kataria DO, Curtis CC, Hsieh KC, Sandel BR, Grande M, Koskinen HEJ, Säles T, McKenna-Lawlor S, Orsini S, Cerulli-Irelli R, Mura A, Milillo A, Maggi M, Roelof E, Brandt P, Russell CT, Szego K, Winningham JD, Frahm RA, Scherrer JR, Sharber JR, Wurz P, Boschler P (2008) The Venusian induced magnetosphere: a case study of plasma and magnetic field measurements on the Venus Express mission. Planet Space Sci 56:796–801

Kivelson MG, Bargatze LF, Khurana KK, Southwood DJ, Walker RJ, Coleman PJ (1993) Magnetic field signatures near Galileo's closest approach to Gaspra. Science 261:331–334

Leinweber HK, Russell CT, Torkar K, Zhang TL, Angelopoulos V (2008) An advanced approach to finding magnetometer zero levels in the interplanetary magnetic field. Meas Sci Technol 19:055104

Lowes FJ (1966) Mean-square values on sphere of spherical harmonic vector fields. J Geophys Res 71:2179

Mansbach EN, Shah J, Williams W, Maurel C, Bryson JFJ, Weiss BP (2022) Size ranges of magnetic domain states in tetrataenite. Geochem Geophys Geosyst 23:e2022GC010631

Matthes M, Van Orman JA, Kleine T (2020) Closure temperature of the Pd-Ag system and the crystallization and cooling history of IIIAB iron meteorites. Geochim Cosmochim Acta 285:193–206

Maurel C, Weiss BP, Bryson JFJ (2019) Meteorite cloudy zone formation as a quantitative indicator of paleomagnetic field intensities and cooling rates on planetesimals. Earth Planet Sci Lett 513:166–175

Maurel C, Bryson JFJ, Lyons RJ, Ball MR, Chopdekar RV, Scholl A, Ciesla FJ, Bottke WF, Weiss BP (2020) Meteorite evidence for partial differentiation and protracted accretion of planetesimals. Sci Adv 6:eaba1303

Maurel C, Bryson JFJ, Shah J, Chopdekar RV, Elkins-Tanton LT, Raymond CA, Weiss BP (2021) A long-lived planetesimal dynamo powered by core crystallization. Geophys Res Lett 48:e2020GL091917

McDonough WF, Sun S-S (1995) The composition of the Earth. Chem Geol 120:223–253
Mehlem K (1978a) Magnetostatic cleanliness analysis by the multiple dipole modelling method. In: Proc. first spacecraft electromagnetic compatibility seminar, pp 165–179
Mehlem K (1978b) Multiple magnetic dipole modeling and prediction of satellites. IEEE Trans Magn MAG-14:1064–1071
Mehlem K, Wiegand A (2010) Magnetostatic cleanliness of spacecraft. In: Proc. 2010 Asia-Pacific international symposium on electromagnetic compatibility, pp 936–944
Merayo JMG, Brauer R, Primdahl F, Peterson JR, Nielsen OV (2000a) Scalar calibration of vector magnetometers. Meas Sci Technol 11:120–132
Merayo JMG, Primdahl F, Brauer P, Risbo T, Olsen N, Sabaka T (2000b) The orthogonalization of magnetic systems. Sens Actuators A, Phys 89:185–196
Merayo JMG, Brauer P, Primdahl F (2005) Triaxial fluxgate gradiometer of high stability and linearity. Sens Actuators A, Phys 120:71–77
Merayo JMG, Jørgensen JK, Friis-Christensen E, Brauer P, Primdahl F, Jørgensen PS, Allin TH, Denver T (2008) The Swarm magnetometry package. In: Sandau R, Röser H-P, Valenzuela A (eds) Small satellites for Earth observation. Springer, Dordrecht, pp 143–151
Merkin VG, Lionello R, Lyon JG, Linker J, Török T, Downs C (2016) Coupling of coronal and heliospheric magnetohydrodynamic models: solution comparisons and verification. Astrophys J 831:23
Moffatt K, Dormy E (2019) Self-exciting fluid dynamos. Cambridge University Press, Cambridge
Moskovitz NA, Walker RJ (2011) Size of the group IVA iron meteorite core: constraints from the age and composition of Muonionalusta. Earth Planet Sci Lett 308:410–416
Muxworthy AR, Bland PA, Davison TM, Moore J, Collins GS, Ciesla FJ (2017) Evidence for an impact-induced magnetic fabric in Allende, and exogenous alternatives to the core dynamo theory for Allende magnetization. Meteorit Planet Sci 52:2132–2146
Ness NF, Behannon KW, Lepping RP, Schatten KH (1971) Use of two magnetometers for magnetic field measurements on a spacecraft. J Geophys Res 76:3564–3573
Neubauer FM, Schatten KH (1974) Theoretical and observational analysis of spacecraft fields. J Geophys Res 79:1550–1554
Neufeld JA, Bryson JFJ, Nimmo F (2019) The top-down solidification of iron asteroids driving dynamo evolution. J Geophys Res 124:1331–1356
Nichols CIO, Bryson JFJ, Herrero-Albillos J, Kronast F, Nimmo F, Harrison RJ (2016) Pallasite paleomagnetism: quiescence of a core dynamo. Earth Planet Sci Lett 441:103–112
Nichols CIO, Krakow R, Herrero-Albillos J, Kronast F, Northwood-Smith G, Harrison RJ (2018) Microstructural and paleomagnetic insight into the cooling history of the IAB parent body. Geochim Cosmochim Acta 229:1–19
Nichols CIO, Bryson JFJ, Cottrell RD, Fu RR, Harrison RJ, Herrero-Albillos J, Kronast F, Tarduno JA, Weiss BP (2021) A time-resolved paleomagnetic record of main group pallasites: evidence for a large-cored, thin-mantled parent body. J Geophys Res 126:e2021JE006900
Nielsen OV, Gutierrez J, Hernando B, Savage HT (1990) A new amorphous ribbon fluxgate sensor based on torsional-creep-induced anisotropy. IEEE Trans Magn 26:276–280
Nielsen OV, Petersen JR, Primdahl F, Brauer P, Hernando B, Fernández A, Merayo JMG, Ripka P (1995) Development, construction and analysis of the 'Ørsted' fluxgate magnetometer. Meas Sci Technol 6:1099–1115
Nielsen OV, Brauer P, Primdahl F, Risbo T, Jørgensen JK, Boe C, Deverler M, Bauereisen S (1997) A high-precision triaxial fluxgate sensor for space applications: layout and choice of materials. Sens Actuators A, Phys 59:168–176
Nimmo F (2009) Energetics of asteroid dynamos and the role of compositional convection. Geophys Res Lett 36:L10201
O'Brien T, Tarduno JA, Anand A, Smirnov AV, Blackman EG, Carroll-Nellenback J, Krot AN (2020) Arrival and magnetization of carbonaceous chondrites in the asteroid belt before 4562 million years ago. Commun Earth Environ 1:54
Odstrcil D (2003) Modeling 3-D solar wind structure. Adv Space Res 32:497–506
Omidi N, Blanco-Cano X, Russell CT, Karimabadi H, Acuna M (2002) Hybrid simulations of solar wind interaction with magnetized asteroids: general characteristics. J Geophys Res 107:1487
Oran R, van der Holst B, Landi E, Jin M, Sokolov IV, Gombosi TI (2013) A global wave-driven magnetohydrodynamic solar model with a unified treatment of open and closed magnetic field topologies. Astrophys J 778:176
Oran R, Weiss BP, Cohen O (2018) Were chondrites magnetized by the early solar wind? Earth Planet Sci Lett 492:222–231
Oran R, Weiss BP, de Soria Santacruz-Pich M, Jun I, Lawrence DJ, Polanskey C, Ratliff M, Ream JB, Russell CT, Shprits Y, Zuber MT, Elkins-Tanton LT, Raymond CA (2022) Maximum energies of trapped particles around magnetized planets and small bodies. Geophys Res Lett 49:e2021GL097014

Pätzold M, Wennmacher A, Häusler B, Eidel W, Morley T, Thomas N, Anderson JD (2001) Mass and density determinations of 140 Siwa and 4979 Otawara as expected from the Rosetta flybys. Astron Astrophys 370:1122–1127
Pesonen LJ, Terho M, Kukkonen IT (1993) Physical properties of 368 meteorites: implications for meteorite magnetism and planetary geophysics. Proc NIPR Symp Antarct Meteor 6:401–416
Petersen JR, Primdahl F, Hernando B, Fernandez A, Nielsen OV (1992) Ther ring core fluxgate sensor null feed-through signal. Meas Sci Technol 3:1149–1154
Polanskey CA, Elkins-Tanton LT, Bell JF III, Alonge NK, Bairstow S, Bass D, Biswas A, Binzel R, Cisneros E, Han D, Jun I, Klipstein WM, Lawrence DJ, McCoy T, Mastrodemos N, Merayo J, Noble SK, Oh D, Oran R, Park RS, Peplowski P, Prettyman T, Ream J, Russell CT, Seal D, Weise T, Wenkert D, Weiss BP, Zuber M (2023) The Psyche mission overview. Space Sci Rev. in preparation
Primdahl F (1979) The fluxgate magnetometer. J Phys E, Sci Instrum 12:241–253
Primdahl F, Jensen PA (1982) Compact spherical coil for fluxgate magnetometer vector feedback. J Phys E, Sci Instrum 15:221–226
Primdahl F, Ripka P, Peterson JR, Nielsen OV (1991) The sensitivity parameters of the short-circuited fluxgate. Meas Sci Technol 2:1039–1045
Primdahl F, Brauer P, Merayo JMG, Petersen JR, Risbo T (2002) Determining the direction of a geometrical/optical reference axis in the coordinate system of a triaxial magnetometer sensor. Meas Sci Technol 13:2094–2098
Ream JB, Weiss BP, Oran R, Raymond CA, Polanskey CA, Wenkert DD, Elkins-Tanton LT, Hart RA, Russell CT, Merayo JMG (2021) Magnetic gradiometry using frequency-domain filtering. Meas Sci Technol 33:015104
Richter I, Brinza DE, Cassel M, Glassmeier K-H, Kuhnke F, Musmann G, Othmer C, Schwingenschuh K, Tsurutani BT (2001) First direct magnetic field measurements of an asteroidal magnetic field: DS1 at Braille. Geophys Res Lett 28:1913–1916
Richter I, Auster HU, Glassmeier K-H, Koenders C, Carr CM, Motschmann U, Müller J, McKenna-Lawlor S (2012) Magnetic field measurements during the ROSETTA flyby at asteroid (21)Lutetia. Planet Space Sci 66:155–164
Runcorn SK (1975) On the interpretation of lunar magnetism. Phys Earth Planet Inter 10:327–335
Russell CT, Raymond CA, Ammannito E, Buczkowski DL, De Sanctis MC, Hiesinger H, Jaumann R, Konopliv AS, McSween HY, Nathues A, Park RS, Pieters CM, Prettyman TH, McCord TB, McFadden LA, Mottola S, Zuber MT, Joy SP, Polanskey C, Rayman MD, Castillo-Rogez JC, Chi PJ, Combe JP, Ermakov A, Fu RR, Hoffmann M, Jia YD, King SD, Lawrence DJ, Li JY, Marchi S, Preusker F, Roatsch T, Ruesch O, Schenk P, Villarreal MN, Yamashita N (2016) Dawn arrives at Ceres: exploration of a small, volatile-rich world. Science 353:1008–1010
Scheinberg A, Elkins-Tanton LT, Schubert G, Bercovici D (2015) Core solidification and dynamo evolution in a mantle-stripped planetesimal. J Geophys Res 121:2–20
Scheinberg A, Fu RR, Elkins-Tanton LT, Weiss BP, Stanley S (2017) Magnetic fields on asteroids and planetesimals. In: Elkins-Tanton LT, Weiss BP (eds) Planetesimals: early differentiation and consequences for planets. Cambridge University Press, Cambridge, pp 180–203
Scholl A (2002) Applications of photoemission electron microscopy (PEEM) in magnetism research. Curr Opin Solid State Mater Sci 7:59–66
Shah J, Williams W, Almeida TP, Nagy L, Muxworthy AR, Kovács A, Vladez-Grijalva M, Fabian K, Russell S, Genge MJ, Dunin-Borkowski RE (2018) The oldest magnetic record in our solar system identified using nanometric imaging and numerical modeling. Nat Commun 9:1173
Shepard MK, de Kleer K, Cambioni S, Taylor PA, Virkki AK, Rívera-Valentin EG, Rodriguez Sanchez-Vahamonde C, Fernanda Zambrano-Marin L, Magri C, Dunham D, Moore J, Camarca M (2021) Asteroid 16 Psyche: shape, features, and global map. Planet Sci J 2:125
Simon SB, Bagdonat T, Motschmann U, Glassmeier KH (2006) Plasma environment of magnetized asteroids: a 3-D hybrid study. Ann Geophys 24:407–414
Srnka LK (1976) Magnetic dipole moment of a spherical shell with TRM acquired in a field of internal origin. Phys Earth Planet Inter 11:184–190
Stephenson A (1976) Crustal remanence and the magnetic moment of Mercury. Earth Planet Sci Lett 28:454–458
Sugiyama T, Kusano K (2007) Multi-scale plasma simulation by the interlocking of magnetohydrodynamic model and particle-in-cell kinetic model. J Comput Phys 227:1340–1352
Tarduno JA, Cottrell RD, Nimmo F, Hopkins J, Voronov J, Erickson A, Blackman E, Scott ERD, McKinley R (2012) Evidence for a dynamo in the main group pallasite parent body. Science 338:939–942
Uehara M, Gattacceca J, Leroux H, Jacob D, van der Beek CJ (2011) Magnetic microstructures of metal grains in equilibrated ordinary chondrites and implications for paleomagnetism of meteorites. Earth Planet Sci Lett 306:241–252

Uehara M, Gattacceca J, Rochette P, Demory F, Valenzuela EM (2012) Magnetic study of meteorites recovered in the Atacama Desert (Chile): implications for meteorite paleomagnetism and the stability of hot desert surfaces. Phys Earth Planet Inter 200–201:113–123
Villarreal MN (2018) Understanding the interiors of Vesta and Ceres through their interactions with the solar wind. University of California, Los Angeles, p 204
Voorhies CV (2008) Thickness of the magnetic crust of Mars. J Geophys Res 113:E04004
Voorhies CV, Sabaka TJ, Purucker M (2002) On magnetic power spectra of Earth and Mars. J Geophys Res 107:1–10
Wang Z, Kivelson MG (1996) Asteroid interaction with solar wind. J Geophys Res 101:24479–24493
Wang H, Weiss BP, Bai X-N, Downey BG, Wang J, Wang J, Suavet C, Fu RR, Zucolotto ME (2017) Lifetime of the solar nebula constrained by meteorite paleomagnetism. Science 355:623–627
Weiss BP, Berdahl S, Elkins-Tanton LT, Stanley S, Lima EA, Carporzen L (2008) Magnetism on the angrite parent body and the early differentiation of planetesimals. Science 322:713–716
Weiss BP, Gattacceca J, Stanley S, Rochette P, Christensen UR (2010) Paleomagnetic records of meteorites and early planetesimal differentiation. Space Sci Rev 152:341–390
Weiss BP, Elkins-Tanton LT, Barucci MA, Sierks H, Snodgrass C Vincent J-B, Marchi S, Weissman PR, Pätzold M, Richter I, Fulchignoni M Binzel RP, Schulz R (2012) Possible evidence for partial differentiation of asteroid Lutetia from Rosetta. Planet Space Sci 66:137–146
Weiss BP, Wang H, Sharp TG, Gatttaceca J, Shuster DL Downey B, Hu J, Fu RR, Kuan AT, Suavet C, Irving AJ Wang J, Wang J (2017) A nonmagnetic differentiated early planetary body. Earth Planet Sci Lett 468:119–132
Weiss BP, Bai X-N, Fu RR (2021) History of the solar nebula from meteorite paleomagnetism. Sci Adv 7:eaba5967
Weiss BP, Elkins-Tanton LT (2013) Differentiated planetesimals and the parent bodies of chondrites. Annu Rev Earth Planet Sci 41:529–560
Wieczorek MA (2018) Strength, depth, and geometry of magnetic sources in the crust of the Moon from localized power spectrum analysis. J Geophys Res 123:291–316
Williams Q (2009) Bottom-up versus top-down solidification of the cores of small solar system bodies: constraints on paradoxical cores. Earth Planet Sci Lett 284:564–569
Winslow RM, Anderson BJ, Johnson CL, Slavin JA, Korth H, Purucker ME, Baker DN, Solomon SC (2013) Mercury's magnetopause and bow shock from MESSENGER magnetometer observations. J Geophys Res 118:2213–2227
Worsham EA, Bermingham KR, Walker RJ (2017) Characterizing cosmochemical materials with genetic affinities to the Earth: genetic and chronological diversity within the IAB iron meteorite complex. Earth Planet Sci Lett 467:157–166
Yang J, Goldstein JI (2006) Metallographic cooling rates of the IIIAB iron meteorites. Geochim Cosmochim Acta 70:3197–3215
Yang J, Goldstein JI, Scott ERD (2008) Metallographic cooling rates and origin of IVA iron meteorites. Geochim Cosmochim Acta 72:3043–3061
Yang J, Goldstein JI, Scott ERD (2010) Main-group pallasites: thermal history, relationship to IIIAB irons, and origin. Geochim Cosmochim Acta 74:4471–4492
Yin F, Lühr H (2011) Recalibration of the CHAMP satellite magnetic field measurements. Meas Sci Technol 22:055101
Zhang TL, Berghofer G, Magnes W, Delva M, Baumjohann W, Biernat H, Lichtenegger H, Nakamura R, Schwingenschuh K, Auster H-U, Fornacon K-H, Richter I, Glassmeier K-H, Carr C, Balogh A, Barabash S, Kudela K, Balikhin M, Russell CT, Motschmann U, Lebreton J-P (2007) MAG: the fluxgate magnetometer of Venus Express. ESA SP 1295:1–10
Zhang Z, Bercovici D, Elkins-Tanton L (2022) Cold compaction and macro-porosity removal in rubble-pile asteroids: 2. Applications. J Geophys Res 127:e2022JE007343
Zuber MT, Park RS, Elkins-Tanton LT, Bell JF, Bruvold KN, Bercovici D, Bills BR, Binzel RP, Jaumann R, Marchi S, Polanskey CA, Raymond CA, Roatsch T, Wang CC, Weiss BP, Wenkert D, Wieczorek MA (2022) The Psyche gravity investigation. Space Sci Rev 218:57

Space Science Reviews (2025) 221:47
https://doi.org/10.1007/s11214-025-01169-3

The Psyche Multispectral Imager Investigation: Characterizing the Geology, Topography, and Multispectral Properties of a Metal-Rich World

J.F. Bell III[1] · M.A. Ravine[2] · M.A. Caplinger[2] · J.A. Schaffner[2] · S.M. Brylow[2] · M.J. Clark[2] · D.A. Peckham[2] · P.T. Otjens[2] · G.J. Price[2] · T. Rowell[2] · J.W. Ravine[2] · J.D. Laramee[3] · R.C. Juergens[4] · W. Morgan[4] · A.G. Parker[4] · D.A. Williams[1] · A. Winhold[1] · S. Dibb[1] · E. Cisneros[1] · M. Walworth[1] · H. Zigo[1] · L. Auchterlonie[1] · N. Warner[5] · H. Bates-Tarasewicz[5] · N. Amiri[5] · C. Polanskey[5] · N. Mastrodemos[5] · R.S. Park[5] · N.K. Alonge[5] · R. Jaumann[6] · R.P. Binzel[7] · T.J. McCoy[8] · M.G. Martin[5] · P.A. Arthur[5]

Received: 10 September 2024 / Accepted: 28 April 2025 / Published online: 21 May 2025

Abstract
The Psyche Multispectral Imager ("the Imager") is a payload system designed to directly achieve or to indirectly enable the key scientific goals and optical navigation requirements of NASA's Psyche mission, which will conduct the first up-close orbital investigation of the metal-rich Main Belt asteroid (16) Psyche. The Imager consists of a pair of block redundant cameras and electronics that are mounted inside the thermally controlled spacecraft body, with a view out the spacecraft −X panel that will be nadir-pointed during nominal asteroid orbital mapping operations. The two identical Camera Heads are connected to a separate Digital Electronics Assembly (DEA) box that interfaces to the spacecraft avionics and that provides power, commanding, data processing, and onboard image storage. The Imager system shares significant heritage with imaging instruments flown on the *Mars Climate Orbiter*, the *Mars Science Laboratory* and *Mars 2020* rovers, and *Juno*. Each camera consists of a 1600×1200 photosensitive pixel charge-coupled device (CCD) detector and its associated electronics, a 9-position filter wheel assembly, a compact catadioptric $f/2.9$ telescope with a fixed focal length of 148 mm, and a sunshade to minimize stray and scattered light. The Imager CCD, filters, and optics enable broadband polychromatic ($\sim$540 $\pm$ 250 nm) imaging plus narrowband imaging in 7 colors centered from 439 to 1015 nm. An additional neutral density filter enables protection of the CCD from direct solar illumination. Each camera has a field of view of $4.6° \times 3.4°$ and an instantaneous field of view of 50 μrad/pixel that enables imaging of the asteroid at scales ranging from $\sim$35 m/pix from 700 km altitude to $\sim$4 m/pix at 75 km altitude. The primary camera ("Imager A") is pointed along the spacecraft −X axis, and the backup camera ("Imager B") is toed-out by 3.7° to potentially enable greater surface area coverage per unit time if both Imagers are operated simultaneously during some mission phases. Stereoscopic mapping is performed by observing the same surface regions with either camera over a range of off-nadir pointing angles.

Keywords Asteroids · (16) Psyche · Psyche mission · Space instrumentation · Space imaging

Extended author information available on the last page of the article

1 Introduction

The NASA Psyche Discovery mission (Elkins-Tanton et al. 2020; Polanskey et al. 2025) will conduct the first detailed, *in situ* exploration of an asteroid interpreted to have a composition significantly dominated by metal. Asteroid (16) Psyche, the largest of the M-class asteroids (∼225 km diameter), is thought to either be the exposed core of a pre-existing larger differentiated body or composed of primordial material having accreted from highly reduced metal-rich nebular condensates (e.g., Elkins-Tanton et al. 2020, 2022). The Psyche orbital mission was selected for flight in 2017 as the 14th of the NASA Discovery series of competed, principal investigator-led missions. The spacecraft launched in October 2023 and, after a Mars flyby gravity assist in 2026, will arrive at Psyche (in the Main Belt at ∼3 AU heliocentric distance) in mid 2029.

A key payload element required for addressing the science goals and objectives of the Psyche mission is a visible to short-wavelength near-infrared multispectral imaging system known as the Psyche Multispectral Imager (or just "the Imager" for short). The Imager will acquire data needed to characterize the shape, topography, geology, and (to a limited extent) surface compositional variations of Psyche. Here, we describe the specific science goals, measurement requirements, design, testing, and planned capabilities of the Imager, as well as the kinds of science products it will generate and the kinds of results that it is expected to obtain.

2 Psyche Multispectral Imager Science Goals and Objectives

2.1 Psyche Mission Science Goals and Objectives Related to Imaging

Table 1 summarizes the high-level science goals of the Psyche mission, as well as the specific more detailed Psyche mission science objectives that drove the choices of payloads and measurements needed to reach these goals (e.g., Polanskey et al. 2025). In addition, the specific goals of the Psyche Multispectral Imager investigation are also listed in Table 1, and described in more detail below. As also described in detail below, the Psyche Multispectral Imager investigation provides critical data to address all three of the mission's primary science goals as well as four of its five driving objectives (Objectives A, B, D, and E in Table 1).

2.2 Imager Scientific Investigation Objectives

2.2.1 Geology/Geomorphology

A primary Psyche Multispectral Imager science objective is to characterize the geology and geomorphology of a never-before-seen kind of object: a metallic world (e.g., Marchi et al. 2022; Jaumann et al. 2022). Is the geology of Psyche similar to that of a stony or icy small body, or is its surface geomorphology radically different from that of other known worlds? How do craters, structural features, and/or erosional landforms scale to a body with potentially significantly different physical and compositional properties? Is there geologic evidence for events (e.g., cratering, brecciation, faulting) that could have disrupted or modified any potential magnetization? The Imager investigation will directly address these kinds of questions by acquiring broadband visible-wavelength, multi-look angle images that will enable mapping at the required few to tens of meters/pixel spatial scales and derivation of

Table 1 *Psyche* Mission and Imager Investigation Science Goals and Objectives

Psyche Mission Science Goals	
Goal 1	Understand an unexplored building block of planet formation: Iron cores
Goal 2	Look inside the terrestrial planets, including Earth, by directly examining the interior of a differentiated body, which otherwise could not be seen
Goal 3	Explore a new type of world for the first time by examining a small body made not of rock or ice, but one that is metal-rich
Psyche Mission Science Objectives	
Objective A	Is Psyche the core of a differentiated planetesimal?
Objective B	What are the relative ages of surface regions on the asteroid?
Objective C	What are the abundances of light elements on the asteroid's surface?
Objective D	Did Psyche form in oxidizing or reducing conditions?
Objective E	What is the nature of Psyche's detailed surface geomorphology and topography?
Psyche Multispectral Imager Investigation Objectives	
(1)	Characterize the geology and geomorphology of (16) Psyche
(2)	Derive the surface topography and shape of (16) Psyche
(3)	Search for and characterize the nature of metal-silicate mixing on (16) Psyche's surface
(4)	Search for and detect any diagnostic sulfide minerals on (16) Psyche's surface
(5)	Acquire images to support optical navigation and orbital spacecraft operations

$\leq$50 m required vertical precision Digital Terrain Models (DTMs). The Imager's pixel scale and DTM requirements are based on experience from previous imaging investigations of planetary and especially small body surfaces, and are intended to enable the detailed analysis of impact cratering and any volcanic, tectonic, and/or gradational processes on this unique world at the appropriate scales (e.g., Sullivan et al. 2002; Marchi et al. 2022, 2023; Jaumann et al. 2022).

Another major mission objective relevant to imaging is characterization of the relative ages of Psyche's surface regions (Marchi et al. 2022). The Imager will contribute to this mission objective by mapping $\geq$50% of the asteroid's surface in the broadband filter at a scale of $\leq$200 m/pix (Polanskey et al. 2025), enabling the counting and characterization of craters with diameters larger than 1 km. This range of crater sizes will also enable assessment of the potential crustal implications of the simple to complex crater transition for an object of this size, as well as the assessment of materials potentially excavated from depth (e.g., Bottke et al. 2020; Marchi et al. 2022). Imager data will also be used to construct a global geologic map of the surface (Polanskey et al. 2025), and potentially to define a global chronostratigraphy as was done, for example, by the Dawn mission at the asteroid Vesta (cf., Williams et al. 2014).

2.2.2 Topography and Shape

The Imager's observations of Psyche will also be designed to enable the derivation of the asteroid's surface topography and shape, using a combination of stereo photogrammetry (SPG; geometrical analysis of the parallax of surface features observed over a range of stereo convergence angles; see e.g., the review in Jaumann et al. 2022), and stereo photoclinometry (SPC; "shape from shading", or the analysis of brightness variations of surface features observed over a range of solar illumination conditions; e.g., reviews in Zuber et al. 2022 and Jaumann et al. 2022).

The Imager's combination of high-resolution ($\leq$20 m/pix) broadband images, mosaics of the landforms, and topographic maps of the surface of Psyche (from digital terrain models with height accuracy of 50 m or smaller), will enable lower spatial resolution gravity field (Zuber et al. 2022) and neutron/gamma-ray compositional measurements (Lawrence et al. 2025) to be correlated with surface geology, providing further insights into the asteroid's internal structure and homogeneity.

2.2.3 Mineralogy and Composition

Because of the inferred relatively high metal abundance of Psyche (based on its visible to near-IR reflectance spectrum, relatively high density, and high radar reflectivity; e.g., Elkins-Tanton et al. 2020), a major goal of the Psyche mission is to test the hypotheses that the asteroid is or is part of a differentiated planetesimal core, or that it is not a core but is instead a different class of object composed of a significant fraction of primordial unmelted metal (e.g., Elkins-Tanton et al. 2022). The Imager's observations can address these hypotheses by using multispectral imaging to search for and characterize the nature of any large-scale metal-silicate mixing on the asteroid's surface.

Specifically, visible to near-infrared multispectral imaging at key wavelengths between 450 to 1000 nm (Table 2; Fig. 1) can enable the detection of re-accreted, exogenically-supplied, or impact-exposed iron-bearing silicate minerals like pyroxene and (to a lesser extent) olivine, and mapping of the asteroid's spectral properties in this wavelength range can quantify the degree of mixing of these kinds of silicates with metals on the surface (e.g., Cloutis and Gaffey 1991; Cloutis et al. 2010; Dibb et al. 2022, 2023). The Imager will observe the surface at key wavelengths and at a spatial resolution $\leq$ 500 m/pix. Modeling of the expected Imager performance [assuming Signal to Noise Ratio (SNR) $\geq$ 50 in all filters] indicates that even relatively weak absorption features seen in laboratory and meteorite metal + silicate mixture spectra with iron-bearing silicate abundances of only $\approx$5–10 wt.% (depending on specific compositions and silicate grain sizes) will be detectable in the Imager's spectroscopic band ratios and band depth maps (e.g., Dibb et al. 2023).

In addition to silicate detection and mapping, the Imager will also address the Psyche mission objective of determining whether Psyche was formed under conditions more oxidizing or more reducing than Earth's core by searching for and detecting diagnostic sulfide minerals like troilite (FeS) and oldhamite [(Ca,Mg,Fe)S] (e.g., Dibb et al. 2021). The presence of such cubic monosulfide phases has been interpreted to be consistent with highly reducing formation conditions (e.g., Clark et al. 2004; Fornasier et al. 2008, 2010; Elkins-Tanton et al. 2020; McCoy et al. 2022). To search for sulfides, the Imager will use narrow-band filters to detect and map a weak cubic monosulfide feature that has been identified in laboratory and telescopic data in the 435-550 nm wavelength region (e.g., Gaffey et al. 1993; Burbine et al. 2002; Clark et al. 2004). Spectroscopic ratio and band depth analyses of these images are expected to enable the search for signatures of S-bearing phases on Psyche with local mixture abundances $\gtrsim$ 10 wt.% (Elkins-Tanton et al. 2022; Dibb et al. 2023).

2.3 Imager Optical Navigation and Orbital Operations Support Objectives

The Imager also serves as the primary Optical Navigation ("OpNav") camera system for the Psyche mission. OpNav is an essential component of mission design and navigation and will be used during the Approach phase and all Science orbits (see Sect. 5 below). OpNav

Table 2 Psyche Multispectral Imager Filter Bandpass Data

Filter #	Imager A (primary)			Imager B (backup)			Justification (see text for details)
	Band Center[1] (nm)	Effective Center at Psyche[2] (nm)	FWHM[3] (nm)	Band Center[1] (nm)	Effective Center at Psyche[2] (nm)	FWHM[3] (nm)	
0	–	–	–	–	–	–	Neutral Density 5 (factor of 10^{-5}) Sun blocking filter
1	539	540	259	538	539	250	Broadband polychromatic bandpass for maximum sensitivity OpNav, topography, and geologic imaging
2	439	442	49	439	441	48	Sulfide detection
3	494	494	24	493	493	24	Sulfide detection
4	548	548	24	548	548	24	Sulfide detection
5	722	722	40	723	722	40	Reflectance continuum for 1000 nm features
6	841	840	48	842	841	48	Detection of Fe-bearing silicates
7	942	942	46	942	941	46	Detection of Fe-bearing silicates
8	1015	1014	40[4]	1015	1014	41[4]	Detection of Fe-bearing silicates

[1]Band Center calculated as the wavelength corresponding to 50% of the integral under the full system's filter profile, which is the measured system-level bandpass that includes the filter, optics transmission, and CCD quantum efficiency.

[2]Effective centers are computed after convolution of each full system filter profile with the solar spectrum and assuming a linearly increasing albedo of (16) Psyche of 0.11–0.15 between 442–1014 nm, based on the summary by Dibb et al. (2023).

[3]Full Width at Half Maximum, calculated using the same effective bandpass approach outlined in Note 2.

[4]Filter 8 is significantly asymmetric on its longer wavelength side, and so the traditional concept of FWHM does not accurately capture its full range of responsivity. See Winhold et al. (2025) for additional details.

data are used in conjunction with radio tracking data in the orbit determination process. The key OpNav functions during Psyche operations are: (1) Continuously supporting real-time orbit determination to assist in updating the science instrument sequences, pointing and data acquisition, in-orbit trajectory correction maneuvers, successful entry into initial orbit, and transfers between science orbits; (2) Assisting in the science orbit reconstruction at the end of each orbit phase; and (3) Characterizing key physical properties of Psyche, such as the spin state and the global shape of the asteroid that are used by both the mission design and science planning teams.

In addition, during the Cruise phase of the mission, OpNav will support imaging of stars and star clusters and data analyses that can be used to assess spacecraft pointing accuracy and jitter, radiometric responsivity and stability, and the nature of any geometric distortion across the Imager field of view (Winhold et al. 2025). The accuracy of the geometric camera models is of particular importance to OpNav to enable high-quality astrometric optical data analysis. Specifically, the goal for the camera geometric model is to be of sufficient accuracy so as not to limit the expected astrometric accuracy of Imager observations in the clear filter,

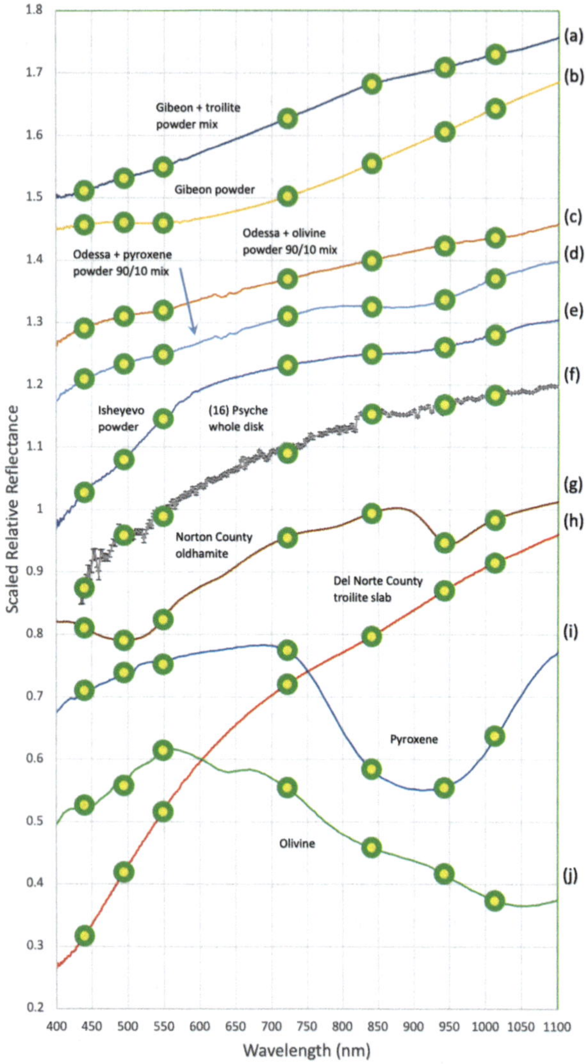

Fig. 1 Examples of relevant high resolution visible to near-IR telescopic, meteorite, and laboratory mineral or mineral mixture spectra (solid lines) resampled over the bandpasses of the Psyche Multispectral Imager's narrowband filters (green/yellow data points; Table 2). These examples include many of the "best matches" to the telescopic spectrum of (16) Psyche compiled in Table 3 of Dibb et al. (2023). All spectra have been scaled to 1.0 at 550 nm and then vertically offset for clarity. Spectra and sources: (a) 80/20 wt.% mixture of < 75 μm grain size powders of the metallic meteorite Gibeon and troilite, from Dibb et al. (2021); (b) Powdered > 75 μm grain size sample of Gibeon, from Dibb et al. (2023): (c) 90/10 wt.% mixture of < 45 μm grain size powders of the metallic meteorite Odessa and olivine, from Cloutis et al. (2009); (d) 90/10 wt.% mixture of < 45 μm grain size powders of the metallic meteorite Odessa and low-Fe pyroxene, from Cloutis et al. (2009); (e) Powdered > 75 μm grain size sample of the CH/CBb chondrite Isheyevo, from Dibb et al. (2023); (f) whole-disk telescopic spectrum of (16) Psyche from the SMASSII survey of Bus and Binzel (2002); (g) Powdered < 1 mm oldhamite-dominated clasts from the Norton County aubrite, from by Dibb et al. (2022); (h) 1200 grit polished troilite slab, from Dibb et al. (2023); (i) Low-Fe < 45 μm pyroxene powder (PYX042) used in mixture (d), from Cloutis (2002): (j) Olivine < 45 umpowder (OLV003) used in mixture (c), from Cloutis et al. (2009)

which from systematic navigation studies from prior small body orbital missions has been shown to be in the range of 0.2-0.1 pixels, in the various stages of orbital operations (e.g., Mastrodemos et al. 2012, 2015).

During the initial approach to the asteroid, the optical data will consist of the estimated center of the body based on limb-fitting and image cross-correlation methods, using an *a priori* shape model. After Psyche exceeds ∼50 pixels in diameter and through to the end of the mission the optical data will include specifically defined landmarks, constructed from DTMs of the surface using the SPC methodology (e.g., Sect. 2.2.2; Jaumann et al. 2022; Zuber et al. 2022). These DTMs will be synthesized into global shape models. Similar landmark-based navigation was carried out with the *Dawn* mission (e.g., Mastrodemos et al. 2012) and to first order the overall process for the Psyche mission will be similar.

2.4 Imager-Specific Design Considerations and Requirements

To help accomplish the scientific goals and objectives of the Psyche mission and the Imager investigation listed in Table 1, the cameras must be able to observe the surface of Psyche from orbit in an approximately nadir-pointed attitude, through a broadband ("clear") filter that enables maximum light collection over minimum exposure times, and through a variety of narrowband color filters optimized to detect specific kinds of diagnostic silicate and sulfide minerals. In addition, the Psyche spacecraft must enable the Imagers to observe the same regions of the surface at multiple off-nadir angles to enable the creation of DTMs using image-based SPG or SPC methods. Stereo imaging will nominally be performed by using just one camera and operationally acquiring ≥4 images of each surface element at nadir and at up to ±20° stereo separation. And as noted above, the Imagers will also be needed for OpNav during cruise, and to provide geologic context for measurements made by the other Psyche mission science investigations.

A detailed set of multi-level Psyche mission and measurement requirements was established and linked to the mission's specific scientific goals and objectives (Table 1) with appropriate margins to guide the design, testing, and eventual operation of all the science instruments on the Psyche spacecraft (e.g., Polanskey et al. 2025). The specific Project requirements levied on the Imager investigation are outlined in Table 3. Imager-specific requirements span the range from contributions to high-level mission-critical science (Level 1 requirements), to measurement requirements needed to meet specific Imager-related mission science objectives (Level 2), to requirements placed on other subsystems within the mission that help enable the Imager to meet specific mission science objectives (Level 3), and finally to detailed requirements on specific individual Imager components, subsystems, or tests that ensure that the as-built Imagers achieve the required level of performance and accuracy to meet Imager investigation goals (Level 4).

3 Instrument Description and Heritage

3.1 Overview

The Psyche Multispectral Imager subsystem is a compact, low mass, and power efficient pair of block-redundant Charge-Coupled Device (CCD) imaging cameras, filter wheels, and electronics that provide a low risk, high heritage way to meet the Psyche mission's imaging-related requirements (Table 3). While the Imagers are not exact copies of any previously-flown instruments, heritage for their CCD sensors, electronics, filters, filter wheel mechanisms, and flight software comes primarily from the *Mars Science Laboratory* (*MSL*) mission's Mastcam, MAHLI, and MARDI imaging systems (Malin et al. 2017) and the *Mars*

Table 3 Psyche Mission and Instrument Requirements Relevant to the Psyche Multispectral Imager

Level[a]	Requirement	Notes and Traceability to Mission Objectives[b]
1	Determine whether there are contiguous silicate regions on scales larger than 1 km, where silicates are defined as having >14% Si by mole.	*Objective A*: Metal/silicate mixing. Multispectral filters used to search for and map regions containing iron-bearing silicate minerals like pyroxene and olivine.
1	Determine the relative ages of regions of the surface of Psyche by counting craters with diameters larger than 1 km over at least 50% of the surface.	*Objective B*: Relative ages. Clear filter used to map the surface between $\approx$35 m/pix (Orbit A) to $\approx$3.5 m/pix (Orbit D), to identify geologic features such as craters.
1	Determine whether Psyche has $>10 \pm 5$ wt.% sulfides like oldhamite (global average), indicating highly reducing conditions.	*Objective D*: Oxidizing vs. reducing conditions. Multispectral filters used to search for and map regions containing sulfides.
1	Characterize Psyche's topography over 50% of the surface at $\leq$200 m/pix horizontal and $\leq$50 m vertical resolution.	*Objective E*: Surface topography. Influences optics and orbital altitude (pixel scale), off-nadir pointing capability of the spacecraft (multiple look angles to produce DTMs using stereo photogrammetry and stereo photoclinometry), and system responsivity (clear filter, sensor, and electronics).
2	The Psyche Project shall map the metal to silicate fraction of at least 80% of (16) Psyche's surface (with filters near 725, 850, 948 and 1041 nm) at a pixel scale $\leq$ 500 m and SNR $\geq$ 50.	*Objective A*: Metal to silicate mixing. Influences optics and orbital altitude (pixel scale), system responsivity, and need for specific near-IR filters.
2	The Psyche Project shall map $\geq$80% of (16) Psyche's surface with the clear filter at a pixel scale of $\leq$20 m and SNR $\geq$ 50.	*Objective E*: Surface geology. Influences optics and orbital altitude (pixel scale), and system responsivity.
2	The Psyche Project shall determine the shape of (16) Psyche by mapping with the clear filter over at least 80% of its surface with a pixel scale $\leq$ 200 m/pix and SNR $\geq$ 50 to obtain $\leq$70 m vertical accuracy using stereo imaging techniques.	*Objective A*: Determine density (mass/volume). Influences optics and orbital altitude (pixel scale), off-nadir pointing capability of the spacecraft, and system responsivity.
2	The Psyche Project shall map (16) Psyche's surface features with the clear filter over $\geq$50% of its surface with a pixel scale $\leq$ 20 m/pix and SNR $\geq$ 50 to obtain $\leq$50 m vertical accuracy using stereo imaging techniques.	*Objective E*: Surface topography. Influences optics and orbital altitude (pixel scale), off-nadir pointing capability of the spacecraft, and system responsivity.
2	The Psyche Project shall map at least 50% of (16) Psyche's surface with the clear filter at a pixel scale $\leq$ 200 m/pix and SNR $\geq$ 50 to support crater age determination.	*Objective B*: Relative ages. Influences optics and orbital altitude (pixel scale) and system responsivity.
2	The Psyche Project shall map at least 80% of (16) Psyche's surface (with filters near 437, 495, and 550 nm) at a pixel scale $\leq$ 200 m/pix and SNR $\geq$ 50 to search for evidence of oldhamite.	*Objective D*: Oxidizing vs. reducing conditions. Influences multispectral filters used to search for and map sulfides, optics and orbital altitude (pixel scale), and system responsivity
3	The Spacecraft shall accommodate operating Imager A and Imager B simultaneously.	*Objective E*: Geology and topography. Simultaneous operation of both Imagers allows more clear filter mapping per unit time in Orbit D, closest to the asteroid.

Table 3 *(Continued)*

Level[a]	Requirement	Notes and Traceability to Mission Objectives[b]
3	Imager A boresight and Imager B boresight are "toed out" relative to each other such that Imager A is mounted nominally nadir (boresight parallel to the spacecraft $-$X axis) and Imager B is mounted with a toe-out angle of $3.7° \pm 0.4°$.	*Objective E*: Geology and topography. Toe-out and simultaneous operation of both Imagers enables $\approx 1.9\times$ surface area coverage per unit time in Orbit D.
4	The images shall have an average signal to noise ratio of greater than or equal to 50:1 in each spectral band.	*Objectives A and D*: Metal/silicate mixing and sulfides. Influences system responsivity (filter bandwidths, sensor, and electronics).
4	Each Imager instrument Instantaneous Field of View (IFOV) shall be smaller than or equal to 50 μrad $\pm$ 2%.	*Objectives A, B, D, E*: Metal/silicate mixing, relative ages, sulfides, geology/topography: Spatial resolution requirements drive optics design.
4	Each Imager shall provide a Field of View (FOV) of $4.4° \times 3.3° \pm 2\%$.	*Objectives A, B, D, E*: Metal/silicate mixing, relative ages, sulfides, geology/topography: Surface coverage requirements drive optics design
4	Each Imager shall have a Modulation Transfer Function (MTF) performance of >0.2 at Nyquist (67.6 lp/mm) over a required scientific operating range of -10 °C to $+25$ °C	*Objectives A, B, D, E*: Metal/silicate mixing, relative ages, sulfides, geology/topography: Spatial resolution requirements drive optics design.
4	The Imager data shall be radiometrically accurate to $\leq 10\%$ absolute and $\leq 2\%$ relative radiance (pixel to pixel and band to band) through each spectral band across the full FOV.	*Objectives A and D*: Metal/silicate mixing and sulfides. Influences system responsivity (filter bandwidths, sensor, and electronics).
4	Distant imaging of representative natural scenes and/or color/reflectance standards shall be used to validate the radiometric calibration of the Imager.	*Objectives A, D, and E*: Metal/silicate mixing, sulfides, topography from stereo photogrammetry. Influences system responsivity (filter bandwidths, sensor, and electronics).

[a]See text for details; [b]See Table 1.

2020 mission's Mastcam-Z imaging system (Bell et al. 2020), which were also built by Malin Space Science Systems, Inc. (MSSS), with only slight changes in filter wavelengths needed to meet Psyche-specific multispectral imaging requirements. Indeed, the Imager uses exactly the same high-heritage interface to the Psyche spacecraft avionics as the *MSL* and *Mars 2020* heritage cameras used with those NASA Jet Propulsion Laboratory (JPL) rover spacecraft; this was part of the justification for selecting this system as part of a spacecraft also to be operated with JPL avionics. Additionally, partial heritage for the Imager optics comes from the *Mars Climate Orbiter* mission's Mars Color Imager medium angle imaging system (Malin et al. 2001), also a previously built MSSS instrument, but with a slightly different lens prescription for the Imager to meet Psyche-specific spatial resolution requirements.

The Imager system consists of four components: Two identical 148-mm focal length $f/2.9$ Camera Heads, each consisting of a camera electronics box with a 1648×1214 total pixel (1600×1200 unmasked, photoactive pixels) focal plane array CCD, a filter wheel assembly including 9 rectangular glass-coated interference filters driven by a CDA InterCorp GZ4 16:1 gear ratio stepper motor actuator, an optics/Sun shade assembly, and an electronics card housing motor drivers, flash memory, and SDRAM controlled by a Xilinx Virtex-2 Field Programmable Gate Array (FPGA). The purpose of the second Camera Head is to provide redundancy primarily for mission-critical OpNav, as was done on the *Dawn*

Fig. 2 Flight Psyche Multispectral Imager Camera Heads (left) and flight Digital Electronics Assembly (right), prior to integration with the spacecraft. For scale, the pocket knife is 3.5 inches (90 mm) long

mission; redundancy for science is a secondary benefit. The two independent Imager electronics cards are housed in a single separate Digital Electronics Assembly (DEA) box. The onboard DEA hardware and software enables lossless predictive or lossy JPEG compression, a limited amount of image processing (e.g., subframing, multi-pixel summing, as well as "companding" – a term for compressing to a smaller bit depth then later expanding back to the original bit depth; e.g., Bell et al. 2017), and storage/manipulation of ~4000 full-frame uncompressed 12-bit images in each camera's flash memory prior to transfer to the spacecraft for downlink. The two flight Imager Camera Heads and the flight DEA are shown in Fig. 2.

3.2 Imager Electronics

As shown in Fig. 3, the Imager electronics are partitioned into two subsystems: the Camera Head (CH) electronics and the Digital Electronics Assembly (DEA). As noted above, both subsystems are very similar to those used previously for cameras on the *MSL* and *Mars 2020* missions, but both subsystems were modified slightly to address issues with parts obsolescence and to improve reliability. Most of the changes involved power circuits: different voltage regulators and passive filters in the CH and a different DC-DC converter architecture in the DEA. Also, the earlier instruments had internal power switching to allow the DEA to operate with the CH turned off, for power management purposes specific to rover operations constraints. Such constraints were not relevant for Psyche mission operations, and thus the internal power switching was removed for the Imager implementation, meaning that both electronics subsystems are powered on when the spacecraft's Imager power switch is enabled. The robustness of the electronics to radiation during the long Psyche mission cruise and operations phases has been demonstrated by the successful performance of the identical electronics used for the *Juno* mission's JunoCam instrument investigation during cruise

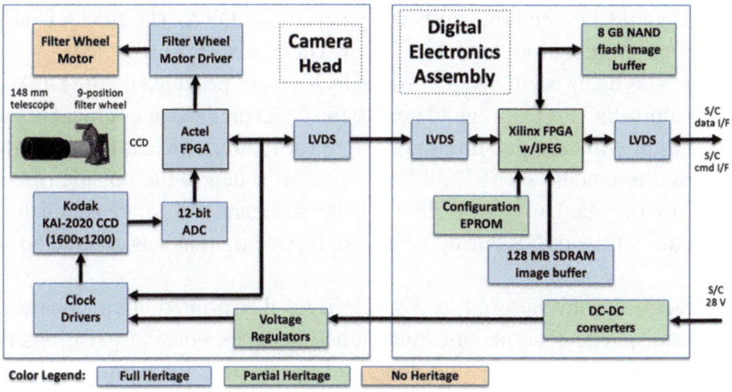

Fig. 3 Block diagram for the two fully redundant Psyche Multispectral Imager Camera Heads and their associated Digital Electronics Assembly (DEA) electronics

and at Jupiter (e.g., Hansen et al. 2017). Specifically, during more than 6 years of Primary and Extended Mission operations in the extremely harsh radiation environment of Jupiter, JunoCam experienced an estimated cumulative dose of ∼28 KRad(Si) behind 1.5 inches of Aluminum shielding before experiencing its first radiation-induced anomaly (Schaffner et al. 2025). In contrast, the mission expected shielded dose for the Imager electronics is estimated to not exceed only ∼3.5 KRad(Si), partly because the electronics are housed relatively deep inside the body of the spacecraft itself (see Sect. 3.6).

3.2.1 Camera Head (CH) Electronics

The CH electronics are designed around the ON Semiconductor (formerly Kodak) KAI-2020M CCD image sensor. This sensor has 1648 × 1214 pixels (1600 × 1200 photoactive), each with a 7.4-µm pitch, and uses interline transfer in single output mode to implement electronic shuttering. Unlike its heritage predecessors on *MSL* and *Mars 2020*, which used the Bayer-pattern color variant, this sensor is monochrome. The sensor has microlenses to improve its quantum efficiency, which has a peak of about 55% at around 500 nm (see Fig. 10 in ON Semiconductor Inc 2015). The "fast-dump" capability of the sensor is used to clear residual charge prior to integration and also allows vertical sub-framing of the output image, if commanded.

The output signal from the CCD is AC-coupled and then amplified. The amplified signal is digitized in an Analog to Digital Converter (ADC) circuit to 12 bits at a maximum rate of 10 Mpix/sec. For each pixel, both reset and video levels are digitized and then subtracted in the digital domain to perform correlated double sampling (CDS) that nominally yields an output signal encompassing the full dynamic range between black and full well within 11 bits, from 0-2047 Data Numbers (DN). The CH readout rate can be selected from values of 2.5, 5, and 10 Mpix/sec. For the Psyche mission, the 5 Mpix/sec rate is used by default, as pre-flight characterization testing showed that this provides the best noise performance. Based on performance of this sensor on previous missions, our expectation for initial exposure time modeling was that it would deliver a gain of ∼16 e^-/pix, read noise ≤15 e^-, and full well ∼30,000 e^-. As described in Sect. 4.2, all of these expectations have been met during actual pre-flight and in-flight performance.

All CH functions are supervised by a single Actel RTSX FPGA. In response to commands from the DEA, the FPGA generates the CCD clocks, reads samples from the ADC and performs digital CDS, and transmits the pixels to the DEA. The FPGA is also responsible for operating the stepper motor for the filter wheel mechanism.

The CH operates using regulated 5 V and ±15 V power provided by the DEA. A platinum resistance thermometer (PRT) mounted next to the CCD on the camera focal plane is read by the spacecraft to provide temperature knowledge for radiometric calibration. An additional pair of PRTs and redundant etched-foil heaters are attached to the outside of each Camera Head and are thermostatically controlled by the spacecraft to remove moisture from the metering structure of the optics during a bakeout procedure that was conducted shortly after launch.

The CH electronics are laid out as a single rigid-flex printed circuit board (PCB) with three rigid sections. The sections are sandwiched between housing partitions that provide mechanical support and radiation shielding, and the flexible interconnects are enclosed in metal covers (Fig. 4). Each Imager CH assembly bolts to the spacecraft via four feet on its base housing.

3.2.2 Digital Electronics Assembly (DEA) Electronics

The DEA interfaces the Camera Heads to the spacecraft avionics. Each redundant interface from the Imager to the Psyche spacecraft is two (one each for command and telemetry) four-signal flow-controlled LVDS synchronous-serial links, running at 8.25 Mbit/s maximum.

The data interface between the DEA and each CH is a three-signal LVDS synchronous serial link transmitting commands from the DEA to the CH at 2 Mbit/sec and an 8-signal synchronous 6-bit parallel interface from the CH to the DEA at selectable rates from 30 to 120 Mbit/sec (5 to 20 MHz). The DEA is powered from the spacecraft 28 V power bus, provides regulated power to the CHs, and contains a PRT for temperature monitoring.

The core functionality of the DEA is implemented in a Xilinx Virtex-II FPGA. All interface, compression, and timing functions are implemented as logic peripherals of a Microblaze soft-processor core in the FPGA. The DEA provides an image-processing pipeline which includes 12-to-8-bit companding of input pixels based on pre-defined onboard lookup tables; horizontal sub-framing; 2×2, 3×3, or 4×4 pixel summing; and optional lossless predictive or lossy JPEG compression. The image-processing pipeline can run at the full speed of the CH input, writing the processed data stream directly into the DEA memory.

The DEA memory subsystem contains 256 MB of DRAM and 8 GB of non-volatile NAND flash memory. The flash is organized as a large image buffer, allowing images to be acquired without the use of spacecraft memory resources at the maximum camera data rate. The DRAM is typically used as scratchpad space and to store file system information, but can also be used as a buffer for a small number of images, if needed.

The DEA electronics for each Imager are laid out on a single rectangular PCB. The DEA boards for both Imagers are integrated into one unit to save mass. Each PCB is sandwiched between housing sections, and the entire assembly bolts onto a mounting plate inside the +Y panel of the spacecraft (Fig. 5).

3.3 Flight Software

DEA hardware functions are coordinated by the DEA flight software (FSW), which runs on the Microblaze. The FSW receives and executes commands from the spacecraft and transmits any resulting data. The FSW can also implement autoexposure algorithms for image acquisition, perform error correction on the contents of the flash, and implement mechanism control and fault protection. The FSW consists of about 10,000 lines of ANSI C code. Nearly all the code was inherited without modification from the earlier heritage instruments. New developments included additional fault protection, control of the nine-position filter wheel and motor (the heritage filter wheels had eight positions), addition of 3 × 3 summing capability, and removal of color processing and other mission-specific functions not needed for Psyche (such as focus and zoom control and focus stack merging).

The Imager FSW also allows considerable flexibility for image acquisition. Images can be acquired in uncompressed form, stored in the DEA flash buffer, and transmitted to the spacecraft multiple times using different compression factors or summing modes. Alternatively, images can be compressed during acquisition and stored as JPEGs in the DEA flash to minimize the amount of buffer space they occupy; the latter capability is useful for video sequences, for example.

3.4 Filter Wheel Mechanism

Each Imager Camera Head includes a nine-position filter wheel (Table 2 and Fig. 4), controlled using a CDA Model GZ4 0.5-inch diameter three-phase stepper motor. The motor drives an integral gear on the outer diameter of the filter wheel, and the wheel itself is supported by a bearing on its inner diameter. Position sensing of the wheel is done using a Hall-effect sensor and magnet pairs at each filter location, with an extra magnet pair for absolute indexing. This mechanism is similar to and based on the ones used for the earlier Mastcam and Mastcam-Z instruments, but using a new motor (because the heritage motors could not be delivered on our required schedule), an additional filter (an opaque Sun-blocking filter to accommodate the need to occasionally "Sun-safe" the optics, a requirement imposed by the spacecraft's solar-electric propulsion system), and a new mounting accommodation for the Imager's specific optics. A flight qualified CDA motor with the same specifications as the flight motors was taken through a life testing program in November 2019. No anomalies were observed during 3 × nominal required lifetime testing of the motor and filter wheel mechanisms.

3.5 Imager Optics and Optomechanical Design

The Imager optics (Fig. 4) consists of a catadioptric telescope with spherical fused silica optical elements providing a fixed focal length of 148 mm, an $f/\#$ of 2.9, and a T/# [defined as $(f/\#) \times (\text{system transmission})^{-0.5}$] of 4.7 based on the combination of optics transmittance and the telescope's large central obscuration. The optics yield an Instantaneous Field of View (IFOV) or ground sampling distance of 50 µrad/pix (yielding 3.5 m/pix sampling from an altitude of 70 km), a field of view (FOV) of 4.6° × 3.4°, and a depth of field from ∼800 m to infinity. The telescope's focus is fixed and set by the use of a precision-machined focus spacer/light shield located just above the image sensor.

In order to be able to achieve excellent image contrast in family with numerous previous deep space imaging systems, the optics are required to deliver a Modulation Transfer Function (MTF) performance > 0.2 at the Nyquist frequency of the sensor (67.6 lp/mm) and over the required operating range for scientific observations of −10 °C to +25 °C. The corresponding Point Spread Function (PSF) requirement was set as a FWHM of < 3 pixels, to provide a good combination between the accuracy of astrometric center finding for OpNav on approach to (16) Psyche and adequate topographic resolution of an extended body during proximity operations (e.g. Mastrodemos et al. 2012, 2015). As described in Winhold et al. (2025), both flight Imagers exhibit PSFs with FWHM ∼2 pixels based on both pre-flight testing using point source targets and in-flight testing using imaging of stars.

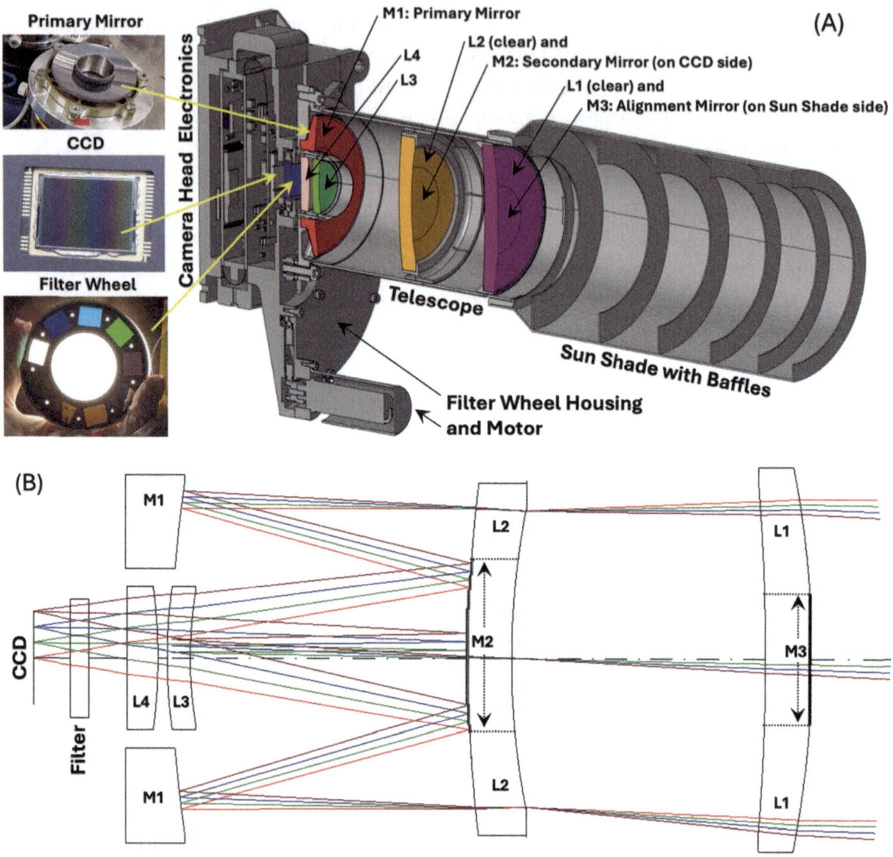

Fig. 4 (A) Cutaway CAD view of the Psyche Multispectral Imager, along with actual photos of some of the key hardware components, including the 57 mm diameter Primary Mirror M1 on a test bench, the $\sim 12 \times 9$ mm CCD sensor, and the 9.1 cm diameter filter wheel. (B) Ray trace of the Psyche Multispectral Imager catadioptric telescope, with lens and mirror elements as identified in (A) above. M1 is the spherical primary mirror, and M2 is the spherical secondary mirror, coated onto the back (CCD side) of lens element L2. The part of the front of L2 corresponding to M2 is coated with Z306 diffuse black paint. M3 is a flat-surface mirror coated onto the front of lens element L1 and used for laser alignment verification of camera boresight pointing during pre-flight testing

All of the Imager lens elements and the primary mirror are made of fused silica. Z306 is used for the black coatings on the back sides of mirrors M2 and M3, and the mirrors themselves are silver coated. The lens cells and the optical bench holding the primary are made of Invar 36, and the metering structure and Sun shade are made of graphite cyanate ester (GR/CE) composite. The optics are attached to the filter wheel housing using titanium flexures, and the electronics housing sections are aluminum (7075 alloy).

Over the course of the optics development, the configuration of the optical bench evolved. Initially, it was highly light-weighted, but this proved detrimental to the performance stability of the system over temperature. This configuration thus appears only in non-flight prototypes and a flight spare. A second, more massive revision of the optical bench was found to be much more athermal and was used for the second flight Imager (Imager B). A third revision, which incorporated additional flexures in the primary mirror mounting interface and which was also measured to be athermal over the required operating temperature range, was used for the primary Imager (Imager A). Analysis, pre-flight testing, and in-flight observations demonstrate that the performance is the same within the uncertainties between the two flight optical benches and meets requirements (Winhold et al. 2025).

Though it is a relatively small effect compared to other parts of the optical system, a slight change of focus was expected between operation under terrestrial ambient conditions and in-flight vacuum conditions, caused by small dimensional changes in the composite metering structure from absorption of water. The magnitude of this effect was modeled and factored into the thickness of the focus spacer used to focus the telescope, and then verified in pre-flight thermal vacuum testing. An external bakeout heater was used to dehydrate the structure by baking it out for ∼2 weeks at +55 °C to +65 °C shortly after launch.

3.6 Imager Integration and Accommodation

For full block redundancy, there are two independent Imager Camera Heads, one designated "primary" (Imager A) and the second designated as the "backup" (Imager B). Both Camera Heads are mounted next to each other on a common plate within the thermally controlled spacecraft interior volume, with just ∼3 cm of each camera's composite Sun shade protruding through holes in the spacecraft's −X panel. The ends of the Sun shades are protected by a "doghouse" enclosure wrapped with multi-layer thermal blanketing (Fig. 5). The DEA is also mounted inside the spacecraft body, but on the inside of the spacecraft's +Y panel. A 2.57 m long roundwire harness connects each independent DEA electronics card to its respective Camera Head.

Both Imager Camera Heads are mounted with the long axis of their CCDs perpendicular to the intended direction of flight over the surface of Psyche. Imager A is boresighted to what will nominally be nadir pointing (parallel to the spacecraft's −X axis), and Imager B is "toed-out" by 3.7° in the −Y direction. Nominally, only Imager A will be operated during the orbital phase of the mission, although both Imagers can be operated simultaneously if desired. The Imager B offset field of view overlaps with Imager A's by ∼10%, enabling greater total surface area coverage per unit time if both Imagers are operated simultaneously during some mission phases. Imager B's boresight is also close enough to nadir to serve as the effectively nadir-viewing primary camera, if needed.

4 Psyche Multispectral Imager Testing and Calibration

4.1 Expected Instrument Performance

During the Imager's design phase, we developed a model of the expected radiance incident on the Imager CCD focal plane from sunlight reflected from Psyche based on the average whole-disk telescopic spectrum of the asteroid (summarized in Table 2 of Dibb et al. 2023), and vendor-supplied data on the expected in-flight performance of the CCD, electronics, optics, and filters. We assumed conservative imaging assumptions: heliocentric distance of 3.3 AU, average surface albedo of 0.12, a 70° solar incidence angle, and a relative ground track speed of 120 m/sec (the latter to constrain exposure times to yield ≤0.5 pixels of ground motion "smear"; Polanskey et al. 2025). Further, because the CCD and telescope are housed within the thermally controlled spacecraft body to a required temperature range of −10 °C to +25 °C and thermally isolated from the Sun Shade, we assumed worst-case expected CCD dark current levels at +25 °C, based on previous in-flight performance of

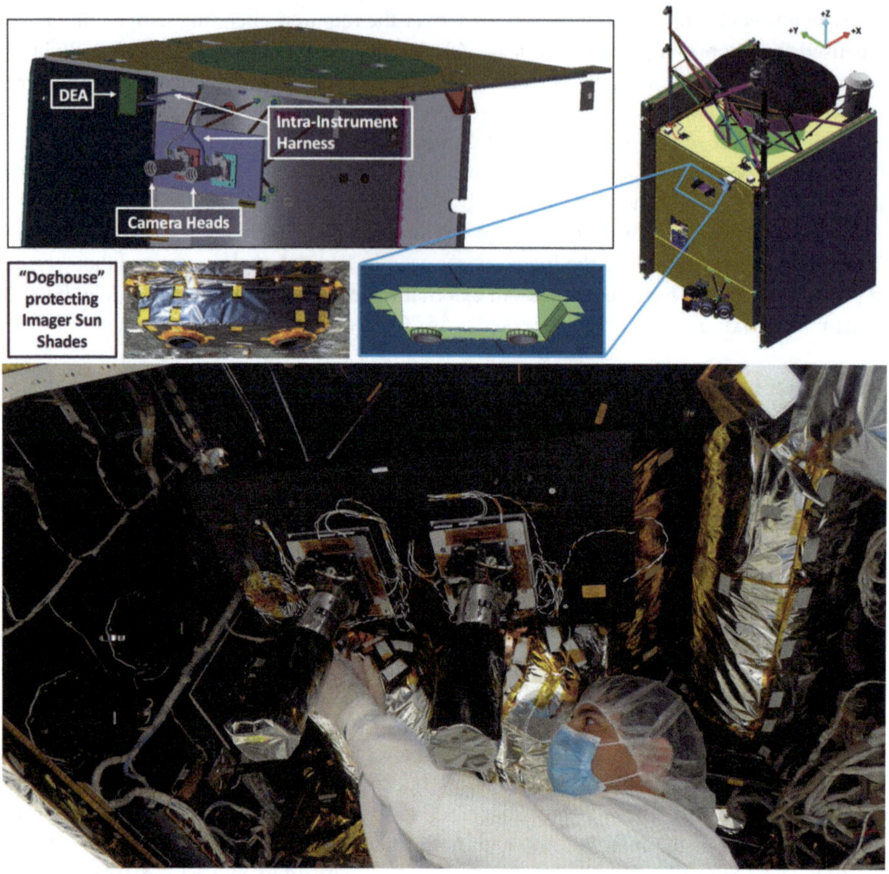

Fig. 5 (Top Right) CAD view of the locations of the Psyche Multispectral Imager Camera Heads, protruding out the −X panel of the Psyche spacecraft. The ends of the Imager Sun Shades are protected by a Mylar-blanketed "doghouse" assembly, shown in the CAD view inset and the actual flight configuration to the inset's left. (Top Left) Cutaway CAD view of the inside of the Psyche spacecraft showing the Imager Camera Heads mounted to a rectangular camera plate that is attached by struts to the spacecraft's central structural cylinder. The Imager Digital Electronics Assembly (DEA) is a separate box mounted to the spacecraft's +Y panel. The DEA and Camera Heads are connected by a harness. (Bottom) JPL Engineer Stephen Partida finalizing the installation of Imager-A (right) and Imager-B (left) on the Psyche spacecraft [photo credit: Sean Howard, JPL]

this specific CCD sensor (e.g., Bell et al. 2017; Hayes et al. 2020). Under such conservative conditions, our model predicted that the Imager design would yield per-pixel SNR $\geq$ 50 for imaging in all filters, but that 2×2 and 3×3 pixel coadding, respectively, would be required to yield SNR $\geq$ 50 for imaging in the two longest wavelength near-IR filters (near 942 and 1014 nm).

4.2 Instrument Performance Verification and Validation: Pre-Flight Through Cruise

To verify that the as-built Imagers meet the SNR, optical performance, and other imaging requirements predicted during design, and to validate their expected performance at Psyche, each Imager underwent extensive independent testing and calibration in ambient

Fig. 6 Examples of "real world" imaging with Imager A in pre-flight testing at MSSS in San Diego. (A) Full frame (1648 × 1200 pixels, including masked reference columns) color composite image of the view out a special cleanroom window of houses on a ridge ∼1 km away. RGB = 722, 548, and 439 nm. (B) Enlarged 548 nm view of power lines ∼2 km away and behind the houses, demonstrating good focus under typical terrestrial imaging conditions. (C) Subset of an Imager A night-time image of the Moon, acquired during relatively clear and dry conditions on Jan. 27, 2022 at 02:26 UTC. The Imager's 50 µrad IFOV (∼10 arcsec on the sky) corresponded to a scale of ∼19 km/pixel on the Moon. Imager B performed similarly under terrestrial conditions

laboratory bench and thermal vacuum chamber conditions at MSSS in San Diego, California, between February 2022 through March 2023, prior to their delivery and installation on the Psyche spacecraft (Winhold et al. 2025). This testing included CCD bias, dark current, read noise, linearity, and full-well characterization; mapping of pixel-to-pixel responsivity ("flatfield") variations; geometric IFOV, FOV, and distortion characterization; assessment of stray and scattered light; quantitative assessment of image quality via measurements of the MTF and PSF; and measurement of the system-level transmission profiles (including out-of-band leakage) and absolute system responsivity for each of the Imager filters. Additional assessment of "real world" image quality was also performed by imaging distant buildings and other structures as well as imaging of the Moon and stars (Fig. 6). Analysis of these pre-flight calibration and characterization images by Winhold et al. (2025) demonstrates that the as-built Imagers meet or exceed their design-based SNR expectations and other requirements (Table 3). A summary of the currently best-known characteristics of the Psyche Multispectral Imagers is provided in Table 4.

Both flight Imagers and the flight DEA underwent individual vibration and pyroshock testing to simulate the vibration and shock environment levels expected during the SpaceX Falcon Heavy launch and Psyche spacecraft deployment processes. No observable changes in optical performance were observed as a result of these tests.

Additional functional testing, boresight determination relative to the spacecraft coordinate system, and basic performance characterization (MTF and relative radiometric responsivity) were also conducted for both Imagers at ambient cleanroom temperature and pressure conditions after the Imagers were installed on and electrically integrated to the Psyche spacecraft at the Astrotech facility in Titusville, Florida, in May and June of 2023. No observable changes in optical performance were observed as a result of spacecraft integration and testing.

Table 4 Psyche Multispectral Imager Instrument Characteristics

Optics	*Description*
Focus	Fixed; Working distances 800 m to ∞
MTF (optics + filters + CCD)	>0.2 at Nyquist (67.6 lp/mm) between −10 °C and +25 °C
PSF (optics + filters + CCD)	1.8 ± 0.3 and 2.1 ± 0.2 pixels FWHM for Imagers A and B, respectively[a]
Filter bandpasses	Two 9 position filter wheels: see Table 2
FOV (1600 × 1200 pix)	$4.6° \times 3.4°$
IFOV (pixel scale)	50 µrad (~35, 15, 9, and 3.5 m/pix from Orbits A, B1/B2, C, & D; Polanskey et al. 2025)
Focal ratio	$f/2.9$
Effective focal length	148 mm (catadioptric)
Entrance pupil aperture	50.8 mm
1st optic (L1) inner/outer diameter, area	30/60 mm, 2148 mm^2
Primary mirror (M1) in./out. diam., area	32.3/56.9 mm, 1716 mm^2
Primary mirror obscuration ratio	0.508
Optics transmission (total system)	0.781
T/# [$(f/\#) \times (\text{sys. transmission})^{-0.5}$]	$= 2.9 \times (0.508 \times 0.781)^{-0.5} = 4.7$
Pixel solid angle	2.5×10^{-9} sr
Throughput ($A_0 \bullet \Omega$)	5.11×10^{-6} mm^2 sr
Optics Distortion	<0.2%
Detector & Electronics	*Description*
CCD	ON Semi (Kodak) KAI-2020M interline transfer, operated in single output mode
Color	Monochrome
Array size	1600 × 1200 unmasked photoactive pixels (1648 × 1214 total)
Pixel size	7.4 µm (square pixels)
Preferred Dark Pixels	Left side: Columns 9-15; Right side: Columns 1633-1640 (where first column is 0)
Gain; Read Noise; Full Well (Imager A/B)[a]	$16.1 \pm 0.3/16.3 \pm 0.2$ e^-/DN; $12.9 \pm 0.1/13.6 \pm 0.1$ e^-; $30{,}352 \pm 35/28{,}202 \pm 21$ e^-
Quantum Efficiency	>46% at 550 nm
Radiometric Coeff. (Filter 1: Broadband)	$\sim 3.0 \pm 0.1 \times 10^{-8}$ [(W m^{-2} sr^{-1} nm^{-1}) / (DN/sec)] for both cameras[a]
Pixel Readout Rate	2.5, 5 (default), or 10 Mpix/sec
Digitization	12 bits/pixel optimized for 11-bit dynamic range (0-2047 DN); single gain, no offset states
Data Interface	Synchronous LVDS: 8 Mbit/sec
Command Interface	2 Mbit/sec serial link
Memory	128 MB SDRAM and 8 GB flash buffer (~4000 raw images) for each camera
Power	5.6 W standby and 8.7 W imaging, per camera

Table 4 (*Continued*)

Exposure	*Description*
Duration	0 to 838.8 sec; commanded in units of 0.1 msec
Auto-exposure	Based on MSL and MER auto-exposure algorithm (Maki et al. 2003)
Onboard Compression	*Description*
Uncompressed	12-bit raw data optimized for 11 bits of dynamic range (0-2047 DN)
Lossless	∼1.7:1 lossless predictive compression
Lossy	Realtime JPEG; compression quality 1–100
Companding	12-bit to 8-bit square-root (default) or linear encoding via lookup tables
Video Group of Pictures (GOP)	JPEG-compressed GOPs, $\leq$2 MB file size and $\leq$16 frames/GOP
Deferred compression	Images can be stored onboard the DEA uncompressed; specified compression can be performed at a later time for transfer to spacecraft and then downlink
Physical Characteristics	*Description*
Data Volume: full raw image	1648 × 1200 × 11 bits = 2.7 MBytes/image
Data Volume: 8-bit companded	1648 × 1200 × 8 bits = 2.0 MBytes/image
Data Volume: 8-bit + lossless comp.	∼1.2 MBytes/image, assuming ∼1.7:1 lossless compression
Dimensions: CCD Sensor	11.84 × 8.88 mm (14.8 mm diagonal)
DEA Dimensions	22 × 12.2 × 6.3 cm
Camera Head	15.5 × 11.5 × 34 cm (each camera)
Camera pointing	Imager A points nadir (along spacecraft −X axis), Imager B toed-out 3.7° along −Y axis
Mass: Camera Head	1.77 kg each
Mass: DEA	1.58 kg

[a] See Winhold et al. (2025) for more details and in-flight updates.

Shortly after launch in October 2023 and following the bakeout out of the Imager composite structures the following month (see Sect. 3.5), the first sets of in-flight images (of star fields) were acquired from both Imagers in December 2023 and January 2024. These images (Fig. 7) included standard stars and star clusters for radiometric and geometric validation, and long-duration images with the solar blocking filter (Filter 0) in place to assess CCD dark current and noise levels. Quantitative analyses of these results, which met expectations based on pre-flight testing, are reported in Winhold et al. (2025).

During the remainder of the Cruise phase of the Psyche mission, the Imagers will perform periodic maintenance and calibration observations of radiometric standard stars, star clusters, and distant multi-pixel-wide solar system objects (e.g., Jupiter) approximately every six months, a process started in mid-2024. In May 2026, the spacecraft will perform a Mars gravity assist flyby at a range of ∼5750 km (Polanskey et al. 2025), enabling the first in-flight imaging of a well-known field-filling extended object at a scale of ∼300 m/pix at closest approach. Mars flyby Imager data will be used to assess radiometric performance, calibration accuracy, and image quality, and to identify any significant changes in flatfield

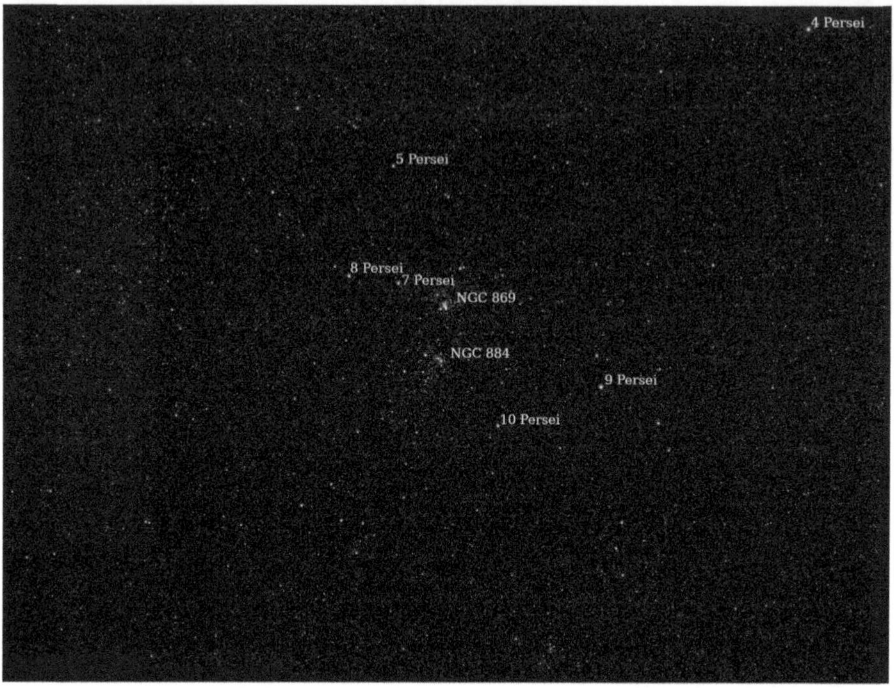

Fig. 7 Example of a 3 × 3 mosaic of images from Imager-B of a star field in the constellation Perseus centered on the famous double cluster NGC869 and NGC884 (also known as h and χ Persei). The images were acquired during initial checkout observations on Jan. 6, 2024. The clusters and some prominent stars are annotated. In-flight focus and performance of both Imagers are comparable based on cruise observations to date

variations (e.g., dust or other particles on the sensors) since pre-flight testing. Additionally, the Mars flyby will serve as a useful practice activity to simulate the later approach to Psyche and to exercise and refine relevant Imager sequencing, data flow, data processing/calibration, and data archiving activities.

5 Prime Mission Operations and Expected Data Products

5.1 Operational Considerations

As described in detail by Polanskey et al. (2025), the Psyche mission's Prime Mission activities are divided into specific phases called Approach, starting shortly after the asteroid begins to be an extended object rather than a point source; Orbits A-D, with nominal altitudes ranging from ~700 km to ~75 km; and Transition Orbit periods between Orbits A-D. Here we provide additional details on Imager-related activities and the estimated number and volume of data products (Table 5) planned to be acquired during each of these mission phases.

Table 5 Psyche Multispectral Imager Prime Mission Data Products Summary

Products	Estimated number of files	Estimated data volume (GB)
Raw (Level 0) binary images, all filters	~85,000	~350
Calibrated (L1A) Radiance and I/F images, all filters	~170,000	~700
Housekeeping data	~2000	~3
Pre-launch calibration files	~1000	<0.5
Calibration Support files	~120	~10
Software Interface Specification (SIS) and other processing documents	~14	negligible
Browse images	~85,000	~250
Derived (L1C) global clear image mosaic (20 m/pix scale)	1	~1
Derived (L1C) global color image mosaics (8 filters, 20 m/pix scale)	8	~20
Derived (L3) global spectral parameters maps (3-5)	3	~10
Derived (L3) global topographic map (DTM), from SPG	1	<0.5
Derived (L3) global topographic map (DTM), from SPC	1	<0.5
Derived global geologic map (from orbits B1 + B2)	1	~3
TOTALS:	**~ 340,000**	**~ 1400**

5.1.1 Approach to (16) Psyche

The Psyche mission's Approach phase to Psyche will begin in early 2029, as the asteroid slowly transitions from a small extended object of ~7 pixels diameter to more than twice the FOV at the end of Approach. During the Approach phase, Imager observations will mostly include OpNav imaging sessions. These sessions consist of imaging the asteroid against fields of background stars, while the asteroid is smaller than the FOV, and a combination of nadir and off-nadir imaging of the asteroid surface during the last week of the Approach phase, when the asteroid greatly exceeds the FOV. The purpose of the OpNav imaging is to assist in the determination of the spacecraft's trajectory relative to the asteroid, to estimate and refine the asteroid's spin state, and to construct the asteroid's shape and topography at increasing resolution, reaching up to 200 m/pix spatial scale at the end of the Approach. Most of the Approach OpNav sessions will span one full asteroid rotational period (~4.2 hr) and images will be acquired with the clear filter. However multispectral observations through all filters will be included in five OpNav sessions with each camera, to search for early evidence of regional albedo or mineralogic variations and to assist in establishing more accurate exposure estimates for imaging once in orbit. Additional imaging opportunities will include deep time-series surveys to cover the asteroid's Hill sphere of gravitational influence to search for any satellites, rings, or surface activity; and imaging on and off the bright limb of the asteroid, to characterize stray and scattered light within the cameras. Approach imaging will be performed with both Imagers separately to reduce mission risk, although simultaneous observations with both cameras could be used to make collecting the satellite survey imaging mosaic more time efficient, if needed.

5.1.2 Orbit A

The Psyche spacecraft is expected to be captured into a high polar orbit (~700 km altitude, 90° inclination) around Psyche in mid 2029. Imager observations during this "Orbit

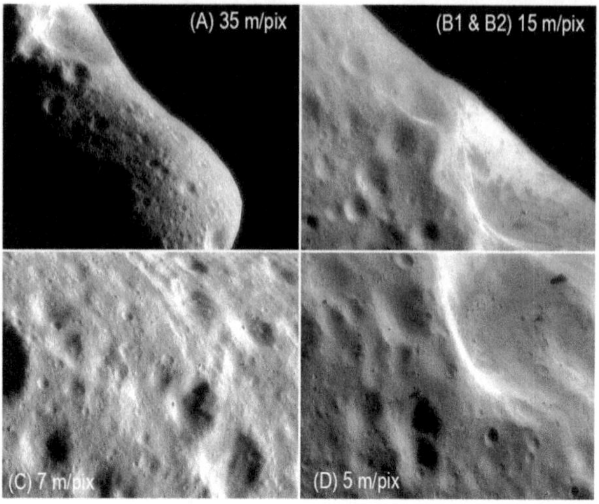

Fig. 8 Images of 433 Eros from the NEAR-Shoemaker mission at the approximate pixel scales representative of those typically expected during the Psyche mission's Orbits A, B1 and B2, C, and D, as labeled

A" phase of the mission, expected to nominally last ∼56 Earth days, will include initial assessment of the asteroid's geology, albedo, and multispectral properties at a relatively low resolution (∼35 m/pixel; e.g., Fig. 8a). The main focus of Orbit A for the Imager investigation will include identifying and characterizing craters and large boulders down to ∼100 m diameter, as well as generating initial global-scale maps of large-scale terrain differences based on albedo, color, morphology, and crater density (Marchi et al. 2022). The majority of Orbit A observations will be conducted at nadir pointing. After the asteroid has been imaged redundantly at nadir, a number of off-nadir observations will be acquired during Orbit A over a total of nine orbital periods. The main purpose of these observations will be to acquire stereo images that enable the estimation of surface topography at close to the image spatial scale, to assist spacecraft navigation for the later science orbits and also to refine the asteroid's spin state that was initially determined during the Approach phase. These observations are planned to be conducted using only Imager A.

5.1.3 Orbit B, Part 1 (Southern Hemisphere)

During an orbit transfer period, the spacecraft will descend to "Orbit B1" (∼300 km altitude, 90° inclination) in late 2029, for approximately three months of global clear filter and multispectral imaging at ∼15 m/pixel scale (e.g., Fig. 8b). Imager observations during this phase of the mission will meet all of the imaging resolution and coverage requirements of the mission (Table 3; see also Polanskey et al. 2025) for the southern hemisphere of Psyche, which is illuminated during this time because of the asteroid's ∼95° obliquity. Meeting these requirements will enable the creation of hemispheric-scale topographic maps for shape modeling, geologic mapping, and topographic mapping (e.g., Jaumann et al. 2022; Zuber et al. 2022), relative age determinations via crater size frequency statistics (e.g., Marchi et al. 2022), as well as assessment of the presence and distribution of both silicates and sulfides on the asteroid's surface (e.g., Dibb et al. 2023).

5.1.4 Orbit D

Then, after a much longer orbit transfer period that includes a transition to a near-equatorial orbit (∼160° inclination), by early 2030 the spacecraft will descend to its lowest Prime Mission altitude of ∼75 km. During this 100-day "Orbit D" phase of the mission, the Imager will obtain its highest resolution images (up to ∼4 m/pixel; e.g., Fig. 8d), focusing on clear filter imaging for characterization and mapping of smaller (20-30 m diameter) craters, boulders, boulder clusters, and fine surface details regarding potential regolith movement and any features suggestive of volcanic activity or tectonic deformation (e.g., Robinson et al. 2002; Sullivan et al. 2002). A limited amount of multispectral imaging could also be conducted in Orbit D, especially if small-scale features with interesting or anomalous multispectral properties are identified in images from Orbits A or B1. However, the high expected ground track motion in the image plane at this low altitude (∼30 pixels/sec) means that only very short exposure times can be used to keep image smear below 0.5 pixels, limiting the SNR of very high resolution multispectral images. Most imaging will likely be performed using only the primary Imager A camera, although simultaneous observations with both cameras during this low orbit phase could be used to cover more of the surface per unit time with high-resolution images. Imaging in Orbit D is limited to nadir-pointed attitudes and by the volume of data that can be returned to the ground. Margin is built into the nominal Orbit D plan to accommodate any extra time needed transitioning from Orbit B1.

5.1.5 Orbit C

The spacecraft will then ascend to "Orbit C," returning to a near-polar orbital inclination at an altitude of ∼190 km in early 2031. Because of the advancing seasons on Psyche since the beginning of Orbit A, Imager data at ∼10 m/pixel (e.g., Fig. 8c) during this ∼100 day phase of the mission will increase geologic and topographic mapping coverage deeper into the asteroid's northern hemisphere, and will enable the first regional high-resolution multispectral imaging of northern equatorial to mid-latitudes.

5.1.6 Orbit B, Part 2 (Northern Hemisphere)

For the final phase of the Prime Mission, the spacecraft will transition back to an altitude of ∼300 km in mid 2031 for "Orbit B2", to conduct the same kind of imaging study as in Orbit B1, but covering more of the northern hemisphere. Thus, by the end of Orbit B2, Imager observations will have met all of the global imaging resolution and coverage requirements of the mission (Table 3; see also Polanskey et al. 2025).

Approximately two months of additional operations margin are built into the Psyche mission plan, to accommodate potential anomalies or surprises that could cause any of the Orbit phases or the transitions between them to require longer than expected to achieve relevant mission or Imager goals or objectives. The nominal Prime Mission is currently scheduled to end on Nov. 1, 2031. Polanskey et al. (2025) provide significantly more detail about the Psyche mission's nominal expected orbital operations and timeline.

5.2 Imager Commanding, Sequencing, Onboard Image Processing, and Data Management

Commanding of the Imager is performed through ground developed command sequences which are uplinked to the spacecraft and executed at their pre-defined epoch time. For orbital phase imaging, the command sequences make use of reusable sequences, stored onboard the spacecraft's Psyche Compute Element (PCE) computer, that perform repeatable sets of command. At specified times, the PCE sends the commands in those sequences to the Imager DEA, which then commands the Imager Camera Head (CH) electronics to acquire the specified images (potentially sub-framing the image within the CH electronics), change the required filter, etc. Images are transferred from the CH to the DEA one at a time after acquisition and are stored in the DEA's flash memory as raw Camera Data Products, each with a unique memory location identifier (CDPID). When an image is acquired, the spacecraft Flight Software (FSW) adds the CDPID to a data transfer queue. After the acquisition of a specified number of images (for example, a swath of images across the sunlit dayside of the asteroid during a particular orbital phase), the imaging sequence commands the transfer of the images in the transfer queue, using the desired companding, compression, and/or summing factor, from the Imager DEA to the PCE for eventual downlink to Earth.

DEA processing of Imager data transferred to the PCE is expected to be relatively minimal. Nominally, this is expected to always include companding of the data from 12-bit to 8-bit using a default square-root lookup table developed during pre-flight calibration (Winhold et al. 2025), to avoid downlinking pixel raw signal level variations that correspond to photon shot noise variations, especially at higher DN levels (a strategy used effectively on almost all MSSS deep space cameras on previous NASA missions; e.g., Bell et al. 2017). As an additional processing step, when images are square-root companded (the default behavior), a user-defined bias level (specified by the DC_OFFSET parameter and set to a 12-bit DN level of 110 by default) is automatically subtracted from the 12-bit data before it is companded to 8-bit, to help maximize the dynamic range of the companded images. The DEA can compress images prior to transfer to the PCE using lossless predictive (the default for Imager operations) or lossy JPEG compression. Optionally, the DEA can also be commanded to sum pixels in existing stored images by factors of 2×2, 3×3, or 4×4 pixels (returning the average pixel value) and transmit those smaller images back to the PCE.

The 8 GB of DEA NVM storage can hold ~4000 raw uncompressed 8-bit images per Imager, so this is a resource that will need to be managed during orbital operations. The NVM used for image storage is allocated using a simple first-fit algorithm without wear leveling. The Imager Operations team at Arizona State University (ASU) will monitor the number of images and their locations in memory for each Imager, using directory and memory dump commands. In general, images will be erased in large groups, with older images being erased before newer ones, and images will only be erased from the DEA after verification of their successful downlink. There must always be enough available memory space (plus margin) to allow for a missed DSN downlink pass without overrunning the DEA NVM.

5.3 Calibration Observations During the Orbital Phases of the Mission

Orbital calibrations will primarily include dark current measurements before or after each swath of imaging of the sunlit surface of the asteroid. Images of standard stars and star clusters for radiometric, pointing, and geometric validation, and imaging on and off the limb of Psyche for stray light characterization, are not planned but could be commanded if required.

5.4 Downlink Assessment and Data Processing/Calibration

Images and other data from the Imagers, DEA, and Imager-relevant spacecraft telemetry channels (e.g., spacecraft-controlled external temperature sensors on/near the Imagers or DEA) are downlinked to Earth through NASA's Deep Space Network and then to the Psyche Mission Operations Center at JPL. The raw images are unpacked from the telemetry stream and converted by JPL into individual 8-bit *.DAT science data and *.EMD telemetry packet files and then into *.IMG files with embedded keyword-value label information that is compliant with NASA Planetary Data System (PDS) 3 Object Description Language (ODL) format.[1] These images are used by the JPL OpNav team and the Psyche Gravity investigation team (Zuber et al. 2022), and are also immediately converted to Portable Network Graphics (PNG) format images that are posted to the NASA/JPL "Psyche Raw Images" public web site[2] for everyone to enjoy.

The ASU Imager Operations team merges additional relevant telemetry from the temperature sensors external to the Camera Heads and DEA as well as updated image timing and other mission/telemetry information into expanded version of the raw image files that are saved in the Flexible Image Transport System (FITS) file format (e.g., Marmo et al. 2018). The team also generates an associated additional Extensible Markup Language (XML) label file that makes these raw Imager team data products compliant with the NASA PDS4 archive format.[3]

Initial downlink assessment by the Imager Operations team focuses on verifying the health and safety of the Imagers and DEA via monitoring for appropriate temperatures, voltages, and currents in the Imager and spacecraft telemetry. Additional assessments include sequence completion status and run times, data downlink completeness, verification of expected spacecraft pointing, and verification of adequate signal level/SNR in each image. Relevant Imager data and telemetry are captured into an SQL database that is maintained in collaboration with the Psyche mission's Science Data Center (SDC), also located at ASU.

In the event of downlink of incomplete (partial) raw data products or telemetry, Imager operations staff work with the JPL Psyche Mission Operations staff to request retransmits of the missing telemetry packets, and then re-creation of the associated raw images or telemetry data.

Individual raw 8-bit images are calibrated by the Imager operations staff into floating point absolute radiance units (e.g., $W/m^2/nm/sr$) via the calibration pipeline described by Winhold et al. (2025). The pipeline also creates images further calibrated to radiance factor, also known as "I/F", where I is the observed radiance and πF is the radiance of sunlight incident on the scene at the time of the observation (e.g., Hapke 1993). Like the raw images, individual radiance- and I/F-calibrated images are saved in a NASA PDS4-compliant format, including an external XML label file that contain additional keyword-value information relevant to the calibration of the images.

I/F is comparable to reflectance, enabling comparisons of the Imager's multispectral data to telescopic reflectance spectra of Psyche and other asteroids as well as laboratory reflectance spectra of analogous minerals and meteorites. For isotropically scattering planetary surfaces, I/F divided by the cosine of the solar incidence angle provides an estimate of the surface geometric albedo. More generally, variations of I/F as a function of illumination and viewing geometry can be used to infer information about the physical properties of the surface of Psyche, including particle size, macroscopic surface roughness, and forward vs. backward scattering properties that can inform estimates of surface composition (e.g., Hapke 1993; Gaffey et al. 1993; Clark et al. 2004).

Calibrated clear and multispectral filter images of Psyche will be mosaicked onto a global latitude/longitude grid using the US Geological Survey's Integrated Software for Imagers and Spectrometers (ISIS) routines (e.g., Anderson et al. 2004) and the camera model information needed for orthorectification of images described in Winhold et al. (2025). Clear and color filter image mosaics will be used to create geologic maps, 3-D anaglyphs, and spectral

[1] https://pds.nasa.gov/datastandards/pds3/standards.

[2] https://solarsystem.nasa.gov/psyche-raw-images.

[3] https://pds.nasa.gov/datastandards/about.

feature maps used to infer surface composition and/or mineralogy (e.g., Dibb et al. 2023; Polanskey et al. 2025).

Calibrated images acquired over a range of viewing and illumination angles will also be processed via SPC and SPG techniques to create the global shape and topographic models as well as the local/regional DTMs required to achieve a number of primary and Imager-related Psyche mission requirements (Tables 1 and 2; e.g., Jaumann et al. 2022; Zuber et al. 2022). These topographic image products will also be used to generate a number of special 3-D products expected to be useful for education and public outreach, including red/blue anaglyphs, "flyover" movies, and other special products.

5.5 Archive Products

Individual raw and calibrated Imager data products will be archived by the Psyche SDC in the NASA Planetary Data System's Small Bodies Node on approximately 3-month intervals during the orbital mission. Archived images will be compliant with FITS and NASA PDS4 data formats; the former allows the Imager team (and the broader planetary science community) to leverage existing/heritage tools and processes, the latter is a newer data archiving format required by NASA for archiving of all Psyche mission data sets.

Additionally, once the global coverage requirements have been met, a number of additional derived Imager data products will also be archived in the PDS, in PDS4 compliant format. These include global clear-filter and multispectral filter image maps at $\leq$20 m/pixel scale; global spectral parameter maps, including pseudo true color ($\leq$20 m/pixel), sulfide distribution ($\leq$200 m/pixel), and silicate distribution ($\leq$500 m/pixel); a global topographic map derived using SPG methods; a tessellated-plates global shape model derived using SPC methods; and a global geologic unit map (e.g., Jaumann et al. 2022; Zuber et al. 2022; Dibb et al. 2023; Polanskey et al. 2025).

6 Summary

This paper describes the science goals, requirements, design, testing, and initial flight performance of the NASA Psyche mission's Psyche Multispectral Imager ("the Imager") instrument and investigation. The Imager consists of a pair of block redundant 148 mm focal length, $f/2.9$ cameras (and their associated electronics) that are nominally pointed in the nadir direction during orbital imaging of the asteroid (16) Psyche. The primary camera ("Imager A") is pointed along the spacecraft $-$X axis, and the backup camera ("Imager B") is toed-out by 3.7° to potentially enable greater surface area coverage per unit time if both Imagers are operated simultaneously during some mission phases. Each camera consists of a 1648 $\times$ 1200 active pixel charge-coupled device (CCD) detector and its associated electronics, a 9-position filter wheel assembly that enables broadband polychromatic ($\sim$540 $\pm$ 125 nm) imaging and narrowband color imaging in 7 wavelengths spanning 400-1000 nm, and a multi-baffle Sun shade to minimize scattered light. Each camera has a FOV of 4.6° $\times$ 3.4° and an IFOV of 50 μrad/pixel that enables imaging of the asteroid at scales ranging from $\sim$35 m/pix from 700 km altitude to $\sim$4 m/pix at 75 km altitude. Off-nadir pointing will be used during orbital operations to acquire stereo imaging coverage that enables the creation of surface topographic maps and DTMs with $\leq$50 m vertical precision (e.g., Jaumann et al. 2022). The two Camera Heads are connected to a separate Digital Electronics Assembly (DEA) box that interfaces to the spacecraft avionics and that provides power, commanding, data processing, and onboard image storage. The Imager system shares significant heritage with imaging instruments flown on the *Mars Climate Orbiter*, *Mars Science Laboratory* and *Mars 2020* rovers, and *Juno*. Additional details about the pre-flight testing and in-flight calibration of the Imagers is provided in a companion paper by Winhold et al. (2025).

Acknowledgements The design, fabrication, testing, integration, launch, and in-flight operation of the Psyche Multispectral Imagers and the Psyche spacecraft could not have occurred without the incredible skill, teamwork, and camaraderie of the extended Psyche mission, payload, and spacecraft teams at JPL, Maxar Technologies, and Astrotech Space Operations, and the launch and administrative support teams at Space Exploration Technologies and at NASA's Kennedy Space Center and Marshall Space Flight Center. We would like to specifically thank Rob Calvet and Brian Monacelli at JPL for their critical help in modifying and testing the ultimate flight Imager optical bench design. We are also grateful to the many commercial vendors and subcontractors who provided components and/or subsystems for the Psyche Multispectral Imagers, including colleagues at Motiv Space Systems, Inc. (filter wheel mechanisms), II-VI, Inc. (optical components), and Materion Corp. (filters). Many of the people involved in Imager development and testing and Psyche mission development, testing, and launch made significant personal sacrifices to continue their work during the unexpected global pandemic of 2020-2022, for which we express additional sincere gratitude. The Imager design, development, and testing, as well as subsequent operations, is supported by a Psyche mission subcontract to Arizona State University from the NASA/MSFC Planetary Missions Program Office.

Declarations

Competing Interests - Authors M.A. Ravine, M.A. Caplinger, J.A. Schaffner, S.M. Brylow, M.J. Clark, D.A. Peckham, P.T. Otjens, G.J. Price, T. Rowell, and J.W. Ravine are employees of Malin Space Science Systems, Inc., which has a financial interest in the subject matter discussed in this manuscript.

- Author J.D. Laramee is an employee of Motiv Space Systems, Inc., which has a financial interest in the subject matter discussed in this manuscript.

- Authors R.C. Juergens, W. Morgan, and A.G. Parker are employees of II-VI Optical Systems, Inc., which has a financial interest in the subject matter discussed in this manuscript.

- All other authors have no conflicts of interest.

References

Anderson JA, Sides SC, Soltesz DL, Sucharski TL, Becker KJ (2004) Modernization of the integrated software for imagers and spectrometers. In: 35th lunar planet sci conf. Abstract #2039

Bell JF III, Godber A, McNair S, Caplinger MC, Maki JN, Lemmon MT, Van Beek J, Malin MC, Wellington D, Kinch KM, Madsen MB, Hardgrove C, Ravine MA, Jensen E, Harker D, Anderson RB, Herkenhoff KE, Morris RV, Cisneros E, Deen RG (2017) The Mars Science Laboratory Curiosity rover mast camera (mastcam) instruments: pre-flight and in-flight calibration, validation, and data archiving. Earth Space Sci 4:396–452. https://doi.org/10.1002/2016EA000219

Bell JF III, Maki JN, Mehall GL, Ravine MA, Caplinger MA, Bailey ZJ, Brylow S, Schaffner JA, Kinch KM, Madsen MB, Winhold A, Hayes A, Corlies P, Barrington M, Deen R, Cisneros E, Jensen E, Paris K, Crawford K, Rojas C, Mehall L, Joseph J, Proton JB, Cluff N, Betts B, Cloutis E, Coates A, Colaprete A, Edgett KS, Ehlmann BL, Fagents S, Grotzinger J, Tate C, Hardgrove C, Herkenhoff K, Horgan B, Jaumann R, Johnson JR, Lemmon M, Paar G, Caballo-Perucha M, Gupta S, Traxler C, Preusker F, Rice M, Robinson MS, Schmitz N, Sullivan R, Wolff MJ (2020) The Mars 2020 rover mast camera zoom (Mastcam-Z) multispectral, stereoscopic imaging investigation. Space Sci Rev 217:24. https://doi.org/10.1007/s11214-020-00755-x

Bottke WF, Vokrouhlický D, Ballouz RL, Barnouin OS, Connolly HC, Elder C, Marchi S, McCoy TJ, Michel P, Nolan MC, Rizk B, Scheeres DJ, Schwartz SR, Walsh KJ, Lauretta DS (2020) Interpreting the cratering histories of Bennu, Ryugu, and other spacecraft-explored asteroids. Astron J 160:14. https://doi.org/10.3847/1538-3881/ab88d3

Burbine TH, McCoy TJ, Nittler LR, Benedix GK, Cloutis EA, Dickinson TL (2002) Spectra of extremely reduced assemblages: implications for Mercury. Meteorit Planet Sci 37:1233–1244. https://doi.org/10.1111/j.1945-5100.2002.tb00892.x

Bus SJ, Binzel RP (2002) Phase II of the small main-belt asteroid spectroscopic survey: the observations. Icarus 158:106–145. https://doi.org/10.1006/icar.2002.6857

Clark BE, Bus SJ, Rivkin AS, McConnochie T, Sanders J, Shah S, Hiroi T, Shepard M (2004) E-type asteroid spectroscopy and compositional modeling. J Geophys Res 109:1–11. https://doi.org/10.1029/2003JE002200

Cloutis EA (2002) Pyroxene reflectance spectra: minor absorption bands and effects of elemental substitutions. J Geophys Res 107:E6. https://doi.org/10.1029/2001JE001590

Cloutis EA, Gaffey MJ (1991) Pyroxene spectroscopy revisited: spectral-compositional correlations and relationship to geothermometry. J Geophys Res 96:22809–22826. https://doi.org/10.1029/91je02512

Cloutis EA, Hardersen PS, Reddy V, Gaffey MJ, Bailey DT, Craig MA (2009) Metal-orthopyroxene and metal-olivine mixtures: spectral reflectance properties and implications for asteroid spectroscopy. In: 40th lunar planet sci conf. Abstract #1332

Cloutis EA, Hardersen PS, Bish DL, Bailey DT, Gaffey MJ, Craig MA (2010) Reflectance spectra of iron meteorites: implications for spectral identification of their parent bodies. Meteorit Planet Sci 45:304–332. https://doi.org/10.1111/j.1945-5100.2010.01033.x

Dibb SD, Bell JF III, Garvie LA (2021) Reflectance spectra of metal-troilite mixtures: implications for m-/x-type asteroid exploration. In: 52nd lunar planet sci conf. Abstract #1543

Dibb SD, Bell JF III, Garvie L (2022) Spectral reflectance variations of aubrites, metal-rich meteorites, and sulfides: implications for exploration of (16) Psyche and other 'spectrally featureless' asteroids. Meteorit Planet Sci 57:1570–1588. https://doi.org/10.1111/maps.13891

Dibb SD, Bell JF III, Elkins-Tanton LT, Williams DA (Psyche Mission Team) (2023) Visible to near-infrared reflectance spectroscopy of asteroid (16) Psyche: implications for the Psyche mission's science investigations. Earth Space Sci 10:1. https://doi.org/10.1029/2022EA002694

Elkins-Tanton LT, Asphaug E, Bell JF III, Bercovici H, Bills B, Binzel RP, Bottke WF, Dibb S, Lawrence DJ, Marchi S, McCoy TJ, Oran R, Park RS, Peplowski PN, Polanskey CA, Prettyman TH, Russell CT, Schaefer L, Weiss BP, Wieczorek MA, Williams DA, Zuber MT (2020) Observations, meteorites, and models: a pre-flight assessment of the composition and formation of (16) Psyche. J Geophys Res 125:e2019JE006296. https://doi.org/10.1029/2019JE006296

Elkins-Tanton LT, Asphaug E, Bell JF III, Bills B, Bottke W, Jun I, Lawrence D, Marchi S, McCoy T, Merayo J, Oran R, Park R, Peplowski P, Prettyman T, Raymond C, Weiss B, Wieczorek M, Zuber M (2022) Distinguishing the origin of asteroid (16) Psyche. Space Sci Rev 218:17. https://doi.org/10.1007/s11214-022-00880-9

Fornasier S, Migliorini A, Dotto E, Barucci MA (2008) Visible and near infrared spectroscopic investigation of E-type asteroids, including 2867 Steins, a target of the Rosetta mission. Icarus 196:119–134. https://doi.org/10.1016/j.icarus.2008.02.015

Fornasier S, Clark BE, Dotto E, Migliorini A, Ockert-Bell M, Barucci MA (2010) Spectroscopic survey of M-type asteroids. Icarus 210:655–673. https://doi.org/10.1016/j.icarus.2010.07.001

Gaffey SJ, McFadden LA, Nash DB (1993) Ultraviolet, visible, and near-infrared reflectance spectroscopy: laboratory spectra of geologic materials. In: Pieters C, Englert P (eds) Remote geochemical analysis: elemental and mineralogical composition. Cambridge University Press, Cambridge, pp 43–71

Hansen CJ, Caplinger MA, Ingersoll A, Ravine MA, Jensen E, Bolton S, Orton G (2017) Junocam: Juno's outreach camera. Space Sci Rev 213:475–506. https://doi.org/10.1007/s11214-014-0079-x

Hapke B (1993) Theory of reflectance and emittance spectroscopy. Cambridge University Press, New York

Hayes AG, Corlies P, Tate C, Bell JF III, Maki JN, Caplinger M, Kinch KM, Herkenhoff K, Horgan B, Johnson J, Parr G, Rice MS, Jensen E, Kubacki TM, Cloutis E, Ehlmann B, Lakdawalla E, Sullivan R, Winhold A, Barrington M, Parkinson A, van Beek J, Caballo-Perucha P, Cisneros E, Dixon D, Donaldson C, Jensen OB, Kuik J, Lapo K, Magee A (2020) Pre-flight calibration of the Mars 2020 rover mastcam zoom (Mastcam-Z) multispectral, stereoscopic imager. Space Sci Rev 217:29. https://doi.org/10.1007/s11214021-00795-x

Jaumann R, Bell JF III, Polanskey CA, Raymond CA, Aspaugh E, Bercovici D, Bills BR, Binzel R, Bottke W, Christoph JM, Marchi S, Neesemann A, Otto K, Park RS, Preusker F, Roatsch T, Williams DA, Wieczorek MA, Zuber MT (2022) The Psyche topography and geomorphology investigation. Space Sci Rev 218:7. https://doi.org/10.1007/s11214-022-00874-7

Lawrence DJ, Goldsten JO, Peplowski PN, Burks MT, Cheng S, Cully MJ, Effron JM, Elkins-Tanton LT, Espiritu RC, Fix SG, Graziano MB, Hoffer EM, Jun I, Kim G-B, Hines NR, LeBlanc MT, Livingstone EM, Marcotte KM, McCoy TJ, Polanskey CA, Sreekantamurthy M, Yokley ZW (2025) The Psyche Gamma-Ray and Neutron Spectrometer. Space Sci Rev 221

Maki JN, Bell JF III, Herkenhoff KE, Squyres SW, Kiely A, Klimesh M, Schwochert M, Litwin T, Willson R, Johnson A, Maimone M, Baumgartner E, Collins A, Wadsworth M, Elliot ST, Dingizian A, Brown D, Hagerott EC, Scherr L, Deen R, Alexander D, Lorre J (2003) The Mars Exploration Rover Engineering Cameras. J Geophys Res 108:8071. https://doi.org/10.1029/2003JE002077

Malin M, Bell JF III, Calvin W, Clancy RT, Haberle RM, James PB, Lee SW, Thomas PC, Caplinger MA (2001) The Mars color imager (MARCI) on the Mars climate orbiter. J Geophys Res 106:17651–17672. https://doi.org/10.1029/1999JE001145

Malin MC, Calvin WM, Cantor BA, Clancy RT, Haberle RM, James PB, Thomas PC, Wolff MJ, Bell JF III, Lee SW (2008) Climate, weather, and north polar observations from the Mars reconnaissance orbiter Mars color imager. Icarus 194:501

Malin MC, Ravine MA, Caplinger MA, Ghaemi FT, Schaffner JA, Maki JN, Bell JF III, Cameron JF, Dietrich WE, Edgett KS, Edwards LJ, Garvin JB, Hallet B, Herkenhoff KE, Heydari E, Kah LC, Lemmon MT, Minitti ME, Olson TS, Parker TJ, Rowland SK, Schieber J, Sletten R, Sullivan RJ, Sumner DY, Yingst RA, Duston BM, McNair S, Jensen EH (2017) The Mars Science Laboratory (MSL) mast cameras and descent imager: I. Investigation and instrument descriptions. Earth Space Sci 4:506–539. https://doi.org/10.1002/2016EA000252

Marchi S, Asphaug E, Bell JF III, Bottke WF, Jaumann R, Park RS, Polanskey CA, Prettyman TH, Williams DA, Binzel R, Oran R, Weiss B, Russell CT (2022) Determining the relative cratering ages of regions of Psyche's surface. Space Sci Rev 218:24. https://doi.org/10.1007/s11214-022-00891-6

Marchi S, Bell JF III, Bierhaus B, Spencer J (2023) Surface geology of Jupiter's Trojan asteroids. Space Sci Rev 219:44. https://doi.org/10.1007/s11214-023-00985-9

Marmo C, Hare TM, Erard S, Minin M, Pineau FX, Zinzi A, Cecconi B, Rossi AP (2018) FITS format for planetary surfaces: definitions, applications, and best practices. Earth Space Sci 5:640–651. https://doi.org/10.1029/2018EA000388

Mastrodemos N, Rush B, Vaughan A, Owen W (2012) Optical navigation for the Dawn mission at Vesta. In: 23rd International Symposium on Space Flight Dynamics, Pasadena, CA. https://ntrs.nasa.gov/citations/20150004573

Mastrodemos N, Rush PB, Owen WM Jr (2015) Optical navigation for the Rosetta mission. In: AAS guidance and control conference, Breckenridge, Colorado. Paper 15-123

McCoy TJ, Dibb SD, Peplowski PN, Maurel C, Bercovici HL, Corrigan CM, Bell JF III, Weiss BP, Lawrence DJ, Wenkert DD, Prettyman TH, Elkins-Tanton LT (2022) Deciphering redox state for a metal-rich world. Space Sci Rev 218:6. https://doi.org/10.1007/s11214-022-00872-9

ON Semiconductor Inc (2015) KAI-2020 1600 (H) x 1200 (V) Interline CCD Image Sensor. Pub. KAI-2020/D. http://www.onsemi.com/pub/Collateral/KAI-2020-D.PDF

Polanskey CA, Elkins-Tanton LT, Bell JF III, Alonge NK, Bairstow S, Bass D, Biswas A, Cisneros E, Han D, Jun I, Klipstein WM, Lawrence DJ, McCoy T, Mastrodemos N, Merayo J, Noble SK, Oh D, Oran R, Park RS, Peplowski P, Prettyman T, Ream J, Russell CT, Seal D, Weise T, Wenkert D, Weiss BP, Zuber M (2025) Psyche mission description and design rationale. Space Sci Rev 221

Robinson MS, Thomas PC, Veverka J, Murchie SL, Wilcox BB (2002) The geology of 433 Eros. Meteorit Planet Sci 37:1651–1684. https://doi.org/10.1111/j.1945-5100.2002.tb01157.x

Schaffner JA, Caplinger MA, Ravine MA, Hansen CJ, Madsen SN, Berkun AC, Rax BG, Lipkaman Vittling L, Krysak DJ, Johnson MB, Sturm EJ, Delavan JA, Yonter H, Bolton SJ (2025) Recovery of JunoCam by annealing in the Jovian radiation environment. In: IEEE nuclear and space radiation effects conference, Nashville, TN

Sullivan RJ, Thomas PC, Murchie SL, Robinson MS (2002) Asteroid geology from Galileo and NEAR Shoemaker data. In: Bottke WF Jr, Cellino A, Paolicchi P, Binzel RP (eds) Asteroids III. University of Arizona Press, Tucson, pp 331–350

Williams DA, Jaumann R, McSween HY Jr, Raymond CA, Marchi S, Schmedemann N, Russell CT (2014) The chronostratigraphy of protoplanet Vesta. Icarus 244:158–165. https://doi.org/10.1016/j.icarus.2014.06.027

Winhold A, Bell JF III, Caplinger MA, Brylow S, Ravine M, Kubacki T, Dahl-Hansen R, Walworth M, Dibb S, Zhao A, Rodea Y, Cisneros E, Williams D, Warner N, Bates-Tarasewicz H, Monacelli B, Mastrodemos N (2025) Pre-flight and in-flight calibration of the Psyche Multispectral Imagers. Earth Space Sci. to be submitted

Zuber MT, Park RS, Elkins-Tanton LT, Bell JF III, Bruvold KN, Bercovici D, Bills BR, Binzel RP, Jaumann R, Marchi S, Polanskey CA, Raymond CA, Roatsch T, Wang CC, Weiss BP, Wenkert D, Wieczorek MA (2022) The Psyche gravity investigation. Space Sci Rev 218:57. https://doi.org/10.1007/s11214-0

Authors and Affiliations

J.F. Bell III[1] · M.A. Ravine[2] · M.A. Caplinger[2] · J.A. Schaffner[2] · S.M. Brylow[2] · M.J. Clark[2] · D.A. Peckham[2] · P.T. Otjens[2] · G.J. Price[2] · T. Rowell[2] · J.W. Ravine[2] · J.D. Laramee[3] · R.C. Juergens[4] · W. Morgan[4] · A.G. Parker[4] · D.A. Williams[1] · A. Winhold[1] · S. Dibb[1] · E. Cisneros[1] · M. Walworth[1] · H. Zigo[1] · L. Auchterlonie[1] · N. Warner[5] · H. Bates-Tarasewicz[5] · N. Amiri[5] · C. Polanskey[5] · N. Mastrodemos[5] · R.S. Park[5] · N.K. Alonge[5] · R. Jaumann[6] · R.P. Binzel[7] · T.J. McCoy[8] · M.G. Martin[5] · P.A. Arthur[5]

✉ J.F. Bell III
Jim.Bell@asu.edu

1 Arizona State University, School of Earth & Space Exploration, Tempe AZ, USA

2 Malin Space Science Systems, Inc., San Diego CA, USA

3 Motiv Space Systems, Pasadena, CA, USA

4 II-VI Optical Systems, Tustin, CA, USA

5 Jet Propulsion Laboratory/California Institute of Technology, Pasadena CA, USA

6 Freie Universität Berlin, Institute of Geological Sciences, Berlin, Germany

7 Massachusetts Institute of Technology, Cambridge, MA, USA

8 Smithsonian Institution, Washington, DC, USA

Space Science Reviews (2025) 221:78
https://doi.org/10.1007/s11214-025-01201-6

The Psyche Gamma-Ray and Neutron Spectrometer

David J. Lawrence[1] · John O. Goldsten[1] · Patrick N. Peplowski[1] · Morgan T. Burks[2] · Shuo Cheng[1] · Michael J. Cully[1] · Jordan M. Effron[1,3] · Linda T. Elkins-Tanton[4,5] · Raymond C. Espiritu[1] · Samuel G. Fix[1] · Milena B. Graziano[1] · Erin M. Hoffer[1] · Insoo Jun[6] · Geon-Bo Kim[2] · Nathan R. Hines[2] · Mark T. LeBlanc[1] · Evan M. Livingstone[1,7] · Kathryn M. Marcotte[1] · Timothy J. McCoy[8] · Carol A. Polanskey[6] · Meena Sreekantamurthy[1] · Zachary W. Yokley[1]

Received: 14 December 2024 / Accepted: 24 July 2025 / Published online: 1 September 2025

Abstract
A Gamma-Ray and Neutron Spectrometer (GRNS) instrument has been developed as part of the science payload for NASA's Discovery Program Psyche mission to the M-class asteroid (16) Psyche. The GRNS instrument is designed to measure the elemental composition of Psyche with the goal to understand the origin of this mysterious, potentially metal-rich planetary body. The GRNS will measure the near-surface abundances for the elements Ni, Fe, Si, K, S, Al, and Ca, as well as the spatial distribution of Psyche's metal-to-silicate fraction (or metal fraction). These measurements address three of the five Psyche mission science objectives: determine if Psyche is a core; determine whether small metal bodies incorporate light elements into the metal phase; and determine whether Psyche was formed under reducing conditions. The Gamma-Ray Spectrometer (GRS) uses a cryocooled, high-purity Ge (HPGe) sensor to detect cosmic-ray generated gamma rays in the 60 to 9000-keV energy range. The HPGe sensor is surrounded by a borated plastic anticoincidence shield that provides three functions: active background rejection from charged particle interactions in the HPGe sensor; fast neutron measurements; and direct measurements of the incident galactic cosmic ray flux. The Neutron Spectrometer (NS) uses three ^{3}He gas proportional sensors, each with different material wraps to measure thermal (<0.4 eV), low-energy epithermal (0.4 eV to 1 keV), and high-energy epithermal (up to 100 keV) neutrons. This paper provides an overview of the Psyche GRNS, including: its science and measurement objectives; the design of the instrument hardware, software, and operation; pre-launch performance measurements and its initial performance in space; and an overview of its data products and expected operation for different Psyche mission phases.

1 Introduction

NASA's Psyche mission will carry out orbital reconnaissance of main-belt asteroid (16) Psyche (hereafter called Psyche), which belongs to the M spectral class, a group of solar system objects whose red-sloped, featureless reflectance spectra are consistent with laboratory measurements of metal-dominated surfaces (Vernazza and Beck 2017). Psyche was chosen for detailed study with the expectation that it might represent the exposed core of

Extended author information available on the last page of the article

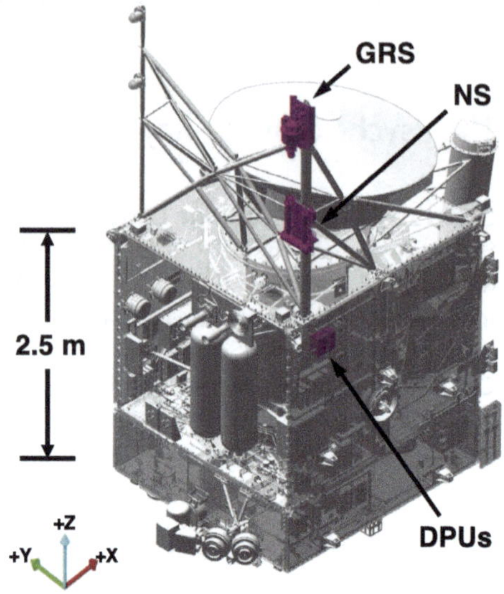

Fig. 1 View of the Psyche spacecraft highlighting the locations and relative dimensions of the GRNS and the spacecraft coordinate system. The GRS and NS are located on -Y boom, with the GRS at the end of the boom and NS at the boom midpoint. The two GRNS DPUs are located on inside the spacecraft on the -Y panel, which is illustrated by the DPU label pointing to their location. Additional components of the Psyche spacecraft are described by other papers in this collection (e.g., Polanskey et al. 2025, this collection)

a proto-planetary body (Elkins-Tanton et al. 2017, 2020). To accomplish its science objectives, the Psyche spacecraft carries a payload of three instruments: a magnetometer (Weiss et al. 2023, this collection), redundant multispectral imagers (Bell et al. 2025, this collection), and a gamma-ray and neutron spectrometer (GRNS) (Fig. 1). The mission will also carry out gravity science using its radio system (Zuber et al. 2022, this collection).

The goal of the Psyche GRNS is to quantify Psyche's surface elemental composition using the established technique of planetary nuclear spectroscopy, where cosmic-ray generated gamma rays and neutrons are passively measured from orbit in close proximity to an airless body (Reedy 1978; Evans et al. 1993; Feldman et al. 1993). Each measured element produces a unique "fingerprint" of gamma-ray energy lines that identifies the element that produced them. The number of gamma rays at these characteristic energies is used to quantify the abundance of that element. Neutrons are measured in broad energy ranges, i.e., not specific to a given element, but where each energy range is sensitive to a different mix of elements such that compositional information can be inferred.

Planetary nuclear spectroscopy is a mature technique, having made compositional measurements of multiple planetary bodies such as the Moon, Mars, Mercury, and the asteroids Eros, Vesta, and Ceres (e.g., Goldsten et al. 1997; Boynton et al. 2004; Feldman et al. 2004; Goldsten et al. 2007; Prettyman et al. 2011, 2019). On the Psyche mission, the gamma-ray spectrometer (GRS) will quantify the elemental abundances for the following elements: Fe, Ni, Si, K, S, Al, and Ca. These abundance measurements provide key information that address three of the five Psyche mission objectives (see Sect. 2). Of particular importance is Psyche's Ni abundance, which is an element that has yet to be remotely measured on a planetary surface with gamma-ray spectroscopy but will be key to determining Psyche's evolutionary history. Because of the importance (and challenge) of obtaining a robust Ni composition measurement, this Ni composition requirement drives much of the GRS design, particularly the need to use a high-purity Ge gamma-ray sensor that has the excellent energy resolution needed for separating the key Ni gamma-ray line from a nearby K line.

The neutron spectrometer (NS), in conjunction with the multispectral imager, will quantify the metal-to-silicate fraction. For Psyche, Ni, C and S are expected to partition with the metallic phase, as would the majority of Fe. Sulfur would be expected to occur in sulfides, primarily troilite (FeS), and carbon as graphite. In contrast, major rock-forming elements (Si, O, Al, Mg, Ca) are expected to partition into the silicates. The relative amount of reduced (metallic) to oxidized Fe can be deduced from the total Fe measured by NS and the composition of FeO-bearing mafic silicates constrained by depth of the absorption bands of olivine and pyroxene from the multispectral imager. Thus, Fe, Ni, and S present in the metallic portion can be distinguished from Si, Mg, Fe, Al and Ca present as oxides in the silicate portion. Psyche might also have spatial variations of H content on its surface (200 to 300 ppm) (Reddy et al. 2018), which can be quantified by the NS. In addition to its own scientific interest, H abundance variations can affect the neutron composition parameters as well as the gamma-ray data, and thus such variations need to be understood. More generally, neutron flux information is useful for understanding and carrying out corrections for neutron capture and inelastic gamma-ray measurements (e.g., Lawrence et al. 2002; Prettyman et al. 2006).

Planetary nuclear spectroscopy measurements have a few foundational top-level requirements that drive the mission and spacecraft design. Most importantly, gamma-ray and neutron signals have inherently low counting rates – fractions of counts per minute per elemental line for gamma rays, and up to a few counts per second for neutrons – and therefore require many days of accumulation to achieve statistically meaningful results. Second, since the sensors are omnidirectional with a full hemispherical view of the planetary body, low altitudes of roughly one planetary radius (or lower) are needed to detect sufficient quantities of planet-originating gamma rays and neutrons (Peplowski 2016). Finally, it is advantageous to place the gamma-ray and neutron sensors on an extension (i.e., fixed boom) away from the primary spacecraft material. The reason is that spacecraft materials generate background gamma rays and neutrons from the same processes that generate planetary gamma rays and neutrons. While some missions have successfully used body mounted instruments due to accommodation and resource constraints (e.g., Goldsten et al. 2007; Prettyman et al. 2011), the high precision Ni measurement needed at Psyche drove the design to require a boom (see Fig. 1). Based on the same "one-body radius" principle, the spacecraft-generated background can be substantially reduced when the sensor is located at least one spacecraft radius away from the spacecraft. The Psyche spacecraft is roughly a cube, 2.5 m on a side, giving an effective "radius" of $\sim$2 m, which sets the 2-m boom length.

The GRNS contains four primary components: the GRS, the NS, and two Data Processing Units (DPUs), one dedicated to each of the two sensors. The GRS inherits its basic functional design from the MESSENGER GRS (Goldsten et al. 2007) and, similarly, uses a cryocooled high-purity germanium (HPGe) gamma-ray sensor to measure gamma rays with optimum energy resolution. A new type of pulse-tube cryocooler will provide longer operational life compared to the rotary cryocooler used for MESSENGER. The gamma-ray sensor is surrounded by a plastic scintillator anticoincidence (AC) shield that serves three primary functions: active background rejection, fast neutron detection, and galactic cosmic-ray (GCR) monitoring. The plastic scintillator also provides an auxiliary function as a gamma-ray burst monitor. Since the GRS is most sensitive to spacecraft backgrounds, it is located at the end of the boom. The NS inherits its basic functional design from the Lunar Prospector NS (Feldman et al. 2004) and uses three ^{3}He gas proportional counters to measure neutrons in three separate energy ranges. The NS is placed halfway up the boom at $\sim$1 m from the spacecraft deck. The DPUs are located within the spacecraft body and are connected to each of the sensor components via harnesses attached along the boom.

The flight model GRNS was delivered to the NASA Jet Propulsion Laboratory (JPL) in August 2021 for integration on the Psyche spacecraft, and the spacecraft was launched on 13 October 2023. Subsequent sections describe the design and performance of the delivered GRNS, including its comprehensive performance test (CPT) on the Psyche spacecraft, as well as its initial performance in space. All portions of the GRNS are presented, including

Table 1 List of elements and composition parameters measured with the GRNS, along with required detection thresholds, precision, and expected accumulation time in Orbit D to meet requirements

Element	Gamma-ray line or neutron energy band	Required detection threshold	Required relative precision		Expected nadir-pointing time at altitudes <1 Psyche radius to meet global requirement (days)
			Global	Mapped on 200 km^2 area	
Al[b]	2211 keV	1.3 wt.%	20%	-	22
C[a]	4438 keV	n/a	-	-	-
Ca[b]	1943 or 3736 keV	0.3 wt.%	20%	-	52
Fe	846 keV	4 wt.%	20%	20%	0.8
H[a]	Low-energy/ high-energy epithermal	n/a	-	-	-
K	1460 keV	200 ppm	20%	20%	4
Mg[a]	1369 keV	n/a	-	-	-
Ni	1454 keV	4 wt.%	20%	-	36
P[a]	1266 keV	n/a	-	-	-
S	2232 keV	3 wt.%	20%		46
Si	1778 keV	2.5 wt.%	20%	20%	16
Th[a]	2615 keV	n/a	-	-	-
U[a]	609 keV	n/a	-	-	-
Metal-to-silicate fraction	Thermal, fast via Σ_a and <A>	n/a	-	0.02	Few days

[a]Not required to meet mission science objectives but detectable if present at greater than ~1 wt.% (C, Mg, P), ~10 ppm (H), ~1 ppm (Th), and ~25 ppb (U). [b]Requirements are met either with the Al or Ca measurement.

the mechanical and electronic aspects of the sensors, as well as an overview of its firmware and software. We also present the end-to-end functional performance of the GRNS components, a description of the GRNS data products, and a summary of the initial cruise operation as well as the expected operation of the GRNS both during the remaining portion of cruise and while in orbit at Psyche.

2 Overview of Elemental Composition Measurements on the Psyche Mission

It has been proposed that Psyche could be the stripped iron core of a differentiated protoplanet, in which case the exploration of Psyche offers an unique opportunity to provide valuable insights into planetary formation processes and test our theories for core formation for terrestrial planets (Elkins-Tanton et al. 2022, this collection). The Psyche mission has five science objectives: A) Determine whether Psyche is a core, or if it is primordial unmelted material; B) determine the relative ages of regions of Psyche's surface; C) determine whether small metal bodies incorporate the same light elements into the metal phase as are expected in the Earth's high pressure core; D) determine whether Psyche was formed under conditions more oxidizing or more reducing than Earth's core; and E) characterize Psyche's morphology (Elkins-Tanton et al. 2022).

The GRS is required to measure the abundances for the elements Fe, Ni, Si, K, S, Al, and Ca. GRNS data will also measure the bulk composition parameters of macroscopic neutron absorption (Σ_a) (Feldman et al. 2000) and average atomic mass (<A>) (Gasnault et al. 2001) from which Psyche's metal-to-silicate fraction can be determined. Table 1 lists the elements and composition parameters that will be measured with the GRNS, along with the required detection limits, precisions, and expected accumulation times needed to make each measurement. The GRNS composition measurements directly address three of the five Psyche mission science objectives (A, C, and D). For objective A, all of the compositional measurements contribute towards the determination if Psyche is a core. Ni concentrations in the metal phase below ~4 wt.% are consistent with Psyche not being a product of core formation processes (Elkins-Tanton et al. 2022). If the Ni concentration is ~4 wt.% or higher, then the measured value provides information about how core-like material may have formed (Elkins-Tanton et al. 2022), as well as the oxidation or reduction state of its material (McCoy et al. 2022). If the Ni concentrations are very low (i.e., $\ll$4 wt.%) along with high metal (e.g., Fe) abundances, such a measurement would indicate that Psyche could be a metal-rich body that may never have melted, and thus represent highly reduced, primordial metal. Alternatively, high Ni concentrations (>12 wt.%) are uncommon in iron meteorites and occur in irons that originated from oxidized cores, where metallic Fe is oxidized to FeO and incorporated in the mantle, leaving a Ni-rich core. Such a high Ni concentration would also indicate a low or very low silicate concentration, as Ni concentrations in silicates are typically at the part per million levels. A primary means for measuring Ni abundances is via a 1454-keV gamma ray produced by an inelastic scatter reaction between a fast neutron and a Ni nucleus. Figure 2 shows simulated gamma-ray spectra, which provides an example of the type of data that can be acquired with the GRS in order to quantify Ni abundances on Psyche's surface.

The elements Si, Al, K, and Fe help discriminate whether any non-metal portions of Psyche's surface are chondritic or achondritic (Elkins-Tanton et al. 2020). Earth-based measurements show evidence that Psyche may have large-scale (hemispherical) metal-to-silicate heterogeneity (Sanchez et al. 2017; Takir et al. 2017; Shepard et al. 2021; Cambioni et al. 2022; Hasegawa et al. 2024), suggesting that GRNS measurements may likewise reveal compositional heterogeneity. Due to better statistical precision, the neutron-based composition parameters will provide the highest spatial-resolution GRNS maps, particularly for Psyche's metal-to-silicate fraction. For objective B, composition measurements of Si, K, and S will provide information about light-element partitioning in the core, if present, or possibly of unmelted (i.e., primordial) metal-rich material. Finally, for objective D, Ni composition measurements will provide information about oxidizing and reducing conditions such that oxidizing conditions are likely required to produce Ni abundances greater than 12 wt.%, and reducing conditions are likely required to produce low Ni abundances (McCoy et al. 2022, this collection).

Recent Earth-based measurements and new interpretations of existing measurements have updated our understanding of the nature of Psyche (Elkins-Tanton et al. 2020). The currently accepted density of Psyche is bounded to be 3.4–4.1 g/cm^3 (Elkins-Tanton et al. 2020) with a recent value of 4.172 ± 0.145 g/cm^3 having been reported (Farnocchia et al. 2024). It is therefore now understood that Psyche can have a broad range of possible compositions. Psyche's Fe abundances could plausibly range from 25 wt.% to upwards of 90 wt.%, and by inference, the silicate fraction could range from less than 0.1 to greater than 0.7. Thus, instead of assuming Psyche may be an analog of iron meteorites, a wider range of possible compositions can be considered that match the density of Psyche, including analogs such as CB and enstatite chondrites, pallasites, and mesosiderites (Peplowski et al. 2019b), although spectral reflectance considerations suggest that enstatite chondrites and mesosiderites may be a less likely match to Psyche (Dibb et al. 2023).

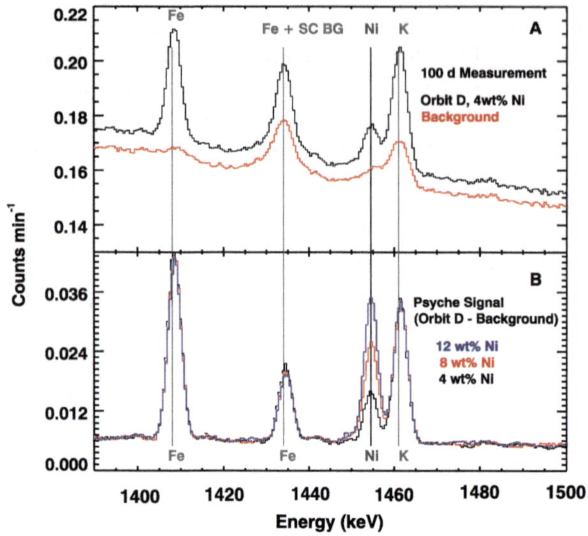

Fig. 2 Simulated gamma-ray spectra for a 100-day-long Orbit D at Psyche. For this orbit, we assumed an altitude of one Psyche radius with a 70% duty cycle; see Sect. 7 for more discussion of duty cycle when in Psyche orbit. Panel A shows combined Psyche-originating (black) and background (red) gamma rays. Psyche-originating gamma rays are calculated for an assumed Psyche composition that includes 85 wt.% Fe, 4 wt.% Ni with the remaining material being Mg-rich orthopyroxene. The Psyche-originating gamma rays are propagated through the GRS sensor using estimated performance parameters for the Psyche HPGe crystal size and an energy resolution of 3.5 keV full-width, half maximum at 1332.5 keV. This energy resolution was set as the pre-launch energy resolution requirement based on the best pre-radiation damage performance of the MESSENGER GRS (see Sect. 4.1). The gamma-ray and continuum background was estimated using the measured background from the MESSENGER GRS (Evans et al. 2012). Panel B shows the Psyche-originating spectra minus the background for different amounts of assumed Ni abundances. Gamma rays for different elemental lines including background (BG) are labeled

For meteorites, the metal-to-silicate fraction is generally defined as the total amount of metal (e.g. metallic iron, nickel) versus total silicates (elements bound as oxides). In this scenario, iron in silicates (e.g. FeO) is "book kept" with silicates. In contrast, GRNS measures elements without regard to their mineralogical state. Since the largest concentration of "metal" elements being considered consist of Fe and Ni, "metal" content is defined as the total concentration of Fe and Ni, and "silicates" are defined as other major rock-forming elements (Si, O, Al, Mg, Ca, etc.)

The lower density value for Psyche detailed in (Elkins-Tanton et al. 2020) requires either less metal than originally thought, and/or significantly higher porosity than is typical for >100-km-diameter asteroids (e.g., Zhang et al. 2022). For the low metal case, measurements of additional elements beyond that originally planned could be important for fully characterizing Psyche's bulk elemental composition. For example, for the meteorite types listed above, Mg concentrations can provide important discrimination between different composition types (Peplowski et al. 2019b). While Mg is not a required element for the baseline Psyche mission science plan, the GRS sensor was fabricated without magnesium materials to enable Mg measurements to be accomplished with no changes to the instrument design or operations plan. Similarly, other elements that could be detected include C and P, if present at high enough concentrations (e.g., greater than 1 wt.%).

Another element that has gained increased importance at Psyche is H. Ground-based measurements using the NASA Infrared Telescope Facility have detected a 3-μm absorption on Psyche that is consistent with hydration features attributed to OH and H_2O bearing phases (Takir et al. 2017). Because Psyche likely has no extensive permanently shaded regions, and thermal conditions do not allow for water ice to be stable at its surface, this detection suggests that, as with the asteroid Vesta (Prettyman et al. 2012), Psyche may also have significant amounts of water-rich carbonaceous-chondritic material accumulated on its surface. Based on a cross correlation of Earth-based data from both Vesta and Psyche with orbital H measurements at Vesta, Reddy et al. (2018) estimated that the bulk hydrogen concentrations could be in the range of 200 – 300 ppm, which is comparable to the range of H variations (0 – 400 ppm) seen at Vesta (Prettyman et al. 2012; Lawrence et al. 2013b). While H measurements are not needed to directly address the mission's science objectives, H abundances in this range can affect the measured neutron composition parameters (Lawrence et al. 2013b; Prettyman et al. 2013), and possibly gamma-ray data when making composition measurements using both capture and inelastic gamma-ray lines (Peplowski et al. 2015a; Wilson et al. 2019). Thus, effects from bulk H concentrations and its possible spatial variability need to be taken into account with the measured GRNS data. The Psyche NS was specifically designed to be sensitive to a large range of possible metal content on Psyche (i.e., Fe and Ni) as well as possible H abundances greater than a few tens of ppm. To illustrate, Fig. 3 shows simulated neutron spectra for different metal-to-silicate content and H abundances. In typical silicate dominated compositions, variations of thermal (neutron energy $E_n < 0.4$ eV) and low-energy epithermal (0.4 eV $< E_n <\sim 100$ eV) neutrons are most sensitive to neutron absorption and H concentrations, respectively (Feldman et al. 2000; Lawrence et al. 2006). In contrast, when there are large concentrations of the neutron absorbers Fe and Ni (e.g., greater than Fe + Ni $\sim$30 wt.%), these elements will strongly suppress thermal neutrons and shift H-dependent variations to higher energies (Fig. 3B). Thus, in addition to sensors that measure thermal and low-energy epithermal neutrons, the NS has a third sensor that is sensitive to higher energy epithermal neutrons (Sect. 3.3).

Finally, Psyche's lower-than-anticipated density means that a wider range of metal and silicate abundances need to be considered. Of primary importance is the relative proportions of Ni and K concentrations, which have closely spaced gamma-ray lines at 1454 keV and 1461 keV, respectively (see Fig. 2). For example, if the metal fraction of Psyche is significantly lower than originally expected, then its Ni concentration will likely be correspondingly lower, and K (which can be enhanced in silicate materials) could be correspondingly higher, which can make separating the two lines more challenging. This possible condition reinforces the need for the excellent energy resolution provided by the HPGe sensor.

3 GRNS Instrument Design

3.1 GRNS Overview

The GRNS was designed and built by the Johns Hopkins University Applied Physics Laboratory (APL) in collaboration with the Lawrence Livermore National Laboratory (LLNL). LLNL designed and built the GRS detector and cryostat assembly. APL designed and built the AC shield, the NS sensor assembly, the GRS and NS front-end electronics, and the GRS and NS DPUs. The cryocooler for the GRS detector was designed and built by the Lockheed Martin Advanced Technology Center. The overall design and functionality of the GRNS follows closely the MESSENGER GRNS, which successfully made compositional measurements of planet Mercury (Peplowski et al. 2011; Evans et al. 2012; Peplowski et al. 2012; Lawrence et al. 2013a; Peplowski et al. 2014; Evans et al. 2015; Peplowski et al.

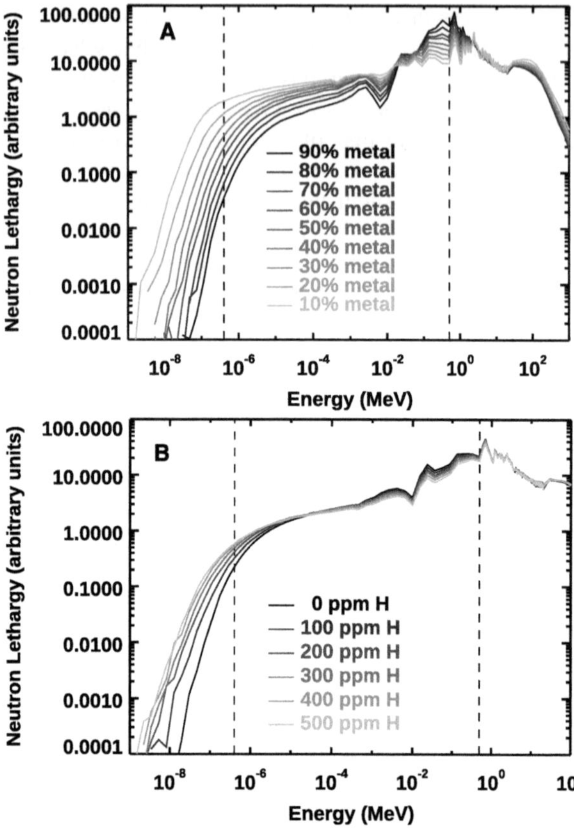

Fig. 3 Calculated neutron lethargy spectra for different assumed Psyche compositions. Lethargy (flux times energy) flattens out the otherwise steep power law spectra. Panel A shows the fluxes for compositions ranging from 10 volume % metal to 90 volume % metal. Panel B shows fluxes for a 90 volume % metal and variable hydrogen abundances from 0 to 500 ppm H. The vertical dashed lines are located at the lower energy boundaries for epithermal (0.4 eV) and fast neutrons (0.7 MeV)

2015b,c, 2016; Lawrence et al. 2017). The similarity with MESSENGER includes an HPGe sensor with plastic scintillator anticoincidence shield, separate neutron spectrometer, and separate DPUs for the two sensor subsystems, where APL "slice" heritage (i.e., modular boards that couple together to form a compact electronics stack) is used for the DPU boards. The AC shield fast-neutron signal processing is largely inherited from the MESSENGER NS fast-neutron measurements.

Despite these similarities, a number of key changes were made to the GRNS design. These changes were made to accommodate different requirements unique to the Psyche mission design as well as incorporate lessons learned from the MESSENGER GRNS. Also notable is that when the Psyche GRNS started its conceptual design phase in 2016, it had been almost 15 years since the initial design work on the MESSENGER GRNS. As a consequence, significant redesign of the electronics and software was needed to bring the instrument up to date (e.g., due to parts obsolesce) and to accommodate new mission assurance requirements.

For the GRS, there were important lessons learned from MESSENGER that resulted in improvements to the Psyche design. In particular, six months into the primary mission, the MESSENGER HPGe detector started to develop leakage currents that degraded its energy resolution performance in the final months of operation (Evans et al. 2017). In addition, the radiation damage that was generated by being exposed to 6.5 years of GCRs during cruise was never fully reversed despite extended annealing at 85 °C, resulting in an unanticipated degradation in energy resolution performance. There were attempts made to mitigate both these effects in flight but, ultimately, the bearings in the rotary-style cryocooler reached end of life and gave out during MESSENGER's first extended mission ending its HPGe sensor operation.

To address the unexpected leakage currents in the MESSENGER HPGe sensor that scaled with total anneal time (Evans et al. 2017), we incorporated two changes to the Psyche cryostat: a passivation layer was added to the HPGe crystal and the detector encapsulation was vented. Passivation is now a standard process for the manufacture of HPGe crystals as a way to stabilize its highly reactive surface, thereby reducing its susceptibility to contamination that can lead to leakage currents. The MESSENGER HPGe crystal was not passivated. Instead, it was hermetically sealed in an aluminum encapsulation filled with two atmospheres of clean dry nitrogen, where the nitrogen gas was intended to occupy surface sites and therefore act as a barrier to contamination. However, the repeated high-temperature annealing attempts of the MESSENGER GRS likely generated an excessive build-up of internal outgassing products within the sealed enclosure itself that became trapped, ultimately overwhelming the intrinsic detector surface to the point of producing significantly elevated detector leakage currents that, in turn, degraded system energy resolution. As a way to avoid this catastrophic scenario, the Psyche design switched to a "vent-to-space" encapsulation scheme. Because the HPGe surface is now passivated, it is relatively insensitive to contaminants, making a hermetically sealed enclosure no longer necessary. Long-term exposure to air has not been shown to be a problem, as was previously demonstrated on INTEGRAL, where their detectors were vented approximately two years prior to launch (Vedrenne et al. 2003). Testing of the Psyche GRS vented design has also shown stable performance after long exposures to air for periods up to 9 months.

To address the incomplete annealing observed on MESSENGER, we carried out a comprehensive study to determine quantitatively the most effective annealing strategy for recovering full performance from radiation-damaged HPGe detectors. The results of this study, reported by Peplowski et al. (2019a), showed that higher temperatures (105 °C) than used by MESSENGER were required to achieve full recovery following irradiation for selected mission scenarios, a result consistent with previous investigations (Albernhe et al. 2002) and verified with our particular detector geometry and predicted radiation dose. We have fully incorporated this revised annealing strategy into our operational plans.

Compared to MESSENGER, the Psyche GRS uses a fully redesigned cryostat that includes several new thermal and mechanical features based on many years of improvements since the MESSENGER design (Burks et al. 2020). The new design is completely separable from the AC shield, which streamlines assembly and facilitates stand-alone testing of each subsystem. Improved energy resolution was achieved by mounting the preamplifier front-end circuit board directly to the detector encapsulation inside the cryostat for better noise performance and less susceptibility to microphonic pick-up of vibrations from the cryocooler.

Psyche also took advantage of advances in electronic parts available for spaceflight missions to improve performance. MESSENGER was limited in the number of digital logic gates available for signal processing, and so relied on a hybrid analog/digital signal processing scheme for pulse height analysis. The Psyche GRS implements a fully digital design, which eliminates analog filtering stages and their associated temperature dependence and adds the capability to adjust filter shaping parameters in flight to optimize energy resolution. The improvement in pre-launch performance – less than 2.0 keV full-width, half-maximum (FWHM) for Psyche (Sect. 4.1) compared to 3.5 keV FWHM for MESSENGER (Goldsten et al. 2007) – is largely due to these improvements.

Table 2 Types and characteristics of measurements from the five GRNS sensors

Measurement Type	Sensor	Characteristics
High-gain gamma-ray spectra (separate accepted & rejected)	HPGe sensor	60 – ∼3000 keV gamma-rays
Low-gain gamma-ray spectra (separate accepted & rejected)	HPGe sensor	60 – ∼9000 keV gamma-rays
Fast neutrons	AC shield, high gain	0.7 – 5 MeV neutrons; double pulse interaction
Broad-band epithermal neutrons	AC shield, high gain	Few hundred to 0.7 MeV neutrons; single pulse interactions within the ∼90 keV $^{10}B(n,\alpha)$ peak
Galactic cosmic-rays	AC shield, low gain	0.5 MeV to ∼35 MeV, mostly protons
Gamma-ray burst detection	AC shield, high gain; after triggering a set threshold	1024-sample, 50-ms resolution profile of integrated counts of gamma-rays falling within an energy window ∼20 keV to ∼100 keV (commandable upper limit)
Slow neutrons	Bare ^{3}He sensor	0 – few hundred keV neutrons
Low-energy epithermal neutrons	Cd-covered ^{3}He sensor	0.4 eV – few hundred keV neutrons
High-energy epithermal neutrons	Polyethylene-covered ^{3}He sensor	∼1 keV – few hundred keV neutrons
Thermal neutrons	Bare and Cd-covered ^{3}He sensor	Count rate difference of bare and Cd-covered sensors

Another significant change from the MESSENGER design is the replacement of its limited-life rotary cryocooler with a long-life pulse-tube cryocooler. While a pulse-tube cryocooler is less efficient than a rotary cryocooler, its lower level of exported vibration allows the signal processing chain to reach the system noise floor even under ambient thermal conditions, achieving laboratory-grade energy resolution performance in a fully qualified spaceflight instrument. The virtually unlimited life of a pulse-tube cooler ensures the GRS will be ready for any exciting extended mission opportunities the Psyche mission may bring.

For the NS, the Psyche design returns to the use of ^{3}He neutron sensors like those used on Lunar Prospector (Feldman et al. 2004) rather than the scintillator-based sensors used for MESSENGER. The MESSENGER scintillator sensors were designed to take advantage of the high orbital velocity of the MESSENGER spacecraft around Mercury (∼few km/s) and use a Doppler filter technique (Feldman and Drake 1986) to discriminate neutron energies. The spacecraft velocity around Psyche is much lower (<200 m/s) and thus the Doppler filter technique is not appropriate at Psyche. Further, ^{3}He neutron sensors are robust and reliable sensors and were easily incorporated into the Psyche design by adopting a straightforward mounting approach.

Separate GRS and NS harness bundles connect the sensor assemblies at their respective boom locations to their DPUs. The DPUs mount to the −Y spacecraft panel on the inside of the spacecraft. Harness cables from each DPU are feed-through bulkhead connectors on the + Z deck to maintain the spacecraft Faraday cage.

Taken together, the GRNS employs five separate sensors to make its measurements: a HPGe gamma-ray sensor, a plastic scintillator AC shield, and three ^{3}He gas proportional counter neutron sensors. These five sensors combine to produce 10 different types of measurements, summarized in Table 2. The rest of Sect. 3 describes further details of each sensor assembly, the DPUs, the software and firmware that controls the GRNS, and a summary of

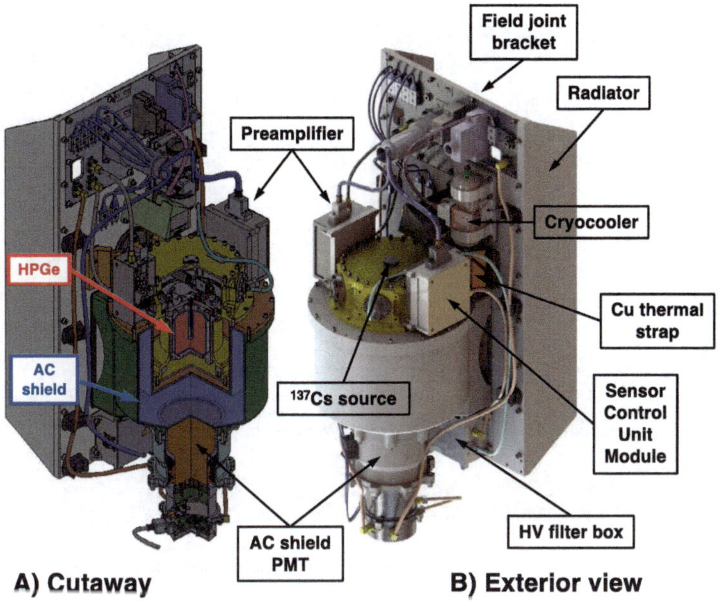

Fig. 4 Cutaway (A) and exterior (B) views of the GRS. Different sensor components as described in the text are labeled. Figure adapted from Peplowski et al. (2025)

the environmental tests that were carried out on the hardware prior to delivery to the spacecraft.

3.2 Gamma-Ray Spectrometer (GRS)

A cutaway view of the GRS is shown in Fig. 4; photographs of the GRS are shown in Fig. 5; a functional block diagram of the GRS is shown in Fig. 6; detector characteristics and resource use are given in Table 3. The GRS consists of multiple components located on a single mounting deck. At the inside of the GRS is the HPGe sensor, which is housed within the cryostat. The cryocooler (Sect. 3.2.3) and three sensor electronics boxes (Sect. 3.2.2) are connected to the cryostat and perform various functions. The cryostat is surrounded by the AC shield, which contains the plastic scintillator sensor and photomultiplier tube (PMT). The combined GRS/AC shield housing is attached to the mounting deck that then attaches to the spacecraft boom structure via three titanium "nodes" with integral flexures (Fig. 5B).

The GRS mounting deck serves multiple purposes. First, it provides the mounting structure for the GRS/AC shield housing along with the other GRS components. These components include: a split-configuration pulse-tube cryocooler that is attached both on, and extends through the deck (Fig. 5B); a dual-armed copper heat strap that thermally connects the cryocooler warm flange to the deck; and a high-voltage filter box, which receives up to −3000 V from the DPU and attenuates any voltage ripple before feeding it to the sensitive HPGe detector (Fig. 4B). Second, the upper end of the deck contains a field joint bracket (labeled in Fig. 4B), which groups together several bulk-head connectors, simplifying installation and connection of the GRS sensor assembly to the boom harness carrying signal and power cables to the DPU. The backside of the field joint bracket contains bulk-head connectors into which the boom harness cables attach. This arrangement allowed the entire

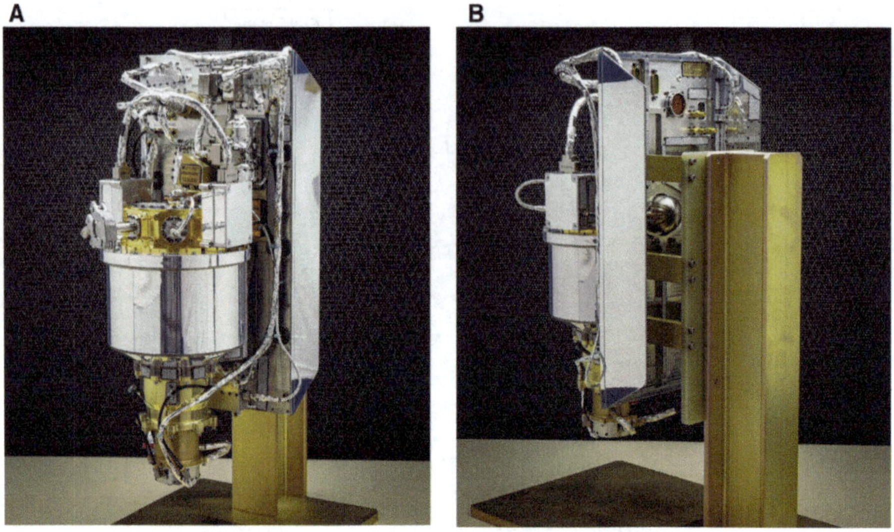

Fig. 5 Photographs of the FM GRS on a test stand. Panel A and B show the front and back of the GRS, respectively

GRS and most of its sensor component connectors to be assembled at APL prior to delivery to the spacecraft. Thus, for spacecraft installation, the only harness mates that needed to be completed were those on the backside of the GRS and those to a spacecraft field joint bracket near the base of the boom that fed signals through to the DPU. Finally, the deck provides the mounting structure for two opposing bolt-on radiator wings that help reject waste heat from the cryocooler to space (exposed portions of the AC-shield housing also provide significant radiator area). The deck, the radiator wings, and the AC-shield housing are all made of AlBeMet, a metal matrix composite material comprised of aluminum and beryllium. AlBeMet has better thermal properties than aluminum and its increased stiffness improves mechanical response under launch vibration, as well as reduces the conducted vibration from the cryocooler to the HPGe sensor. The lower density and effective atomic number of AlBeMet also helps to minimize attenuation of gamma rays that pass through the AC shield. The next subsections describe details for each GRS component.

3.2.1 GRS Sensor and Cryostat

The central gamma-ray detector is a 5-cm-diameter, 5-cm-long coaxial HPGe crystal (Fig. 4A) that was manufactured by Mirion Technologies. The crystal is N-type germanium with reverse-electrode configuration, which offers better resistance to radiation damage (Pehl et al. 1979), and is specified with an impurity level that targets an initial depletion voltage of 2000 V. The bare Ge crystal is encapsulated within a gold-plated aluminum encapsulation that is highly polished to achieve low emissivity in the infrared. The encapsulated detector is then suspended using a Kevlar suspension inside a similarly plated and polished cryostat, forming a separable unit that can then be lowered into the well of the AC shield. The GRS cryostat (Fig. 7) is optimized for three goals: thermal isolation from the environment, mechanical protection of the germanium crystal against launch vibration loads, and high-resolution readout of the gamma-ray signal. More details regarding the development of the cryostat are given in (Burks et al. 2020).

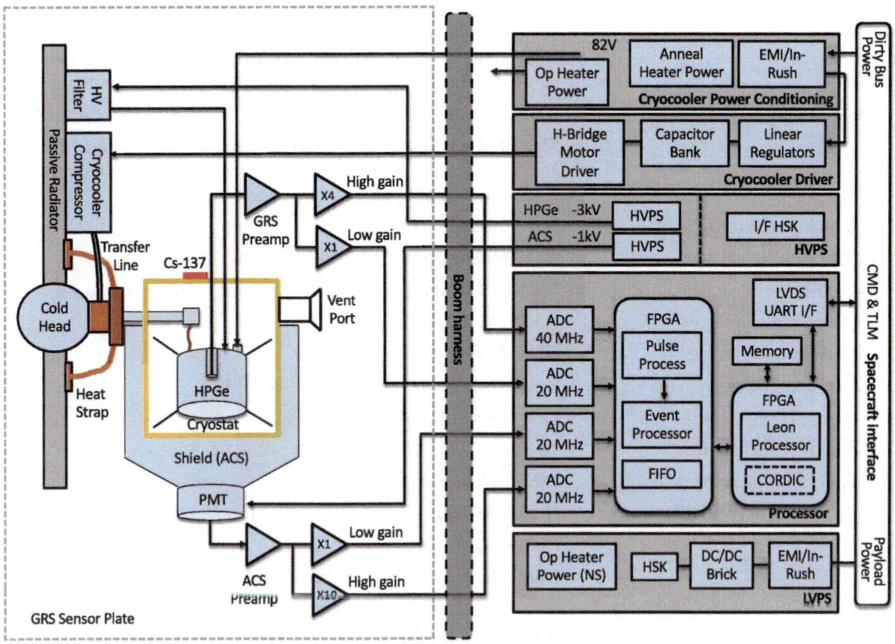

Fig. 6 GRS functional block diagram. All instrument components that are located on the boom are shown within the large dashed rectangle. The components for the five-slice DPU are shown on the right side of the block diagram. Individual components within each DPU slice are labeled. Additional details are given in the text

Thermal isolation is required to operate the germanium crystal at cryogenic temperature, typically 85 K to 95 K, while protecting against the ambient environment. In space, where liquid nitrogen is not an option, sufficient cooling can be achieved using a mechanical cryocooler. For terrestrial applications these coolers can have a mass of several kg and consume considerable power. The Psyche GRS uses a compact, low-mass, low-power, pulse-tube cryocooler (Sect. 3.2.3). The tradeoff is that its heat lift capacity can be as low as ∼500 mW at cryogenic temperatures. The design of the cryostat must therefore constrain the heat loss to within the capacity of the cooler. This requirement is achieved by combining two methods. First, the conductive heat loss is minimized by mounting the encapsulation on a low-conductivity Kevlar-29 suspension system. Second, infrared heat transfer is minimized by employing infrared (IR) shielding comprised of highly reflective, low-emissivity surfaces. The cryostat is vented directly to space so conductive heat transfer is negligible. The total heat load on the detector was measured at 380 mW ± 20 mW at 23 °C ambient temperature, well within the capacity of the cryocooler.

Mounted onto the detector encapsulation are two power Zener diodes serving as heaters. These are used to bake the system at ∼50 °C to desorb water from the low-emissivity surfaces. They are also used to anneal the detector at 105 °C to reverse accumulated radiation damage (Peplowski et al. 2019a). Zener diodes were selected as heaters because they operate at high voltage and low current (∼80 V × ∼20 mA), allowing for the use of low-conductivity, 0.25-mm diameter, NbTi wires, with minimal conductive heat loss. Two temperature-sensing diodes are mounted onto the detector encapsulation and cryocooler cold-tip, respectively, to allow for temperature feedback and control. By having temperature sensors on each end of the short, braided copper heat strap that makes the thermal connec-

Table 3 GRS characteristics and resource use

Primary sensor	High-purity germanium (HPGe) coaxial detector, 5 cm × 5 cm beveled cylinder; inner hole is 8 mm diameter; 3.8 cm deep
Anticoincidence shield	Borated plastic scintillator (BC454) cup coupled to a 7.6-cm PMT;
	Measured scintillator mass: 1.63 kg annulus 12.9 cm inner diameter; 16.5 cm outer diameter; 1.8 cm thick puck: 4.92 cm tall; diameter is 7.7 cm at the PMT and 12.9 cm at the annulus
Photomultiplier tube (PMT)	Hamamatsu R6233-01; 76-mm diameter
Effective field-of-view	4π omnidirectional; full view of Psyche asteroid
Cryocooler	Lockheed Martin pulsed-tube cryocooler, ~500 mW heat lift
HPGe operating voltage	<3000 V, leakage current <10 pA
Gamma-ray energy range	60 keV to 9000 keV
Neutron energy range	Broad-band epithermal neutron: <0.7 MeV
	Fast neutron: 0.7 MeV to ~10 MeV
Number spectral channels	16,384 HPGe; 512 AC high gain; 1024 AC low gain
Energy resolution (FWHM)	1.92–2.09 keV @ 1332 keV
Pulser resolution (FWHM)	Hardware high gain: 1.12 keV; rate: 1 to 10,000 Hz
	Hardware low gain: 1.32 keV; rate: 1 to 10,000 Hz
	Digital high gain: 1.12 keV; rate: any
	Digital low gain: 1.32 keV; rate: any
Intrinsic efficiency	0.212 @ 511 keV, 0.102 @ 1332 keV, 0.020 @ 6130 keV
Maximum throughput	>5,000 events per second
Background rejection ratio	Varies from 2:1 @ ~1000 keV to >10:1 @ ~8000 keV
Differential nonlinearity (DNL)	<1%
Integral nonlinearity (INL)	<3 channels (high gain); <2 channels (low gain)
Electronics gain drift	0.2 channels per °C
Preamplifier gain drift	0.1 channels per °C
Sensor thermal environment	$-25\ ^\circ\text{C} < T_{\text{radiator}} < 30\ ^\circ\text{C}$
Cool-down time to 90 K	<12 hrs, T_{radiator} @ 0 °C
HPGe crystal temperature regulation	<0.25 °C, $-25\ ^\circ\text{C} < T_{\text{radiator}} < 65\ ^\circ\text{C}$ (transient conditions)
HPGe crystal annealing temperature	105 °C, regulated
GRS sensor mass	9.06 kg (including mounting deck and passive radiator)
GRS DPU mass	2.64 kg
GRS harness mass	4.44 kg (includes harness along boom and DPU to + z deck harness)
GRS total mass	16.14 kg
Power use	57.7 W peak (DPU: 10.6 W; cryocooler, sensor dissipation: 21.0 W; cryocooler, heater dissipation: 25.4 W; harness dissipation: 0.7 W)
Data rate	6000 bits per second Orbit D allocation (combined with NS)

tion between the cold finger and the encapsulation, one can directly measure and monitor the heat flow in or out of the detector once the conductance of the cold braid is calibrated.

The Kevlar suspension also serves an important secondary purpose: it holds the detector encapsulation firmly in place during launch. A BeCu wave spring maintains a preload on the encapsulation/Kevlar suspension system. The detector encapsulation contains an internal BeCu wave spring as well to hold the germanium crystal in place within the encapsulation. BeCu springs were selected over stainless steel to minimize the amount of Fe and Ni close to the detector, so as to reduce background gamma rays specifically from Ni. The elimination

Fig. 7 Photograph of the GRS cryostat prior to delivery from LLNL to APL. The bottom lid (on left) shows initials of LLNL persons who contributed to the build of the cryostat. The iron meteorite Odessa is shown on the right side of the photograph

of Ni in these springs was part of a more general effort to reduce Ni in the Psyche spacecraft with the goal of minimizing background gamma rays from Ni (Bradford et al. 2022). Initial results of this effort are described in Sect. 5.1.

Of particular importance for the Psyche cryostat is optimizing the gamma-ray energy resolution to distinguish the important ^{58}Ni gamma-ray line at 1454 keV from potential interference from ^{40}K at 1461 keV. In the MESSENGER GRS, the three biggest limitations to resolution were caused by radiation damage, microphonic pickup from the cryocooler, and the pulse-processing electronics. Radiation damage is addressed in the Psyche GRS through an improved annealing procedure (Sect. 3.1). Microphonic pickup arises when vibrations in the system induce electrical noise in the sensitive front-end electronics. The pulse tube cryocooler was chosen (Sect. 3.2.3) primarily for its long life but also benefits from its reduced exported vibration. The most sensitive components of the preamplifier – the front-end junction-gate field-effect transistor (JFET) and feedback components – were mounted directly onto the detector encapsulation. This approach reduces the JFET gate-lead length and thus its susceptibility to microphonics. There is a slight noise penalty for operating the JFET at a temperature below optimum of $\sim$115 K, but the improved end-to-end system energy resolution with the cooler running justifies this trade.

The cryostat has several vacuum feed throughs that allow for the connection of the cryocooler, high voltage (HV), Zener and temperature sensors, and the preamplifier signal. Another port is used to mount a light shield (Hines et al. 2022). This light shield has been designed to allow the cryostat to vent to space while blocking sunlight from penetrating into the cryostat. Direct sunlight is a concern because the photons cause thermal excitation in the crystal resulting in leakage current. The light filter uses a torturous path, painted with a layer of low-reflectance Stycast (Loctite 2850FT), to reduce the ambient light by several orders of magnitude.

A low-activity ^{137}Cs source (5 nCi) is attached to the cryostat lid (Fig. 4B). This source enables energy and efficiency calibration of the detector in flight, which allows flight monitoring of the detector efficiency as it changes with each radiation damage/annealing cycle (Peplowski et al. 2019a). The low activity of the source ensures that it does not interfere with the measured signal from Psyche.

3.2.2 GRS Electronics Modules

The GRS preamplifier module attaches to one side port of the cryostat. Its overall topology and functionality are similar to that flown on MESSENGER (Goldsten et al. 2007), but with a few significant improvements. The front-end JFET and feedback components now reside inside the cryostat and operate at cryogenic temperature, improving noise performance. A BF862 JFET provides improved noise performance at lower power. Two low-noise amplifier stages were added to split the signal into low- and high-gain outputs, where the high-gain channel provides better energy resolution and finer bin spacing for the low-energy region spanning ~100 keV to ~3000 keV, a region that includes the critical Fe and Ni lines, specific to the Psyche mission. Since the charge-amplifier is direct-current-coupled to the detector, a buffered version of its output allows monitoring the detector leakage current, which is nominally <10 pA but can measure up to ~4,000 pA, if necessary.

The Sensor Control Unit Module (SCUM) attaches to the opposite side port of the cryostat. It contains the readout circuitry for the cryogenic diode temperature sensors along with two mechanical relays that short the anneal heater lines to chassis when not in use to prevent external interference from entering the cryostat. During annealing operations, the relays are opened, allowing up to ~2 W at ~100 V to power the Zener diodes used as heaters. The SCUM also contains a digitally activated analog pulser that injects charge onto the high-voltage side of the detector, providing a way to measure the depletion voltage of the detector during voltage ramp-up or ramp-down. The detector depletion voltage generally moves down with increasing radiation damage and is then restored following high-temperature annealing, so provides an additional measure of detector health.

The High Voltage Filter Box (HVFB) contains a two-stage ripple filter with ~120 dB rejection. It was too large to mount directly to a port on the cryostat, so it is attached to the mounting deck. It is located opposite the cryocooler to distance it from any exported vibrations generated by the cryocooler compressor, as high voltage components can be microphonic, injecting unwanted interference into the signal chain. Only capacitors with Negative-Positive-Zero (NPO) dielectric were used as added precaution. Shielded coaxial cables carry the high voltage up the boom from the DPU to the HVFB and also for the short distance from the HVFB to the sensor. Versions of the obsolete Reynolds Century Plus series were resurrected by special build. For the final mating of the connectors, a thin layer of Braycote high vacuum grease was applied to the mating surfaces to fill any trapped volumes, avoiding the possibility of a latent high-voltage breakdown after many months in space.

3.2.3 Cryocooler

The cryocooler for the GRS sensor (Fig. 8) is a pulse tube microcryocooler developed and manufactured by Lockheed Martin's Advanced Technology Center (Olson et al. 2015). Their Micro-1 model was chosen for its low mass (~450 g), cooling performance (~500 mW at 85K with ~20 W input and 300 K reject temperature), and long life. Oxford-style pulse tube coolers of this type are expected to last 10 + years, far exceeding the ~100-day science requirement. Two engineering model units were built prior to the flight unit, with one designated as a Life Cycle Test Unit (LCTU) that underwent accelerated life testing. The LCTU exceeded the equivalent of three times the total expected number of operating hours, including extensive thermal cycling and characterization, with no sign of degradation.

The cryocooler consists of three separate components: a compressor module, a cold head with cold finger, and a transfer line between them. The cold head mounts directly to the cryostat with a flange and O-ring interface. A short, flexible copper braid thermally links a tab on the end of the cold finger to the detector encapsulation. The copper braid has also been shown to provide significant damping of radial loads acting on the cold finger. A cutout in the mounting deck allows the bulbous end of the cold head to protrude through it so as to minimize the length of a two-armed copper heat strap that conducts heat from the cold head to the deck (Fig. 5B). The compressor mounts directly to the radiator deck via a thermal

Fig. 8 Photograph of the GRS cryocooler on its delivery test stand. The compressor (left side) is connected to the cold head with cold finger (right side) through the transfer tube

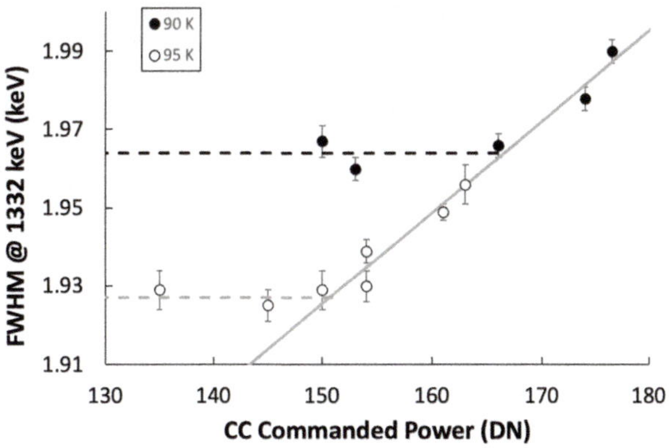

Fig. 9 Energy resolution of the 1332.5-keV gamma-ray peak, as measured via the peak full-width at half maximum (FWHM), as a function of cryocooler power level as measured in digital units (DN). These measurements were taken during GRS ground calibration

gasket for efficient transfer of heat. Cooler power dissipation is split roughly evenly between the cold head and the compressor. The connecting transfer line between them is looped to provide strain relief on the pressure seals, given the separate mounting surfaces for the two components. The fill pressure of the system is 625 PSIA at room temperature.

The cryocooler has been tested regularly to ~40 W input power but will be limited in software to a maximum of ~30 W in flight to prevent overheating under hot conditions or excessive piston stroke under cold conditions. Once in space, the temperature of the entire sensor is tightly regulated to −10 °C using operational heaters, where the expected input power to the cryocooler will be ~20 W or less. See Sect. 5.1 for details regarding the flight performance of the cryocooler.

Pulse tube coolers contain no moving parts in the cold head and no contact bearings in the compressor, so they produce low exported vibrations. The GRS HPGe detector system is "microphonic" and therefore sensitive to cooler vibrations, which can degrade overall system energy resolution slightly when operating at high power. Figure 9 shows GRS energy resolution versus cryocooler (CC) commanded power. Cryocooler power is shown in digital numbers (DN), where the calibration to watts, W, is: $W = 47(DN/186)^{2.2}$. Data are shown for detector temperatures of 90 K (closed circles) and 95 K (open circles). These data show that for power levels above ~150 DN (~29 W) higher power contributes to poorer energy resolution. These data further show that better energy resolution performance is achieved for the warmer detector temperature of 95 K, due to the detector JFET, which performs better at slightly warmer temperatures (see Sect. 3.2.1). In the context of full system performance, when radiation damage is being accumulated (as is the case during space-based operation), performance degradation manifests more quickly at higher detector temperatures, which offsets the improvement from a warmer JFET operating temperature. In flight, the cryocooler operates at much lower power levels (~110 DN or ~14 W; see Sect. 5.1), which is well below the range where energy resolution is impacted by cryocooler power level.

3.2.4 GRS Thermal Subsystem

The GRS thermal subsystem is designed to accommodate the heat produced by the cryocooler. While the cryostat isolates the detector, the AC shield provides additional isolation from the thermal environment. In order to reject the heat produced by the cryocooler, the instrument not only utilizes existing areas on the cryostat and the AC shield housing, but also has dedicated radiator wings bolted onto the instrument deck (Figs. 4, 5, and 10). Thermal management on instrument surfaces is achieved with the use of fluorinated ethylene propylene (FEP) tapes backed with silver lining. Due to its exceptional optical properties (Fig. 10), silvered FEP tapes can efficiently emit heat in the IR spectrum while minimizing solar absorptance. Additionally, these silvered FEP tapes are adhered to existing surfaces, reducing the size of the dedicated radiators and added mass impact.

The cryocooler distributes its dissipated heat approximately equally in two places: the compressor and the cold head interface to the cryostat (the "warm flange"). The cryocooler compressor is directly coupled to the instrument deck thermally, but the warm flange is thermally coupled to the cryostat, so a copper thermal strap was added to provide a heat pathway from the warm flange to the radiator wings as well (Fig. 4B). Other key components, such as the PMT, on-sensor electronics boxes, and AC shield are all thermally tied to the sensor housing and generally have a similar temperature, with the PMT temperature running slightly cooler being the furthest away from the operational heaters.

The GRS has operational heaters located on the mounting deck, while the survival heaters are distributed on the AC shield housing and radiator wings. The operational heaters are controlled by the DPU, while the survival heaters are controlled by the spacecraft. Surfaces not being used as radiative area were covered in multilayer insulation (MLI) blankets. These blankets limit heat loss from undesired areas, as well as provide shielding from solar ultraviolet (UV) radiation. The vast majority of the on-instrument harnessing was also wrapped in MLI. This was especially important as the high-voltage harnessing had more restrictive temperature requirements.

3.2.5 GRS Anticoincidence Shield

A gamma-ray spectrometer system operating in space measures a significant background from charged-particle interactions in the gamma-ray sensor. For the Psyche GRS, GCR protons can deposit sufficiently low energy in the HPGe sensor to create a continuum that is in the same energy range as the gamma-ray signals of interest. This increased background lowers the sensitivity of the HPGe measurements. Following NEAR, Lunar Prospector, and MESSENGER flight heritage (Goldsten et al. 1997; Feldman et al. 2004; Goldsten et al. 2007), the Psyche GRS therefore includes an anti-coincidence (AC) shield.

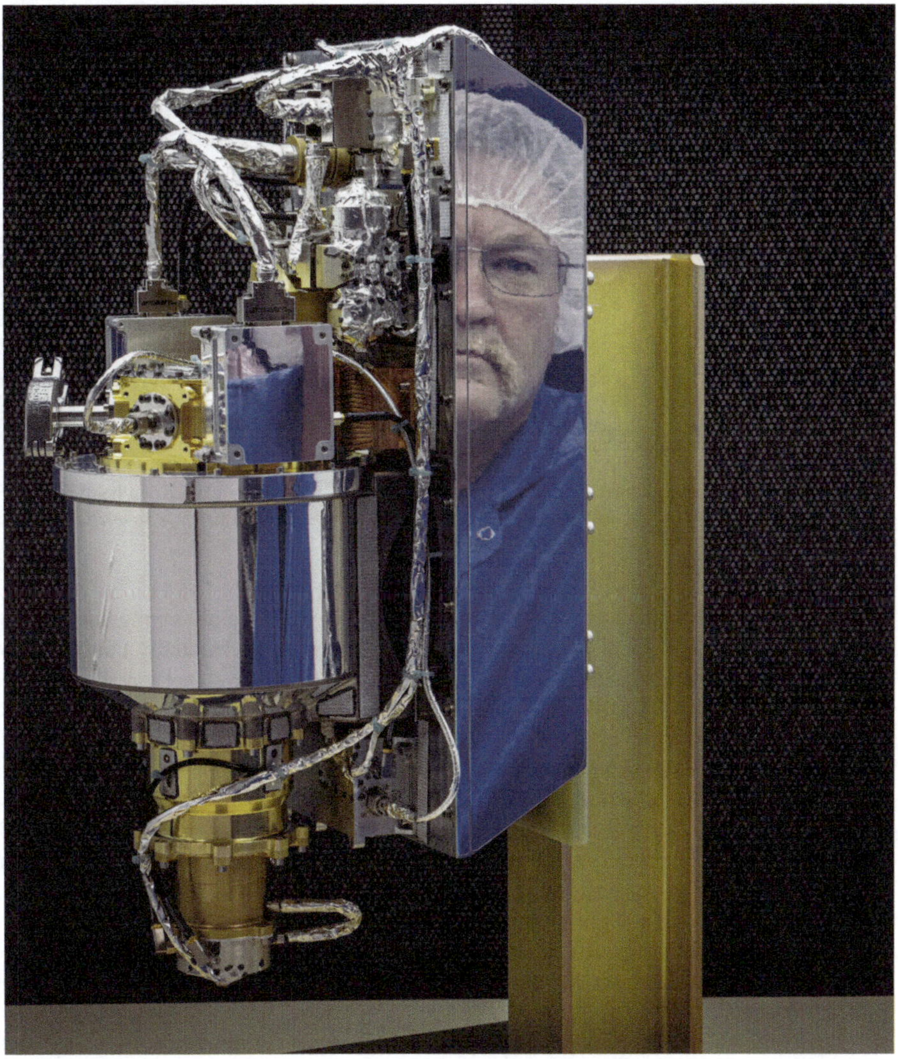

Fig. 10 Photograph of the GRS illustrating the high degree of reflectivity of one of the two radiators. The reflection of technician Mark Hoff (who carried out much of the mechanical assembly of the GRS) is clearly seen

The Psyche AC Shield is a cup-shaped, monolithic piece of borated plastic scintillator (EJ-254), read out by a 7.6-cm-diameter photomultiplier tube (Hamamatsu R6233-01) operating at approximately 400 V. The geometry of the scintillator (Figs. 4A, 11) was chosen to surround the HPGe sensor such that any highly penetrating charged particle reaching the interior gamma-ray detector must also interact with the AC shield. The scintillator is wrapped with several layers of polytetrafluoroethylene (PTFE) reflective tape that in combination with the conical sides, guides the scintillation light produced in the detector toward the bottom where it is read out by the PMT via an optical coupling pad. A charge-sensitive preamplifier contained in the end cap of the PMT housing converts the PMT charge pulses to voltage pulses, suitable for transmission down the boom to the DPU. The preamplifier

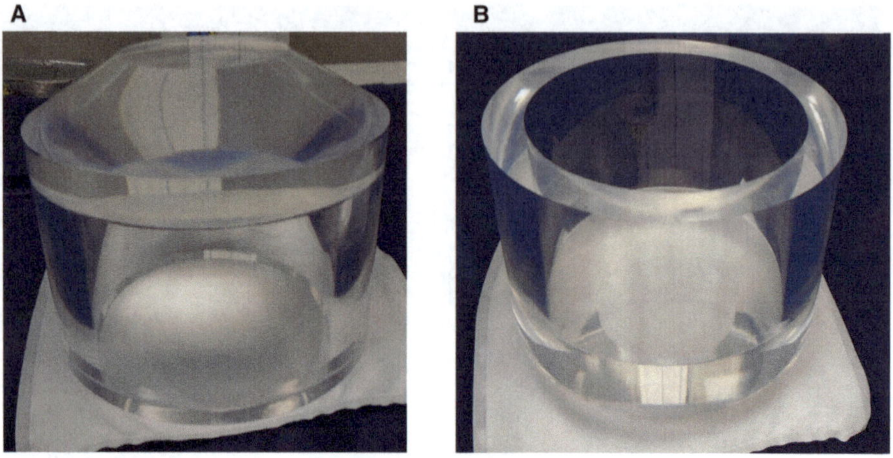

Fig. 11 Photographs of the of the AC shield scintillator. Panels A and B show the top and bottom, respectively, of the scintillator. Prior to installation, the scintillator is wrapped in reflective tape to internally reflect the scintillator light, and foil to provide light tight cover

contains two signal outputs that transmit the PMT signal with two different gains (see below). Because the AC shield scintillator (low density, low Z) is essentially transparent to gamma rays, AC shield signals that are in coincidence with HPGe signals are classified as background events and electronically removed (vetoed) from the HPGe AC spectral products.

In addition to the HPGe anticoincidence function, the AC shield also measures neutrons from two different energy bands, as well as GCR protons through energy deposition spectra that are recorded by the GRS DPU. Two spectral products (low and high gain) are recorded. The low-gain "GCR" spectrum extends to a maximum energy of 35 MeV, sufficiently high to record the characteristic peaks resulting from minimum-ionizing particle (MIP) interactions in the scintillator. Cosmic-ray protons produce characteristic MIP peaks at 4 and 17 MeV, resulting from interactions in the annular and base portions of the AC shield, respectively (see Sect. 5.1). The GCR spectrum therefore provides a means to measure and monitor the GCR proton environment in flight. The high-gain spectrum extends to a maximum energy of approximately 0.5 MeV, and includes the ~90 keV peak resulting from the $^{10}B + n \rightarrow {}^{7}Li + \alpha$ neutron capture reaction. Section 3.5.2 describes how the neutron and energetic proton measurements are carried out within the GRS DPU. Sections 4 and 5 provide examples of pre-launch and in-flight AC shield measurements, respectively.

3.3 Neutron Spectrometer (NS)

A graphic view of the NS is shown in Fig. 12; photographs of the NS are shown in Fig. 13; a functional block diagram of the NS is shown in Fig. 14; detector characteristics and resource use for the NS are given in Table 4. The NS consists of three ^{3}He neutron sensors that are sensitive to three different neutron energy ranges via different surrounding materials. Each cylindrical sensor is identical and is filled with 10 atm of ^{3}He gas. A thin wire attached at each end runs through the center of each sensor and is set to 1360 V relative to the grounded sensor housing. Neutrons are detected via the ^{3}He neutron-capture reaction:

$$^{3}He + n \rightarrow H + t + 764\ keV, \tag{1}$$

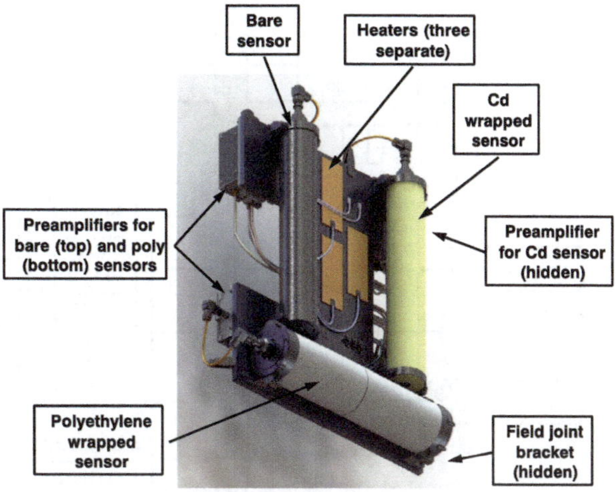

Fig. 12 Graphical view of the NS. The different sensor components described in the text are labeled. Figure adapted from Peplowski et al. (2025)

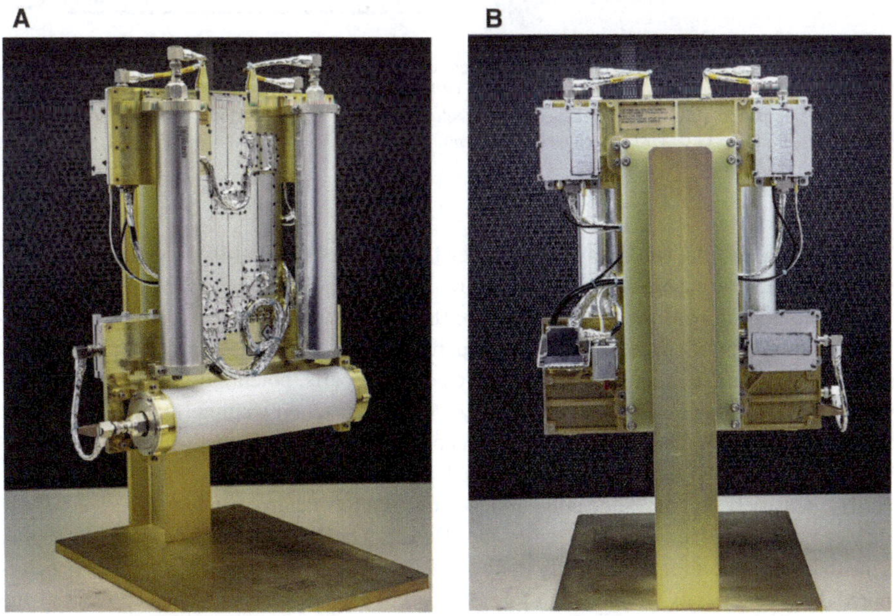

Fig. 13 Photographs of the NS on a test stand. Panels A and B show the front and back of the NS, respectively

where the combined kinetic energy (764 keV) from the proton (H) and triton (t) is detected as a charge pulse on the high-voltage line. Each sensor has dimensions of 5.08-cm diameter by 30.6-cm-long, and "dead volume" lengths of 2.07 cm (non-connector end) and 3.13 cm (connector end) that are not sensitive to neutron detection. The sensor length sensitive to neutron detection is 25.41 cm. Further details of the sensors and their calibration are given in Peplowski et al. (2020) and (Peplowski et al. 2025).

Neutron energy discrimination is achieved by surrounding each sensor with different types of materials. The bare sensor (no surrounding material) is sensitive to thermal and low-energy epithermal neutrons. The Cd-covered sensor is surrounded by a 0.5-mm layer

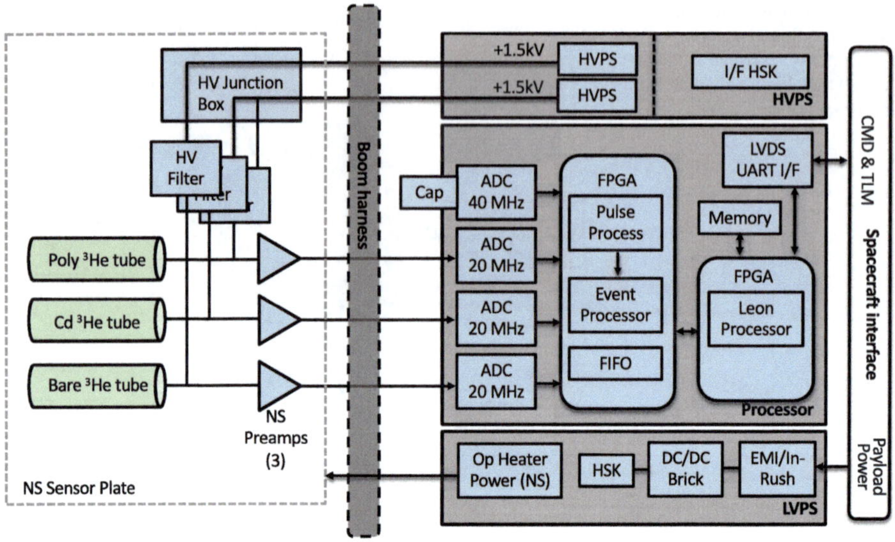

Fig. 14 NS functional block diagram. All instrument components that are located on the boom are shown within the large dashed rectangle. The three-slice DPU is shown on the right side of the block diagram. Individual components within each DPU slice are labeled. The 40 MHz ADC, which is used for the high-gain HPGe channel in the GRS DPU, is capped off and not used in the NS DPU. Additional details are given in the text

of Cd, which absorbs neutrons with energies less than 0.4 eV, and is therefore sensitive to low-energy epithermal neutrons. The count-rate difference between the bare and Cd-covered sensor provides a measure of thermal ($E_n < 0.4$ eV) neutrons. For applications in the vacuum environment of space, Cd has the property that it is easily evaporative and fragments of Cd can migrate to other mechanical and electronics parts, potentially affecting their performance. To mitigate this effect, the Cd layer on the low-energy epithermal neutron sensor was plated with a 3–12 μm layer of Ni to encapsulate the Cd to ensure there would be no Cd migration. The third sensor is covered in a 1-cm-thick layer of high-density polyethylene (C_2H_4, density 0.9 g/cm^3). The polyethylene downscatters neutrons to lower energies where they can be more efficiently detected by the ^{3}He gas. The polyethylene thus provides higher sensitivity detection for high-energy epithermal ($\sim$1 keV $< E_n <\sim$100 keV) neutrons. It is noted that while the AC shield is not formally part of the NS, its broad-band epithermal and fast-neutron measurements ($E_n > 500$ keV), which are similar to measurements carried out on the Lunar Prospector GRS (Gasnault et al. 2001; Genetay et al. 2003), complete the full GRNS five energy-band neutron detection capability. The modeled effective area for all five neutron energy bands is shown in Fig. 15.

All three neutron sensors are mounted to a single aluminum plate (1 mm thick with 8 mm thick reinforced ribs) that is attached to the boom at two attachment points (Fig. 13B). In contrast to the end-to-end sensor arrangement for the Lunar Prospector NS, the Psyche bare and Cd-covered sensors are oriented parallel to each other. Sensor "self-shielding" effects are minimized by separating the sensors by multiple sensor diameters. The polyethylene-covered sensor is oriented perpendicular to the bare and Cd-covered sensors and is located at the "dead volume" ends of the bare and Cd sensors, thus minimizing its shielding effects on the other two.

Table 4 NS characteristics and resource use

Neutron sensors	^{3}He gas proportional sensors; 5.08-cm diameter; 25.4-cm active length; aluminum housing; all sensors identical;
	Thermal: bare (no covering)
	Low-energy epithermal: 0.5 mm Cd covered; 3–12 μm Ni plating
	High-energy epithermal: 1 cm polyethylene covered
Energy range	Thermal: ~0.025 eV to 0.4 eV
	Low-energy epithermal: > 0.4 eV
	High-energy epithermal: ~1 keV to ~100 keV
Energy resolution (FWHM/mean)	<4% for 764 keV neutron capture peak
Effective field-of-view	4π omnidirectional; full view of Psyche asteroid
Number of spectral channels	256 channels for each ^{3}He sensor
Lower-level discriminators	Separate adjustable discriminator settings for each sensor
Maximum throughput	>5000 events per second
Operating voltage	1360 V
High-voltage stability	≪1 V
Electronics gain drift	None
Sensor gain drift	0.05 channels per °C
NS sensor mass	4.54 kg (including mounting plate)
NS DPU mass	1.56 kg
NS harness mass	1.94 kg (includes harness along boom and DPU to + z deck harness)
NS total mass	8.04 kg
Power use	11.4 W (DPU: 7.9 W; operational heater: 3.2 W; harness dissipation: 0.3 W)
Data rate	6000 bits per second Orbit D allocation (combined with GRS)

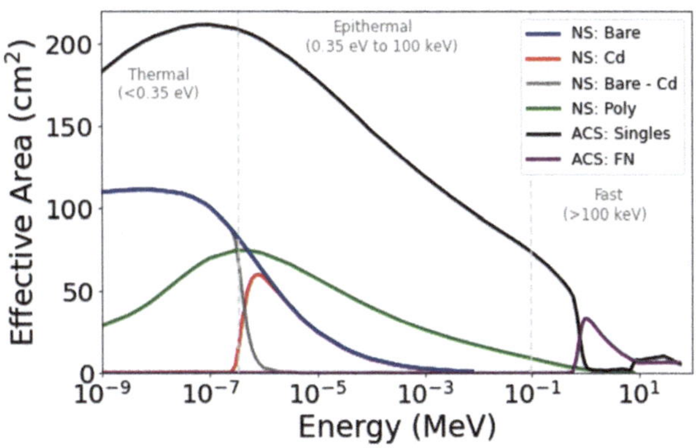

Fig. 15 Effective area of all GRNS neutron sensors and derived thermal neutron measurement as a function of incident neutron energy. Effective area is the sensor efficiency times its cross-sectional area, and were derived as part of the GRNS calibration activities (Peplowski et al. 2025). ACS singles is the AC shield broad-band epithermal neutrons; ACS FN is the AC shield fast neutrons

The backside of the NS mounting plate contains three separate but identical preamplifier boxes (Figs. 12 and 13) that feed high-voltage lines to each sensor, and receive and condition each signal line before sending the signals through the boom harness to the NS DPU. The fourth backside corner of the mounting plate contains a high-voltage junction box and field-joint bracket (lower left in Fig. 13B). The high-voltage junction box receives the two high-voltage lines from the DPU and distributes them to the three sensors, with a single high-voltage line going to the bare sensor, and two lines (from a single high-voltage supply)

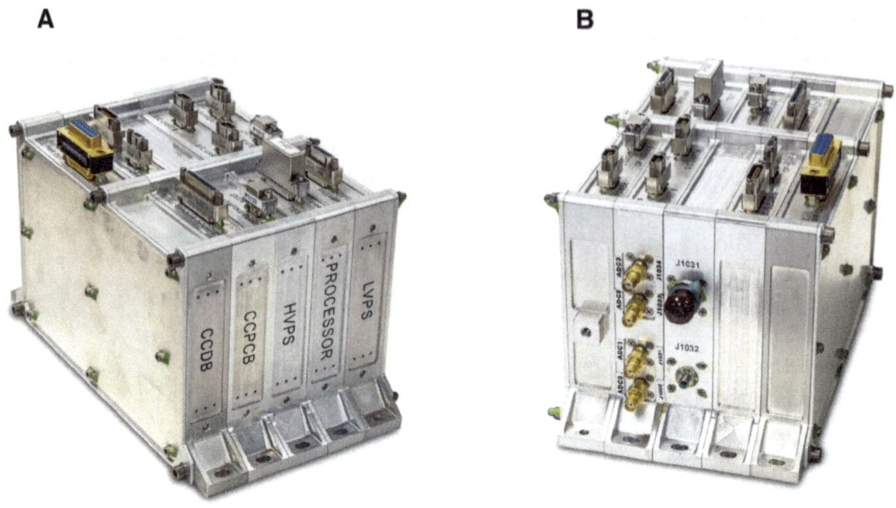

Fig. 16 Photographs for two views of the GRS DPU show the top connectors and slice names (A) and side connectors (B)

going to the Cd- and polyethylene-covered sensors, respectively (see Fig. 14). The field-joint bracket combines all the individual cables from the sensors into a single harness that runs from the sensor plate to the NS DPU. In its final configuration on the spacecraft, the sensors and plate are covered with a thermal blanket of MLI, and are kept thermally stable with thermal straps on the front side of the plate. The NS is enclosed by another layer of thermal blanketing that encompasses most of the GRS/NS boom structure.

3.4 Data Processing Units (DPUs)

3.4.1 GRS DPU

The GRS DPU consists of five boards (Figs. 6 and 16). Three of the boards are common to the GRS and NS DPUs: a Low Voltage Power Supply (LVPS), Processor, and High Voltage Power Supply (HVPS). The other two boards, the Cryocooler Power Conditioning Board (CCPCB) and Cryocooler Driver Board (CCDB) are part of the GRS DPU only. Each board in the DPU is a slice about 10.2 cm × 16.5 cm with an internal stacking connector providing the connections required between boards. They are each housed in an aluminum frame that is stacked together to form a complete housing (the end slices receive end covers). External connections to the spacecraft and to the sensor assemblies are provided by connectors on the top of each slice and on the side on the Processor and HVPS slices.

The Processor board contains several Analog-to-Digital Converters (ADCs), spacecraft communications interfaces, memories, and two Field Programmable Gate Arrays (FPGAs). The Pulse-Process FPGAs perform pulse processing algorithms on the incoming signals via the science ADCs and provides the processed signals to the Event-Processor FPGA. The Event-Processor FPGA contains an embedded LEON3-FT processor along with custom logic for instrument control. The LEON3-FT processor core runs instrument flight software, which is responsible for processing commands, generating telemetry, operating the instrument, and handling science data from the Pulse Process FPGA.

The LVPS is a Printed Circuit Board (PCB) that powers the DPU electronics. Both the GRS and NS DPUs have nearly identical LVPS boards. The LVPS converts spacecraft input voltage, which ranges from 27.2 VDC to 33.6 VDC, to isolated conditioned voltages for the electronics in the DPU and for the NS and AC shield preamplifier boards which are external electrical PCBs to the DPU stack. The LVPS includes telemetry circuits to monitor all voltages and currents to other DPU slices, primary current on the LVPS, NS operational heater current, temperatures on the AC shield preamplifier board and board temperature on the LVPS. ADCs convert these analog-sensed housekeeping signals into serial digital signals that are sent via a Serial Peripheral Interface bus to the Processor board. The LVPS also has circuity to turn on and off the NS operational heaters, which are external to the DPU. The LVPS has an inrush current limiter and an electromagnetic interference (EMI) filter to properly handle power-on transients and prevent noise from going back onto the spacecraft payload power bus. The LVPS includes a DC/DC converter that steps down the input voltage from the spacecraft to ± 12 VDC and + 3.3 VDC. Other voltages required for the instrument such as ± 6 VDC and ± 5 VDC are generated by linear regulators and sent to the stacking connector, the sensor electronics, or utilized on the LVPS itself.

The CCDB is a low-distortion 50-W Class-D-type amplifier that supplies the cryocooler with sinusoidal power of variable amplitude and frequency. The CCDB circuits are powered directly from spacecraft primary power to maximize efficiency, made possible because the cryocooler motor coils are completely isolated from chassis. The CCDB draws filtered input power via the CCPCB, which interfaces to a dedicated spacecraft "dirty" bus to ensure that any conducted emissions emanating from the cryocooler electronics will not reach and potentially interfere with other sensitive instruments or spacecraft subsystems. Communication with secondary-referenced electronics is achieved via high-speed digital isolators. Several protection circuits are incorporated into the design including an overcurrent trip, shoot-through prevention, and "launch lock" motor coil shorting relays.

All CCDB drive signals are digitally synthesized. The Event Processor FPGA generates a numerical sinusoidal waveform of variable frequency and amplitude using a Coordinate Rotation Digital Computer (CORDIC) algorithm. The frequency is selected to match the optimum drive frequency of the cryocooler, a value that varies linearly with temperature over the range 117-132 Hz. The resulting sinusoid is then compared against a digitally-generated sawtooth carrier that converts the signal to pulse-width modulation (PWM) format, where the carrier frequency is also selectable. This feature is used to set the carrier to a value not harmonically related to the magnetometer signal processor. The PWM signals are sent to the CCDB over the stacking connector. "3-level" modulation is used to minimize total harmonic distortion (THD), a scheme that requires independent control of each half of the Class-D H-bridge. Dead time can also introduce a significant source of distortion and can limit maximum output power capability. Gate drivers that offer independent high and low side control were chosen to allow selectable dead time. After extensive testing and timing analysis, a dead time of 50 ns was chosen, yielding a THD of $\sim 1\%$. The output is low-pass filtered with balanced two-stage Butterworth LC filters that provide >60 dB differential-mode attenuation at switching frequencies above 100 kHz. The PWM frequency is nominally set to ~ 107 kHz.

The primary bus voltage is nominally ~ 31 V but can range $\pm 5\%$ so is regulated down further to produce a "stiff" 27-V cryocooler power bus based on a bank of eight paralleled linear regulators mounted to the frame to dissipate heat. Each regulator is capable of delivering up to 1 A, providing ample margin and a degree of fault tolerance. Small ballast resistors associated with each regulator ensure equal distribution of the load current to within ~ 50 mA. An additional two linear regulators create a 12.5-V bus for the H-bridge gate drivers and a 5-V bus that powers local digital circuits, housekeeping monitors, and the Hall Effect position sensors. The Hall Effect sensors provide piston stroke displacement information, an important cryocooler health diagnostic.

The CCDB incorporates several protection features. An overcurrent-monitor circuit trips if the peak current exceeds 8 A, a value chosen to be nearly twice the maximum operating current of the cryocooler, so as not to trip unnecessarily, but sufficient to interrupt a serious fault condition at the output. When the associated comparator circuit triggers, the circuit quickly disables the gate drivers (<10 μs), sets a flag to the Event Processor FPGA, and then remains latched in a disabled state until cleared by ground command or power-on reset. Another protection feature is an H-bridge shoot-through prevention circuit that rejects erroneous digital PWM signals that could otherwise short the cryocooler bus to ground. Lastly, magnetic latching relays are included to short the cryocooler motor coils during launch. Shorting the coils heavily dampens vibration-induced motion of the cryocooler pistons with an opposing back electromotive force. Without this feature, the pistons could potentially impact the end-stops with sufficient force to cause permanent damage. A grouping of four relays provides full series-parallel redundancy, i.e., if one relay opens during launch, the short is maintained by a parallel branch, and if one relay fails closed during launch, then a series relay ensures that the short can be opened to allow the cryocooler to operate.

The HVPS slice consists of two boards: the digital interface board and the analog high voltage board. The digital interface board is the low-voltage portion of the slice that communicates between the processor board and the analog high voltage board. The analog high voltage board amplifies the command voltage (0-5 V) to the voltage levels required by the GRS and NS. The digital interface board was designed and manufactured at APL, while the analog high voltage board was designed and manufactured at University of California, Berkeley. APL integrated both boards into one slice and tested the integrated slice. The digital interface board consists of signal chains that convert and scale commanding and housekeeping signals between the processor board and the analog high voltage board, a safing circuit that prevents unintended high voltage turn-on during ground testing, and a sequence circuit that ensures that the digital board is powered before the analog board to prevent board damage. The analog high voltage board consists of self-resonant power converters that amplify the commanding voltages, and housekeeping circuits that measure voltages and currents.

3.4.2 NS DPU

The NS DPU (Fig. 17) is a similar design to the GRS DPU with the exception of the cryocooler board set. The Processor board runs an NS-specific flight software application and the HVPS outputs are tailored to provide the correct high voltages for the NS ^{3}He sensors. Both DPUs have identical LVPS boards.

3.5 GRNS Flight Software and Operation

3.5.1 Software Overview

The GRNS flight software is based heavily on the APL-generated "common core" library used for APL space instruments. Most software tasks utilize routines from this common library, which is actively used and maintained for multiple on-going missions. The common core library includes modules to handle the following types of functionality: command buffering, parsing, and execution; telemetry packet generation, compression, formatting,

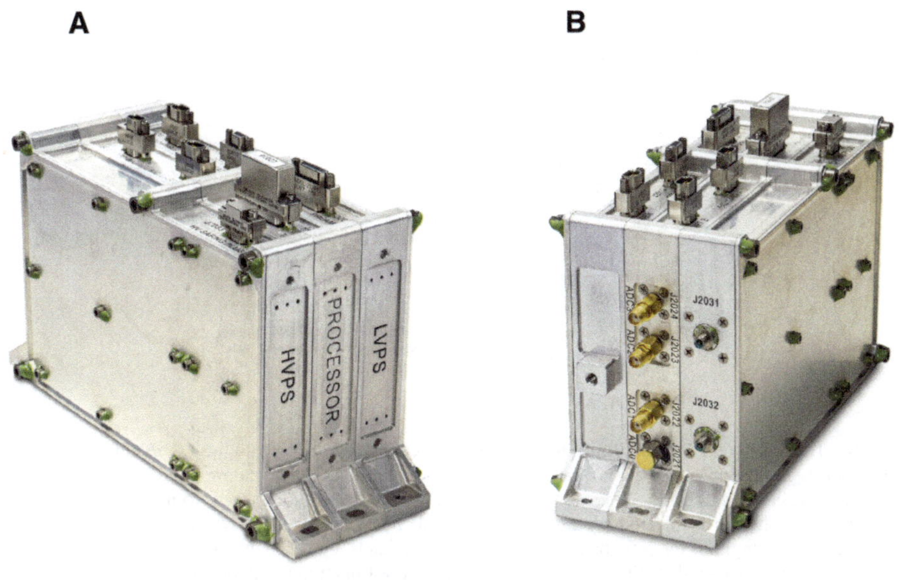

Fig. 17 Photographs for two views of the NS DPU show the top connectors and slice names (A) and side connectors (B)

and output buffer management; alarm monitoring, response execution, and reporting; on-board definition, execution, and maintenance of stored sequences (macros); memory management including load, dump, copy, and checksum; stored parameter management; housekeeping, command echo, and alarm packet generation; and variable telemetry collection intervals.

The common core library provides hooks to include custom functionality where necessary. The GRS and NS flight software applications include hardware device drivers for the GRS and NS Pulse Process FPGAs, respectively. There is a device driver for each major hardware function in the FPGA design. In the GRS flight software, there is a driver for each of the following functions: housekeeping ADCs; cryocooler functionality; anneal heater functionality; high voltage functionality; pulse processing functionality; miscellaneous instrument interfaces. In the NS flight software, there is a driver for each of the following functions: housekeeping ADCs; high voltage functionality; pulse processing functionality; miscellaneous instrument interfaces. Each flight software application also includes a science processing task which is responsible for controlling the cadence of science data collection.

3.5.2 GRS and NS Data Processing

The underlying framework for retrieving and histogramming GRS and NS products is similar for the two different software applications on each DPU. Each DPU processor treats the Pulse Process FPGA as an input/output memory interface, with registers to read and write. The software configures the pulse processing channels by writing to control registers and retrieves pulse processing event data and status by reading data and status registers. The Pulse Process FPGA multiplexes event data from all channels into a single first-in, first-out (FIFO) buffer and presents it to software as a single data source. The processor is interrupted at a "half full" mark of the FIFO to trigger the driver to read out all available events in the

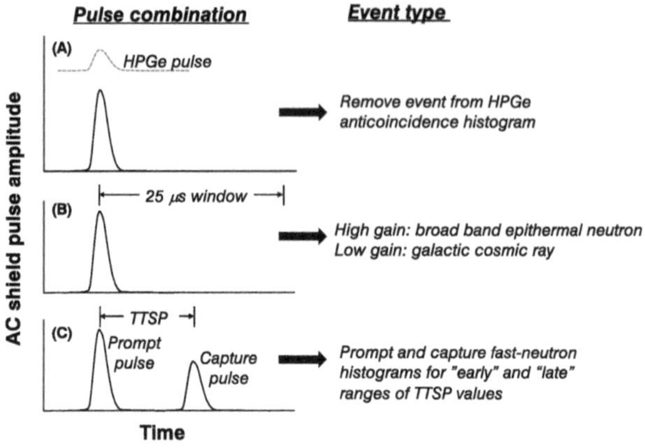

Fig. 18 This schematic diagram illustrates the pulse processing associated with the AC shield. A) When an AC shield high-gain pulse (solid line) arrives in coincidence (within a 0.5 μs window; window size not shown) with a HPGe pulse (dashed line) this event is kept in the raw spectrum but removed from the anticoincidence spectrum. B) When an AC shield high-gain pulse is the only pulse within a 25 μs window, the pulse amplitude is recorded in histograms for both the high- and low-gain channels. C) When two AC shield high-gain pulses arrive within a 25-μs window, this event is categorized as a fast-neutron candidate event. The first (prompt) and second (capture) events are recorded in early and late histograms based on the time between the two pulses (i.e., the TTSP). More details for fast-neutron events are provided in the text

FIFO. Based on which data products are enabled and the histogram configuration, the software will histogram valid events into the appropriate data product or drop invalid events. All histogram data produced by the GRS and NS are compressed within the DPU using a lossless fast compression algorithm. Non-histogrammed data from both the GRS and NS are uncompressed.

HPGe spectra The HPGe spectra are the primary data product of the GRS. The GRS sensor has two parallel signal processing chains from the HPGe pre-amplifier. One is the low-gain channel that characterizes gamma-ray energies up to 9000 keV and the other is the high-resolution high-gain channel that characterizes energies up to 3000 keV. The high-gain channel digitizes the analog signals from the sensor with a 40 mega-samples per second (MSPS) ADC (AD9246S) on the processor board (Fig. 6), which is integrated down to 10 MSPS in the Pulse Process FPGA for the signal processing chain. The low-gain channel digitizes incoming signals with a 20 MSPS ADC (RHF1401) (Fig. 6) integrated down to 10 MSPS by the Pulse Process FPGA. After the integration stage, both channels have a three stage (unipolar, bipolar, tripolar) shaping algorithm for detection of HPGe events. The processor bins tripolar events into appropriate histograms based on the amplitude reported by the Pulse Process FPGA. Each channel has two 16,384-bin histograms where the highest 5 V signal goes into the highest bin. The "raw" histogram contains every valid energy event detected and the "anticoincident" histogram only contains events that also saw no coincident activity on the AC shield during a 0.5 μs analysis window of the event (see below).

AC shield neutron and GCR spectra Fig. 18 illustrates how different AC shield pulse combinations are categorized into different event types. When an AC shield high-gain pulse is seen in coincidence with an HPGe pulse (Fig. 18A), the event is categorized as an HPGe coincidence event, and the HPGe pulse is removed from the HPGe AC histogram; the AC shield pulse for this event type is not placed in any histogram. When an AC shield high-gain pulse is detected without a coincident HPGe pulse, this opens a 25-μs analysis window. If no second pulse is detected with this window (Fig. 18B), the high- and low-gain pulse amplitudes are added to separate high- and low-gain histograms, respectively. The high-gain histogram (512 bins) provides a measure of broad-band epithermal neutrons by detecting the energy deposition from a single ^{10}B(n,α) capture in the scintillator. The low-gain histogram (1024 bins) measures the energy deposition from energetic GCRs. If a second pulse is detected within the 25-μs-long window, then this double-pulse event is categorized as a fast-neutron candidate event (Fig. 18C). For these events, the first pulse in the analysis window is the "prompt" pulse, and the second pulse is the "capture" pulse. The time difference between the two pulses is the time-to-second-pulse (TTSP).

The physical process of the fast-neutron detection is the following. When a fast neutron enters the AC scintillator with an energy above the nominal threshold of ~0.7 MeV (Peplowski et al. 2025), it quickly slows down through interactions with the hydrogen atoms in the scintillator. The energy deposition from this slowing down process produces the prompt pulse; the amplitude of the prompt pulse provides a measure of the initial neutron energy. After the neutron has sufficiently slowed down in the scintillator, it can be captured by a ^{10}B atom in the scintillator with an exponential TTSP distribution. The mean time, τ, for capture is ~2 μs, which is determined by the 5% ^{10}B in the scintillator (Feldman et al. 1991). The capture pulse identifies the 90 keV energy deposition from the ^{10}B(n,α) reaction. The size of the analysis window is set to cover multiple e-folding lengths such that most fast neutrons are dominantly detected in the early portion of the window (0–5 μs) and accidental coincidences dominate the late portion of the window (20.6 – 25 μs). The nominal fast-neutron data products are four separate 256-bin histograms of early and late histograms of the prompt and capture pulses, respectively. The net fast neutron histograms are derived by taking the differences between the early and late histograms for the respective prompt and capture pulses. More details about the AC shield fast-neutron energy threshold and detection calibration are provided by (Peplowski et al. 2025). Fast neutron data acquired in flight are discussed in Sect. 5.1.

Gamma-ray burst The AC shield scintillator is sensitive to the energy deposition of gamma rays of energies above ~20 keV. A gamma-ray burst (GRB) product is generated using the full energy range of the AC shield high-gain channel. While this product is not part of the science requirements, it is included for additional science investigation to help triangulate galactic gamma-ray burst events as part of an ongoing interplanetary burst detection network (Hurley et al. 2010); the MESSENGER NS included a similar capability (Goldsten et al. 2007). Qualified above-threshold signals are counted (not histogrammed) at a 20 Hz cadence (i.e., 50 ms time resolution) and stored in a 1024-sample ring buffer. A trigger is detected if the count rate exceeds a threshold above a moving average across the time samples, indicating the detection of a gamma-ray burst. The collection of count rates will continue until the ring buffer has been filled post-trigger, leaving 256 samples of pre-trigger count rates for analysis. Once the buffer has been filled, it will be queued for downlink and the system will continue to look for the next trigger. The trigger threshold is nominally set at 20 keV, but will be tuned in flight to minimize the number of false positive triggers.

^{3}He neutron sensor spectra Each NS sensor generates a 256-bin histogram using a dedicated signal processing chain that run signals in parallel. The processor board digitizes signals with a 20 MSPS ADC (Fig. 14) and the Pulse Process FPGA integrates the samples down to 10 MSPS for shaping using a bipolar shaper for pulse detection. The amplitudes of the pulses from each neutron sensor are histogrammed in the flight software. Additionally, the time between the positive and negative peaks of each bipolar shaped pulse in each sensor is reported on an event-by event basis. This value, called 'dt' (i.e., 'delta time') can be used for background rejection. More information about the use of dt for background rejection is given by Peplowski et al. (2025) and in Sect. 5.1.

Additional data products The flight software has other data products that are produced either nominally in flight and/or upon command. During nominal operation, both the GRS and NS produce raw engineering and counter data packets. The engineering packets contain various instrument health and safety parameters (e.g., voltages, currents, temperatures), as well as instrument operational settings. Each of the counter packets contains a collection of various sensor and software count rates such as trigger rates, out-of-range rates, and other values. The accumulation period for these engineering and counter packets is commandable. The flight software also provides for the collection of GRS and NS raw event data where individual pulse parameter values (e.g., pulse amplitudes, time tags) are provided. For raw GRS and NS events, these products contain up to 50 events per second. The fast-neutron raw-event packet provides up to 255 events per second. The generation of these raw event packets is enabled and disabled via command. Detailed descriptions of these and all other GRS and NS data packets are given by Espiritu et al. (2022a) and Espiritu et al. (2022b).

3.5.3 Other Subsystem Controls

Operational heaters Both the GRS and the NS have two operational heaters under software control. On each sensor, the two operational heater controls are independent. Parameters define a temperature range for the operational heaters to control with hysteresis. If the temperature sensor dips below the lower temperature bound, the operational heater will turn on. Likewise, if the temperature goes above the upper bound, the operational heater turns off. The purpose of the operational heaters is to more tightly regulate the temperature of the sensors as a way to minimize temperature variations of the sensor that could result in gain variations producing "smear" as well as to improve long-term reliability of the sensor electronics. The nominal setpoint is -10 °C $\pm$ 1 °C.

HVPS control Both the GRS and the NS have two high voltage power supplies (Figs. 6 and 14) under software control. On each sensor, the two high voltage supplies are independently controlled. High voltage is ramped slowly to a commandable goal using a parameterized three-stage ramping pattern. While high voltage is ramping, the software monitors the housekeeping for unsafe conditions. If any such conditions are seen during a ramp, the high voltage will be brought back to a safe level and the safing level will be updated as appropriate. If at safing level 1, the high voltage can automatically re-attempt to ramp for a commandable number of retries after a waiting period. If the high voltage cannot be safely ramped after exhausting all retries, that high voltage supply will be set to safing level 2 and requires ground intervention before attempting to ramp again.

Sensor safing The flight software employs a number of autonomous safing mechanisms to keep the instrument safe in case of anomalous behavior due to either internal instrument and/or environmental conditions.

On the GRS HPGe high voltage supply, the detector leakage current is monitored during different stages of the ramp. The safing behavior is dependent on which stage of the ramp and the leakage current seen. Stage 1 of the HPGe high voltage ramp is a blind ramp and the leakage current is not monitored. While ramping above the bias voltage (Segment 1

Hold Voltage) in stages 2 and 3, the bias current is checked against non-extreme leakage current threshold. If the leakage current is persistently above this threshold, the software will trigger a safing level 1 event. Persistence in the leakage current check is required to avoid instantaneous spikes in the leakage current due to GCR hits. If this safing level 1 event occurs below the depletion voltage (Segment 2 Hold Voltage), the high voltage is held until the retry interval expires or the safing level is reset. If this safing level 1 event occurs above the depletion voltage, the high voltage is ramped back down to the depletion voltage and held there until the retry interval expires or the safing level is reset. At any time that the high voltage is above the depletion voltage, including when at goal, if the leakage current is persistently above an extreme leakage current threshold, the software will automatically trigger a safing level 2 event and ramp back to the depletion voltage. The thresholds for non-extreme and extreme leakage current, as well as the persistence of both checks, are configurable by command. A second safing condition for the HPGe sensor is to bring high voltage back to 0 V and disable the high voltage control if the temperature at the detector is detected to be out of range persistently for three seconds. The leakage current thresholds for the HPGe detector are set well above nominal expected values but below those that would indicate a more serious condition, i.e., ~1000 pA.

The NS sensors and AC shield PMT have a count-rate safing feature where the flight software monitors the count rates in each of these sensors. If the count rates on any of these sensor channels exceeds unique thresholds for each sensor, the software triggers as safing level 1 event and brings the high voltage back to 0 V. Typical scenarios for triggering this type of safing are when large solar particle events hit the spacecraft and induce count rates that are significantly above nominal background levels. When a level 1 safing event occurs, the software waits a specified amount of time (e.g., a few hours) and then autonomously raises the high voltage. If the count rate has gone down, then the sensors resume their normal operation. However, if the count rate remains high after a set number of attempts to raise the high voltage, then the flight software enters safing level 2 and will keep the high voltage at 0 V and wait for ground intervention. The count rate thresholds were chosen to be above the expected nominal science counting rates for the sensors yet comfortably below any rate that could accelerate aging (even for an extended period of time). Current values are 3 kHz for the NS and 20 kHz for the ACS. Autonomous recovery times are based on typical profiles for large solar particle events. Attempts at recovery are typically made every 12 hours for up to a few days.

Temperature safing and control for all subsystems use a pair of redundant temperature sensors. If the primary sensor were to fail, the software automatically switches control to the secondary sensor to main instrument operation and safety without intervention. If both sensors fail, the software configures the subsystem in a safe state until the sensors can be recovered.

Cryocooler Using a three-level PWM scheme (see Sect. 3.4.1), the event processor controls the power to the cryocooler to manage the temperature. The power output is nominally set automatically by a proportional integral controller implemented in software but can also be set to fixed values manually via command. Control loop parameters are configurable, as is the temperature set point of the detector and the maximum output power allowed. To avoid damage to the sensor, the cryocooler will be disabled automatically if the temperature at the compressor or the warm flange is detected to be out of limit persistently for three seconds.

Anneal heaters The GRS sensor has two anneal heaters to heat the Ge sensor during bake-out or annealing operations. The processor generates a pulse train to the anneal heater to provide power. The software uses "on-off" control to reach its temperature set point, with parameterized upper and lower dead band limits to reduce switching. A power switch must be enabled for each anneal heater for the pulse train to generate any power. Only one anneal heater may be powered at a time. The anneal heater control will be disabled if the Zener diode is detected to be in an "open" or "shorted" condition persistently for three seconds. The thresholds for open and short are parameterized. The anneal heater control will also be disabled if the temperature at the detector is over 115.5 °C persistently for three seconds. The anneal heaters use the same redundant sensors as the high voltage and cryocooler for detector temperature sensing. If both primary and secondary sensors have failed, anneal heater control is disabled.

3.5.4 GRNS Commanding

The GRNS has an extensive set of available commands for instrument turn on, configuration, maintenance, debugging, and data collection. The set of GRS commands is more extensive than the NS, and the NS commands are similar or identical to subsets of the GRS commands. The GRNS is able to generate and execute command macros, which are a sequence of commands that can be stored and then executed later.

The GRNS commands can be divided into the following categories: instrument power on/off, software core commands, GRS bake out/anneal, GRS cryocooler, operational heaters, GRS pulser, and data accumulation. The power on/off commands are formally spacecraft commands that provide the spacecraft voltage to each of the DPUs, which then for power-on initiates the boot process for each of the DPUs. Software core commands are an extensive set of commands that carry out various flight software tasks, such as memory check loads, and clears, and setting up of macros. The GRS bake out/anneal commands carry out the processes for operating the GRS anneal heaters that are used both for the initial GRS bake out following launch, and subsequent GRS anneal activities. The GRS cryocooler has a set of commands used to turn on and configure the cryocooler for initial and nominal operation. Operational heater commands are used to configure the autonomy for the operational thermal heating as well as set various temperature limits and thresholds. Pulser commands set up and configure parameters (e.g., rate, width) of the SCUM and digital pulsers. Finally, data accumulation commands enable and disable the various data products, as well as set up various parameters such as thresholds, accumulation times, and pulse processing parameters.

3.6 Summary of Instrument Builds and Environmental Testing

As part of the GRNS development program, a full set of engineering model (EM) hardware was built for development, debugging, and instrument characterization. A second complete EM DPU was built and delivered to JPL for ongoing development and testing in the Psyche mission testbed. Environmental tests carried out on the EM included vibration, shock, and electromagnetic interference and compatibility (EMI/EMC) qualification tests. Following the EM, the full flight model was built, tested, qualified, and delivered to JPL and installed on the Psyche spacecraft. Qualification testing included standard tests of vibration, shock, and thermal vacuum testing.

Photographs of the final installed GRS and NS are shown in Fig. 19. Figures 19A and 19B show the Psyche spacecraft following final GRS instrument closeout in August 2023, which included the installation of the light shield and calibration source. Figure 19C shows each sensor component with only the NS thermal blanket installed. The NS is covered with a secondary blanket that is a portion of the boom thermal blanketing (Fig. 19A). The radiator surfaces on the GRS are exposed, whereas all other portions of the GRS (e.g., cryocooler, harness, PMT, etc.) are covered in thermal blankets.

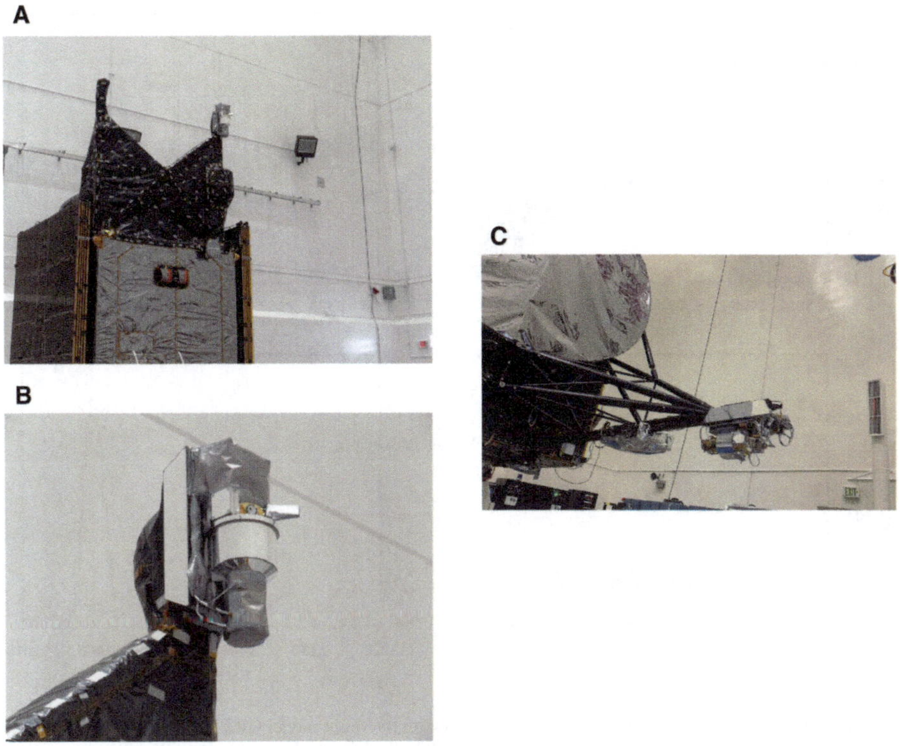

Fig. 19 (A) Photograph of the GRS and NS on the Psyche spacecraft with final thermal blanketing installed after the GRNS closeout. (B) Focused picture of the GRS after its closeout. (C) Photograph of installed GRS and NS on the Psyche spacecraft prior to the installation of the final thermal blankets. The thermal blanketing for the NS is seen

A full set of characterization and calibration measurements were carried out on both the NS and GRS. Initial NS calibration measurements are described by Peplowski et al. (2020). Comprehensive GRS and additional NS performance and calibration measurements were carried out prior to delivery and are described by Peplowski et al. (2025).

4 Ground Performance Measurements

4.1 GRS Performance Measurements

The performance of the GRS was characterized during development, final instrument calibration, and post-spacecraft installation. Here we present instrument checkout measurements performed on the flight model GRS prior to launch; Sect. 5 describes the GRNS performance post launch.

Figure 20 shows the full gamma-ray spectrum measured with an experimental setup designed to illuminate the GRS with gamma rays up to ~9000 keV (Peplowski et al. 2025). This setup used an AmBe neutron source surrounded by various types of materials (polyethylene, stainless steel, potassium chloride) to generate neutron capture and inelastic scatter gamma rays from elements such as K, H, Ni, Cl, and Fe. Carbon gamma

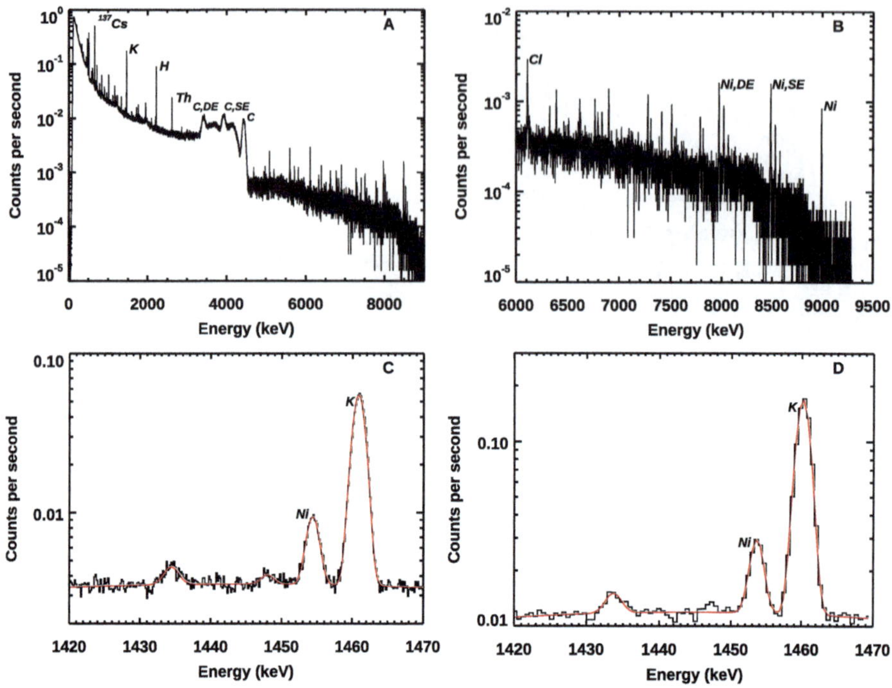

Fig. 20 Measured laboratory gamma-ray spectra taken with the flight model GRS as described by Peplowski et al. (2025). Part A shows the energy range from 0 to 9000 keV for the low-gain channel, with some prominent gamma-ray peaks labeled. Part B shows the high-energy portion of the low-gain channel with Cl and Ni gamma-ray peaks labeled. Part C shows the high-gain spectra in the energy range that highlights the 1454 and 1461 keV peaks from Ni and K, respectively. Fits to the peaks in this energy range are shown in red. Part D shows the same energy range shown in Part C but with the low-gain channel

rays from the AmBe source were also seen. The full low-gain spectrum (Fig. 20A) shows gamma rays from ∼90 keV to >9000 keV with 16,384 energy channels. Figure 20B shows the high-energy range from 6000 keV to 9500 keV, including capture lines from Cl and Ni. Figures 20C and 20D show the energy range around the key 1454 keV ^{58}Ni gamma-ray line for both the high- and low-gain spectra, respectively. The high-gain spectrum provides measurements of gamma rays with energies of <3000 keV, again with 16,384 energy channels. The oversampling of the high-gain data (Fig. 20C) provides better energy resolution than the low-gain spectrum (Fig. 20D). For this experiment, the count rate of K gamma rays at 1461 keV was sufficiently large to dwarf the measured count rate of Ni gamma rays from the stainless steel. Even with this significant count-rate difference, the Ni line is cleanly resolved and thus demonstrates that even with a similar count-rate difference at Psyche between these two gamma-ray lines (see Fig. 2), the GRS will be able to robustly quantify the flux of 1454-keV gamma rays.

The performance of the GRS was further verified by checking the energy resolution using a ^{60}Co gamma-ray source. Following ANSI standard N42.14, which characterizes the FWHM energy resolution of gamma-ray spectrometers at 1332.5 keV, we performed spectral analysis of the HPGe measurements. Figures 21 and 22 show gamma-ray spectra collected during the first CPT in September 2021 after the GRS was installed on the Psyche spacecraft in August 2021. These spectra were obtained with full processing of the data

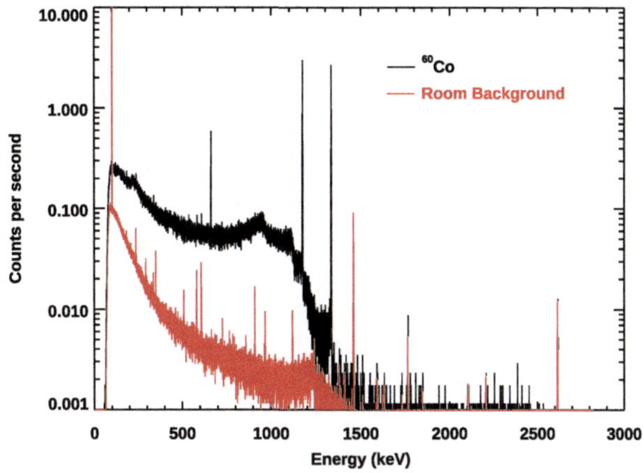

Fig. 21 Measured gamma-ray spectra taken from the high-gain channel during the September 2021 CPT after the GRS was installed on the Psyche spacecraft. Black shows data taken using the ^{60}Co calibration source, and red shows room background data

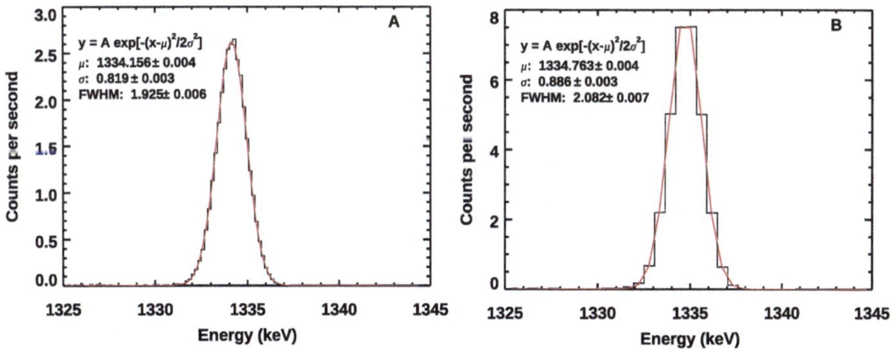

Fig. 22 High-gain (A) and low-gain (B) data taken using the ^{60}Co calibration source during the September 2021 GRS CPT while on the Psyche spacecraft. Gaussian fits to the peaks are shown in red, and the fit parameters are shown in the plots. We note that the best-fit peak centroid in both cases has a higher energy than the standard 1332.5 keV; however, this fit value is consistent with the standard value when the uncertainties in the pre-launch calibration parameters are taken into account (Peplowski et al. 2025)

through the spacecraft flight software and ground data system. The measured spectra include gamma-ray peaks from room backgrounds (primordial radionuclides ^{40}K, ^{232}Th, and U isotopes). A fit to the 1332.5-keV peak from the ^{60}Co source is shown in Fig. 22. The optimum high-gain measurement demonstrates 1.93 ± 0.01 keV FWHM energy resolution (Fig. 22A). This value is typical of laboratory-quality measurements and exceeds the 3.5 keV FWHM pre-launch requirement. As expected, the energy-resolution performance of the low-gain channel is about 8% poorer than the high-gain channel with a resolution of 2.08 ± 0.01 keV (Fig. 22B). The GRS performance was characterized through a range of operating environments during four separate pre-launch tests (e.g., thermal vacuum, spin up of all reaction wheels, etc.), and showed consistent performance during all tests.

The neutron detection performance of the AC shield was characterized with a moderated and unmoderated AmBe neutron source as well as a ^{137}Cs source. The high-gain spectrum is shown in Fig. 23. The 60-keV X-ray and ^{137}Cs Compton edges were used to establish the energy scale of the spectrum. The moderated AmBe measurements (red) show a clear spectral peak at ~100 keV that is not present when a moderated neutron source is absent. This

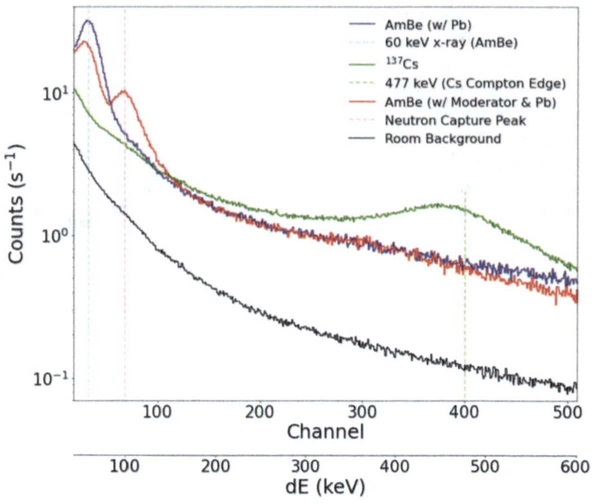

Fig. 23 Neutron events observed in the high-gain spectrum of the GRS AC shield. An estimated channel-to-energy calibration is included as the second horizontal axis. The 60-keV X-ray from AmBe (blue), neutron capture peak (red), and 661-keV Compton edge (477 keV; green) features are marked with vertical dashed lines

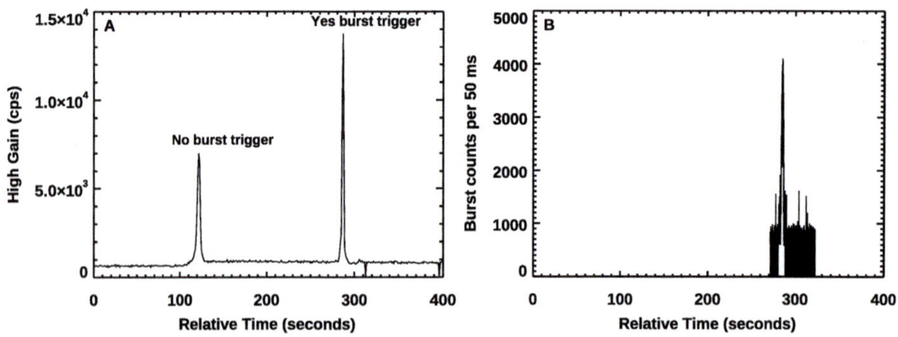

Fig. 24 Example gamma-ray burst data from pre-launch ground-based testing of the AC shield taken on 23 September 2021. A radioactive gamma-ray source was twice moved rapidly towards the AC shield to induce a count-rate enhancement. The high-gain count rate with a time resolution of one second is shown in (A) as a function of time relative to a fiducial time at 0 seconds. The first time the source was moved, the source did not get too close to the AC shield and so did not initiate a burst trigger (left peak at 120 seconds); the second time the source was moved, it came closer to the AC shield and initiated a burst trigger (right peak at a time of 290 seconds). The burst trigger with 50-ms time resolution (B) is shown at the time of 290 seconds

spectral feature is the $^{10}B(n,\alpha)^{7}Li$ neutron capture peak. Its presence is indicative of neutrons, and its amplitude provides a measure of the count rate for broad-band neutron energies less than 0.7 MeV. Note that double-pulse neutron events are not classified as single-pulsed events and therefore do not appear in this spectrum.

Ground-based fast neutron data and calibration are described in detail by Peplowski et al. (2025). Flight-based fast neutron data are presented in Sect. 5.

The low-gain AC shield spectrum provides a means for monitoring the flux of GCR protons. Laboratory testing of cosmic-ray monitoring is challenging, as cosmic-ray protons do not reach Earth's surface due to the presence of the thick atmosphere. GCR-generated muons do reach Earth's surface, however, and can serve as a proxy for testing the GCR measurement capability. Peplowski et al. (2025) describe how this ground-based muon measurement was carried out with the AC shield.

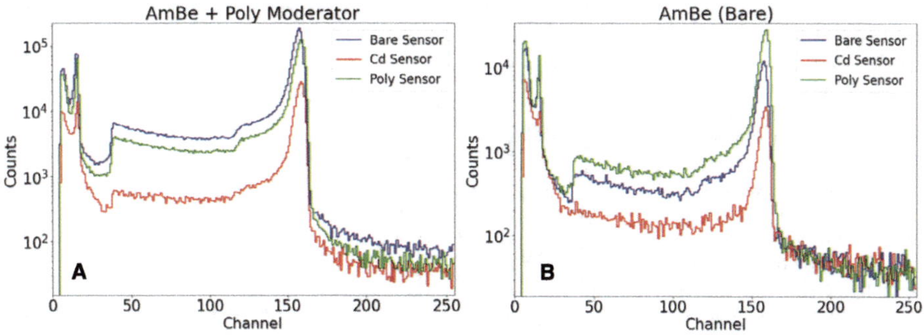

Fig. 25 Pre-launch energy deposition spectra measured by the three NS sensors. Relative energy deposition is given in channel numbers. (A) Data shown for an AmBe neutron source surrounded by a polyethylene shield that moderates the neutron energies to lower energies than the few MeV neutrons naturally emitted from the AmBe source. Compared to the bare sensor, the polyethylene covered sensor (green) shows a lower relative count rate indicating fewer higher energy neutrons compared to thermal neutrons as measured by the bare sensor (blue). (B) Data shown for a bare AmBe neutron source that has a high proportion of higher energy neutrons than the shielded source. Here, the polyethylene covered sensor shows the highest count rate of the three sensors indicating the detection of a larger proportion of higher energy neutrons than for the case of the shielded source

The performance of the GRB mode was demonstrated by manually moving the ^{60}Co source near the GRS. The source was moved toward the GRS twice (Fig. 24A). The first time, the peak rate in the AC shield was insufficient to trigger the GRB mode. The second time, the threshold was met and a 50-ms-time-resolution spectrum showing the counting rates before, during, and after the "burst" was produced as seen in Fig. 24B.

4.2 NS Performance Measurements

The performance of the NS was characterized during development, final instrument calibration, and post-spacecraft installation. Detailed NS calibration data were presented by Peplowski et al. (2020) based on measurements taken at the National Institute of Standards and Technology Center for Neutron Research.

Figure 25 shows measured pre-launch neutron spectra taken with the NS and DPU using an AmBe neutron source close to the sensors. The data in Fig. 25A were taken with the AmBe source surrounded by polyethylene in order to moderate the neutrons to include a range of lower energy neutrons. In contrast, the data in Fig. 25B used a bare AmBe source, which consequently had a higher fraction of higher energy neutrons. This difference in neutron energies is reflected in the data such that the bare sensor shows the highest count rate for the moderated source whereas the polyethylene covered sensor – which has the highest sensitivity to higher energy neutrons – shows the highest count rate with the bare source. Additional quantitative characterization of the NS (e.g., temperature dependence, gain characteristics, flight performance) is presented by Peplowski et al. (2025).

5 GRNS in-Flight Performance

5.1 Initial Checkout

The Psyche spacecraft launched on 13 October 2023 from Cape Canaveral, Florida on a Falcon Heavy launch vehicle. As part of the multi-month spacecraft initial checkout (ICO)

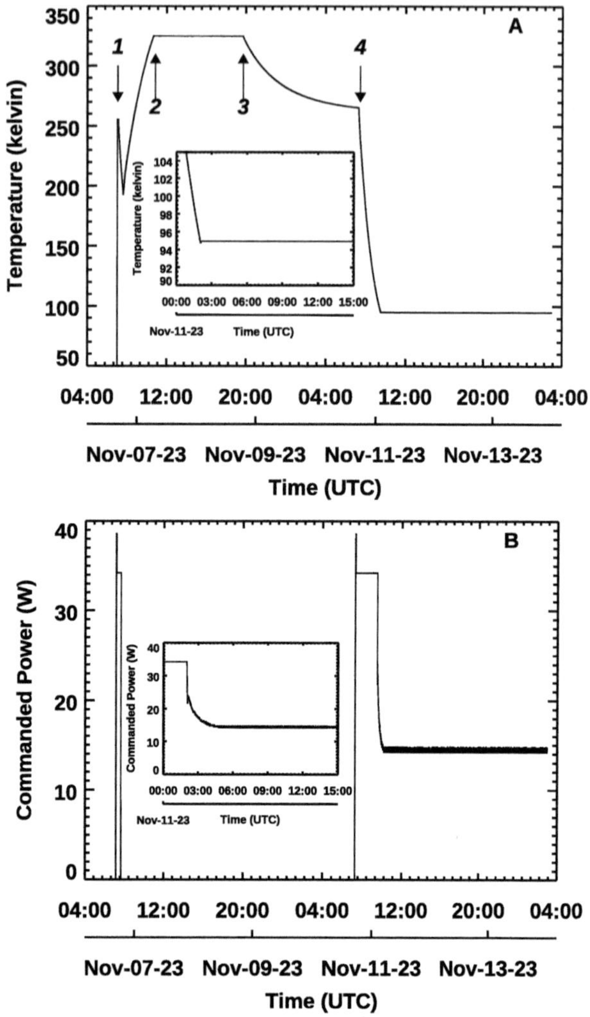

Fig. 26 GRS cryocooler temperature (A) and power (B) profiles during the initial turn on and first full cooldown of the cryocooler. Arrow labels indicate the following: 1): initial two-hour cryocooler operation; 2) start of 36-hour cryostat bakeout at 50 °C; 3) start of passive cool down of GRS cryostat; 4) start of full GRS cooldown. The insets in each figure show the temperature and power profiles around the approach to and regulation at the equilibrium temperature

activities, the GRS ICO took place over three weeks from 6 November 2023 to 27 November 2023. The NS ICO took place on 11–12 December 2023. This section describes the results from these first in-space operations of all GRNS components.

The first turn on of the GRS DPU occurred on 6 November. There was a one-time operation to disable launch locks on the cryocooler. This was followed by a two-hour cryocooler operation to ensure that the cryocooler survived launch and operated correctly. GRS detector temperature and command power profiles for this and the subsequent full-cooldown operation are shown in Fig. 26. The initial two-hour cool down is shown as arrow '1' in Fig. 26A. During this quick cooldown, the cryocooler demonstrated good operation, and the GRS reached 163 K (- 110 °C). Following this cooldown, the GRS anneal heater was turned on for a bake out to 50 °C (323 K) in order to remove any volatiles that may have accumulated inside the cryostat prior to launch. The heater took ~12 hours to reach the bake out temperature, and the total time at bake out temperature (arrow '2' to arrow '3' in Fig. 26A) was 36 hours.

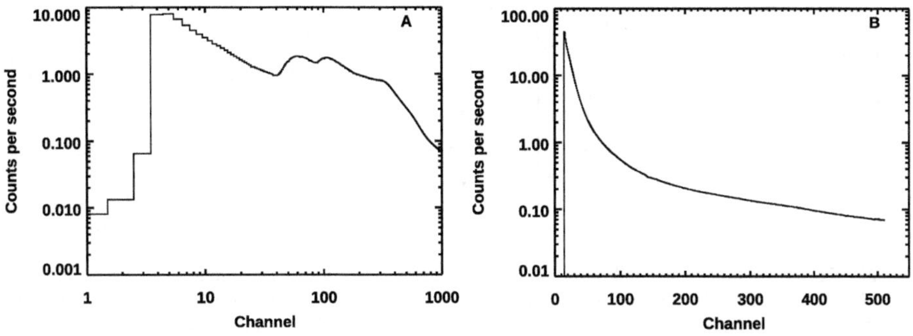

Fig. 27 Initial checkout data from the GRS AC shield. (A) Energy deposition spectrum of low-gain, AC shield channel. (B) Energy deposition spectrum of the high-gain AC shield channel. Both spectra are averaged from the dates of 12 November to 27 November 2023

At the end of the GRS bake out (arrow '3' in Fig. 26A), the GRS was passively cooled, and the full cooldown was initiated on 10 November 2023 (arrow '4' in Fig. 26A). The cooler took roughly 12 hours to reach its commanded temperature of 95 K. The temperature of 95 K was chosen to ensure sufficient thermal margin when operating relatively close to the Sun. While in orbit at Psyche, the cryocooler will operate at a temperature of 85 K to 90 K. It is preferred to operate at the lower 85 K temperature, as this temperature slows down the process of radiation damage, thus enabling longer accumulation periods without the need to anneal the detector (Jourdain 2011). The commanded cryocooler power for these operations is shown in Fig. 26B; the equilibrium power draw after reaching temperature was 14.5 W. The insets in each figure show the detector temperature and cryocooler power for the approach to, and at equilibrium. The temperature control loop held the temperature constant to less than 0.5 K.

Soon after the bake out command was sent on the first day of operation, the AC shield high voltage was raised for the first time. Figure 27 shows AC-shield data from the low-gain (Fig. 27A) and high-gain (Fig. 27B) spectra, respectively. The low-gain spectrum shows various features indicative of charged particle energy deposition in the AC shield. Specifically, the peaks around channels 70, 120, and 400 indicate energy deposition in different geometrical portions of the AC shield. Details of the low-gain response to charged particles are given by Peplowski et al. (2025). The high-gain spectrum shows a featureless spectrum with no indication of lower energy neutrons. This lack of lower energy neutrons while in cruise is consistent with prior space-based measurements (e.g., Lawrence et al. 2013a), and indicates that the amount of material in typical spacecraft is insufficient to downscatter energetic neutrons to lower energies.

Figures 28A and 28B show the four fast-neutron spectral products (Sect. 3.5.2) from data acquired on November 12–27, 2023. The two early spectra (black lines in Figs. 28A and 28B) were accumulated from early TTSP values (0 – 5 μs) using onboard processing; similarly, the two late spectra (red lines in Figs. 28A and 28B) were accumulated from late TTSP values (20.6 – 25 μs), also using onboard processing. The net fast-neutron energy spectrum (blue line in Fig. 28A) is consistent with similar feature-less profiles seen in prior space-based fast-neutron measurements (e.g., Lawrence et al. 2014). The $^{10}B(n,\alpha)$ neutron capture peak is seen in Fig. 28B near channel 70, and indicates a clean and unambiguous detection of neutrons. The average fast neutron count rate (e.g., the sum of the net peak counts in Fig. 28B) is ~2 cps, which is broadly consistent with that measured with the

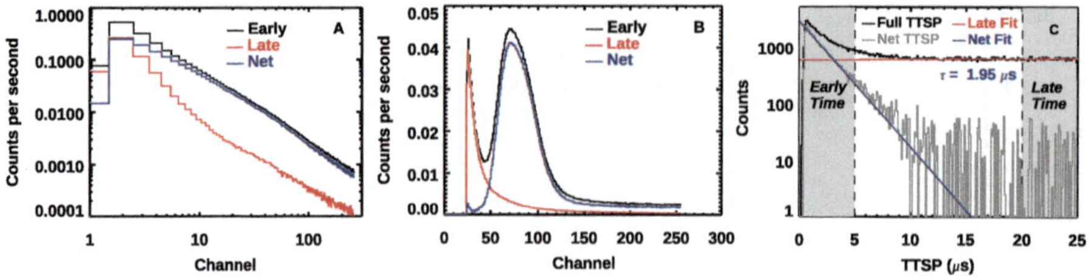

Fig. 28 Fast neutron data collected during ICO. (A) and (B) show prompt and capture spectra as generated by onboard DPU processing as described in Sect. 3.5.2. (C) shows TTSP spectra acquired during four-hours of fast-neutron event mode collected on November 6, 2023. The left shaded region shows early TTSP times (0 to 5 μs); the right shaded region shows late TTSP times (20.6 – 25 μs). Additional details are described in the text

MESSENGER NS (1.8 to 2.5 cps for variable GCR conditions) (Rodgers et al. 2015) using a similar sized borated plastic scintillator.

Figure 28C shows TTSP data that were acquired during a four-hour accumulation of fast-neutron event-mode data early in ICO on November 6, 2023. This data mode is not usually turned on during nominal operation, but was turned on during ICO to demonstrate that the fast-neutron processing is working as expected. The spectrum of TTSP values (black line in Fig. 28C) is a combination of real neutron detections at early times of 0 to 5 μs (left shaded region in Fig. 28C), and accidental coincidences at late times of 20.6 to 25 μs (right shaded region in Fig. 28C). The net TTSP histogram (gray line in Fig. 28C) is derived by subtracting an exponential fit to the late TTSP values (red line in Fig. 28C) from the measured TTSP spectrum. This net TTSP spectrum has a mean decay time of 1.95 μs, which is consistent with that expected for the 5% ^{10}B in the type of scintillator material used in the AC shield (Feldman et al. 1991).

On 13 November 2023, the HPGe high voltage was raised in a stepwise fashion from 0 V to 2500 V with ground-in-the-loop monitoring for each step. The HPGe sensor was then operated for nearly 14 days (333.5 hours, or 13.9 days). There was no significant solar energetic particle activity during this time period. In addition, there were no activation lines apparent in the data that would indicate significant solar activity in the weeks or months prior to operation. Therefore, this dataset provides a clean accumulation of GCR-generated background gamma rays. In addition, prior to its initial power on, the GRS was only in space for 30 days of GCR radiation. Thus, the accumulated radiation damage was relatively low for the initial GRS operation.

Summed high-gain and low-gain spectra for the 13.9-day period are shown in Fig. 29. Raw spectra are shown in red, and anticoincidence spectra are shown in black. The ratio of the raw to anticoincidence spectra (see Peplowski et al. 2025) shows that the background rejection ranges from a factor of three for the high-gain channel, to greater than a factor of 10 for the low-gain channel. Over 100 gamma-ray peaks are identified in the high- and low-gain spectra, and many of these are listed in Table 7. The Appendix describes the channel-to-energy calibration derived for this analysis.

Figure 30 shows the high-gain spectrum in an energy range that contains the key 1454 keV Ni gamma-ray peak. The six strongest peaks in the region are labeled along with the location of the ^{58}Ni peak. A fit to all peaks – including the putative ^{58}Ni peak – is shown by the solid red line. To obtain a good fit for the ^{58}Ni peak region, the peak centroid and width needed to be fixed, with its magnitude the only variable. In the other peaks, all peak parameters were allowed to converge on their best-fit values. The net total

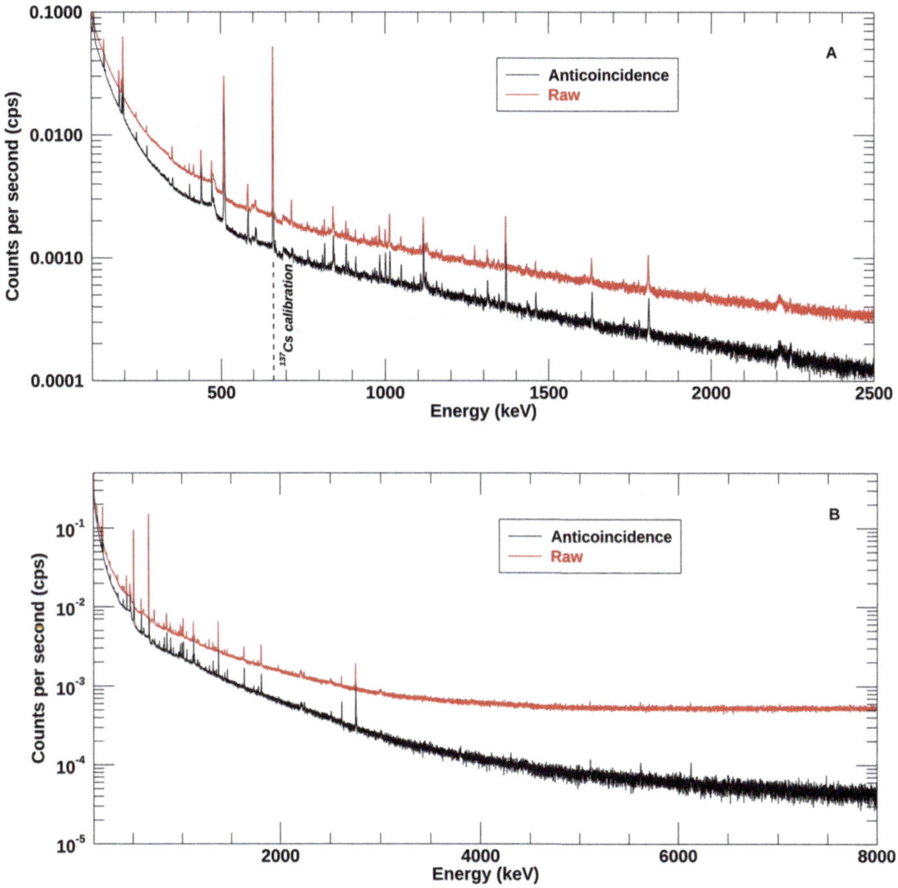

Fig. 29 High-gain (A) and low-gain (B) gamma-ray spectra for the 333 hours of accumulation time during the GRS initial checkout. Raw spectra are shown in red, and anticoincidence spectra are shown in black. In the high-gain spectra, the location of the 662 keV ^{137}Cs calibration peak is labeled

counts for the ^{58}Ni peak, and its neighboring ^{40}K peak at 1460.8 keV are 141.9 ± 170.7 and 2629.0 ± 195.2, respectively (or equivalent count rates of $1.19 \times 10^{-4} \pm 1.43 \times 10^{-4}$ cps and $2.20 \times 10^{-3} \pm 1.63 \times 10^{-4}$ cps). The derived uncertainties are one standard deviation, and assume Poisson statistics in the peak and background. The measured counts in the ^{58}Ni peak are consistent with zero. This result demonstrates that the effort to minimize nickel-bearing materials on the spacecraft was largely effective (Bradford et al. 2022). Further, when the low nickel background is combined with the excellent energy resolution of the GRS (see below), we conclude the GRS should meet its nickel measurement requirement with substantial margin, perhaps lowering our detection threshold, which may prove useful if Psyche turns out to have less exposed metal-rich surface than currently expected.

For the high-energy portion of the low-gain spectrum (Fig. 31), the 6129 keV oxygen gamma-ray peak is seen in both the raw and anticoincidence spectra. Here, the anticoincidence spectrum shows roughly a factor of eight background reduction compared to the raw spectrum. Fits to both peaks are shown in the plot. The net counts for the raw and anticoincidence peaks are 709.5 ± 336.0 and 403.5 ± 113.3, respectively (or $5.94 \times 10^{-4} \pm 2.8 \times 10^{-4}$

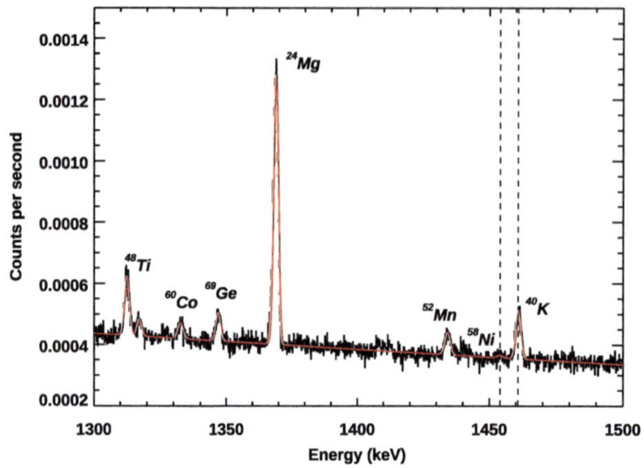

Fig. 30 High-gain anticoincidence spectra for the energy region from 1300 to 1500 keV. The locations of prominent peaks are labeled, and the locations of the 1454 keV ^{58}Ni and 1460 ^{40}K peaks are identified by vertical dashed lines. The red line shows a fit using a gaussian plus a second order polynomial where all clear peaks are fit along with the region around the 1454 keV peak. In the peak fits for six identified peaks (plus an unidentified peak close to the ^{48}Ti peak), all peak parameters (energy position, width, height) are allowed to vary and converge to a best fit. For the 1454 keV region, the energy position and width were fixed, with the height allowed to vary. The net counts in the 1454 keV region are 141.9 ± 170.7, which is consistent with zero at a one standard deviation uncertainty

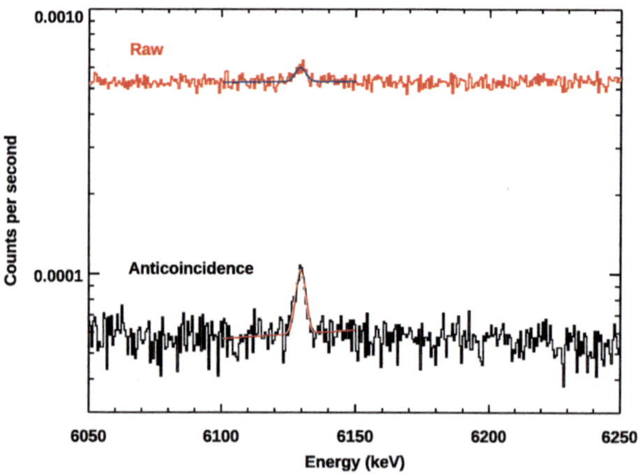

Fig. 31 Low-gain spectra in the energy region from 6050 to 6250 keV region. The raw and anticoincidence spectra are labeled. The 6129 keV oxygen peak is clearly seen in both spectra, and peak fits for each are shown by the blue trace for the raw spectrum, and the red trace for the anticoincidence spectrum. The net counts for the raw and anticoincidence peaks are 709.5 ± 336.0 and 403.5 ± 113.3, respectively

cps and $3.38 \times 10^{-4} \pm 9.49 \times 10^{-5}$ cps). The count rates are consistent to within the one-standard-deviation uncertainty. The uncertainty for the anticoincidence measurement is nearly a factor of three improved compared to the raw spectrum measurement. Since Poisson counting uncertainties scale as the square root of time, this implies that in this energy range, the anticoincidence shield provides a signal-to-background improvement roughly equivalent to almost a factor of nine increase in accumulation time. Further, when assessing the degree to which Psyche is a metal or silicate body, oxygen could be a key measurement since oxygen will only be present in silicate materials and not in pure iron-nickel metal.

Comparing the GRS energy resolution, $\sigma_{E,FWHM}$, as measured in space, with the prelaunch measurements is not straightforward. The reason is that the statistical precision of the canonical 1332.5 keV ^{60}Co peak is significantly worse in the space measurements compared to the pre-launch data with the ^{60}Co source (Fig. 22). The flight data show a small 1332.5 keV peak (Fig. 30) from GCR-induced spallation reactions on the copper heat strap that is part of the GRS housing. The energy resolution of this peak is 2.41 ± 0.40 keV. This large uncertainty allows a range of $\sigma_{E,FWHM}$ values from 2.0 to 2.8 keV. Thus, we use an alternate way of establishing the energy resolution at 1332.5 keV. To achieve a higher precision estimate of energy resolution, we measure the energy resolution for a number of peaks across the full-energy range of both the high- and low-gain spectra, fit a functional form to the $\sigma_{E,FWHM}$ versus energy (E) data, and then use this functional form to infer the energy resolution at 1332.5 keV. These data are shown in Fig. 32. For the functional form, we use a general relation given by Knoll (2000):

$$\sigma_{E,FWHM} = \sqrt{b_0 + b_1 E + b_2 E^2}, \tag{2}$$

where b_0 represents the intrinsic electronic noise, b_1 represents the statistical fluctuations inherent in the physics of detection in the HPGe sensor, and b_2 is a term that broadly accounts for effects such as incomplete charge collection. Initially, all peaks in Table 7 were available as candidate data points for this analysis. However, prior to fitting the σ_E versus E values, some selections were made to the data to obtain a more robust fit. First, some peaks (e.g., 511 keV, 1808 keV) have an intrinsic broadening that is larger than the optimum width of the HPGe detector. If included in the fit to Eq. (2), these peaks would bias the fit to high σ_E values. Second, other peaks have sufficiently large counts that non-Gaussian effects are present in the specific peak fits and can affect the measured peak width and uncertainty. This kind of effect is identified via a large reduced chi-squared value (χ^2/ν) from each of the peak fits (see Table 7 for the χ^2/ν values). Finally, to ensure sufficiently small $\sigma_{E,FWHM}$ uncertainties, a lower limit was placed on the total counts in a given peak. For the high-gain spectra, the χ^2/ν and peak count limits were $\leq$1.5 and $\geq$2500 counts, respectively. For the low-gain spectra, these limits were relaxed to ensure that the highest energy peaks were used, and were $\leq$2.5 and $\geq$300 counts, respectively. The red data points in Fig. 32 are those that passed these selections, and are used in the high- and low-gain fits to Eq. (2).

For the final high-gain fit, the parameter b_2 was set to zero, for the reason that when it was included, it resulted in a poorer χ^2/ν value (1.11 versus 1.04), the overall uncertainty for $\sigma_{E,FWHM}$ was significantly larger (0.13 keV versus 0.03 keV), and the b_2 value was quite small (5.9×10^{-8}). In short, including the b_2 parameter resulted in a poorer fit. When b_2 is set to zero, the resulting fit for the high-gain spectra shows a reduced χ^2/ν of 1.04 indicating a robust fit. Using the fitted parameters and their associated uncertainties, the inferred energy resolution at 1332.5 keV is 2.09 ± 0.03 keV. This performance is slightly poorer than what was measured just after spacecraft delivery (Fig. 22) and may reflect multiple factors. For example, the ^{137}Cs peak shows a drift of 0.038 keV from the beginning to the end of the measurement period (Fig. 45), so correcting this variation could reduce the measured energy resolution to as low as 2.05 keV. There may also be a modestly different electronic noise environment that could affect the energy resolution. Finally, there may also be a slight amount of radiation damage due to the 30 days that the GRS spent in space prior to its initial operation (see Sect. 5.2).

Even so, this performance is significantly better than previous space-based HPGe detectors, which generally have energy resolution values at 1332.5 keV in the 3 to 4 keV range (Smith et al. 2002; Goldsten et al. 2007; Diehl et al. 2018; Roques et al. 2022). The prior best-in-class energy resolution for a space-based HPGe detector was 2.27 keV at 1117 keV for one of the nine detectors on the RHESSI solar gamma-ray spectrometer (Smith et al.

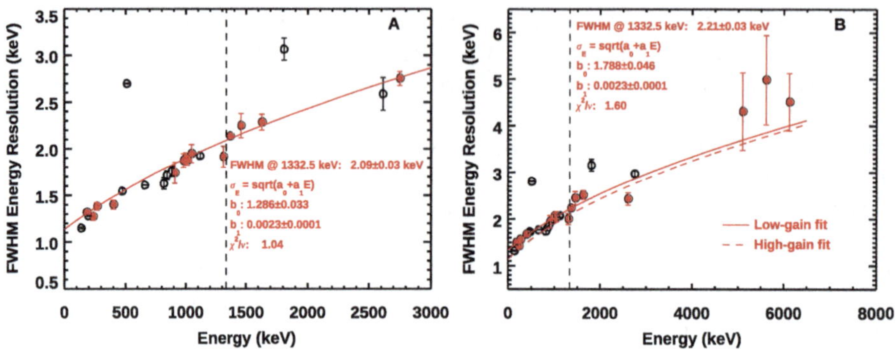

Fig. 32 GRS FWHM energy resolution, $\sigma_{E,FWHM}$, versus gamma-ray energy, E, for the high-gain (A) and low-gain (B) channels. The data points are those given in Table 7 from 28 prominent peaks seen in the GRS ICO measurements. The $\sigma_{E,FWHM}$ values are those obtained by carrying out the peak fits described in the Appendix. The error bars represent the full propagated one-standard-deviation uncertainties from each peak fit. Red data points indicate those values that passed total counts and χ^2/ν selection criteria. For the high-gain spectra, the criteria were >2500 counts and $\chi^2/\nu < 1.5$; for the low-gain spectra, the criteria were >300 counts and $\chi^2/\nu < 2.5$. The selected data points were fit to Eq. (2). The fit parameters and associated uncertainties are shown in the plot along with the goodness-of-fit value (χ^2/ν) for each fit. The best-fit functions are shown by the solid red lines. The best-fit function for the high-gain spectra is shown by the dashed red line in panel B

2002). For the low-gain spectra, the inferred energy resolution is 2.21 ± 0.03 keV, which is about 6% wider than the high-gain measurement, and is broadly consistent with the increase seen in the pre-launch data for the low-gain spectra (Fig. 22).

The fit parameters for Eq. (2) can provide information about the detector performance characteristics. The parameter b_0 provides a measure of the baseline energy resolution due to energy-independent factors, such as electronics noise and broadening due to different energy per channel factors from the high- and low-gain spectra. The b_0 value of 1.29 keV for the high-gain spectra is slightly larger than the width of the electronic pulser peak of 1.11 keV (which occurs at an equivalent energy of 107 keV; the spectral fit for this peak is not shown). For the low-gain spectra, b_0 has a larger value than can be attributed, at least in part, to the larger energy-per-channel covered by the low-gain spectra. The b_1 parameter provides information about the statistical fluctuations in the Ge sensor (Knoll 2000), where

$$b_1^2 = (2.35)^2 \, F \epsilon E. \tag{3}$$

Here, $\epsilon = 2.96$ eV is the energy necessary to create an electron-hole pair in the Ge sensor, and F is a factor known as the Fano factor, which quantifies the reduction in statistical fluctuations from Poisson statistics seen with various types of detectors. For both the high- and low-gain spectra, it is inferred that $F = 0.14$. According to Knoll (2000) and other studies (e.g., Pehl and Goulding 1970; Croft and Bond 1991; Devanathan et al. 2006), for Ge sensors F is typically in the range of 0.06 to 0.15, with indications that values in the lower range are preferred. Thus, while 0.14 might be considered "acceptable", the fact that it is in the upper range indicates that for ICO there are some residual broadening effects that have not been fully quantified. Finally, since this analysis shows that $b_2 = 0$ is a best-fit value, it indicates that incomplete charge trapping is, at worst, a small effect for the ICO operation of the GRS. As cruise and orbital operations continue, and radiation damage and subsequent annealing operations occur, it will be instructive to observe how the Ge sensor performance changes from this well characterized in-space baseline performance.

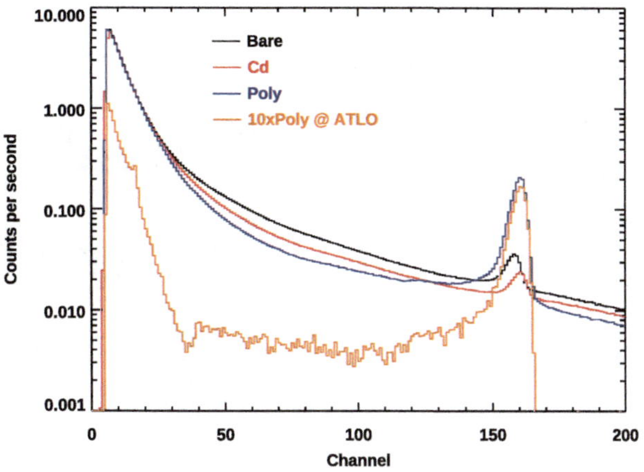

Fig. 33 Energy deposition spectra from the NS ^{3}He sensors. Data from the bare, Cd-covered, and polyethylene-covered sensors are shown in the black, red, and blue traces, respectively. The orange trace shows ground-based data from the polyethylene-covered sensor, where it has been multiplied by a factor of 10 to match the peak values of the flight polyethylene-covered sensor data

On 11 December 2023, the high voltages on the NS were raised and the three ^{3}He sensors started to collect data. Figure 33 shows space-based energy deposition spectra for the three sensors compared to the ground-based measurements with the polyethylene sensor. A significant difference between the space- and ground-based data is that the background at the lower channels is dramatically higher for the space-based measurements compared to the ground-based data. These higher count rates are due to charged particle energy deposition present in space from GCRs that is not present with the ground-based data. The highest count rate of the three sensors is seen in the polyethylene sensor. After subtracting a non-neutron continuum, the net neutron background count rates are 0.23 cps, 0.14 cps, and 1.61 cps for the bare, Cd-covered, and polyethylene-covered sensors, respectively. This observation confirms that the dominant neutron energy range at the Psyche spacecraft is higher energy neutrons, and that the flux of lower energy epithermal and thermal neutrons is low.

Figure 34 shows a summary of NS event-mode data that were collected during the initial turn on operation. Here, *dt* versus peak-to-peak data show that the population of charged particle energy deposition extends over a wider range of *dt* values than the true neutron counts (the peak-to-peak value is proportional to the energy deposition in the sensor). When a *dt* cut is applied to the data (which can be done via a commandable parameter in flight software), this cut can reduce the background by nearly an order of magnitude, thus increasing the signal-to-background in a manner that is similar to the improvement seen in the GRS anticoincidence spectra shown in Fig. 31. Additional analyses of these event-mode data are given by Peplowski et al. (2025) where further background reduction can be achieved.

5.2 Periodic Maintenance Calibration #1

The HPGe sensor was operated a second time during a spacecraft operations period known as the Periodic Maintenance Calibration (PMC) #1. During this time, the HPGe sensor was cooled to 90 K, the high voltage was raised to 2000 V, and data were collected for four days. Figure 35 compares data from the ^{137}Cs calibration source taken during three time periods: during the initial pre-launch CPT in September 2021 (Fig. 35A); during ICO in November 2023, 35 days after being in space (Fig. 35B); and during PMC #1, 277 days after being in space (Fig. 35C). The trending of peak widths (both FWHM and full-width, tenth maximum, FWTM) show an increasing broadening due to space-based radiation damage (Fig. 35D). The source of radiation damage is due not only to nominal GCRs hitting the HPGe sensor,

Fig. 34 NS in-space event-mode data from the Cd-covered sensor. (A) shows a two-dimensional plot of dt values versus peak-to-peak values. The scale bar is given in log units of 10^{counts}. (B) shows the dt-summed peak-to-peak values when accepting all dt values (black), and when only allowing dt values in the range from 250 to 275

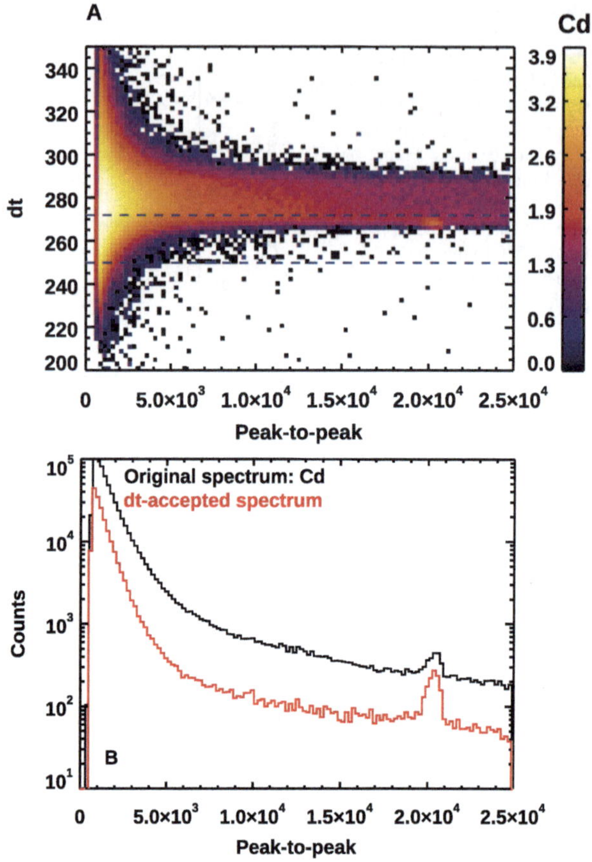

but there were a number of high intensity solar energetic particle events that hit the Psyche spacecraft during the time between ICO and PMC #1 (see Sect. 5.3) that possibly contributed to the radiation damage. Based on prior operation with space-based HPGe detectors, the amount of energy resolution degradation is consistent with prior experience (e.g., Evans et al. 2017). Subsequent PMCs (roughly every six months) will be used characterize the energy resolution performance versus time (see Sect. 7.1). This information will then be used as input, in part, for determining the annealing duration needed prior to the science measurements at Psyche.

5.3 Early Cruise Operation

Figures 36 and 37 provide a summary of the operation of the GRS AC shield and NS during the first year after initial checkout. A number of different space environmental and operational aspects are illustrated with these data.

Most prominently, these data show numerous instances of solar particle events (SPEs) hitting the spacecraft. These events, which are mostly made up of protons with energies greater than tens of MeV, typically last many hours to many days. These events occur more frequently during solar maximum, where the current solar maximum is expected to occur in the 2024 to 2025 time period. Figure 38 shows GRNS count-rate measurements along

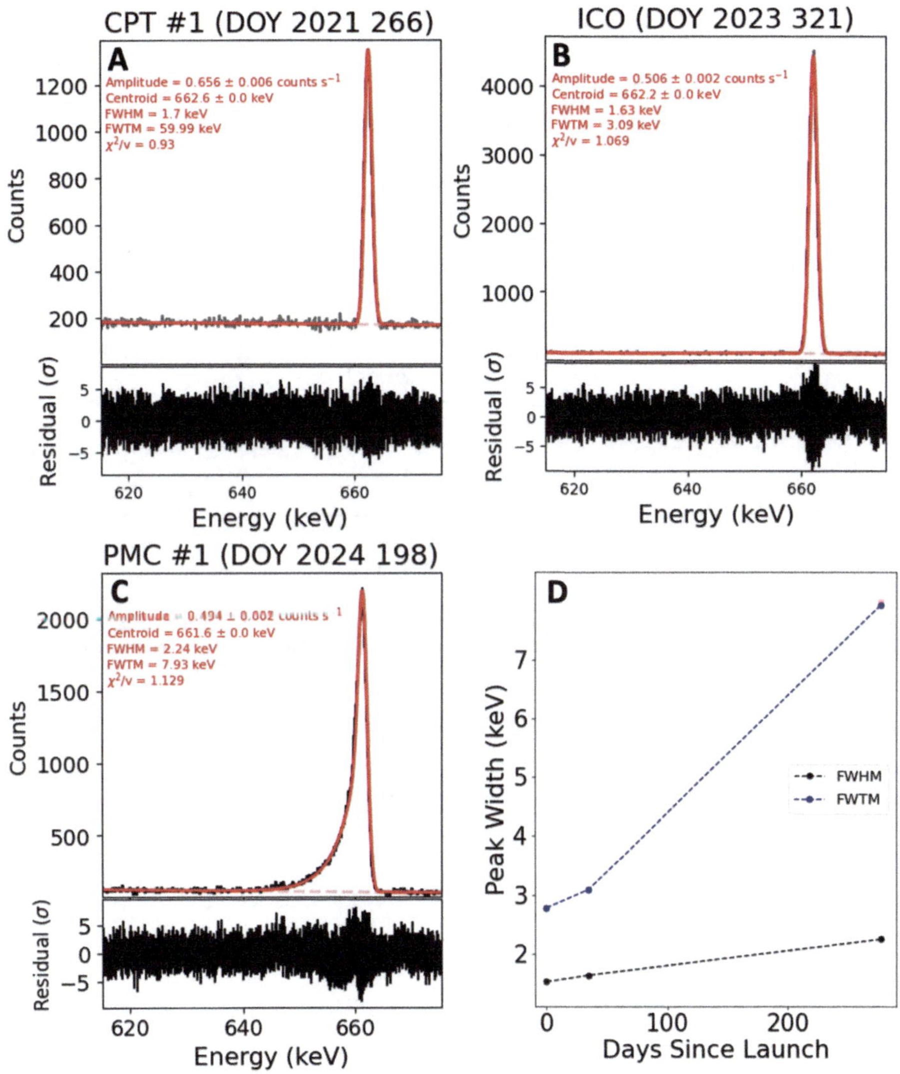

Fig. 35 Comparison of energy resolution performance for three different time periods using the 661 keV ^{137}Cs peak. (A) This shows the performance during the September 2021 pre-launch CPT at JPL where the energy resolution is 1.52 keV. (B) The energy resolution of the 661 keV peak was 1.61 keV during ICO. (C) The energy resolution of the 661 keV peak is significantly broadened at 2.25 keV during the first PMC due to radiation damage from GCRs and solar particle events. (D) Time dependence of the FWHM and full-width, tenth-maximum (FWTM) for the three time periods

with GOES Earth-orbiting energetic proton data (University of Colorado Boulder and SWx TREC 2019). These data, which were acquired soon after launch when the Psyche spacecraft and Earth were relatively closely aligned in solar longitude, indicate that the same SEP events detected by GOES were also detected by the GRNS. As the Psyche spacecraft travels away from Earth alignment, this agreement will lessen, since Psyche will be located in a different space environment than at Earth. Also notable from these events is that the GRNS

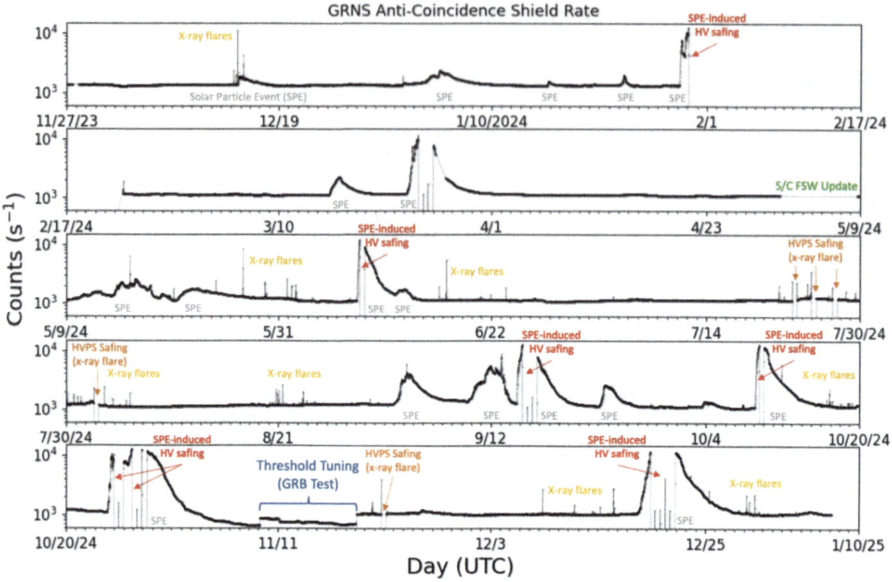

Fig. 36 Time variation of the AC shield high-gain count rate during the first seven months of the orbital cruise mission. Various environmental and operational events that took place during this time period are labeled

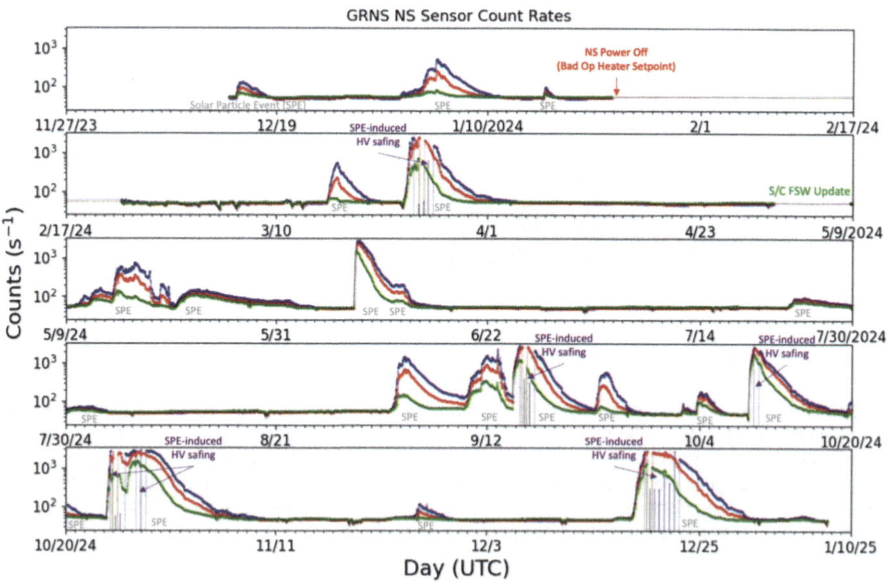

Fig. 37 Time variation of the three NS sensor count rates during the first seven months of the orbital cruise mission. Blue, red, and green show data from the bare, Cd-covered, and polyethylene-covered sensors, respectively. Various environmental and operational events that took place during this time period are labeled

data not only carries information about the existence of SPE events, but likely has information about the proton energy spectra, where lower energy protons show higher count rates

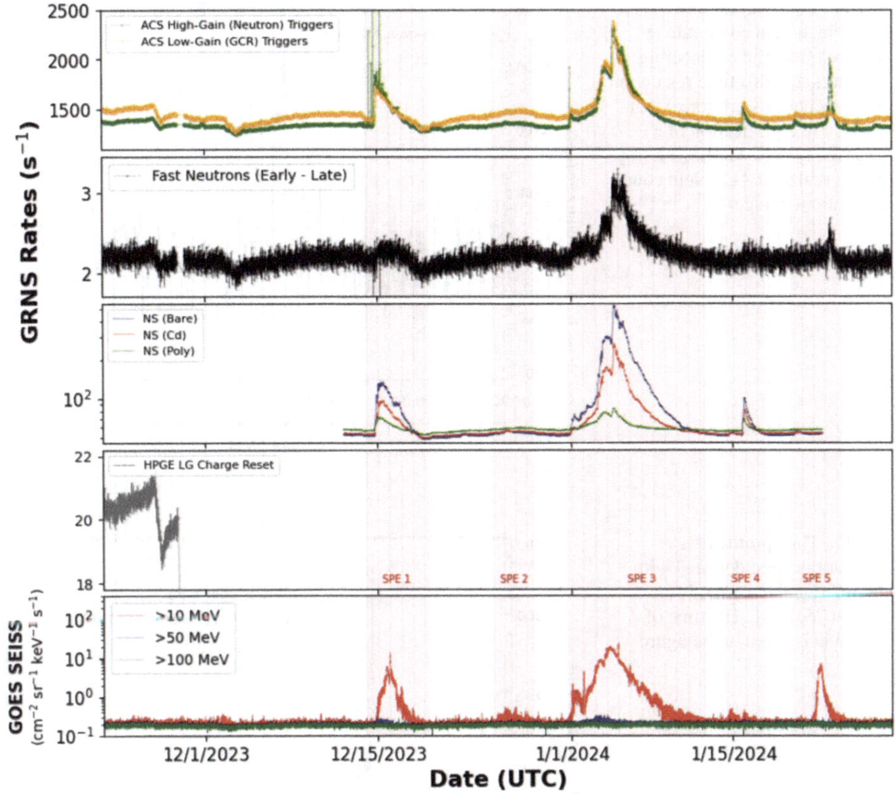

Fig. 38 Time variation of various GRNS count rate values are shown in the top four panels (A, B, C, D) during the first two months of orbital operation. The bottom panel (E) shows the time variation of three proton energy bands from the Earth-orbiting GOES spacecraft (University of Colorado Boulder and SWx TREC 2019)

than higher energy protons. This is particularly seen with the NS count rates, where the three different ^{3}He sensor coverings (bare, Cd, polyethylene) stop increasingly higher proton energies. Short, few-minute-long count-rate spikes are seen in the high-gain AC shield data, but not in the low-gain, fast neutron or ^{3}He sensor data. Because these enhancements are only seen in the high-gain AC shield data, which has a photon energy threshold of ~20 keV, it is likely they are due to X-ray energy deposition in the AC shield. A number of these events align in time with known X-ray flares as identified by GOES X-ray data (University of Colorado Boulder and SWx TREC 2019), an example of which is shown in Fig. 39. We therefore conclude that most, if not all, of these spikes are due to short duration (few tens of minutes) X-ray flares.

Operational aspects of the AC shield and NS cruise data include the stochastically timed safing of these sensors. As described in Sect. 3.5.3, all the GRNS sensors have a safing feature such that when the count rates rise above a threshold count rate, the HV is lowered for a set period of time, and then re-raised to determine if the count rate is reduced below that same threshold. As seen, this safing has been triggered a number of times during the first year of operation. Interestingly, there are some events that cause the AC shield to safe,

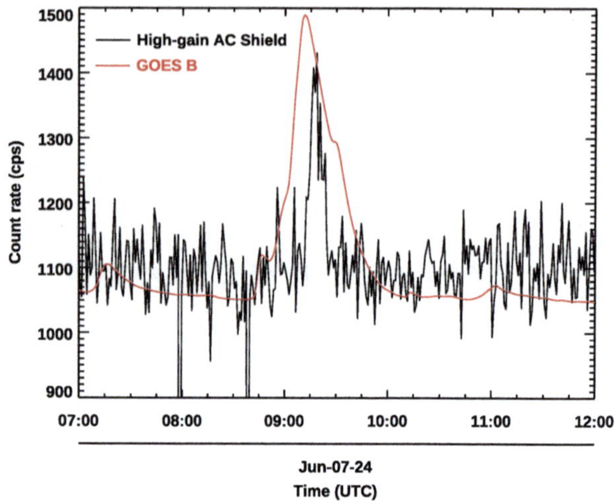

Fig. 39 Time variation of the AC shield high-gain count rate as acquired from the engineering data packet with a time resolution of 60 seconds (black trace) that shows a likely detection of a solar X-ray flare. GOES B X-ray data is scaled to AC-shield count rate data (red trace)

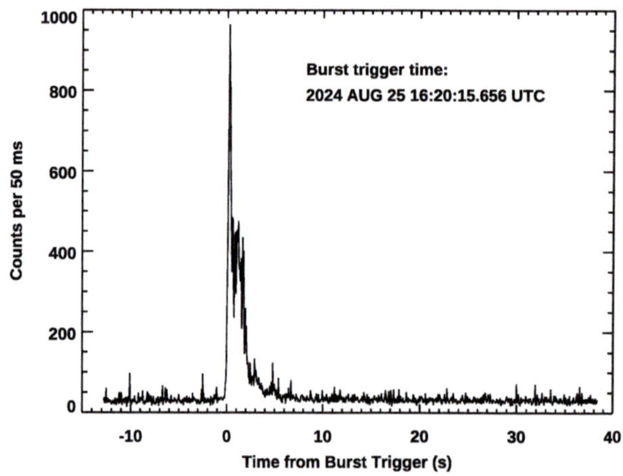

Fig. 40 Time profile of a gamma-ray burst detected with the AC shield burst mode on August 25, 2024. The time of detection is given in the figure

but not the NS sensors (e.g., see event between 5/31 and 6/22/2024), which indicates that the AC shield is likely more sensitive to charged particle deposition than the NS.

Figure 40 shows a gamma-ray burst (GRB) that was detected on 25 August 2024, and corresponds to GRB240825A (https://user-web.icecube.wisc.edu/~grbweb_public/Individual_GRB_webpages/GRB240825A.html). While a full analysis of this GRB is yet to be carried out, we can check the consistency of this measurement with known parameters for this GRB. As viewed from the Earth, this burst was located at right ascension (RA) 344.5°, declination (DEC) 1.03°. For this date, the Psyche spacecraft as viewed from the Earth was at RA 174.2°, DEC 3.0°, and at a distance of 4.98×10^8 km, which is almost directly opposite in the sky compared to the GRB. The light travel time difference between Psyche and Earth was 1662.1 seconds. At Earth, the GRB was detected at a time of 15:52:59 UTC; at Psyche the detection time was 16:20:15.656 UTC, for a measured time difference of 1636.7 seconds. The difference between the light travel and detected time differences is 25.5 seconds, which is consistent with the GRB occurring at a location roughly opposite on the sky from the Psyche spacecraft location. Resolving the remaining time difference will require

Table 5 Summary of GRS raw data products

GRS Product	Product Type	Description
HPGe Low Gain	16,384 bin spectra	Raw spectra collected from the high purity germanium (HPGe) detector. Covers 0 to 9200 keV energy range.
HPGe High Gain	16,384 bin spectra	Raw spectra collected from the high purity germanium (HPGe) detector. Covers 0 to 2800 keV energy range.
HPGe AC Low Gain	16,384 bin spectra	Anticoincidence (AC) spectra collected from the high purity germanium (HPGe) detector. Covers 0 to 9200 keV energy range.
HPGe AC High Gain	16,384 bin spectra	Anticoincidence (AC) spectra collected from the high purity germanium (HPGe) detector. Covers 0 to 2800 keV energy range.
AC Shield Low Gain	1024 bin spectra	AC spectra collected from the shield detector. Used for cosmic-ray proton monitoring. Sensitive to energy deposition events up to ~35 MeV.
AC Shield High gain	512 bin spectra	AC spectra collected from the shield detector. Used for neutron science. Sensitive to energy-deposition events up to ~500 keV.
AC Shield Fast Neutron	4 × 256 bin spectra	AC Shield spectra capturing double-pulse events, grouped into four separate categories, based on their time to second pulse values.
Gamma-Ray Burst	1024 time-binned counts	Counts from a triggered gamma-ray burst event.
Gamma-Ray Counters	binary table	Counters of events measured across the integration period. Used for diagnostic purposes.
Gamma-Ray Events	binary table	Events read from the hardware. Includes events not considered valid to be included in a spectrum. Used for diagnostic purposes.
Fast Neutron Events	binary table	Fast neutron events and metadata read from the hardware. Includes events not considered valid to be included in a spectrum. Used for diagnostic purposes.
GRS Engineering	binary table	Measurements of the health and status of the GRS instrument.

an analysis of the exact position locations of the spacecraft, Earth, and the GRB, and their relative timings.

6 Data Products and Processing

The GRNS data are distributed to the science community via NASA's Planetary Data System (PDS) through the Psyche mission Science Data Center at Arizona State University. Per PDS convention, the GRNS data categories include raw, calibrated, and derived data. There are 12 GRS raw data product types (Table 5): high- and low-gain HPGe detector raw spectra, high- and low-gain HPGe AC spectra, high- and low-gain AC shield spectra, shield fast-neutron spectrum, gamma-ray burst counts, gamma-ray counters, gamma-ray events, fast neutron events, and engineering data.

The NS raw product pipeline generates four raw data product types (Table 6): NS spectra, NS counter, raw event, and engineering data products. The raw NS spectra product consists

Table 6 Summary of NS raw data products

NS Product	Product Type	Description
NS Spectra	3 × 256 bin spectra	Spectra histogram arrays collected from the Bare, Cadmium-, and Polyethylene-wrapped sensors.
NS Events	binary table	Event time series data collected by the NS instrument. Events are collected even if they are not considered valid to be included in an NS spectra packet.
NS Counters	binary table	Contains counters of events across the integration period.
NS Engineering	binary table	Measurements of the health and status of the NS instrument

of three 256 bin histograms, from the polyethylene, cadmium, and bare wrapped detectors. As with the GRS, the NS histogram data are compressed using a fast lossless compression. More details for the raw and calibrated data are described by Espiritu et al. (2022a,b).

7 Expected Measurements During Cruise, Mars Flyby, and at Psyche

7.1 Summary of Expected Measurements for Each Mission Phase

The major GRNS operational phases throughout the Psyche mission include:

- Instrument Checkout Operations (ICO): November – December 2023
- Interplanetary Cruise 1: January 2024 – March/April 2026
- Mars gravity assist: March/May 2026
- Interplanetary Cruise 2: May 2026 – May 2029
- Psyche Approach: May 2029 – August 2029
- Psyche Orbit Operations (Orbits A through B1): August 2029 – February 2030
- Psyche Orbit Transfer (Orbit B1-to-D transfer): February 2030 – May 2030
- Psyche Orbit D: May 2030 – August 2030
- Psyche Orbit D margin: August 2030 – October 2030
- Psyche Orbit Transfer (Orbit D-to-C transfer): October 2030 – January 2031
- Psyche Orbit C: January 2031 – April 2031
- Psyche Orbit Transfer (Orbit C-to-B2 transfer): April/May 2031
- Psyche Orbit B2: May 2031 – August 2031

Full GRNS operations are performed during ICO, Mars gravity assist, Psyche Orbit B1-to-D transfer, Orbit D, and Psyche Orbit D-to-C transfer. During interplanetary cruise and early Psyche orbit operations (Orbits A and B), the GRNS operates in a reduced science mode where neutron and AC shield measurements are made but the HPGe sensor is not generally operated.

The cruise measurements, during interplanetary cruise periods 1 and 2, include the near-continuous operation of the AC shield and all three NS sensors, as well as GRS HPGe operation roughly every six months during spacecraft PMCs. The entire GRNS will operate during the Mars gravity assist in the spring of 2026, where the Psyche spacecraft will approach the surface of Mars at an altitude of ∼3500 km. There are multiple purposes for these cruise operations. Most importantly, there are aspects of GRNS operation that cannot simulated or replicated on the ground. For example, the GCR flux was measured at 1 AU, contemporaneously with Earth-orbiting space weather assets (see Fig. 38). The continued operation of the AC shield and NS sensors enables the GCR flux variation to be monitored throughout the mission, which facilitates precise knowledge of the asteroid-incident GCR flux once reaching (16) Psyche. Knowledge of this flux is an important input for interpreting the GRNS measurements. Continuous monitoring of the energetic particle environment is important not only to fully understand the GCR flux while at Psyche, but provides information about the energetic particle radiation encountered by the HPGe sensor. This information is important for optimizing the HPGe annealing procedure(s) prior to science data acquisition while in orbit about Psyche (Peplowski et al. 2019a). We finally note that the data acquired during cruise is critical not only for achieving Psyche science objectives, but can be useful for other studies of the space environment, such as understanding the time and spatial dependent GCR flux (Rodgers et al. 2015; Lawrence et al. 2016), as well as carrying out space weather and heliophysics studies with multi-spacecraft correlation analyses (Battarbee et al. 2018; Winslow et al. 2018). Additional studies could combine the GRNS data with cruise data from other instruments (e.g., interplanetary magnetic field data from the magnetometers) and from engineering subsystems (e.g., single event effect data from onboard computer memories), which in part can help inform radiation effects on the Psyche spacecraft.

During inbound and outbound Orbit D transfer phases, the spacecraft is at >3.5-Psyche-radii altitude, and neutron and gamma-ray measurements provide a measure of background rates and spectra. During Orbit D, the spacecraft is within 1 Psyche-radius altitude and gamma-ray and neutron measurements are sensitive to the composition of Psyche's surface. The average planned altitude during this period is 76 km (0.68 Psyche radius), with a range of 59 to 91 km (0.53 to 0.83 Psyche radius) (Fig. 41). These are the measurements that contribute to Psyche science requirement closure. Orbit D is planned to last for 100 days, approximately 75% of which will be nadir-pointed and thus contribute to GRNS science observations. The remaining 25% of time includes off-nadir pointing to downlink data to Earth. An additional $\sim$60 days of margin are included in the Orbit D phase. Based on the current orbit-plan estimates and including the 75% duty cycle for nadir-pointed attitudes, the total Orbit D times including margin for altitudes less than 0.6, 0.8, and 1.0 body radius are $\sim$19 days, $\sim$119 days, and $\sim$122 days, respectively. The total times less than 1.0 body radius for the B1-to-D and D-to-C transfer orbits are $\sim$53 and $\sim$2.5 days, respectively. It is unknown how much nadir-pointing time will be available during these transfer orbits.

Due to Psyche's irregular shape, a polar orbit is not expected to be stable at an orbit radius of <1-Psyche-radius for a sufficiently long time to meet mission safety guidelines. Thus, during Orbit D the spacecraft will be in an equatorial orbit that samples the surface composition equatorial of $\pm 15°$ latitude. The total surface coverage is 86%, which is enabled by the broad size of the NS and GRS spatial footprints. The polar regions will not be measured or mapped by Psyche GRNS during Orbit D. However, during Orbit C, which is currently planned to take place after Orbit D, the spacecraft will have full latitude coverage. The altitude range during Orbit C is 1.3 to 2.1 body radii (Fig. 41). Neutron data from the Lunar Prospector mission can be used to estimate the order-of-magnitude NS count rate for Psyche at this altitude range. Prior to going into a circular orbit, the Lunar Prospector spacecraft obtained data during multiple elliptical orbits about the Moon (Maurice et al. 2004). The Lunar Prospector and Psyche neutron measurements are similar such that both use the same sized neutron sensors, and the Lunar Prospector measurements were made around solar minimum conditions, as will be Psyche during Orbit C in 2031. The Lunar Prospector count rate of its Cd-covered neutron sensor for 1.3-to-2.1 body radii altitude range at the Moon is 1.91 ± 0.54 cps. The measured Psyche cruise count rate is 0.14 cps. Because of

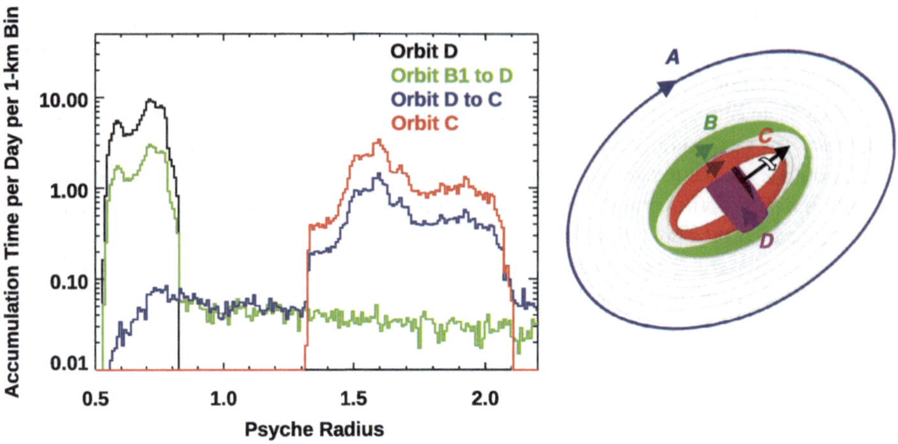

Fig. 41 (Left) Histograms of altitude occurrences for the Psyche mission Orbits D and C for the current orbital mission plan (black and red, respectively), along with the transfer orbits B1 to D (green) and D to C (blue). The altitude bins are 1-km wide, and the occurrences indicate the number of days per 1-km-wide altitude bin. For Orbit D, altitudes range from ~0.53 to 0.83 Psyche radius, and for Orbit C, altitudes range from 1.3 to 2.1 Psyche radii. (Right) For reference, graphical representation of the four Psyche orbits during the primary orbital mission (Polanskey et al. 2025, this collection). The average altitudes for Orbits A and B (not shown at left) are 6.4 and 2.7 Psyche radii, respectively

this low background level, there is therefore a good opportunity to obtain additional neutron data to characterize global compositional variabilities during Orbit C, albeit with broader spatial resolution than for the lower altitude Orbit D.

7.2 Science Closure

During the early development of the Psyche mission, the asteroid (16) Psyche was thought to be the exposed core of a protoplanetary object. Initial theories suggested that the asteroid would be ~90% metal (≥4 wt% Ni) and ~10% Mg-rich silicates. On the basis of these expectations, sensitivity simulations and calculations were carried out to determine what elements would be measured by the GRNS, including required measurement times, detection limits, and statistical plus systematic uncertainties. Figure 2 shows an example of simulated spectra in the region around the 1454 keV Ni gamma-ray line. While our expectations of (16) Psyche's composition have broadened significantly since the original studies, our sensitivity studies and mission requirements are still based on this assumed composition. These sensitivity calculations, which were finalized after the GRNS design was completed, included the following details. For GCR-generated gamma-ray lines, gamma-ray production was calculated using the particle transport code MCNPX (Pelowitz 2005). The gamma-rays were transported to the assumed spacecraft altitude with standard geometric relations (e.g., Prettyman et al. 2006; Chabot et al. 2021). The GRS efficiency was determined using particle transport calculations (Agostinelli et al. 2003), as was done for the MESSENGER (Peplowski et al. 2012), and used a full accounting of the Psyche GRS assembly. Gamma-ray background (line and continuum) was taken into account using the measured MESSENGER GRS background. A number of factors were given conservative estimates to provide margin in the sensitivity calculation. These margins guard against unanticipated performance degradations – either in the instrument, mission operations, or both – that would prevent the accomplishment of the primary measurement requirements. These factors include: 1) an assumed energy resolution of 3.5 keV (MESSENGER GRS optimum performance); 2) a constant altitude of one Psyche radius; 3) a lower limit value of the incident GCR flux at the start of the original orbit phase; 4) a partial background reduction to account for the expected improvement from the boom. Given the detection thresholds and required precisions listed in Table 1, the output values of the sensitivity calculations are the times listed in the rightmost column of Table 1. For some elements (e.g., Fe, K), the time needed to meet the requirements is fairly short (less than a week). The elements that take the longest to meet the required threshold and precision are S and Ca. Post-calibration measurements of the Psyche GRNS showed that the instrument meets or exceeds all performance requirements (Peplowski et al. 2025), and the current Psyche mission operations plan includes at least 75 days of nadir pointing (not including time margin) at an altitude well below 1-Psyche-radius. We therefore conclude that the GRNS will exceed the required performance, providing margin against the now less-constrained range of compositions expected for the asteroid.

Appendix: GRS Energy, Total Counts Check, and Gain Temperature Dependence

This first part of the appendix describes the channel-to-energy calibration used for the GRS ICO dataset. The gain of the system is dependent on the temperature, T_{Ge}, of the Ge crystal. Thus, the energy calibration derived here for the crystal at $T_{\mathrm{Ge}} = 95$ K will be different than that derived for either the pre-flight calibration data ($T_{\mathrm{Ge}} = 90$ K) and future flight data, for which T_{Ge} is expected to be in the range of 85 K to 90 K.

Table 7 lists 28 of the most prominent peaks that are used for the channel-to-energy calibration. Each peak in Table 7 was fit with a gaussian plus a second-order polynomial function, $f(x)$, within a ~7-keV window around each peak:

$$f(x) = A_0 e^{\frac{-[z(x)]^2}{2}} + A_3 + A_4 x + A_5 x^2, \tag{4}$$

where,

$$z(x) = \frac{x - A_1}{A_2}. \tag{5}$$

These initial fits were carried out using spectra channels (i.e., $x = C$) that ranged in value from 0 to 16,303 (i.e., a total of 16,384 channels). The returned parameters (including parameter uncertainties) for each fit are the peak height (A_0), centroid (A_1), width (A_2) and three polynomial parameters (A_3, A_4, A_5). Note that for the high-gain data, only peaks with energies up to the 2754-keV peak are included. Figure 42 shows the peak centroid values ('centroid channel' column in Table 7) plotted versus the reference peak energy values ('energy' column in Table 7) for both the high- and low-gain spectra. A second-order polynomial was fit to the values in Fig. 42 with the fit parameters shown in the figure. To demonstrate the robustness of the channel-to-energy calibration, a calibrated energy scale, E_{cal}, was generated using the fit parameters of Fig. 42, and each peak was re-fit using these energy values (i.e., $\mathrm{x} = E_{\mathrm{cal}}$). The residual between the reference peak energies, E_{ref}, and newly fit energy centroids is shown in Fig. 43. The error bars in Fig. 43 are the one-sigma uncertainties for the fitted energy centroid values. As seen, within the uncertainties, the residuals show no obvious bias or energy dependent pattern. This indicates that the energy calibration is robust across all energies for both the high- and low-gain spectra.

Table 7 List of prominent gamma-ray lines in the ICO dataset. The reference energies ('Energy' column) are taken from Evans et al. (2017) and Chadwick et al. (2006). The 'Centroid channel' is the fit centroid channel value when a single gaussian plus second-order polynomial function is fit to each peak. The FWHM value is determined from a fit using a gaussian plus a second-order polynomial function using the calibrated energy scale. The FWHM is a factor of 2.35 times each peak width. The * symbol denotes a gamma ray that can be created by more than one nuclear process.

Reference Energy (keV)	Source	Centroid channel	Centroid energy (keV)	FWHM (keV)	χ^2/ν	Counts	Counts per second	Centroid channel	Centroid energy (keV)	FWHM (keV)	χ^2/ν	Counts	Counts per second
		High Gain						Low Gain					
139.69	^{75m}Ge	809.0	139.8 ± 0.004	1.150 ± 0.0094	11.11	125170.6 ± 909.4	0.1048	247.9	139.8 ± 0.005	1.324 ± 0.0124	32.17	121670.9 ± 966.5	0.1019
185.72	^{235}U	1076.1	185.9 ± 0.005	1.319 ± 0.0108	1.17	94959.2 ± 68.9	0.0795	329.3	185.9 ± 0.005	1.497 ± 0.0130	2.34	95695.6 ± 739.1	0.0801
198.40	^{71m}Ge	1148.4	198.3 ± 0.001	1.280 ± 0.0029	134.48	399256.8 ± 857.3	0.3342	351.4	198.3 ± 0.002	1.455 ± 0.0035	445.58	397447.6 ± 862.1	0.3327
238.63	^{232}Th	1382.7	238.7 ± 0.017	1.275 ± 0.0409	1.20	16400.1 ± 466.2	0.0137	422.8	238.8 ± 0.021	1.425 ± 0.0511	0.72	15698.1 ± 508.2	0.0131
271.25	^{44}Sc	1571.5	271.2 ± 0.017	1.385 ± 0.0400	1.18	16583.8 ± 415.5	0.0139	480.3	271.4 ± 0.019	1.561 ± 0.0472	1.38	16754.8 ± 422.4	0.0140
403.20	^{67}Ga + K	2336.8	403.1 ± 0.020	1.403 ± 0.0485	1.15	9597.0 ± 292.0	0.0080	713.1	403.1 ± 0.025	1.675 ± 0.0604	0.96	10033.6 ± 304.2	0.0084
472.20	^{24m}Na	2737.5	472.2 ± 0.012	1.546 ± 0.0275	3.13	19734.5 ± 304.7	0.0165	834.9	472.1 ± 0.013	1.727 ± 0.0319	7.10	20045.0 ± 310.4	0.0168
511.00	Annih.	2962.3	510.9 ± 0.003	2.698 ± 0.0057	6.19	312183.1 ± 638.0	0.2613	903.3	510.8 ± 0.003	2.802 ± 0.0060	15.73	312249.2 ± 643.6	0.2614
661.66	^{137}Cs	3836.7	661.6 ± 0.001	1.610 ± 0.0018	6.54	597944.4 ± 795.4	0.5005	1169.6	661.6 ± 0.001	1.768 ± 0.0020	14.46	599456.6 ± 798.3	0.5018
817.90	^{58}Co + K	4743.0	817.8 ± 0.025	1.627 ± 0.0606	2.39	5289.9 ± 175.4	0.0044	1445.5	817.8 ± 0.028	1.740 ± 0.0668	6.17	5309.1 ± 179.3	0.0044
843.70	^{27}Al	4893.8	843.8 ± 0.019	1.714 ± 0.0458	5.57	7938.5 ± 185.5	0.0066	1491.4	843.8 ± 0.021	1.800 ± 0.0504	17.72	7762.1 ± 188.9	0.0065
882.50	^{69}Ge + K	5118.3	882.4 ± 0.022	1.756 ± 0.0533	2.70	6629.7 ± 176.3	0.0055	1559.8	882.5 ± 0.025	1.869 ± 0.0589	5.79	6557.2 ± 185.8	0.0055
911.20	^{232}Th	5284.6	911.1 ± 0.046	1.743 ± 0.1117	1.04	2878.2 ± 160.2	0.0024	1610.6	911.2 ± 0.054	1.945 ± 0.1333	0.58	2860.8 ± 161.5	0.0024
983.50	^{48}Ti	5705.0	983.5 ± 0.030	1.875 ± 0.0732	1.37	4884.8 ± 163.4	0.0041	1738.4	983.6 ± 0.035	2.002 ± 0.0853	2.55	4754.4 ± 174.2	0.0040
1001.03	^{238}U	5806.6	1001.0 ± 0.025	1.898 ± 0.0611	0.95	5978.9 ± 166.6	0.0050	1769.2	1001.1 ± 0.029	2.049 ± 0.0705	0.68	5918.1 ± 174.1	0.0050
1014.50	^{27}Al	5884.4	1014.4 ± 0.023	1.879 ± 0.0546	0.97	6572.4 ± 167.9	0.0055	1792.9	1014.5 ± 0.027	2.073 ± 0.0646	1.13	6578.7 ± 174.9	0.0055
1048.90	^{66}Ga + K	6083.8	1048.8 ± 0.040	1.949 ± 0.0969	0.93	3670.4 ± 158.0	0.0031	1853.6	1048.9 ± 0.043	2.055 ± 0.1062	1.10	3682.0 ± 169.0	0.0031
1117.10	^{69}Ge + K	6480.6	1117.1 ± 0.013	1.924 ± 0.0291	2.29	13778.7 ± 185.1	0.0115	1974.2	1117.1 ± 0.014	2.071 ± 0.0327	4.64	13776.8 ± 190.0	0.0115
1312.10	^{48}Ti	7613.6	1312.3 ± 0.046	1.917 ± 0.1130	1.20	2613.0 ± 132.2	0.0022	2318.8	1312.2 ± 0.049	2.007 ± 0.1192	1.70	2661.7 ± 135.4	0.0022
1368.67	^{24}Mg	7940.6	1368.7 ± 0.013	2.137 ± 0.0292	1.23	13922.4 ± 170.9	0.0117	2418.1	1368.5 ± 0.014	2.238 ± 0.0316	1.73	13720.0 ± 171.8	0.0115

Table 7 (*Continued*)

Reference Energy (keV)	Source	Centroid channel	Centroid energy (keV)	FWHM (keV)	χ^2/ν	Counts	Counts per second	Centroid channel	Centroid energy (keV)	FWHM (keV)	χ^2/ν	Counts	Counts per second
		High Gain						Low Gain					
1460.82	^{40}K	8475.7	1460.8 ± 0.053	2.250 ± 0.1305	0.99	2642.2 ± 128.7	0.0022	2530.8	1460.6 ± 0.058	2.451 ± 0.1429	2.05	2720.3 ± 133.8	0.0023
1633.60	^{20}Ne*	9479.0	1633.7 ± 0.035	2.287 ± 0.0856	1.09	3924.2 ± 124.3	0.0033	2836.0	1633.5 ± 0.039	2.519 ± 0.0953	1.42	4061.0 ± 128.9	0.0034
1808.74	^{26}Mg*	10495.9	1808.8 ± 0.046	3.066 ± 0.1175	1.02	4253.3 ± 132.2	0.0036	3195.3	1808.6 ± 0.049	3.154 ± 0.1249	1.95	4215.1 ± 133.2	0.0035
2614.51	^{232}Th	15175.4	2614.6 ± 0.069	2.587 ± 0.1745	1.00	1840.5 ± 103.6	0.0015	46[illegible]8.7	2614.7 ± 0.055	2.436 ± 0.1333	0.66	1672.9 ± 78.4	0.0014
2754.02	^{24}Na, ^{24}Mg	15984.9	2753.9 ± 0.031	2.755 ± 0.0749	1.27	5643.7 ± 129.7	0.0047	4864.8	2754.1 ± 0.026	3.001 ± 0.0616	2.38	5914.2 ± 103.8	0.0050
5107.30	^{16}O, DE							9019.0	5107.6 ± 0.300	4.312 ± 0.8318	0.93	345.4 ± 45.3	0.0003
5618.30	^{16}O, SE							9919.4	5617.9 ± 0.346	4.997 ± 0.9594	0.99	371.5 ± 46.2	0.0003
6129.30	^{16}O							10821.9	6129.4 ± 0.201	4.518 ± 0.6140	0.68	456.3 ± 43.4	0.0004

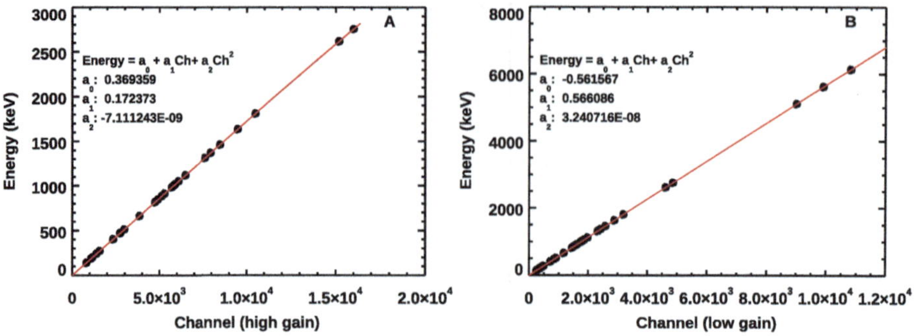

Fig. 42 Channel-to-energy calibration of high-gain (A) and low-gain (B) spectra. The data points show the data values given in Table 7; the red solid line shows the fit of a second-order polynomial function. The best-fit parameters for each fit are shown in each figure

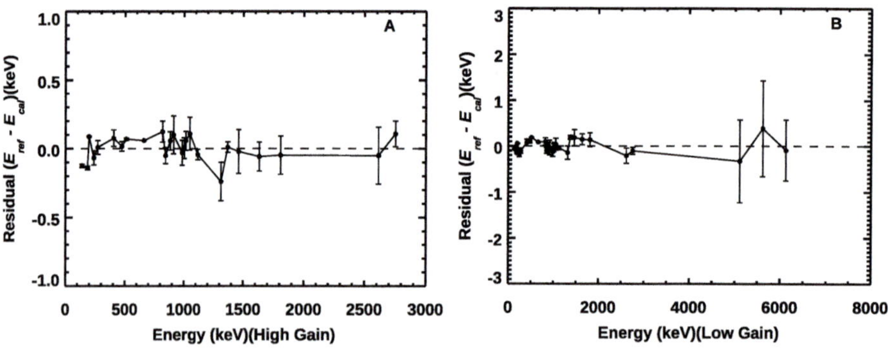

Fig. 43 Residuals from the channel-to-energy calibration where the differences are shown between the reference and calibrated energy values versus the calibrated energies. The error bars show the one-standard-deviation statistical uncertainty in the calibrated energy peak centroids. Panel A shows the high-gain data; panel B show the low-gain data

Another data robustness test is to compare the total counts within each peak for the high- and low-gain data. The derived total counts should be the same for each type of spectra since they both originate from the same detector signal. The total counts for each peak were determined by subtracting the fitted polynomial background and summing the net peak counts. The count uncertainty is assumed to be based on Poisson counting statistics, and is the square root of the sum of peak and background counts within a window of three FWHM around the peak centroid. Here, the FWHM value is 2.35 times the peak width A_2. Figure 44 shows the difference between the high- and low-gain peak counts as a function of the high-gain peak counts. As expected, the count difference is consistent with zero within the one-standard-deviation uncertainty. This demonstrates that both high- and low-gain spectra individually return the same counts within the limits of the available statistics.

Finally, to understand the effects of possible temperature variations on the gain of the GRS, Fig. 45 shows the variation of the peak centroid of the ^{137}Cs peak at ~662 keV. As seen, there is a roughly monotonic variation of the peak centroid throughout the 13.9-day GRS background measurement. A linear fit to this relation is shown by the solid red line. The energy difference between the beginning and end times is 0.038 keV. Additional detail

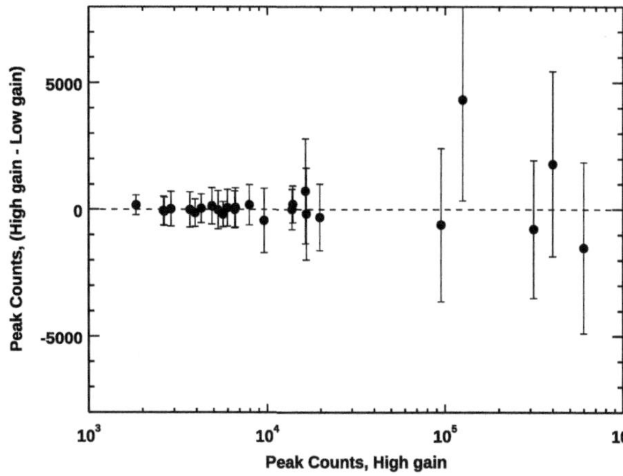

Fig. 44 Total count difference between the high- and low-gain data for all the peaks given in Table 7. The errors bars show the combined one-standard-deviation statistical uncertainties for both peak fits

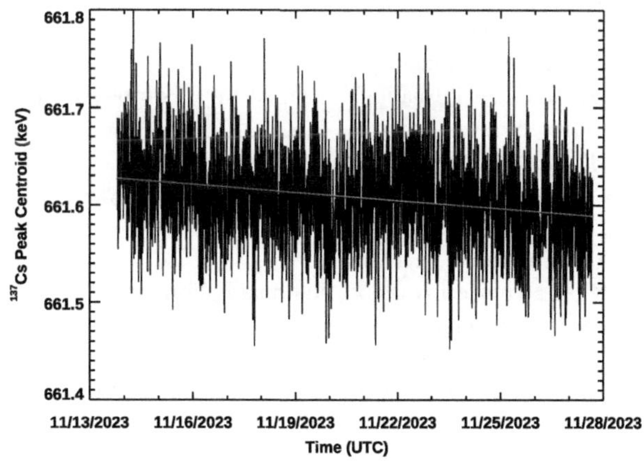

Fig. 45 Time-dependent variation of the ^{137}Cs peak centroid throughout the GRS ICO data collection. A linear fit to the data is shown by the solid red line

about the temperature dependent gain variation of the GRS is given by Peplowski et al. (2025).

List of acronyms

AC Anticoincidence
ADC Analog-to-Digital Converter
ANSI American National Standards Institute
APL Applied Physics Laboratory
BG Background
CC Cryocooler
CCDB Cryocooler Driver Board
CCPCB Cryocooler Power Conditioning Board
CORDIC Coordinate Rotation Digital Computer
CPT Comprehensive performance test
DN Digital number
DNL Differential nonlinearity
DPU Data Processing Unit
EM Engineering model

EMC Electromagnetic compatibility
EMI Electromagnetic interference
FEP fluorinated ethylene propylene
FIFO First-in, first-out
FPGA Field Programmable Gate Array
FWHM Full-width, half-maximum
FWTM Full-width, tenth-maximum
GCR Galactic cosmic ray
GOES Geostationary Operational Environmental Satellite
GRB Gamma-ray burst
GRNS Gamma-Ray and Neutron Spectrometer
GRS Gamma-Ray Spectrometer
HPGe High-purity Germanium
HV High voltage
HVFB High Voltage Filter Box
HVPS High Voltage Power Supply
ICO Initial checkout
INL Integral nonlinearity
INTEGRAL INTErnational Gamma-Ray Astrophysics Laboratory
IR Infrared
JFET Junction-gate field-effect transistor
JHU/APL Johns Hopkins University Applied Physics Laboratory
JPL Jet Propulsion Laboratory
LCTU Life Cycle Test Unit
LLNL Lawrence Livermore National Laboratory
LVPS Low Voltage Power Supply
MCNPX Monte Carlo N-Particle eXtended
MESSENGER Mercury Surface, Space Environment, Geochemistry, and Ranging
MIP Minimum-ionizing particle
MLI Multilayer insulation
MSPS Mega-samples per second
NEAR Near Earth Asteroid Rendezvous
NPO Negative-Positive-Zero
NS Neutron Spectrometer
PCB Printed Circuit Board
PDS Planetary Data System
PMC Periodic Maintenance Calibration
PMT Photomultiplier tube
PTFE Polytetrafluoroethylene
PWM Pulse-width modulation
SCUM Sensor Control Unit Module
SPE Solar particle event
THD Total harmonic distortion
TTSP Time-to-second pulse
UV Ultraviolet
VDC Volt direct current

Acknowledgements This work was funded by NASA's Discovery Program, under the Psyche: Journey to a Metal World mission. Psyche funding is provided to the Johns Hopkins University Applied Physics Laboratory (JHU/APL) through contracts 1569206 from NASA Jet Propulsion Laboratory and 17–256 from Arizona

State University. Co-authors from institutions other than JHU/APL are funded by respective contracts or subcontracts to their institutions. The design, development, and delivery of the Psyche GRNS was a significant effort that was only made possible by the dedicated effort of a large number of talented people. Acknowledgement and grateful thanks go to the following people: Jeff Olson from Lockheed Martin Advanced Technology Center for leading the design and building of the GRS cryocooler; Peter Berg from University of California Berkeley for leading the design and building of the analog portion of the high-voltage power supplies; Mark Hoff from JHU/APL (Fig. 10), who as the primary technician for the GRNS completed incredibly complicated work to help assemble many portions of the GRNS; Lena Heffern, currently from Laboratory for Atmospheric and Space Physics at the University of Colorado, who assisted with design work for the GRS while she was at Lawrence Livermore National Laboratory; colleagues from JHU/APL who worked on various aspects of the GRNS development (S. Begley, D. Bonner, C. Delmar, S. Disque, E. Dustin, C. Henry, C. Kim, S. Krupa, M. Lawrence, N. Lopez, L. Mehr, T. Palmer, A. Pergosky, S. Price, K. Runkles, B. Schratz, A. Shah, D. Street); colleagues either previously at APL and/or JPL who helped with the GRNS development (E. Barlow, M. DeSoria-Santacruz Pich, D. Nuding, Z. Staniszewski, H. Stone, K. Sukhatme, N. Warner). The authors are grateful for the detailed reviews by two reviewers, which resulted in an improved manuscript. The Psyche GRNS spaceflight data used in this paper are located at: https://doi.org/10.26033/24s8-7q30 (for the GRS) and https://doi.org/10.26033/10xg-da31 (for the NS).

Declarations

Competing Interests The authors declare no conflict of interest.

References

Agostinelli S, et al (2003) GEANT4-a simulation toolkit. Nucl Instrum Methods Phys Res A 506(3):250–303. https://doi.org/10.1016/S0168-9002(03)01368-8

Albernhe F, Borrel V, Frabel P, Vedrenne G, Coszach R, Denis JM, Leleux P (2002) Degradation and recovery of Ge detectors: tests prior to a space mission. Nucl Instrum Methods Phys Res A 492(1–2):91–96. https://doi.org/10.1016/s0168-9002(02)01289-5

Battarbee M, Guo J, Dalla S, Wimmer-Schweingruber R, Swalwell B, Lawrence DJ (2018) Multi-spacecraft observations and transport simulations of solar energetic particles for the May 17th 2012 event. Astron Astrophys 612. https://doi.org/10.1051/0004-6361/201731451

Bell JF, et al (2025) The Psyche multispectral imager investigation: characterizing the geology, topography, and multispectral properties of a metal-rich world. Space Sci Rev 221:47. https://doi.org/10.1007/s11214-025-01169-3

Boynton WV, et al (2004) The Mars Odyssey Gamma-Ray Spectrometer instrument suite. Space Sci Rev 110(1–2):37–83. https://doi.org/10.1023/b:spac.0000021007.76126.15

Bradford E, Kokorowski M, Lawrence D, Cullinan J, Ieee (2022) Nickel Accounting for the Psyche Spacecraft. Paper presented at IEEE Aerospace Conference (AERO), Big Sky, MT, Mar 05-12

Burks MT, et al (2020) GeMini: a high-resolution, low-resource, gamma-ray spectrometer for planetary science applications. Space Sci Rev 216:115. https://doi.org/10.1007/s11214-020-00746-y

Cambioni S, de Kleer K, Shepard M (2022) The heterogeneous surface of asteroid (16) Psyche. J Geophys Res Planets 127(6):e2021JE007091. https://doi.org/10.1029/2021je007091

Chabot NL, Peplowski PN, Ernst CM, Nair H, Lucks M, Steele RJ, Lawrence DJ (2021) MEGANE investigations of Phobos and the Small Body Mapping Tool. Earth Planets Space 73(1):217. https://doi.org/10.1186/s40623-021-01509-x

Chadwick MB, Oblozinsky P, Herman M et al (2006) ENDF/B-VII.0: Next Generation Evaluated Nuclear Data Library for Nuclear Science and Technology. Nuclear Data Sheets 107:2931–3060.

Croft S, Bond DS (1991) A determination of the Fano factor for germanium at 77.4-K from measurements of the energy resolution of a 113 cm^3 HPGe gamma-ray spectrometer taken over the energy-range from 14 to 6129 keV. Appl Radiat Isot 42(11):1009–1014. https://doi.org/10.1016/0883-2889(91)90002-I

Devanathan R, Corrales LR, Gao F, Weber WJ (2006) Signal variance in gamma-ray detectors - a review. Nucl Instrum Methods Phys Res A 565(2):637–649. https://doi.org/10.1016/j.nima.2006.05.085

Dibb S, Bell J, Elkins-Tanton L, Williams D (2023) Visible to near-infrared reflectance spectroscopy of asteroid (16) Psyche: implications for the Psyche Mission's science investigations. Earth Space Sci 10(1):e2022EA002694. https://doi.org/10.1029/2022EA002694

Diehl R, Siegert T, Greiner J, Krause M, Kretschmer K, Lang M, Pleintinger M, Strong AW, Weinberger C, Zhang XL (2018) INTEGRAL/SPI γ-ray line spectroscopy: response and background characteristics. Astron Astrophys 611. https://doi.org/10.1051/0004-6361/201731815

Elkins-Tanton LT, et al (2017) Asteroid (16) Psyche: visiting a metal world. In: 48th Lunar and Planetary Science Conference, Houston, Texas. Abstract #2614

Elkins-Tanton LT, et al (2020) Observations, Meteorites, and Models: a Preflight Assessment of the Composition and Formation of (16) Psyche. J Geophys Res Planets 125(3):e2019JE006296. https://doi.org/10.1029/2019je006296

Elkins-Tanton LT, et al (2022) Distinguishing the origin of asteroid (16) Psyche. Space Sci Rev 218:17. https://doi.org/10.1007/s11214-022-00880-9

Espiritu R, Peplowski P, LeBlanc M, Goldsten J, Lawrence D, Cully M, Marcotte K, Fix S (2022a), Psyche Mission Gamma Ray Neutron Spectrometer (GRNS) NS Uncalibrated to Calibrated Data Product Software Interface Specification Rep. 7506-9075. https://doi.org/10.26033/10xg-da31

Espiritu R, Peplowski P, LeBlanc M, Goldsten J, Lawrence D, Cully M, Marcotte K, Fix S (2022b), Psyche Mission Gamma Ray Neutron Spectrometer (GRNS) GRS Uncalibrated to Calibrated Data Product Software Interface Specification Rep. 7506-9074. https://doi.org/10.26033/24s8-7q30

Evans LG, Reedy RC, Trombka JI (1993) Introduction to planetary remote sensing gamma ray spectroscopy. In: Pieters CM, Englert PAJ (eds) Remote geochemical analysis: elemental and mineralogical composition. Cambridge University Press, Cambridge, pp 167–198

Evans LG, et al (2012) Major-element abundances on the surface of Mercury: Results from the MESSENGER Gamma-Ray Spectrometer. J Geophys Res Planets 117(E12). https://doi.org/10.1029/2012JE004178

Evans LG, et al (2015) Chlorine on the surface of Mercury: MESSENGER gamma-ray measurements and implications for the planet's formation and evolution. Icarus 257:417–427. https://doi.org/10.1016/j.icarus.2015.04.039

Evans LG, Peplowski PN, Rhodes EA, Goldsten JO, Starr RD, Solomon SC (2017) The MESSENGER gamma-ray spectrometer: calibration and operations. Icarus 288:186–200. https://doi.org/10.1016/j.icarus.2017.01.022

Farnocchia D, Fuentes-Muñoz O, Park R, Baer J, Chesley S (2024) Mass, density, and radius of asteroid (16) Psyche from high-precision astrometry. Astron J 168:21. https://doi.org/10.3847/1538-3881/ad50ca

Feldman WC, Drake DM (1986) A Doppler filter technique to measure the hydrogen content of planetary surfaces. Nucl Instrum Methods Phys Res A 245(1):182–190. https://doi.org/10.1016/0168-9002(86)90274-3

Feldman W, Auchampaugh G, Byrd R (1991) A novel fast-neutron detector for space applications. Nucl Instrum Methods Phys Res A 306:350–365

Feldman WC, Boynton WV, Drake DM (1993) Planetary neutron spectroscopy from orbit. In: Pieters CM, Englert PAJ (eds) Remote geochemical analysis: elemental and mineralogical composition. Cambridge University Press, Cambridge, pp 213–234

Feldman W, Lawrence D, Elphic R, Vaniman D, Thomsen D, Barraclough B, Maurice S, Binder A (2000) Chemical information content of lunar thermal and epithermal neutrons. J Geophys Res Planets 105(E8):20347–20363. https://doi.org/10.1029/1999JE001183

Feldman W, et al (2004) Gamma-Ray, Neutron, and Alpha-Particle Spectrometers for the Lunar Prospector mission. J Geophys Res Planets 109:E07S06. https://doi.org/10.1029/2003JE002207

Gasnault O, Feldman WC, Maurice S, Genetay I, d'Uston C, Prettyman TH, Moore KR (2001) Composition from fast neutrons: application to the Moon. Geophys Res Lett 28(19):3797–3800. https://doi.org/10.1029/2001GL013072

Genetay I, Maurice S, Feldman W, Gasnault O, Lawrence D, Elphic R, d'Uston C, Binder A (2003) Elemental content from 0 to 500 keV neutrons: Lunar Prospector results. Planet Space Sci 51(3):271–280. https://doi.org/10.1016/S0032-0633(02)00152-6

Goldsten JO, et al (1997) The X-ray/gamma-ray spectrometer on the near Earth Asteroid Rendezvous mission. Space Sci Rev 82(1–2):169–216. https://doi.org/10.1023/a:1005052720796

Goldsten JO, et al (2007) The MESSENGER gamma-ray and neutron spectrometer. Space Sci Rev 131(1–4):339–391. https://doi.org/10.1007/s11214-007-9262-7
Hasegawa S, et al (2024) Candidate main-belt asteroids for surface heterogeneity. Astron J 167:224. https://doi.org/10.3847/1538-3881/ad3045
Hines NR, Burks MT, Kim GB, Lawrence DJ, McNeil WJ (2022) A compact light shield for the Psyche gamma-ray spectrometer. Nucl Instrum Methods Phys Res A 1043:167491. https://doi.org/10.1016/j.nima.2022.167491
Hurley K, et al (2010) The Third Interplanetary Network, paper presented at Conference of the Gamma Ray Bursts, Annapolis, MD, Nov 01-04
Jourdain E (2011) Status of the INTEGRAL/SPI calibration. In: 6th international astronomical consortium for high-energy calibration meeting, Grottaferrata, Italy
Knoll GF (2000) Radiation detection and measurement, 3rd edn. Wiley, New York
Lawrence DJ, Feldman WC, Elphic RC, Little RC, Prettyman TH, Maurice S, Lucey PG, Binder AB (2002) Iron abundances on the lunar surface as measured by the Lunar Prospector gamma-ray and neutron spectrometers. J Geophys Res 107(E12):5130. https://doi.org/10.1029/2001JE001530
Lawrence DJ, Feldman WC, Elphic RC, Hagerty JJ, Maurice S, McKinney GW, Prettyman TH (2006) Improved modeling of Lunar Prospector neutron spectrometer data: Implications for hydrogen deposits at the lunar poles. J Geophys Res 111(E8):E08001. https://doi.org/10.1029/2005JE002637
Lawrence DJ, et al (2013a) Evidence for water ice near Mercury's North Pole from MESSENGER neutron spectrometer measurements. Science 339(6117):292–296. https://doi.org/10.1126/science.1229953
Lawrence DJ, Peplowski PN, Prettyman TH, Feldman WC, Bazell D, Mittlefehldt DW, Reedy RC, Yamashita N (2013b) Constraints on Vesta's elemental composition: fast neutron measurements by Dawn's gamma ray and neutron detector. Meteorit Planet Sci 48(11):2271–2288. https://doi.org/10.1111/maps.12187
Lawrence DJ, Feldman WC, Goldsten JO, Peplowski PN, Rodgers DJ, Solomon SC (2014) Detection and characterization of 0.5-8 MeV neutrons near Mercury: evidence for a solar origin. J Geophys Res Space Phys 119(7):5150–5171. https://doi.org/10.1002/2013JA019037
Lawrence DJ, Peplowski PN, Feldman WC, Schwadron NA, Spence HE (2016) Galactic cosmic ray variations in the inner heliosphere from solar distances less than 0.5AU: measurements from the MESSENGER neutron spectrometer. J Geophys Res Space Phys 121(8):7398–7406. https://doi.org/10.1002/2016JA022962
Lawrence DJ, Peplowski PN, Beck AW, Feldman WC, Frank EA, McCoy TJ, Nittler LR, Solomon SC (2017) Compositional terranes on Mercury: information from fast neutrons. Icarus 281:32–45. https://doi.org/10.1016/j.icarus.2016.07.018
Maurice S, Lawrence DJ, Feldman WC, Elphic RC, Gasnault O (2004) Reduction of neutron data from Lunar Prospector. J Geophys Res 109(E7):E07S04. https://doi.org/10.1029/2003JE002208
McCoy T, et al (2022) Deciphering redox state for a metal-rich world. Space Sci Rev 218:6. https://doi.org/10.1007/s11214-022-00872-9
Olson JR, Kaldas G, Champagne P, Roth E, Nast T (2015) MatISSE microcryocooler. In: Weisend Ii JG, et al (eds) Advances in cryogenic engineering. https://doi.org/10.1088/1757-899x/101/1/012025
Pehl RH, Goulding FS (1970) Recent observations on Fano factor in germanium. Nucl Instrum Methods 81(2):329–330. https://doi.org/10.1016/0029-554x(70)90569-0
Pehl RH, Madden NW, Elliott JH, Raudorf TW, Trammel RC, Darken LS (1979) Radiation-damage resistance of reverse electrode ge coaxial detectors. IEEE Trans Nucl Sci 26(1):321–323
Pelowitz DB (2005) MCNPX™ User's Manual, Version 2.5.0. Los Alamos National Laboratory Report LA-CP-05-0369
Peplowski PN (2016) The global elemental composition of 433 Eros: first results from the NEAR gamma-ray spectrometer orbital dataset. Planet Space Sci 134:36–51. https://doi.org/10.1016/j.pss.2016.10.006
Peplowski PN, et al (2011) Radioactive elements on Mercury's surface from MESSENGER: implications for the planet's formation and evolution. Science 333(6051):1850–1852. https://doi.org/10.1126/science.1211576
Peplowski PN, et al (2012) Variations in the abundances of potassium and thorium on the surface of Mercury: results from the MESSENGER gamma-ray spectrometer. J Geophys Res Planets 117(E12):E00L04. https://doi.org/10.1029/2012JE004141
Peplowski PN, et al (2014) Enhanced sodium abundance in Mercury's North polar region revealed by the MESSENGER gamma-ray spectrometer. Icarus 228:86–95. https://doi.org/10.1016/j.icarus.2013.09.007
Peplowski PN, Bazell D, Evans LG, Goldsten JO, Lawrence DJ, Nittler LR (2015a) Hydrogen and major element concentrations on 433 eros: evidence for an L- or LL-chondrite-like surface composition. Meteorit Planet Sci 50(3):353–367. https://doi.org/10.1111/maps.12434
Peplowski PN, et al (2015b) Constraints on the abundance of carbon in near-surface materials on Mercury: results from the MESSENGER gamma-ray spectrometer. Planet Space Sci 108:98–107. https://doi.org/10.1016/j.pss.2015.01.008

Peplowski PN, Lawrence DJ, Feldman WC, Goldsten JO, Bazell D, Evans LG, Head JW, Nittler LR, Solomon SC, Weider SZ (2015c) Geochemical terranes of Mercury's northern hemisphere as revealed by MESSENGER neutron measurements. Icarus 253:346–363. https://doi.org/10.1016/j.icarus.2015.02.002

Peplowski PN, Klima RL, Lawrence DJ, Ernst CM, Denevi BW, Frank EA, Goldsten JO, Murchie SL, Nittler LR, Solomon SC (2016) Remote sensing evidence for an ancient carbon-bearing crust on Mercury. Nat Geosci 9(4):273–276. https://doi.org/10.1038/ngeo2669

Peplowski PN, Burks M, Goldsten JO, Fix S, Heffern LE, Lawrence DJ, Yokley ZW (2019a) Radiation damage and annealing of three coaxial n-type germanium detectors: Preparation for spaceflight missions to asteroid 16 Psyche and Mars' moon Phobos. Nucl Instrum Methods Phys Res A 942:162409. https://doi.org/10.1016/j.nima.2019.162409

Peplowski PN, McCoy TJ, Liebel M, Beck AW, Burks MT, Elkins-Tanton LT, Lawrence DJ, Prettyman TH (2019b) Gamma-ray spectroscopy of metal-rich meteorites: preparations for the Psyche gamma-ray spectroscopy investigation. In: 50th Lunar and Planetary Science Conference, p #1731

Peplowski PN, Yokley ZW, Liebel M, Cheng S, Elphic RC, Hoogerheide SF, Lawrence DJ, Nico JS (2020) Position-dependent neutron detection efficiency loss in ^{3}He gas proportional counters. Nucl Instrum Methods Phys Res A 982:164574. https://doi.org/10.1016/j.nima.2020.164574

Peplowski P, et al (2025) Ground calibration and in-flight commissioning of the Gamma-Ray and Neutron Spectrometer (GRNS) for NASA's Psyche mission. Nucl Instrum Methods Phys Res A 1077:170542. https://doi.org/10.1016/j.nima.2025.170542

Polanskey CA, et al (2025) Psyche mission description and design rationale. Space Sci Rev 221

Prettyman TH, Hagerty JJ, Elphic RC, Feldman WC, Lawrence DJ, McKinney GW, Vaniman DT (2006) Elemental composition of the lunar surface: analysis of gamma ray spectroscopy data from lunar prospector. J Geophys Res 111(E12):E12007–E12007. https://doi.org/10.1029/2005JE002656

Prettyman TH, et al (2011) Dawn's gamma ray and neutron detector. Space Sci Rev 163(1–4):371–459. https://doi.org/10.1007/s11214-011-9862-0

Prettyman TH, et al (2012) Elemental mapping by dawn reveals exogenic H in Vesta's regolith. Science 338(6104):242–246. https://doi.org/10.1126/science.1225354

Prettyman TH, et al (2013) Neutron absorption constraints on the composition of 4 Vesta. Meteorit Planet Sci 48(11):2211–2236. https://doi.org/10.1111/maps.12244

Prettyman T, et al (2019) Elemental composition and mineralogy of Vesta and Ceres: distribution and origins of hydrogen-bearing species. Icarus 318:42–55. https://doi.org/10.1016/j.icarus.2018.04.032

Reddy V, Lawrence DJ, Elkins-Tanton LT, Takir D (2018) Constraining hydrogen abundance on asteroid (16) Psyche. In: 49th Lunar and Planetary Science Conference, edited, Houston, Texas. Abstract #1344

Reedy RC (1978) Planetary gamma-ray spectroscopy. In: 9th Lunar and Planetary Science Conference. Pergamon Press, New York, pp 2961–2984

Rodgers DJ, Lawrence DJ, Feldman WC, Peplowski PN (2015) Neutrons and energetic charged particles in the inner heliosphere: measurements of the MESSENGER neutron spectrometer from 0.3 to 0.85 AU. J Geophys Res Space Phys 120(2):841–854. https://doi.org/10.1002/2014JA020263

Roques JP, Teegarden BJ, Lawrence DJ, Jourdain E (2022) The use of germanium detectors in space. In: Bambi C, Santangelo A (eds) Handbook of X-ray and gamma-ray astrophysics. Springer, Singapore, pp 1–44. https://doi.org/10.1007/978-981-16-4544-0_163-1

Sanchez JA, Reddy V, Shepard MK, Thomas C, Cloutis EA, Takir D, Conrad A, Kiddell C, Applin D (2017) Detection of Rotational Spectral Variation on the M-Type Asteroid (16) Psyche. Astron J 153(1). https://doi.org/10.3847/1538-3881/153/1/29

Shepard MK, de Kleer K, Cambioni S, Taylor PA, Virkki AK, Rivera-Valentin EG, Rodriguez Sanchez-Vahamonde C, Fernanda Zambrano-Marin L, Magri C, Dunham D (2021) Asteroid 16 Psyche: Shape, features, and Global Map. Planet Sci J 2(4). https://doi.org/10.3847/PSJ/abfdba

Smith DM, et al (2002) The RHESSI spectrometer. Sol Phys 210(1–2):33–60. https://doi.org/10.1023/a:1022400716414

Takir D, Reddy V, Sanchez JA, Shepard MK, Emery JP (2017) Detection of water and/or hydroxyl on asteroid (16) Psyche. Astron J 153(1):31. https://doi.org/10.3847/1538-3881/153/1/31

University of Colorado Boulder and Space Weather Technology, Research, and Education Center (2019) SWx TREC Space Weather Data Portal. Laboratory for Atmospheric and Space Physics. https://doi.org/10.25980/NMFX-XX89

Vedrenne G, et al (2003) SPI: the spectrometer aboard INTEGRAL. Astron Astrophys 411(1):L63–L70. https://doi.org/10.1051/0004-6361:20031482

Vernazza P, Beck P (2017) Composition of Solar System small bodies. In: Elkins-Tanton LT, Weiss BP (eds) Planetesimals: early differentiation and consequences for planets. Cambridge University Press, Cambridge, pp 269–297. https://doi.org/10.1017/9781316339794.013

Weiss BP et al (2023) The Psyche magnetometry investigation. Space Sci Rev 219:22. https://doi.org/10.1007/s11214-023-00965-z

Wilson JT, Lawrence DJ, Peplowski PN, Feldman WC (2019) MESSENGER gamma ray spectrometer and epithermal neutron hydrogen data reveal compositional differences between Mercury's hot and cold poles. J Geophys Res Planets 124(3):721–733. https://doi.org/10.1029/2018je005871

Winslow RM, et al (2018) Opening a window on ICME-driven GCR modulation in the inner solar system. Astrophys J 856(2):139. https://doi.org/10.3847/1538-4357/aab098

Zhang ZT, Bercovici D, Elkins-Tanton L (2022) Cold Compaction and Macro-Porosity Removal in Rubble-Pile Asteroids: 2. Applications. J Geophys Res Planets 127:e2022JE007343. https://doi.org/10.1029/2022je007343

Zuber M, et al(2022) The Psyche gravity investigation. Space Sci Rev 218:57. https://doi.org/10.1007/s11214-022-00905-3

Publisher's Note Springer Nature remains neutral with regard to jurisdictional claims in published maps and institutional affiliations.

Authors and Affiliations

David J. Lawrence[1] · John O. Goldsten[1] · Patrick N. Peplowski[1] · Morgan T. Burks[2] · Shuo Cheng[1] · Michael J. Cully[1] · Jordan M. Effron[1,3] · Linda T. Elkins-Tanton[4,5] · Raymond C. Espiritu[1] · Samuel G. Fix[1] · Milena B. Graziano[1] · Erin M. Hoffer[1] · Insoo Jun[6] · Geon-Bo Kim[2] · Nathan R. Hines[2] · Mark T. LeBlanc[1] · Evan M. Livingstone[1,7] · Kathryn M. Marcotte[1] · Timothy J. McCoy[8] · Carol A. Polanskey[6] · Meena Sreekantamurthy[1] · Zachary W. Yokley[1]

✉ D.J. Lawrence
David.J.Lawrence@jhuapl.edu

1 Johns Hopkins University Applied Physics Laboratory, Laurel, MD, USA

2 Lawrence Livermore National Laboratory, Livermore, CA, USA

3 Present address: NASA Goddard Spaceflight Center, Greenbelt, MD, USA

4 School of Earth and Space Exploration, Arizona State University, Tempe, AZ, USA

5 Present address: Space Sciences Laboratory, University of California, Berkeley, Berkeley, CA, USA

6 Jet Propulsion Laboratory/California Institute of Technology, Pasadena, CA, USA

7 Present address: Northrop Grumman – AMS, Linthicum Heights, MD, USA

8 National Museum of Natural History, Smithsonian Institution, Washington DC, USA

Space Science Reviews (2022) 218:57
https://doi.org/10.1007/s11214-022-00905-3

The Psyche Gravity Investigation

Maria T. Zuber[1] · Ryan S. Park[2] · Linda T. Elkins-Tanton[3] · J.F. Bell III[3] · Kristoffer N. Bruvold[2] · David Bercovici[4] · Bruce G. Bills[2] · Richard P. Binzel[1] · R. Jaumann[5] · Simone Marchi[6] · Carol A. Raymond[2] · T. Roatsch[5] · Charles C. Wang[2] · Benjamin P. Weiss[1] · Daniel Wenkert[2] · Mark A. Wieczorek[7]

Received: 31 May 2021 / Accepted: 5 May 2022 / Published online: 18 October 2022

Abstract
The objective of the NASA Psyche mission gravity science investigation is to map the mass distribution within asteroid (16) Psyche to elucidate interior structure and to resolve the question of whether this metal-rich asteroid represents a remnant metal core or whether it is a primordial body that never melted. Measurements of gravity will be obtained via the X-band telecommunication system on the Psyche spacecraft, collected from progressively lower mapping altitudes. Orbital gravity will allow an estimate of GM to better than 0.001 $km^3\ s^{-2}$. A spherical harmonic model of gravity to degree and order 10 will be achievable and, in concert with spherical harmonic data sets from topography and magnetometry, as well as surface composition data, will provide information regarding the spatial and radial distribution of mass that will be used to constrain the origin and evolution of (16) Psyche.

Keywords Gravity · Asteroid · Remote sensing · Spacecraft

1 Introduction

Critical to achieving the scientific objectives of NASA's Psyche mission is characterization of the asteroid's internal structure. Gravitational mapping of (16) Psyche will add to the

The NASA Psyche Mission: Science Instruments and Investigations
Edited by James F. Bell III, Carol Polanskey and Lindy Elkins-Tanton

✉ M.T. Zuber

1 Department of Earth, Atmospheric and Planetary Sciences, Massachusetts Institute of Technology, Cambridge, MA 02139-4307, USA

2 Jet Propulsion Laboratory, Pasadena, CA 91109, USA

3 School of Earth and Space Exploration, Arizona State University, Tempe, AZ 86387-2001, USA

4 Department of Earth and Planetary Sciences, Yale University, New Haven, CT 06520, USA

5 Institute of Geological Sciences, Free University of Berlin, 12249 Berlin, Germany

6 Southwest Research Institute, Boulder, CO 80302, USA

7 Observatoire de la Côte d'Azur, CNRS, Laboratoire Lagrange, Université Côte d'Azur, Nice, France

quantitative hypothesis testing as to whether the asteroid represents an iron core, a fundamental building block of terrestrial planets (Elkins-Tanton 2016; Elkins-Tanton et al. 2022), or if it consists of unmelted metal-rich material. Gravity will also contribute towards interpretation of spatially large structural features of the asteroid, and provide linkage between surface and subsurface processes. Secondarily, the gravity investigation, in coordination with topography (Jaumann et al. 2022), improves global positioning and the reference frame, increasing the ease of integration of all the Psyche mission data sets.

2 Tracking the Psyche Orbiter

The Psyche spacecraft (Hart et al. 2018) contains an X-band telecom system that provides uplink (7.2 GHz) and downlink (8.4 GHz) radio signal and is used to track the spacecraft via range rate and range observations. Spacecraft Psyche's telecom system features a 2-m diameter high-gain antenna and three low-gain antennae, and radio tracking is accomplished using NASA's Deep Space Network (DSN) using 34-m and 70-m antennae at Goldstone, California, Madrid, Spain and Canberra, Australia. The Psyche spacecraft (Brown et al. 2021) also includes an experimental Deep Space Optical Communications (DSOC) laser communication link that will downlink to the Palomar Observatory.

The Doppler shift of the radio signal detected on Earth is sensitive to the changes in the motion of a spacecraft due to gravitational variations, which can be used to recover the gravity field of the body being orbited. The accuracy is defined by the X-band telecom system performance that is typically of order 0.1 $\mathrm{mm\,s^{-1}}$ at 60-s integration time. Radiometric Doppler tracking occurs as two-way, with the same transmit and receive stations, and sometimes three-way, with a different receiver station.

The Psyche mapping plan (Polanskey et al. 2018) is operationally similar to that of the Dawn mission (Russell et al. 2007; Park et al. 2016), on which a number of members of the Psyche Gravity Team participated, providing important experience. Specifically, the mapping operations plan consists of an approach phase and progression to lower altitude orbits as knowledge of the asteroid's gravity field progressively improves (see Fig. 1). During Orbit C, which is the prime for gravity science, radio tracking data will be acquired regularly over the 100-day orbit phase for a total of at least 35 days (i.e., $\geq 35\%$), which is expected to be sufficient for a global sampling of Psyche's gravity field. Gravity observations will also be acquired in the lower-altitude Orbit D, during which time measurements of surface composition are prioritized.

Figure 2 shows the latest shape of Psyche (Shepard et al. 2021) from different viewing geometries. The shape, developed using a wide range of observations (Arecibo, ALMA, Keck, VLT, stellar occultation), is used in mission planning. Psyche's mean radius is $\sim$113 km, and compared to previously orbited asteroids and dwarf planets, Psyche is smaller than 1 Ceres ($\sim$470 km) and 4 Vesta ($\sim$265 km), but much larger than 433 Eros (8 km), 26143 Itokawa ($<$1 km), 162173 Ryugu ($<$1 km), and 101955 Bennu ($<$1 km).

DSN range measurements to the Psyche spacecraft, which have an accuracy of 1-2 m, are included in inversions for the gravitational field and associated parameters, though experience dictates that this data type does not add particular strength in the determination of the gravity field. However, even if observations are infrequent ($\sim$1 hour over several days), ranging is valuable in improving (16) Psyche's heliocentric orbital ephemeris (cf. Konopliv et al. 2018).

Corrections for Earth media fluctuations are applied to Doppler and range data at each DSN site. These include daily wet and dry terrestrial troposphere corrections and terrestrial

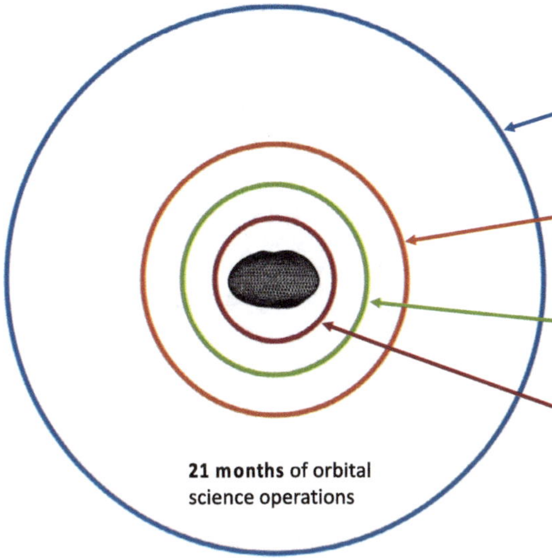

Fig. 1 Schematic depicting the progressive lowering of science mapping orbits, designed based on experience with the Dawn mission. Gravity science is prioritized in Orbit C

GPS-based ionosphere calibrations. The three-way Doppler data are corrected for tracking station timing errors to a level $<0.1\ \mathrm{mm\,s^{-1}}$ at 60-s integration time. The range measurements are corrected for spacecraft clock delay and a range calibration correction of <10 m is made at each DSN site (Asmar et al. 2005).

Solar plasma represents an additional source of data noise, with X-band undergoing considerable degradation at Sun-Earth-Psyche (SEP) angles $<20°$, close to solar conjunction (Asmar et al. 2005). Such observations are typically de-weighted in gravity field solutions (see Sect. 3.3). During the entire science mapping phase, the SEP angle will vary essentially from 0° to 180°, so there will be some periods in which tracking is interrupted. However, mission design simulation (Oh et al. 2019) indicates that enough data will be collected during Orbit C with SEP$>40°$ that achieving an $n = 10$ gravity field should not be a problem (cf. Fig. 5).

Finally, correction for non-gravitational forces, including angular momentum desaturations and solar radiation pressure modeling of the Psyche spacecraft, is essential. The angular momentum desaturations, often known as small forces, are required to unload large cyclic gravity gradient torques. For Psyche, there may even be non-trivial differences in the gravity gradient torques between a point-mass model and a model with high degree harmonics. During science orbits, environmental torques are compounded much more rapidly when the spacecraft bus is pointing off nadir. Perturbations by small forces are typically a few minutes in duration and will be modeled in 60-s increments. The Psyche spacecraft solar panels will be pointed mostly toward the sun, making the solar radiation pressure fairly smooth and easy to model. However, the spacecraft bus will be rotating relative to the sun to point the instrument deck towards (16) Psyche. The orientation of both the spacecraft bus and panels will be modeled using the attitude of the Psyche spacecraft derived from onboard star trackers and the acceleration due to solar radiation pressure can be typically modeled with accuracy better than $10^{-12}\ \mathrm{km\,s^{-2}}$.

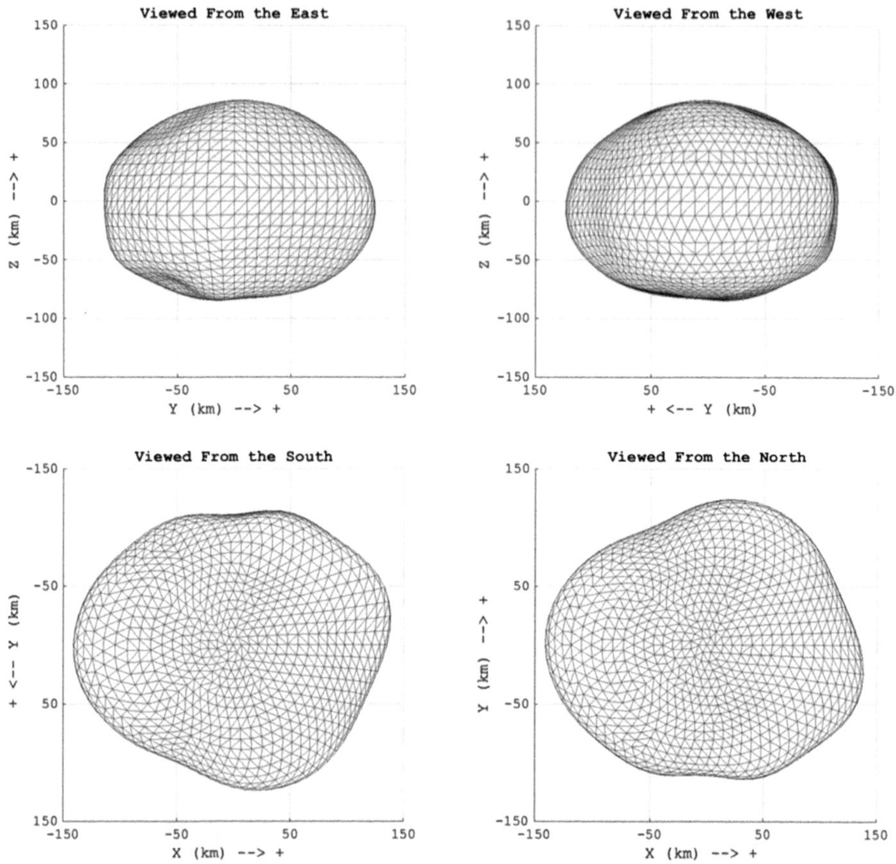

Fig. 2 Shape of Psyche (Shepard et al. 2021) shown from different viewing angles

2.1 Psyche Telecommunication Subsystem

The Psyche Telecom Subsystem operates at X-band for nominal two-way communications (command uplink and telemetry downlink) and two-way tracking (Doppler and ranging). These are the high-level functions, with the details explained in the following sections. The Telecom subsystem block diagram is presented in Fig. 3. This subsystem is block redundant, consisting of two Small Deep Space Transponders (SDSTs), two traveling-wave tube amplifiers (TWTAs), two diplexer-isolators (DXIs), seven waveguide transfer switches (WTS), one high-gain antenna (HGA), and three low-gain antennas (LGAs). Right-hand circular polarization (RHCP) is used for all antennas for nominal operations. Figure 3 also shows how the SDSTs, TWTAs, diplexers-isolators, and two of the waveguide switches (WTS-1 and WTS-2) form a primary and a redundant telecom block. Only one block will be powered on at any given time in flight for nominal operations. Waveguide transfer switches (WTS-3 and WTS-4) are used to route the signals between the antennas and the two telecom blocks. WTS-5, WTS-6, and WTS-7 are used to provide redundant paths to the mission-critical antennas using left-hand circular polarization (LHCP) in case WTS-3 or WTS-4 fails. These three switches will only be actuated in case of a switch failure. The waveguide runs to the mission critical antennas, HGA, +X-LGA, and –Z-LGA, are shown in red in Fig. 3.

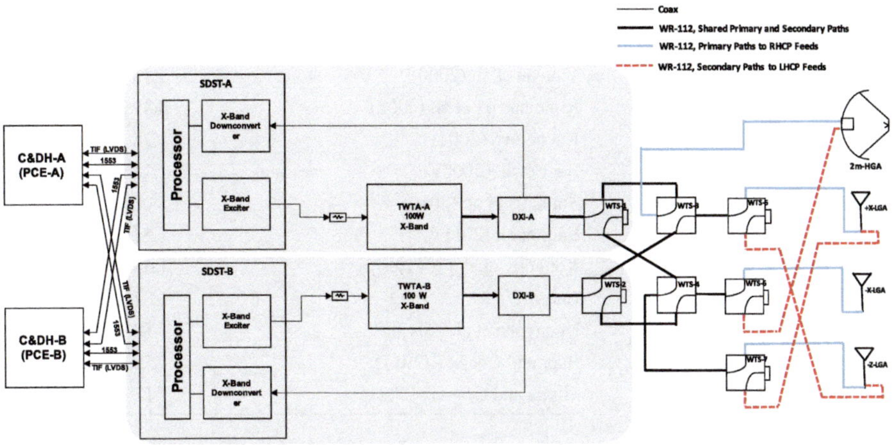

Fig. 3 Block diagram of the Psyche telecommunication system

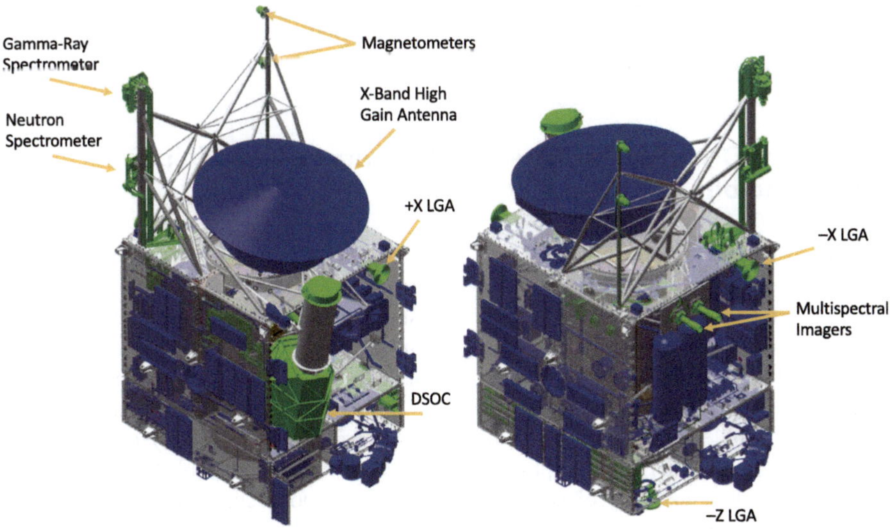

Fig. 4 Schematics showing key components of the Psyche telecommunication system

The locations of the HGA and the three identical LGAs are shown in Fig. 4. The HGA is mounted on the +Z deck of the spacecraft. It supports high-data rate communications, as well as Delta-Differential One-way Ranging (Δ-DOR), and Doppler measurements. In practice, Δ-DOR measurements will be acquired during cruise, but probably not during the orbital phase as Doppler data are much more sensitive. The -Z-LGA, +X-LGA, and -X-LGA are mounted at the -Z, +X, and -X decks of the spacecraft, respectively. The -Z-LGA will provide most of the two-way Doppler measurements for gravity science. It is also used for safe mode communication during the early part of the mission. The +X-LGA is used for safe mode communication for most of the mission. In safe mode, the spacecraft will point the +X-axis of the spacecraft to the sun to charge the solar arrays to ensure the

Table 1 Psyche GM Values

Source	Value $\pm$ σ (km^3 s^{-2})
Viateau et al. (2000)	1.15 $\pm$ 0.35
Kuzmanoski et al. (2002)	4.49 $\pm$ 0.37
Kochetova (2004)	1.78 $\pm$ 0.29
Baer et al. (2008)	1.71 $\pm$ 0.23
Somenzi et al. (2010)	3.06 $\pm$ 1.29
Baer et al. (2011)[a]	1.51 $\pm$ 0.06
Konopliv et al. (2011)	1.65 $\pm$ 0.46
Carry (2012)	1.82 $\pm$ 0.50
Kuchynka et al. (2013)	1.18 $\pm$ 0.28
Baer and Chelsey (2017)	1.53 $\pm$ 0.05
Siltala and Granvik (2021)	1.12 $\pm$ 0.20

[a]The study of Baer et al. (2011) is considered to be the best determination and is used for mission planning purposes

spacecraft is power-positive, with the Earth staying within 40° of the +X-LGA boresight. Both +X-LGA and -X-LGA are used to support two-way Doppler measurements in support of gravity science during the orbit phases of the mission. The LGAs have been designed to have broad patterns to support both safe mode telecommunications and two-way Doppler measurements for gravity science.

3 Precision Orbit Determination and Gravity Field Modeling Methodology

3.1 Precision Orbit Determination

Determination of precise orbits of the Psyche spacecraft about the asteroid requires full integration of the spacecraft trajectory from an initial state, accompanied by the application of informed models to account for non-conservative forces. Models include an *a priori* for Psyche based on astronomical estimations of GM (see Table 1), rotation parameters for Psyche (see Table 2), an initial ephemeris obtained from the JPL Solar System Dynamics group, and various other *a priori* physical parameters. In terms of published GM values, the team considers the study of Baer et al. (2011) to be most reliable because: it has the smallest uncertainty, and checking predictions for Vesta and Ceres by these authors (Baer et al. 2008) showed robust estimations. In addition we note that Baer et al. (2011) was the best value available at the time orbit design simulations were being performed (2018 time frame). Other values are listed in Table 1 to provide a compendium of published estimations.

DSN Doppler and range measurements, and optical positions acquired from the approach and Psyche mission orbital phases are co-processed to yield solutions for the spacecraft orbit, gravitational field, pole orientation and rotation rate, and heliocentric orbit. These observables are processed and filtered using JPLs MONTE (Mission Analysis, Operations, and Navigation Toolkit Environment) MONTE, which employs a least-squares approach to converge orbital arcs (Evans et al. 2018). MONTE, which has been utilized in a similar capacity in numerous other planetary gravity investigations, is capable of processing Doppler, range, and landmark observations. Processing of surface landmarks via optical imaging is accomplished using a stereophotoclinometric (SPC) process (Owen et al. 2001; Konopliv et al. 2002; Park et al. 2019, 2020).

Table 2 Psyche Physical Parameters

Parameter	Value
Spin-pole Ecliptic Longitude (λ) (deg)	35 ± 5 (Kaasalainen et al. 2002)
	32 ± 5 (Hanus et al. 2013)
	34 ± 5 (Shepard et al. 2017)
	32 ± 3 (Drummond et al. 2018)
	36 ± 2 (Shepard et al. 2021)
Spin-pole Ecliptic Longitude (β) (deg)	9 ± 5 (Kaasalainen et al. 2002)
	7 ±5 (Hanus et al. 2013)
	7 ±5 (Shepard et al. 2017)
	8 ± 3 (Drummond et al. 2018)
	8 ± 2 (Shepard et al. 2021)
Spin period (hour)	4.195947 ± 0.000001 (Kaasalainen et al. 2002)
	4.195948 ± 0.000001 (Hanus et al. 2013)
	4.195948 ± 0.000001 (Shepard et al. 2017)
	4.195948 ± 0.000001 (Drummond et al. 2018)
	4.195948 ± 0.000001 (Shepard et al. 2021)
Obliquity (deg)	94. (Bills and Scott 2021)
Maximum Dimensions (km)	279×232×189 (Shepard et al. 2021)

3.2 Data Calibration

Calibration of the tracking data is accomplished by routinely analyzing the noise level, range biases, and media effects of the Doppler and range data. Attention is focused on assuring that the quality of tracking data is at the level required to meet measurement requirements.

3.3 Gravitational Field Estimation

The gravitational field potential, $U(r, \lambda, \theta)$, associated with (16) Psyche is expressed as a spherical harmonic expansion (Kaula 1966; Heiskanen and Moritz 1967)

$$U(r, \lambda, \theta) = \frac{GM}{r}\left\{1 + \sum_{l=2}^{\infty}\left(\frac{R}{r}\right)^{l}\sum_{m=0}^{l}\left[\underline{C}_{lm}cos(m\lambda) + \underline{S}_{lm}sin(m\lambda)\right]\underline{P}_{lm}sin(\theta)\right\} \quad (1)$$

where GM is the mass parameter, l is the spherical harmonic degree, m is the angular order, $\underline{P}_{lm}$ are the fully normalized associated Legendre polynomials, $\underline{C}_{lm}$ and $\underline{S}_{lm}$ are the normalized spherical harmonic coefficients (*cf.* Wieczorek 2015), R is the reference radius of Psyche (113 km), λ is longitude, θ is latitude, and r is the radius at the evaluation point of the Psyche spacecraft relative to the (16) Psyche body-fixed frame. Following Kaula (1966) and Lambeck (1990), the normalization of the gravity coefficients is defined such that the integral of the harmonic squared is equated to the area of a unit sphere. Normalized coefficients are related to unnormalized coefficients by

$$(C_{lm}, S_{lm}) = \left[\frac{(l-m)!\,(2l+1)\,(2-\delta_{0m})}{(l+m)!}\right]^{1/2}\left(\underline{C}_{lm}, \underline{S}_{lm}\right) \quad (2)$$

where δ is the Kronecker delta. In this formulation zonal coefficients are defined as $\underline{J}_l = -\underline{C}_{l0}$. Because the origin of the planetary system is defined to be (16) Psyche's center of mass, the degree-one coefficients are identically zero.

In practice, the potential in Eq. (1) is obtained from the inversion of normal equations in a sparse matrix of tracking observations that are weighted according to quality and spatial and temporal distribution (Balmino et al. 1982; Smith et al. 1993). A single solution yields the potential field as well as estimates of dynamical parameters that include the asteroid pole position (obliquity) and rotation rate.

4 Science Investigation

4.1 Pre-Mission Reference Field and Gravity from Shape

In order to design the Psyche mission mapping scenario and to plan for data acquisition, it is necessary to develop a reference gravity field of Psyche. Consistent with X-band tracking, the simulations assume a Doppler accuracy of 0.1 mm s^{-1} and 60-s sample time. As detailed in Sect. 3.3, we develop a notional spherical harmonic expansion of the gravitational potential, defined by a reference radius, R, mass, M, and normalized spherical harmonic coefficients $\underline{C}_{lm}$ and $\underline{S}_{lm}$ (Kaula 1966; Heiskanen and Moritz 1967).

The reference gravity of (16) Psyche was computed by scaling the harmonic coefficients using the Kaula rule (Kaula 1966), an empirical power law for estimating gravitational power on solid planetary bodies. All terrestrial bodies that have measured gravitational field follow such a relationship in their degree variance spectrum (Kaula 1966; Konopliv et al. 2011; Zuber et al. 2013)

$$\sigma_l^2 = \frac{\left[\sum_{m=0}^{l}\left(\underline{C}_{lm}^2 + \underline{S}_{lm}^2\right)\right]}{(2l+1)} \tag{3}$$

Assuming a Kaula rule of 0.05 n^{-2}, the variance spectrum is shown in Fig. 5.

We also computed a gravity model associated with the 3-dimensional Psyche shape models from Shepard et al. (2017), which is the baseline model being used by the Psyche project. Additional shape models are also being investigated (Hanus et al. 2017; Viikinkoski et al. 2018) and the mission plan is robust with these additional models as well. The spherical harmonics of the shape model were computed using a quadrature method (Werner and Scheeres 1997) assuming the GM of 1.53 km^3 s^{-2}, where $G = 6.674 \times 10^{-11}$ m^3 kg^{-1} s^{-2}. The power spectrum of the baseline gravity field from the Shepard shape is shown in Fig. 5, as well as the expected accuracy of the recovered Psyche gravity field (*i.e.*, error spectrum) from Orbit C. The result shows that a $l = 10$ gravity field should be easily achievable (i.e., the intersection of the power spectrum and the error spectrum shows the recoverable resolution of the gravity field). The result also indicates that the spin pole direction can be recovered with accuracy better than 0.001° while the current estimate of Psyche's spin rate is sufficiently accurate (Shepard et al. 2021).

4.2 Implications for Psyche's Structure, Origin, Evolution and Current State

Psyche's gravity investigation is designed to yield a spherical harmonic representation of the field to better than degree and order 10 (spatial block size, or half-wavelength, average resolution of ~36 km; since Psyche is ellipsoidal the range of resolved block sizes from 30

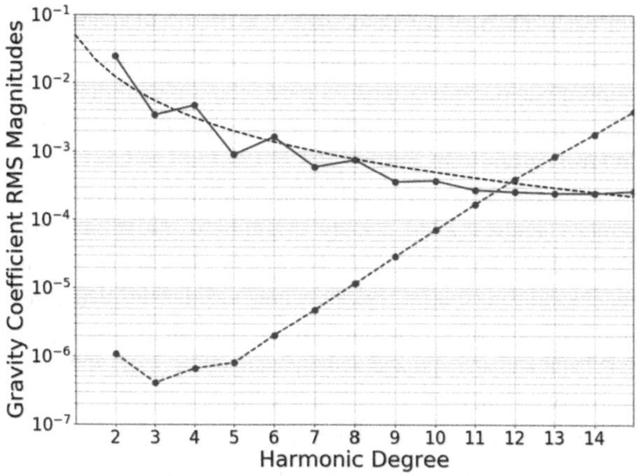

Fig. 5 Gravitational power spectra for an assumed Kaula rule (black-dashed) and for the Shepard et al. (2021) Psyche shape model (blue solid; cf. Fig. 2) assuming a constant density interior. The reference radius of Psyche was assumed to be 113 km. The power spectrum of the expected accuracy of the recovered gravity field from Orbit C is shown in the dashed-blue line. Expected accuracy is defined by the degree at which the formal error spectrum intersects the power spectrum

to 44 km), using data from Orbits A-C. This measurement depends upon the spacecraft and uplink transponder stability, and downlink signal-to-noise ratio. Each of these requirements is met with a substantial margin in performance. If we consider Orbit D in addition to Orbits A-C, we expect a degree and order 12 could be recovered (i.e., spatial resolution of ~30 km). A spherical harmonic model of free-air anomalies, derived from the differentiating the gravitational potential (Eq. (1)), will be converted to Bouguer gravity by subtracting the gravitational attraction of topography. Bouguer gravity provides information on the internal distribution of mass.

First, the Gravity Science investigation will produce a much improved estimate of the asteroid mass (cf. Table 1), reducing the uncertainty in GM to better than 0.001 $km^3\,s^{-2}$. That measurement, combined with the volume provided from the shape determination (Jaumann et al. 2022), will provide a greatly improved estimation of bulk density. A summary of bulk density measurements of Psyche based on recent measurements of GM and volume are summarized by Elkins-Tanton et al. (2020), who notes that the bulk density is one of the highest of the main belt asteroids. The range of measured values falls in the range for other asteroids that display metal-rich surface spectra (X- and M-class; Hanus et al. 2017). The preferred current Psyche value of $3780 \pm 340\ kg\,m^{-3}$ (Elkins-Tanton et al. 2020) is inconsistent with Psyche being a solid metallic core. Indeed, if Psyche is metallic the bulk density implies a porosity of ~50%, a void structure that would be difficult to maintain if developed in its earliest evolution given Psyche's size and estimated early cooling rate (Nichols-Fleming et al. 2022).

Higher-degree harmonics of gravity and shape will be analyzed to discern mass variations in spatial location and depth. The gravitational flattening, $C_{2,0}$, combined with any observed spin axis precession, if it occurs and is measurable, will permit estimation of C/MR^2, the moment of inertia factor that provides a measure of the radial distribution of mass. If Psyche's structure predominantly reflects solidification from a liquid, it should exhibit radial variation in its density structure and approximate azimuthal symmetry after accounting for effects of impacts, which will have greatest effect at shallow depths. As a case in point, a model of Vesta's gravity field with an assumed three-layer model showed nearly perfect correlation for all measurable degrees (Konopliv et al. 2014; Park et al. 2014), indicating no substantial density variations in the thickness or density of Vesta's crust. Azimuthal symmetry, at least at crustal levels, can also be achieved by impact homogenization, as has been observed for the Moon (Zuber et al. 2013). Estimates of the effective surface density for each spherical harmonic degree will provide constraints on how the density varies with depth (Wieczorek 2013; Besserer et al. 2014). Gravity, like all potential field data, provides non-unique interpretations. In particular, internal mass variations can be associated with either compositional variations or porosity. The addition of other data allows the range of interpretations to be narrowed.

If Psyche was instead substantially disrupted and/or was never molten, it will likely have a more heterogeneous density structure. Correlation of its gravity field with surface topography (Jaumann et al. 2022) via transfer function techniques of admittance and coherence, would indicate spatial regions where the density structure is not consistent with simple azimuthally symmetric models. Gravity and shape combined with magnetometry (Weiss et al. 2022) and surface compositional data (Bell et al. 2022; Lawrence et al. 2022), will be collectively required to constrain the structure and degree of azimuthal symmetry, including the possible disruption of the body into a rubble pile (Elkins-Tanton et al. 2022). If Psyche does turn out to be largely metallic, it almost certainly will have acquired its porosity subsequent to cooling.

The spatial resolution of gravity noted above, combined with that of topography (50 m pixel^{-1}; Jaumann et al. 2022) dictates that regional density variations across Psyche's shallow subsurface, including in association with large craters (Marchi et al. 2022) can be mapped. And combined with surface observations to provide geophysical information to support geological context based on observations from other Psyche sensors (Jaumann et al. 2022; Lawrence et al. 2022; Marchi et al. 2022).

5 Data Archiving and Distribution

Standard data products from the Psyche Gravity Investigation will be archived in the Small Bodies Node of the NASA Planetary Data System on a schedule set by the Psyche Project, as per its contract with NASA. Level 0 data products, in TRK-34 format, will be delivered to the PDS for verification and archiving within 6 months of collection and will be disseminated to the scientific community immediately upon verification by PDS reviewers.

Level 3 data, including spherical harmonic models of gravitational potential, free air gravity and Bouguer gravity, and determinations of other physical parameters (*e.g.*, planetocentric reference frame, spin-pole axis, rotation rate, etc.) will be archived at a Planetary Data System (PDS) Small Bodies Node as part of the Psyche archival plan.

6 Summary

The Psyche mission affords the unprecedented opportunity to map a possible protoplanetary core. Gravity observations, combined with other orbital observations of topography, magnetometry and compositional measurements, will collectively provide a quantitative view of the internal mass distribution and implications for origin and composition.

Acknowledgements This investigation is supported by the NASA Psyche Mission. A portion of this research was carried out at the Jet Propulsion Laboratory, California Institute of Technology, under a contract with the National Aeronautics and Space Administration (80NM0018D0004).

References

S.W. Asmar et al., Spacecraft Doppler tracking: noise budget and accuracy achievable in precision radio science observations. Radio Sci. **40** (2005). https://doi.org/10.1029/2004RS003101

J. Baer et al., An observational error model, and application to asteroid mass determination. Bull. Am. Astron. Soc. **40**, 493 (2008)

J. Baer et al., Astrometric masses of 26 asteroids and observations on asteroid porosity. Astron. J. **141**, 1–12 (2011)

J. Baer, S.R. Chelsey, Simultaneous mass determination for gravitationally coupled asteroids. Astron. J. **154**, 76 (2017). https://doi.org/10.3847/1538-3881/AA7DE8

G. Balmino et al., Gravity field of Mars in spherical harmonics up to degree and order eighteen. J. Geophys. Res. **87**, 9735–9746 (1982)

J.F. Bell et al., The multispectral imaging investigation on the NASA Psyche discovery mission. Space Sci. Rev. (2022, this journal). https://doi.org/10.1007/s11214-022-00874-7

J. Besserer et al., GRAIL gravity constraints on the vertical and lateral density structure of the lunar crust. Geophys. Res. Lett. **41**, 5771–5777 (2014). https://doi.org/10.1002/2014GL060240

B. Bills, B.R. Scott (2021). https://ui.adsabs.harvard.edu/abs/2019AGUFM.P51E3412B/abstract

M. Brown et al., Psyche spacecraft design. Space Sci. Rev. (2021, this journal). https://doi.org/10.1109/AERO.2018.8396444

B. Carry et al., Density of asteroids. Planet. Space Sci. **73**, 98–118 (2012)

J.D. Drummond et al., The triaxial ellipsoid size, density, and rotational pole of asteroid (16) Psyche from Keck and Gemini AO observations 2004–2015. Icarus **305**, 174–185 (2018). https://doi.org/10.1016/j.icarus.2018.01.010

L.T. Elkins-Tanton, The taxonomy of planetesimals: consequences for planets, in *Planetesimals: Early Differentiation and Consequences for Planets*, ed. by L.T. Elkins-Tanton, B.P. Weiss (Cambridge University Press, Cambridge, 2016), pp. 365–375

L.T. Elkins-Tanton et al., Observations, meteorites, and models: a preflight assessment of the composition and formation of (16) Psyche. J. Geophys. Res., Planets **125**, e2019JE006296 (2020). https://doi.org/10.1029/2019/JE006296

L.T. Elkins-Tanton et al., Distinguishing the origin of (16) Psyche. Space Sci. Rev. (2022, this journal). https://doi.org/10.1007/s11214-022-00880-9

S. Evans et al., MONTE: the next generation of mission design and navigation software. CEAS Space J. **10**, 79–86 (2018). https://doi.org/10.1007/s12567-017-0171-7

J. Hanus et al., Sizes of main-belt asteroids by combining shape models and Keck adaptive optics observations. Icarus **226**, 1045–1057 (2013). https://doi.org/10.1016/j.icarus.2013.07.023

J. Hanus et al., Volumes and bulk densities of forty asteroids from ADAM shape modeling. Astron. Astrophys. **601**, A114 (2017). https://doi.org/10.1051/0004-6361/201629956

W. Hart et al., Overview of the spacecraft design for the Psyche mission concept, in *IEEE Aerospace Conf.*, Big Sky, MT (2018), pp. 1–20

W.A. Heiskanen, H. Moritz, *Physical Geodesy* (W.H. Freeman, San Francisco, 1967), 404 pp.

R. Jaumann et al., The Psyche topography and geomorphology investigation. Space Sci. Rev. (2022, this journal). https://doi.org/10.1007/s11214-022-00874-7

M. Kaasalainen et al., Models of twenty asteroids from photometric data. Icarus **159**, 369–395 (2002). https://doi.org/10.1006/icar.2002.6907

W.M. Kaula, in *Theory of Satellite Geodesy*, Waltham, Blaisdell (1966), 124 pp.

O.M. Kochetova et al., Determination of large asteroid masses by the dynamical method. Sol. Syst. Res. **38**, 66–75 (2004)

A.S. Konopliv et al., A global solution for the gravity field, rotation, landmarks, and ephemeris of Eros. Icarus **166**, 289–299 (2002)

A.S. Konopliv et al., The dawn gravity investigation at Vesta and Ceres. Space Sci. Rev. **163**, 461–486 (2011). https://doi.org/10.1007/s11214-011-9794-8 Also in *The Dawn Mission to Minor Planets 4 Vesta and 1 Ceres*, ed. by C.T. Russell, C.A. Raymond

A.S. Konopliv et al., The Vesta gravity field, spin pole and rotation period, landmark positions, and ephemeris from the Dawn tracking and optical data. Icarus **240**, 103–117 (2014). https://doi.org/10.1016/j.icarus.2013.09.005

A.S. Konopliv et al., A global solution for the gravity field, rotation, landmarks, and ephemeris of Eros. Icarus **299**, 411–429 (2018). https://doi.org/10.1016/j.icarus.2017.08.005

M. Kuzmanoski, A. Kovačević, Motion of the asteroid (13206) 1997GC22 and the mass of (16) Psyche. Astron. Astrophys. **395**, L17–L19 (2002)

P. Kuchynka, W.M. Folkner, A new approach to determining asteroid masses from planetary range measurements. Icarus **222**, 243–253 (2013).

K. Lambeck, *Geophysical Geodesy* (Cambridge University Press, New York, 1990)

D.J. Lawrence et al., The Psyche gamma ray % neutron spectrometer investigation. Space Sci. Rev. (2022, this journal)

S. Marchi et al., Determining the relative cratering ages of regions of Psyche's surface. Space Sci. Rev. (2022, this journal). https://doi.org/10.1007/s11214-022-00891-6

F. Nichols-Fleming et al., Porosity evolution in metallic asteroids: implications for the origin and thermal history of asteroid 16 Psyche. J. Geophys. Res., Planets **127**, e2021JE007063 (2022). https://doi.org/10.1029/2021JE007063

D.Y. Oh et al., Development of the Psyche mission for NASA's discovery program, in *The 36th Electric Propulsion Conf., IEPC-2019-192*, 12-20 September 2019 (Univ. Vienna, Austria, 2019), 29 pp.

W.M. Owen Jr. et al., in *NEAR Optical Navigation at Eros. AAS/AIAA Astrodynamics Specialists Conference*, Quebec City, Quebec, CA (2001)

R.S. Park et al., Gravity field expansion in ellipsoidal harmonic and polyhedral internal representations applied to Vesta. Icarus **240**, 118–132 (2014). https://doi.org/10.1016/j.icarus.2018.10.024

R.S. Park et al., A partially differentiated interior for (1) Ceres deduced from its gravity field and shape. Nature **537**, 515–517 (2016). https://doi.org/10.1038/nature18955

R.S. Park et al., High-resolution shape model of Ceres from stereophotoclinometry using Dawn imaging data. Icarus **319**, 812–827 (2019). https://doi.org/10.1016/j.icarus.2018.10.024

R.S. Park et al., Evidence of non-uniform crust of Ceres from Dawn's high-resolution gravity data. Nat. Astron. **4**, 748–755 (2020). https://doi.org/10.1038/s41550-020-1019-1

C.A. Polanskey et al., Psyche science operations concept: maximize reuse to minimize risk, in *SpaceOps Conf.*, Marseille, France (2018)

C.T. Russell et al., Dawn mission to Vesta and Ceres: symbiosis between terrestrial observations and robotic exploration. Earth Moon Planets **101**, 65–91 (2007). https://doi.org/10.1007/s11038-007-9151-9

M.K. Shepard et al., Radar observations and shape model of asteroid 16 Psyche. Icarus **281**, 388–403 (2017). https://doi.org/10.1018/j.icarus.2014.09.016

M.K. Shepard et al., Asteroid 16 Psyche: shape, features, and global map. Planet. Sci. J. **2**, 125 (2021). https://doi.org/10.3847/PSJ/abfdba

L. Siltala, M. Granvik, Mass and density of asteroid (16) Psyche. Astrophys. J. Lett. **909**, L14 (2021). https://doi.org/10.3847/2041-8213/abe948

D.E. Smith et al., An improved gravity model for Mars: Goddard Mars Model-1 (GMM-1). J. Geophys. Res. **98**, 20871–20889 (1993)

L. Somenzi et al., etermination of asteroid masses from their close encounters with Mars. Planet. Space Sci. **58**, 858–863 (2010)

M. Viikinkoski et al., (16) Psyche: a mesosiderite-like asteroid. Astron. Astrophys. **619**, L3 (2018) (1993). https://doi.org/10.1051/0004-6361/201834091

B. Viateau et al., Mass and density of asteroids (16) Psyche and (121) Hermione. Astron. Astrophys. **354**, 725–731 (2000)

B.P. Weiss et al., The Psyche magnetometry investigation. Space Sci. Rev. (2022, this journal)

R.A. Werner, D.J. Scheeres, Exterior gravitation of a polyhedron derived and compared with spherical harmonic and mascon gravitation representations of asteroid 4769 Castalia. Celest. Mech. Dyn. Astron. **65**, 313–344 (1997)

M.A. Wieczorek, The crust of the Moon as seen by GRAIL. Science **339**, 671–675 (2013). https://doi.org/10.1126/science.1231530

M.A. Wieczorek, *Gravity and Topography of the Terrestrial Planets, Treatise on Geophysics*, 2nd edn. (Elsevier, Oxford, 2015), pp. 153–193. https://doi.org/10.1016/B978-0-444-53802-4.00169-X

M.T. Zuber et al., Gravity field of the Moon from the Gravity Recovery and Interior Laboratory (GRAIL) mission. Science **339**, 668–671 (2013). https://doi.org/10.1126/science.1231507

Space Science Reviews (2025) 221:96
https://doi.org/10.1007/s11214-025-01222-1

Design and Development of the Psyche Spacecraft for NASA's Discovery Program

David Y. Oh[1] · Dan M. Goebel[1] · G. Mark Brown[1] · Steve Collins[1] · Tracy Drain[1] · William Hart[1] · Travis Imken[1] · Kristina Larson[1] · Jennifer Maxwell[1] · Danielle Marsh[1] · Max Martin[1,2] · John Steven Snyder[1] · Denis Trofimov[1,3] · Tim Weise[1] · Ian Johnson[4] · Peter Lord[4] · Zack Pirkl[4]

Received: 26 April 2025 / Accepted: 21 September 2025 / Published online: 14 October 2025

Abstract
The NASA Psyche mission represents a milestone in the application of solar electric propulsion for deep-space exploration. Developed collaboratively by the Jet Propulsion Laboratory (JPL) and Maxar Space Systems, the Psyche spacecraft integrates a commercial high-power solar electric propulsion chassis with JPL's heritage in autonomous deep-space flight systems. The spacecraft uses no chemical propulsion – it employs a fully electric Hall thruster based propulsion system augmented by nitrogen cold-gas thrusters. The design adapts commercial hardware for thermal, power, and fault protection to the deep space environment. Following significant challenges in the verification and validation of its guidance and navigation software, the successful launch of Psyche in October 2023 validated its hybrid commercial-NASA development model for planetary missions. The Psyche experience offers key lessons for future cost-capped, SEP-enabled deep-space missions.

1 Spacecraft Description

1.1 Overview

The NASA Discovery Announcement of Opportunity in 2014 solicited proposals for space science missions under a cost-cap of US$500M for Phases A through D (NASA 2014). The

The NASA Psyche Mission: Science Instruments and Investigations
Edited by James F. Bell III, Carol Polanskey and Lindy Elkins-Tanton

In October 2025, the organization known as "Maxar Space Solutions" changed its name to "Lanteris Space Systems". We refer to this organization as "Maxar" throughout this paper.

✉ D.Y. Oh
david.y.oh@jpl.nasa.gov

1 Jet Propulsion Laboratory, California Institute of Technology, Pasadena, CA, 91109, USA

2 Present address: Aegis Aerospace, Webster, TX, USA

3 Present address: SpaceX, Hawthorne, CA, USA

4 Lanteris Space Systems, Palo Alto, CA, 94303, USA

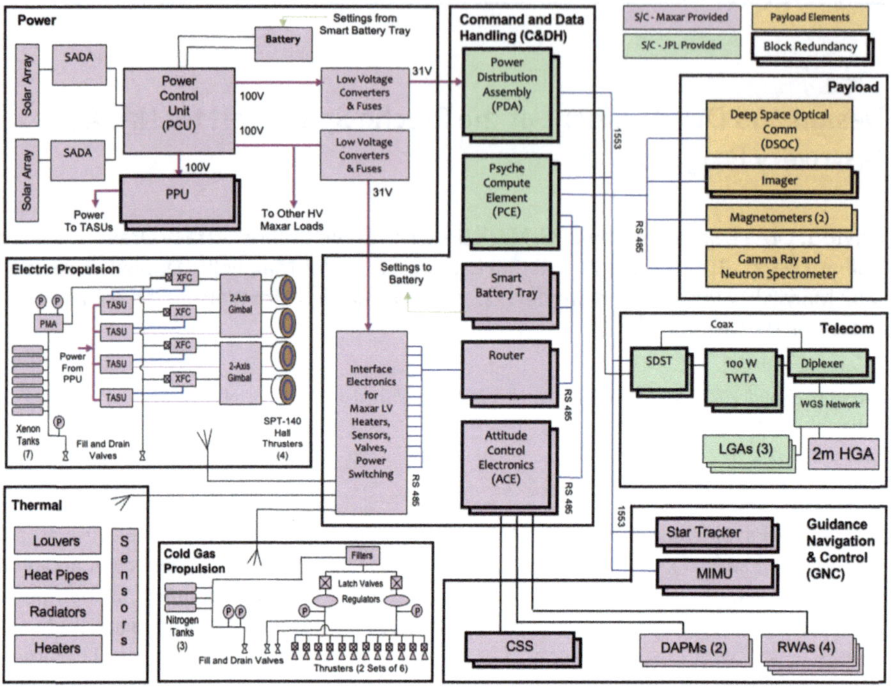

Fig. 1 Psyche spacecraft block diagram

earlier Dawn Discovery Mission had demonstrated the value of solar electric propulsion (SEP) for enabling planetary science missions that were too expensive and impractical or impossible to conduct using other propulsion methods (Garner and Rayman 2018). The Psyche mission was likewise formulated to use SEP to reach and orbit a Type-M (metal) asteroid in the asteroid belt while remaining under the Discovery cost cap. However, at the time of Psyche's development, Dawn's electric propulsion (EP) system, based on the NSTAR gridded-ion thruster, was found to be obsolete and not available for use. For all SEP missions following Dawn, other EP thrusters and power systems that form the heart of the spacecraft needed to be found. Early in mission concept development of the Psyche design, the team prioritized finding an EP system with proven heritage and high Technology Readiness Level (Oh et al. 2014b). It quickly became apparent that Psyche would benefit from an EP system provided by a commercial satellite manufacturer (Oh et al. 2017).

The proposal team chose to partner with Space Systems/Loral (now Maxar Space Systems), a company that has built over 40 Earth orbiting spacecraft that have accumulated over 250,000 hours of active EP operation in space. The team developed a spacecraft design that combines Maxar's experience in high-power electric propulsion systems with JPL's experience building highly autonomous spacecraft for deep space. The architecture is built around a solar electric propulsion (SEP) chassis, derived from Maxar's GEO product line, that is coupled with JPL's deep-space heritage Command & Data Handling (C&DH) systems and flight software to form the Psyche spacecraft bus (Oh et al. 2019). The SEP chassis incorporates EP, power, thermal, and structure subsystems with attitude control system (ACS) sensors from Maxar's 1300 modular spacecraft platform (see Fig. 1 and Fig. 2). JPL provides the spacecraft's main computer and flight software, X-band telecommunications systems (with a High Gain Antenna provided by Maxar), and a low voltage power distribution

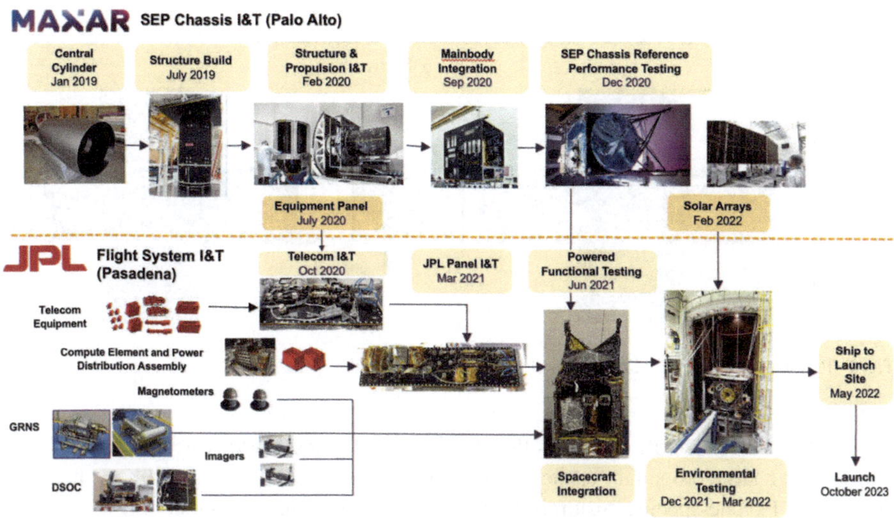

Fig. 2 The Psyche spacecraft is built around a solar electric propulsion chassis provided by Maxar

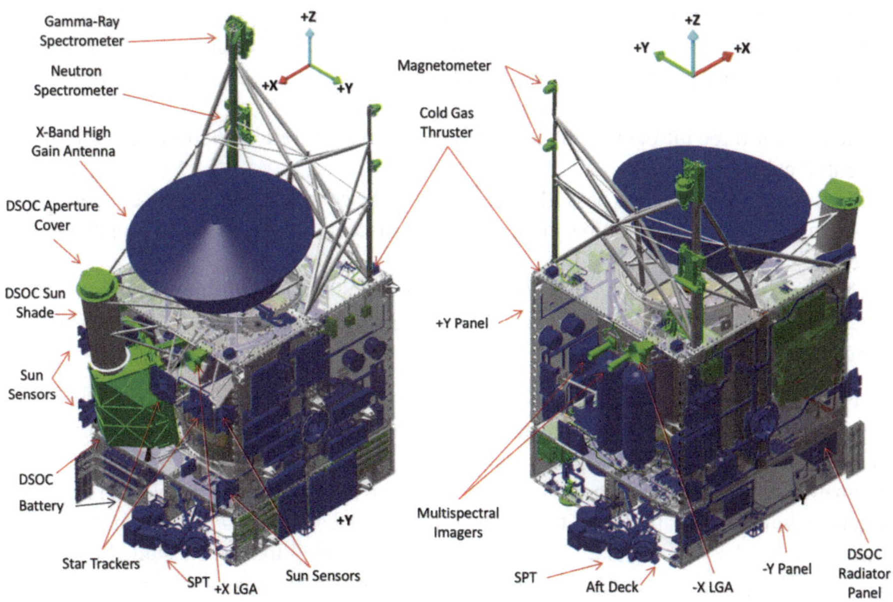

Fig. 3 Psyche spacecraft without the solar arrays or thermal blanketing, showing the various components installed on the bus

assembly that complete the spacecraft bus. The Command and Data Handling System is a joint effort, with JPL and Maxar both providing components for Psyche. The spacecraft has no chemical propulsion system, only electric and cold gas systems. The overall spacecraft configuration is shown in Fig. 3 and Fig. 4.

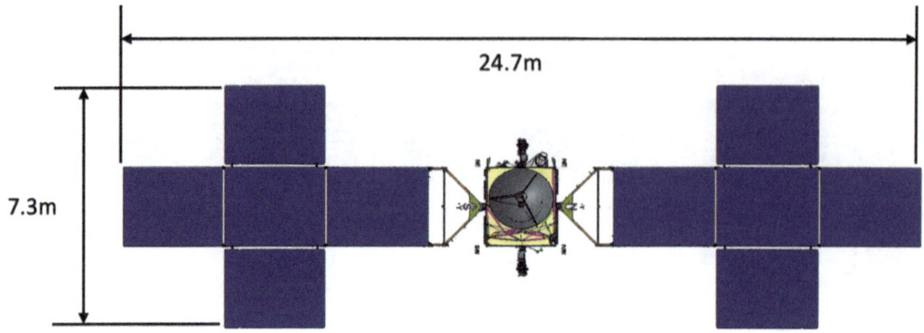

Fig. 4 Psyche spacecraft with large 5-panel solar arrays deployed

The Psyche spacecraft combines the smallest and lightest 1300 structure available that can support a large propellant-capacity xenon electric propulsion system with a medium size solar array. Psyche-specific customizations made to the structure, include:

- Two 2-meter-long booms and a custom support structure on the +Z face to support the magnetometer and gamma-ray spectrometer (GRNS) instruments and the High Gain Antenna (HGA)
- Stable mounting points on the +X side of the central support cylinder for the Deep Space Optical Communications (DSOC) payload and its associated sunshade
- Stable mounting points on the −X side of the central cylinder for the multi-spectral imagers
- Support for three nitrogen cold gas propellant tanks interior to the spacecraft and off the central cylinder
- Placement of seven xenon propellant tanks inside the central cylinder, replacing the bipropellant fuel and oxidizer tanks present in 1300 spacecraft with chemical propulsion systems.
- Custom mounting points on the +/−X faces for the Stationary Plasma Thrusters (SPTs)
- Customized locations and support bracketry around the spacecraft for Cold Gas Thrusters and Sun Sensors

Figure 5 shows the as-built High Gain Antenna and boom structure (under thermal blanketing) holding the GRNS and magnetometer instruments. Also visible are the aluminum louvers (on the left) covering the thermal radiators placed over the internally mounted high dissipation components. Figure 6 shows the Psyche spacecraft fully assembled on the vibration table with DSOC and its white sunshade that protects the detector from sun irradiation. Also seen are the stowed solar arrays and two SPT-140 Hall thrusters installed on the deployable SPT gimbal-modules. The HGA is seen on top of the spacecraft and a Low Gain Antenna (LGA) and two star-trackers are seen on the front.

Primary propulsion is provided by four Stationary Plasma Thruster (SPT)-140 Hall effect thrusters. These thrusters provide higher thrust levels and lifetime than the SPT-100 system originally flown on the Maxar 1300 bus and were first flown by Maxar in 2018 (Hart et al. 2018). The propulsion system is further discussed in Sect. 1.2. The power system is split between Maxar and JPL, with most of the hardware supplied by Maxar using elements of its EPS 2.0 power subsystem. More detail is provided in Sect. 1.4.

For guidance and control, the Psyche spacecraft is three-axis stabilized and uses star trackers and ring laser gyros for inertial attitude and rate measurements. Four reaction

Fig. 5 Psyche spacecraft undergoing solar array deployment testing

Fig. 6 Psyche spacecraft fully assembled on the vibration table

wheels, mounted in a pyramid configuration, are the primary means of attitude control. Sun sensors are distributed to minimize Sun search durations during safe mode. Nitrogen cold gas thrusters provide support for tip-off from the launch vehicle and safe mode scenarios when the wheels cannot be used. The cold gas thrusters are described in Sect. 1.3 and methodology of attitude control for Psyche is described in Sect. 1.5.

Psyche's fully redundant C&DH system includes components built by both JPL and Maxar. The Rad750 version 3 processor in the Psyche Compute Element is the computer

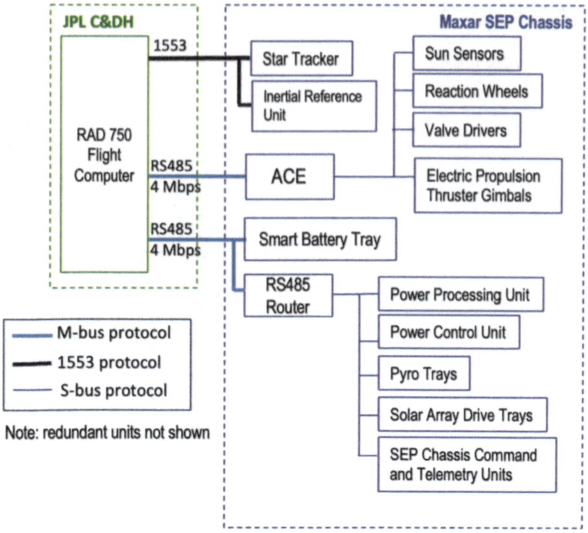

Fig. 7 Maxar command and data handling units control all SEP chassis hardware except for the star tracker and Inertial Reference Unit, which are controlled directly from the JPL provided flight computer

that hosts the flight software that controls the C&DH system. The compute element collects and stores engineering and science data in non-volatile memory, and formats the data for downlink to Earth. It incorporates many heritage interfaces from the compute element in the Curiosity Rover, including high speed serial interfaces for the science instruments and a 1553 data bus. A clean interface is provided between JPL and Maxar provided avionics hardware. As shown in Fig. 7, the compute element controls the SEP chassis via two RS-485 serial data lines that use Maxar's proprietary M-Bus protocol to connect to three Maxar provided units: the Attitude Control Electronics (ACE) (which in turn provides command and telemetry interfaces for ACS sensors and actuators), the router, and the Smart Battery Tray. The 1553 bus provides the interface to the inertial measurement units and star trackers. Maxar's proprietary S-Bus protocol is used internal to the SEP Chassis to interface to the thermal, propulsion, pyro and power components.

JPL's Flight Software Core Product Line forms the basis for Psyche's flight software. This architecture allows software to be organized into modules which are allocated to time partitions and run in the Green Hills Software INTEGRITY Real Time Operating System environment. Core software components based on Curiosity Rover/SMAP spacecraft heritage include basic infrastructure (e.g. initialization, timekeeping, inter and intra component messaging, etc.), input/output (IO) interfaces (e.g. 1553 bus management, serial IO, etc.), data storage services and management of uplink and downlink functions. Psyche's mission-specific software modules include power management, fault protection and attitude control functionality.

The thermal system architecture is similar to Maxar's geostationary orbit (GEO) communication satellites and uses heaters to maintain minimum allowable flight temperatures and passive radiators with heat pipes to dissipate excess heat – with sizes and placement adjusted to accommodate solar ranges from 1 AU to 3.33 AU. The addition of louvers to the main radiator panels allows the system to control temperatures across a wide range of solar distances extending all the way out to the main asteroid belt. Psyche's louvers incorporate mechanically controlled vanes that automatically open to expose the radiators to space when warm and close to insulate them when cold, thus preserving heat when the spacecraft is far away from the Sun. The location of louvers and optical surfaces on the exterior of the

Fig. 8 SPT-140 thruster operating in vacuum chamber at JPL (Image credit: NASA/JPL-Caltech)

radiator panels maintains allowable flight limit temperatures for the units installed on the interior of the spacecraft (see Fig. 5). A thermally isolated cold zone, separated from the rest of the heat pipe matrix, keeps DSOC's temperature sensitive hardware below its maximum operating temperature.

Figure 2 shows the overall spacecraft integration and test (I&T) flow. Maxar produced the SEP chassis on their spacecraft production line in Palo Alto. A test computer and test software taken from Maxar's standard 1300 product line were used to conduct end-to-end functional testing on the SEP chassis before it was shipped to JPL. In parallel with SEP chassis integration, Maxar delivered the spacecraft's +Y equipment support panel to Pasadena, where JPL integrated the main computer, flight software, and telecom subsystems onto the equipment panel, and then tested these elements in a tabletop environment. When the SEP chassis was delivered to Pasadena, the +Y panel was integrated onto the spacecraft, then the spacecraft proceeded through system-level functional and environmental test and delivery for launch. JPL was responsible for final spacecraft assembly, test, and launch operations, all of which occurred at JPL's facilities in Pasadena. The I&T process and flow are further described in Sect. 2.

1.2 Electric Propulsion

1.2.1 Design Description

The electric propulsion system for Psyche is based on Maxar's GEO heritage SPT-140 subsystem (Jameson-Silva et al. 2017), with focused modifications made to enable deep-space operation. The system includes four SPT-140 thrusters (Fig. 8), each paired with its own Xenon Flow Controller (XFC) (Fig. 9). Two Power Processing Units (PPUs) are cross-strapped to the four thrusters through Thruster Auxiliary Support Units (TASUs). The TASUs incorporate cross-strapping relays, the discharge filter capacitor, electrostatic discharge protection, and ground-test features. Either of the PPUs can power any of the four thrusters. Two thruster/XFC pairs are mounted on each of two Deployable SPT Modules (DSMs). The DSMs are mounted on opposite sides of the spacecraft and include independent dual axis positioning mechanisms (DAPMs) for thruster articulation in two axes.

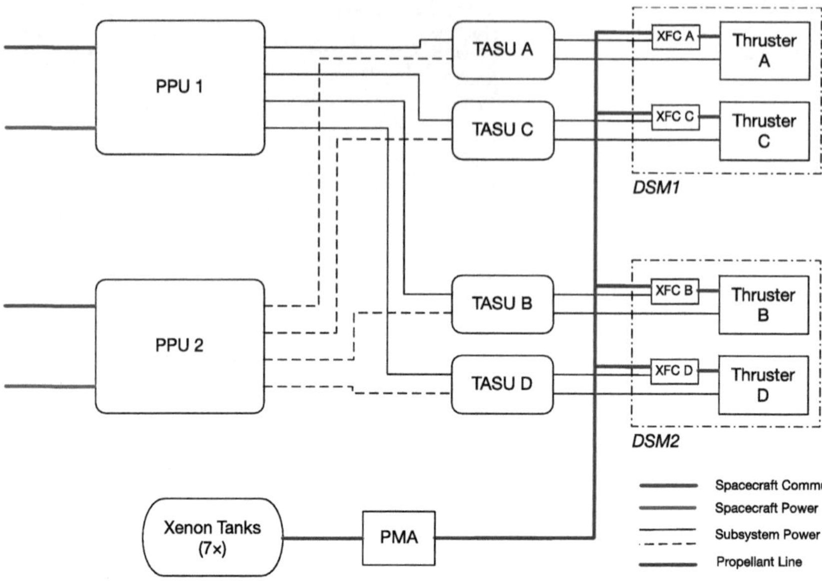

Fig. 9 Electric propulsion subsystem block diagram

Propellant storage is provided by seven composite-overwrapped pressure vessels that are ganged together through a common plenum, each holding up to 155 kg of xenon for a total storage capacity of 1085 kg. The Propellant Management Assembly (PMA) contains parallel-redundant pyrotechnic valves, latch valves, and mechanical regulators for propellant isolation and distribution to the XFCs, as well as four pressure transducers to monitor tank storage, and downstream pressure regulators. The entire system is single-fault tolerant, apart from the tanks, propellant lines, and mechanical portions of the DSM actuators. Further details of the propulsion system architecture, including the xenon propellant budget, are provided in Snyder et al. (2019).

The most significant changes to the heritage Maxar system are in the XFC and PPU designs to enable the wider propellant flow range for the thruster throttling required for Psyche's outbound trajectory. The mission requires power throttling down to 0.9 kW discharge power, significantly lower than the 3.0 kW low-power limit used for GEO station keeping. The heritage XFC uses a thermothrottle for flow control, which does not have the required flow throttling range, so Psyche uses a new XFC based on proportional flow control valves (PFCV) (Lenguito et al. 2019). The Psyche XFC provides the same functionality as the heritage unit but with a wider range of flows.

Modifications to the PPU and XFC designs are also made to accommodate the unique low-power operating conditions at Psyche (Malone et al. 2019). Thruster testing shows that the cathode-to-ground voltages at lower powers are significantly more negative than observed in thruster operation at 3.0 to 4.5 kW discharge powers. Those cathode-to-ground voltages are mitigated by flowing additional xenon to the cathode and applying a modest current of 1.15 A to the cathode keeper at low power (Snyder et al. 2018). Hence, the XFC includes an additional flow path to provide higher flow rates to the cathode at low power. The XFC includes a solenoid valve for flow isolation and a proportional flow control valve for flow control upstream of the cathode-leg and anode-leg orifices. The cathode flow fraction can be changed from the nominal 5% used for high-power conditions to 9% for low-power conditions by actuating a latch valve. Additional details regarding the XFC design and testing are found in Lenguito et al. (2019). The PPU cathode ignitor supply is modified to provide a steady-state keeper current at discharge powers less than 1.5 kW (Snyder et al. 2019).

The PPU design changes for Psyche were incorporated into an existing engineering-model unit that was sent to JPL for system integrated testing with the development-model XFC and the qualification-model thruster (Malone et al. 2019). The primary goal of the integrated testing was to validate the PPU design changes in a system environment, e.g. the PFCV control circuitry, new low-power operating mode, and the associated new logic and control functions. Testing included startups, steady-state operation, and throttling between power levels across the entire required operating power range for Psyche from 0.9 to 4.5 kW thruster discharge power. It also included forced off-nominal and stressing operating conditions to test system responses; transient system responses; and system fault protection functions. These tests were successfully completed in 2019 (Malone et al. 2019).

The SPT-140 thrust and propellant flow rate performance models used for trajectory analysis, called the throttle curves, evolved over the course of subsystem development program done for Psyche. Initial models were based on development-model-thruster test data and a simple model of the effects of facility pressure on thruster performance. The throttle curves used for trajectory analyses today improved significantly with the use of flight thruster acceptance test data, improved knowledge of the effects of background pressure as a function of power, and the comparisons to Maxar SPT-140 flight data collected in Earth orbit. Pressure-effects test results provided information on the magnitude and behavior of the effects of pressure on thruster performance at different power levels (Snyder et al. 2018). The first Maxar flight results provided quantitative information on in-space performance at the single throttle condition used for that mission (Johnson et al. 2020). Together those results, along with the Psyche thruster acceptance test data, were used to create beginning-of-life throttle curves for the mission (Snyder et al. 2019). Thruster performance changes with operating time (i.e. aging effects) were also incorporated to predict the performance at any given point in the mission.

The Psyche mission requires the largest total xenon throughput of any space mission flown to date. At the project's preliminary design review, the total required xenon load was projected to be 1030 kg, of which 1007 kg would be (or could be, in the case of propellant margin) processed by three of the SPT-140s to allow for single-fault tolerance. The main life test of the SPT-140 was designed for commercial missions at Earth (Delgado et al. 2014) and it was later extended to demonstrate low-power operation and additional throughput for Psyche (Jameson-Silva et al. 2017). At completion the full life test demonstrated 10,371 hours of thruster operation with a total propellant throughput of 500 kg and a total impulse of 8.79 MN-sec (Jameson-Silva et al. 2017), the latter of which is 160% of the mission-required total impulse per thruster (Snyder et al. 2019). Although this validates the total impulse required for the Psyche mission, there are three notable differences between the life test operating conditions and those for the Psyche mission: 1) the background pressure, 2) thruster magnetic field strength, and 3) the throttle profile. To bridge this gap, model simulations were performed to assess thruster erosion for Psyche operating conditions and environments. The simulation results showed that thruster usage requirements, including an additional 50% margin, were well within the wear capability of the thruster. Discussion of the models, their validation, and the Psyche wear simulations can be found in Lopez Ortega et al. (2019).

1.2.2 Operational Differences Between Psyche's Hall Thruster and Dawn's Gridded Ion Thruster Systems

While thousands of Hall thrusters are flying on Earth orbiting spacecraft in a variety of different applications varying from orbit maintenance to primary propulsion, Psyche is the first mission to use Hall thrusters beyond the Earth-Moon system. Unlike Earth or Lunar orbiting spacecraft, which constantly vary their thrust direction in inertial space, Psyche operates Hall thrusters for days to weeks at a time with the thrust vector in a nearly-fixed orientation in inertial space during interplanetary cruise. This causes the spacecraft to accumulate transverse "swirl torque" momentum in a single axis at a relatively rapid rate. In addition, the need to rendezvous with planetary bodies within scheduled windows of time (for example, to achieve the Mars gravity assist) places a premium on the controllability and predictability of thruster performance over long periods of time which is not present in most Earth orbit applications.

NASA's Dawn Mission utilized gridded ion thrusters which were throttled about a factor of five in input power. This throttle range is similar to Psyche. Although both gridded ion thrusters and Hall effect thrusters ionize xenon and accelerate ions to provide thrust, a key difference is that an ion thruster uses an electrostatic field to accelerate the ions, while a Hall thruster uses a crossed electric and magnetic field for the same purpose. In operational terms, the differences between the two designs can be grouped into two main categories: control and performance.

In terms of control, the Dawn ion system used seven setpoints to control the thruster: four electrical (currents and voltages) and three flow rates. These were stored in a table which assigned a specific combination of setpoints to 112 discrete throttle levels. These setpoints provided the operations team with flexibility if a change in performance was observed or was necessary to meet mission objectives, but with this flexibility came the need to configuration manage and track those values over the life of the mission. Because the system controlled the xenon flow rate to a commanded setpoint, flow control was implemented via a calibrated metering system that provided the flight operations team with precise knowledge of the xenon mass flow rate.

In contrast, Psyche's Hall thrusters have 235 discrete throttle levels and no operator commanded flow rate set points. This is because the Hall thruster power processing unit incorporates autonomous feedback flow control logic that adjusts the xenon flow rate, which changes the discharge current, to achieve the desired discharge current setpoint. Although the hardware is capable of operating at any setpoint, mission design requirements drive the discretization into throttle levels. The requirements associated with deep space navigation prioritize thrust predictability and repeatability over thrust magnitude, so from an overall mission perspective it is not necessary or desirable to use a continuously-throttling EP system. Psyche chose to implement 235 thrust levels in software based on a balance between operations complexity and power-level increment. Consequently, on Psyche, determining the precise mass flow rate during operations in space requires analysis and interpolation/extrapolation from ground-test data. Finally, Dawn's ion thrusters had an operating range from about 0.51 kW to 2.48 kW input power, while the SPT-140s support a range from about 1.0 kW to 4.9 kW input power. When these ranges are broken down into 112 or 235, discrete throttle levels respectively, the difference between levels are similar – $\sim$20 W for Dawn and $\sim$17 W for Psyche.

Both systems control thrust using commanded setpoints that specify parameters to operate the thruster for a given input power, rather than the thrust itself. The resulting thrust and mass flow rate at a given setpoint were predictable with high accuracy for Dawn as compared to Psyche. For Dawn, the mass flow rate knowledge was used to book keep propellant consumed and compared with a tank model using tank pressure and temperature to determine propellant remained as a cross check. For Psyche, the tank model is the primary reference for propellant consumption. In the absence of telemetered mass flow rate, a bookkeeping method derived from the thruster life test is also used. This results in the Psyche mission design carrying larger uncertainties on mass flow and thus a larger xenon allocation.

Another key difference is that for the ion thruster system on Dawn, the thrust was well known and didn't change significantly over time – for example the full power thrust uncertainty was +/− 0.25 mN (0.27%). For the Hall thrusters on Psyche, the thruster life test showed a thrust decrease of ∼10% over the first ∼2000 hours of operation at 4.5 kW, after which it stabilized. The Psyche operations team relies on timely in-space observational data to plan near term maneuvers, and carries larger uncertainties on the models for longer term planning.

Finally, both types of thrusters produce a swirl torque as the xenon ions exit the thruster. These torques are small and extremely difficult to measure on the ground. Dawn observed swirl torques on all three thrusters, with the largest being ∼55 μN-m. Psyche's SPT-140 thrusters have an analytically estimated swirl torque of 740 μN-m at full power from a hybrid Hall2De code simulation. The values measured in flight on two thrusters at the full-power 4.5 kW operating point were 0.83 and 0.86 mN-m, which is about 15% higher than the value calculated using an analytic model and by the hybrid Hall2De code (Snyder et al. 2025). Dawn used hydrazine reaction control system thrusters to unload the swirl torque momentum buildup, while Psyche uses the electric propulsion system itself for this task. The Psyche implementation is to pause mission thrusting, reorient the spacecraft and fire a different electric propulsion thruster positioned to provide a counter torque to the accumulated momentum. The total execution time for these momentum unloads is about 40 minutes (the actual burn time from the thruster is lower). The frequency of the unloads is a factor of the reaction wheel assembly (RWA) momentum storage capability and the actual swirl torque, and just after launch, unloads were required as often as every 7 hours at full power. The full operational implications and the adaptations made by the operations team to accommodate these maneuvers will be described in future publications describing the launch and initial checkout of the Psyche spacecraft (Oh et al. 2025). The frequency of the unload maneuvers drops substantially as the spacecraft moves away from the Sun and the EP system is operated at lower thrust and power levels.

1.2.3 Plume Erosion and Contamination Control

The erosion of material from surfaces exposed to ionized xenon plumes and the possibility for redeposition of this material onto sensitive areas of the spacecraft have long been recognized as issues of concern for spacecraft using Hall thrusters. The Psyche team used plume models to analyze the amount of erosion and deposition which would occur on the spacecraft. The amount of surface erosion caused by plume impingement is dependent on several factors including thruster voltage, firing time, material erosion yield, angle from thruster centerline and distance from thruster centerline. On Psyche, the hardware located closest to the thruster is the XFC, which is mounted on the DSM just behind the exit plane of the thrusters. The XFC is protected by a cover which is in turn covered by a thermal blanketing material (primarily fused quartz with internal aluminized polyimide layers). Plume simulations were used to assess two primary erosion cases for the flight thruster-XFC geometry and are summarized in Table 1.

Once the erosion levels for each case were calculated, transport of the eroded products, primarily sputtered quartz and polyimide material, was simulated using CERN's MolFlow+

Table 1 Psyche Plume Erosion Case Descriptions

Case	Case-unique Assumptions (Note 1)	Predicted Erosion Depth (mm)	Eroded Material Volume (cm^3)	Erosion Mechanical Impact
#1 Updated XFC Cover Design – Worst Case	4.5 kW Power 9400 hrs firing time	0.44 Maximum	19.4	Penetration of top 3 blanket layers
#2 Updated XFC Cover Design – Flight Prediction Case	3.5 kW Power 6800 hrs firing time	0.26 Maximum 0.10 Average (Note 2)	5.6	Penetration of top 2 blanket layers

Note 1: All cases use the same material erosion yield, angle from thruster centerline, and minimum distance from thruster centerline.

Note 2: The JPL contamination control erosion and transport model was updated to allow for variable erosion rates across the XFC cover so that an average erosion value could be determined for case 2.

Table 2 Summary of Contamination Deposition Results

Sensitive Surface	Case 1 Deposition (angstroms)	Case 2 Deposition X-Y SA Orientation (angstroms)	Case 3 Deposition Y-Z SA Orientation (angstroms)
Solar Arrays	<200	30	40
DSOC	Negligible	Negligible	Negligible
Psyche Multispectral Imagers	<10	Negligible	Negligible
Star Trackers	<20	Negligible	Negligible

raytracing software. Eroded products were released from the XFC cover surface according to a cosine distribution and the eroded particles were tracked to determine what fraction of particles would impact various spacecraft surfaces. Multiple Psyche thruster orientations were evaluated in this investigation, but results were shown to be insensitive to the small thruster rotations associated with the changing spacecraft center of gravity throughout the mission. The maximum levels of contamination deposition are tabulated in Table 2. Case 3 was assessed with solar arrays (SAs) in both 90-degree and 180-degree orientations, but both orientations showed acceptable results (negligible effect on array performance) without a significant difference between them. Based on the results from Case 3, the Maxar solar array team determined that the predicted level of erosion and deposition of contaminants was acceptable for flight. A graphical representation of the Case 2 X-Y deposition results is shown in Fig. 10.

1.3 Cold Gas Propulsion

While Hall thrusters use propellant very efficiently to produce thrust, an additional small propulsive attitude control system was determined to be needed to accomplish short duration, power-critical, and time-critical maneuvers. A comparison of the propulsive maneuvers conducted by the EP vs. cold gas thruster subsystems is shown in Table 3, and CGT requirements are summarized in Table 4

Psyche uses a regulated nitrogen CGT system based on previously qualified hardware used on Maxar spacecraft. Figure 11 shows a block diagram of the nitrogen cold gas system.

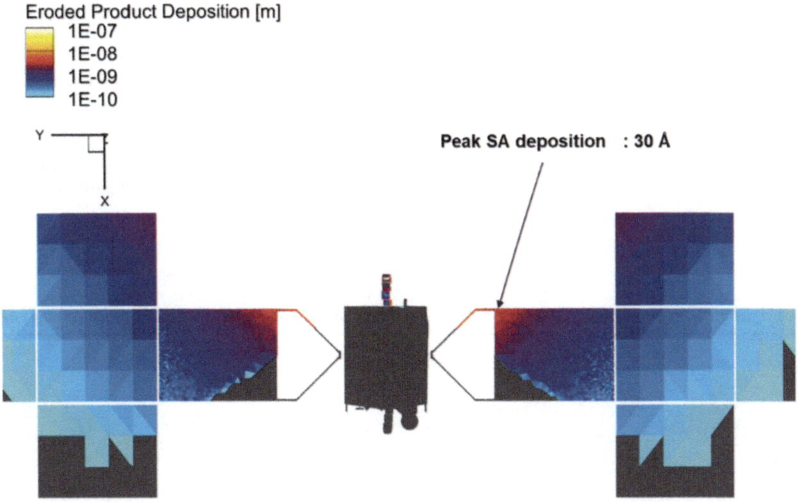

Fig. 10 Case 2 X-Y solar array orientation results show a peak contaminant deposition of 30 angstroms, which was deemed acceptable for flight

Table 3 Usage of Electric Propulsion and Cold Gas Propulsion Systems for Psyche

	Electric Propulsion	Cold Gas
Launch		
Post-Launch Detumble and Sun Acquisition		X
Cruise		
Primary Propulsion	X	
Momentum Management	X	
Asteroid Orbital Operations		
Orbit Transfers	X	
Orbit Maintenance	X	
Momentum Management		X
Off-Nominal Scenarios		
Safe Mode Turns		X
Safe Mode Wheel Desaturation		X

There are two independent, fully redundant, cold-gas thruster strings. Three 82-liter nitrogen tanks are isolated from the rest of the system during launch via latch valves, while series redundant regulators limit thruster inlet pressures to about 360 psi. Pressure transducers, a filter, and fill/drain valves complete the system. As shown in Fig. 12, thrusters 5/6 and 11/12 operate in pairs to provide rotation about the X axis. The remaining eight thrusters provide rotation about the Y and Z axis. Each single-seat thruster is rated at 0.45 N thrust and 70 seconds specific impulse at 360 psi inlet pressure and operates at a maximum 30% duty cycle to avoid overheating. All six thrusters firing on a single string result in a 4.5 g/sec flow rate, which is accommodated by the system. The thrust vs. tank pressure curve for the

Table 4 Overview of Psyche cold gas subsystem requirements

Requirement		Rationale
Fault tolerance	Single	Allows for a single hardware failure without impacting mission
Tank capacity	>45 kg	Allows for multiple 3-sigma contingency scenarios if fully loaded. Final decision on loaded mass is not required until late in the build.
Torque across any axis	>0.3 N-m	Supports safe mode recovery timeline
Thrust	>0.25 N	
Minimum impulse bit	<8 mN-s	Allows for dead-banding and precise momentum management

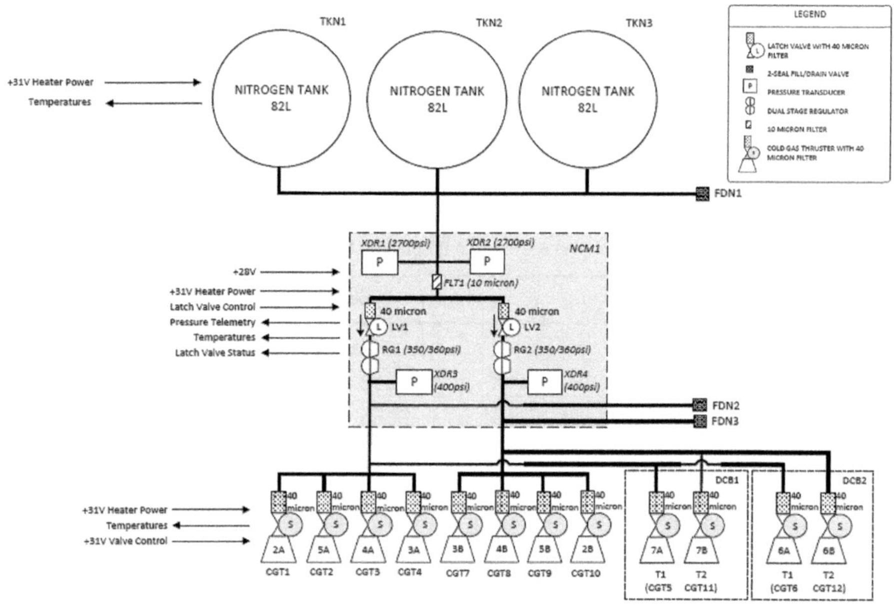

Fig. 11 Block diagram of the cold gas propulsion system for Psyche

Moog 58E151 thruster is shown in Fig. 13. Over 40 kg of the nitrogen propellant can be used at the rated performance of the system. The cold gas propellant budget is shown in Table 5.

1.4 Power

A major achievement for Psyche's power system is the successful adaptation of hardware originally designed for use in Earth orbit to a deep space mission that travels to solar distances as high as 3.33 AU. Conducting a deep space mission poses unique challenges for a system originally designed to operate in the traditional operational setting of GEO spacecraft. In particular, the solar array's electrical behavior changes substantially since the current and voltage characteristics of photovoltaic cells vary as a function of temperature and solar intensity (Fig. 14). The power management electronics must accommodate the full

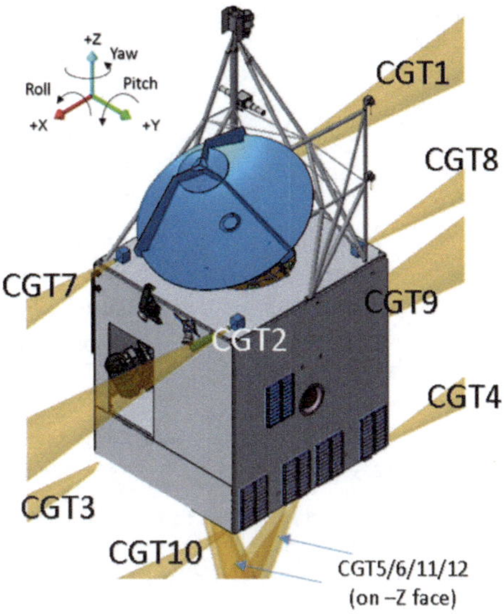

Fig. 12 Placement of cold gas thrusters on the Psyche spacecraft

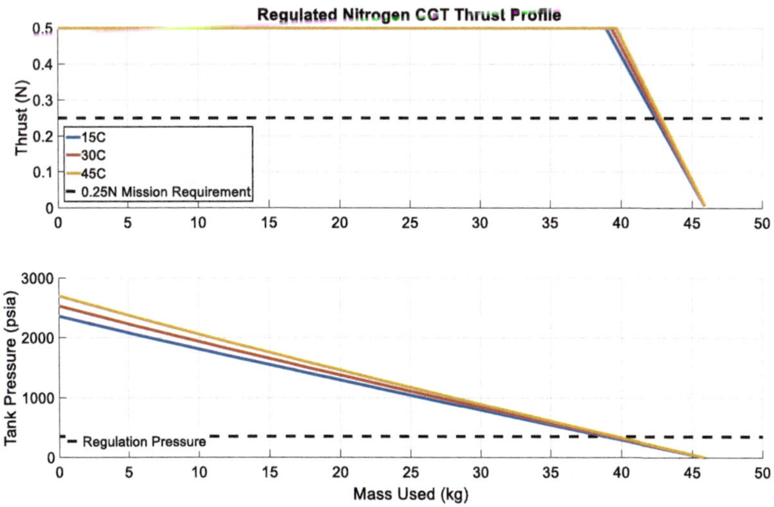

Fig. 13 Cold gas thruster subsystem thrust and pressure curves vs. mass usage

range of solar array operating conditions as the spacecraft travels away from the sun over the course of the mission.

1.4.1 Power Subsystem Heritage and Architecture

Maxar has over 14 years of flight heritage with space power systems using Lithium-ion batteries. Its first-generation lithium-ion space power system, developed in the early 2000's, has successfully flown on over 60 missions in Earth orbit. Its second-generation system, EPS 2.0, launched into space for the first time in 2019. The Psyche mission uses a version of EPS 2.0 that has been modified to account for its deep space mission profile. Maxar's GEO

Table 5 Nitrogen cold gas propellant budget

	Mass (kg)
Maximum Potential Load Mass (2700 psi, 45 C)	**45.9**
Launch Detumble	0.9
Initial Checkout/Commissioning	0.2
Momentum Management	2.4
Allocated Leak	2.1
Tank Residuals	6.9
Minimum Required Load for Nominal Mission	**12.5**
Available Nitrogen Propellant for Contingencies	**33.4**

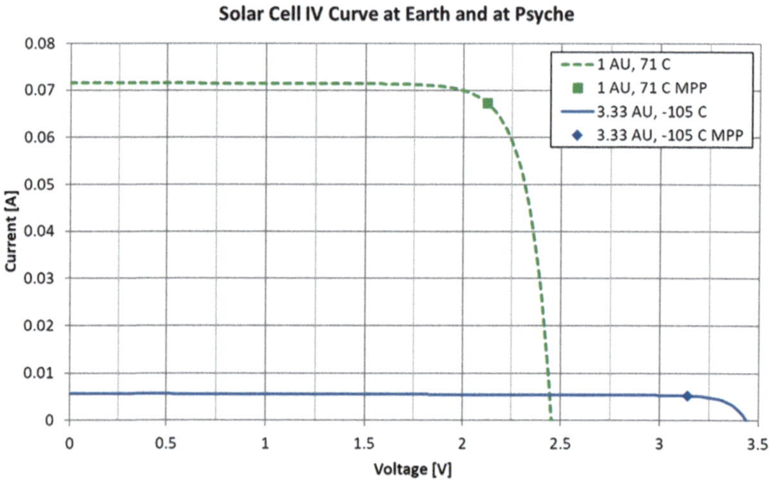

Fig. 14 Solar cell I-V curves differ substantially as solar distance varies from 1 AU to 3.33 AU. MPP indicates the Maximum Power Producting point on the I-V curve

spacecraft use a shunt regulated system to provide power from the array to the spacecraft bus at a nominal regulated voltage of 100 V. Shunt regulated systems work efficiently when the array's peak-power operating voltage is close to the target voltage for the power bus, which is easily obtainable in Earth orbit. In the case of Psyche, the current-voltage (I-V) curve for its solar cells varies substantially with distance from the Sun (Fig. 14). Performance modeling of the array as a whole indicated that the resulting peak-power voltage could vary from 60 to 105 V. The PPU for the SPT-140 is qualified to receive power from a 100 V regulated bus and is therefore incompatible with this wide voltage range.

The solution to this incompatibility was found in the power system technology already used in Maxar's Earth orbiting spacecraft. When spacecraft in GEO enter eclipse, discharge converters in the Power Control Unit (PCU) boost the voltage from batteries (which can vary from 40 V-to-100 V) to create the regulated 100 V power bus. Because the solar array voltage range to be accommodated for Psyche is enveloped by this GEO application, the battery discharge converters were repurposed without modification to directly boost solar array voltage to create a 100 V regulated power bus, regardless of distance from the sun. Any bus voltage transients are damped by a capacitor bank on the power bus that also provides surge current capability during transient start-ups of loads like the PPU.

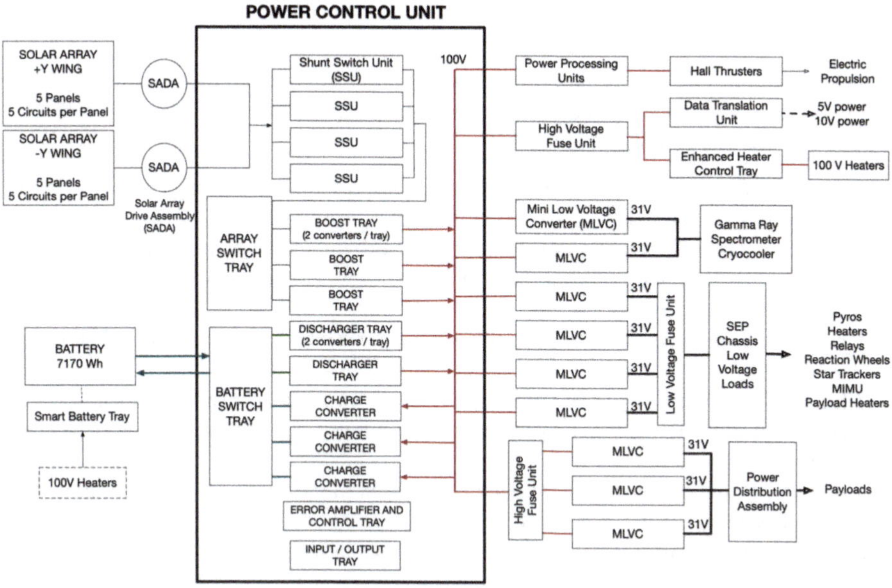

Fig. 15 Psyche spacecraft power system block diagram

An enabling feature of the power system design is the selection of a solar array voltage which is less than 100 V at the start of the mission, when the spacecraft is closest to the sun. At beginning of life, when solar fluence and array output power are highest, the array voltage is at its lowest due to the high cell temperature and discharge converters boost the array output voltage to the bus voltage of 100 V. The resulting power conversion losses are easily tolerated by the spacecraft, since kilowatts of excess power are available at beginning of mission. As the spacecraft moves away from the sun, the optimal solar array voltage rises as the arrays cool. The solar array is designed so that the array's peak power voltage, including solar array degradation and low intensity illumination effects, reaches 100 V on approach to Psyche. At this point, because the array voltage matches the bus voltage, the power conversion boost stage is no longer needed and is replaced with a diode pass-through circuit from the solar array to the PPU. This eliminates boost stage conversion losses, thus conserving power when array power is at its minimum and all available power is required for spacecraft operation. By design, the conversion efficiency effectively increases from 97% to 99% as the spacecraft moves away from the Sun, maximizing power available for science operations. If the solar array voltages exceeds 103 V, any excess power is shunted back to the array by connecting unused solar array strings to ground using the shunt electronics within the Power Control Unit (PCU). This effectively leaves excess power on the array, providing efficient bus voltage regulation and overvoltage protection.

This SEP power train design trades higher losses when solar power is abundantly available for high efficiency when solar power availability is low. It also uses flight proven hardware that is economical vs overly sophisticated, dedicated deep space-specific peak power tracking approaches that strive to minimize power train losses over the entire mission.

1.4.2 Power Electronics

Figure 15 provides a detailed view of the Psyche power subsystem. The PCU is powered by the solar array and batteries and performs the core function of creating a regulated 100 V spacecraft bus. It is comprised of the following cards:

- **Sequential Shunt Unit (SSU) (x4):** Shunts excess solar array power to ground when the solar array is operating above the regulation voltage setpoint
- **Boost Converter Tray (x3):** Boosts solar array voltage to 100 V when the array is operating below 100 V, passes power through to the bus when the array is operating at 100 V or higher
- **Discharge Converter Tray (x2):** Extracts power from the battery as required and boosts it to 100 V
- **Charge Converter Tray (x3):** Charges the battery on flight software command
- **Error Amplifier Tray:** Provides control logic and error signals for the power bus regulation control loops
- **I/O Tray:** Provides key input and output interfaces to the Psyche Compute Element and other subsystems via the router (see Fig. 7)
- **Capacitor Tray:** Adds capacitance to the intermediate solar array power bus to ensure regulation stability
- **Battery Switch Tray:** Provides an interface between the battery and converters and includes relays to isolate failed converters, if necessary
- **Array Switch Tray:** Provides an interface between the solar array and converters and includes relays to isolate a failed converter, if necessary

1.4.3 Low Voltage Power Topology

While the PPUs operate directly off fused outputs from the 100 V power bus, other loads on the spacecraft, including the science instruments and telecom systems, operate at lower input voltages. The spacecraft has three separately isolated 31 V buses: The Maxar Bus, JPL Bus, and GRNS Cryocooler Bus (see Fig. 15). These busses are powered by Mini Low Voltage Converters (MLVCs) that down convert power to a regulated 32 V +/− 0.6 V. Each MLVC can provide up to 350 W of power. The MLVCs are arranged in parallel in an N + 1 redundancy scheme, where the system can lose an entire unit and still meet its performance requirements.

Downstream of the MLVCs, the spacecraft's low voltage loads are provided switched power by either Maxar power distribution hardware or by the JPL-provided Power Distribution Assembly (PDA). The loads are generally allocated based on the delivering organization, so SEP chassis equipment is primarily powered by the Maxar power distribution hardware while JPL provided equipment is powered by JPL-provided Power Distribution Assembly (PDA), though some Maxar loads are capable of being switched on/off through the JPL PDA during contingency scenarios. The internally redundant PDA is comprised of three Power Switch Slices (PSSs), two Remote Engineering Units (REUs), and two Housekeeping Power Converter Units (HPCUs). The PSS is an internally redundant power distribution slice that provides 32 channels of high and low side switched power on the low voltage bus. The REU is a block redundant power management card with extensive JPL heritage, including the Curiosity and SMAP missions, that contains autonomous control logic that can execute fault management functions independent of the main computer. The HPCU is a block redundant card that provides secondary voltages to the REU as well as some auxiliary PDA functions, such as communication bus cross-strapping. The key advantage of the

PDA is that it provides communication interfaces and fault protection functions that have shared heritage with the compute elements on previous JPL missions.

1.4.4 Power Generation

Power is generated by a solar array composed of two wings, each having five separate panels with 65 solar cell strings per panel. The strings are predominately 35 cells in length, which was designed to achieve a nominal peak power voltage of 100 V at 3.33 AU solar distance. There are also a few 34-cell strings present due to layout constraints. The 65 strings are wired into five circuits per panel (25 per wing) and sent through Solar Array Drive Assembly (SADAs) to the SSUs. The solar array uses SolAero ZTJ triple junction cells, which have a BOL AM0/28 °C efficiency of 29.5%. Half-wafer cells of 27.55 cm^2 area are used, resulting in a total array active area of 62.6 m^2. At 1 AU and 71 °C, the array is capable of producing over 21 kW, although only 9.5 kW is available to the spacecraft due to the current limits in the boost converters. At Psyche conditions of 3.33 AU and −105 °C, the array is designed to produce 2.3 kW. One of the Psyche mission's key science investigations involves detection and measurement of the Psyche asteroid's magnetic field; in support of this investigation, the flow of currents between the panels and the spacecraft was carefully designed and modeled to show that the solar arrays would provide a sufficiently clean magnetic environment at the location of the spacecraft's boom-mounted magnetometer.

ZTJ cells have extensive heritage in GEO and low Earth orbit, and are currently in use for the NASA LUCY, MAVEN and Insight missions. They have undergone extensive test campaigns for use at Jupiter/Europa (5.5 AU) and Saturn (9.1 AU), and the results have demonstrated strong performance from the cells even in high radiation, low intensity/low temperature (LILT) conditions (Boca et al. 2017 and Boca 2019). While LILT conditions are not as severe at 3.33 AU, some deleterious effects are observed. A critical step in mitigating these effects for Psyche was screening the flight cells for a threshold fill factor limit in low intensity/room temperature (LIRT) conditions. While screening the flight lot in LILT test conditions would be ideal, LILT testing in high volumes is cost prohibitive. LIRT screening has shown to be a sufficient for mitigating poor LILT performance (Dawson et al. 2012).

1.4.5 Energy Storage

Energy storage on Psyche is implemented using a single 7170 Watt-hour Li-Ion battery, which provides power during eclipses or when the flight system power requirements exceed the solar array capability. The battery is a Maxar GEO heritage unit composed of a single string of 13 Li-ion cells, with an EOL rating of 144 Ah after 15 years in a GEO environment. To charge the battery during sunlight operations, the charge converters in the PCU down-convert 100 V bus power to provide charging in a constant current charge mode. Charge control is provided by battery management software in the compute element, with the charge algorithm specifically tailored for Psyche. The battery design includes a battery management unit, referred to as the Smart Battery Tray that provides telemetry for every cell in the entire battery, controls battery temperature, and supports cell balancing to preserve battery health.

1.5 Telecommunications

The Psyche spacecraft uses an X-band Small Deep Space Transponder (SDST) to communicate with the Deep Space Network (DSN). The X-band system block diagram is shown in Fig. 16. The telecom system provides uplink at rates of up to 2 kbps via a 2 m fixed HGA

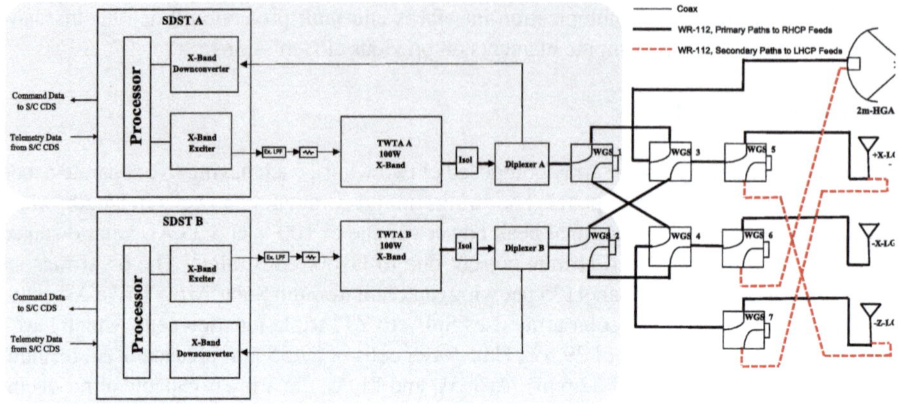

Fig. 16 The X-band telecom system block diagram showing the dual redundant SDSTs and TWTAs coupled through seven waveguide switches to the high gain and low gain antennas to provide single fault tolerance

mounted on the +Z deck or via a network of three body-mounted LGAs oriented to provide wide field of view coverage. The spacecraft's 100 W Traveling Wave Tube Amplifier (TWTA) enables downlink rates of up to 180 kbps via the HGA at over 4 AU from Earth. The spacecraft can also downlink via a selected LGA, ensuring a downlink data rate of at least 10 bps in safe-mode at over 4 AU Earth-Probe distance to a 70 m DSN antenna. A series of waveguide switches (WGSs) direct RF energy from the TWTA to the selected antenna. The graphite composite HGA, provided by Maxar, leverages their substantial experience manufacturing antennas for communications satellites. The remainder of the telecommunications equipment is provided by JPL and draws heritage from previous missions such as Juno and Dawn. A photograph of the telecommunications panel is shown in Fig. 17 prior to its integration onto the spacecraft.

1.6 Guidance, Navigation and Control

Psyche includes a Guidance, Navigation and Control system (GN&C) for use throughout the mission. The spacecraft contains no chemical propulsion system, so all control functions are conducted using a combination of reaction wheels, electric propulsion, and cold gas thrusters. Psyche's GN&C system supports detumble following launch vehicle separation, followed by solar array deployment and acquisition of inertial reference. GN&C also provides Earth pointing of the body-fixed HGA and the DSOC tech-demo for communications, implements flexible inertial pointing of the science instruments for checkouts and calibrations, and allows inertial pointing of the EP thrust vector to support cruise mission thrusting.

Along with providing functions for managing the stored Reaction Wheel Assembly (RWA) momentum, the GN&C flight software implements control laws for pointing the solar arrays and positioning the EP thruster gimbals. After arrival in orbit around Psyche, GN&C provides nadir and off-nadir pointing control for the science instruments, along with time-profiled pointing of the EP thrust vector to accomplish orbit transfers and orbit maintenance.

GN&C additionally supports spacecraft "safe mode", providing fault protection with functions for detumble, sun acquisition and sun-referenced attitude control using either RWAs or Cold-Gas System (CGS) thrusters. The GN&C software also implements an RWA

Fig. 17 The X-band telecom system mounted to a dedicated panel, prior to installation on the SEP Chassis. The dual redundant TWTAs are on the left and the waveguide and switches for routing to the four antennas is on the right

based Earth-pointed "standby" mode. Attitude sensors on the Psyche spacecraft include Honeywell Miniature Inertial Reference Units (MIMUs), Jena-Optronik star trackers and an eight detector "fan" configuration of Adcole coarse sun sensors (CSSs). Each sensor is redundant. For nominal inertial attitude estimation, star tracker attitude fixes are combined with a bias-corrected MIMU-based attitude estimate using a "fast observer" algorithm. For safe mode, attitude estimation is sun-referenced, using only the uncorrected MIMU and sun sensor. A 360-degree, single axis sun search achieves full-sky sun coverage.

Nominal 3-axis attitude control is accomplished using a pyramid of four Honeywell HR 16-100 reaction wheels operated near zero momentum. To increase torque authority and disturbance momentum capacity, the nominal configuration uses four wheels, but 3-wheel operation is supported, providing three-for-four redundancy. Backup attitude control for use during initial detumble and CGS safe mode is provided by the CGS system. The CGS also provides an alternate RWA momentum unloading capability to supplement the use of EP-based momentum control.

Unlike Deep Space 1 and Dawn, which used the EP gimbals for direct attitude control of two axes during thruster operations, the Psyche GN&C uses RWAs for full 3-axis control even while thrusting. GN&C performs nominal RWA unloading and momentum control using torques generated by the EP system. The approach is similar to that used for Maxar's GEO communications satellites. The RWAs provide 3-axis control of the spacecraft attitude while a low-bandwidth momentum control algorithm positions the EP gimbal to direct the thrust vector near the vehicle center of mass. The GNC system actuates the active thruster's DAPMs to create small offsets to the EP thrust vector that produce torques that allow near continuous unloading of two transverse axes of RWA momentum.

As discussed in Sect. 1.2.2, Hall effect thrusters create a "swirl" disturbance torque that must be accounted for in operation. Since swirl torque is aligned with the thrust axis, it cannot be directly unloaded using a two-axis gimbal and continuous unloading. Instead, an alternate method must be used (Oh et al. 2014a). Psyche implemented a method that minimized the need for new capabilities onboard the spacecraft and maximized the predictability of disturbances to the trajectory. During cruise, when the inertial thrust vector changes only slowly, the remaining axis of momentum (along the thrust vector) is unloaded periodically using short (∼15 min) EP-unload burns using a thruster on the opposite side of the spacecraft. The timing and direction of these burns is determined by ground operators and is optimized to maximize the efficiency of these maneuvers. The burn duration is predicted by the ground, but ultimately determined autonomously onboard the spacecraft. Because the time of the burn is deterministic and the magnitude is predictable, spacecraft navigators are able to predict and model the velocity change created by these maneuvers and account for them in the trajectory design. Once in orbit around Psyche, where the inertial thrust direction will change more quickly, continuous unloading can be used to accomplish most of the required momentum control, so dedicated EP-unload burns are expected to be rare. If required, EP unloading can be supplemented with momentum adjustment burns using the CGS thrusters.

1.7 Autonomous Operations

A major challenge for the operations team is to maintain the flexibility and responsiveness needed to conduct the science mission while operating at a previously unexplored body at distances over 30 light-minutes from Earth. Psyche's concept of operations is based on strategies that were developed for the Dawn mission, which successfully conducted low thrust orbital science campaigns at Vesta and Ceres in 2012 and 2016 respectively (Garner and Rayman 2018). The operations team utilizes onboard command sequences to minimize the ground-spacecraft interactions needed to control the spacecraft. Command sequences contain lists of commands, including simple logic checks and directives, that are uploaded to the spacecraft and execute autonomously at pre-designated times or under pre-designated conditions.

During cruise, command sequences are created using a four-week build process, uplinked to the spacecraft, and then executed over a four-week period. In addition to the command sequences, the operation team also uplinks the latest navigation data to the spacecraft in the form of ephemeris polynomials and thrust pointing polynomials used for thrusting and pointing the spacecraft. During Approach and Orbital Operations, the commencement of science activities and need for frequent navigational updates lead to an accelerated operational pace and the concurrent development of command sequences packages. As the spacecraft enters Psyche's sphere of gravitation influence, the accuracy of the orbital determination (OD) solutions used for navigation degrade more rapidly than during interplanetary cruise, and the thrust plan becomes more sensitive to EP thruster performance. To compensate, the schedule for incorporation of OD data into thrust sequences is compressed so the latest thrust performance data and OD solution available can be used, and sequence packages are uploaded to the spacecraft on a more frequent basis (Oh et al. 2019).

The spacecraft does not perform autonomous guidance along its trajectory. Ground-provided command sequences specify when EP thrust arcs begin and end, the desired thrust vector orientation, and the desired throttle level to be used. To allow ground operators to choose the highest throttle level possible without concern for short-term variation in the power demands by the rest of the spacecraft (e.g. heater cycling), the flight software autonomously monitors the battery state of charge. The flight software will tolerate short-term, low-level depletion of the battery, but if the battery state of charge drops below a prescribed threshold that is indicative of a longer-term, more significant energy imbalance, the flight software autonomously responds by reducing the EP throttle level. This autonomous behavior will repeat up to three times, with successively lower state of charge thresholds triggering additional throttle level reductions. After the third (lowest) state of charge threshold is crossed, the flight software halts the EP thrusting and triggers a fault protection response that is aligned with other directly triggered EP fault protection responses (see Sect. 1.8 below). In either situation - autonomous reduction of the EP throttle level or autonomous suspension of thrusting – the net velocity change (ΔV) achieved by the spacecraft will be less than originally planned. The trajectory is designed accommodate this reduction and the operations team can compensate for this shortfall as follows.

When the nominal thrust plan is first developed, it is designed to tolerate periods of unexpected missed thrust with minimal impact to the mission. The plan is required to accommodate at least 12 consecutive days of missed thrust at any single time during the mission, and at least 14 non-consecutive days of missed thrust per year while still arriving at the asteroid on schedule. In practice, the plan is often tolerant to much longer outage periods. When a deviation from the nominal thrust plan occurs, the impact is observed afterwards during downstream orbit determination activities. The downstream thrusting plan is then adjusted to accommodate the deviation. New radiometric and/or optical measurements are obtained and incorporated into the orbit determination solution at least once per week and the downstream thrusting plan is elaborated in ground-managed sequences that are typically designed and uplinked to the spacecraft once per month. While the spacecraft typically cannot "make up" for past thrusting shortfalls by thrusting later at higher throttle levels, the spacecraft can "make up" for past thrusting shortfalls by thrusting on days that were previously planned as "coasting" (non-thrusting) days in downstream thrust arcs. In some cases, thrusting on those make up days will contribute to an increase in the total mission propellant expenditure, and sufficient xenon propellant has been allocated in the overall propellant budget for this purpose.

1.8 Fault Protection for Deep Space Low Thrust Missions

The Psyche spacecraft is single-fault tolerant, meaning that no credible fault can result in failure to achieve mission success. Single fault tolerance is implemented primarily via cold-spare block redundancy, with cross-strapped interfaces between most hardware (including all interfaces between JPL and Maxar hardware) that provide small fault containment regions and multiple fault tolerance across critical functions.

Fault protection on-board the Psyche spacecraft is implemented in hardware and software. Hardware-based fault protection provides mitigation for low-level faults isolated to a particular hardware unit (e.g., error detection and correction for memory upsets). Software-based fault protection provides mitigation for component-, function-, and behavioral-level faults that require autonomous hardware swaps, span multiple hardware units, require coordination of multiple actions across subsystems, and/or trigger system safe mode (e.g., excess attitude control errors, command loss timer expiration, low battery state of charge). All software-based fault protection is delivered by JPL and implemented via centralized fault protection engine in the compute element.

Because the spacecraft is located many light-minutes from Earth, long communication outages with the ground are an expected condition. The Psyche spacecraft autonomously detects and responds to faults and can autonomously achieve a sustainable power-safe, thermal-safe, and communicative safe mode when the spacecraft's health is threatened. If safe mode is triggered, the spacecraft will stop execution of on-board command sequences, power-off non-critical hardware, switch to LGA communication, and point its solar arrays to the Sun. In order to minimize cold gas propellant usage, Psyche primarily uses a reaction wheel based safe mode, but Psyche also has a cold gas thruster based safe mode to deal with RWA-faults and low power faults. Aside from launch, there are no time-critical/mission-critical events and the Psyche spacecraft is therefore designed to fail-safe and remain in safe mode for up to 34 days, until the flight operations team can diagnose the fault and return the spacecraft back to nominal science operations. The mission's missed-thrust margin accounts for safe mode occurrences during cruise and science operations (Oh et al. 2019).

Because Psyche uses a cold-spare architecture, careful consideration is taken to ensure safe recovery in the event of a computer fault. The PDA contains the hardware, logic and interfaces necessary to independently detect a computer fault, and will swap the main computer to the backup unit if a serious fault is detected. The swap can take several minutes, and during this time there is no software control onboard the spacecraft. All spacecraft hardware, including heaters and actuators, must be put into a safe state during this potential flight software control outage, and the REU's personality PROM (PPROM) provides a mechanism to place devices onboard the spacecraft into a safe state when the compute element is not operational.

The time of EP thrusting is a resource that should be preserved, so the Psyche fault protection design incorporates features to limit the amount of missed thrust that occurs due to a fault. One example of this is a "light-touch" fault response mode that maintains star tracker- and reaction wheel-based pointing and HGA communications, for quick ground-based recovery from low severity faults. Another example of this is limited graceful recovery from some hardware faults before demoting to safe mode, such as resetting the star tracker so the spacecraft can "fly-through" radiation-induced single event upsets rather than immediately entering safe mode.

1.9 Heritage Hardware Assessment and Delta-Qualification

During the spring of 2018, as part of the project's preparations for its Preliminary Mission and Systems Review, the spacecraft management team conducted a series of subsystem-by-subsystem Inheritance Reviews to confirm the compatibility of inherited hardware designs with the Psyche mission requirements, and to identify any incompatibilities that would need to be addressed via design changes, additional testing, or additional analysis. As described earlier (Sects. 1.2 and 1.4), the project knew at its conception that certain elements of the power subsystem and the electric propulsion subsystem would require design modifications to extend the spacecraft's propulsive capabilities over the full range of solar distances required for the Psyche mission. The Inheritance Reviews served as a "backcheck" on the compatibility of elements that had not been explicitly modified as part of the early conceptual design.

The results of these early Inheritance Reviews were very valuable to the project, leading to a variety of crucial changes and additions spanning several subsystem implementation plans. These modifications included:

- A change to a different flight IMU package due to a fundamental mismatch that was discovered between the Psyche GNC performance requirements and the performance of the originally baselined Maxar gyro.
- A delta-qualification test for the inherited solar array panel design to encompass the colder temperatures that the panels were to experience at 3.33 AU.
- A delta-qualification test for the inherited SADA design to demonstrate its ability to stop and then reverse direction thousands of times because the Psyche mission requires frequent changes in spacecraft orientation (e.g. turns between thrusting attitudes and communications attitudes).

- A unique thermal characterization test of a previously flown Cold Gas Thruster design to determine the maximum duty cycle at which these thrusters could be continuously operated by the Maxar-heritage Attitude Control Electronics (ACE) without overheating the thruster valve coils.

2 Spacecraft Manufacturing and Testing

2.1 Spacecraft Assembly Flow

Psyche's primary spacecraft I&T flow, shown in Fig. 2, was conducted jointly by JPL and Maxar with integration and testing (I&T) of the SEP chassis conducted at Maxar, and spacecraft integration and environmental testing conducted at JPL. Maxar was responsible for all mechanical integration and testing on the spacecraft throughout the campaign and used Maxar-provided ground support equipment, regardless of location. Construction of the SEP chassis began at Maxar in January 2019 with the manufacturing of the composite central cylinder that serves as the spacecraft's core structure. Maxar delivered a structural support/thermal radiator panel for the spacecraft's telecom and core avionics equipment to JPL in July 2020. This allowed JPL's telecom team to conduct integration and testing of the X-band telecommunication system, including the SDST and TWTAs, in parallel with integration of the SEP chassis through early 2021.

Full flight system integration began in March 2021 when the SEP chassis was delivered to a JPL cleanroom in Pasadena, CA and the telecommunications equipment panel was integrated with the SEP Chassis from Maxar (Maxwell et al. 2022). In April of that year, the core avionics hardware began to arrive, and the magnetometers were the first instrument hardware to show up in May. Integration of the other science instruments and the DSOC technology demonstration payload took place throughout summer 2021, interspersed with spacecraft subsystem functional checkouts. During this time period, the SEP chassis and the rest of the spacecraft were exercised using the JPL-provided avionics and flight software, interfacing with a JPL-provided Ground Data System (GDS). The GDS used during system testing at JPL was the same GDS that was ultimately used for JPL-led mission operations.

By fall of 2021, flight system integration was nearly complete, and the spacecraft finished its initial system-level functional checkouts in preparation for the system environmental test campaign. The environmental test campaign began in December 2021 with Electromagnetic Interference and Compatibility testing followed by a Thermal-Vacuum test campaign. Once that completed in January 2022, the team proceeded with solar array installation and solar array deployment testing, followed by Dynamics testing and Launch Vehicle Separation testing. The stowed solar arrays were then removed from the spacecraft and returned to Maxar for post-dynamics inspections and deployment tests. The rest of the spacecraft was shipped to Kennedy Space Center in May 2022. Final Deep Space Network compatibility testing and final System Scenario testing were conducted at the launch site to close out the testing with the flight vehicle. During the summer of 2023, the stowed solar arrays were shipped from Maxar to launch base and re-installed onto the spacecraft, and both xenon and nitrogen propellants were loaded into the spacecraft tanks. The spacecraft was successfully launched on a SpaceX provided Falcon Heavy launch vehicle on October 13, 2023.

2.2 Spacecraft Testbeds

The project built and operated a total of three hardware-resident testbeds at JPL. These three testbeds were used throughout spacecraft development, and have remained at JPL for use during mission operations. Although the testbeds were built, maintained, and operated by JPL, they incorporated Maxar-provided elements of the SEP chassis, via combinations of engineering model hardware and simulation software.

In each of the three Psyche testbeds, a dedicated set of Simulation Support Equipment (SSE) provides realistic stimulus, via a collection of dynamics simulation, hardware drivers, and subsystem simulators, that makes it appear to the Flight Software as if the entire spacecraft is present. The SSE also provides power switch monitoring, analog sensor stimulation, and interface connections for device simulations, signal monitoring, and fault injection.

The highest-fidelity testbed, known as the Flight System Testbed (FSTB), employs two engineering model Psyche Compute Elements operating in a full dual-string configuration. The FSTB also hosts flight-like data buses and at least one hardware instance of each type of Maxar peripheral electronics, with which the compute element interact. This allows the FSTB to provide realistic onboard spacecraft data communications timing. The FSTB also hosts science instrument electronics of engineering model or better quality, sufficient to test flight-like commanding and sequencing of the science instruments during science data gathering scenarios involving high-cadence imaging and concurrent payload operations.

There are two lower-fidelity testbeds, known collectively as the Flight Software Testbeds (FSWTB1 and FSWTB2), that each host a single engineering model compute element. In each FSWTB, everything outside of the compute element is simulated via the SSE. During development, one FSWTB was almost exclusively dedicated to the FSW team, for FSW pre-release testing. The other FSWTB was shared by a variety of other users.

Closed-loop dynamics simulation was not implemented in support equipment used for the flight vehicle. Therefore, the testbeds were crucial to validating the functional behavior of the GNC flight software. The testbeds were also the primary venues used for fault protection testing, stress testing, and risk reduction testing. For additional information on the Psyche testbeds, their fidelity and their uses, see Maxwell et al. (2022 and 2024).

2.3 Spacecraft System Verification and Validation

Psyche's verification and validation (V&V) program was conducted across multiple venues including the software simulation-based Work Station Test Set (WSTS) venues, the three hardware-based testbeds described in Sect. 2.2 above, and the spacecraft itself in the spacecraft assembly, integration and test (AIT) venue. The spacecraft system V&V program made use of these venues to execute a comprehensive test program that ensured that Psyche meets design requirements and will achieve its mission objectives.

The spacecraft system V&V program began when the first flight software release was formally delivered to the system testbeds in Fall 2020. This software build included the core software functionality necessary to integrate and test subsystem hardware as it was integrated to the flight vehicle: basic uplink, downlink, file management, communications, power management, propulsion, and payload support. The systems engineering team performed testbed and WSTS testing of these capabilities over a period of a few months to identify critical issues to be fixed before the flight software was delivered to the flight vehicle.

In Summer 2021, the systems engineering team began to test the more complex mission capabilities needed to support launch, cruise, and orbital operations. Mission Scenario Tests (MSTs) were created to validate mission capabilities by executing in both the system testbed and AIT venues (Imken et al. 2022). The MSTs included multiple system-level tests of nominal and off-nominal in-flight scenarios across each phase of the mission. These tests ensured that the flight system correctly executed key mission capabilities using flight software behaviors and flight-like Mission Operations System (MOS) and Ground Data System

(GDS) generated sequences. A key limitation of the AIT venue was its lack of closed-loop guidance, navigation, and control (GNC) simulation, so all system-level closed-loop GNC testing was performed using closed-loop guidance simulations in the testbeds. Careful analysis of testbed fidelity differences and all other Test As You Fly (TAYF) exceptions were identified and mitigated by additional analysis and demonstrations to mitigate the risk associated with this approach.

The first test of nominal launch behaviors was conducted in AIT in Fall 2021. The launch behavior was one of the most critical elements of the flight software as it autonomously conducted all of the critical events that brought the spacecraft to a power positive, communicative, and thermally stable state after spacecraft separation. The nominal scenario exercised launch file and parameter load, the day-of-launch countdown procedure, a simulated ascent leading into the autonomous solar array deployment behavior, entry into spacecraft safe-mode after deployment, and the initial steps of the safe mode transition into nominal operations. The off-nominal launch behaviors were tested in AIT in Spring 2022. The off-nominal scenario exercised the spacecraft's autonomous response in faulted scenarios by injecting simulated faults into key hardware used during spacecraft separation and solar array deployment. This served to validate the system's fault protection responses and showed that the vehicle could ultimately recover and safely complete solar array deployment.

A sample set of data from the launch nominal mission scenario test is shown in Fig. 18. The figure shows a graphical representation of selected MST 1 Launch Nominal engineering health and accountability data from the execution on the spacecraft in AIT. Each row represents the state of an individual telemetry point as a function of time, with time on the X-axis and the data value on the Y-axis. The name of the telemetry channel, both the short channel designation and full name, are listed at the top middle of each row. The spacecraft's software operating mode is shown in the telemetry channel at the top of the figure. The figure's selected time range shows the final minutes before simulated separation from the launch vehicle. The step in SM-6008 (third row) reflects separation detection and drives the change from Launch Mode to Solar Array Deploy Mode.

Once in solar array deployment mode (SA Deploy) mode, the autonomous solar array deployment behavior stepped through actions to detumble from tip off rates and release both +Y and −Y solar arrays. Nine flight software behavior milestones are incremented in SM-0004 (second row) so that the flight operations team can quickly determine progress and reconstruct behavior actions. The commanded deploy progress channel SM-6009 communicates the parameterized onboard expected state of the arrays. The final two rows (PYRO-6119, PYRO-6219) highlight two channels indicating FIRE counts of the release mechanisms – one for each array. This telemetry shows each release mechanism is fired twice, as expected. The yellow coloring of the dots at the steps indicates the telemetry channel triggered an operational alert (alarm) at that point in time. The alarms shown in this figure are expected during ground test and are implemented for personnel safety. The spacecraft operated nominally throughout this period. Once the solar array deployment behavior completes, channel SM-0001 shows the transition into safe mode, which was the intended end state of the Launch Behavior.

The first test of science operations sequences and behaviors was an Orbit D functional scenario conducted during the system Thermal-Vacuum test. This test brought the spacecraft into a flight-like cold environment and operated all available payloads concurrently while generating high-volume GNC data products. The test performed a short simulated electric propulsion system thrust arc, simulated a solar eclipse to discharge the battery, configured the battery to recharge while performing a subsequent short thrust arc, and finished with continued science data acquisition. This was the most flight-like scenario possible in the

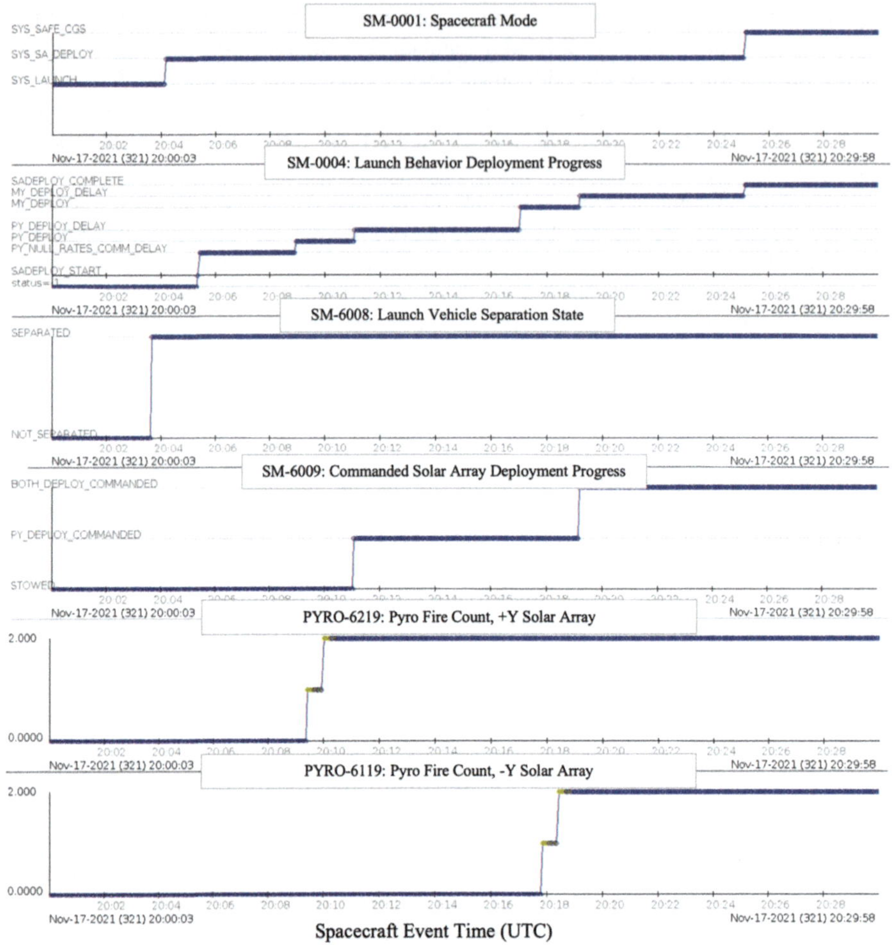

Fig. 18 Telemetry data from Psyche's launch nominal mission scenario test

thermal-vacuum venue and was the most stressful scenario for power and thermal – two subsystems that are only simulated in the Psyche testbed venues.

Two MSTs were completed in summer 2022 for the Cruise Phase, one nominal and one off-nominal. The nominal scenario was an accelerated run-through of activities that occur over the course of a week in Cruise operations. This test started with a weekly spacecraft to Earth communications pass and then proceeded through a transition to a cruise thrust maneuver. The cruise thrust maneuver had two primary thrust arcs separated by a shorter momentum management (swirl torque) maneuver. During this maneuver, the spacecraft was oriented to offload momentum stored in the reaction wheels using its electric propulsion system. The cruise off-nominal scenario incorporated simulated faults that are injected into the spacecraft during what have been identified as critical activities, such as during a cruise thrust maneuver. These faults triggered fault protection software responses that led to thrust termination and transition to safe mode. The spacecraft was then transitioned back to nominal operating configuration via ground command.

In addition to the previously mentioned Orbit D functional scenario that occurred during thermal-vacuum testing, an Orbit B nominal operations scenario was executed during functional testing in AIT in Spring 2022. Orbit B operations consist of the most driving imaging cadences seen in the mission, which stress the onboard data management capability since the spacecraft must store large amounts of image data between downlink opportunities. This test used flight-like command sequences and was conducted after all science instruments were integrated onto the spacecraft. The scenario includes a single lit-side orbit of imaging consisting of 58 image stations, each acquiring an image through all eight filters per station with the primary imager (a total of 464 images over approximately six hours), followed by a dark-side downlink pass to return all data to the ground. The magnetometer and GRNS were also commanded to flight-like configurations using standard sequences. Following the Orbit B scenario, a small session of concurrent imaging with both imagers was added to represent the most challenging aspect of Orbit D for the instruments. Because of limitations in the AIT venue for closed-loop simulation of the guidance, navigation and control subsystem, the spacecraft turns were omitted from the AIT run, although they were included in a later run on the system testbeds.

Orbital transfer maneuvers conducted at Psyche using the electric propulsion system are more challenging than those conducted during interplanetary cruise. Orbital transfer maneuvers require the navigation and spacecraft teams to design and produce updated thrust designs and sequences at a much more aggressive cadence than other times in the mission, and challenge the spacecraft to follow aggressive thrust pointing profiles. Additional orbital operations scenario tests are occurring in the testbeds post-launch, including an Orbit B to C transfer thrusting scenario test, to support post-launch validation of these scenarios.

2.4 Launch Delay and Completion of Spacecraft System V&V

Spacecraft integration and test activities were completed on a satisfactory schedule, and the flight vehicle was shipped to Kennedy Space Center in May 2022, in preparation for an October 2022 launch opportunity (see Fig. 2). Shortly after arrival at the launch site, a project-sponsored assessment of the spacecraft system Verification and Validation (V&V) program, much of which involved thoroughly exercising the flight software (FSW) on the JPL-resident testbeds, concluded that the project would be unable to complete all required elements of the V&V program in time to launch with acceptable risk in the fall of 2022.

In June 2022, the project and its sponsors together decided to delay the launch by a year and target an alternate October 2023 launch opportunity. The move to October was driven by mission design constraints - the mission requires a Mars Gravity Assist, and appropriate Mars rendezvous trajectories are only available every $\sim$ 12-13 months. A move to October also provided enough time to complete a robust spacecraft system V&V program.

Following the decision to delay the launch, NASA and JPL jointly established a Psyche Independent Review Board (IRB), to fully investigate the reasons for the delay and to make recommendations for corrective actions needed to ensure a successful launch in October 2023. From July to September 2022, the IRB conducted a series of targeted stand-up reviews and informal personnel interviews. These activities were centered at JPL, and the presenters and interviewees were primarily Psyche JPL project personnel and JPL institutional leaders, though the interviewees also included some Maxar personnel. The IRB concluded their activities in October 2022 with a review of the Psyche Project's replan for achieving the October 2023 Launch Readiness Date. The Psyche IRB published their final report on 4 November 2022, in which they endorsed the Psyche Project's replan and recommended that the project proceed forward with the new launch readiness date.

While the IRB agreed that the most pressing reason for the launch delay was a lack of readiness for the GNC FSW (including the readiness of the testbeds to confidently V&V the GNC FSW), the IRB also identified several other launch readiness concerns. As stated in their final report, "During the push to prepare for a 2022 launch, it became clear that although the GNC FSW and its V&V were the most apparent problems, there were other serious issues that would have led to the launch delay. These issues included a large number of open FSW issues, incomplete definition of fault protection behavior, incomplete V&V, and insufficient plans and preparation for mission operations (Psyche IRB Report 2022). The report cited the following deficiencies as contributors to the project's inability to complete its preparations for the Fall 2022 launch opportunity:

- Inadequate staffing
- Inadequate metrics for the tracking of spacecraft system V&V progress
- Communication breakdowns within the Psyche Project team
- Inadequate penetration of Psyche development challenges by JPL Institutional Management

The report went on to cite both the large number of flight projects that JPL was managing in the 2020-2022 timeframe, as well as the drastic reduction in face-to-face workplace interactions during this same time frame due to the COVID-19 pandemic as cross-cutting contributors to these deficiencies.

The Psyche IRB report contained a large set of path forward recommendations, both for the Psyche Project and for JPL/Caltech as an institution. The Psyche-specific recommendations that were successfully implemented and incorporated into the spacecraft system V&V program for the October 2023 launch opportunity were:

- Significant staffing augmentations, across several elements of the project. These augmentations enabled parallel completion of the spacecraft system V&V activities and the flight operations preparations.
- Restructuring of several distributed systems engineering teams into a single, centralized systems engineering organization. This restructuring enabled centralized management and prioritization of all the remaining systems engineering activities.
- Development and reporting of more comprehensive and informative V&V metrics. These improved metrics allowed for more optimal measurement and tasking of the systems engineering staff, by the newly centralized systems engineering management team.

The complete set of path forward recommendations are beyond the scope of this paper. The Psyche IRB Report (2022) provides further detail on the membership, methods, findings and recommendations of the Psyche IRB.

3 Post-Launch Reflections

The Psyche spacecraft successfully launched from Launch Complex 39A at Kennedy Space Center on October 13, 2023 (Fig. 19). A series of In-Orbit Checkout activities were successfully completed during the first 120 days of the mission, during which no serious problems were found, and no new significant constraints against future operations were identified. A Post-Launch Assessment Review (PLAR) was successfully completed in March 2024, and purposeful SEP thrusting towards the Psyche asteroid rendezvous started on schedule in April 2024.

Fig. 19 Psyche launch from pad LC-39A at Kennedy Space Center (image credit: NASA)

During In-Orbit Checkout and subsequent cruise thrusting, the flight team encountered several idiosyncrasies in the spacecraft's behavior and performance and implemented several operational efficiency improvements. A full listing and discussion of operational discoveries and improvements is beyond the scope of this paper. Instead, we close with a few reflections on the Psyche spacecraft development effort, informed by its successful post-launch performance to date.

3.1 Psyche's Hybrid Spacecraft Architecture

The spacecraft hybrid architecture (a Maxar-provided SEP Chassis with deep space avionics and flight software provided by JPL) proved to be a successful, implementable architecture. It leveraged the high heritage, high reliability, demonstrated capabilities of the Maxar 1300 bus married with JPL telecommunications and compute capabilities that enabled the sort of communications protocols and autonomous behaviors that are needed for deep space exploration missions. During Phase B, the Psyche spacecraft design experienced a series of sizing changes (e.g. total solar array area, total propellant tank capacity, total number of temperature sensors). But the fundamental split of Maxar vs. JPL responsibilities, as outlined in the original Phase A Concept Study Report, remained true throughout the spacecraft implementation. We believe this sort of hybrid spacecraft implementation, leveraging existing industry buses with particular power generation and ΔV capabilities, is a good model for future JPL-managed missions. More generally, the success of Psyche's hybrid architecture shows how the capabilities of commercial spacecraft providers can be combined with specialized mission-unique capabilities from NASA to enable missions which would otherwise be unachievable or impractical to achieve using commercial spacecraft. This model will become increasingly relevant as policy makers navigate the rapid emergence of new commercial space companies and capabilities referred to collectively as "New Space" and their implications for NASA science missions.

3.2 Spacecraft Technical Margins

As mentioned above, we chose to "up-size" a few subsystems in Phase B, in order to have robust technical margins at Project PDR. Those decisions paid off handsomely in Phase

C/D, as we were able to complete development without having to do take drastic measures to manage our technical system resources (e.g. mass, power). The high heritage subsystems from Maxar experienced relatively little mass/power growth, and we were able to solve some other technical challenges along the way, by consciously spending mass or power from our ample reserves.

3.3 Flight Software Development

Within the spacecraft hybrid architecture, JPL assumed 100% of the Psyche Flight Software (FSW) development responsibility, and did not attempt to directly transfer any Maxar source code into the JPL code base. This allowed all FSW, including the FSW that interfaced with the Maxar-provided hardware, to be developed in a consistent style to a consistent coding standard. This also yielded a consistent level of expertise among the JPL software developers and testers with respect to all of the FSW modules; in other words, there were no Maxar-centric FSW modules that were "black boxes" to the JPL team that had to integrate and test the FSW.

While we found this to be a solid implementation approach that was ultimately successful due to the efforts of a very dedicated and talented FSW team, we also found that the team had a difficult time obtaining accurate and complete documentation of the Maxar hardware. Because some of the original Maxar hardware designs were more than 10 years old, and some designs had been modified over the years, the existing documentation was sometimes of questionable accuracy. Because some of the legacy hardware had been used in the Maxar product line for so long, it was often difficult for Maxar to answer the sorts of questions that naturally arise when a FSW developer is asked to create a brand new interface module (for example, what is its response time if we do X instead of Y?). Furthermore, FSW development was hindered by the inherently different approaches to fault protection, between the two organizations. Maxar Spacecraft feature hot spares and extensive cross strapping designed to allow the spacecraft to fail-operational with ground in the loop. JPL missions, on the other hand, fly cold spares within fault containment zones that are carefully designed to allow the spacecraft to fail-safe autonomously. These distinctly different paradigms have profound implications on the desired behavior of the FSW, in both nominal and off-nominal scenarios.

Ultimately, the project underestimated both the total effort and the calendar time required to fully understand the Maxar hardware behavior and document the desired FSW interface behavior in a manner that allowed the JPL FSW development to proceed.

3.4 Spacecraft Integration & Test

Psyche's primary spacecraft I&T flow, shown in Fig. 2 and described in Sect. 2.1, was conducted in three distinct phases:

- SEP Chassis I&T at Maxar – this phase was managed by Maxar and staffed solely by Maxar personnel with some JPL quality assurance oversight.
- Full Spacecraft I&T at JPL – this phase was managed by JPL and was staffed by both Maxar and JPL personnel, with quality assurance oversight provided by both JPL and Maxar. This phase included all system-level environmental testing.
- Full Spacecraft I&T at the Launch Site – this phase was also managed by JPL and was also staffed by both Maxar and JPL personnel, with quality assurance oversight by both JPL and Maxar. This phase included final solar array installation, propellant loading, and fairing encapsulation.

The first I&T phase (SEP Chassis I&T) followed a typical paradigm for both JPL and Maxar. Maxar is accustomed to managing their own I&T in their own facilities, and JPL is accustomed to providing some level of onsite QA presence at our industry partners.

The second and third I&T phases (the full spacecraft I&T campaigns at JPL and the Launch Site) were certainly not business as usual for either JPL or Maxar. Maxar was not accustomed to performing I&T activities under the management of another organization and was not accustomed to having its overall I&T schedule controlled by its customer. JPL was unaccustomed to hosting I&T activities that were led by experts from outside organizations, and was not accustomed to accommodating outside quality assurance representatives. Because the second and third phases were unconventional, the project suffered schedule delays as we ran into unforeseen obstacles and communication challenges. These challenges were exacerbated by the fact that the bulk of the activity in these phases took place in 2021 and 2022, under the umbrella of constantly evolving and changing COVID workplace restrictions. Ultimately these phases required significant extra management attention across both organizations to ensure that all parties were clear on their roles and responsibilities for each I&T activity and were clear on the paperwork required as entry and exit criteria for various types of I&T activities.

Reflecting on the I&T campaign as a whole, we conclude that our overall I&T approach, even though it involved a rather unusual split of responsibilities, was the least unacceptable of the available alternatives. It would have been extremely risky to perform ongoing I&T of the spacecraft at JPL without Maxar's mechanical I&T team. The entire spacecraft primary structure was a Maxar structure, and the I&T campaign at JPL included numerous additional mechanical integrations. Furthermore, all the mechanical ground support equipment (GSE) that was used to move, support, and pose the spacecraft, was owned and operated by Maxar. But if we had attempted to leave the spacecraft at Maxar for the balance of the I&T campaign, a huge effort would have been required to deploy the JPL ground data system and all the JPL personnel who were responsible for developing and executing all of the system-level functional testing to Maxar's facility in Palo Alto. Furthermore, Maxar had no experience at allowing customers to conduct tests at Maxar-managed facilities. These challenges would have been partially mitigated if the project had been able to follow its original plan of sending key leaders from the JPL I&T team to Maxar in Palo Alto during SEP chassis I&T. That would have allowed the JPL I&T team to become more familiar with Maxar I&T processes and constraints, and to foresee some of the gaps and overlaps that might later occur during the I&T campaign at JPL. Unfortunately, the project was unable to follow that plan during SEP Chassis I&T in 2020 and 2021, due to COVID-19 restrictions.

3.5 Unique V&V Challenges

Psyche had several V&V challenges, some of which were key contributors to our ultimate decision to delay the launch from October 2022 to October 2023. Some of these challenges are traceable to the hybrid spacecraft architecture, especially for those V&V activities that involved the synthesis of Maxar-provided V&V evidence and JPL-provided V&V evidence. This synthesis was uniquely challenging for two subsystems that are difficult to test exhaustively at the system level using true "test as you fly" (TAYF) methods: GNC and Thermal. What was unique about these two subsystems on Psyche was the intersection of the testing challenges with the cross-organizational work required to complete the V&V program using analysis and simulations.

It is generally impractical for JPL missions to do true TAYF closed-loop testing of GNC functions on the actual flight vehicle, and basically impossible to "fly" a Psyche-sized spacecraft in a test chamber at 1 g. For this reason, GNC V&V is highly reliant on a mix of simulation environments, including a dynamics simulation environment that can execute in real-time, in parallel with the real-time FSW. On Psyche, we chose to use our JPL-resident FSW testbed to run a simulation environment in parallel with the real-time FSW. Since the flight GNC sensors and actuators were part of the Maxar-provided SEP Chassis, we ported several Maxar simulations (both GNC hardware simulations and dynamics simulations) into the JPL-resident FSW testbeds. This porting of the Maxar simulations into the FSW testbed created exactly the sort of mixed source software environment that we had avoided in the FSW itself. Numerous problems were experienced getting the Maxar simulations to run reliably and predictably in the JPL-resident FSW testbeds. Furthermore, because the Maxar simulations had been designed to support different testing objectives in support of the Maxar satellite product line, they did not possess some of the GNC fidelity that JPL had expected (but had failed to formally specify) at the start of GNC development. What followed was a lengthy, iterative cross-organizational campaign of GNC simulation requirements definition and GNC simulation "upgrades". Our challenges in implementing a reliable real-time GNC simulation environment ultimately impacted the readiness of our JPL-resident FSW testbeds, and our ability to complete GNC V&V on schedule. In retrospect, we should have used 100% JPL-provided simulation code in the FSW testbeds for the same reason that we chose to use 100% JPL-provided FSW. We ultimately completed a comprehensive GNC V&V program in time to launch in October 2023.

JPL missions do closed-loop thermal testing in a thermal vacuum chamber, and we typically classify this thermal testing as TAYF testing. However, thermal testing is typically limited to a few extreme thermal environments and a few representative power configurations due to the cost and time required to run such tests. Ensuring that the thermal subsystem works within the heater power allocations, across all the anticipated mission environments and all the anticipated mission power configurations is handled via a system-level analysis that requires a test-correlated spacecraft thermal model. Maxar had clearly assumed responsibility for producing a test-correlated spacecraft thermal model, but had no experience tuning their model to generate the large set of additional power/thermal analyses required to close the thermal V&V for a deep space mission. The project ultimately completed a comprehensive set of power/thermal analyses in time to launch in October 2023, but in retrospect, we should have more carefully probed the fidelity and operability of the typical Maxar thermal models, and we should have had more workforce allocated to exercising those models and completing the thermal V&V.

4 Conclusions

The reflections presented in the previous section encompass both successes and challenges, including the significant obstacles presented by the COVID-19 pandemic during spacecraft assembly, integration and test. Nevertheless, the successful design and development of the Psyche spacecraft unequivocally demonstrate the unrelenting commitment and outstanding teamwork that defined the entire project. Launched successfully in October 2023, the spacecraft is now nominally operating in deep space, steadily progressing towards its 2029 rendezvous with the asteroid Psyche.

Acknowledgements The authors thank the entire ASU, APL, DTU, JPL, Maxar, and MIT teams for getting the Psyche spacecraft ready for launch and we would like to particularly acknowledge the contributions of Dr. Lindy Elkins-Tanton and Dhack Muthulingam as well as many others to the work presented here. Portions of this work were previously presented at the International Electric Propulsion Conference in 2017 and 2019 (References Oh et al. 2017 and 2019). The authors would also like to thank NASA's Planetary

Science Division, NASA's Discovery Program Office, and JPL's Solar System Exploration Program Office for their support of this work. During the preparation of this work the authors used Chat GPT to edit the abstract. After using this tool, the authors reviewed and edited the content and take full responsibility for the content of the publication.

Author Contribution All authors wrote parts of the paper and/or commented on previous versions of the manuscript. All authors read and approved the final manuscript.

Funding Information This work was carried out at the Jet Propulsion Laboratory, California Institute of Technology (80NM0018D0004), under a contract with the National Aeronautics and Space Administration, and at Maxar Space Systems.

Declarations

Competing Interests The authors have no competing interests to declare.

References

Boca A (2019) Solar cell testing for the Jupiter environment: low irradiance, low temperature and high radiation. In: 46th IEEE photovoltaic specialists conference, pp 2787–2791. https://doi.org/10.1109/PVSC40753.2019.8981342

Boca A, Warwick R, White B, Evel R (2017) A data-driven evaluation of the viability of solar arrays at Saturn. IEEE J Photovolt 7(4):1159–1164. https://doi.org/10.1109/JPHOTOV.2017.2698499

Dawson SF, Stella P, McAlpine W, Smith B (2012) JUNO photovoltaic power at Jupiter. In: 10th International Energy Conversion Engineering Conference, p AIAA-2012-3833. https://doi.org/10.2514/6.2012-3833

Delgado JJ, Corey RL, Murashko VM, Koryakin AI, Pridannikov SY (2014) Qualification of the SPT-140 for use on Western Spacecraft. In: 50th AIAA/ASME/SAE/ASEE Joint Propulsion Conference, p AIAA 2014-3606. https://doi.org/10.2514/6.2014-3606

Garner C, Rayman M (2018) Operation of the Dawn ion propulsion system through the final orbit around dwarf planet Ceres. In: 2018 Joint Propulsion Conference, p AIAA-2018-4641. https://doi.org/10.2514/6.2018-4641

Hart W, Brown GM, Collins S, De Soria-Santacruz Pich M, Fieseler P, Goebel D, Marsh D, Snyder JS, Warner N, Whiffen G, Elkins-Tanton L, Bell J, Lawrence D, Lord P, Pirkl Z (2018) Overview of the spacecraft design for the Psyche mission concept. In: IEEE Aerospace Conference. https://doi.org/10.1109/AERO.2018.8396444

Imken T, Hart W, Larson K, Bairstow S (2022) Preparation for the launch, cruise, and orbital operations phases of the Psyche mission. In: 2022 IEEE Aerospace Conference. https://doi.org/10.1109/AERO53065.2022.9843808

Jameson-Silva K, Delgado JJ, Liang R, Lord PW, Rotlisburger L, Torres M, Tomescu B, Malone SP, Werner E, Waranauskas J (2017) Adaptability of the SSL electric propulsion-140 subsystem for use on a NASA discovery class missions: Psyche. In: 35th international electric propulsion conference, p IEPC-2017-181

Johnson I, Santiago G, Li J, Baldwin JA (2020) 100,000hrs of on-orbit electric propulsion and Maxar's first electric orbit raising. In: AIAA SciTech 2020 Forum, Orlando, FL, Jan. 6–10, 2020, p AIAA-2020-0189. https://doi.org/10.2514/6.2020-0189

Lenguito G, Neff K, Barbarits J, Snyder JS, Chaplin V (2019) Versatile xenon flow controller for extended Hall effect thruster power range. In: 36th international electric propulsion conference, p IEPC-2019-303

Lopez Ortega A, Mikellides IG, Chaplin VH, Snyder JS, Lenguito G (2019) Numerical investigations of background pressure effects and channel erosion in the SPT-140 Hall thruster for the Psyche mission. In: 36th international electric propulsion conference, p IEPC-2019-263

Malone S, Aghazadeh F, Lenguito G, Staley M, Tomescu B, Snyder JS (2019) Deep space power processing unit for the Psyche mission. In: 36th international electric propulsion conference, p IEPC-2019-280

Maxwell J, Imken T, Mase R, Oh R, Safavizadeh M, Solish B, Warner N, Elkins-Tanton L, Lord P (2022) Psyche project implementation during the COVID pandemic. In: 2022 IEEE Aerospace Conference. https://doi.org/10.1109/AERO53065.2022.9843595

Maxwell J, Hart W, Hernandez C, Oh D, Sanchez S, Snyder JS, Weisner F, Zorn T, Johnson I (2024) Achieving mission success: a comprehensive look at the verification and validation approach for the Psyche spacecraft for NASA's discovery program. In: 2024 IEEE Aerospace Conference

NASA (2014) Discovery 2014 announcement of opportunity. NASA Solicitation and Proposal Integrated Review and Evaluation System. https://nspires.nasaprs.com/external/solicitations/summary.do?method=init&solId=%7BFE7B4C63-873D-63C1-4D15-1D46E2FEA949%7D&path=open

Oh D, Collins S, Randolph T, Vanelli C, Tilley S (2014a) Feasibility of all-electric three axis momentum management for deep space small body rendezvous. In: 50th AIAA/ASME/SAE/ASEE Joint Propulsion Conference, p AIAA 2014-3906. https://doi.org/10.2514/6.2014-3906

Oh D, Goebel D, Hofer R, Snyder JS (2014b) Solar electric propulsion for discovery class missions. J Spacecr Rockets 51(6):1822–1835. https://doi.org/10.2514/1.A32889

Oh D, Collins S, Goebel D, Hart B, Lantoine G, Snyder JS, Whiffen G, Elkins-Tanton L, Lord P, Pirkl Z, Rotlisburger L (2017) Development of the Psyche mission for NASA's discovery program. In: 35th international electric propulsion conference, p IEPC-2017-153

Oh D, Collins S, Drain T, Hart W, Imken T, Larson K, Marsh D, Muthulingham D, Snyder J, Trofimov D, Elkins-Tanton L, Johnson I, Lord P, Pirkl Z (2019) Development of the Psyche mission for NASA's discovery program. In: 36th international electric propulsion conference, p IEPC-2019-192

Oh D, Alonge N, Bairstow S, Biswas A, Chesko C, Goebel D, Imken T, Na-Nakornpanom A, Nicholas A, Snyder JS, Johnson I, Linggi P, Lord P, Malone S, Pirkl Z, Ream J, Weiss B, Lawrence D, Peplowski P, Goldsten J, Burks M, Bell JF (2025) Launch and initial checkout of the Psyche spacecraft. J Spacecr Rockets (submitted for publication)

Psyche Independent Review Board Report (IRB) (2022). https://www.nasa.gov/wp-content/uploads/2022/11/psyche_irb_report_and_response_nov_2022.pdf?emrc=cca37d

Snyder JS, Lenguito G, Frieman JD, Haag TW, Mackey JA (2018) The effects of background pressure on SPT-140 Hall thruster performance. In: 2018 Joint Propulsion Conference, p AIAA 2018-4421. https://doi.org/10.2514/6.2018-4421

Snyder JS, Goebel DM, Chaplin V Lopez Ortega A, Mikellides IG, Aghazadeh F, Johnson I, Kerl T, Lenguito G (2019) Electric propulsion for the Psyche mission. In: 36th international electric propulsion conference, p IEPC-2019-244

Snyder JS, Nicholas A, Collins S, Ortega A (2025) Measurement of SPT-140 Hall thruster swirl torque during the Psyche mission. AIAA SCITECH 2025 Forum, p AIAA 2025-0062. https://doi.org/10.2514/6.2025-0062

Space Science Reviews (2023) 219:25
https://doi.org/10.1007/s11214-023-00967-x

Mission to Psyche: Including Undergraduates and the Public on the Journey to a Metal World

Catherine D. D. Bowman[1] · Linda T. Elkins-Tanton[1] · Adriana Talamante[1] · James F. Bell III[1] · Ernest Cisneros[1] · Alexandra Cook[1] · Jason D. Frieman[2] · Danya Gainor[1] · Jamie Hunziker[1] · Shaheer Khan[3] · Christopher R. Lawler[3] · Jessica Maschino[1] · Timothy J. McCoy[4] · Kaxandra Nessi[1] · Rona Oran[5] · David Seal[3] · Amber Simon[1] · Rohit Singh[1] · Carol M. Tolbert[2] · Karin Valentine[1] · Benjamin Weiss[5] · Daniel D. Wenkert[3] · David A. Williams[1]

Received: 14 June 2021 / Accepted: 1 March 2023 / Published online: 5 April 2023

Abstract
The NASA Psyche mission's program to engage university undergraduates and the public in the mission is inspired by and built upon the extensive foundation of public engagement, educational outreach activities, and expertise of NASA and mission partner institutions. The program leverages the enthusiasm and contributions of undergraduates nationwide to the benefit of the mission, the students and their institutions and communities, and the broader public. Psyche Student Collaborations consists of four main programs, two (Psyche Capstone and Psyche Inspired) are available solely to undergraduates enrolled at universities or community colleges in the United States and its territories and two (Innovation Toolkit free online courses and Science Outreach Interns and Docents) invite broader participation by engaging the talents and creativity of undergraduate interns to help create content and events to reach the public and lifelong learners. Together, these offerings provide multiple entry points and a spectrum of intensity of experiences, numbers of participants, disciplinary diversity, and mode of delivery. Involving undergraduates in all phases of the program supports the development of the next generation of explorers, contributes to the nation's workforce preparation, and complements NASA's existing undergraduate offerings by providing long-term opportunities for students to participate with the mission through established postsecondary education structures like capstone courses.

Keywords Psyche · Asteroid · Undergraduate students · Public engagement · STEAM · Interdisciplinary

---

The NASA Psyche Mission: Science Instruments and Investigations
Edited by James F. Bell III, Carol Polanskey and Lindy Elkins-Tanton

---

Extended author information available on the last page of the article

1 Introduction

NASA's Discovery Program (https://www.nasa.gov/planetarymissions/discovery.html) aims to "achieve outstanding results by launching more smaller missions using fewer resources and shorter development times" than larger flagship missions. Discovery missions are led by a principal investigator and are selected through peer review during a multi-stage selection process (NASA 2014). In January 2017, NASA selected the Psyche mission to investigate the asteroid (16) Psyche, led by Arizona State University (ASU), as its 14th Discovery mission.

Among the goals of the Discovery Program, as stated in the 2014 Discovery Announcement of Opportunity (AO), is to "announce scientific progress and results in the peer-reviewed literature, popular media, scholastic curricula, and materials that can be used to inspire and motivate students to pursue careers in science, technology, engineering, and mathematics" (NASA 2014). Although the 2014 AO precluded an *Education and Communication (E&C) Program*, the mission team submitted a plan for the optional *Student Collaboration* opportunity, defined in the AO as "a separate part of the proposed investigation" that "may not increase the mission development risk" and "must include appropriate plans for the mentoring and oversight of students to maximize the opportunity for teaching, learning, and success in contributing to the mission" (NASA 2014). This paper outlines the history, development, implementation, and future plans for including undergraduate students and the public in the Psyche mission.

1.1 Psyche Student Collaborations Program History

NASA has a decades-long legacy of planetary science educational and public outreach efforts (Table 1; see also agency-wide program summaries in e.g., Kaminski 2021; Smith et al. 2014) at the programmatic (e.g., Mars Public Engagement), flagship mission (e.g., Cassini), and principal investigator-led mission (e.g., Dawn, Lucy, MAVEN) levels. In developing the Psyche Student Collaborations program, the team was informed by many previous and ongoing efforts within specific NASA divisions and missions (e.g., Cassini Inspires, Send Your Name to Mars, JWST Art, Dawn Journal blogs) as well as agency-wide initiatives (e.g., Science Activation Program, Museum Alliance [now the NASA Museum and Informal Education Alliance], Solar System Ambassadors, NASA Solve, NASA Space Place, Night Sky Network, and NASA Citizen Science).

Through the Discovery proposal process, the program design was refined with feedback from NASA's Science Mission Directorate, culminating in the present program, which leverages the enthusiasm and contributions of undergraduates nationwide to the benefit of the mission, the students and their institutions and communities, and the broader public. The Psyche Student Collaborations logic model (Appendix A) shows the alignment of the present program to the latest strategic goals outlined in the NASA Strategy for STEM (Science, Technology, Engineering, and Math) Engagement 2020-2023: "Create unique opportunities for a diverse set of students to contribute to NASA's work in exploration and discovery"; "Build a diverse future STEM workforce by engaging students in authentic learning experiences with NASA's people, content and facilities"; "Attract diverse groups of students to STEM through learning opportunities that spark interest and provide connections to NASA's mission and work" (NASA 2020).

Table 1 Selection of NASA program exemplars

Program	Description	URL
Cassini Inspires	A campaign that invited the public to create and post their artistic creations inspired by NASA's Cassini mission on social media with the tag #CassiniInspires.	https://solarsystem.nasa.gov/resources/17598/cassini-inspires/
Send Your Name to Mars	A campaign program that encourages space enthusiasts to submit their name to be etched onto a microchip that will be carried by a Mars rover.	https://mars.nasa.gov/participate/send-your-name/
JWST Art	An artistic program collaborating with interns and the public for the creation of art about the James Webb Space Telescope.	https://webb.nasa.gov/content/features/jwstArt/publicArt.html
Dawn Journal Blogs	A NASA blog that provided journal entry descriptions discussing Dawn mission milestones with the public.	https://solarsystem.nasa.gov/missions/dawn/mission/dawn-journal/
Science Activation	A cooperative network connecting NASA science experts, content, and experiences with community leaders.	https://science.nasa.gov/learners
Museum & Informal Edu. Alliance	An education program providing free NASA educational resources and services to educators and the general public.	https://informal.jpl.nasa.gov/museum/
Solar System Ambassadors	A public engagement program using nationwide volunteers to communicate the science and excitement of NASA's missions and discoveries.	https://solarsystem.nasa.gov/solar-system-ambassadors/
NASA Solve	A website inviting the public to solve problems on NASA projects and participate in research and prize competitions.	https://www.nasa.gov/solve/
NASA Space Place	A website featuring games, articles, videos, and hands-on activities meant to educate children on space and Earth science.	https://spaceplace.nasa.gov/
Night Sky Network	A nationwide coalition of astronomy clubs sharing information about NASA missions and telescopic observations with the general public at museums, observatories, and classrooms.	https://nightsky.jpl.nasa.gov/
NASA Citizen Science	Scientific project collaborations between NASA scientists and public volunteers interested in participating in research.	https://science.nasa.gov/citizenscience

1.2 The Discovery 2014 Announcement of Opportunity: The "Charge" for Student Collaborations

The Discovery 2014 Announcement of Opportunity (AO) reiterated that "among NASA's strategic goals is to communicate the results of its efforts to the American public and to enhance the science and technical education of the next generation of Americans" (NASA 2014). Though an Education and Communication plan was not requested in the AO, proposers were invited to submit an optional Student Collaboration plan. The language in the AO suggested that an expected Student Collaboration opportunity could "take the form of an instrument development, an investigation of scientific questions, analysis and display of data, development of supporting hardware or software, or other aspects of the investigation", likely something like a camera, requiring that the proposal "provide details of the development schedule of the SC [Student Collaboration], including decision points for determining SC readiness for flight" (NASA 2014). In Discovery 2014, optional proposals for Student Collaboration projects were expected to be aimed at university undergraduates only, and with a budget up to 1% of the proposal cost cap, which for that call for Phases A-D was $450M. Psyche proposed for the full amount spread over the full life of the mission, equaling slightly over $400,000 per year (the majority of the budget is dedicated to one full-time lead, supported by up to 6 student interns, with the remainder dedicated to supporting the program elements detailed below).

During the Step 1 proposal process, the Psyche mission proposal team proposed a cubesat as a Technology Demonstration Option with associated Student Collaboration opportunities. Upon updated NASA guidance indicating that cubesats would count against total mission risk, the Psyche mission proposal team removed the cubesat element from the proposal. The team instead outlined activities for Student Collaboration related to instrument development and science operation opportunities, as well as research projects. Upon selection in 2017, however, NASA challenged the mission team to produce a more ambitious Student Collaboration program than previously proposed. The mission team entered a few months of conceptualization and iteration to finalize the current program with NASA (see Appendix B for a description of additional projects proposed after selection but not included in the final program). The goal was to include a strong element of interdisciplinarity in the program components to reflect the reality of working in teams at NASA and other large, complex endeavors, rather than the single-discipline experiences undergraduate students tend to encounter in their coursework and degree programs (NASEM 2018).

1.3 Psyche Student Collaborations Guiding Concepts

The guiding concepts for the Psyche Student Collaborations program reflect NASA's strategic goals as well as the charter of Arizona State University (ASU), which states, "ASU is a comprehensive public research university, measured not by whom we exclude, but rather by whom we include and how they succeed" (ASU 2014). The program is designed to:

- Attract and train future instrument and mission leads, both science and engineering;
- Facilitate innovation by including people from fields outside science and engineering, including the arts and humanities;
- Offer a portfolio of programs that range in the extent of direct engagement with mission team members and the numbers of participants who can participate (e.g., from small numbers of internships working directly with team members to thousands able to take the asynchronous free online courses available worldwide);
- Extend through the life of the mission, starting in Phase B, and;
- Include ways for people at any level of education or nationality to participate through opportunities developed and facilitated by undergraduate science outreach interns.

Additionally, the mission team holds as a core value that there are space exploration-related jobs in all disciplines and that learning to value all disciplines within interdisciplinary teams is a key skill for exploration and for everyday life. These guiding concepts are addressed by four main program components and two cross-cutting efforts (evaluation; publication and dissemination).

1.4 Psyche Student Collaborations Program Components and Timeline

Psyche Student Collaborations consists of four main programs, two of which are available solely to undergraduates enrolled at universities or community colleges in the United States and its territories (Psyche capstone projects and Psyche Inspired, outlined in Sect. 4) and two of which invite broader participation and leverage the talents and creativity of undergraduate student workers to help create content to engage the public and lifelong learners (Innovation Toolkit online courses and Psyche Outreach Interns and Docents, outlined in Sect. 6). These programs are managed by one full-time research faculty member at ASU with a team of six part-time ASU undergraduate student workers. The program benefits from voluntary in-kind contributions (such as STEM expertise, presentations, content review, social media amplification, etc.) from Psyche mission team members at NASA, JPL, mission partner institutions, and members of the ASU community and the School of Earth and Space Exploration.

- Psyche Capstone Projects (https://psyche.asu.edu/get-involved/capstone-projects/): Capstone courses (also called Senior Design) are culminating, project-based courses undertaken by university students in any major in the final (senior) year of university. Psyche develops mission-aligned capstone projects undertaken by teams of university capstone students nationwide in engineering, the sciences, the humanities, and the arts.
- Psyche Inspired (https://psyche.asu.edu/get-involved/psyche-inspired/): Psyche Inspired is a virtual "creatives-in-residence" program that brings undergraduate students from any discipline or major together to share the excitement, innovation, and scientific and engineering content of NASA's Psyche mission with the public in new ways through artistic and creative works.
- Innovation Toolkit Online Courses (https://psyche.asu.edu/get-involved/innovation-toolkit/): A growing series of free, asynchronous online courses based on the real-world challenges and skills associated with the mission's science, engineering, technology, and teamwork.
- Psyche Science Outreach Interns and Docents to Support Public Engagement (https://psyche.asu.edu/get-involved/): These student workers form the majority of the Psyche Student Collaborations team and support the program components and public engagement efforts while gaining valuable work experience.

Additionally, working with Psyche public engagement, public affairs, and media relations team members at NASA, the Jet Propulsion Laboratory (JPL), and partner institutions, the program supports the development of outreach products and public engagement opportunities, including the mission website, social media, video products, and public events (outlined in Sect. 4). These offerings provide multiple entry points and present experiences of varying intensity and time commitment, accommodate differing numbers of participants, and offer in person and remote modes of delivery (Fig. 1).

These program components have strong support and extensive voluntary participation from mission team members at all levels and from many disciplines, including mission leadership, management, scientists, engineers, communications professionals, and student interns. The results are documented and disseminated through ongoing evaluation and publication.

The program components were piloted at ASU in the early part of the mission's Phase B (2017-2018 academic year) and expanded nationally beginning with the 2018-2019 academic year. Program participation and reach continues to grow and active participation by

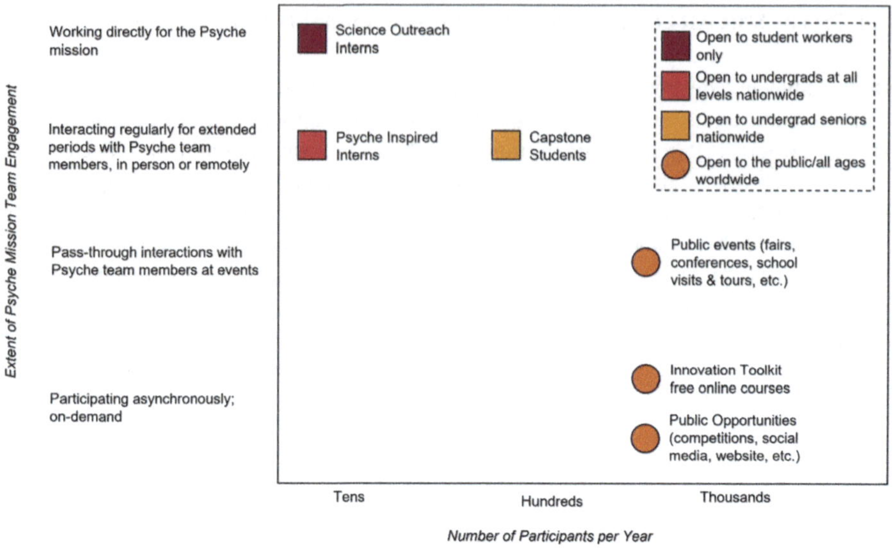

Fig. 1 The portfolio of Psyche Student Collaborations programs balances intensity of direct engagement of mission team members with numbers of participants and mode of delivery (e.g., in person; remote but synchronous; remote but asynchronous). In this figure, programs indicated by a square are for undergraduate participants and programs indicated by a circle are for the general public of all ages

students and the public is planned through Phase E of the mission. Phase F of the mission will focus on final documentation and publication.

1.5 Mission Team Member Involvement

All members of the broader mission team (from NASA, JPL, and partner universities and institutions) are made aware of opportunities to contribute to Psyche Student Collaborations efforts. As with the programs for undergraduates, opportunities for mission team members to be involved cover a range of offerings that vary in terms of intensity and commitment so they may choose what works best for them given their interests, preferences, and available time. Activities with greater time commitment or intensity of engagement include providing scientific or technical advising for capstone projects, crafting capstone project briefs, reviewing Psyche Inspired applications, creating and reviewing content for Psyche online courses, and writing abstracts and journal articles (such as the present article). Activities requiring less intensity or time commitment include being guest speakers for capstone teams or Psyche Inspired, providing suggestions for social media content, giving public talks or participating in classroom visits, assisting in writing or reviewing Frequently Asked Questions (FAQs) or content for the website, testing new student products, or answering questions that come through the public contact form or social media. At any given time, about 10% of the Psyche team (along with non-Psyche team members from NASA, industry, universities, and the arts) is actively participating in one or more of these activities, ranging from 30 minutes to 50 hours per year per team member, depending on the activity. Additionally, dozens of Psyche capstone, Psyche Inspired, and science outreach intern and docent alumni stay involved through offering job-seeking advice to current students, supporting networking events, or advising on their projects' next steps on an "on-call" basis.

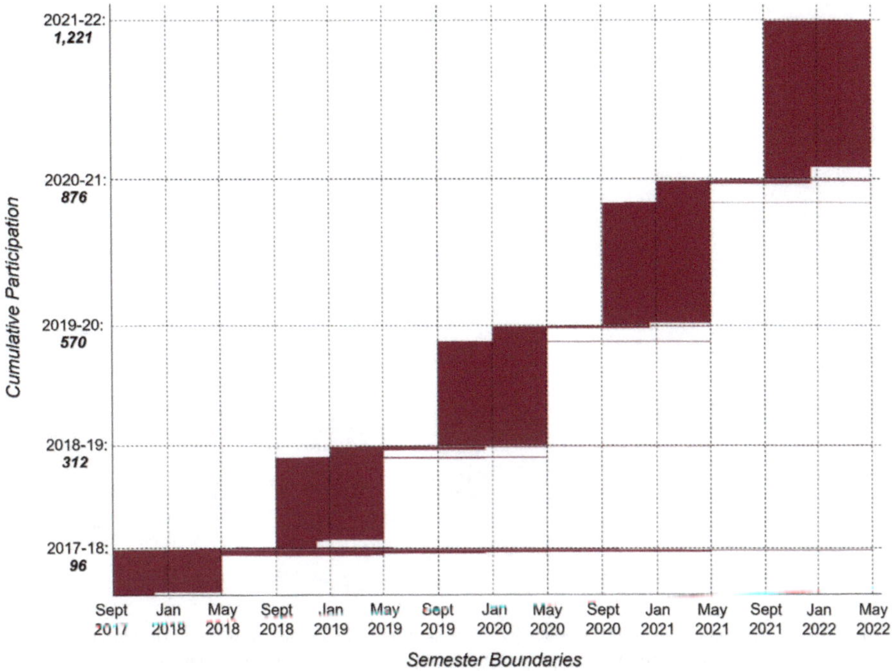

Fig. 2 Cumulative undergraduate participation (n = 1,221) September 2017 - May 2022 (chart concept credit: R. Binzel). The bulk of the participants every year are part of the Psyche capstone program (n = 1,108) followed by Psyche Inspired (n = 75) and science outreach interns (n = 24) and docents (n = 14). Some participants take part in multiple programs or participate for multiple years (such as interns and docents) but they are only counted once horizontally

2 Undergraduate Participation

The design of the three main programs open specifically to undergraduate students (Psyche capstone projects, Psyche Inspired, and the Psyche outreach interns and docents) seeks to strike a balance between depth of engagement with mission personnel and the reach of the programs and ability to scale to include greater numbers of students.

2.1 Cumulative Undergraduate Participation to Date

Over the course of five academic years (from fall 2017 to spring 2022), more than 1,200 undergraduates have participated in Psyche Student Collaborations through the Psyche Capstone program, Psyche Inspired, and as Psyche Outreach Interns and Docents. Figure 2 shows cumulative participation. The programs provide deep and extended engagement; at a minimum, students participate for one semester, but most participate for two semesters. Some students continue over time and some have participated in multiple programs (in this figure, if a student participates in multiple programs or continues over time, they are not "added" vertically but may be seen extending horizontally).

2.1.1 Mission Contact and Participant Effort

The estimated minimum average contact hours and effort hours of each type of participant in the undergraduate programs is shown in Fig. 3. "Contact hours" in the figure refer to direct

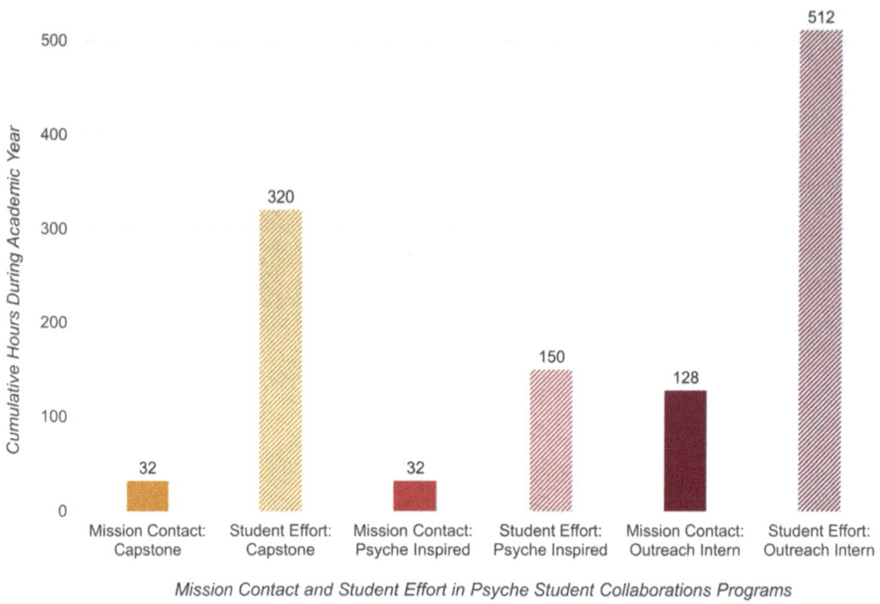

Fig. 3 Estimated cumulative contact hours (with mission personnel) and effort hours (minimum time *each* participant is expected to work on their project [capstone and Psyche Inspired] or paid to work [outreach interns]) during the academic year quantify the substantial contribution that undergraduate participants make to the mission through Psyche Student Collaborations beyond the direct contact they have with mission personnel. For example, each Psyche Inspired intern has a minimum of 32 hours of contact with mission personnel and contributes a minimum of 150 hours of work on their project

interaction between mission personnel and the participant, whether synchronous (such as in meetings) or asynchronous (emails, messaging). Capstone program participants meet in teams with the project sponsor (and other technical and scientific advisors as appropriate) 30 minutes biweekly, with the remaining contact hours with either the team or individual team members through communication on an online messaging platform or emails, feedback on deliverables, connecting them with resources, and attendance at mission presentations and question-and-answer sessions. Psyche Inspired participants meet as a group with program staff and special speakers one hour per week, with incidental communication individually through an online messaging platform as needed to support their progress. Each Science Outreach Intern has contact (individually or in groups) with mission personnel approximately an hour per day when they are working (an average of four days per week). "Effort hours" in the figure refer to expected individual participant effort or contribution related to the mission. Capstone program participants, who are receiving course credit, are each expected by their course faculty to each contribute ten hours per week to their project. To accomplish their creative works (from ideation to proposal to execution), each Psyche Inspired participant is expected to spend five hours per week on their projects at a minimum, though for some it may be much more, depending on the project scope and medium. Science Outreach Interns, who are paid by the mission, each work an average of 16 hours per week (beyond contact time with mission personnel, which is approximately 4 hours per week).

The undergraduate programs, initially piloted at ASU during the 2017-2018 academic year, expanded nationally at the start of the 2018-2019 academic year and have involved undergraduate participants at 50 universities or community colleges through the 2021-2022 academic year (see Appendix C, Table 4) in 25 states with students majoring in dozens of fields and disciplines across the arts, humanities, sciences, and engineering (see Appendix D, Table 5). The programs continue to expand and diversify in terms of geographic location (of students and institutions), institution types (university, college, community college), and participating major or discipline.

2.1.2 Alumni Engagement

The hundreds of undergraduates that work with Psyche Student Collaborations each year will culminate in thousands of former participants by the end of the mission. As part of their final capstone project, in fall semester 2017 a Public Relations Psyche capstone team recommended the creation of an alumni program and suggested that the classes be designated with names, rather than years, since not all participants are seniors when engaging with Psyche Student Collaborations (for example, as science interns and docents or in Psyche Inspired). The naming convention adopted was a list of metals that may be found at Psyche (Elkins-Tanton et al. 2020, 2022), ordered by their number on the periodic (Appendix E, Table 6). The Psyche Student Collaborations team maintains private LinkedIn and Slack groups to connect with alumni and post internship and employment opportunities in all disciplines, highlight upcoming events, and share mission updates. A recent scan of current status of the alumni connected with the Psyche LinkedIn alumni group include 38 that are working for government agencies or non-profits, 53 that are still students (either finishing undergraduate degrees or in graduate school), and 380 that are working in industry. A periodic email newsletter is sent containing similar information. Alumni are able to reach out to Psyche Student Collaborations team members to request references for jobs, letters of recommendation for graduate school, and advice on postgraduate studies and careers.

Alumni have also proven to be a valuable source of advice and contacts for current participants. For example, Psyche capstone alumni were contacted in August 2020 to share advice (anonymously) they would like to give to incoming capstone students. Their responses were aggregated and became part of a presentation, "Use Capstone to Your Advantage," presented to ASU's Computer Science (CS) and Computer Systems Engineering (CSE) capstone students at the start of the 2020 fall semester. Themes of their advice included: Focus on teamwork and communication; make sure to document your work; network and build long-term relationships (as one noted, "Your teammates will be your long-term connections and possibly your references in job applications so try to foster a closely knit team dynamic"); be practical; and stretch yourself and have fun. Many alumni have also been willing to connect with current students to discuss their workplace, share tips on the job search, and even advise on continuing projects they worked on previously. This has been especially useful for students participating in the Psyche capstone program since most are graduating at the end of the program.

2.2 Psyche Capstone

Capstone courses (also known as Senior Design) are culminating, project-based courses undertaken by university students in any major in the final (senior) year of university. Although there is some variation in terms of logistics and implementation of different capstone or senior design programs (see e.g., Howe et al. 2017), a common strength of capstone courses is that they involve students in applying their knowledge and skills to real-world, team-based projects and "[offer] closure and a focus for the sense of achievement that comes with completion" (Lee and Loton 2015). With over 70% of colleges and universities offering capstone courses, Hauhart and Grahe (2015) estimate that approximately a million U.S. college students participate in capstone projects annually.

Although specific logistics and structures may vary by course, department, and university, in general they involve teams of students, course faculty, project sponsors or mentors, and the projects themselves (see Howe et al. 2017). Students form or are assigned by their professor to teams of an average of three to five students (or more, depending on the course or project need). The team selects or is assigned to a project based on a project brief, which is often written by the project sponsor (most commonly a company or faculty member), though sometimes formulated by the students themselves. The project brief outlines the motivation for the project, the desired outcome, and other relevant background information. In undertaking the project, the student team meets with the sponsor regularly, follows the schedule of deliverables required by the capstone course, and engages in independent and group work in pursuit of the project outcome. The ubiquity and diversity of capstone and senior design courses at colleges and universities around the country offers an accessible and sustainable way for missions to engage with undergraduates in small groups or at scale.

2.2.1 Program History and Structure

The Psyche capstone program started in the 2017-2018 academic year. To facilitate broad adoption, the program focuses on developing standalone capstone projects that may be undertaken by students in any capstone or senior design course (or across capstone courses in multiple disciplines) with relevant interests and abilities. Psyche projects range from the practical (a digital publication repository and archive) to the exploratory (automated meteorite image analysis) to the futuristic (robotic explorers designed to accommodate all potential hypothesized surface characteristics of the asteroid for as-yet unplanned future exploration). Some projects are designed to inform and enhance the Student Collaborations program directly. Examples include the creation of an online interdisciplinary capstone "marketplace" to match students with projects across disciplines, evaluation of the Psyche Inspired program, and an analysis of ways to optimize the capstone program itself to enable scalability. Other projects may inform the decision making and development of future mission public engagement opportunities, such as communications campaigns and comparative analyses of previous mission opportunities (discussed in Sect. 5). The process of capstone project formulation is discussed in Sect. 2.2.3.

Given the large variation across institutions in the structure, content, and administration of capstone courses (Howe et al. 2017), Psyche projects are intentionally open and flexible to being modified and adapted to accommodate the needs of participating faculty and courses. As an example, during the 2020-2021 academic year three teams from different departments and universities each worked on the project brief "Planetary Geologic Mapping of a Hypothesized Surface," which was proposed by a Psyche Co-Investigator. Each team chose to approach the project in a different way based on their experience, career interests, and requirements of their capstone courses. One, an interdisciplinary team made up of astronomy, geology, and engineering systems design students, focused on forecasting and implementing a feasible geologic map and developing methods for identifying interesting as well as safe landing sites for hypothesized surface explorers. A computer science team from a capstone course that is open to research-style projects chose to try to implement code capable of automatically detecting surface features. A computer science team from a different university whose course requires product development focused on creating a webpage capable of displaying ArcGIS maps of planetary surfaces (including the artist's rendition of the Psyche surface) and explaining core concepts of geologic mapping for users.

Being flexible to these types of variables has been a key element in the development of project briefs from the start of the program, which was piloted during the academic

Table 2 Capstone team compositions

	All On-Campus Students	All Online Students	On-Campus & Online Students
Single University, Single Discipline	•	•	•
Single University, Multiple Disciplines	•		•
Multiple Universities, Multiple Disciplines	•		•

year 2017-2018. The pilot year involved 10 teams of students at ASU working on four project briefs (with an additional two teams working on capstone-style projects but in a non-capstone course), including five interdisciplinary teams made up of computer science, graphic design, and engineering management students who worked to create a mobile app for the mission. From the start in 2017 through the 2021-2022 academic year, the program has engaged 19 university systems (see Appendix C) and worked with over 1,100 capstone students from majors in various fields of engineering, science, communications, and the arts (see Appendix D). At present capacity, the program engages approximately 300 capstone students in over 60 capstone teams each academic year.

Since exploration endeavors such as the Psyche mission, or any other complex, real-world problem, require the skills of a wide range of disciplines working together, the mission strongly encourages interdisciplinary capstone teams where possible. As few universities offer interdisciplinary or multidisciplinary capstone courses, the program works with capstone partners to offer interdisciplinary projects that can incorporate and accommodate students enrolled in separate single-discipline capstone courses. Through these projects, students from different disciplines work together on the project as an integrated interdisciplinary team while still meeting the deliverables of their individual capstone courses. In some cases, students from multiple universities work together as a team, providing an opportunity to prepare them to engage with a diversity of disciplines and collaborate in remote teams in the workplace, as is typical of large-scale projects like space missions. Engineering students have seldom worked on project teams with graphic design students, for example, and graphic design have seldom worked on projects guided by project managers, but such scenarios are common in the workplace and this experience can help provide them with important preparation. Additionally, through such capstone projects, universities may gain access to non-local, specialized technical mentors and to disciplines not offered at their institutions.

Aligned with the diversity of modern universities and the modern workplace, the mission has formed seven different team composition types to date (Table 2). ASU offers many online degree programs (https://asuonline.asu.edu/online-degree-programs/) and to-date the Psyche capstone program has worked with online students in engineering management (who often join other hybrid teams as project managers), software engineering, and electrical engineering. As online degree programs with online capstone course sequences become more common (both at ASU and at other partner universities), there may be opportunities in the future for online students from multiple disciplines and from multiple universities to team together in capstone.

Regardless of the type of team, the students receive some measure of structure from their capstone course itself. The Psyche capstone program also provides all participating students with a secondary, overarching structure to facilitate communication, professional development, and exposure to the mission. Key elements include onboarding (an introductory presentation about the mission, the Psyche team guidelines, team members and contact information; an Intellectual Property [IP] agreement to be signed; and other administrative tasks), a range of program communication and information options (messaging platforms [such as Slack, Ryver, and Basecamp], a shared Google drive, bi-weekly meetings with the program lead and relevant subject matter experts), and resources (mission speakers, optional one-on-one meetings with the program lead, technical support, training opportunities, relevant publications, and an alumni program). This program-level structure allows the mission to work with many universities and styles of capstone and senior design courses in a way that is sustainable in terms of staffing and resources and could be passed on to other missions or university entities interested in replicating or continuing the model following the conclusion of the Psyche mission.

2.2.2 Leveraging University Partnerships

Arizona State University (ASU), with more than 100,000 students spanning multiple campuses as well as a large online program, provided the opportunity to expand rapidly even in the pilot year of the program. Though there was no centralized office for capstone courses university-wide, a science outreach intern created an internal database of potential faculty contacts by searching the digital course catalog for senior-level courses with keywords such as "capstone", "senior design", or "project-based", as well as reviewing online undergraduate major maps. The Student Collaborations team approached faculty to arrange individual introductory meetings and then hosted a kick-off meeting in August 2017 for capstone faculty from multiple disciplines and ASU campuses to meet. The outcome was a pilot year involving 12 capstone (or capstone-style) teams with members from computer science, computer systems engineering, communications, graphic design, and engineering management, and the creation of an ASU-wide capstone email listserv with over 100 members. As planned, the following year the program expanded nationally, with partners solicited via conference participation (e.g., 2018 Capstone Design Conference), word-of-mouth (within ASU and beyond), social media posts directing interested participants (whether faculty or students) to submit their details via the online interest form, and, importantly, direct referrals from mission team members. Referrals from mission team members may be in the form of a direct contact at their alma mater, a colleague working at a university, or a previous institutional partnership.

As an example, NASA's Glenn Research Center (GRC) is located in Cleveland, OH and has a long history of providing capstone projects and mentorship to student teams at Cleveland State University (CSU) in the Washkewicz College of Engineering. Through their prior connection, Psyche team members at GRC offered an introduction to the CSU capstone coordinator, as well as applied their subject matter expertise in solar electric propulsion (SEP) to the development of capstone project briefs and subsequent technical and scientific mentoring (both of CSU teams and others, see Sect. 2.2.3). The direct referral from GRC served as a "stamp of approval" for the emergent Psyche capstone program, providing a strong foundation from which the partnership with CSU has grown and a model for other direct referrals that may be possible in the future. Through the 2021-2022 academic year, the Psyche capstone program has worked with CSU capstone teams each year since national expansion, including with teams focused on GRC-generated projects related to SEP (such as Xenon Flow Controller; Hall Thruster Database; and Hall Thruster Visualization), with teams receiving technical and scientific mentorship from GRC mission team members.

2.2.3 Project Formulation and Mentoring

Capstone project briefs are solicited from mission team members annually and result in 15-25 distinct project briefs each year. The criteria for projects is that they are relevant

Table 3 Summary of Psyche capstone project formulation approaches

Approach	Description	Example Project(s)
1) Self-contained challenge	A general space mission development challenge is isolated for student teams to explore areas of engineering that may have been unexplored or underexplored during spacecraft development and may prove fruitful in future applications	• Xenon Flow Controller
2) Harness fresh perspectives for existing challenges	Fresh perspectives & new skills of students are used to further a field of inquiry faced as part of the Psyche development that may not have been fully solved or that resulted in the opening of additional avenues for research.	• Machine Learning Analysis of HET Facility Effects • Solar Wind Propagation • Meteorite Image Analysis
2) Mission team standalone projects	Standalone projects inspired by the Psyche mission team. Deliverables provide tools that can be leveraged by the mentor and the mission to further educational and public outreach goals while also providing the student teams the opportunity to directly contribute to an active NASA mission.	• Hall Thruster Visualization and Animation Project • 3-D Asteroid Viewer • Science Activity Planner
4) Future-looking projects	Future-looking projects that push students, as well as science and technical mentors, to create solutions to challenging "what if" scenarios	• Hypothesized Surface Exploration (Landers, Robotics, Sampling, etc.)

to the mission but not in the critical path—ideally projects that the mission team member thinks would be a good idea to try or would be nice to have completed but has not had the time or resources to pursue themselves. Projects may be intended to be completed in a semester, an academic year, or across multiple academic years by multiple teams. As appropriate, projects may be modified to accommodate student interests, skills, or previous coursework, and new projects or extensions to previous projects may be added. Mission team members submitting projects are not required to provide scientific or technical mentoring of the projects but are welcome to participate at the level they choose, and this varies by team member and project. Below are examples from mission team members involved in capstone, which illustrate the range of approaches taken in formulating projects and providing scientific or technical guidance and mentoring. Formulation of Psyche capstone projects generally follows four approaches (see summary in Table 3): 1) a specific, self-contained challenge faced during mission development; 2) the opportunity to re-think or refine existing solutions through the fresh perspectives and skills of students in interdisciplinary teams; 3) a standalone project inspired by, or of use to, the Psyche team; and, 4) future-looking projects that push students, as well as science and technical mentors, to explore solutions to challenging "what if" scenarios.

One project formulation approach is to isolate a self-contained challenge faced during general space mission development and present it to the students. The Xenon Flow Controller project from the 2018-2019 academic year is an example of this approach. In this project, student teams from Cleveland State University (CSU), Arizona State University (ASU), and Florida A&M University-Florida State University (FAMU-FSU) were asked to design a system to control the flow of xenon propellant to the electric propulsion system used by the spacecraft and described in detail in Snyder et al. (2020a). The students were given similar requirements to those imposed on the system actually developed for the spacecraft and provided an opportunity to apply their classroom knowledge to replicate the real-world engineering processes that lead to closure of that design challenge on the actual spacecraft. From the mentor's perspective, this approach provides the unique opportunity to leverage the creativity of student teams to explore areas of the engineering trade space that may have been unexplored or underexplored during spacecraft development but may prove fruitful for further development or future applications. Given that a known engineering solution exists for the projects developed in this project formulation approach, the challenge for the mentor is to avoid reducing the project to a rederivation of that solution. Instead, the mentor is challenged to support and guide the student teams towards a technically viable solution without stifling creative alternative solutions.

A second formulation approach is to harness the fresh perspectives or new skills of student teams from different technical disciplines in order to further a field of inquiry faced by the mission that may not have been fully solved or that resulted in the opening of additional avenues for research. An example of this approach is the Machine Learning Analysis of Hall Thruster Facility Effects Data project from the 2019-2020 academic year. One of the challenges facing the spacecraft is the difficulty in predicting on-orbit performance of the spacecraft's electric propulsion system due to the known sensitivity of these devices to their operational environment and the inability to perfectly replicate orbital conditions on the ground (Snyder et al. 2020b). For the mission, detailed performance measurements were taken in the lab pre-launch and used to create a model for approximating on-orbit performance. In this project, computer science student teams were presented with the same data and asked to apply advanced machine learning techniques in order to determine if these more advanced techniques could improve the performance predictions and uncover previously unexplored sensitivities and trends. For the students, this approach provides an opportunity to apply their classroom knowledge to a real-world problem while, for the mentor, it provides an opportunity to explore new approaches and methodologies to further their technical inquiries. Here the mentor is challenged with distilling their central research problem so that it is understandable to student teams outside their core discipline and navigating that difference in disciplines in order to achieve a mutually beneficial outcome.

The Solar Wind Machine Learning Propagation Model capstone project is another example of such an approach. This project is directly related to the magnetometry investigation, which is tasked with determining whether the asteroid is magnetized and to what extent. Since the asteroid's field would be distorted by the solar wind flow around it, knowledge of the solar wind conditions at the spacecraft can assist in interpreting the measurements. Since the flight system does not carry instrumentation that directly measures the wind, the magnetometry investigation will use concurrent observations of the solar wind measured by other spacecraft near Earth to estimate the wind condition at the spacecraft. However, accurate predictions can only be achieved for certain alignments of the spacecraft, Earth, and the Sun.

The solar wind propagation model capstone project explored the possibility that a machine-learning based model can out-perform existing methods, designed as an exploratory project that complements efforts within the Magnetometry Investigation team to build a non-machine learning based model. In the first iteration of this project brief (during the 2019-2020 academic year), the students were undergraduates attending ASU, with technical advising from Magnetometry Investigation members from Massachusetts Institute of Technology (MIT). Two teams worked in parallel on the project. For the students, this project offered the opportunity to learn about a real-world problem facing the science team: how to remove uncertainty from the observations and make a more accurate interpretation of the data. Furthermore, this problem is open-ended and stands at the forefront of current research efforts. As such, it allowed the students to carve their own path and explore different methods. The student teams also self-organized and divided the project into subtasks between different participants. Some tasks were carried out by several students independently, which provided a means for comparing different machine learning approaches.

For the mentors, this project provided the advantage of taking on a wholly different approach to a problem already in the works within the science team, without diverting team resources to exploring this avenue. This was particularly suitable to a machine-learning based project: the field of machine learning is a growing field and new approaches and software platforms are being constantly introduced. At the same time, the scientific return from machine learning models, specifically in fields like space physics, is still uncertain. It is therefore more suitable to use such methods for exploratory projects, rather than putting them on the critical path of mission design and planning. The outcome of the first iteration of the project was that several students succeeded in producing software tools for processing data from existing missions and designing preliminary predictive models. These were later used to inform the magnetometry investigation team in carrying on further projects. Lessons learned were also implemented in the second iteration of this capstone project during a subsequent academic year.

Challenges encountered during this project will help to more clearly define and manage future projects. For example, while the project was mostly aimed at computer science students familiar with machine learning, it still required students to develop an understanding of what the quantities in their machine learning models represent in the real world, so they could make meaningful inferences and be able to test the model. To achieve this, students were given written materials and short "crash course" presentations by the scientific advisor. However, retention and integration of such information by students from a different discipline is challenging. Another hurdle to student performance was clear and well-defined instructions. While mission scientists are accustomed to exploring open-ended questions with no "right answer", undergraduate students are not often required to engage in such activities. However, while a more narrowly defined project can produce better performance from the students, the open-ended nature of the "real world" capstone project has the advantage of exposing the students to the challenges and frustrations of real discovery work, which can be a unique and inspiring experience.

Another project was formulated from an underappreciated challenge in planetary science missions–ensuring a robust set of data for comparison to mission results. The Meteorite Imaging Project addressed a fundamental weakness in current knowledge of the abundance of sulfur, phosphorus, and silica-bearing minerals within iron meteorites. Although the bulk of Psyche's material was expected to be metal per the mission team's pre-flight assessment, the light elements are important constituents of the core of the Earth and other terrestrial planets and are known from iron meteorites. The existing methods for quantifying these elements' abundance in large slabs of iron meteorites is, however, tedious and underutilized.

The challenge put forth to the capstone teams was to harness the power of imaging, image processing, and feature recognition, with only initial user training of the system, to determine the abundances of these light elements in all available large slabs of iron meteorites in the collections of Arizona State University and the Smithsonian's National Museum of Natural History. Over multiple years, teams have combined expertise from students in project management, engineering, photography, image analyses, and automated image processing. Faced with real world challenges, including the need to provide uniform illumination and viewing geometry, the ability to distinguish different minerals that contain these light elements, challenges in edge detection of these phases, and automated processing, the teams have developed a wide range of solutions to date, both from the engineering perspective and software development.

Teams over five academic years have benefited from this real-world experience and made, most notably, significant advances in the systems needed for image capture and software development. A side benefit is the growing relationship between the mentors and the institutions involved, including a number of virtual and in person consultations with experts in meteoritics and planetary science to whom a largely engineering-focused group might not otherwise be exposed. Though the project was put on hold due the pandemic, but continued starting with the 2021-2022 academic year.

A third approach is to present students with a standalone project inspired by, or of use to, the Psyche mission team specifically. The Hall Thruster Visualization and Animation Project from the 2019-2020 academic year is an example of this approach. In this project, computer science capstone teams were challenged to create an entertaining and interactive animation to educate a general audience on the electric propulsion system used on the spacecraft. The final project deliverables provided tools that could be leveraged by the mentor and the mission to further their educational and public outreach goals while simultaneously providing the student teams the unique opportunity to directly contribute to an active NASA mission.

Another example is a need identified by Psyche team members (and of use to researchers and the general public) to have an accessible, quick-glance way to determine what sorts of data are available for a given target of interest on a planetary object. A seemingly straightforward solution to this problem is a graphical presentation of the target (in this case, eventually the Psyche asteroid) with easy-to-find information on what data have been taken for each point on that target. There are some detailed and powerful tools to do this, in particular those developed by NASA's Solar System Treks project (https://trek.nasa.gov) and JMARS (https://jmars.mars.asu.edu/). However, a use case exists for a less feature-rich tool to quickly identify the data sets acquired to date. The Psyche capstone 3-D Asteroid Viewer project was proposed with the aim to present to the user, through a web browser, a simple graphic of a target in the form of a globe which can be rotated on all axes, zoomed in and out quickly, etc. The graphic is normally presented in shaded relief. The user is able to zoom into an area of interest and query the tool to see what data have been acquired. This is provided in the form of outlines of data takes (e.g., image outlines) on the globe or the centers of such data takes. By pointing and clicking on a data take location, a list of what data are in the Planetary Data System becomes available. (These functions are similar to what was once provided by the browser-based Google Mars website prior to it being incorporated into the Google Earth computer program [see e.g., Google Earth Blog 2009; NASA 2009) As a recurring project since the 2018-2019 academic year, teams of students at several universities have been designing and developing this tool, learning to use specialized NASA products such as the Planetary Data System (https://pds.nasa.gov/) and NASA's SPICE, "an information system… to assist NASA scientists in planning and interpreting scientific observations from space-borne instruments" (https://naif.jpl.nasa.gov/naif/), strengthening and building upon each other's work each year. The Psyche team mentor at JPL (and other team members as appropriate) attends monthly or bimonthly videoconference meetings during the academic year. At each meeting, the tool is demonstrated online to the mentor, using a target with existing datasets from flyby or orbital missions (e.g., the Dawn mission at Vesta). Each step in the design and development process is thereby made clear and the mentor shares in the excitement of the project with the students. In addition, the mentor is able to provide advice on how to improve current characteristics of the tool and suggest future modifications or additional features prior to sharing the tool with the Psyche team or the public.

The Activity Planner capstone project, pursued by capstone teams at two different institutions (ASU and Penn State Behrend) in 2020-2021, is an example of a capstone project with significant positive impact to the mission. This project idea was inspired by the nature of the Psyche operations concept, where reference plans are generated during the Integrated Sequence Build process before the start of orbit operations and where activity absolute start times will change due to geometric epoch updates late in the uplink process. These concepts together mean that quickly understanding distinct classes of differences between the approved reference plans and the current baseline is essential to having confidence while approving the uplink products. Though not in the critical path for operations, the desire for a plan "diff" (difference) software tool on Psyche was a lesson learned from NASA's Dawn mission. In addition to operations, the tool will be useful in development as well since many notional or preliminary baseline plans are already being created and understanding differences between them can speed analysis and validation.

Both teams designed tools that essentially consisted of two parts: one part which evaluates changes to activity plans and one part which displays those differences. The first part analyzes activity plan JSON files and determines which and how many activities were added or removed from one plan to another, as well as which activities have modified start times, durations, or parameters. Because activities can be timed relative to geometric epochs, the tools need to be able to distinguish between activity start times that have shifted due to a change in geometric epoch timing and activity start times that have shifted due to manual changes. The second part of the programs display the results of this analysis in an interactive format. Users can see timelines of activity plans side by side, see differences between the plans highlighted by color, and hover over activities in the timeline to display tooltips with more detail.

Both capstone teams that worked on the Activity Planner project produced tools that the Psyche operations team could use as-is, although there will be ongoing work at JPL to take the best ideas from both tools and create one final version in time for use in Operational Readiness Tests. The availability of this tool is such a large quality-of-life increase that current Psyche MPST and SOST members are excited to use it as soon as it is available to them. The project was designed to have positive outcomes for the students as well. For many, this was their first experience with a medium-scale software project that was less structured than a typical class project, so they got experience talking to customers and interpreting their responses to form use cases, as well as making wide-ranging decisions about implementation to best meet them. They were also coached by JPL engineers about best practices like unit testing, git flow, encapsulation, and more. Technology-wise, the teams gained experience with PyQT, dash, plotly, numpy/pandas, and JPL open-source activity- and time-related libraries.

A fourth approach is to formulate future-looking projects that push students, as well as science and technical mentors, to create solutions to challenging "what if" scenarios. For Psyche, these projects have centered on teams proposing, designing, simulating and/or prototyping solutions for mapping, landing, exploring, and sampling the range of hypothesized surfaces of Psyche. These projects are similar to the types of exploratory projects undertaken by researchers and mission formulation teams at NASA centers and universities and funded by programs like NASA Innovative Advanced Concepts (e.g., https://www.nasa.gov/directorates/spacetech/niac/2018_Phase_I_Phase_II/Shapeshifters_from_Science_Fiction_to_Science_Fact/).

2.2.4 Capstone Products

Capstone courses in all disciplines are structured so that the end of the course (if one semester) or course sequence (if two semesters) results in a final product from each team

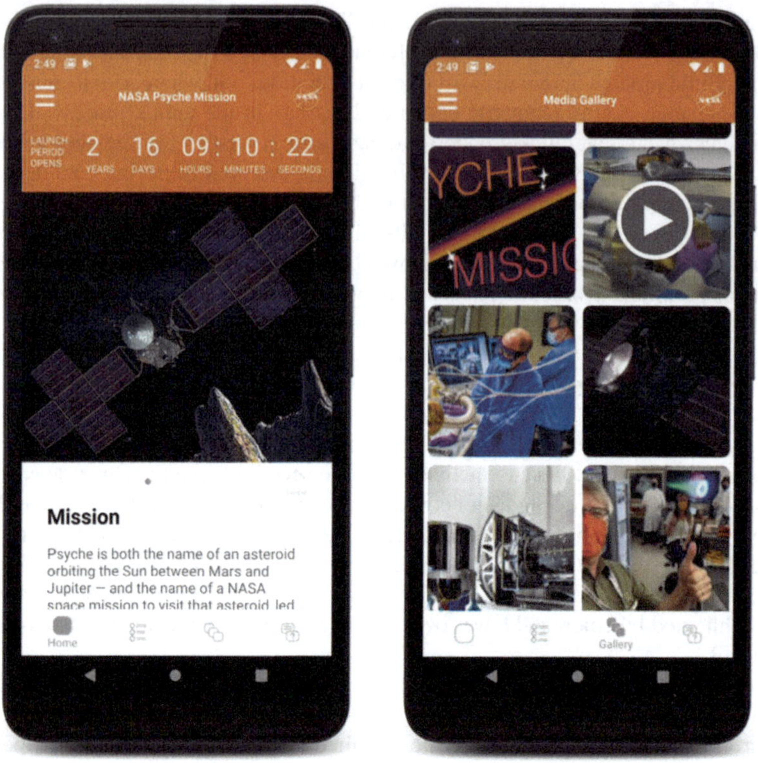

Fig. 4 Capstone team-built mobile app

that is shared with the capstone project sponsor and showcased to their school and the public through a culminating event, presentation, or web-based summary. The course faculty structure the interim deliverables and assignments to help the teams reach this conclusion. Each team provides the sponsor with a final report and, if appropriate to their project, a final deliverable such as software, a prototype, or implementation plan. In the case of some Psyche projects that are especially challenging and long-term, requiring multiple years of teams to complete, the final deliverable includes handover materials (such as a github repository, equipment, instructions, recommendations, etc.) to guide a subsequent team. As projects are finalized, they are made available to the project advisors (such as with the Activity Planner) or shared with the public (as in the case of a public mobile app: https://psyche.asu.edu/mobile-app; Fig. 4). Regardless of the destination of the final product, each team has a dedicated page on the Psyche capstone program page with a URL that they may include on their resume to document their participation.

Beyond products from the capstone students themselves, the Psyche capstone program has resulted in presentations and posters at the *Capstone Design Conference* and the *Undergraduate Research Programs* conference, the *Council on Undergraduate Research* conference, as well as an article in the *International Journal of Engineering Education*. Additional dissemination and professional development efforts take place through a summer capstone faculty journal club organized by the Psyche capstone program team and a workshop on interdisciplinary, inter-university capstone teaming as part of the *Capstone Design Conference* virtual offerings in summer 2021. It is also hoped that the Psyche capstone program can provide a feasible model for other PI-lead missions to consider that are developing undergraduate-focused Student Collaborations plans.

2.2.5 Feedback and Program Improvement

The individual capstone courses and instructors conduct student- and team-level assessment, including soliciting capstone project sponsor feedback, and assign grades as directed by their department and institution. Additionally, the Psyche capstone program lead and the capstone student manager conduct an IRB-approved internal program-level evaluation using an adapted version of empowerment evaluation (Fetterman 2001) focused on stakeholder (program staff and participant) engagement. It is conducted electronically (via survey) and spread across the full academic year, in which students participate in "(a) developing a mission, vision, or unifying purpose [for their year of the program]; (b) taking stock or determining where the program stands, including strengths and weaknesses; and (c) planning for the future by establishing goals and [determining] strategies to accomplish program goals and objectives" (Fetterman 2001). Used with a previous NASA student program with distributed participants (Fetterman and Bowman 2002), the modified, survey-based empowerment evaluation process, repeated annually, provides an efficient way to engage participants in identifying the salient elements of the program, guiding iterative corrections and enabling continuous program improvement, and informing future adaptations as the needs of participants and other stakeholders change over the lifetime of the program (see discussion of the Psyche capstone evaluation in Bowman et al. 2019 and Talamante et al. 2022).

2.3 Psyche Inspired

Space exploration is often thought to be dominated by engineers, scientists, and the imaginations of science fiction writers. But as in any large human endeavor, there is need and room for many kinds of expertise, including artists and other creators, as NASA has long recognized through groups and projects like The Studio at JPL (https://www.jpl.nasa.gov/thestudio), the Mars as Art exhibits (http://mars.nasa.gov/multimedia/marsasart/), the JWST Art campaign (https://www.jwst.nasa.gov/content/features/jwstArt/index.html), and many others). Psyche Inspired addresses this by bringing undergraduate students together, virtually, to spend an academic year as "creatives-in-residence" to share the excitement, innovation, and scientific and engineering content of the mission with the public through artistic and creative works. The Psyche asteroid, the mission's destination, remains unique and mysterious, both because it will be the first metal-rich world ever explored and because it is not yet known what it looks like. Through a telescope, it looks like a bright spot; with radar, scientists have learned about its approximate shape and size, but not its appearance. This provides a special opportunity for creative exploration and for inspiring the public (and the mission team, e.g., Appendix F) and encouraging people worldwide to dream, innovate, and explore in their own lives.

2.3.1 Program History and Structure

Each year the interns are selected through an online application process that includes submission of sample works and written responses to four questions (example application details available at https://psyche.asu.edu/get-involved/psyche-inspired/psyche-inspired-application-2022/). The application is advertised as widely as possible through direct emails and email newsletters (e.g., NASA's Museum & Informal Education Alliance email newsletter; NASA STEM Express), listservs, social media (both the mission's social media accounts and those of collaborators and other NASA accounts), and national organizations that reach undergraduates at universities and community colleges. The program selects up to 16 students annually (approximately 10-30% of applicants, depending on the application cycle), in honor of Psyche's place as the 16th asteroid discovered. The selected interns are supplied with a stipend to support the creation of 4 (or more if they wish) original creative works to excite and educate the public about Psyche. In addition to weekly virtual meetings, interns create connections with speakers brought in from various disciplines and roles related to the mission, partner institutions, or professional artists and other creators.

Psyche mission team members support the Psyche Inspired program by volunteering their time to review the submitted sample works and accompanying interpretive text during the application period through a streamlined process enabled by a custom web application developed by Psyche software engineering interns. Approximately 10 Psyche team members from science, engineering, management, and public engagement participate in this phase each year. Written responses to the application questions are reviewed and scored by the Psyche Inspired management team based at ASU (the Student Collaborations lead plus 2-4 interns, depending on the year), who review and score each application separately, following the selection process guidelines that are outlined for applicants in the application (see Appendix G).

The Psyche Inspired program was piloted during the 2017-2018 academic year with 13 competitively selected interns from ASU. This inaugural group was referred to as the Titanium Class, as part of the naming convention of the Psyche Alumni program (see Sect. 2.1.2). Since all students in the pilot program were on-campus ASU students, they met in-person weekly with mission team members. At these weekly meetings, interns proposed and produced artworks about the Psyche mission, and they were exposed to subject matter experts working on the mission, to provide context and content. Produced artworks were shared through the mission social media accounts, exhibited to the public via an on-campus Showcase, and compiled into a coffee table book, which was made freely available via download to the public with printed copies sent to interns and mission leadership for outreach purposes.

Following its inaugural year, the Psyche Inspired program expanded nationally. The second cohort of Psyche Inspired (2018-2019), referred to as the Iron Class, was comprised of students from nine different universities in pursuit of 12 different majors, followed by the third year, referred to as the Cobalt Class, with 16 interns from 12 different universities in 18 different majors (including double majors). The Nickel Class, representing the fourth year of Psyche Inspired, included 15 interns from 8 different universities across the nation in pursuit of 15 different majors (including double majors). The Copper Class, representing the fifth year of Psyche Inspired, included 15 interns from 14 different universities in 13 different majors (including double majors). Through these first years of Psyche Inspired, participants from 38 colleges and universities (see Appendix C) have created more than 260 artworks to engage the public in the mission. These pieces and any subsequent works may be found in the Psyche Inspired gallery: https://psyche.asu.edu/galleries/artwork/.

2.3.2 Psyche Inspired Products

The primary products of Psyche Inspired are the artworks themselves (see e.g., Fig. 5, Fig. 8). Over the course of an academic year (September to May), each Psyche Inspired intern creates 4 creative works meant to communicate Psyche mission concepts to the public.

Fig. 5 Psyche Inspired projects range from posters (L: Daniel Zepeda-Cuba) to a handmade dress (R: Siena Smania)

In total, each cohort is responsible for the creation of more than 60 art pieces. Each piece is displayed, with accompanying interpretive text, on the interns' individual webpages on the Psyche mission website, as well as highlighted on the @MissionToPsyche social media accounts throughout the year. They are further archived and disseminated through a coffee table book that displays each piece along with introductory front matter such as letters from the program's student manager and the mission principal investigator and short descriptions of relevant mission content. The covers of these coffee table books are designed by either a member of the Psyche Inspired class or by a member of the Psyche Student Collaborations team. The books are made available to the public via a downloadable PDF on the Psyche mission website.

At the end of the academic year, Psyche interns have their works exhibited in an annual showcase. Originally, the showcase was held entirely in person but, as the Psyche Inspired program expanded to include national participants, it has evolved to incorporate a virtual component as well. This hybrid approach allows local communities to see a subset of works in person as well as appreciate the full catalog of art through online exhibits.

In response to the COVID-19 pandemic, the Psyche Inspired program transitioned its showcase to an entirely virtual format. Interns were interviewed in pre-recorded sessions to discuss their participation in the program. These interviews were made available on their respective web pages along with an opening introduction video by mission leadership. In addition to these recordings, the public was encouraged on social media or through a comment form to send in questions to the interns about their experiences within the program and their inspiration and interpretation of their art pieces. As appropriate, questions were distributed to the interns and intern responses were posted and used along with the pre-recorded interviews to create short videos to help promote the application for the next Psyche Inspired cohort.

The virtual showcase experience was enhanced by the use of a 3D gallery created on the free *ArtSteps* platform. With this platform, the public was able to walk around the virtual gallery and enjoy the pieces created by the interns by interacting with objects to read about both the pieces and the interns that created them. As possible, capstone or other undergraduates contribute to the showcase, such as an additional virtual experience created for the Cobalt Class Showcase through the involvement of a mobile app development class at ASU. Through this partnership, an app was designed to showcase the Psyche Inspired pieces, as well as highlight the nationwide distribution of interns within the Cobalt Class.

For the Nickel Class, an ASU Public Relations capstone team aided in the development of promotional content and marketing plans for Psyche Inspired, as well as showcase design and implementation.

Psyche Inspired art pieces are also exhibited beyond their cohort year at public events (such as The Art of Planetary Science at the University of Arizona, ASU Earth & Space Exploration Day and Homecoming, and Phoenix Fan Fusion) and museums (such as the i.d.e.a. Museum in Mesa, AZ, the MOXI in Santa Barbara, CA, and the Phoenix Airport Museum in Phoenix, AZ). The works are cataloged and stored at ASU when not on exhibit at local, regional, and national events and spaces.

2.3.3 Feedback and Program Improvement

As the Psyche Inspired program has grown in its reach, structured vetting efforts of proposed artwork concepts have been implemented to ensure that Psyche mission content is accurately (while still creatively) represented in the artworks and accompanying interpretive text. This is implemented through three formal checkpoints during the development of each artwork: Once when the artwork concept is proposed, once at the halfway mark, and once at completion. At these checkpoints, the artworks are reviewed by program coordinators and the mission principal investigator. The Psyche mission content portrayed in the works and the written descriptions provided as interpretive text are reviewed for potential misconceptions, and feedback or editing is provided to the interns as appropriate. Art pieces are also reviewed and discussed informally during Psyche Inspired group meetings when interns are encouraged to share the status of their project with their cohort, further ensuring accurate portrayal and discussion of Psyche mission information throughout the creative process.

Beyond the feedback cycle between the Psyche Inspired interns and the program coordinators about the artwork, the coordinators implement program-wide improvements based on student feedback gathered through one-on-one check-ins throughout the year and a final group debrief following the Psyche Inspired Showcase. Additionally, Psyche Inspired has leveraged the talents of capstone teams participating in the Psyche capstone program. This interdisciplinary, inter-program approach allows the two programs and their participants to inform one another's experiences. For example, in 2019, an applied engineering capstone team from Michigan State University (MSU) conducted a formal analysis of Psyche Inspired. They assessed the program timeline, joined Psyche Inspired meetings to engage with interns and observe the synchronous component of the internship, and conducted surveys and interviews with the interns. Their resulting recommendations (such as streamlined due dates, thematic project assignments, greater involvement by science, engineering, and arts experts, and requiring interns to complete the introductory *Process and Lifetime of a Space Mission* online course [see Sect. 3.1]) were implemented by the program with the subsequent cohort. In spring 2021, a public relations capstone team from ASU devised a marketing plan with creative methods for advertising and augmenting the year-end showcase, leading to the creation of a special Psyche mission Snapchat filter to boost social media presence, virtual gift bag items, and future plans for a Unity experience for users to view the art pieces in a more elaborate virtual setting. Other projects relevant to Psyche Inspired undertaken by Psyche capstone teams include an industrial engineering team that built an online artwork archiving system to keep track of the artworks, including temporary loans to exhibit spaces, and multiple computer science and software engineering capstone teams that have incorporated the art into their web-based game projects.

3 Public Programs

Though all Psyche Student Collaborations programs are university- or undergraduate-based, many also include components that are available to the public, including lifelong learners worldwide and K-12 teachers and students, supported by undergraduate student participants.

3.1 Innovation Toolkit – Online Courses

The NASA Psyche Mission Innovation Toolkit (https://psyche.asu.edu/get-involved/innovation-toolkit/) includes a series of free online courses based on the real-world challenges and skills associated with the mission's science, engineering, technology, and teamwork. The short courses are self-paced, may be started at any time, and may be taken as stand-alone courses or be completed as part of a sequence. Each course results in a certificate of completion and may be included on a participant's resume or CV as "professional development." Though targeted at early college level, through these courses lifelong learners worldwide can learn and practice a range of new information and skills critical for innovation - both technical and "soft" skills sought after by employers in diverse work environments.

Developed in collaboration with ASU's EdPlus (digital education group) and student worker science writers, with the expertise of subject matter experts from the mission and other collaborators, the courses incorporate new and pre-existing content and materials, with the goal of presenting them in ways that are engaging and comprehensible to a wide audience. Topics alternate between science or engineering content (e.g., space mission life cycle and small solar system bodies) and general workplace skill or competency content (e.g., inclusive teamwork or troubleshooting strategies). Current and upcoming courses descriptions are outlined in Appendix H.

The courses are advertised via traditional and social media, flyers at public events, and via team member presentations. As of the end of the 2021-2022 academic year, more than 2,500 learners have enrolled in *The Process and Lifetime of a Space Mission* and almost 700 learners have enrolled in *Inclusive Mindset.* Upcoming courses include *Countless Worlds in Our Solar System: Asteroids, Comets, and Meteorites* and *Strategies for Troubleshooting Problems in Everyday Life and Work.* Additional courses will be developed throughout the lifetime of the mission.

3.2 Science Outreach Interns and Docents

As with other university-based principal investigator-led missions, student workers, both undergraduate and master's students (as well as summer students and other interns at partner institutions as described below) contribute significantly to Psyche Student Collaborations efforts. These students are employed as either science outreach interns (24 as of the 2021-2022 academic year) or docents (14 as of the 2021-2022 academic year), with associated responsibilities and opportunities. Doctoral-level science and engineering students and postdocs are employed by individual science co-investigators. Some participate with Psyche Student Collaborations as volunteers, but it is not a requirement of their involvement with the mission.

3.2.1 Intern and Docent Overview

Psyche science outreach interns directly support and co-manage the programs within Psyche Student Collaborations and associated public engagement efforts. They are hired for specific program needs and, with a wide range of majors and previous experience, they provide essential work and a fresh perspective. With mission team member guidance, the interns co-manage the Psyche capstone program, Psyche Inspired, the mission website, @MissionToPsyche social media, and school visits and public events. They develop and edit content for the Innovation Toolkit online courses, public outreach materials, and conference and journal publications. They provide support to Psyche capstone students and Psyche Inspired interns, helping with technical questions, programming, access to resources, review of materials, and administrative documentation.

Docents work for the ASU School of Earth and Space Exploration (SESE) and are partially supported by the mission to engage more than 10,000 K-12 students and teachers annually in giving tours of the Gallery of Exploration at ASU, including an introduction to the mission (this number was accurate as of the 2019-2020 academic year, prior to the pandemic, and is expected to return to pre-pandemic levels). The docents receive training from the Community Outreach, Education, and Public Engagement staff for SESE at ASU covering a variety of subjects, including robotic exploration, geologic and planetary sciences, and instrument assembly and testing. Their majors range from music and performing arts to communications to various engineering disciplines and the sciences, which enables them to connect in unique ways with diverse groups of K-12 students and teachers.

Psyche Student Collaborations has also benefited through the participation of NASA Space Grant Interns (as volunteers at events) and summer students (such as students participating in a Research Experiences for Undergraduates [REU] program or interns working with Co-Investigators at other institutions). Depending on their role, these students are onboarded through the standard Psyche Student Collaborations process and integrated into the team.

3.2.2 School Tours

Interested Phoenix-area schools can schedule an onsite guided tour of SESE's Gallery of Scientific Exploration. During these tours, K-12 students and their teachers engage with a university student docent to learn about space exploration and the range of people contributing to it. Docents (predominantly undergraduates majoring in the sciences or engineering) are trained by educational, science, and engineering subject matter experts in ASU's School of Earth and Space Exploration.

Participating school groups learn about the Psyche mission through discussions about differentiated bodies, meteorites, cratering, and the simulated experience of flying by the asteroid as a part of the 3-D astronomy show they attend. Each year since the mission's selection, the reach of the school tours has grown by approximately 10%. The program reported its highest reach to-date in 2019 (pre-pandemic), engaging with over 10,000 students. The tour groups represent over 100 schools and 40 districts within Arizona. To date, the bulk of participating classes (95%) are between 5th and 8th grade. Those in grades 9-12 and homeschool groups make up an additional 4% and the remaining (1%) are from grades K-4th.

3.2.3 School Visits

For K-5th grade classes in the greater Phoenix area unable to visit the ASU campus, the Psyche outreach interns offer on-site classroom visits with the goal of exciting students about space exploration and the mission and allowing them to meet college students and ask questions about their education and career aspirations. The content of the classroom visits is tailored to the age group and focused on what the mission entails, why the team is interested

Fig. 6 JPL Ticket to Explore (shown) and other public events locally and nationally are supported by Psyche mission team members, including science outreach interns

in studying Psyche, and the Student Collaborations opportunities of the mission. During the 2019-2020 academic year, the intern team visited 10 schools and presented to approximately 400 students and teachers. As a result of the COVID-19 pandemic, school visits for the 2020-2021 and 2021-2022 school years were canceled, however, materials from the school visits are also shared on the mission website by grade band (https://psyche.asu.edu/get-involved/models-and-materials/) and were used for a number of virtual school-related events during the pandemic.

3.2.4 Public Events and Presentations

The science outreach interns, docents, and many members of the mission team from universities, NASA centers, and partner institutions participate in public events and presentations each year, and efforts are made to reach and inform a wide range of communities and stakeholders about the mission. This includes scientific and technical peers at major conferences (e.g., American Geophysical Union [AGU] Fall Meeting, American Institute of Aeronautics and Astronautics [AIAA], Institute of Electrical and Electronics Engineers [IEEE], Capstone Design Conference, Undergraduate Research Programs conference), large local public events (e.g., USA Science & Engineering Festival, Phoenix Fan Fusion, JPL Ticket to Explore, ASU Earth and Space Exploration Day, and the Cambridge Science Festival in Massachusetts), and smaller community organizations (e.g., astronomy clubs, school science nights). In the first four years since mission selection, the mission team has participated in hundreds of events reaching more than 100,000 attendees. Mission team members coordinate with the Student Collaborations team to report their participation details and to request materials or student staffing if needed.

4 Outreach Products and Efforts

Psyche Student Collaborations interns and program participants formulate, create, implement, maintain, and disseminate many mission-related outreach products and efforts (e.g., the mission website, social media, videos, and public events and presentations) in close coordination with the Psyche Student Collaborations lead and Psyche mission team members. This helps extend the reach of the mission and the Student Collaborations program beyond the program participants themselves to many new audiences.

4.1 Mission Website

The Psyche mission is represented by NASA (https://www.nasa.gov/psyche/), JPL (https://www.jpl.nasa.gov/missions/psyche/), and NASA Solar System Exploration (https://solarsystem.nasa.gov/missions/psyche/in-depth/). Although NASA's webpage is considered the primary page, as with other principal investigator-led missions (e.g. Lucy: http://lucy.swri.edu/, OSIRIS-Rex: https://www.asteroidmission.org/, MAVEN: https://lasp.colorado.edu/home/maven/) the mission team also maintains https://psyche.asu.edu/ to provide extensive in-depth information about the science, engineering, timeline, and public opportunities related to Psyche and, once at the asteroid, will host processed and calibrated Imager data (see Sect. 5.3.4; raw uncalibrated images will be hosted at JPL). The website and other outreach products align to the Psyche Brand Guidelines, which serve "to establish boundaries and best practices in the use of the Psyche Mission badge, visual system, and appropriate copy [to]... ensure consistency and quality in all Psyche-related designs" (https://psyche.asu.edu/wp-content/uploads/2018/03/20200528_Psyche_BrandGuide-v2_6.1_20_rev-.pdf). Details about the Psyche mission website are provided in Appendix I.

4.2 Social Media

Major mission milestones and highlights are shared via @NASASolarSystem on Twitter, Facebook, and Instagram (and occasionally amplified by other NASA accounts), which is managed by JPL. To share daily mission content, opportunities to get involved, and an inside look at the team, the mission uses @MissionToPsyche (formerly @NASAPsyche) on Twitter, Facebook, and Instagram, which is managed by a trained outreach intern with oversight from the Student Collaborations lead and the mission principal investigator. When appropriate, upcoming posts from @MissionToPsyche are coordinated with other mission partner accounts (such as @Maxar) to help amplify the content. As possible, the @MissionToPsyche account implements suggestions and content from communications-related capstone teams (such as recommendations from a Public Relations capstone team to make regular use of certain hashtags and platform-specific content, or recommendations for images and animations from a graphic design capstone team), mission personnel, alumni, and other stakeholders.

In particular, the @MissionToPsyche accounts allow the mission to acknowledge the many people and institutions involved. Special themed campaigns, such as #PsychePartner and #PetsofPsyche, highlight those elements. Additionally, on January 1, 2020, the Psyche principal investigator, with the encouragement of NASA Headquarters, started a new series of regular tweets with the tag #PI_Daily from her personal account (@ltelkins) and invited other NASA principal investigators and deputy principal investigators to do the same, in order to share the "chores, challenges, tasks, assignments, and duties" and "share the experience." @MissionToPsyche regularly retweets her #PI_Daily posts (and occasionally reposts on other platforms) and maintains an archive on the mission website (https://psyche.asu.edu/mission/faq/pi_daily/). Details about the brand strategy and engagement for the @MissionToPsyche accounts are provided in Appendix J).

4.3 Video Products

Videos constitute another series of projects created to build public awareness of, and engagement with, the Mission, as well as providing important introductory material for new Student Collaborations program participants being onboarded each fall. From the time of mission selection, the goal of the videos has been to tell the story of the mission and the people behind it in a series of short vignettes, which can be combined into a more extensive video documentary post-launch, after arrival at the asteroid, and when the mission is complete. An Arizona-based production company, True Story Films, was selected to be the official videographers for the mission, in addition to in-house videographers from NASA, JPL, and Maxar and cinematic illustrator Peter Rubin.

The Psyche mission team works together to determine the focus of each video, ensuring a diversity of team members, locations, subject areas, and perspectives. The ultimate goal of the collection of videos, beyond public awareness and engagement, is to highlight the diversity of background, knowledge, and skill needed for a successful space mission. Each video is relatively short, most are less than five minutes, making them easy to watch on YouTube, for public presentations, to share on social media platforms and press releases, for events, and other public outreach venues. Details about the Psyche mission video products are provided in Appendix K.

5 Future Plans and Expanded Opportunities

As described above, the Psyche Student Collaborations program is inspired by and built upon the extensive foundation of public engagement, education, and outreach efforts and expertise of NASA and partner institutions. Below is presented a summary of existing (and potential) plans for Psyche Student Collaborations post-launch.

5.1 Capstone

The Psyche capstone program will continue to grow in diversity of participating institutions, disciplines, and projects, if not in total number of teams participating annually (currently approximately 60 teams per academic year engaging more than 300 students). Goals informed by feedback from participants and participating institutions include engaging more interdisciplinary teams, developing targeted professional development to help students improve teamwork and communication, creating more casual networking opportunities between students and alumni and Psyche team members, providing archived lectures as podcasts to facilitate access for students who cannot attend synchronously, and increasing interim opportunities for students to get direct feedback on their projects from subject matter experts. Additionally, engaging capstone teams (such as those in applied engineering, engineering management, graphic design, and public relations) in helping to improve aspects of other Psyche programs (described below) has been fruitful and is an area planned for expansion, as possible.

5.2 Psyche Inspired

In the future, Psyche Inspired will continue supporting undergraduates in creating artworks about Psyche in collaborative ways that invite the public to participate with the pieces made by the interns. Through displays on their campuses, exhibits at regional museums, and opportunities to engage with members of existing NASA programs such as the Museum & Informal Education Alliance and Solar System Ambassadors, the Psyche Inspired interns and their artworks will be able to reach a wide audience interested in STEAM. Artworks may continue to be displayed at venues such as the MOXI STEAM Museum in Santa Barbara, CA, the Sky Harbor Airport Museum in Phoenix, AZ, the Visitor Center at NASA's Jet Propulsion Laboratory, and at other NASA venues. Expansion of social media presence is planned through the integration of strategic promotional timelines and more interactive means of spreading the word (e.g., filters or lenses on Instagram and Snapchat, Instagram and Facebook Live events, swipe up options for informational links, etc.). Further development and implementation of the showcase gallery enhancements as recommended by capstone team collaborators will strengthen the virtual component of the showcase. These include offering immersive, virtual campus backgrounds from each of the institutions the interns attend, expansion of the virtual galleries to include interactive 3D YouTube videos and gamified experiences, and a forum for guests to ask interns questions about their art pieces and experiences in the program.

5.3 Public Engagement and Partnership Opportunities

The energy, enthusiasm, creativity, and relatability of undergraduate students engaged in Psyche Student Collaborations as interns, participants, and program alumni offers a renewable resource that facilitates the program's efforts to bring as many people as possible along on the journey to a metal world. With their help, and building on the history and experience of NASA's extensive portfolio of programs, subject matter experts, and public engagement, public affairs, and media relations professionals, the mission is able to efficiently implement a number of public opportunities based on successful prior models. The implementation of many of these opportunities also benefits from analyses and reports produced by Psyche capstone teams, as described below.

5.3.1 Pre-Launch and Launch Coverage

NASA and JPL provide support to missions for public engagement and media coverage leading up to and including launch, through efforts such as the development of fact sheets and press kits, coordination of press briefings and NASA Socials (https://www.nasa.gov/connect/social/index.html), and live launch coverage on NASA TV and NASA social media channels. Recently, commercial spaceflight companies have conducted similar efforts with their own launches and may offer other ideas for the mission to consider. To help prepare for the most comprehensive and compelling launch lead-up and coverage possible, during the 2021 spring semester a team of capstone students in applied engineering at Michigan State University (MSU) conducted research and a comparative analyses of lead-up events, coverage, promotion, and public participation opportunities of recent launches by NASA, SpaceX, and other commercial space entities and created a customizable cause and effect matrix to help the mission team explore what would be likely to be the most effective in terms of engaging wide audiences. They also provided recommendations for opportunities the mission could incorporate to help gain substantial public attention and generate excitement and participation. Results from this capstone project, particularly the cause-and-effect matrix, may be used to inform the design and implementation of mission launch lead-up.

5.3.2 Beam Your Name to Psyche

Many NASA missions have conducted successful "Send Your Name" public campaigns in which members of the public worldwide submit their names via websites, receive a certificate, "boarding pass," or other acknowledgement of their participation, and have their name included on a chip or plaque on a spacecraft (e.g., https://mars.nasa.gov/participate/send-your-name/, http://parkersolarprobe.jhuapl.edu/The-Mission/Name-to-Sun/, http://lro.jhuapl.edu/NameToMoon/, http://lasp.colorado.edu/maven/goingtomars/send-your-name/). The mission is investigating a similar opportunity, *Beam Your Name to Psyche*, with the goal highlighting the way the mission team communicates with the flight system via the Deep Space Network and, potentially, through the Deep Space Optical Communication (DSOC) technology demonstration (see e.g., NASA's Laser Communications Relay Demonstration public campaign: https://www.nasa.gov/mission_pages/tdm/lcrd/participate-in-lcrd-first-light.html). As an early step in determining the preparation and resources needed to support the collection of names, during the 2019 spring semester an applied engineering capstone team from MSU researched the processes, work flows, data management needs, and public interest of previous opportunities, including interviewing subject matter experts from Mars Public Engagement (*Send Your Name to Mars*) and surveying members of the public about preferences related to a potential "Send Your Name" registration experience. In the 2020 fall semester, a public relations capstone team from ASU conducted an audit of previous "Send Your Name" campaigns and provided suggestions for language, graphics, and promotion of a similar Psyche opportunity. The results from these two capstone projects will help guide the development of "Beam Your Name to Psyche."

5.3.3 Asteroid Appearance Competition

One of the challenges of engaging the public in the Psyche mission is that there are no images of the Psyche asteroid, and the earliest images of it will be obtained as the flight system arrives at its target. A series of scientifically informed artist's renderings (see Fig. 7) and the imaginings of the Psyche Inspired interns (see e.g., Fig. 8) provide tantalizing possibilities.

The truth, however, is that no one can know what Psyche will look like, and that presents an opportunity to invite the public to imagine Psyche along with the team. Informed by successful NASA projects such as *Cassini Inspires* (https://saturn.jpl.nasa.gov/mission/cassiniinspires/) and *Imagine Ceres* (http://dawn.jpl.nasa.gov/DawnCommunity/imagine_Ceres_gallery.asp), during cruise the public would be invited to submit their creative ideas in different media categories (such as drawing, painting, ceramic, Lego, Minecraft, animation) as well as different categories for students, for professional artists, for enthusiasts, etc. After inspection, the creative works would be put online for general review, and then winners announced after the flight system has arrived at Psyche and images are released. To be informed of best practices for such opportunities, during the 2021 spring semester a team of capstone students in applied engineering at MSU conducted research and analysis of the various creative contests and competitions NASA has hosted in the past to compare the procedures and rules, how they have been promoted, what the outcomes have been, and what has been effective in engaging the public in order to make recommendations about what the mission could do to make the effort stand out and reach a wide audience. The findings from this capstone project will be used to finalize the design of the asteroid appearance competition.

Fig. 7 Scientifically-informed artist's illustrations. Based on data obtained from Earth, scientists believe Psyche is a mixture of metal and rock. The rock and metal may be in large provinces (L) or intimately mixed on a scale too small to detect from orbit (R). Image credit: NASA/JPL-Caltech/ASU

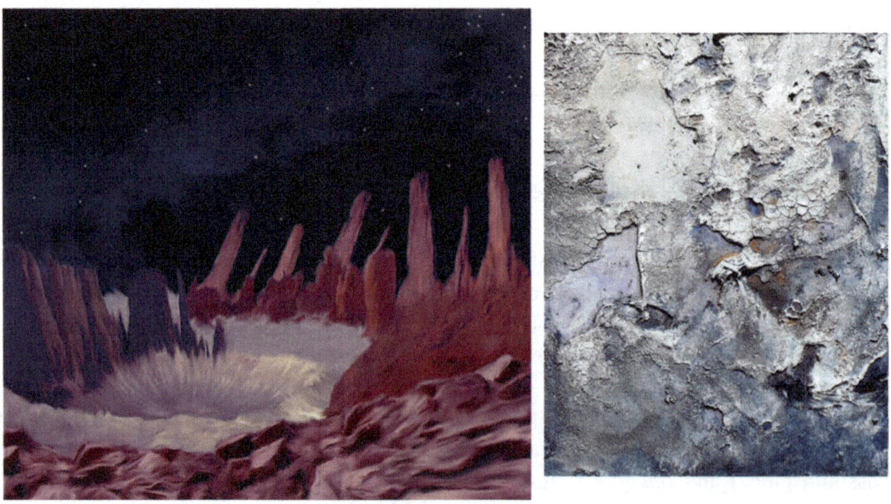

Fig. 8 Creative interpretations by Psyche Inspired interns Finn Witt (L) and Zarina Karapetyan (R)

5.3.4 Multispectral Imagers Data Direct to Public Websites

Following the tradition established by many previous NASA Solar System exploration missions, including principal investigator-led missions (e.g., Dawn: https://solarsystem.nasa.gov/missions/dawn/galleries/images/, InSight: https://mars.nasa.gov/insight/multimedia/raw-images/, Lucy: https://svs.gsfc.nasa.gov/Gallery/Lucy.html, and OSIRIS-Rex: https://www.asteroidmission.org/galleries/spacecraft-imagery/), the Psyche team plans to share raw and derived data products with the world as quickly and openly as possible. The aim is to increase awareness and excitement about the mission by inviting the public to be among the first to see each new spectacular scene coming from the mission's exploration of this enigmatic asteroid. Specifically, raw images from the Psyche Multispectral Imagers ("the Imagers") will be automatically created after each downlink and then rapidly and automatically posted as PNG-format image files on a publicly accessible web site hosted by JPL. Based on experience with similar JPL image distribution systems used for images from the Spirit, Opportunity, Curiosity, and Perseverance rovers (e.g., https://mars.nasa.gov/mars2020/multimedia/raw-images/), for example, the Psyche team expects to be able to post raw images for public viewing within only seconds to perhaps minutes after each set is received and processed by JPL and NASA's Deep Space Network.

With oversight from the Imager lead and his team, science outreach interns will help manage a second publicly accessible web site, hosted at ASU and part of ASU's overall mission web page structure, that will host additional processed and calibrated Imager data. These additional images will include calibrated individual monochrome images, natural color and false color images made by blending data from various combinations of the Imager's seven different color filters, monochrome and color mosaics of specific regions of geologic interest, and eventually, global monochrome and color mosaics of the entire surface. The team envisions that these products will be made rapidly available when it is possible to do so (for example, within a few hours of downlink for calibrated individual image products), but other products (such as large regional mosaics or global maps) might take days, weeks, or months to complete for posting. As with other NASA missions, all of the Imager calibrated and derived image products will also be available through the Planetary Data System (PDS), which is the official repository for all science data products. Nonetheless, the intention is that the ASU Imager website will be a user-friendly long-term source for highest-resolution versions of the many kinds of derived image products that the team plans to share with the world from this exciting mission of exploration. This resource may also be a source of inspiration for new capstone projects and for cohorts of Psyche Inspired interns in the future.

5.3.5 Engagement with NASA's Solar System Ambassadors and Museum & Informal Education Alliance

NASA has existing and robust programs involving enthusiasts, public engagement professionals, and informal learning venues in communicating with the public about space exploration, for example the Solar System Ambassadors program (https://solarsystem.nasa.gov/solar-system-ambassadors/directory/) and the Museum & Informal Education Alliance (https://informal.jpl.nasa.gov/museum/). In Phase B and early in Phase D of the mission, team members gave professional development presentations about the mission to NASA's Solar System Ambassadors and members of the Museum & Informal Education Alliance (formerly the Museum Alliance) and additional presentations will take place pre- and post-launch and arrival. Additionally, relevant materials developed by the team, including the Psyche science outreach interns, Psyche Inspired interns, and capstone teams, are available to be shared with these groups. In particular, capstone-developed printable models, WebXR experiences for mobile phones, Virtual Reality (VR) experiences, and web-based games are available to disseminate to interested members of both groups. Other materials (classroom activities, videos, demonstrations, etc.) are also available for members of those groups who wish to present about Psyche to their local communities.

5.4 Other Possible Public Opportunities

A benefit of Psyche's status as a principal investigator-led mission based at a large university is that the talents, ideas, and enthusiasm of large numbers of undergraduate and graduate students, staff, and faculty inspire and enable the implementation of a wide range of public opportunities. Ideas are also shared by Psyche team members at other institutions. Possible student and public opportunities include:

- **"Hi"-ku from Psyche**
  In 2013, NASA's Mars Atmosphere and Volatile EvolutioN (MAVEN) orbiter mission invited the public to submit a haiku about Mars for possible inclusion on the spacecraft https://lasp.colorado.edu/maven/goingtomars/send-your-name/haiku-with-maven/ In a nod to that previous competition, the "Hi"-ku from Psyche opportunity would give members of the public the chance to submit a haiku to say "hi" to the MAVEN orbiter during the Psyche spacecraft's Mars fly-by during interplanetary cruise.
- **Deep Space Optical Communications (DSOC) Observing Opportunity**
  The mission will be carrying the Deep Space Optical Communication (DSOC) technology demonstration, which will demonstrate high-rate data transmission to Earth by shining a laser to a telescope at the Palomar Observatory in California. This signal also will be observable in the southwestern U.S. and, at times, beyond. With partners at NASA's Deep Space Network, the mission is assessing the feasibility of engaging observers (whether other observatories or serious amateurs) in helping to track the signal to provide data to aid in pinpointing the position of the flight system and assessing the performance of the laser. A related effort could involve public viewing parties and "Psyche Nights" coordinated with amateur astronomy clubs, universities, and NASA's Night Sky Network (https://nightsky.jpl.nasa.gov/).
- **Citizen Science: Crater/Surface Characterization**
  NASA has significant citizen science resources and infrastructure available to the public to engage with space missions (https://science.nasa.gov/citizenscience). As the Psyche mission is the first flight system to visit a body made largely of metal, there is significant uncertainty about the composition and appearance of the asteroid, including what geologic features may be found on the surface. Although the Psyche team has undertaken analyses and experiments to explore the possibilities of cratering on a metal-rich body (described in https://psyche.asu.edu/2019/09/05/simulating-asteroid-collisions-and-making-craters/ and in Marchi et al. 2020), upon arrival there will be extensive characterization activities needed. One possibility is to engage "citizen scientists" in characterizing craters (and possibly other surface features) in images of the surface, as has been done for other planetary bodies (e.g., NASA Citizen Science projects listed at https://science.nasa.gov/citizenscience, such as Planet Hunters TESS and Disk Detective, projects listed on the Planetary Science Institute's CosmoQuest site https://cosmoquest.org/x/ and on Zooniverse https://www.zooniverse.org/; see also Fortson 2021). During the 2020-2021 academic year, an interdisciplinary team of ASU capstone students from computer science and engineering management, along with an astronomy and planetary science undergraduate from Northern Arizona University, investigated new automated crater identification techniques as a first step to planning for potential hybrid (computer/human) crater and surface characterization citizen science activities related to Psyche.
- **Crochet Psyche**
  A community with the potential to be engaged directly with planetary science is the extensive network of crafters around the country. The pre-flight conception of Psyche, with the potential for spires or flaps of metal, intricately mixed materials of different color and texture, and possibility of large fractures due to cooling of a metallic core yields visions of a stark landscape unlike anything on Earth. Yet the potentially angular forms are not dissimilar to those seen on the seafloor, particularly where reefs rise above the shallow

ocean floor, teaming with spiky coral of every imaginable shape and color. The Institute of Figuring developed the *Crochet Coral Reef Project* (https://ocean.si.edu/oceanlife/invertebrates/when-art-meets-science-hyberbolic-crochet-coral-reef) to allow crafters to contribute to a reef-scale complex that has been exhibited at, among other places, the Smithsonian's National Museum of Natural History. The myriad of shapes, textures and colors on Psyche could provide a comparable opportunity to allow crafters to engage in the science of Psyche by crocheting individual features (e.g., craters, metal spires, flaps, offset cracks) that, when combined, could be exhibited together in an interesting and creative expression of Psyche's unique surface.

6 Conclusions

In FY20, NASA had 1,861 total undergraduate interns (NASA OSTEM 2020), an opportunity open only to U.S. Citizens (https://intern.nasa.gov/) or, in the case of NASA's I^2 internship program, open to citizens of the 15 countries that have an agreement with NASA (https://www.nasa.gov/stem/international-internships-for-students.html). The 5% acceptance rate for NASA internships (Anderson 2022) and roughly 400,000 STEM bachelor's degrees conferred annually in the United States (Hussar et al. 2020) (even if only a subset of them are interested in working in space exploration) indicates that the demand for available NASA internships exceeds supply. With more than 300 undergraduates participating actively with the Psyche mission for a full academic year through capstone projects and arts and science outreach internships, and with no citizenship requirement (as long as the student is enrolled full-time in a U.S. institution), Psyche Student Collaborations demonstrates that individual mission programs, whether PI-led or flagship, have the potential to significantly increase this capacity to help meet demand.

By leveraging extensive engagement by undergraduates at all levels of Psyche Student Collaborations, and working within existing academic structures such as senior capstone courses, the component programs benefit from a cost-effective, renewable, and regenerative source of creativity, talent, and enthusiasm. Involving undergraduates in all phases of the program also supports the development of the next generation of explorers, contributes to the nation's workforce preparation, and complements NASA's existing undergraduate offerings by providing long-term opportunities for students to participate with the mission through established postsecondary education structures like capstone courses. PI-lead missions not headquartered at a university may wish to consider engaging university partners in their Student Collaborations planning in order to facilitate these opportunities (such as the case with the Lucy mission's partnership between the Southwest Research Institute and ASU). Additionally, when selecting opportunities to implement, missions may wish to consider the intensity of resources, such as required mission team member involvement and the number of people who can participate in each opportunity (such as in Fig. 1 above), in order to develop a diverse program that balances breadth and depth. This includes designing opportunities so that interested mission team members, whatever their subject matter expertise, may contribute at a level that is comfortable and realistic for them. Finally, in selecting opportunities for inclusion in a comprehensive Student Collaborations program, remaining grounded in the core values and vision of the mission team and the participating institutions can help ensure that the program is considered, and supported as, an integral component of the mission as a whole.

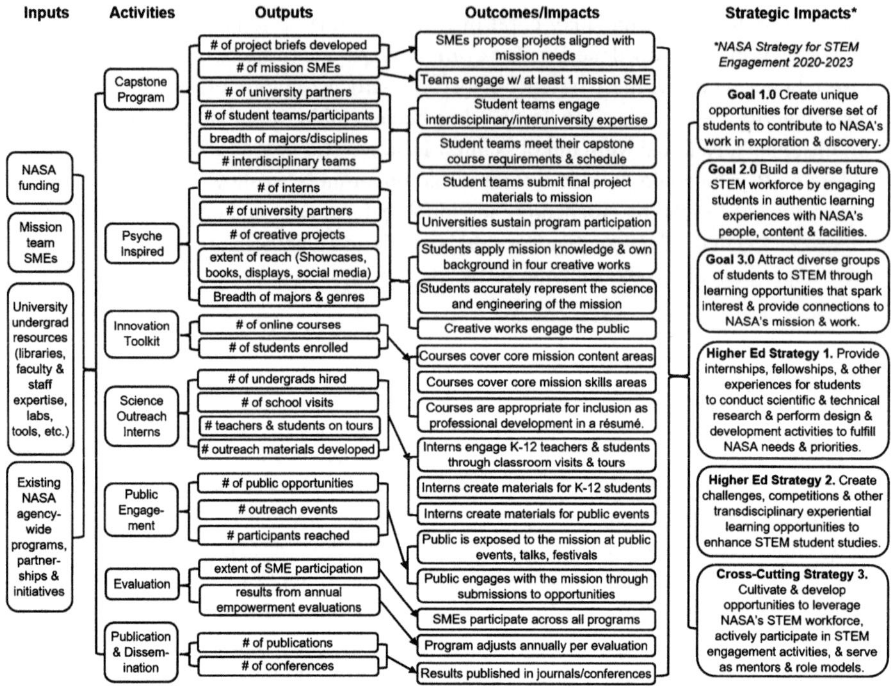

Fig. 9 Psyche Student Collaborations logic model

Appendix A: Logic Model

The logic model (Fig. 9) depicts the relationships between the program inputs (the "funding sources and resource streams that provide support to the project"), activities (the "services, materials, and actions that characterize the project's thrusts"), outputs ("products of these activities or a count that describes the activity"), and outcomes and impacts (the "the changes that occur as a result of the activities; they may be short or longer term") (Fretchling 2010, p. 17; see also Friedman 2008). The final column, strategic impact, lists the specific goals and strategies from the NASA Strategy for STEM Engagement 2020-2023 are addressed by the Psyche programs (NASA 2020).

Appendix B: Additional Proposed Student Collaborations Projects (Not Selected for Implementation)

As described in Sect. 1.1, NASA's long history of educational and public engagement programs provided significant inspiration for the development of the Psyche Student Collaborations program. The list below provides an archive of additional proposed elements of Psyche Student Collaborations that were not selected by NASA for implementation.

Planetary Science Summer Seminar (PSSS) Feeder Program

To develop a program to complement PSSS (https://pscischool.jpl.nasa.gov/), acting as an entry point for students from community colleges and non-Research One institutions.

Tracking and Gravity Science

To partner with the NASA/JPL Goldstone Apple Valley Radio Telescope Project (GAVRT) to engage high school, community college, and undergraduate students in two Psyche-related observing campaigns to compare with Psyche team data (tracking the spacecraft trajectory and modeling the asteroid's gravity field), modeled on similar previous campaigns (http://www.lewiscenter.org/Global-Programs/GAVRT/About/).

CubeSat Summer School

To create and implement a new annual summer school at LANL, JPL, and ASU on CubeSat mission design or space hardware design and development to be planned and executed with members of the ASU School of Earth and Space Exploration, the Psyche team, JPL, and LANL.

Remote Team Building and Competition

To carry out student competitions open to teams assembled anywhere in the world, modeled on NASA's Space Apps Challenge (https://2017.spaceappschallenge.org/), with opportunities to conduct substantive work related to their expertise that can be listed on a resume or CV.

Participating Early Career Scientist Program[1].

As a complement to NASA's existing participating scientist opportunities, an open call to scientists within five years of their PhD to join the science team for periods of one or two years each, to learn the running of the team, the process of a mission, to produce some research, and to meet and become integrated into the mission community.

Appendix C: Psyche Student Collaborations Participating Institutions

Table 4 Psyche Student Collaborations Participating Institutions (through 2021-2022 academic year)

	Capstone	Psyche inspired
Albion College		•
Arizona State University - Downtown Campus	•	
Arizona State University - Main Campus (Tempe)	•	•
Arizona State University - Online	•	•
Arizona State University - Polytechnic Campus	•	
Boston University		•
Bowdoin College		•
Brown University		•
California College of the Arts		•
California State University - Los Angeles	•	
Caltech		•
Chaffey College		•
Cleveland State University	•	
Columbia University		•

[1] Though this program component was not incorporated into Psyche Student Collaborations, the Psyche mission participated in NASA's *Mission Science Team Meeting Opportunities for Early Career Scientists* in 2020

Table 4 (*continued*)

	Capstone	Psyche inspired
Columbus College of Art and Design		•
Creighton University		•
Emory University		•
Florida A&M U-Florida State University College of Engineering	•	
Florida State University - Panama City	•	
Georgia Tech		•
Illinois State University		•
Indiana University-Purdue University Indianapolis (IUPUI)		•
Kansas City Art Institute		•
Lawrence University		•
Lehigh University		•
Michigan State University	•	
Michigan Technological University	•	
Morgan State University		•
Northern Arizona University	•	
Pasadena City College		•
Penn State	•	
Penn State - Behrend	•	
Phoenix College		•
Rochester Institute of Technology	•	•
Seattle University	•	•
Temple University		•
The Citadel		•
Trinity University		•
University of Arkansas	•	
University of California, Berkeley		•
University of California, Santa Barbara		•
University of Colorado – Colorado Springs		•
University of Florida		•
University of Illinois – Chicago	•	
University of Maryland		•
University of Michigan		•
University of Minnesota		•
University of Pittsburgh		•
University of Southern California		•
University of Texas – Tyler	•	
Utah State University		•
Virginia Commonwealth University	•	•
Virginia Tech		•

Appendix D: Participant Undergraduate Majors

Table 5 Undergraduate majors (n = 1221)

	Students
Arts	(n = 61)
Graphic Design	35
Other (Animation, Illustration, Music, etc)	26
Communications & Journalism	(n = 23)
Engineering	(n = 947)
Aerospace Engineering	35
Applied Engineering	29
Biomedical, Chemical, Materials Science	21
Computers/Programming	
Computer Science	343
Computer Systems Engineering	59
Informatics	20
Information Technology	14
Software Engineering	92
Electrical Engineering	82
Engineering Management	61
Industrial Engineering	10
Mechanical (including Robotics)	180
Natural Sciences (Astrobiology, Astrophysics, Biology, Geoscience, Physics)	(n = 61)
Other (e.g. business, forensics, interdisciplinary studies, public policy, etc.)	(n = 129)

Appendix E: Alumni Class Names

Table 6 Alumni class names to date

Academic year	Class name
2017-2018	Titanium
2018-2019	Iron
2019-2020	Cobalt
2020-2021	Nickel
2021-2022	Copper

Appendix F: Student Artist Contributions to Project Development

The work of Psyche Inspired interns engages the mission team as well as the public. While the most memorable end results of NASA's space program certainly include footprints on the Moon and breathtaking images of celestial objects, NASA's successes begin with blank pages and screens upon which tough technical analyses must be conducted by both scientists and engineers. Though there may seem to be little room for art in serious space exploration,

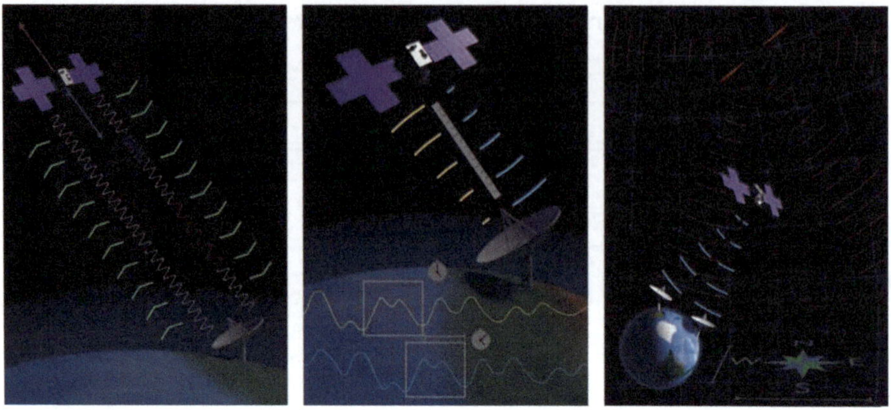

Fig. 10 "Navigation Data Types Triptych," a digital media piece created by Psyche Inspired alumnus Ben Conway for the Psyche project team during a JPL internship. Image Credit: NASA/JPL-Caltech

none of NASA's successes could have come about without the imagination of its team members. The imagination is the genesis of all flight projects, and imagination is made into reality through visualization, which can and often must include art. Likewise, some of NASA's most difficult moments can and have been traced to what might be categorized as a failure of imagination, where a key piece of data, presented and visualized in the right form, could have made the difference. No space fan who was alive in 1986 will forget Nobel Laureate and Caltech Professor Richard Feynman dropping a sample of the Space Shuttle O-ring into ice water to demonstrate its loss of resilience at cold temperatures. Though not necessarily a work of art, the principle is the same: showing, not telling, has power.

There is a place in space exploration for artistry, which can go beyond inspiration and make real contributions to the advancement of space missions. Mission project members spend a great deal of time plotting data and showing the geometry of their spacecraft and its environment in ways that can easily be understood. It is critically important to create visualizations of how operations may work and accurately show these results to other project members who are trying to solve tough technical problems and need to understand what they are viewing. Effective visualization techniques can contribute to efficiently communicating complex topics.

As an example of the importance of visualizations, Psyche (the asteroid) rotates at an angle of nearly 90° to the plane of the planets, spinning "on its side." For the mission to succeed, a fast-orbiting spacecraft must fly in a variety of orbits and make measurements of a fast-spinning asteroid, all while avoiding too-long eclipses (when the spacecraft is behind the irregular-shaped spinning body) while both the spacecraft and the asteroid orbit around the Sun resulting in changes in surface lighting and eclipse zones over time. Even the most experienced rocket scientists and engineers may struggle to visualize this scenario in their minds, but it can be depicted- including portraying critical estimations of the required science coverage to meet mission goals - through visualization and art.

The Psyche project has already directly employed, as JPL interns, members of the Psyche Inspired program to produce graphic designs to effectively describe its mission. These designs have helped provide quick-glance depictions of the mission goals, telecommunications, launch and orbit geometry, navigation techniques, physics of electric propulsion, and optical communication for team members that might not be specialists in a specific area

(Fig. 10). Furthermore, the Psyche Mission Plan - arguably the most widely read project-level document - is graced with over a dozen works from the Psyche Inspired collection to date (with more coming in future releases).

Appendix G: Psyche Inspired Application FAQs Provided to Applicants

From https://psyche.asu.edu/get-involved/psyche-inspired/psyche-inspired-application-2022/

What Will Psyche Inspired Interns do?

Psyche Inspired interns will use their creative talents to share, explain, and highlight the excitement, innovation, and scientific and engineering content of NASA's Psyche mission. They will become part of the Psyche team, with the opportunity to interact with Psyche scientists and engineers from NASA, ASU, and other partner institutions, and with other Psyche Inspired interns and staff.

What do We Expect from Our Psyche Inspired Interns?

Psyche Inspired interns will *meet once a week via videoconferencing* with staff from Psyche Inspired and the mission team. The meetings are required and will be held on Fridays from 12 p.m. to 1 p.m. Pacific Time.[2]

Psyche Inspired interns will *create at least one* original *creative work every six weeks* during the 2022-2023 school year, which will amount to four works throughout the school year. These works will be guided by the mission themes of science, technology, engineering, mathematics, exploration, inspiration, teams, and inclusion. Proposals, drafts, and final products will be provided to the Psyche Inspired staff for review and then will be posted to, and highlighted on, the Psyche mission website hosted by ASU (https://psyche.asu.edu), blog, and/or social media accounts. In late spring 2023, these works will be curated into an online showcase with distributed campus participation (if possible) and then incorporated into a coffee table book.

Psyche Inspired interns will take the Psyche online course, *The Process and Lifetime of a Space Mission.*

Psyche Inspired interns will follow NASA's Image Use policies and other guidelines (which will be distributed by the Psyche Inspired staff upon selection).

Psyche Inspired interns will sign an intellectual property agreement with ASU prior to joining the program.

Psyche Inspired interns will be enthusiastic ambassadors for the Psyche mission, striving to inspire and excite the public through their creative work!

How do Psyche Inspired Interns Benefit?

Psyche Inspired interns will be supplied with a stipend of $300 per intern to purchase materials needed to create their works (art supplies, computer software, crafting materials, etc.). The program will pay for completed physical works to be shipped to the project office at Arizona State University.

[2]Note for Arizona applicants: Since Arizona does not change with daylight savings time, you need to be aware that the time of the meeting will be 12 p.m. – 1 p.m. for you during daylight savings time, and 1 p.m. – 2 p.m. for you during standard time (starting with our meeting on November 18th and reverting back starting with our meeting on March 17th), so you need to make sure you are available for both timeslots.

Psyche Inspired interns will have their work highlighted on the official Psyche mission website, blog, and/or social media accounts at least once per project.

Psyche Inspired interns will have their work exhibited in an online showcase with distributed campus participation (if possible) in late spring 2023.

Psyche Inspired interns will have their work showcased in a printed coffee table book and receive a copy of the book.

Psyche Inspired interns will have their work archived in an online portfolio.

Psyche Inspired interns may have their work presented and highlighted in Psyche presentations and exhibited at conferences, libraries, Psyche partner universities, NASA Centers, and at public events.

Psyche Inspired interns will gain knowledge and skills that may benefit them in their education and future employment. They will learn about the science and engineering of a space mission as well as gain skills and experience developing and promoting their own creative work locally and virtually.

Psyche Inspired interns will meet other creative students from around the country and will have the opportunity to share ideas and feedback and to collaborate.

What Are the Eligibility Requirements of Psyche Inspired?

To be eligible, an applicant must:

- Be a current full-time enrolled undergraduate student at a university or community college in the United States and its territories and must have completed at least two semesters/terms at your current university or community college prior to applying. This will be confirmed with your institution before you are admitted to the program.
- Not be graduating any earlier than the end of the spring semester 2023 (nominally no earlier than May 2023).
- Commit to creating four original creative works over the period of the program (approximately one work per six weeks through May 2023).
- Be willing to sign an intellectual property agreement with Arizona State University.
- Commit to sending any physical works to the Psyche Inspired project office at Arizona State University (prepaid shipping labels will be provided).
- Commit to attending the weekly required meeting via videoconferencing on Fridays at 12 p.m. (noon) Pacific Time through the entirety of the program (through May 2023).
- Commit to responding to all communications from the program via email, text, or Slack (or similar) in a timely manner (within 48 hours).

Applicants are **not** required to be U.S. citizens.

What Creative Genres May Apply?

We encourage any eligible undergraduate to apply who thinks that his/her creative medium would be effective at sharing the Psyche mission with the public. Examples include (but are not limited to):

3D physical works (such as ceramics, welding, glass blowing, fiber, jewelry, sculpture, etc.)

3D virtual works

Ads/posters

Comics/Animation

Choreography

Dance

Design

Drawing/sketching
Fashion/textiles
Film (including documentary)
Folk arts/crafts
Game design
Graphic art/design
Infographics
Media art
Music composition/performance
Painting
Performance arts
Photography
Poetry
Short story (fiction, nonfiction, science fiction, etc.)
Sonification (take information and turn it into sound)
Spoken word
Tattoo artistry
Theater
Other ways of creatively expressing mission information

Is This Program Only for Art Majors?

No. Psyche Inspired is open to all talented, creative full-time enrolled undergraduate students at universities and community colleges in the United States or its territories, regardless of major.

Previous Psyche Inspired interns have come from a variety of majors, including aerospace engineering, animation, art & design, art history, astrobiology, astronomy, biochemistry, biogeoscience, biology, biomedical engineering, ceramics, chemical engineering, civil engineering, communication, computer graphics technology, computer science, creative technologies, digital culture, drawing, electrical engineering, engineering management, English, environmental sustainability management, forensics, geology, graphic design, graphic information technology, illustration, industrial design, interdisciplinary studies, jewelry/metalworking, kinetic imaging, mass media, materials science, mathematics, mechanical engineering, molecular environmental biology, music, new media design, painting, physics, public service and public policy, and sculpture.

How Many Participants Will Be Selected?

We will select up to 16 interns for the 2022-2023 academic year: Up to 15 will be selected from institutions nationwide. At least 1 will be selected from Arizona State University.

Who Reviews the Applications?

A committee comprised of Psyche mission team members and Psyche Inspired staff will review the applications and make the final selection.

What Is the Process for Selection?

The goal of Psyche Inspired is to communicate the excitement, innovation, and scientific and engineering content of the Psyche mission through creative/artistic expression. To this end, we aim to include the best and widest range of creative genres and presentation styles. Because we invite an open range of creative pursuits, the criteria for selection are based on qualities that all selected interns should share:

- *Quality of example works* (Is the work executed with high quality and care? How creative/original is the work? Is the reviewer excited at the thought of seeing more work like this? Does the reviewer think this type of work would be appropriate for communicating the Psyche mission?)
- *Stated reason(s) for wanting to be part of Psyche Inspired.*
- *Stated reason(s) applicant believes he/she is a good candidate for Psyche Inspired.*
- *Demonstrated ability to follow through with projects.*
- *Demonstrated ability to engage with their local community or general public.*
- *Contact of references* (If deemed necessary, the selection committee will contact the applicant's reference to discuss the qualifications of the applicant.)
- *Confirmation of eligibility* (If deemed necessary, the selection committee will contact the registrar or major department of the applicant's university or community college to confirm enrollment.)

Appendix H: Innovation Toolkit: Free Online Course Descriptions

From: https://psyche.asu.edu/get-involved/innovation-toolkit/

- *Course 1: The Process and Lifetime of a Space Mission*: A space mission encompasses so much more than reaching a destination and sending back discoveries! The Process and Lifetime of a Space Mission course shares the story of a team creating a NASA robotic space mission. As a participant in this course, you will learn about and practice elements of the process and lifetime of a space mission, from idea to flight. Aspects of the preparation and submission of a proposal, team building, design, construction, modeling, testing, launching, tracking, and data collection and analysis processes will be covered.
- *Course 2: Inclusive Mindset: Tools for Building Positive Team Culture*: Organizations, universities, and businesses (and missions!) benefit most when people work together smoothly as a team. An inclusive mindset, collaborative team culture, teamwork, and positive communication are among the skills such organizations value most to achieve their goals. In this course, you will learn about team culture, strategies for working as part of a diverse team, and techniques for developing a positive, collaborative inclusive mindset.
- *Course 3: Countless Worlds in our Solar System: Asteroids, Comets, and Meteorites:* Our solar system is made up of much more than the Sun and the major planets and their moons. Other small objects, including asteroids and comets, along with the pieces of them that have fallen to the Earth as meteorites, provide important insights into the formation of our solar system. In this course you will learn about these objects' origin, categorization, composition, location, and features, along with their importance to understanding the history of the solar system. In addition, you will explore NASA missions, as well as jobs and even hobbies, that gather information about these objects.
- *Course 4: Strategies for Troubleshooting Problems in Everyday Life and Work:* Troubleshooting is a type of problem-solving skill needed by everyone, from students programming a toy robot, to families at home trying to get their garage door to open, to NASA engineers tracking down why their spacecraft has stopped sending data to Earth. If you have ever tried figuring out why something is not behaving as it should (*why won't my lamp turn on?*), you have had a troubleshooting experience! In this course, you will learn about the five global troubleshooting strategies and explore how to choose and use the best strategy for different kinds of problems you face.

Fig. 11 The Psyche mission website (https://psyche.asu.edu)

Appendix I: The Psyche Mission Website

Both the Psyche "badge" ("mission identifier"), in multiple approved file formats and color variations, and the corresponding brand guidelines are available on the mission website (https://psyche.asu.edu/galleries/). The mission team worked with designer Michael Taylor to develop the brand guidelines at the onset of the mission. The guidelines are intended to encourage consistent use of the corresponding fonts, colors, sizes, aspect ratios, and positioning. While brand guidelines are often used only internally in many organizational settings, providing the guidelines on the website ensures consistent use of the mission brand standard.

The mission website (Fig. 11) includes a mission overview section with details about the target asteroid, the spacecraft, the launch, the instruments and science investigations, the team, and frequently asked questions (in English and Spanish); the mission timeline; an overview of the relevant science and mission goals and objectives; a news section with articles and blogs written by or about the mission; a list of future and past events; and an image and video gallery. The "Get Involved" section (https://psyche.asu.edu/get-involved/) highlights the offerings of Psyche Student Collaborations, including a public arts opportunity (based on previous NASA efforts such as #JWSTArt and Cassini Inspires), known as #PsycheSpaceCRAFTY (https://psyche.asu.edu/get-involved/psyche-space-crafty/). There is also a page with downloadable or 3D printable models and activities and educational resources (https://psyche.asu.edu/get-involved/models-and-materials/), and links to opportunities to get involved with NASA and mission partners. To make relevant content easier to find in search engines, the Student Collaborations team has implemented search engine optimization (SEO) best practices (keyphrases, meta descriptions, etc.) and a date on each page lets users know the last time content was updated.

As the mission nears major milestones such as launch, Mars gravity assist, orbit insertion, and data collection, the websites at NASA, JPL, and ASU are augmented and expanded to meet communications and public engagement needs. In particular, post-launch the mission website gallery is updated to accommodate processed and calibrated Imager data (see Sect. 5.3.4. Multispectral Imagers Data Direct to Public Websites).

Appendix J: Mission Social Media (@MissionToPsyche) Brand Strategy and Engagement

The Psyche mission's social media was initiated in 2016, nearly a year before NASA's final selection of the mission. The accounts were managed under the @NASAPsyche username across Twitter, Facebook, and Instagram until a rebranding in June 2020 as part of an agency-wide NASA effort to consolidate individual mission social media pages under larger recognizable umbrella accounts. @MissionToPsyche was adopted, along with the general hashtag #MissionToPsyche, in order to post daily about Psyche team featurettes, daily mission insights, student collaboration highlights, and public outreach programs. As noted above, @NASASolarSystem is the designated umbrella account and posts noteworthy mission milestones as well as partnering with @MissionToPsyche to amplify public opportunities when appropriate. All content on @MissionToPsyche social media accounts is organic only, following NASA guidelines.

In October 2019, the team revamped the mission's social media strategy after identifying three key digital content pillars to focus on: mission and Psyche science-related content (that is not related to mission milestones), student-related content, and public engagement and events. The social media strategy works to incorporate all three elements in the content creation process in a cohesive way to underscore the multifaceted and engaging nature of the mission. From initial selection, the team worked to nurture public interest and encourage involvement through employing a consistent "voice" intended to be welcoming, inspiring, and often playful, employing emojis, puns, relatable GIFs, and friendly responses to mentions. In particular, community management is central to @MissionToPsyche's digital strategy. The social media team receives a number of social media comments and queries each day and ensures that all responses are equally thoughtful and scientifically accurate by cross-checking with internal and external teams.

@MissionToPsyche has engaged the public through various digital campaigns, with the most prominent campaigns being #PsycheSpaceCRAFTY, #PetsofPsyche, and #PsychePartners. Similar to the Psyche Inspired internship program (see Sect. 2.3), #PsycheSpaceCRAFTY encourages the public to participate in creating and submitting original work inspired by the mission. #PetsofPsyche, developed during the start of the COVID-19 pandemic, aims to share photos of Psyche team members' pets through global quarantine and beyond. #PsychePartners shares the contributions of the major mission partners (institutions, universities, and other contractors) through their research, engineering, and manufacturing.

Instagram Theme Experiment A particular benefit of having the accounts managed by a Psyche outreach intern majoring in marketing or communications is that the student is concurrently taking coursework focused on best practices in engaging digital audiences. As an example, originally the Instagram account (which, with 64% of followers reporting as 34 years old or younger, skews younger than the Twitter and Facebook accounts) posted content in the same form as shared on the mission Twitter and Facebook accounts, which left the account with a lack of uniform visible branding and consistent color scheme, important on a visual platform like Instagram. Additionally, complex concepts such as data models proved difficult to interpret without reading accompanying text, which is structured differently on Instagram compared to Twitter or Facebook. Guided by best practices that the social media student interns have learned in their courses, in January 2021 the social media team developed and implemented a "theme" with content-specific templates to give the @MissionToPsyche Instagram page a consistent, cohesive, brand-aligned appearance (Fig. 12).

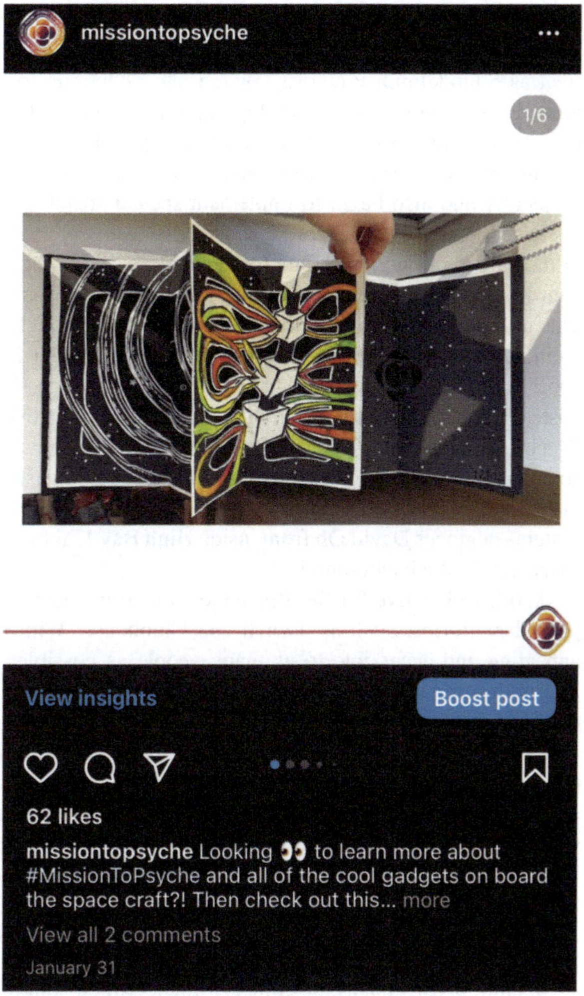

Fig. 12 Example of the new @MissionToPsyche Instagram theme

The content was changed to follow a minimal, streamlined standard on a white frame template with a stronger concentration on imagery than text in the photo to appeal to the average user and consistent mission identifier integration.

The theme and associated templates align with the overall Psyche brand guidelines and the appearance of the website, using the mission colors along with a minimalistic white aesthetic (see Sect. 4.1). The content-specific templates (for example, one style for Psyche Inspired artwork, another for upcoming events, and a distinct layout for science and engineering content) were integrated to encourage follower engagement and retention of content by indicating to them, at a glance, what content they can expect from a given post. In the first thirty days of implementing the new visual design, @MissionToPsyche experienced a 300% increase in Instagram Story interactions, 150% increase in website clicks, an almost 40% increase in impressions, and a more than 35% increase in content interactions. The three months following the implementation of theme and its templates resulted in almost double the increase in followers compared to the preceding three months.

Expanded Social Media Opportunities
The mission's @MissionToPsyche social media accounts (existing, and expanded platforms such as Soundcloud, Pinterest, Twitch, etc.) continue to provide day-to-day mission and Student Collaborations content, highlight opportunities to get involved, and share an inside look at the team. Surrounding launch and post-launch, as mission milestones and newsworthy events increase, @NASASolarSystem accounts are designed to feature Psyche more frequently and also begin to implement special social media opportunities such as NASA Socials, NASA Live content, social media features, and other activities. Additionally, as in the past, there will continue to be efforts to highlight various Psyche team members in their roles pre- and post-launch through open question-and-answer formats and interviews. This provides a closer look into the behind-the-scenes of Psyche and allows members of the public to pose questions to mission experts. In October 2019, the Psyche outreach interns coordinated Psyche's first Reddit "Ask Me Anything" (AMA), which was verified and cross-posted to the subreddit r/space with over 15 million active members, and Ernest Cisneros, Psyche's Science Data Center Manager, fielded questions from the public. Similarly, in December 2020, Dr. David Williams, science team co-investigator held a live Zoom Q&A and in March 2021, NASA's Jet Propulsion Laboratory led a Facebook Live Q&A with project systems engineer David Oh from inside High Bay 1, showcasing the newly delivered Psyche spacecraft in the background.

Another objective for the Psyche team is to increase participation across emerging, approved platforms, such as Twitch and Clubhouse. With the goal of making the mission appealing and accessible to as many people as possible, the mission plans to contribute to a multi-channel approach in expanding the mission's messaging, public education, and outreach. NASA currently operates an official Twitch account with over 900,000 followers, where it highlights rocket launches, spacewalks, and other major occasions. In the future, Psyche hopes to help highlight team members, public engagement events, and NASA livestreams related to pre- and post-launch milestones to share with the public. Additionally, Clubhouse presents as a unique opportunity for the Psyche mission and its team to host informative virtual panels and events; Psyche's principal investigator was a guest speaker at a space-themed Clubhouse event in March 2021. The team seeks to build and grow followings on emerging platforms such as these to reach new audiences and spread the mission's message.

Finally, Psyche's current Student Collaborations alumni program (Sect. 5.3.3) includes many hundreds of current students and graduates from institutions across the United States. Through LinkedIn, the messaging app Slack, and a periodic email newsletter, the mission remains connected to program alumni by sharing mission updates, opportunities to stay involved, and relevant job posts.

Appendix K: Psyche Mission Videos

Mission Trailer Video
The mission trailer video, "Journey to a Metal World" (https://www.youtube.com/watch?v=aExTQGcIGKo&t=1s) was launched in 2018 and filmed at Arizona State University and at NASA's Jet Propulsion Laboratory (with an update released in early 2022). In addition to the principal investigator, Lindy Elkins-Tanton and project manager Henry Stone, the video features Psyche team engineers, scientists, students, and staff. Throughout this five-minute award-winning video (2018 Rocky Mountain Emmy Award for Best Director, Short Form), which airs on the NASA YouTube channel, viewers are taken on a journey to explore why

this mission was selected for NASA's Discovery Program, how the spacecraft will get to the asteroid, what the team hopes to learn from Psyche, and the importance of scientific discovery. It includes a call to action for the public to "Join the Journey" via the NASA and Psyche websites. The link to this video is shared with all incoming Student Collaborations participants as part of their onboarding process.

Instrument and Investigations Highlight Videos

The Psyche team produces video vignettes on each of Psyche's instruments, the Gamma Ray and Neutron Spectrometer (GRNS), the Magnetometer, and the Multispectral Imager, as well as a video featuring the gravity science investigation. Other videos related to technology (such as the Deep Space Optical Communications [DSOC] technology experiment), engineering, or science may also be undertaken throughout the mission. The goal of the videos is to introduce each instrument or investigation, its purpose on the mission from a science and engineering perspective, and the team members involved. As with the Mission Trailer, the instrument videos tell the story of the mission and the teams involved for a general audience.

Animated Mission Videos

In addition to interviews with team members about various aspects of the mission, several animated videos were developed to illustrate the arrival of the spacecraft at the asteroid, to help explain the asteroid's potential violent past, and to simulate what a metallic world might look like based on the latest research. Artist and cinematic illustrator Peter Rubin created these videos in a series which includes an introduction to the mission with a voice over by mission principal investigator Linda Elkins-Tanton (https://www.youtube.com/watch?v=cgtiwqVWfgc), an illustrated video of Psyche's story and the violent impacts that may have stripped away its mantle (https://psyche.asu.edu/galleries/videos/), and an animation of the Psyche spacecraft exploring and charting the metallic world.

Science and Engineering Team Highlights Videos

Throughout the mission, science and engineering team members are selected for video features to highlight their role on the mission, their background, and their area of expertise. To date, science team members Dr. Simone Marchi from the Southwest Research Institute and Dr. Tim McCoy from the Smithsonian Museum of Natural History have been featured. Additional science and engineering team members will be featured as the mission progresses, both by the Psyche mission and by the NASA 360 crew, to showcase the science behind the mission and the breadth of skills and abilities needed to compile a successful space mission team.

Student Collaborations Videos

For promotion to and recruitment of future participants and stakeholders, short videos were made about the Psyche capstone program (https://psyche.asu.edu/gallery/nasa-psyche-mission-undergraduate-capstone-projects/) and the Innovation Toolkit Online Courses (https://psyche.asu.edu/gallery/asu-psyche-online-course/) by the Psyche mission website design team, Fervor Creative. The Capstone program video features testimonials from Psyche capstone alumni and showcases a number of their projects as well as explaining for potential capstone faculty or students the ways to get involved. The Innovation Toolkit Online Courses video emphasizes the opportunity to join the courses at any time from any place and encourages lifelong learners worldwide to build their skills for innovation and exploration.

Videos created by Student Collaborations participants themselves (outreach interns, capstone students, and Psyche Inspired interns) augment the content for the website, social media, and school visits and other outreach events. For example, in response to a request from 4th grade teachers during a school visit, a group of outreach interns researched, wrote the script and had it approved by the Psyche team, recorded, and animated a two-part video series about the engineering process: (https://psyche.asu.edu/2020/12/04/engineering-process-part-1/; https://psyche.asu.edu/2020/02/07/engineering-process-part-2/). Capstone students have made quick videos about Psyche for posting on #PsycheSpaceCRAFTY (e.g., https://psyche.asu.edu/psyche-space-crafty/christina-webb-stephanie-bookout/) and Psyche Inspired interns frequently create videos as a means of capturing their work (https://psyche.asu.edu/gallery/sixteen/) or sharing the mission through animation and song (https://psyche.asu.edu/gallery/an-asteroid-named-psyche-animated-typography/). They are also interviewed for their annual showcase and relevant excerpts from those interviews are compiled into videos used for promoting the annual Psyche Inspired application (e.g., https://www.youtube.com/watch?v=nr2DEmzeSuw).

Acknowledgements Psyche Student Collaborations acknowledges the contributions of Psyche mission team members and participating undergraduate students and faculty and members of the public to the Psyche Student Collaborations programs (https://psyche.asu.edu/get-involved/). This work, and Open Access funding, is supported by NASA contract NNM16AA09, "Psyche: Journey to a Metal World."

Funding This work, and Open Access funding, is supported by NASA contract NNM16AA09, "Psyche: Journey to a Metal World."

Data Availability All Psyche Student Collaborations public material is available at https://psyche.asu.edu.

Declarations

Competing Interests The authors have no conflicts of interest to declare that are relevant to the content of this article.

Ethics Approval The ASU IRB determined that the protocol for Psyche Student Collaborations: Psyche Capstone Program is considered exempt pursuant to Federal Regulations 45CFR46 (1) Educational settings, (2) Tests, surveys, interviews, or observation.

Consent to Participate A consent form is included at the beginning of each survey and subsequent participation in that survey represents consent.

Consent for Publication Not applicable.

References

Anderson M (March 2022) NASA accepts just 5% of applicants to its ultra-competitive internship program. The program's leader says here's how to snag a spot. Business Insider. https://www.businessinsider.com/nasa-internship-intern-summer-engineering-aerospace-astronaut-2022-3. Accessed 17 August 2022

Arizona State University Charter (ASU) (2014) https://www.asu.edu/about/charter-mission-and-values. Accessed 9 December 2020

Bowman CDD, Elkins-Tanton LT, Bacalzo D, Howell P, Montgomery EJ, Sanft A, Briggs D, Chen Y, McCarville D, Meuth R, Zhao M (2019) Coordinating opportunistic interdisciplinary projects across single-discipline capstone courses. Int J Eng Educ 35(6): 1983–1992

Elkins-Tanton LT, Asphaug E, Bell JF III, Bercovici H, Bills B, Binzel R, Bottke WF, Dibb S, Lawrence DJ, Marchi S, McCoy TJ, Oran R, Park RS, Peplowski PN, Polanskey CA, Prettyman TH, Russell CT, Schaefer L, Weiss BP, Wieczorek MA, Williams DA, Zuber MT (2020). Observations, meteorites, and models: a preflight assessment of the composition and formation of (16) Psyche. JGR Planets 125(3):e2019JE006296. https://doi.org/10.1029/2019JE006296

Elkins-Tanton LT, Asphaug E, Bell JF III, Bierson CJ, Bills BG, Bottke WF, Courville SW, Dibb S, Jun I, Lawrence DJ, Marchi S, McCoy TJ, Merayo JMG, Oran R, O'Rourke JG, Park RS, Peplowski PN, Prettyman TH, Raymond CA, Weiss BP, Wieczorek MA, Zuber MT (2022) Distinguishing the origin of asteroid (16) Psyche. Space Sci Rev 218:17. https://doi.org/10.1007/s11214-022-00880-9

Fetterman D (2001) Foundations of empowerment evaluation. Sage Publications, Thousand Oaks, CA

Fetterman D, Bowman CD (2002) Experiential education and empowerment evaluation: Mars Rover educational program case example. J Exp Educ 25(2):286–295. https://doi.org/10.1177/105382590202500207

Fortson L (2021) From Green Peas to STEVE: citizen science engagement in space science. In: Kaminski AP (ed) Space science and public engagement: 21st century perspectives and opportunities. Elsevier, Amsterdam, pp 185–219. https://doi.org/10.1016/B978-0-12-817390-9.00009-9

Fretchling J (2010) The 2010 user-friendly handbook for project evaluation. National Science Foundation, Arlington

Friedman AJ (2008) Framework for evaluating impacts of informal science education projects. National Science Foundation, Washington

Google Earth Blog (2009) Google Earth 5 – The New Google Mars. https://www.gearthblog.com/blog/archives/2009/02/google_earth_5_the_new_google_mars.html. Accessed 24 May 2021

Hauhart RC, Grahe JE (2015) Designing and teaching undergraduate capstone courses. Jossey-Bass, San Francisco

Howe S, Rosenbauer L, Poulos S (2017) The 2015 capstone design survey results: current practices and changes over time. Int J Eng Educ 33(5):1393–1421

Hussar B, Zhang J, Hein S, Wang K, Roberts A, Cui J, Smith M, Bullock Mann F, Barmer A, Dilig R (2020) The condition of education 2020 (NCES 2020-144). (U.S. Department of Education, Washington, DC: National Center for Education Statistics)

Kaminski AP (ed) (2021) Space science and public engagement: 21st century perspectives and opportunities. Elsevier, Amsterdam, pp xv–xxiii. https://doi.org/10.1016/B978-0-12-817390-9.02001-7

Lee N, Loton DJ (2015) Capstone curriculum across disciplines: synthesising theory, practice and policy to provide practical tools for curriculum design. Australian Government Office for Learning and Teaching, Sydney, p 1

Marchi S, Durda DD, Polanskey CA, Asphaug E, Bottke WF, Elkins-Tanton LT, Garvie LAJ, Ray S, Chocron S, Williams DA (2020) Hypervelocity impact experiments in iron-nickel ingots and iron meteorites: implications for the NASA Psyche mission. JGR Planets 125:e2019JE005927. https://doi.org/10.1029/2019JE005927

NASA (2009) Mars, then and now: Google Mars update. https://www.nasa.gov/mission_pages/mars/news/google-mars-update-3-13.html. Accessed 24 May 2021

NASA (2014) Announcement of opportunity: Discovery 2014, NSPIRES 2014. https://nspires.nasaprs.com/external/solicitations/summary.do?method=init&solId={FE7B4C63-873D-63C1-4D15-1D46E2FEA949}&path=open. Accessed 8 December 2020

NASA (2020) NASA strategy for science, technology, engineering and mathematics (STEM) engagement 2020-2023, NASA 2020. https://www.nasa.gov/sites/default/files/atoms/files/nasa-strategy-for-stem-2020-23-508.pdf. Accessed 8 December 2020

NASA Office of STEM Engagement (NASA OSTEM) (2020). NASA STEM engagement highlights 2020, NASA, 2020. https://www.nasa.gov/sites/default/files/atoms/files/nasa_ostem_highlights_2020_0.pdf. Accessed 5 April 2021

National Academies of Sciences, Engineering, and Medicine (NASEM) (2018) The integration of the humanities and arts with sciences, engineering, and medicine in higher education: branches from the same tree. The National Academies Press, Washington. https://doi.org/10.17226/24988

Smith DA, Peticolas L, Schwerin T, Shipp S, Manning JG (2014) Science and science education go hand-in-hand: the impact of the NASA science mission directorate education and public outreach program. In: Manning JG et al (eds) Ensuring STEM literacy: a national conference on STEM education and public outreach. ASP Conference Series, vol 483. Astronomical Society of the Pacific, San Francisco, pp 9–28

Snyder JS et al (2020a) AIAA propulsion and energy forum. https://doi.org/10.2514/6.2020-3607
Snyder JS, Lenguito G, Frieman JD, Haag TW, Mackey JA (2020b) Effects of background pressure on SPT-140 hall thruster performance, J Propuls Power 36(5):668–676. https://doi.org/10.2514/1.B37702
Talamante A, Bowman CDD, Elkins-Tanton LT, Oran R, Prakash R 2022 Too late or just in time? Industry partner networking and professional development opportunities for capstone students. In: Capstone design conference, June 6–8, 2022

Authors and Affiliations

Catherine D. D. Bowman[1] · Linda T. Elkins-Tanton[1] · Adriana Talamante[1] · James F. Bell III[1] · Ernest Cisneros[1] · Alexandra Cook[1] · Jason D. Frieman[2] · Danya Gainor[1] · Jamie Hunziker[1] · Shaheer Khan[3] · Christopher R. Lawler[3] · Jessica Maschino[1] · Timothy J. McCoy[4] · Kaxandra Nessi[1] · Rona Oran[5] · David Seal[3] · Amber Simon[1] · Rohit Singh[1] · Carol M. Tolbert[2] · Karin Valentine[1] · Benjamin Weiss[5] · Daniel D. Wenkert[3] · David A. Williams[1]

✉ C.D. D. Bowman
c.bowman@asu.edu

1 School of Earth and Space Exploration, Arizona State University, Tempe, AZ, USA

2 NASA Glenn Research Center, Cleveland, OH, USA

3 Jet Propulsion Laboratory, Pasadena, CA, USA

4 Smithsonian Institution, Washington, DC, USA

5 Massachusetts Institute of Technology, Cambridge, MA, USA

GPSR Compliance

The European Union's (EU) General Product Safety Regulation (GPSR) is a set of rules that requires consumer products to be safe and our obligations to ensure this.

If you have any concerns about our products, you can contact us on ProductSafety@springernature.com

In case Publisher is established outside the EU, the EU authorized representative is:

Springer Nature Customer Service Center GmbH
Europaplatz 3
69115 Heidelberg, Germany

Batch number: 10628659

Printed by Printforce, the Netherlands